WITHDRAWN

Evolution of Early Earth's Atmosphere, Hydrosphere, and Biosphere—Constraints from Ore Deposits

edited by

Stephen E. Kesler
Department of Geological Sciences
University of Michigan
Ann Arbor, Michigan 48109
USA

Hiroshi Ohmoto
Penn State Astrobiology Research Center
of the NASA Astrobiology Institute
and
Department of Geosciences
The Pennsylvania State University
University Park, Pennsylvania 16802
USA

THE
GEOLOGICAL
SOCIETY
OF AMERICA

Memoir 198

3300 Penrose Place, P.O. Box 9140 ▪ Boulder, Colorado 80301-9140, USA

2006

Copyright © 2006, The Geological Society of America, Inc. (GSA). All rights reserved. GSA grants permission to individual scientists to make unlimited photocopies of one or more items from this volume for noncommercial purposes advancing science or education, including classroom use. For permission to make photocopies of any item in this volume for other noncommercial, nonprofit purposes, contact the Geological Society of America. Written permission is required from GSA for all other forms of capture or reproduction of any item in the volume including, but not limited to, all types of electronic or digital scanning or other digital or manual transformation of articles or any portion thereof, such as abstracts, into computer-readable and/or transmittable form for personal or corporate use, either noncommercial or commercial, for-profit or otherwise. Send permission requests to GSA Copyright Permissions, 3300 Penrose Place, P.O. Box 9140, Boulder, Colorado 80301-9140, USA.

Copyright is not claimed on any material prepared wholly by government employees within the scope of their employment.

Published by The Geological Society of America, Inc.
3300 Penrose Place, P.O. Box 9140, Boulder, Colorado 80301-9140, USA
www.geosociety.org

Printed in U.S.A.

GSA Books Science Editor: Abhijit Basu

Library of Congress Cataloging-in-Publication Data

Evolution of early earth's atmosphere, hydrosphere, and biosphere : constraints
   from ore deposits / edited by Stephen E. Kesler, Hiroshi Ohmoto.
      p. cm. – (Memoir ; 198)
   Includes bibliographical references and index..
   ISBN-13 978-0-8137-1198-0
   ISBN-10 0-8137-1198-3
      1. Geochemistry. 2. Atmosphere. 3. Evolution (Biology). I. Kesler, Stephen E.
   II. Ohmoto, Hiroshi. Memoir (Geological Society of America) ; 198.

QE515.E86 2006
551.5 22 -- dc22

2006044413

Cover: The 2.5 Ga Mount McRae Shale (black color) and iron ores (brown color) developed from the Brockman Iron Formation at the Tom Price Mine, Hamersley, Western Australia.

10 9 8 7 6 5 4 3 2 1

Contents

Preface v

Origin of Life

1. *The onset and early evolution of life* 1
   M.J. Russell and A.J. Hall

2. *Early life signatures in sulfur and carbon isotopes from Isua, Barberton,*
   Wabigoon (Steep Rock), and Belingwe Greenstone Belts (3.8 to 2.7 Ga) 33
   N.V. Grassineau, P. Abell, P.W.U. Appel, D. Lowry, and E.G. Nisbet

Evolution of the Early Continents

3. *Fingerprinting the metal endowment of early continental crust to test for secular changes*
   in global mineralization 53
   C. Thiart and M.J. de Wit

4. *Discovery of the oldest oxidized granitoids in the Kaapvaal Craton and its implications*
   for the redox evolution of early Earth 67
   S. Ishihara, H. Ohmoto, C.R. Anhaeusser, A. Imai, and L.J. Robb

5. *Secular variations of N-isotopes in terrestrial reservoirs and ore deposits* 81
   R. Kerrich, Y. Jia, C. Manikyamba, and S.M. Naqvi

Evolution of Uranium Deposits and Atmospheric Evolution

6. *The sedimentary setting of Witwatersrand placer mineral deposits*
   in an Archean atmosphere 105
   W.E.L. Minter

7. *Witwatersrand gold-pyrite-uraninite deposits do not support a reducing*
   Archean atmosphere 121
   J. Law and N. Phillips

8. *Evidence from sulfur isotope and trace elements in pyrites for their multiple*
   post-depositional processes in uranium ores at the Stanleigh mine, Elliot Lake,
   Ontario, Canada 143
   K.E. Yamaguchi and H. Ohmoto

9. *Time constraint for the occurrence of uranium deposits and natural nuclear fission*
   reactors in the Paleoproterozoic Franceville Basin (Gabon) 157
   F. Gauthier-Lafaye

Evolution of Seawater and Basinal Sulfur Geochemistry

10. *Proterozoic sedimentary exhalative (SEDEX) deposits and links to evolving global ocean chemistry* 169
    T.W. Lyons, A.M. Gellatly, P.J. McGoldrick, and L.C. Kah

11. *Precambrian Mississippi Valley–type deposits: Relation to changes in composition of the hydrosphere and atmosphere* 185
    S.E. Kesler and M.H. Reich

12. *Superheavy S isotopes from glacier-associated sediments of the Neoproterozoic of south China: Oceanic anoxia or sulfate limitation?* 205
    Liu Tie-bing, J.B. Maynard, and J. Alten

Evolution of Seawater Iron and Oxygen Geochemistry and Banded Iron Formations

13. *An evaluation of diagenetic recycling as a source of iron for banded iron formations* 223
    R. Raiswell

14. *Microbially mediated iron mobilization and deposition in iron formations since the early Precambrian* 239
    D.A. Brown

15. *Oxygen isotope composition of hematite and genesis of high-grade BIF-hosted iron ores* 257
    J. Gutzmer, J. Mukhopadhyay, N.J. Beukes, A. Pack, K. Hayashi, and Z.D. Sharp

16. *Rare-earth elements in Precambrian banded iron formations: Secular changes of Ce and Eu anomalies and evolution of atmospheric oxygen* 269
    Y. Kato, K.E. Yamaguchi, and H. Ohmoto

17. *Chemical and biological evolution of early Earth: Constraints from banded iron formations* 291
    H. Ohmoto, Y. Watanabe, K.E. Yamaguchi, H. Naraoka, M. Haruna, T. Kakegawa, K. Hayashi, and Y. Kato

Index 333

Preface

INTRODUCTION

This volume was stimulated by a Pardee Symposium titled "Evolution of the Early Atmosphere, Hydrosphere, and Biosphere: Constraints from Ore Deposits," which we convened in 2002 at the national meeting of the Geological Society of America. The history of Earth's early atmosphere, hydrosphere, and biosphere, from Hadean through Archean and into Proterozoic time, is one of the enduring puzzles in the geological sciences. When did the oceans appear, and did they remain liquid throughout Earth's history? What was the composition of the early atmosphere, and how did it affect climate? How did the atmosphere and ocean compositions change through time, and why? When, where, and how did life emerge on Earth? When did cyanobacteria, sulfate reducers, methanogens, and eukarya appear, and how did they affect their geologic environments? How did changes in the atmosphere, hydrosphere, and biosphere affect the lithosphere, and vice versa?

One of the most prominent aspects of this puzzle is when Earth's atmosphere became oxic. Controversy today focuses on two possibilities: that the atmosphere has been oxic since early Archean time (ca. 3.8 Ga) or that it gained oxygen between about 2.3 and 2.1 Ga, an event recently termed the Great Oxidation Event (GOE). The availability of free oxygen would have had a strong effect on the (bio)geochemical cycles of elements that exist in more than one oxidation state in nature. Of the major rock-forming elements, only iron does this. In contrast, multiple oxidation states are common in nature for many of the trace elements that are concentrated in some ore deposits, such as manganese, molybdenum, uranium, and vanadium. The atmospheric concentrations of CO_2 and CH_4 are also of concern as potential greenhouse gases to resolve the faint young sun problem, as the sources and products of biological activities, and as the source of acid rain for weathering the early continents. Finally, the concentrations of various forms of sulfur species in the oceans, especially H_2S and SO_4^{2-}, are of interest because they are linked to the atmospheric pO_2 history, the evolution of a variety of sulfur-utilizing microbes (e.g., sulfate-reducing bacteria and sulfide-oxidizing bacteria), and the origins of a variety of mineral deposits, including banded iron-formations, volcanogenic massive sulfide, sedimentary exhalative, and Mississippi Valley–type (MVT) deposits. The abundance ratios of many of the elements in ore deposits respond to the oxygen-carbon-sulfur geochemistry of the atmosphere and oceans, making ore deposits particularly good indicators of their geochemical environment.

Historically, the first ore deposit types linked to atmosphere-hydrosphere compositions were uranium paleoplacers and banded iron formations. Later work has expanded the list to include other types of uranium deposits, sedimentary manganese deposits, laterites, and sedimentary exhalative and MVT lead-zinc deposits. Papers in this volume deal with most of these deposit types and include new research results, as well as summaries of the current opinions on how they relate to proposed histories of Earth's early atmosphere, hydrosphere, and biosphere.

SUMMARY OF THE VOLUME

This volume includes papers based on some of the presentations at the symposium, as well as additional papers. It starts with a section dealing with the biosphere and the origin of life. Here, Russell and Hall show that low-temperature hydrothermal systems could have constituted simple reactors in which H_2 and CO_2 formed acetate, complex organic molecules and even cells (i.e., the emergence of life) using a froth-like substrate of iron-sulfides. Next, Grassineau et al. use sulfur and carbon isotope geochemistry of Archean rocks and ore deposits to trace the metabolic evolution of life from 3.8 to 2.7 Ga. The second section focuses on development of the early continents and begins with the study by Thiart and de Wit demonstrating that extraction of ore elements from the mantle has become less efficient through time. This is followed by a study by Ishihara et al. showing that the range in oxidation states of Archean granitoid intrusions is essentially the same as that of younger granitoids, and a final study by Kerrich et al. indicating that atmospheric N_2 has been drawn into ore deposits and other parts of the lithosphere through geologic time by N-fixing microorganisms.

The third section of the volume, which focuses on uranium deposits and their relation to atmospheric evolution, starts with two contrasting views of the famous Mesoarchean Witwatersrand Supergroup in South Africa that contains Earth's largest gold-uranium deposits. These deposits are hosted by quartz-pebble conglomerates, and controversy centers on whether they formed as paleoplacers. In the first paper, Minter shows that the sedimentary setting of gold, pyrite, and uraninite favors a detrital origin and their deposition under a reducing atmosphere. In the second paper, Law and Phillips describe geologic and geochemical features indicating that the gold, uranium, and sulfur were deposited by hydrothermal fluids after deposition of the conglomerates and conclude that these deposits do not provide evidence for a reducing Archean atmosphere. In a third paper, Yamaguchi and Ohmoto argue against a paleoplacer origin for pyrite in the slightly younger Paleoproterozoic Huronian Supergroup in Canada, which also contains quartz-pebble conglomerates that host only uraninite and pyrite, but no gold. In the final paper, Gauthier-Lafaye shows that the well-known Oklo sandstone-type uranium deposits of Gabon, which formed at ca. 2.1 Ga, are best explained if they formed when oxygenated water was available to transport dissolved uranium.

The fourth section of the volume concerns lead-zinc and manganese deposits and the information that they provide about the evolution of sulfur in seawater. The first two papers by Lyons et al. and Kesler and Reich show that the temporal history of sedimentary exhalative (sedex) and MVT lead-zinc deposits, respectively, is best explained by an ocean that lacked widespread, abundant sulfate until at least middle Proterozoic time. Liu et al. show that Neoproterozoic Mn-carbonate deposits in China resulted from low levels of sulfate in the ocean caused by an anomalous influx of Fe from lateritic soils rather than from a "snowball Earth."

The final section of the volume deals with banded iron formation (BIF) deposits and the insights that they provide about the geochemistry of iron, oxygen, sulfur, and carbon and the nature of organisms in the oceans. The paper by Raiswell deals with the relative importance of weathering, diagenetic, and exhalative processes as sources of iron. Brown discusses the problem of depositing iron and concludes from experiments and energy considerations that microbes could have accounted for large volumes of iron carbonate and oxide precipitates, and Gutzmer et al. show that high-grade hematite BIF deposits formed under conditions similar to those of today. In the final two papers, Kato et al. and Ohmoto et al. suggest that BIFs formed throughout geologic history by the mixing of locally discharged hydrothermal fluids with ambient seawater and show that this is consistent with an ocean that has been oxygen- and sulfate rich since ca. 3.8 Ga.

REMAINING CONTROVERSIES

It will be apparent to even the most casual reader that authors who have contributed to this volume hold a wide range of views about the composition of the early Earth atmosphere, hydrosphere, and biosphere and the constraints offered by ore deposits, and we regard this as one of the major contributions of the volume. Our sincere hope is that the range of views propounded in this volume will help students and researchers to realize the nature of the remaining controversies and the sorts of data that are required to resolve them. For example, why do Lyons et al. and Kesler and Reich conclude that the early ocean was low in sulfate whereas Ohmoto et al. conclude that it was high in sulfate? What sort of evidence do they have, how did they interpret this evidence, and what sorts of new studies are necessary to test their assumptions and conclusions? What new methods can be used to resolve these controversies?

Our understanding of early Earth conditions has evolved tremendously over the past few decades, and we expect that this will continue in the future, spurred in part by studies inspired by this volume. As this research continues, three points are apparent. First, in the final analysis, our conclusions are based largely on the rock record, and that record is complex. Mineralogical and textural features of undisturbed rocks are difficult enough to understand, but things get much harder with older rocks because in general, they have been subjected to higher degrees of metamorphism and/or to longer periods of weathering. As geochemical studies become more and more sophisticated, it is increasingly important that the utmost care be taken to assure that samples truly represent the event or feature of interest and to document the textures and features that prove it. Second, the abundance of Precambrian rocks, especially those formed in near-surface and active tectonic settings, decreases with increasing geologic age because most of them were eroded or destroyed by later tectonic events, but also perhaps because the earlier continents were smaller. Therefore, more accurate reconstructions of paleogeographic and tectonic settings of the major Archean and Proterozoic terrains are necessary in order to relate the temporal change in the preserved number/size of a specific type of ore deposit to the chemical and biological evolution of early Earth. Third, there is growing evidence that Earth's ore deposit history rarely reflects a single dominant control and instead is result of complex interaction among many factors. These can be classified in many ways, but are perhaps best viewed as the tectonic, hydrologic, and biologic cycles. Our growing understanding of earth system science shows how intimately these cycles are related and it comes as no surprise that ore deposits reflect this interplay. Hopefully, when the successor to this volume is compiled, continuing research will have brought us closer to an understanding of the constraints that ore deposits provide on early Earth's atmosphere, hydrosphere, and biosphere.

ACKNOWLEDGMENTS

The symposium that stimulated this volume was supported generously by the Geological Society of America, the Society of Economic Geologists and the NASA Astrobiology Institute. We are grateful to all of these for their interest in this important subject.

Stephen E. Kesler
Hiroshi Ohmoto

The onset and early evolution of life

Michael J. Russell*

Scottish Universities Environmental Research Centre, Rankine Avenue, East Kilbride, Glasgow G75 0QF, Scotland, and Géosciences, LGGA, Université de Grenoble 1, 1381, rue de la Piscine, 38400 St. Martin d'Heres, France

Allan J. Hall*

Department of Archaeology, University of Glasgow, G12 8QQ, Scotland

ABSTRACT

The tension between CO_2 dissolved at relatively high atmospheric pressure in the Hadean ocean, and H_2 generated as ocean water oxidized ferrous iron during convection in the oceanic crust, was resolved by the onset of life. We suggest that this chemosynthetic life emerged within hydrothermal mounds produced by alkaline solutions of moderate temperature in the relative safety of the deep ocean floor. Exothermic reaction between hydrothermal H_2, $HCOO^-$ and CH_3S^- with CO_2 was catalyzed in inorganic membranes near the mound's surface by mackinawite (FeS) nanocrysts and "ready-made" clusters corresponding to the greigite (Fe_5NiS_8) structure. Such clusters were precursors to the active centers (e.g., the C-cluster, Fe_4NiS_5) of a metalloenzyme that today catalyzes acetate synthesis, viz., the bifunctional dehydrogenase enzyme (ACS/CODH). The water, and some of the acetate ($H_3C.COO^-$), produced in this way were exhaled into the ocean together as fluid waste. Glycine ($^+H_3N.CH_2.COO^-$) and other amino acids, as well as tiny quantities of RNA, generated in the same milieu were trapped within tiny iron sulfide cavities.

Energy from the acetate reaction, augmented by a proton gradient operating through the membrane, was spent polymerizing glycine and other amino acids into short peptides upon the phosphorylated mineral surface. In turn these peptides sequestered, and thereby protected, the catalytically and electrochemically active pyrophosphate and iron/nickel sulfide clusters, from dissolution or crystallization.

Intervention of RNA as a polymerizing agent for amino acids also led to an adventitious, though crude, process of regulating metabolism—a process that was also to provide genetic information to offspring. The fluxes of energy and nutrient available in the hydrothermal mound—commensurate with the requirements of life—encouraged differentiation of the first microbes into two separate domains. At the bifurcation the Bacteria were to specialize in acetogenesis and the Archaea into methanogenesis. Representatives of both these domains left the mound by way of the ocean floor and crust to colonize the deep biosphere.

Once life had emerged and evolved to the extent of being able to reduce nitrogen for use in peptides and nucleic acids, light could have been used directly as an energy source for biosynthesis. Certain bacteria may have been able to do this, where protected from hard UV by a thin coating of chemical sediment produced at a subaerial hot spring operating in an obducted and uplifted portion of the deep biosphere.

---

*michaelr@chem.gla.ac.uk; A.Hall@archaeology.arts.gla.ac.uk

Embedded in fresh manganiferous exhalites, early photosynthetic bacteria could further protect themselves from radiation by adsorbing manganese on the membrane. Organization of the manganese with calcium, within a membrane protein, happened to result in a $CaMn_3O_4$ cluster. In Mn(IV) mode this structure could oxidize two molecules of water, evolve waste oxygen, and gain four electrons and four protons in the process to fix CO_2 for biosynthesis. All these biosynthetic pathways had probably evolved before 3.7 Ga, though the reduced nature of the planet prevented a buildup of free atmospheric oxygen until the early Proterozoic.

Keywords: CODH/ACS, greigite, origin of life, oxygenic photosynthesis,ranciéite.

By *autogeny* we understand the origin of a most simple organic individual in an *inorganic formative fluid,* that is, in a fluid which contains the fundamental substances for the composition of the organism dissolved in simple and loose combinations (for example, carbonic acid, ammonia, binary salts, etc.).

—Ernst Haeckel 1892, p. 414

INTRODUCTION

In a posthumous paper published in 1952, Goldschmidt presented three principles to be adhered to in origin-of-life studies, principles derived from his geochemical and mineralogical experience:

> The **first** principle is that an a-biotic environment, poor in elementary oxygen, is suitable for the preservation and accumulation of ... organic molecules.
> The **second** principle is the collection, concentration and ordering of such molecules on free rectilinear planes, crystal faces, of minerals, giving them possibilities for further mutual interaction between themselves and the "basement" crystal.
> The **third** principle is the hypothesis that carbon dioxide (and its nearest derivatives) may be the primary material.

These principles, consistent with early geological interpretations of the moderate redox state of the early atmosphere, stand in stark contrast to the assumptions of Oparin (1924, 1938), Haldane (1929), Urey (1952), Oró (1961), Deamer (1985), Joyce (1989), Miller (1992), and Bada (2004) that life originated from the plethora of organic molecules supposedly delivered from space or generated in a putative reduced atmosphere. Goldschmidt, like the evolutionist Haeckel (1892) before him, inferred life to have emerged autogenically (i.e., from the simplest of inorganic substances), whereas Oparin assumed, and his followers assume still, a plasmogenic or organotrophic inheritance to explain their RNA and lipid worlds (Bada, 2004).

In this contribution we develop Goldschmidt's autogenic principles to show that evolutionary steps may be traced, though uncharted in places, mechanistically from aqueous geochemistry and mineralogy, through chemosynthetic biochemistry to oxygenic photosynthesis. The *abiotic environment* we favor for the accumulation and preservation of organic molecules is within FeS microcavities in a submarine hydrothermal mound. The *basement crystals* collecting, concentrating, ordering, and promoting mutual further interactions are the metastable iron sulfides, mackinawite and greigite—sulfides which accommodate that effective and common catalytic metal, nickel. The primary *carbon dioxide*, which composed a proportion of the atmosphere/ocean system (the volatisphere), is fixed by reaction with activated hydrothermal H_2 emanating from the highly reduced Earth to provide the basic organic molecules of life. Hydrogen, as a carrier and donor of high-energy electrons, is the first fuel of life. And as soon as organic molecules are generated they can inhibit crystal growth. Indeed growth of inorganic clusters may be arrested at a very early stage by certain charged or polar organic molecules (Rickard et al., 2001). In some cases these clusters can act as catalysts for further organic synthesis.

Although the potential energy available for reaction between the highly reduced Earth and its moderately oxidized volatisphere is substantial, the kinetic barriers are formidable (Shock, 1992). When hydrothermal solutions first titrated with a sterile prebiotic ocean, much of the thermal energy was effectively dispersed. Not so the chemical energy. Here we attempt to show how a hydrothermal mound at moderate temperature focused and catalyzed the reaction between the main molecules fundamental to life, H_2 and CO_2, and then fractionated, concentrated, and contained the longer charged products. In so doing we rely on the thermodynamic calculations and kinetic considerations of Shock and his collaborators in predicting the likely products of the earliest metabolism (Shock, 1990, 1992; Amend and Shock, 1998; Shock and Schulte 1998; Shock et al., 1998). We also appreciate the operations of Geochemist's Workbench in the presentation of relative stability fields in Pourbaix (Eh/pH) diagrams interpretable by both geochemists and biochemists (Bethke, 1996). Even these geochemical considerations must be given geological context. The emergence of life must be seen as a geological issue, as the first stage in the "evolution of species" and not some separate conception to be examined merely by dismantling a bacterial cell and looking for the oldest bits (Leduc, 1911). And the geochemical reactions must be translatable to an early plausible biochemistry. For example, reactions that make organic polymers only at temperatures above 200 °C have little direct bearing on the problem. Life exists as the energy trap and catalyst for reactions between reduced and oxidized components. Like a mineral exploration geologist searching for a lithochemical aureole to an orebody as an indicator of "spent" ore solutions (Russell, 1974), in order to comprehend the process of emergence we have to examine life's waste products, the entropic sinks. Amongst these are the fluid wastes, water, acetate, methane, hydrogen sulfide, and oxygen. These waste products continued to be dispersed, then and now, by convection and advection, albeit at differing rates. The solid mineral wastes, especially the sulfides, are generally deposited close to the source.

So what were the geological and geochemical conditions in the Hadean that gave rise to life and its waste products?

INITIAL CONDITIONS

As soon as the first ocean condensed and cooled around 4.4 Ga Earth was primed for life (Wilde et al., 2001; Russell and Hall, 1997). But where on Earth could life have begun? Conditions were anything but equable. The temperature of the oceans fluctuated wildly. Large meteorites that partially vaporized the ocean increased atmospheric pressure so that the remaining water might have reached 300 °C or so. In lulls in the bombardment, high CO_2 pressures could induce a 100 °C greenhouse (Kasting, 1993), yet meteorite-induced dust clouds might, on occasion, have masked radiation from the weak young sun. If so, a short-lived icehouse could have ensued (Nisbet and Sleep, 2001). However, conditions in this "water world" were generally tempestuous; the ocean surface was no place to organize the first cell, and shorelines, where they existed, would have suffered continual storms and huge tides, a response to the closer moon rapidly orbiting about an Earth whose day lasted a mere five hours. Darwin's "warm little pond," if it was not swamped, would have been subject to deleterious hard UV at eight times the present flux (Canuto et al., 1982; Bahcall et al., 2001; Abe and Ooe, 2001). The only safe place to be was on the ocean floor.

Can we imagine a window of opportunity for life to onset? After all, there was no shortage of the appropriate energies, both physical and chemical. Hydrothermal convection currents within the thick, fractured, and permeable Hadean crust focused a chemical disequilibrium between reduced iron (in ferrous minerals and as minor native iron) and the relatively oxidized volatisphere. Although there was chemical potential for reaction between the H_2, continuously emanating from Earth's interior and CO_2 in the atmosphere and ocean, an electrochemical potential between redox couples H^+/H_2 (the hydrothermal state) and Fe^{3+}/Fe^{2+} (the state in the ocean) also obtained.

A geological phenomenon to excite an earth scientist imagining the onset of life is the black smoker (Corliss et al., 1981). Such a very hot spring, emanating at close to the critical point of seawater, is acidic as a result of the serpentinization of pyroxenes at high temperature. In these conditions Mg^{2+} from ocean water in the convective downdrafts is precipitated as brucite $(Mg(OH)_2$ or serpentinite and protons are returned in its stead (Seyfried and Bischoff, 1981; Douville et al., 2002):

$$2MgSiO_3(\text{enstatite}) + Mg^{2+} + 3H_2O \rightarrow Mg_3Si_2O_5(OH)_4 \text{ (serpentine)} + 2H^+ \quad (1)$$

At the same time any sulfate is scrubbed out as magnesium hydroxide sulfate hydrate is produced within the crust, further contributing to the low pH of high-temperature hydrothermal solutions (Bischoff and Seyfried, 1978; Janecky and Seyfried, 1983):

$$2Mg^{2+} + 3H_2O + SO_4^{2-} \rightarrow Mg(OH)_2 \cdot MgSO_4 \cdot H_2O + 2H^+ \quad (2)$$

In the near absence of sulfate (equation 2) but with springs exhaling into an aciduous ocean there would have been no anhydrite and sulfide chimneys, and no black smokers in the Hadean. Free sulfide concentrations in the hot fluids were relatively low. The little there was would have reacted with zinc to produce stable ZnS and $Zn_2S_3^-$ clusters (Walker and Brimblecombe, 1985; Luther et al., 1999). Minor to trace quantities of other "biophile" elements such as Mn, Zn, Ni, Co, Mo, Se, and W (Goldschmidt, 1937) would also have been delivered, with iron, to the Hadean ocean through these high temperature springs at oceanic spreading centers (Von Damm, 1990; Hemley et al., 1992). The iron concentrations contributed in this way probably approached 20 mmol. We base this estimate on analyses of the Rainbow hydrothermal system operating in ultramafic rock at the slow spreading Mid-Atlantic Ridge (Douville et al., 2002) (Table 1). Ocean floor spreading in the Hadean was unlikely to have been fast because of the inhibiting effects of a 30-km-thick crust produced by the voluminous melting of a very hot and dry mantle, not to men-

TABLE 1. COMPARISON BETWEEN HIGH AND MODERATE TEMPERATURE SUBMARINE SPRINGS

Parameter	Juan da Fuca	Rainbow	Lost City	Eyjafjordur
T °C	224 °C	365 °C	40°–90 °C	71.4 °C
pH	3.2	2.8	≤11	10.03 (at 24 °C)
H_2	na	13	≤15	na
H_2S	3.5	1.0	0.064	0.01
Fe	18.74	24	na	0.00014
Mn	3.58	2.25	na	0.0000018
Mg	0	0	9–19	0.01
Ca	96.4	67	22	0.061
Na	796	553	482	3.4
K	51.6	20	na	4.2
SiO_2	23.3	6.9	na	1.6
CO_2	δ4.46	na	na	0.57
SO_4	0	(0)	5.9–12.9	0.2
Cl	1087	380	548	1.26
Co	na	0.013	na	na
Ni	na	0.003	na	na
Zn	0.9	0.16	na	na
Mo	na	0.000002	na	na
Duration yr	>1000	>1000	>30,000	~11,000

Note: Data from Von Damm, (1990), Douville et al., (2002), Kelley et al., (2001, 2005), Früh-Green et al. (2003), and Marteinsson et al., (2001). Elemental and molecular concentrations are in millimoles. Note that most submarine springs probably last for at least 100,000 years (e.g., Lalou et al., 1993; Früh-Green et al., 2003).

tion the additional tens of kilometers of oceanic volcanic plateaus produced from mantle plumes (Arndt and Chauvel, 1990; Arndt, 1999; Russell and Arndt, 2005) (Fig. 1). And hydrostatic pressures, and thereby temperatures of the hydrothermal fluid, may have been higher in the Hadean because of the deep penetration of fluids to the margins of the crystallizing magma chambers under a deeper ocean. As a consequence, iron concentrations also may have been correspondingly even higher (Von Damm, 2000; Bounama et al., 2001; Allen and Seyfried, 2003).

Although it has been pointed out that temperatures of such hot acidic springs at spreading centers in the Hadean were so high as to destroy organic molecules (Miller and Bada, 1988), they would have provided phosphate, as well as the trace elements that were to help energize and catalyze life, directly to the ocean (Kakegawa et al., 2002).

Given these hydrothermal contributions, what was the state of ocean chemistry? As atmospheric CO_2 of mainly volcanic derivation was at a pressure of anywhere between 0.2 and 10 bars, oceanic pH probably varied between 5 and 6. It could have been higher following major meteorite impacts when large quantities of rock dust were raised to the atmosphere (Nisbet and Sleep, 2001). Rainfall was likely to have been high, but there was little if any atmospheric weathering and runoff because, though there probably were continents, radioactive heating made them plastic and they rarely emerged above the surface of the relatively deep ocean (Sandiford and McLaren, 2002; Russell and Arndt, 2005). However, the iron contributions from the magmatically driven hot springs were diluted by the iron-free submarine alkaline springs exhaling from the ridge flanks and on the deep ocean floor. There were also quantities of ferric oxyhydroxide (FeOOH) particles (denoted by Fe^{III}) produced by photolysis (Cairns-Smith et al., 1992; Russell and Hall, 2002) (Fig. 2A):

$$2Fe^{2+} + 2H^+ + light \rightarrow 2Fe^{III}\downarrow + H_2\uparrow \qquad (3)$$

Taking account of dilution, photo-oxidation and precipitation, we speculate that the carbonic ocean carried up to 10 mmol of ferrous iron.

The hydrothermal convection cells feeding the alkaline springs were partly driven by exothermic serpentinization, a process that begins slowly at around 85 °C (Martin and Fyfe, 1970; Wenner and Taylor, 1971). Flow in fractures within the oceanic crust was facilitated partly by geodynamic stresses and partly by the tidal stresses induced by the close and rapidly orbiting moon

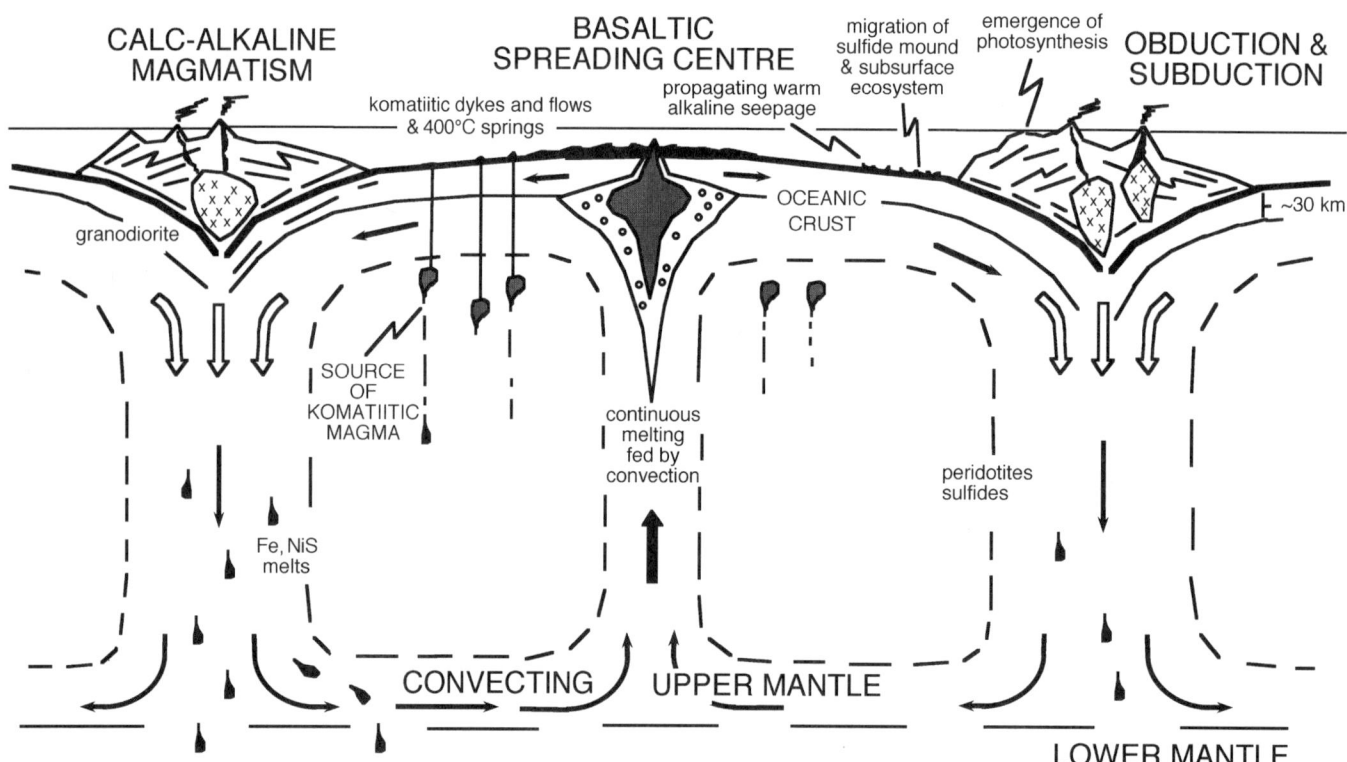

Figure 1. Cross-section of mantle convection cell for Earth at >4 Ga (Smith, 1981; Campbell et al., 1989; Davies, 1992; Karsten et al., 1996; Foley et al., 2003; Russell and Arndt, 2005). Chemosynthetic life emerged at a warm alkaline seepage and expanded into the surrounding sediments and crust, and was conveyed by ocean floor spreading toward a constructive margin produced largely by obduction. Once uplifted at the margin, a proportion of cells invaded sediments in the photic zone where, at a sulfurous spring, some evolved to exploit solar photons. Oxygenic photosynthesis was a further evolutionary development.

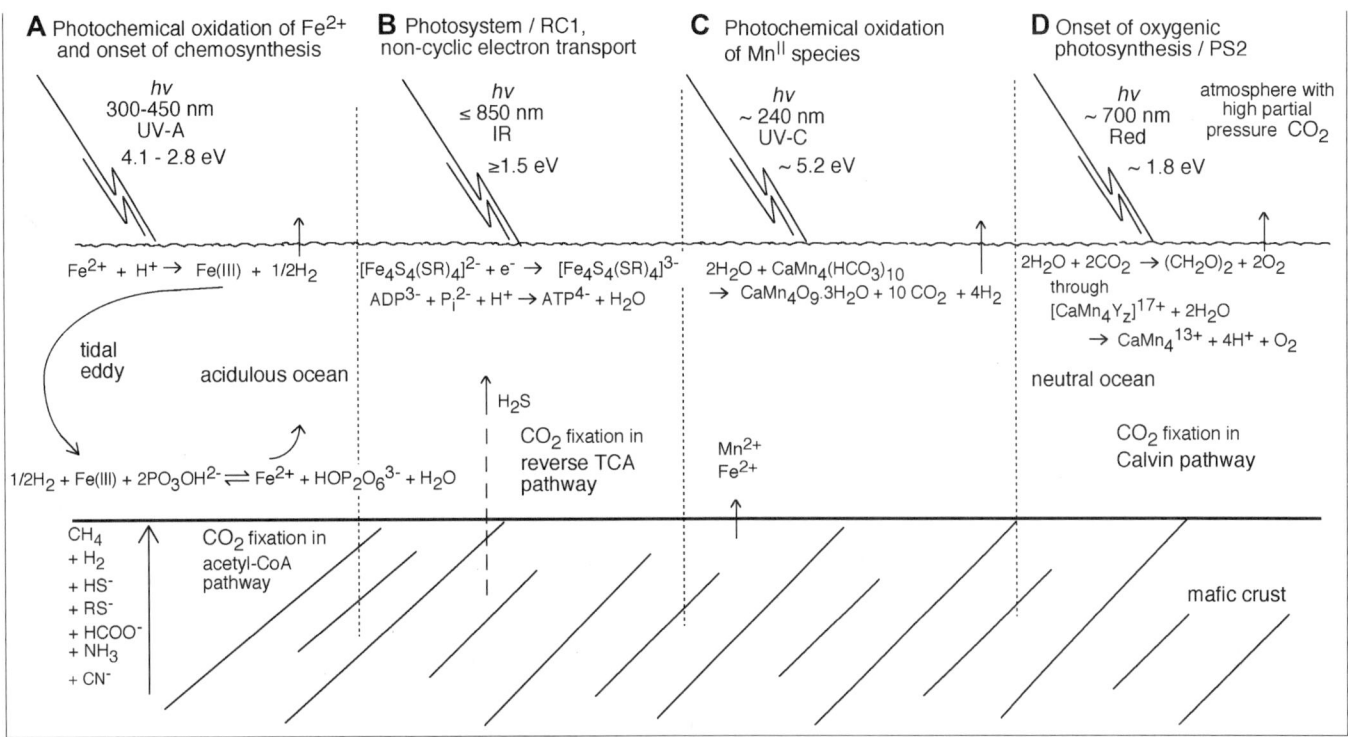

Figure 2. The focusing of solar energy to produce (A) photolytic iron oxidation and the potentiation of chemosynthetic life (Cairns-Smith et al., 1992); (B) reduction of ferredoxin and the onset of photo-induced non-cyclic electron transport (Blankenship, 2002); (C) photo-oxidation of Ca-Mn bicarbonate and generation of a precursor to the water oxidizing complex (Anbar and Holland, 1992; Dismukes et al., 2001; Russell and Hall, 2001, 2002); and (D) oxygenic photosynthesis through reduction of Mn_4^{IV} (Blankenship, 2002). Iron and manganese are exhaled from hot springs at ocean floor spreading centers and at island chains.

(Gaffey, 1997). Nevertheless serpentinization and carbonation may have blocked most fractures at temperatures above ~115 °C (Wenner and Taylor, 1971).

Although it is well understood that high temperature springs have a pH of between 2 and 3, a consequence of the loss and fixation of Mg^{2+} and the concomitant release of two protons (equations 1 and 2) (Von Damm, 1990; Allen and Seyfried, 2003), less well known is the fact that magnesium is rendered more soluble during the exothermic serpentinization of olivine and pyroxene below 200 °C (Fig. 3) (Macleod et al., 1994; Palandri and Reed, 2004). Because of this, hydroxyl rather than H^+ is generated as a byproduct of the serpentinization:

$$3MgFeSiO_3 + 7H_2O \rightarrow 3SiO_2 + Fe_3O_4 + 3Mg^{2+} + 6OH^- + 4H_2\uparrow \qquad (4)$$

Thus these moderate temperature springs are buffered to a pH of 10–11 by the precipitation of brucite $[Mg(OH)_2]$ and their temperature may be controlled at ~115 °C (Wenner and Taylor, 1971).

Hydrogen is produced in both the low and high temperature systems on the oxidation, by water, of ferrous-iron-bearing minerals to magnetite (equation 4). Hydrogen will also have been generated on the oxidation of the vestiges of native iron in the Hadean crust (Righter et al., 1997).

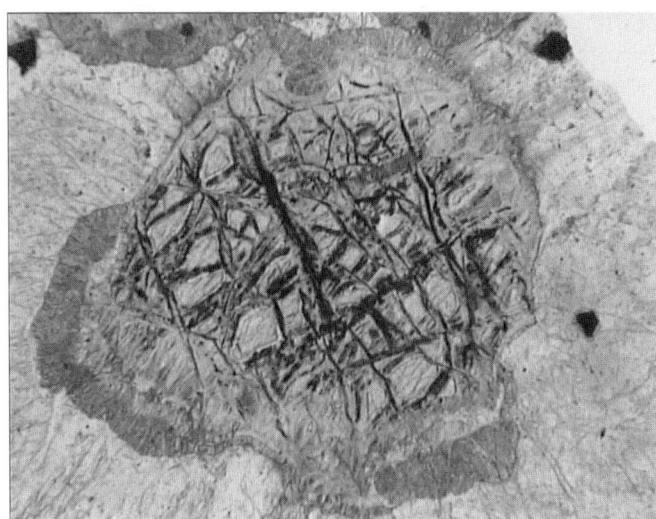

Figure 3. Photomicrograph showing typical texture of a millimetric grain of serpentinized olivine as found in diverse rock types. The process only takes place below 350 °C (Allen and Seyfried, 2003).

$$3Fe^0 + 4H_2O \rightarrow Fe_3O_4 + 4H_2\uparrow \qquad (5)$$

Furthermore, reaction with the nickel-iron alloys produced during serpentinization will have released hydrogen to the hydrothermal fluid (Krishnarao, 1964). Other reduced entities likely to be produced in the low temperature alkaline system are ammonia (NH_3), methane thiol (CH_3S^-), hydrosulfide (HS^-), formate ($HCOO^-$), and minor cyanide (CN^-) (Muller, 1995; 995; Russell and Hall, 1997; Shock et al., 1998; McCollom and Seewald, 2003). These molecules provide most of the basic nutrient and

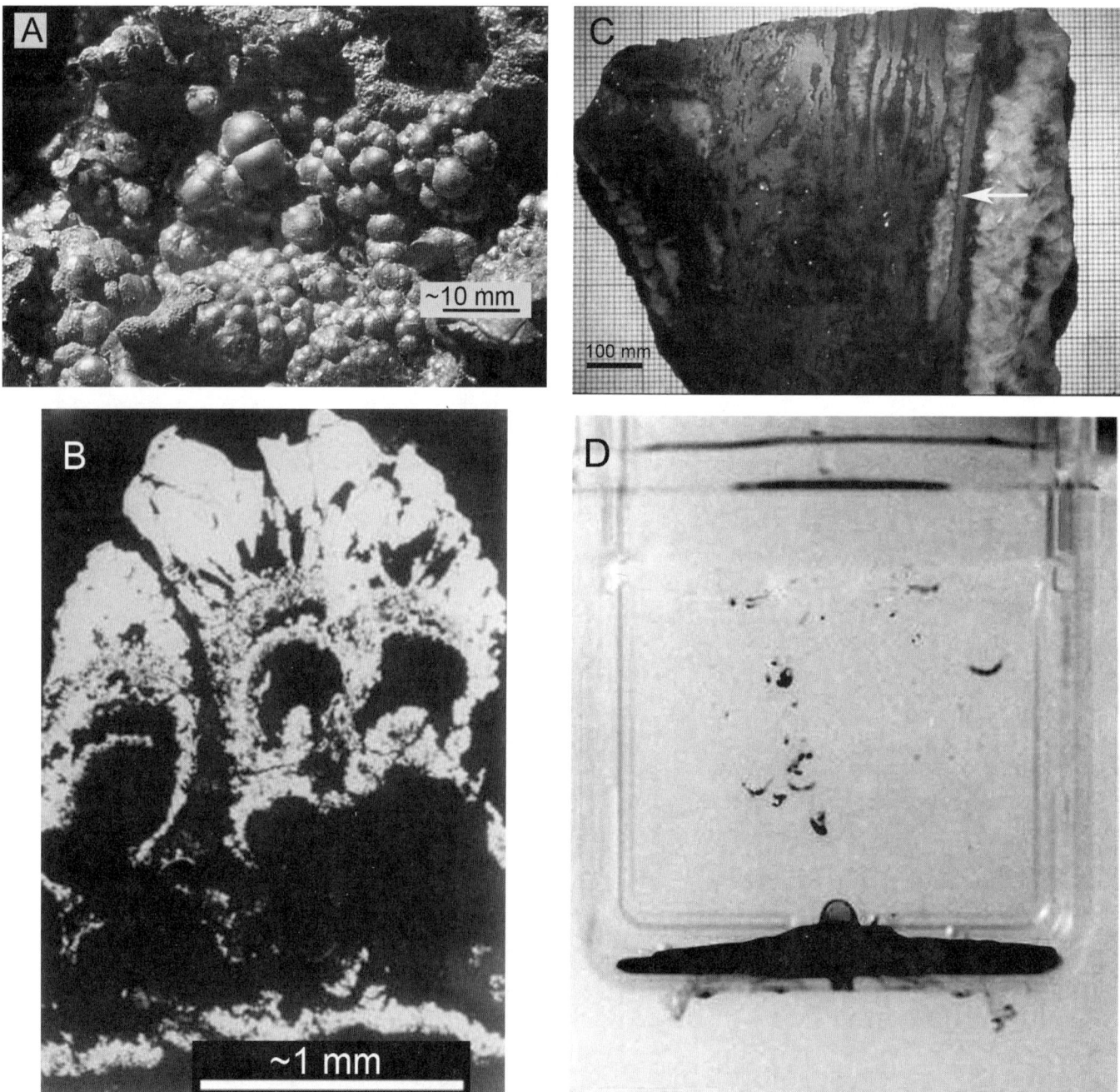

Figure 4. Pyrite botryoids at a 350 Ma fossil warm spring at the Tynagh mine, Ireland: (A) The top surface (field of view measures 2 cm across). (B) Cross-section through pyrite botryoids, revealing bubbles. (C) A natural chemical garden from Tynagh comprising pyrite spires embedded in barite (photographed on centimetric background). (D) Photograph by Martin Beinhorn of a sulfide structure produced as 250 mmol of Na_2S is injected into a 25 mmol $FeCl_2$ solution (field of view measures 4 cm across). (Pyrite is presumed to have replaced iron monosulfide membranes in cases A to C [Banks, 1985; Russell, 1988].)

energy requirements of life. On meeting and mixing with the Hadean ocean, hydrothermal mounds would be formed that seem to us the likely hatcheries of life (Russell and Hall, 1997).

EVIDENCE FOR A LOW TEMPERATURE MOUND

We first assumed (Russell et al., 1988, 1989) that hydrothermal precipitates at an alkaline spring or seepage would result in a porous mound somewhat comparable to the submarine exhalative sulfide constructs at the Tynagh and Silvermines orebodies in Ireland (Fig. 4A–C) (Boyce et al., 1983; Banks, 1985; Banks and Russell, 1992; Samson and Russell, 1987). These were the deposits that first excited our interests in life's emergence (Russell et al., 1988, 1989, 1994). Although the iron sulfide mounds at Tynagh and Silvermines resulted from acidic solutions, originally at ~250 °C and emanating into ~60 °C alkaline brine pools in faulted basins at the bottom of the Mississippian sea ca. 350 Ma, we reasoned that the chemical reactions would just as well result in similar precipitates if the two solutions were inverted (Russell et al., 1989). Observations of the kind of alkaline moderate temperature hydrothermal spring we envisaged (Russell et al., 1989, 1998; Shock, 1992) have been made in 1.5 m.y. old ultramafic oceanic crust, 15 km from the Mid-Atlantic Ridge at the so-called "Lost City" field (Kelley et al., 2001; Früh-Green et al., 2003) (Table 1). Nevertheless, although the solutions here are alkaline as we expected, and carried some H_2, the mounds contrasted significantly with our predictions (Russell et al., 1994; Russell and Hall, 1997).

The large edifices at the Lost City spring are composed mainly of carbonate and brucite [$Mg(OH)_2$], though Kelley et al., (2001) deduce a preoxidation sulfide concentration of ~5 mmol kg^{-1} in the hydrothermal solution. The former presence of a similar <100 °C alkaline spring at a transform fault in the Indian Ocean is also indicated by deposits of finely layered hydrated magnesium silicate (sepiolite) mixed with poorly crystalline Fe-Mn hydroxides (Bonatti et al., 1983). Any original carbonate has been redissolved. Another rather similar deposit, though precipitated from fresh water, has been discovered off the north coast of Iceland (Marteinsson et al., 2001; Geptner et al., 2002) (Table 1). Here cones of Mg-rich clay (saponite) tens of meters high are forming where warm (72 °C) alkaline (pH 10) submarine spring waters exhale into a fjord. Although the cones do offer the kind of porous morphology we expected, no sulfides are recorded (Geptner et al., 2002; Martin and Russell, 2003). Some of the differences between our expectations and the modern submarine springs can be ascribed to contrasting conditions in the Hadean when the crust was more reduced, Fe^{2+} concentrations in the ocean were high, and O_2 was negligible or absent.

In the light of these present-day discoveries how then might we refine our model for the emergence of life? Where the convective up-drafts were vigorous, the alkaline spring waters (pH 10–11, ~100 °C) would have exhaled directly into the acidulous Hadean ocean (pH ~5.5) (Russell et al., 1989; Shock, 1992; Macleod et al., 1994; Russell, 2003). We have assumed that at times when solar radiation was masked, this ocean was ~20 °C or less (Fig. 5). Precipitation at the exhalative center was rapid, but

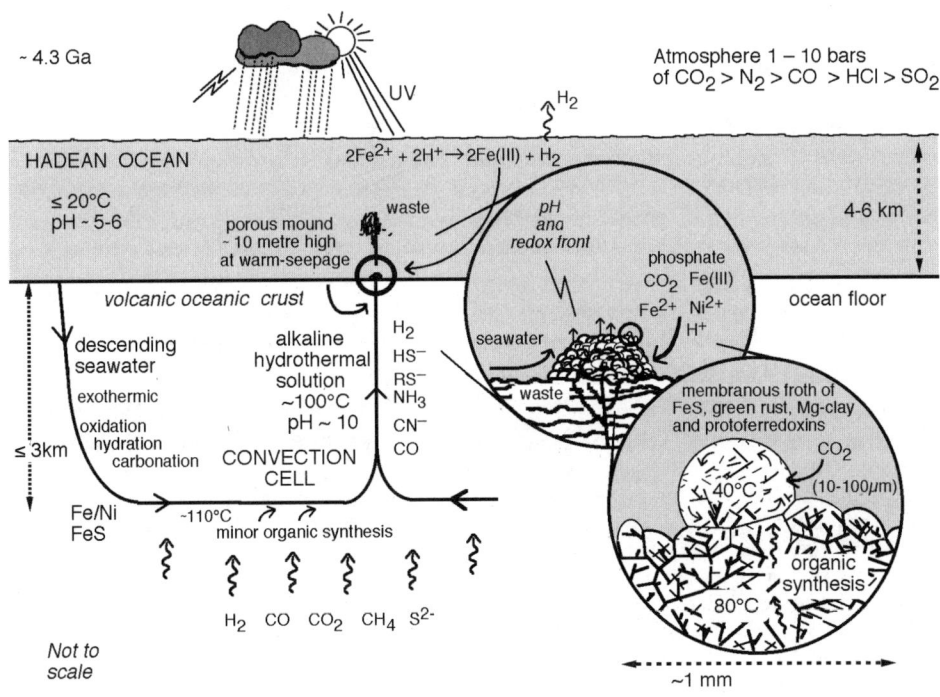

Figure 5. Model environment for the emergence of life at a submarine seepage on the ocean floor (after Russell and Hall, 1997).

as the fracture conduits in the mounds became fouled with carbonate, and with gels and microcrysts composed of silica, sepiolite, saponite, brucite, green rust, and iron sulfide, the fluid egress became diffuse and seepages replaced springs. Such restraint in the seepage mound favored—depending on solutes and local pH—development of siderite, ferrous hydroxide, and/or iron (nickel) monosulfides. What might the structure of these precipitates have looked like?

We originally imagined iron sulfide structures to have precipitated spontaneously at the interface of the hot alkaline seepage waters containing millimoles of HS^- and H_2 with the cool carbonic ocean water bearing millimoles of Fe^{2+} and particulate FeOOH (Figs. 6A, 6B). Our attempts to reproduce similar structures in the lab were relatively successful, though rather high concentrations of HS^- (250 mmol) were required to produce bubbles (Fig. 4D) (Russell, 1988; Russell et al., 1989). Depend-

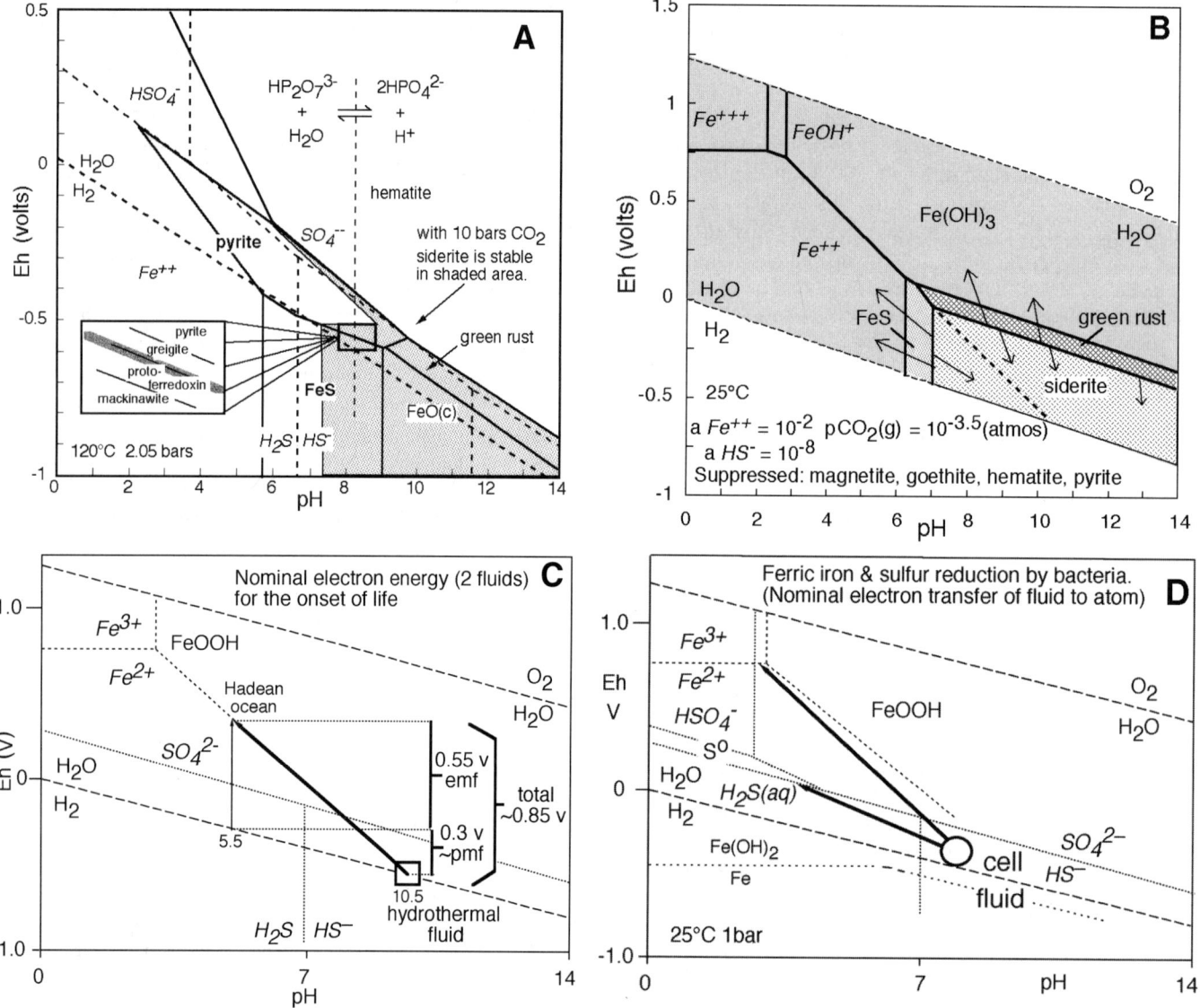

Figure 6. (A) A Pourbaix (Eh/pH) diagram illustrating the stabilities of siderite, mackinawite (as FeS), protoferredoxin, greigite, pyrite, green rust, and hematite, produced for activities of $H_2S(aq) = 10^{-3}$, and $Fe^{2+} = 10^{-6}$, using GWB (Bethke, 1996). The inset shows notional phase relations emphasizing the intermediate oxidation state of the FeS component of membrane protoferredoxins and is positioned to indicate the Eh-pH conditions pertaining to alkaline hydrothermal fluid as it enters Hadean seawater. Note that the pH boundary of monophosphate/polyphosphate intersects this redox position (and see Fig. 16). (B) An Eh-pH diagram computed for modern atmospheric CO_2. At higher pCO_2 the siderite field would expand as indicated by arrows, at the expense of FeS. Such a release of HS^- from pyrrhotite accumulations in the crust to the hydrothermal solution would, on meeting Fe^{2+} within the growing mound, reprecipitate as FeS (Hall et al., 1994). The $Fe^{2+}/Fe(OH)_3$ boundary is projected (dashed line to show its approximate position at very low pCO_2. Calculated using GWB (Bethke, 1996). (C) An Eh/pH diagram illustrating the electrochemical energy potentiating the onset of life and the first microbe. (D) The electrochemical energy available to modern iron reducing bacteria (Zachara et al., 2002) compared with that available from the reduction of native sulfur. After Russell and Hall (1997).

ing on the pH, precipitation of mackinawite [(Fe>Ni)$_{1+x}$S] and green rust [Fe$^{II}_4$Fe$^{III}_2$(OH)$_{12}$.CO$_3^{2-}$.2H$_2$O] produced chemical gardens comparable to those found at Tynagh (Fig. 4C) (Russell, 1988). Greigite [Fe$_3$S$_4$] also occurred, as did violarite [Fe$_2$Ni$_4$S$_8$] if Ni^{2+}:Fe^{2+} ratios were high (Russell, 1988; Russell et al., 1998). Mackinawite nanocrysts were probably the main components of the chemical garden and of the membranous walls to individual compartments. We suggest bubbles like these, or at least microcavities within a mackinawite precipitate, acted as the original catalytic culture chambers for early metabolism and embryonic life. Described thus, the hydrothermal mound begins to take on the attributes of a natural catalytic flow reactor and fractionation column and we now examine it in this light (Russell et al., 2003; Stone and Goldstein, 2004; Russell and Martin, 2004).

MACKINAWITE—MEMBRANE MINERAL, PREBIOTIC CATALYST, AND ELECTRON TRANSFER AGENT

Mackinawite provided the inorganic structure and reaction surfaces of the first membrane. At the molecular level mackinawite (Fe$_{1+x}$S) comprises layers of offset Fe$_2$S$_2$ rhombs (Wolthers et al., 2003). At the atomic level it can be seen that the iron layers in mackinawite are semiconducting (Vaughan and Ridout, 1971) (Fig. 7), yet across the layers the van der Waals bonding of the sulfurs confers an insulating capacity to the mineral. The particle sizes of the mackinawite precipitates are bimodal—one is 2 × 2 × 1.5 nm (at pH 8), the other 7 × 7 × 3 nm (at pH 6) (Wolthers et al., 2003).

Aided by the electrochemical gradients obtaining near the mound's surface, one of the effects of the inorganic membrane would have been to split hydrogen into electrons, protons, and transient activated hydrogen atoms. We imagine electrons transferring from one semiconducting nanocrystal to the next through the membrane as they were drawn toward external Fe^{III} and/or HCO$_3^-$ within the membrane (Fig. 8). To maintain charge balance the protons were forced to follow by rotational/translational diffusion of water/hydronium molecules that adhered to the crystallite surfaces (da Silva and Williams, 1991, p. 103). As we shall see, the addition of further protons to the exterior of the FeS compartments would have augmented the natural protonmotive force acting on the membrane.

Divalent metal ions can also invade the sulfur layer (Fig. 7), and nickel and minor cobalt as well as other metals can replace iron in the metallic layer. Also mackinawite is potentially a major temporary sink for many trace metals in anoxic conditions, even calcium (Morse and Arakaki, 1993). Morse and Arakaki (1993) demonstrate that the surface affinity of mackinawite during adsorption of Mn^{2+}, Cr^{3+}, Co^{2+}, Ni^{2+}, Zn^{2+}, Cd^{2+} and Cu^{2+} is in the order of their decreasing solubility as sulfides. When taken in conjunction with their high surface-to-volume ratios, mackinawite nanocrysts have excellent catalytic properties (Cody, 2004; Cody et al., 2004). And as we would expect of a catalyst, mackinawite is a highly reactive mineral prone to oxidation. In an anaerobic environment it can be oxidized in two ways. Oxidized by Fe^{3+}, it converts to greigite by loss of electrons from, and reorganization and even some dissolution of, the iron (Krupp, 1994) (Figs. 7, 9A). So we expect greigite to be a minor phase,

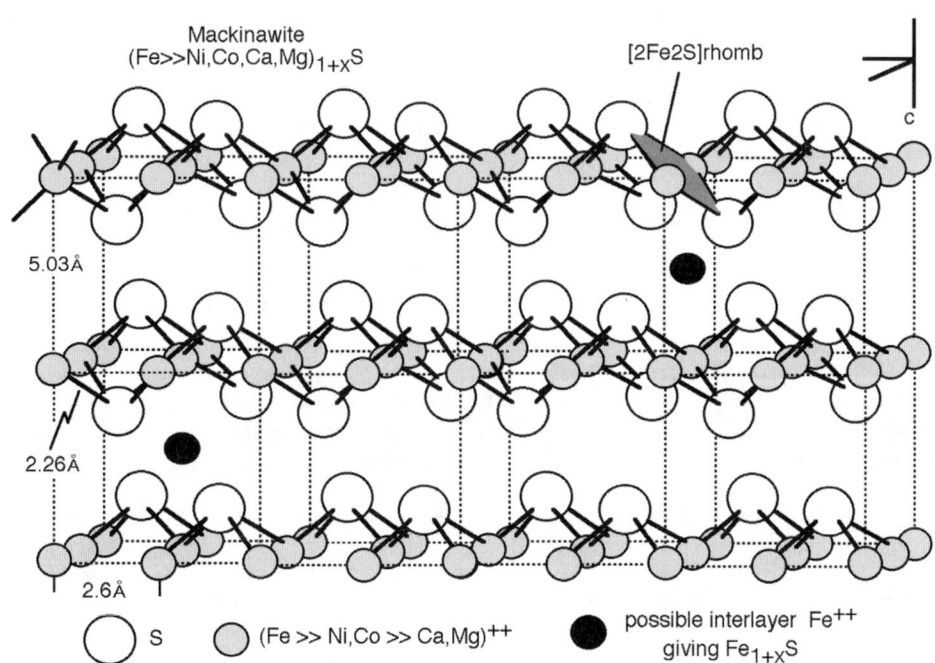

Figure 7. Structure of mackinawite Fe$_{1+x}$S. Mackinawite consists of an assemblage of [2Fe2S] rhombs (Wolthers et al., 2003) arranged in such a way that it acts as a semiconductor in the bc plane and an insulator through the c axis. Comprising the membrane, mackinawite nanocrysts may have acted as the first electron transfer agents from the interior of the protocells to Fe^{III}, the exterior electron acceptor (Fig. 8) (Russell and Hall, 1997; Russell et al., 1998; cf. Ferris et al., 1992).

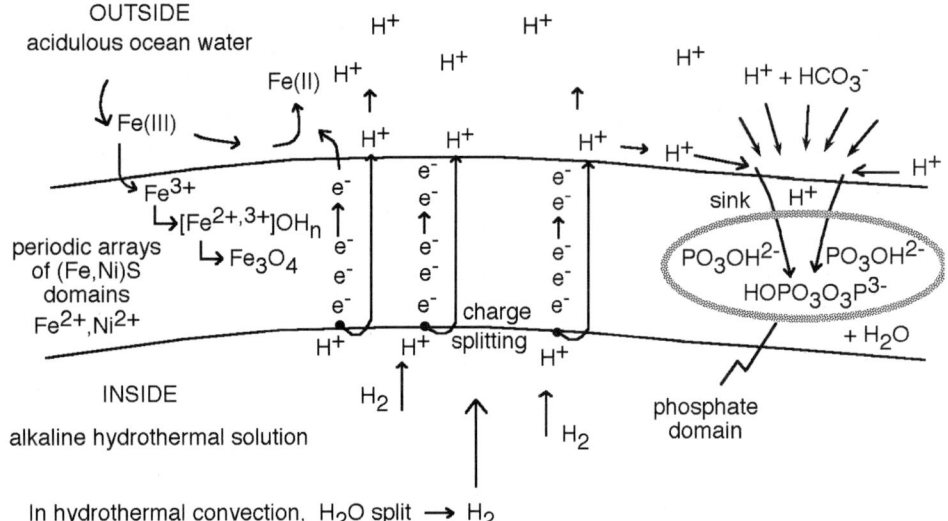

Figure 8. Supposed emergence of chemiosmosis driven by reduction of Fe^{III} on the exterior of the FeS membrane. Electrons are conducted through mackinawite nanocrysts from H_2 oxidized on the interior (cf. Ferris et al., 1992). Protons track electrons through aqueous films by rotational/translational diffusion of H_3O^+/H_2O molecules adhering to the crystallite surfaces (da Silva and Williams, 1991, p. 103) to conserve charge balance. Elsewhere mackinawite may act as an insulator (Fig. 7). The membrane potential is augmented by protons in the acidulous ocean—an ambient protonmotive force.

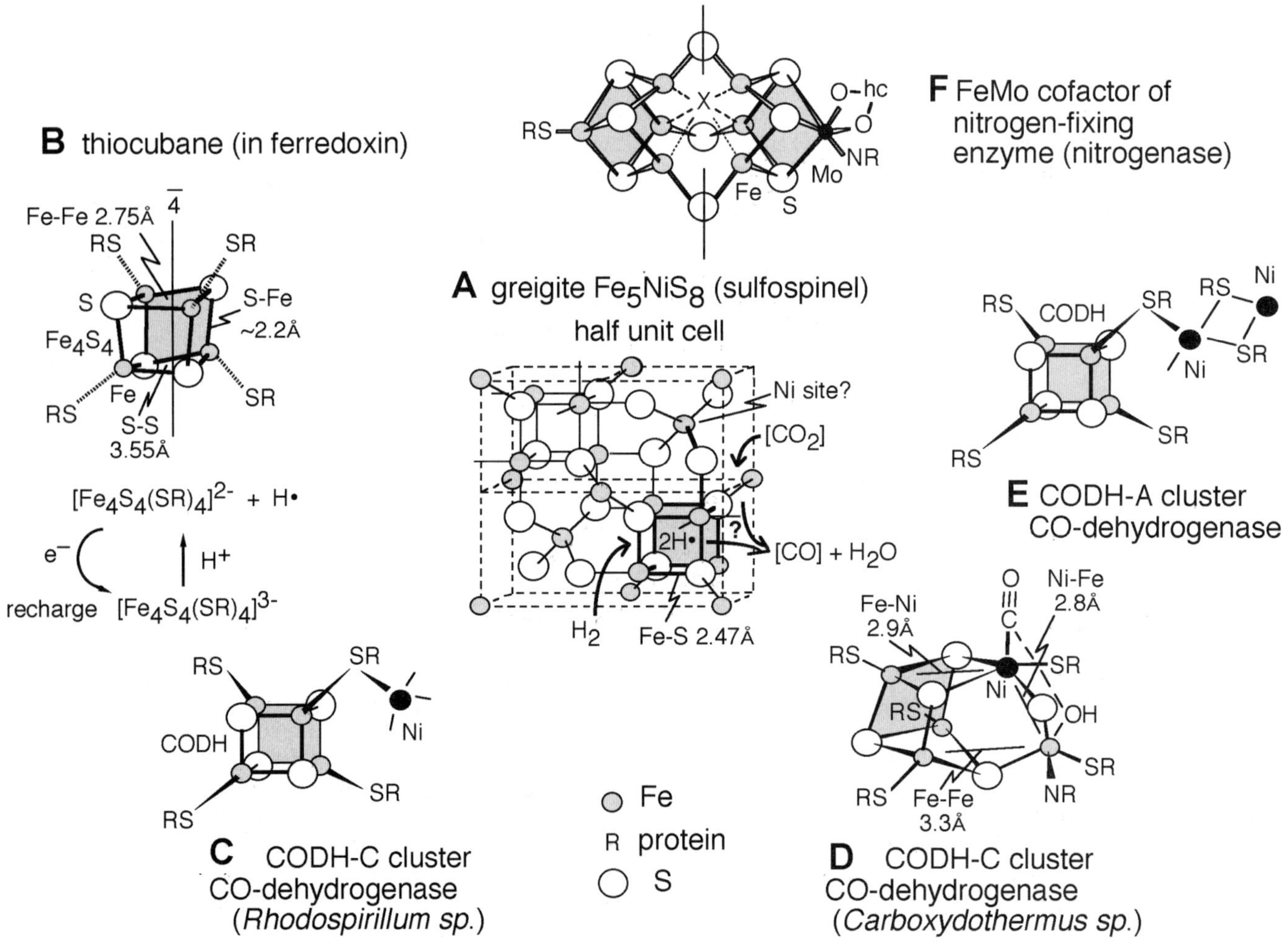

Figure 9. Structural relatedness of (A) greigite Fe_5NiS_8; (B) the thiocubane $[Fe_4S_4]$ unit in protoferredoxins and ferredoxins; (C–E) the $[Fe_4NiS_5]$ open cuboidal complexes in CO-dehydrogenase; and (F) the twinned center to nitrogenase. Affine sulfur sub-lattices, cubic close-packed in A, are distorted in the metalloenzyme centers. The presence or absence of $Fe^{III/3+}$, Mo, Ni, and organic ligands may dictate which of these entities formed in the first cells. RS denotes a link to the protein through the sulfur of cysteine, HN involves a link through a nitrogen ligand of histidine to the same protein, while hc is homocitrate ($^-OOC.CH_2.COH.COO^-.CH_2.COO^-$). Structure (A) from Vaughan and Craig (1978), Krupp, 1994; (B) Hall et al. (1971), Beinert et al. (1997); (C) Drennan et al. (2001); (D) Dobbek et al. (2001); (E) Doukov et al. (2002), Darnault et al. (2003); Svetlitchnyi et al. (2004); (F) Helz et al. (1996), Einsle et al. (2002), Seefeldt et al. (2004).

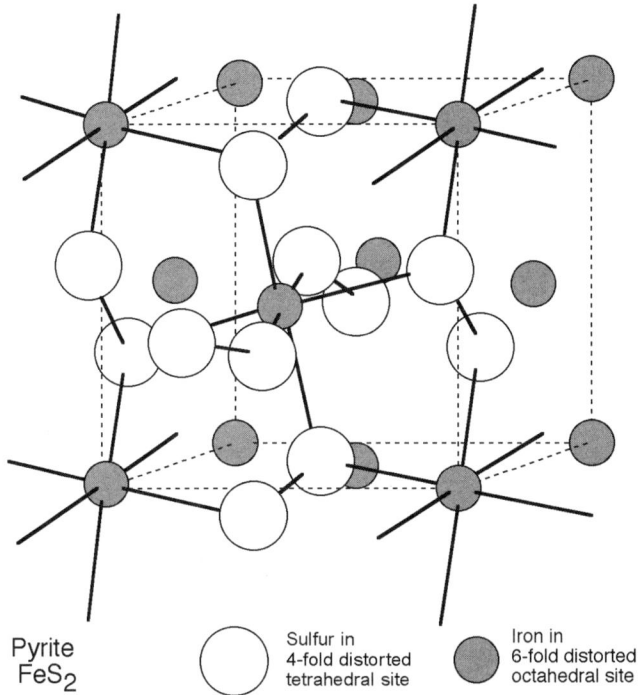

Figure 10. Atomic structure of pyrite drawn to show the ferrous iron ligated to six sulfur-pairs, $Fe^{2+}(S_2^{2-})_6$. The complex structure makes the mineral difficult to nucleate, and difficult to reduce (Finklea et al., 1976) unlike the iron sulfide clusters of metal enzymes (cf. Fig. 9).

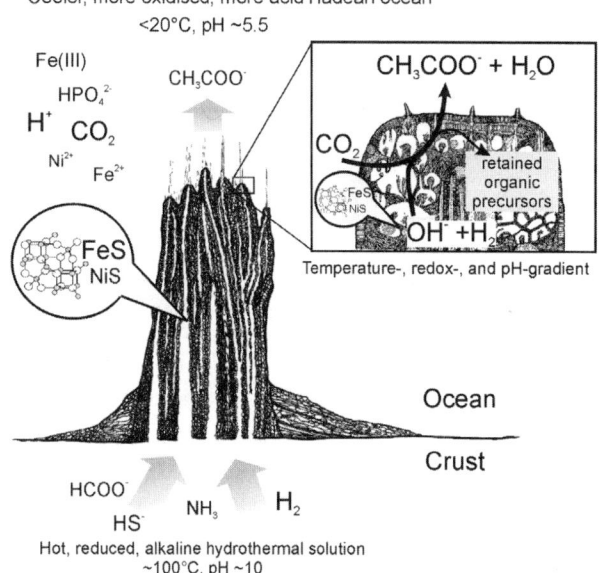

Figure 11. The hydrothermal mound as an acetate generator. The detailed cross-section of the surface illustrates the sites where organic ions are produced, retained, react, and self-organize to emerge as protolife (from Russell and Martin, 2004).

particularly toward the outer margins of the membrane (see next section). Further oxidation converts it to the relatively inert pyrite (Fig. 10). In this latter case it is the sulfur that is oxidized (to S_2^{2-}). Rickard et al. (2001) have shown that this stage of oxidation is prevented by formaldehyde, significant because, as we shall see below, greigite has a structural affinity to ancient metalloenzymes (Fig. 9).

HYDROTHERMAL MOUND AS REACTOR AND ACETATE WASTE GENERATOR

Acting as a natural flow reactor and fractionation column, the hydrothermal mound stood vertically and was composed essentially of brucite, clay, minor sulfides, green rust, and ephemeral carbonate (Fig. 11). Hydrothermal fluid entered through the permeable and porous base. The bubbles and pores stemmed the flow of "electron-rich" molecules such as H_2, NH_3, $HCOO^-$, CN^-, CH_3S^-, and HS^-. The H_2, HS^-, $HCOO^-$ and NH_3 were the most concentrated at 10 or so mmol. This hydrothermal solution mixed with about 100 mmol of HCO_3^- and several millimoles of Fe^{2+} and $HP_2O_7^{3-}/HPO_3^{2-}$ in ocean water that was entrained through the sides of the mound. The HS^- reacted with Fe^{2+} and Fe^{II} (i.e., fixed ferrous iron) to precipitate nickeliferous mackinawite:

$$Fe(Ni)^{2+} + HS^- \rightarrow Fe(Ni)S + H^+ \quad (6)$$

Fresh sulfide nanoparticles acted as sites of adsorption, absorption, and catalysis. Ferric iron particles attracted to the outside of the membrane may have oxidized some of the mackinawite to greigite and even pyrite. Cyanide would have fractionated from the formaldehyde, the former adsorbed on the pyrite, the latter upon the mackinawite (Leja, 1982; Rickard et al., 2001). In places membranes and barriers composed of mackinawite and minor greigite acted as solid phases for further chemical interactions between the reactive solutes (Russell et al., 1994, 2003; Schoofs et al., 2000). What might these reactions have been?

The reaction expected to have released the most energy, i.e., with the greatest thermodynamic drive, was the production of methane and water from the carbon dioxide and the hydrogen (Amend and Shock, 2001):

$$CO_2 + 4H_2 \rightarrow CH_4 + 2H_2O \quad (7)$$

However, there is a major kinetic barrier that faces this reaction, which takes place spontaneously only above 500 °C (Shock et al., 1998). The reaction to produce acetate dissipates less energy, but the kinetic barrier is also lower, though not low enough for H_2 to react spontaneously with the CO_2 (Shock et al., 1998; Amend and Shock, 2001). Shock et al. (1998) calculate that carbon in metastable equilibrium states obtained by mixing hydrothermal fluids with anoxygenic seawater below 110 °C would be mainly as acetate, with subsidiary propanate and dodecanoate (Fig. 12). Therefore, we expect the hydrothermal mound and its compartments to have catalyzed the production of acetate (CH_3COO^-) and water from CO_2 (as bicarbonate, Fig. 13) and H_2 below 50 °C, degrading energy in the process (Russell and Martin, 2004):

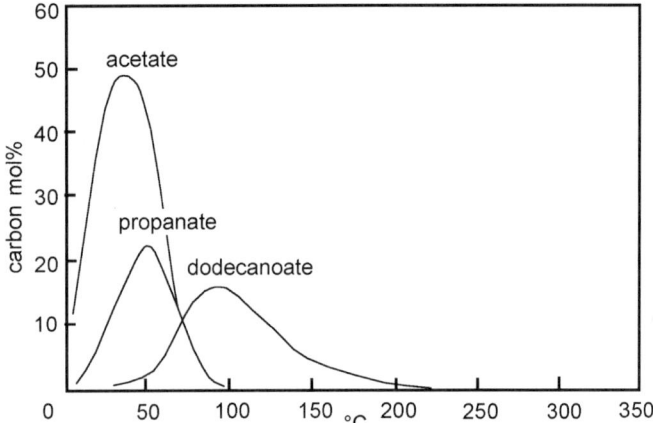

Figure 12. Mole percent distribution of the three most "metastable" carboxylic acids at low to moderate temperatures theoretically obtained through the mixing of hydrothermal fluids with anoxygenic seawater. They are acetate ($CH_3.COO^-$), propanate ($CH_3.CH_2.COO^-$), and dodecanoate ($CH_3.(CH_2)_{10}.COO^-$). (Minor species have been neglected in this redrawing of part of Figure 5.7 in Shock et al., 1998.)

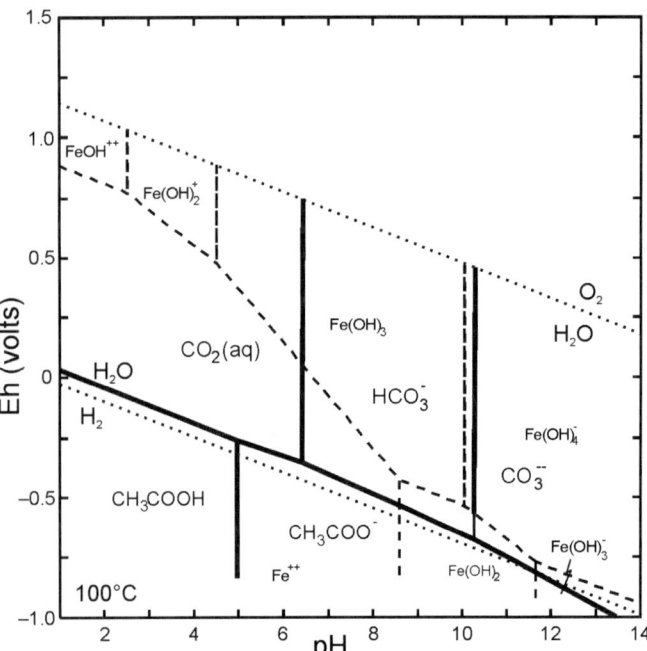

Figure 13. Pourbaix diagram using Geochemist's Workbench ACT2 (Bethke, 1996) showing the acetate and carbonate fields (thick lines) with respect to dissolved and solid iron phases (dashed lines). Conditions: 100 °C, p = 1.013 bars, $CO_2(g)$ log f-3, Fe^{2+} log a-20, fields extended below stability field of water (dotted lines).

$$4HCO_3^- + 8H_2 + 2H^+ \rightarrow 2CH_3.COO^- + 8H_2O \quad (8)$$

There is some experimental evidence to support the conclusions of Shock et al. (1998). Reacting 100 mmol CH_3SH (methane thiol) and 4.5 mmol CO at 100 °C at normal pressure, Huber and Wächtershäuser (1997) produced micromolar amounts of methyl thioacetate (CH_3COSCH_3) in the presence of an FeS/NiS slurry optimally at pH 6.4:

$$CO + 2CH_3SH \rightarrow CH_3.COSCH_3 + H_2S \quad (9)$$

Methane thiol has been synthesized from CO_2 at 100 °C in the presence of FeS and H_2S (Heinen and Lauwers, 1996, 1997). At the same time, the FeS was oxidized to pyrite, as might be expected from the "pyrite-pulled" model of Wächtershäuser, 1988). The yield with respect to H_2S was ~0.25%. In theory methane thiol activities would rise a thousandfold when generated from H_2 and CO (rather than CO_2) in the crust or the hydrothermal mound (Schulte and Rogers, 2004). These conditions would have been met within the mackinawitic membrane that separated ocean water at about pH 5.5 from the more alkaline hydrothermal fluids that contained CO and organic sulfides, i.e., the thiols (Russell and Hall, 1997). Hydrolysis would have liberated a proportion of the acetate while the methane thiol catalyst was returned to the solution:

$$CH_3.COSCH_3 + OH^- \rightarrow CH_3.COO^- + CH_3SH \quad (10)$$

The acetate reaction, augmented by the proton gradient operating across the membrane, may also have been coupled to pyrophosphate generation. (Currently this process is driven by a sodium gradient.) The surviving methyl acetate became involved in further biosynthetic reactions such as the generation of amino acids and lipids. Can we generalize the first evolutionary step?

The notional hydrothermal reactor works to the rule of the second law of thermodynamics. For the most part, we might expect that geochemicals far from equilibrium would react irreversibly, degrading energy in the process (Shock et al., 1998). Minerals such as mackinawite form exothermically on reaction between hydrothermal HS^- and the Fe^{2+} in the entrained ocean water (equation 6), so we can consider the onset of life as similarly exothermic overall and suggest a simplified equation as a demonstration:

$$\{407H_2 + 10NH_3 + HS^-\}_{hydrothermal} + \{210CO_2 + H_2PO_4^-$$
$$+ Fe, Ni, Co, Zn, Mo^{2+}\}_{ocean}$$
$$\rightarrow \{C_{70}H_{129}O_{65}N_{10}P(Fe,Ni,Co,Zn,Mo)S\}_{proto-life}$$
$$+ \{70H_3C.COOH + 219H_2O\}_{waste} \quad (11)$$

The ratio of waste acetate plus water to "proto-life" in this conceptual reaction is high. Most of the output from the reacting monomers elutes to the ocean and entropy thus increases, but the iron sulfide botryoids, bubbles, and pores could act, albeit inefficiently at first, as tiny electrically powered compartments or "tureens" in which a "warm organic soup" could have been synthesized, constrained and concentrated to a critical mass that encouraged further interactions (Russell et al., 1988, 1994; Braun and Libchaber, 2004). Thus, within the compartments processes were reversible, and though entropy was exported, it decreased within the compartments themselves (Prigogine, 1978).

This approach does correspond to what is understood of early life from microbiology: many microbes, including those near the base of the evolutionary tree, can gain energy by generating acetate using the enzyme carbon monoxide dehydrogenase with acetyl-coenzyme-A (which is also an organic sulfide or thiol, co-A.SH), through the acetyl-coenzyme-A pathway, that is in part homologous with the Huber-Wächtershäuser reaction (equation 9) (Schink, 1997; Peretó et al., 1999; Amend and Shock, 2001; Russell and Martin, 2004):

$$2CO_2 + 8[H] + co\text{-}A.SH \rightarrow co\text{-}A.S.OC.CH_3 + 3H_2O \quad (12)$$

A portion of the acetate and the energy released in these exergonic reactions would have gone to waste. But waste, the generation of entropy, is life's raison d'être. We might think of the mound as optimizing the generation of acetate over time while side reactions, including many involving the activated thioacetate ($CH_3.COS^-$), synthesized the more complex molecules that interacted to produce life. This non-vivocentric view is now examined in the context of a notional reactor that produces the acetate and water (Figs. 11–13).

Bubbles comprising the iron sulfide membrane could have been hydraulically inflated over warm seepages, where they encapsulated the reduced alkaline hydrothermal solution (Fig. 4D) (Russell et al., 1989, 1993). As the bubbles became distended they weakened, failed, and daughter bubbles were generated above the punctures (Fig. 4B). Thus the redox and pH front remained at the growing surface of the mound (Figs. 4D, 5, 11). Bubbles farthest from the feeder veins would have been disadvantaged unless the structure of their membranes particularly disposed them to supporting an osmotic pressure. This osmotic pressure would have been induced by the generation of abiotic charged organic molecules. Contiguous compartments (Fig. 5), generated by budding of the iron monosulfide membrane (Fig. 4B), would have contained fluid mixes at slightly different Eh and pH conditions and therefore would have harbored different reactants and products, as energy cascaded from one chemical and electrochemical level to another. These possibilities have been elegantly considered for other types of inorganic membrane by Cairns-Smith (1982, p. 327 and 351–356). As discussed below, organic synthesis would have been catalyzed by the iron (nickel) monosulfide, which, unlike fine metal and oxide/hydroxide catalysts, cannot be poisoned by sulfidation.

Although acetate and water were the main fluid products, the hydrothermal NH_3 and minor CN^- would have reacted with bicarbonate on mineral surfaces and produced amino acids, especially glycine ($^+H_3N.CH_2.COO^-$) (Hennet et al., 1992):

$$3CO_2 + 5H_2 + NH_3 + HCN \rightarrow$$
$$2(^+H_3N.CH_2.COO^-) + 2H_2O \quad (13)$$

The amino acids were mostly adsorbed within compartments in the mound. With the addition of formaldehyde (HCHO), minor concentrations of RNA would also have been produced from these components (Ferris and Hagan, 1984).

Apart from the fluid wastes, mainly water and acetate, there is also the solid waste product to consider.

PYRITE—A SOLID WASTE PRODUCT

When mackinawite or greigite is oxidized to pyrite, for example during the generation of methane thiol, the rhombic [Fe_2S_2] building block is lost; instead, the ferrous iron is ligated to six partially oxidized sulfur-pairs (S_2^{2-}) (Fig. 10). The resulting $Fe(S_2^{2-})_6$ is a complex structure which makes the mineral difficult to nucleate and very difficult to reduce back to FeS (Finklea et al., 1976). Wächtershäuser (1988) has argued for the formation of pyrite on oxidation of FeS by hydrogen sulfide, with the generation of hydrogen, as the primeval energy source for the origin of a (surface) metabolist. Although we consider hydrogen to be delivered to the mound in the alkaline solution, the Wächtershäuser reaction works in acidic and neutral conditions (Taylor et al., 1979; Drobner et al., 1990; Rickard, 1997). Moreover, the hydrophobic pyrite precipitate recorded at the gas-water boundary in the experiments of Heinen and Lauwers (1996) demonstrates that this irreversible (i.e., non-catalytic) oxidation can drive reductions of CO_2 as suggested by Wächtershäuser (1988). Although this reaction produces less energy in alkaline conditions, were pyrite to have been formed it could have taken no further part in protometabolism and must, therefore, be considered as a waste product. Nevertheless, its hydrophobic surfaces would have rendered it a surficial trap for organic molecules and cyanide gleaned from the hydrothermal fluid (Leja, 1982; Russell et al., 1988).

ENERGY FOR POLYMERIZATION

Although we note empirically that the thermal gradient responsible for convection lies between 115° and <20 °C, the electrochemical potential to drive biosynthesis and polymerization could be subscribed by the H^+/H_2 couple (effectively the hydrogen electrode) representing the hydrothermal emanations, as against the Fe^{3+}/Fe^{2+} couple representing the photolytic ferric iron (representing an initially dispersed positive electrode). Theoretically the hydrothermal hydrogen could have reduced ferric oxyhydroxide (signified as Fe^{III}) to ferrous ions, and produced protons in a reversal of equation 3 in which the redox potential was influenced by light energy:

$$H_2 + 2Fe^{III} \rightarrow 2H^+ + 2Fe^{2+} \quad (14)$$

Pourbaix diagrams idealize the thermodynamic potentials on proton activity (pH) and electron activity (Eh) coordinates (Fig. 6). Oxidation of H_2 on one side of the membrane by reduction, through electron transfer, of Fe^{III} on the other side of the membrane, theoretically generates a potential of 770 mV (Fig. 6C)

(Russell et al., 2003). Other species such a Fe(OH)$_3$ have lower redox potentials, for example Fe(OH)$_3$/Fe(OH)$_2$ = 270 mV (Garrels and Christ, 1965, p. 183). Even this potential is more than the 250 mV (−11.5 kcal per mole) required to drive polymerization, especially when the pH gradient is taken into account (Thauer et al., 1977; and see Kell, 1988, Van Walraven et al., 1997). This chemiosmotic proton potential is augmented by proton potential of the acidulous ocean (pH ~5.5 at ~20 °C) acting across the inorganic membrane on the alkaline hydrothermal solution (pH ~10.5 at ~100 °C), a potential that results in a further 300 mV (Fig. 6C). *Geobacter metallireducens*, as well as many other bacteria in the lowest branches of the evolutionary tree, can reduce Fe^{III} using much the same electrochemical potential (Fig. 6D, 14) (Liu et al., 1997; Vargas et al., 1998; Zachara et al., 2002).

In sum, the overall "protonic" potential approximates 1 V, enough geochemical energy to have engendered metabolism across an inorganic membrane on early Earth (Russell et al., 2003). Using a "Beutner rig" (Beutner, 1913, Fig. 1), we have found that a spontaneously generated FeS membrane, formed on reaction between 10 mmol solutions of FeSO$_4$ and Na$_2$S (at pH 4.12 and 10.0 respectively), creates a tension of ~600 mV, which is held for several hours (Fig. 15). Filtness et al. (2003), using more fastidious conditions obtain a total tension closer to 700 mV, differences maintained for more than 4 h. That the pH-dependent boundary between the mono- and di-phosphate fields intersects the (primarily) Eh-dependent iron sulfide fields demonstrates how "proticity" (a proton current) could have driven the condensation of inorganic phosphate (Pi) to pyrophosphate (PPi) (Fig. 16). So in theory the redox reaction outlined in equation 3 could have been simply coupled through the membrane to

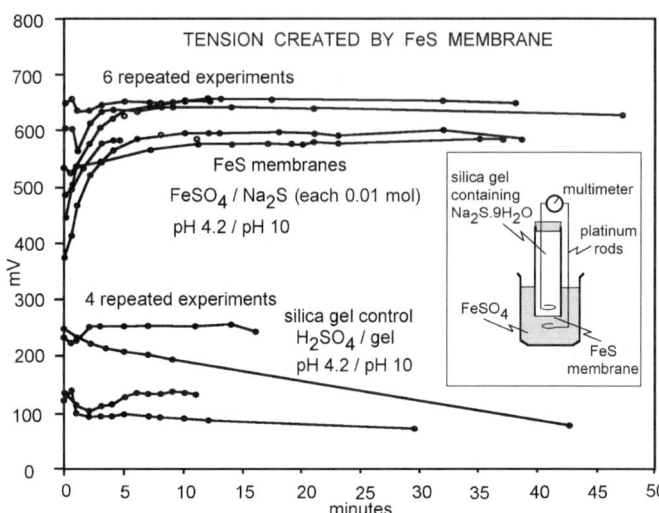

Figure 15. Plot of mV developed across an FeS membrane with time. Preliminary results with conditions as defined in the inset. Schematic illustration of the notional photoelectrochemical cell assumed to have obtained on the Hadean Earth and ocean, and of a preliminary experimental emf/pmf.

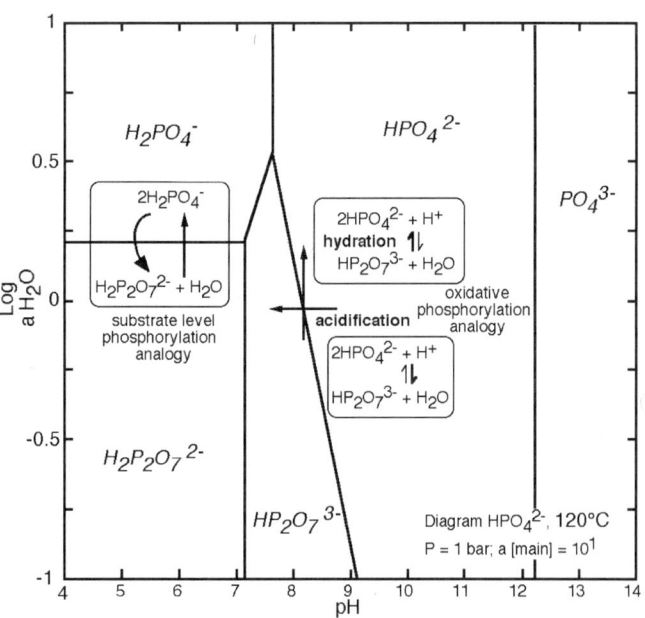

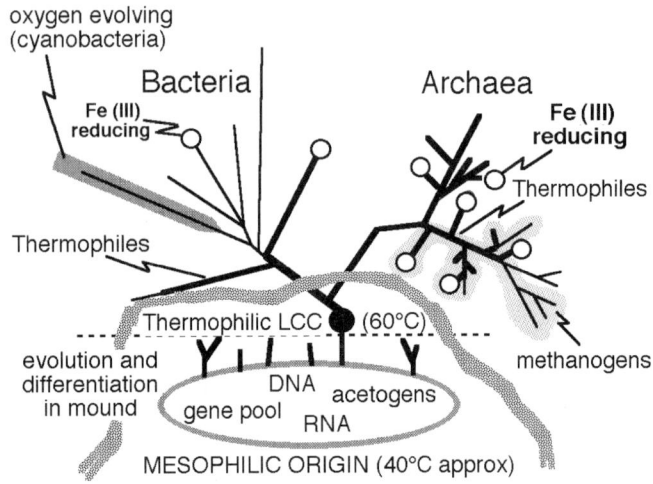

Figure 14. Evolutionary tree (after Woese et al., 1990; Stetter, 1996; Martin and Russell, 2003). The last common community (the LCC) occupied the mound in which life had emerged. Many prokaryotes in the lowest branches of the tree can use Fe^{III} as an electron acceptor (Liu et al., 1997; Vargas et al., 1998; Kashefi et al., 2002). Note that methanogens are found only in the Archaeal domain and that oxygenic photosynthesis is a property only of the cyanobacteria.

Figure 16. GWB diagram (Bethke, 1996) illustrates how *high* energy phosphoanhydride, stable at lower pH and water activity and higher T, forms from *low* energy monophosphate (HOP$_2$O$_6^{3-}$) and can drive dehydration polymerizations on its hydration (cf. the ATP^{4-}/ADP^{3-}/or AMP^{2-} energy cycle of life). In oxidative phosphorylation the dehydrating power of ATP is renewed by pmf (acidification), whereas in substrate level phosphorylation ATP is renewed by removal of H$^+$ and OH$^-$ by a NAD-associated enzyme.

this dehydration or condensation of monophosphate (Russell and Hall, 2002):

$$H^+ + 2HPO_4^{2-} \rightarrow HP_2O_7^{3-} + H_2O \quad (15)$$

In biochemical energetics the process is known as "oxidative phosphorylation," a reference to the fact that protons are initially driven to the outside of the membrane to maintain their balance with electrons flowing outward to an electron acceptor before returning to recharge the phosphate (Mitchell, 1967) (Fig. 8). The protonmotive force is the power behind metabolism and is therefore indispensable to life, so the fact that this ambient force is a feature of the redox and alkaline-to-acid interface separated by a semiconducting inorganic membrane is a compelling aspect of the hypothesis. Nevertheless, the "ATPase" responsible for the conversion in today's organisms is a complex rotating turbo-motor (Elston et al., 1998). We assume, after Baltscheffsky et al. (1999), that the original process used a static prototype H$^+$-PPase, basically an enriched domain of phosphate upon mineral surfaces within the membrane (Fig. 8).

What may be surprising is the assumption that dehydrating reactions can take place in hydrothermal conditions. What to consider is the fact that anions, generated and concentrated in the membrane, would have competed successfully for a place on a growing mineral surface against the minor negative charges on the oxygens of polar water molecules. Given the centrality of acetate, the first organic phosphate may have been acetyl phosphate (CH$_3$.CO.PO$_4^{2-}$), perhaps produced by phosphorylation of the acetyl thioester (from equation 9) (de Duve, 1991; Russell et al., 1994):

$$CH_3COSCH_3 + HPO_4^{2-} \rightarrow CH_3SH + CH_3COPO_4^{2-} \quad (16)$$

Reacting acetyl phosphate with inorganic phosphate (HPO$_4^{2-}$) in the presence of Fe^{II} minerals, de Zwart et al. (2004) have generated pyrophosphate with a yield of 25% at ~40 °C. FeS was found to strongly retard hydrolysis of, and thus preserve, the pyrophosphate:

$$HPO_4^{2-} + CH_3.CO.PO_4^{2-} \rightarrow HP_2O_7^{3-} + CH_3.COO^- \quad (17)$$

The resulting pyrophosphate (PPi) would have had the energy to polymerize amino acids (Romero et al., 1991; Baltscheffsky et al., 1999):

$$^+H_3N.CH_2.COO^- + HP_2O_7^{3-} + {}^+H_3N.CH_2.COO^- \rightarrow$$
$$^+H_3N.CH_2.CO.NH.CH_2.COO^- + 2HPO_4^{2-} + H^+ \quad (18)$$

The monophosphate might then have been repolymerized with protons as shown in equation 15.

Thus, before the advent of the ribosome, random amino acid polymerization could have been driven by pyrophosphate bonded through to sulfide on the surface of mackinawite in a -Fe-S-O-P-motif as suggested by EXAFS (Pattrick, 2001, personal commun.). Such a conformation is congruent with the relationship Wolthers (2003, chapter 4, Fig. 5) finds between mackinawite and arsenate. Whatever the mechanism, these first amino acid polymers or peptides probably consisted mainly of glycine with occasional alanine, aspartate, serine, and valine (Hennet et al., 1992). The last four amino acids have particular stereochemistries; i.e., they can be right-handed (dextro) or left-handed (levo). Polymerization of a racemic mix of dextro and levo amino acids would result in a heterochiral peptide; i.e., it would have had a mixed chirality. A lack of chirality is thought to be a major problem for theorists of the emergence of life (Cairns-Smith, 1982). We show later that a lack of chirality in the first peptides, far from being a problem, was a positive advantage with regard to self-organization of the constituents of the first compartments, though we must emphasize that this unregulated polymerization of amino acids, while a necessary step, was an evolutionary dead end. However, before we examine the unregulated peptides we need to take a careful look at the structure of another key metal sulfide mineral precipitated in the reactor, viz., greigite.

GREIGITE—PRE-ENZYME AND ELECTRON DOCKING SITE

The structure of greigite tends to an inverse spinel, written notionally as SFe^{3+}S[Fe$_4^{2.75+}$S$_4$]SNi^{2+}S (Fig. 9A). In fact the electrons are delocalized, so there is a certain amount of valence resonance. If further nickel is introduced, the inverse spinel verges toward the true thiospinel violarite (as SNi^{2+}S[Fe$_2^{3+}$Ni$_2^{3+}$S$_4$]SNi^{2+}S). We assume that some of the hydrogen dissolved under high pressure in the hydrothermal fluid was adsorbed on, or absorbed within, the [Fe$_4$S$_4$] cube. Once there, hydrogen was split to an electron (reducing the iron) and a proton (protonating one of the four sulfides), leaving a hydrogen radical. This nascent hydrogen [H•] was highly active and might have attacked a bicarbonate ion, the ultimate electron acceptor, at a tetragonally coordinated nickel site (Fig. 9A). Because of the likely contribution of greigite (NiFe$_5$S$_8$) to inorganic membranes developed at alkaline submarine seepages (Russell, 1988), and its similarity to the structure of the active centers to the most ancient proteins, the ferredoxins, we have suggested that molecular constructs of the mineral were, at a later stage of emergence, incorporated by peptides as the first electron transfer agents, redox enzymes and synthases (Russell and Hall, 1997, 2002; Milner-White and Russell, 2005). They are the ready-made, modular mineral clusters (Beinert et al., 1997) that, when combined with other metals and/or organic structures, constitute components of what Baymann et al. (2003) have termed "the redox protein construction kit." The kit is partly based on the (inverse) spinel or greigite structure, which contains an [Fe$_4$S$_4$]$^{-2+}$ cage or cubane in which the electrons are delocalized and in which the iron atoms have a nominal positive charge of 2.5 and have the tendency to switch valence.

FROM CATALYSTS TO ENZYMES

We have seen that iron sulfides and iron nickel sulfides have the catalytic propensity to produce some of the simple molecular modules of life from inorganic constituents. Examples of organic products are methane thiol, the simple amino acids, and acetate. As acetate is the likely first major product of the hydrothermal reactor or mound, it is instructive to consider how modern homoacetogens (i.e., acetogenic bacteria that use only inorganic nutrients and fuel) synthesize acetate (Müller, 2003). The overall reaction of the acetyl-coenzyme-A pathway is shown in equation 12. The key enzyme of the pathway is carbon monoxide dehydrogenase (CODH) (Peretó et al., 1999; Dobbek et al., 2001). As a single unit or homodimer, the enzyme employs five $[Fe_4S_4]$ clusters including a unique $[Fe_4NiS_5]$ cluster where CO_2 is reduced to CO (Figs. 9B, 9D). A more complex form of the enzyme possesses additional FeS clusters and reduces not only CO_2 to CO, but also condenses Ni-bound methyl and CO, yielding a Ni-bound acetyl moiety that is transferred to co-ASH in the acetyl-coenzyme-A synthase reaction (Fig. 9E) (Lindahl, 2002). The formula of the $[Fe_4NiS_5]$ C-cluster in the enzyme is also comparable to that of greigite $[Fe_5NiS_8]$ (Fig. 9C–E). Indeed, the $[Fe_4S_4]$ cube of greigite is also congruent, or nearly so, with the cubanes in the most ancient proteins, the ferredoxins (Eck and Dayhoff, 1966) (Fig. 9B). That there is some play in the placing of the Fe and Ni ions and of their charge in greigite is also echoed by the way Ni and Fe are variously but characteristically sited in the dehydrogenases (redox enzymes) and synthases (biosynthesis enzymes) (Fig. 9C–E). We have considered the component building blocks of the greigite structure to be the likely "ready-made" molecules co-opted by early life before they could be interred in sulfides or oxides (Russell and Hall, 1997; Russell and Martin, 2004; and see Bonomi et al., 1985).

THE ORGANIC TAKEOVER

Recalling our inorganic predilection, we use the way metals are coordinated in minerals and their natural cluster precursors as a heuristic device to inform us as to the likely, or at least possible, coordination chemistries adopted by early biology. Metals now constitute the active centers of ~50% of all protein types (Jernigan et al., 1994). This percentage will be optimal for the present-day geochemical environment where metals are harder to come by. Mono-, di- and tri-phosphates are also vital to coenzymes. At the onset of life amino acid polymers—the peptides—would have sequestered and protected the ubiquitous active inorganic centers in what was the first stage in an organic takeover. The peptides are the mediators of the takeover.

Amino acids are zwitterions, amphoteric molecules that have a negative and a positive charge ("zwitter" is German for hermaphrodite) as well as an organic side chain R—except for glycine ($^+H_3N.CH_2.COO^-$), which only caries a hydrogen atom. In normal conditions amino acids have a negatively charged carboxyl group (-COO^-) and a positively charged amino group (-NH_3^+) e.g., $^+H_3N.HCR.COO^-$. Before the advent of RNA, amino acids could have been polymerized by the loss of the constituents of water, OH^- from the carboxyl of one amino acid and of H^+ from the next. The reaction may have been driven by pyrophosphate hydration where water activity was low (Baltscheffsky et al., 1999) (equation 18). Alternatively, where sulfide concentrations were high, the more reactive thiocarboxyl group (-CSO^-) might have been attacked by the amino group with the loss of a thiol (RSH) (Wächtershäuser, 1992). There are also experimentally inspired suggestions for peptide formation (Ferris et al., 1996; Huber and Wächtershäuser, 1998; Huber et al., 2003; Leman et al., 2004), but a mechanism for prebiotic polymerization has yet to be agreed.

PEPTIDE NESTS FOR SULFIDES AND PHOSPHATE

Potentially the two structures involved in energy transfer are the metal sulfide clusters and the phosphates. The early sulfide clusters are likely to have been sequestered by thiolate (e.g., $[Fe_4S_4][CH_3S]_4^{2-}$) in aqueous solution (Bonomi et al., 1985). Thus both the phosphates (e.g., $HP_2O_7^{3-}$) and the thiolated iron(nickel) sulfide clusters were anionic. As the nitrogens of the amino groups carry a δ^+ charge even when part of a peptide, the inorganic anions would have been drawn to these δ^+ charges on the peptide chain (Fig. 17). At the same time the chain would have bowed to satisfy the negatively charged clusters. Encased in such

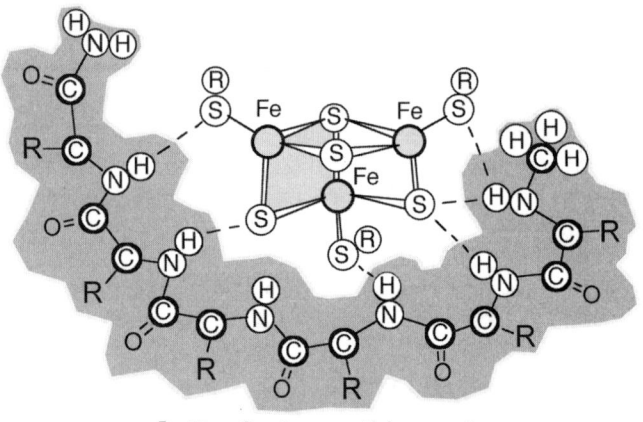

A Fe_3S_4 in peptide nest

B Phosphate in peptide nest

Figure 17. Ball and stick sketches of early nests of short peptides constituting alternating glycines with other abiotic racemic amino acids binding (A) an $Fe_3S_4(RS^{3-})_3$ to produce a protoferredoxin, and (B) an inorganic phosphate in a primitive P-loop. R represents side chains. After Milner-White and Russell (2005).

"peptide nests" the active centers were partially protected from dissolution, or nucleation and crystallization, and would have remained active (Milner-White and Russell, 2005). The cosseting of active centers in this way would have been optimal when the peptide was composed either of glycine, the one achiral amino acid, or of amino acids with random stereochemistries. In other words, for efficient self-organization of protoenzymes (i.e., the ferredoxins, dehydrogenases, and synthases) and the proto-coenzymes (e.g., PPi) it is better for the peptide to have been heterochiral and have consisted of racemic amino acid residues. Chiral peptides would tend to have formed helices or sheets rather than nests. The nest configuration was a likely early outcome because the easiest amino acid to make abiotically is glycine and because the other abiotic amino acids would have been racemic (Hennet et al., 1992; Huber and Wächtershäuser, 2003).

We have seen how electrons, carried first as constituents of the hydrogen molecule in hydrothermal solution, seek out electron acceptors. These are ferric iron and the more recalcitrant carbon dioxide or bicarbonate (Figs. 6D, 13, 18). Electron transfer agents and catalysts are required for the redox reactions. Later in evolution sulfate, nitrogen, and nitrate also came to be used as electron acceptors (Fig. 18). Depending on what oxidation or reduction takes place, electrons had to be transported singly, two at a time, three at a time, or even four at a time. Simple and multiple twinning to dimers and multidimers of the [4Fe-4S]-bearing ferredoxin, results of gene duplications, facilitates multiple electron flow (Adman et al., 1973; Steigerwald et al., 1990) (Fig. 19). But these complex proteins could only have formed when peptides were relatively long and had particular and regulated amino acid sequences to facilitate their folding and the sequestration of the inorganic active clusters.

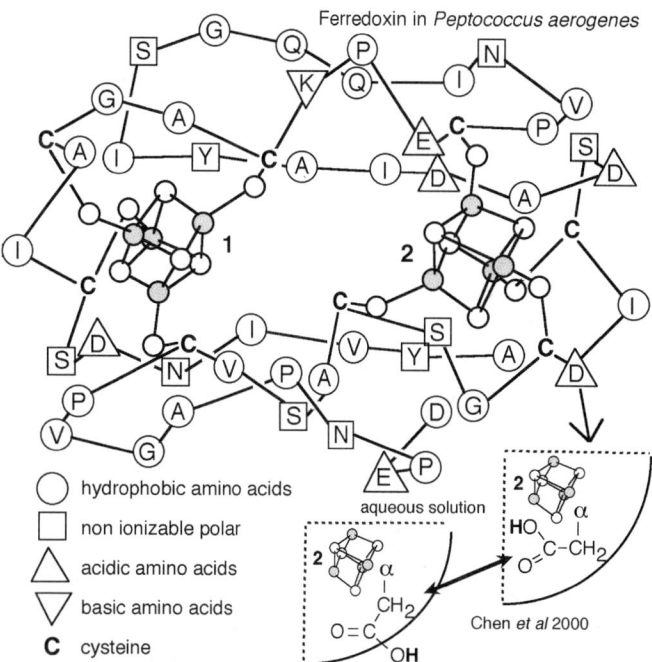

Figure 19. Ferredoxin in *Peptococcus aerogenes* as a comparison to a nested [4Fe-4S] cubane shown in Figure 17A. This is a typical ferredoxin of the type that probably formed by the twinning of a single cubane ligated through cysteines. This ferredoxin has a total of 54 simple amino acids (redrawn form Adman et al., 1973). Inset shown is the mechanism of the transfer of a solvent proton to the buried redox center via the side chain of aspartate (Chen et al., 2000). (A—alanine, Y—tyrosine, V—valine, I—isoleucine, N—asparagine, D—aspartate, S—serine, C—cysteine, G—glycine, K—lysine, P—proline, E—glutamate, Q—glutamine.)

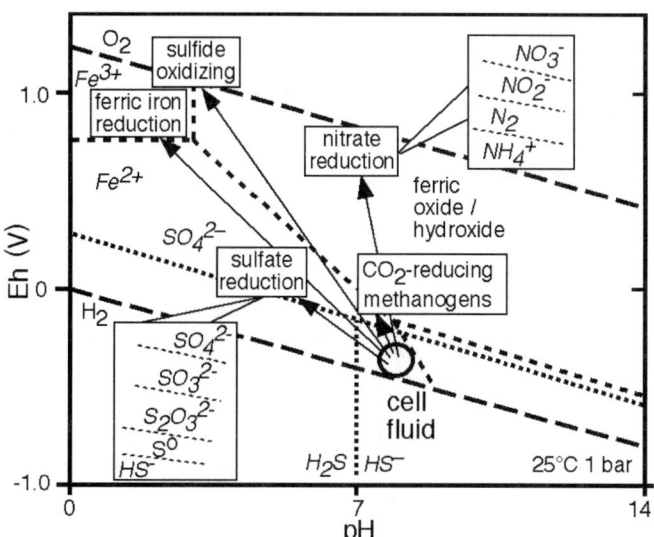

Figure 18. Redox potential/pH vectors between the cellular fluid of prokaryotes and that of the living environment represented by electron acceptors. After Russell and Hall (1997).

COENZYMES WITH ORGANIC RING COMPONENTS

One of the first requirements for organic molecules during early evolution was to cooperate with the redox tasks of the ready-made metal-bearing clusters and their associated peptides. Carbon/nitrogen rings containing three conjugated double bonds and unpaired and delocalized electrons are the coenzymes that carry out some of these processes (Pullman, 1972). Furthermore they have the advantage of being bonded to aliphatic compounds involved in structuring the cell. All the nucleic acid bases and the three aromatic amino acids have such rings as, or on, their side chains (Fig. 20). In terms of the organic takeover, the ferredoxins originally involved in the oxidation of glucose were largely replaced by nicotinamide adenine dinucleotide (NAD) (Daniel and Danson, 1995). And, when iron is in short supply, flavodoxins can substitute as electron carriers for ferredoxins.

Although the heterocyclic rings do not grow to mineral proportions, some of them polymerize to macrocyclic compounds. These compounds can sequester a variety of single metal ions coordinated through four nitrogen atoms (Eschenmoser, 1988). Different metal ions can bestow remarkably different properties

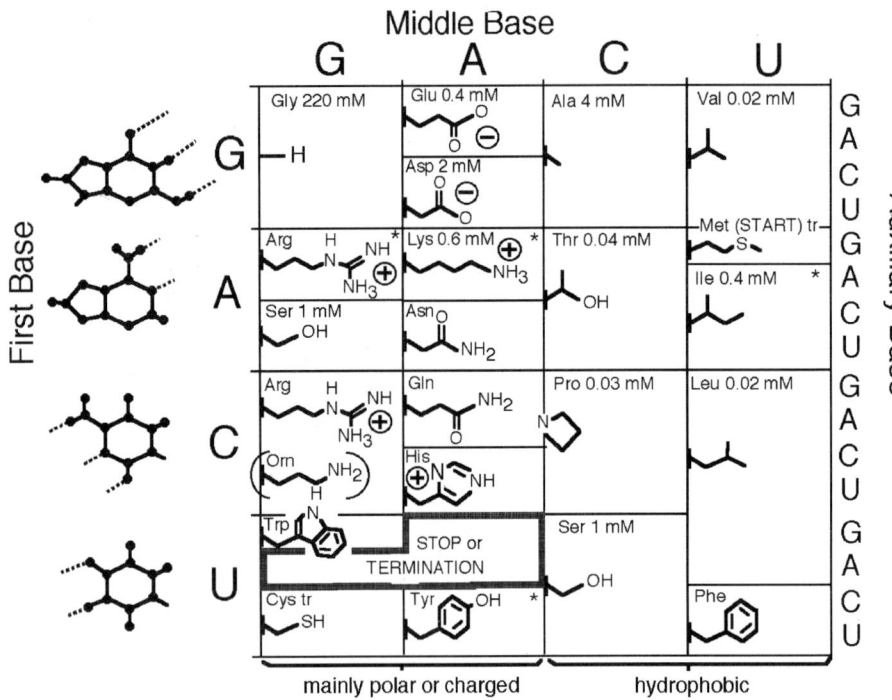

Figure 20. Table of the "universal" genetic code. Side chains of amino acids are shown to be mainly distinctive for three-letter codons, which consist of First Base, Middle Base, and Auxiliary Base. Hence, AAG and AAA are codons for lysine but the only codon for methionine is AUG (Fig. 21). Concentrations indicate amino acids that have been obtained in "prebiotic" syntheses (Hennet et al., 1992; and see Marshall, 1994) and are therefore assumed to be the commonest on the early Earth. Ornithine, not analyzed for in Hennet et al. (1992), has been tentatively assigned to the arginine codons because of the similar nature of their side chains and because arginine has not been recorded in abiotic experiments. The four starred amino acids have been shown to attach to RNA strands that contain their codons. (From Russell et al., 2003 and references therein.)

on these organometallic compounds. Cobalt and nickel corrinoids are involved in biosynthesis; magnesium (and in acid solutions, zinc) produces the chlorophylls used in photosynthesis; and iron produces the heme groups used for electron transfer and in oxygen chemistry (Eschenmoser, 1988). Pratt (1993) considers the structures based on the macrocyclic corrin ligand to date back four billion years. They certainly must be older than 3.8 Ga, for photosynthesis employs a myriad of such structural variants (Blankenship, 2002).

THE PARTICULAR PROBLEMS OF RNA SYNTHESIS

No special place in this "unintentional" world is given to RNA in our decentralized system beyond its being metabolically useful and therefore a surviving molecule. Although an unstable entity, once formed RNA would have been less mutable when secured upon a mineral surface, especially in the presence of highly reduced fluids. Nevertheless, the synthesis of nucleic acids, composed of a phosphorylated ribose sugar attached to one of four possible bases, is a problem more daunting than that of the amino and carboxylic acids. As a start we note that pyrophosphate, introduced through volcanoes to the early oceans, would have remained in solution in the relatively acidic ocean, although some would have been precipitated as vivianite ($Fe_2(PO_4)_2 \cdot 8H_2O$) and as a condensed pyrophosphate on mixing with moderate-temperature alkaline fluids at the hydrothermal mound (Rouse et al., 1988; Yamagata et al., 1991; Russell and Hall, 1997; de Zwart et al., 2004). There are plausible hydrothermal sources of NH_3 and HCN to explain the synthesis of some of the bases of RNA (Schulte and Shock, 1995; Shock et al., 1998). As noted by many others, both bases and a variety of sugars can be formed at low temperature by the condensation of hydrogen cyanide and phosphoglycerate, respectively.

The carbon nitrogen ring compounds constituting some of the nucleic acid bases may have formed by the condensation of HCN on a sulfide surface such as pyrite (Leja, 1982; cf. Sowerby and Heckl, 1998). Adenine $(HCN)_5$, the most common of the bases, vital to energy storage as well as one of the components of RNA, may have formed this way (Oró and Kimball, 1961). And guanine might also have been synthesized in hydrothermal conditions, but at low yields (Ferris et al., 1978). Uracil is a hydrolysis product of HCN oligomers (Voet and Schwartz, 1982). But how or whether unstable cytosine formed at this early stage is not known.

The synthesis of ribose phosphate, the particular pentose sugar attached to the bases, is also difficult to understand but may have been the stable end product, assembled upon a mackinawite surface, from a reaction between phosphorylated chiral glyceraldehyde (GA3P) and the achiral dihydroxy acetone phosphate (DHAP), themselves derived by condensation of formaldehyde adsorbed from the alkaline fluids upon FeS (Quayle and Ferenci, 1978; Schulte and Shock 1995; Pontes-Buarques et al., 2001; Rickard et al., 2001; Russell et al., 2003; Ricardo et al., 2004). If so a feedback or autocatalytic cycle may have been initiated, with a contribution from activated hydrogen that acted as another sink for carbon dioxide (Russell et al., 2003). Phosphogluconate formed in this way decomposes exothermically to GA3P and DHAP again. Reaction between the ribose phosphate adhering to the mackinawite surface and the bases would then produce RNA.

Of course it may be that the particular nucleic acids used before life had fully developed were different, and/or that there were two rather than four bases (e.g., Reader and Joyce, 2002), perhaps adenine and uracil (Jimenez-Sanchez, 1995). We are forced to step over this period for lack of knowledge. Anyway, apart from the easily synthesized adenine, only small concentrations of the rest were needed because, as we shall see, RNA polymers were the "molds" that may have produced a myriad of peptide "casts."

THE ORIGIN OF THE CODE AND THE FIRST CODED POLYMERS OF PARTICULAR CHIRALITY

Although the analogue of the acetyl-coenzyme-A pathway provides a sink for carbon dioxide in the early atmosphere and ocean, and it is possible to envisage how such a growing system would bud and reproduce, this is insufficient for replication and evolution. For this a code was required, probably reliant on a replicating and evolving RNA, to generate and order functionally useful peptides. In living cells amino acids are sequenced and polymerized in a process centered on the ribosome, composed essentially of RNA, that ratchets along messenger RNA (Ban et al., 2000). Such a complex process must have evolved from a simpler system. A hint of what this was is provided by the shape of the RNA codons and the characteristics of the amino acid side chains (Fig. 21).

Geologists are less familiar with the structures of the carbon-, nitrogen-, and oxygen-bearing main-chain polymers and their various and characteristic side chains that constitute the amino and nucleic acids than they are with the internal structure of minerals. Nevertheless, the way the side chains of the organic polymers might have been packed together on a mineral surface, if only ephemerally, takes some of the mystery out of how life works or worked—how, for example, nucleic acid polymers may have first adventitiously coded for peptides and proteins. Looking at the origin of the code from a mineralogical perspective leads us to follow Woese (1967) in his view that genetic information was first transferred directly by selection through a somewhat indiscriminate "codon-amino acid pairing," which relied upon the affinity of the shape and charge of the codon (a triplet of three nucleic acids) to the shape and charge of the amino acid and especially of its side chain (Woese et al., 1966: Woese, 1967, p. 174–175). Thus, what is known as the peptidyl transferase reaction of an RNA molecule probably evolved via direct translation on a protoribosome (Woese, 1967). This relationship happened to provide a rudimentary but direct coding to the polymerizing amino acid sequence.

Developing this idea, Mellersh (1993) emphasized that RNA triplets would only offer a cleftlike (tridentate) conformation to attract amino acids when adhering to a solid phase. For such a solid phase we favor mackinawite (Russell et al., 2003). We have already noted that phosphates may coat a mackinawite surface and there act as a random polymerizing agent for peptide formation. The phosphates of RNA also could have been bonded through to sulfide on the surface of mackinawite in the same way (Fig. 21). First and foremost we should see the affinity between the RNA triplets as offering a mechanism of polymerization more efficient than the chance condensation of amino acids on a simple phosphorylated mineral surface. The rows of RNA triplets could have gripped and juxtaposed amino acid monomers in such a way as to offer the carboxyl group of one to the amino group of the next for

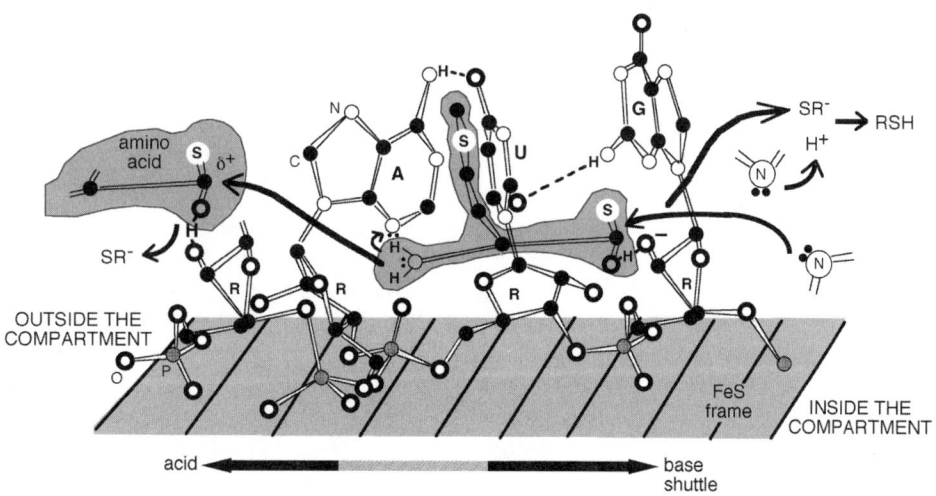

Figure 21. Sketch to show how an amino acid may have been gripped by an RNA triplet (AUG) and offered its nucleophilic amino group to the electrophilic carbon of the thiocarboxyl group of an adjacent amino acid to dimerize (Mellersh, 1993; cf. Muth et al., 2000, fig. 2). A nitrogen atom acts as the basic binding site for the amino group on adenine (A), and oxygen on the next ribose (R) acts as the binding site of the matching thiocarboxyl group. A peptide chain built incrementally in this way would be released by acid-base catalysis (Muth et al., 2000). There would have been a tendency for the AUG triplet to act as the codon for methionine as shown here, if available in the FeS membrane (Mellersh, 1993; Russell et al., 2003). If not, another hydrophobic amino acid at relatively high concentration was likely to have occupied the site.

bonding (Mellersh, 1993). At the same time the affinity between the clefts of RNA and the side chains of the amino acids would happen to effect crude selection by codon-amino acid pairing as envisaged by Woese (1967). For example, clefts in which uracil was the central base would tend to attract only amino acids with hydrophobic side chains such as that of methionine (NH_3^+. $CH_2.(CH_2)_2.SH.CH_3.CSO^-$), whereas those in which adenine was central would show affinity for the hydrophilic charged or polar amino acids (Figs. 20, 21).

We assume amino and nucleic acids would have occupied the iron sulfide cells, though the former would have vastly outweighed the latter, and the crude coding of, or translation to, peptides would have been a feed-forward process. RNAs on the surface of nanocrystals may themselves have replicated, where water activity was low, by Watson-Crick hydrogen bonding in which A bonded to U and G bonded to C (Béland and Allen, 1994). The "protoribosome" would then have operated as a replicase, via the replication of triplets (the "triplicase" of Poole et al., 1999). The first cycle of replication provided an antisense codon, so that a GCC triplet (the codon for alanine and the most likely first triplet) produced the antisense codon "read" in the opposite direction as GGC (glycine) (Trifonov, 2000). Therefore, as touched on above, if the initial triplet coded for a hydrophilic amino acid (with adenine occupying the central position), then its opposite would have coded for a hydrophobic residue (with uracil as the central base) (Béland and Allen, 1994; Konecny et al., 1995).

Point (single base) mutations on these triplets would have tended to code for the amino acids synthesized by Hennet at al. (1991), i.e., glycine (GGC), alanine (GCC), aspartate (GAC), and serine (UCC). More astonishing is the fact that these amino acids constitute the common sequence in that group of metalloproteins acknowledged to have the longest pedigree, the ferredoxins (Eck and Dayhoff, 1966; Trifonov et al., 2001; Russell et al., 2003). This system of direct coding would have been relatively robust, in that mutations not involving the central RNA monomer would have attracted amino acids with similar side chains and thereby similar properties.

However, during the organic takeover the protoribosome would have required another surface in place of mackinawite. This might have been supplied by a peptide sequence rich in positively charged side chains. Such a peptide would have attracted the phosphates of RNA that they might polymerize and still offer the triplet clefts. Lysine, arginine, and ornithine would have been equally useful in such a peptide (Fig. 20). Mellersh and Wilkinson (2000) have demonstrated that poly-adenosine, which includes the clefts AAA expected to have affinity for lysine, does stereoselectively bind L-lysine from dilute aqueous solution of L-amino acids (Mellersh and Wilkinson 2000). Moreover, about half the amount of L-arginine and L-ornithine also was found to bind with poly-adenosine. As adenosine was likely the most common of the nucleic acids, and lysine and probably ornithine can be made abiotically, then we have the makings of a feedback cycle that involves the transfer of information.

The chance stereochemistry of the short RNA polymer would determine whether it catalyzed the polymerization of D- or L-amino acids into peptides (Mellersh, 1993). To achieve a low-energy state, as with mineral growth, we might expect RNA to tend to lengthen while preserving either left or right chiralities, i.e., a favorable packing arrangement (Joyce et al., 1984). Were a monomer with the opposite stereochemistry to be added to a growing chain, growth would be thwarted (Sandars, 2003). That filter would have been sufficient to tip the holochirality scale because, despite the presence of a racemic mixture of amino acids in the microcavities, only amino acids of the same α-carbon configuration (similar stereochemistry) would preferentially have ended up in peptides, to yield a population of distinctly handed peptides. Some of these peptides would eventually feed back in a hypercyclic manner to favor the syntheses of their "stereochemically appropriate" polymerizing template.

Eventually the more robust but less reactive DNA (deoxyribonucleic acid) molecules took over from RNA and thence survived. Braun and Libchaber (2004) have demonstrated that secondary convection and thermophoresis driven by temperature gradients within microcavities in a hydrothermal mound could have concentrated, elongated, and driven the replication of DNA. It remains to be seen if RNA could be elongated and replicated by the same process.

Although the codon-amino acid affinity concept explains why the chiralities of polynucleotides and peptides in life are opposite, a relationship not required by Crick's (1968) frozen accident hypothesis, it does not explain why, on our planet, dextro-DNA and RNA code for levo-proteins. As the energetic differences between right- and left-handed chiral molecules, even for thiosubstituted DNA analogues, are negligible (MacDermott et al., 1992) and would have been "lost in the noise" within a natural hydrothermal reactor, we can only conclude that both chiral forms emerged separately, but that at some unknown stage the present pairing survived, either by fortuitously stealing a march on the other, perhaps through the chance development of a better ribosome, or later in a chirality war between the rival prokaryotes within the biosphere.

A HYDROPHOBIC ORGANIC MEMBRANE

As abiotic lipids of equal carbon chain lengths are unlikely to have been delivered to, and survived on, early Earth in quantities large enough to have allowed continued reproduction (Cairns-Smith, 1982), we have been left to consider FeS bounded "cells" as the hatcheries of life. Short noncoded peptides generated in hydrothermal conditions (Russell et al., 1994; Huber and Wächtershäuser, 1998, 2003; Ferris et al., 1996) could have played an important role in improving membrane characteristics. These peptides and other polymers produced in the mound would have coated the inside of inorganic compartments and partially plugged pore spaces. Excess organic sulfides and nitrides also could have been "entropy driven" into this, the first organic membrane (Cole et al., 1994). We imagine these polymers to have cohered to form proteinaceous membranes and walls to protocells—organic structures that offered several advantages. Eventually, genetically controlled proteinaceous cell envelopes composed substantially of hydrophobic heterochiral peptides, for example, would have the advantage of including metal clusters such as the [Fe_4S_4] centers within their structure as stabilizers and electron transfer agents. At a later stage lipids would have been generated from acetyl-coenzyme-A by continued addition of C_2 components, perhaps until the stable C_{12} fatty acid (dodecanoate) was realized (Fig. 12) (Shock et al., 1998). Once this happened, lipids would have more efficiently denied protons an uncontrolled short circuit back into the cellular interiors.

EARLY EVOLUTION AND THE COMMON ANCESTRAL COMMUNITY

The universal ancestor of life probably comprised a community of single-celled organisms still housed within its hydrothermal hatchery that possessed all of the attributes common to all Bacteria and Archaea: the genetic code; the ribosome; DNA; a supporting core and intermediate metabolism needed to supply the constituents of its reproduction; replication; compartmentation from the environment; redox chemistry; and the use of a proton gradient. This last common community (the LCC of Woese, 1998; Macalady and Banfield, 2003) existed in the hydrothermal mound at the dawn of the biochemical revolution where genes and proteins were diversifying into a myriad of functions, where metal sulfide catalysts were being replaced by proteins, where new pathways and cofactors were being invented to augment and substitute their mineral and RNA precursors, where FeS was being incorporated into proteins as Fe(Ni)S clusters, an imprint of which would be reflected in the FeS centers of ancient protein, and where biochemistry started to diversify into the forms that were both possible and useful (Eck and Dayhoff, 1966; Hall et al., 1971; Martin and Russell, 2003) (Figs. 14, 19).

From the standpoint of protein structure, this age of invention would have witnessed the origin of basic building blocks of biochemical function that (i) are conserved at the level of 3D structure among Bacteria and Archaea and (ii) are recognizable as functional modules in various electron transporting proteins (Beinert et al., 1997; Baymann et al., 2003). From the standpoint of amino acid sequence complexification, this age of invention would have been a phase of molecular evolution where proteins were diversifying and improving their efficiency (Baymann et al., 2003).

It has been argued that the first acetogens were the forerunners of the Bacteria (Fig. 14) (Russell and Martin, 2004). We suggest that a minority of these cells, derived from those that emerged at around 40 °C, exploited the potential offered at higher temperature deeper in the mound where the kinetic energy was greater and the activation energy required for reduction, through acetate (equation 8), all the way to methane (equation 7), was lower. These first so-called methanogens may have evolved while still in the mound, as there is even more energy to be had in the full reduction of carbon dioxide (Amend and Shock, 2001). Moreover, the catalytic/enzymatic machinery required is similar (Thauer, 1998; Fontecilla-Camps and Ragsdale, 1999).

Before their release from the mound the first cells would have responded to environmental differences deterministically in a manner more comparable to an ecosystem than to an individuated cell. Compartments near the edge of the hydrothermal mound, those most distant from the "inorganic formative fluid" (cf. Haeckel, 1892), lie in the steepest chemical, electrochemical, and thermal gradients but at lower temperature. It is here that H^+, FeOOH, PPi, and CO_2 were most concentrated, and this was where we assume the onset of life to have taken place at around 40 °C (Fig. 14). At this early stage, mineral circumscribed compartments in the more restricted environment below this distal group and closer to the hydrothermal fluid were hotter, and though the gradients would have been lower and the immediate environment would have been more depleted in the constituents mentioned above, concentrations of CO, H_2, thiols, and the abiotic amino acids would have been higher. In this proximal zone elemental sulfur polymers, generated in the atmosphere by photolysis of H_2S and SO_2 (Pavlov and Kasting, 2002) and sedimented within the mounds, could stand in as an electron acceptor in place of ferric iron, albeit at a lower potential (Fig. 6D) (Stetter and Gaag, 1983). Alternatively, if cells deeper in the mound were beyond the reach of external electron acceptors, another way of ridding the system of excess reductant was through the discharging of electrons to the oceanic and atmospheric sink in CH_4 (de Duve, 1991).

Evolution in the mound extended beyond mere optimization of the acetate and methane reactions (Martin and Russell, 2003). A next step was adaptation that exploited the reduced carbon and energy to be found in waste products and dead cells:

$$CH_3.COO^- + 8Fe^{III} + 4H_2O \rightarrow 8Fe^{2+} + 2HCO_3^- + 9H^+ \quad (19)$$

The prior use of Fe^{III} as an electron acceptor during autotrophic biosynthesis (equation 14) provided a means of such respiration (oxidative metabolism) (Vargas et al., 1998). Other potential electron acceptors were photolytic S^0 and Mn^{IV} (Figs. 6D, 18) (Nealson and Stahl, 1997; Bahcall et al., 2001; Baymann et al., 2003).

We conclude that the last common ancestral community occupied the very hydrothermal hatchery in which life first emerged. The proto-Bacteria were initially suited to low to moderate temperatures, and the proto-Archaea originally evolved to withstand the shock of relatively high temperatures (i.e., ~60 °C) (Fig. 14). But the propensity to live well above 40 °C was passed back to the nascent Bacteria through genetic transfer. A period of high ambient temperature, caused either by a meteorite impact or by a carbon dioxide greenhouse (Kasting and Ackerman, 1986; Kasting and Brown, 1998; Nisbet and Sleep, 2001) could explain why the last common community may have been thermophilic, perhaps living at 50–60 °C (Gaucher et al., 2003 but see Brochier and Philippe, 2002).

DIFFERENTIATION INTO TWO DOMAINS

Although the thermodynamic drives would be lower in proximal compartments of the mound (Schoonen et al., 1999), the lower kinetic barriers to reductions and condensations at these higher temperatures would encourage reaction. In these hotter compartments at first there would have been no RNA regulation of peptides, and no replication or evolution. This is because RNA is unstable at high temperature (Poole et al., 1999). But early biochemical evolution in the outermost compartments could have produced some "thermotolerant" DNA from RNA that could have invaded the metabolic husks of those below and co-opted this poorer, "thermochallenging" environment (Fig. 22) (Forterre, 1995, 2002). Glansdorf has argued that cotranslation of functionally related proteins from integrated anabolic genes "facilitated the formation of multienzyme complexes" that channeled thermolabile substrates that could invade hotter environments (Glansdorff, 1999, p. 432). Here inherently thermolabile proteins acted to stabilize and protect the whole ensemble (Forterre, 1995). Operons—linear sequences of genes transcribable as a single unit together with a regulatory operator—emerged as a response to these increasing temperatures (Glansdorff, 1999). Such operons would have facilitated the production of multienzyme complexes capable of reducing the deleterious effects of toxic intermediates produced by thermodegradation at high temperature. Srere (1987) points out that the clustering of the functionally related genes responsible for these complexes would also have conferred evolutionary advantage when returned to mesothermal conditions.

Let us recall the importance of a regulated dynamic system. We could show that the hydrothermal system was both thermostat and chemostat. This took the onus off the living system to be a thermostat but it would have needed its own control system for governing the internal state of the protocells. Such a regulatory power is known as homeostasis. It probably emerged as newly generated protons, driven out of the cell by electron transfer, kept the cell neutral to alkaline. This process was augmented by the oxidative formation of disulfides such as the amino acid dimer cystine, from the monomer cysteine, by protons (Russell et al., 1994):

$$2RS^- + 2H^+ \Leftrightarrow H_2 + RSSR \qquad (20)$$

At the point of differentiation or bifurcation into the two prokaryotic domains, we see the precursors to the Bacteria occupying the broad front of the growing hydrothermal mound at its interface with the ocean. The precursors to the Archaea, the sturdy but slowly evolving second domain of the prokaryotes,

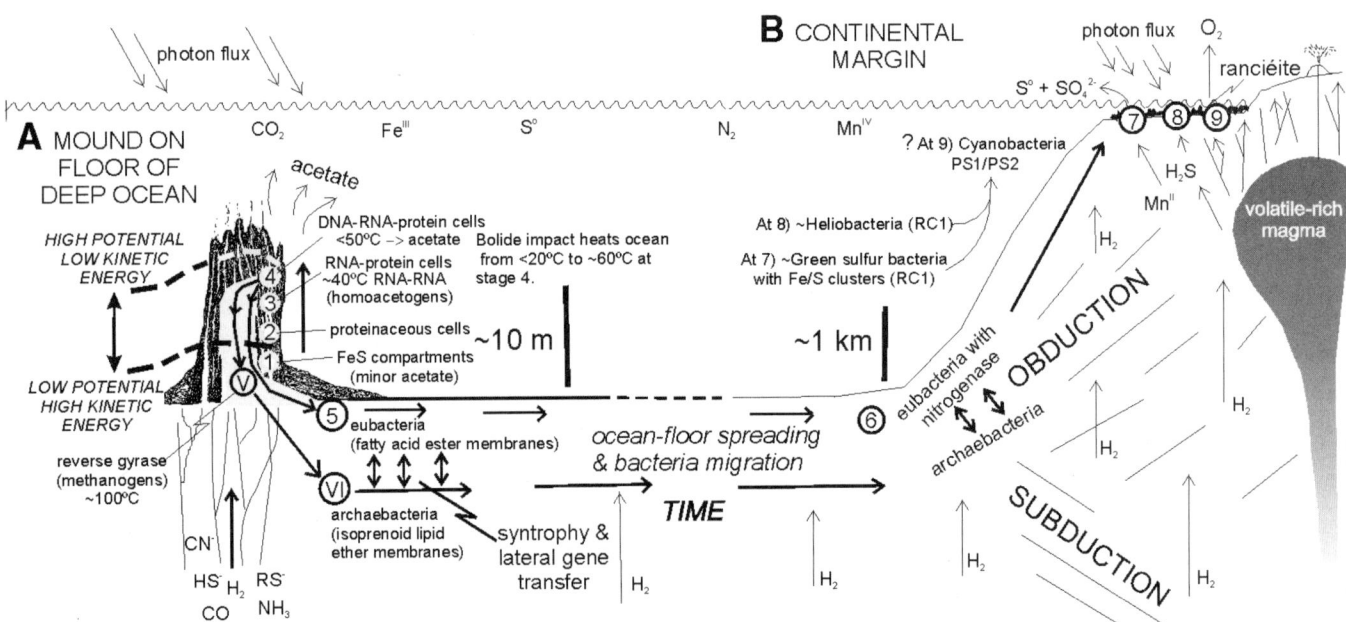

Figure 22. Chemosynthetic life emerges at a warm alkaline seepage and at (A) differentiates into the precursors of the Bacteria and Archaea, and expands downward into the surrounding sediments and crust (Martin and Russell, 2003; Russell and Martin, 2004). From here a proportion is conveyed by ocean floor spreading toward a constructive margin (B) produced by obduction. Once uplifted at the margin, some of the cells happen to invade sediments in the photic zone where, at a sulfurous spring, some evolve to exploit solar photons. Numbers 1–3 relate to life's emergence, and 4 marks the point of differentiation of the Archaea from the Bacteria. Roman numerals V and VI mark evolutionary stages of the Archaea, and 5 and 6 indicate stages of evolution of the Bacteria in the deep biosphere. Photon energy was first mastered by the green sulfur bacteria (7), followed by the heliobacteria (8). These photosynthesizing bacteria had appeared at least by the early Archaean (Westall et al., 2001). Oxygenic photosynthesis (9) is a further development that may have evolved at a manganiferous hot spring by 3.75 Ga. (Various scales.) (From Russell and Arndt, 2005.)

bring up the rear (Figs. 14, 22) (Woese et al., 1990; Russell and Hall, 2002). Both the proto-Bacteria and the proto-Archaea lived up to the opposite edges of reproductive viability, the former at a distance from fuel and challenged by kinetics at low temperature, the latter subject to the dangers of pyrolysis. Certainly the Archaea appear to be the conservative cousins of the Bacteria, as though they have had to hoard resources internally and defend themselves against untoward perturbations. In the anaerobic environment that obtained in the mound at the dawn of life, some of the proto-Archaea probably lived off redox reactions that Bacteria have never mastered, relying on organic sulfides in a series of electron donations and generating methane waste (an electron carrier) from carbon dioxide without a metallic electron acceptor (Schäfer et al., 1999; Amend and Shock, 2001).

Even then, the proto-Bacteria and the proto-Archaea found it advantageous to live syntrophically both within and across domains. Cells would have interacted with their neighbors by swapping genes, providing some of the nutrients, and removing some of the waste. Unfortunates that were entrained within the hydrothermal solution and dispersed to the relative desert of the uncertain ocean could not have survived such vicissitudes and dilution of nutrient (e.g., Bjerrum and Canfield, 2002). The only safe migration route was down onto the ocean floor and into the warm sediments and permeable basalts below, where the essentials—H_2 and CO_2—were assured.

Thus we conclude that the most significant of all cellular differentiations, that between the Bacteria and the Archaea, probably took place before the mound was evacuated (Koga et al., 1998; Martin and Russell, 2003), although up till this time of divergence, genes were shared in what may be called a cellular cooperative (Fig. 14). This differentiation of the precursors of the Archaea from those of the Bacteria was expedited by entropy and the random changes in genes it caused. It eventually produced the two mutually exclusive genotypes (Wicken, 1987).

ESTABLISHMENT OF THE DEEP BIOSPHERE

As the proto-Archaea and the proto-Bacteria began to colonize their surrounds, they eventually found themselves expanding into the sediments and volcanic horizons at the base of the mound (Fig. 22). Here conditions were not so different excepting the much lower flux of nutrient and fuel (Wolin, 1982). Thus the deep biosphere was inaugurated (Parkes et al., 1990, 1994; Pedersen 1993). Once in the deep biosphere SO_4^{2-} and N_2 are likely to have joined Fe^{III}, S^0, and Mn^{IV} as electron acceptors (Fig. 18). Although this probably happened quickly (Pinti, 2002; Shen and Buick, 2004; Raymond et al., 2004) the age of these innovations is not known. We take the view that once chemical energy potentials are available then their exploitation is relatively rapid. Yet at these early stages mineral-like clusters would again have played a critical role in evolution. The metal center responsible for N_2 reduction is comparable to a greigite twin along the sulfur plane, excepting the presence of one proximal Mo atom comparable to a naturally occurring $MoFe_3S_4$ cluster produced from aqueous MoS_4^{2-} and FeS (Russell et al., 1994; Helz et al., 1996). This "Siamese-like twin" possibly contains a nitrogen atom in the central site (X in Fig. 9F) (Einsle et al., 2002; Smith, 2002). Exactly where and how nitrogen is reduced to ammonia via the nitride N^{3-} on this metal sulfide center is as yet unresolved (Seefeldt et al., 2004).

OBDUCTION AND PHOTOSYNTHESIS

We have noted that conditions for the earliest cells in the open sea were inhospitable and periodically impossible. How then do we explain the emergence of photosynthetic organisms in the full glare of hard UV from the young sun (Canuto et al., 1982)? The first step was for organisms to have approached the photic zone with the "safety of numbers." Because of the particular geometry of Hadean oceanic crust, buoyant sediment and hydrated basaltic crust piled up over the subducting, delaminated, eclogitized lower parts of the slab (Russell and Arndt, 2005). There were no deep ocean trenches. Obduction and uplift of oceanic sediments, and of the hydrated basalt beneath, passively transported some bacterial colonies into shallow water and into the photic zone on the margins of volcanic chains (cf. Margolis et al., 1978) (Fig. 22). Cells were protected from deleterious solar radiation beneath a mineral coating (Cockell and Knowland, 1999). Opportunistic protection by superposed minerals and mineral excretions are well-known survival gambits (Phoenix et al., 2001). In these conditions some Bacteria near the surface augmented their protective shield by developing a UV pigment protector from a ring of organic bases. Pigments comprising macrocyclic aromatic rings probably date back to at least 4 Ga (Pratt, 1993). Single ions of Fe, Mg, Zn, Co, and Ni could have been sequestered in variants of what is known as the corrin ring, itself comprising four C/N rings (Eschenmoser, 1988). Pigments developed for photoprotection could then have been adapted as electron transfer agents, as photosynthetic reaction centers and antenna proteins (Fig. 2B) (Mulkidjanian and Junge, 1997; Allen, 2004).

Sediments in and overlying an obducted pile are likely, in places, to have been subjected to hydrothermal H_2S of magmatic or metasomatic derivation (Fig. 22). The first photosynthesist may have been a precursor of the green sulfur bacteria. As the name suggests, like many "primitive" bacteria, they relied on hydrogen sulfide as an electron donor (Baymann et al., 2001; Blankenship, 2002). In these conditions, a reaction center (RC1) was developed that could generate elemental sulfur as waste, and gain electrons for biosynthesis in the process:

$$H_2S + CO_2 + light \rightarrow (CH_2O)_{life} + H_2O + 2S^0 \quad (21)$$

As we might expect of relatively gradualistic evolution, the green sulfur bacteria retain a reliance on iron sulfide clusters as electron transfer agents (Vermaas, 1994; Blankenship, 2002). An evolutionary variant of the fermenting bacteria, a photosynthetic precursor of the heliobacteria, could fix carbon dioxide by the concomitant oxidation of organic waste and detritus. At the same time they used a similar reaction center (RC1) (Baymann et al., 2001; Blankenship, 2002). These two reaction centers hybridized to form what is known as photosystem 2 (PS2), the photosystem invariably employed by all cyanobacteria and the chloroplasts derived from them in plants (Michel and Deisenhofer, 1988; Allen, 2005). PS2 works in conjunction with the first reaction center (RC1), which evolved into photosystem 1 (Baymann et al., 2001). Photosystem 2 is capable, in conjunction with photosystem 1, of oxidizing two molecules of H_2O (cf. H_2S in the green sulfur bacteria and $(CH_2O)_x$ in the halobacteria) for the generation of every one molecule of O_2, while gaining four electrons and four protons for the fixation of carbon from CO_2 for biosynthesis in the process (Hansson and Wydrzynski, 1990; Allen, 2005):

$$2H_2O + light + PS2_{oxidized} \rightarrow 4H^+ + 4e^- + PS2_{excited} + O_2 \quad (22)$$

To generalize:

$$H_2O + CO_2 + light \rightarrow (CH_2O)_{life} + O_2 \quad (23)$$

Reaction 22 invariably involves the oxygen evolving center (OEC) in PS2. The OEC is a [$CaMn_4$]-structure consisting of a trigonal pyramid with calcium at the apex, 3.4Å from each of the three manganese atoms at its base: a distal manganese lies in the same plane as the other three and 3.3Å from its nearest neighbor (Loll et al., 2005). One of the manganese atoms constituting the base of the pyramid is also 3.3Å from another, whereas the other two are 2.7Å apart. To explain such an extraordinary innovation Russell and Hall (2001, 2002) assumed that anoxygenic photosynthesizers in subaerial hot spring carbonate pools adsorbed exhalative manganese precipitates onto their membranes, perhaps while using photolytic Mn^{IV} as an electron acceptor (Fig. 2C) (Burns and Burns, 1979; Myers and Nealson, 1988; Chafetz et al., 1998). In reduced form a manganese coating could protect the cells from hard UV injury (cf. Daly et al., 2004). Photo-oxidation would generate ranciéite (the calcium-bearing birnessite, $CaMn_4O_9 \cdot 3H_2O$) (Anbar and Holland, 1992; Russell and Hall, 2001; 2002, fig. 6). Thus, a cluster of ranciéite may have contributed the "ready-made" [$CaMn_4$] structure that was co-opted by one of the reaction centers, though if so the Mn-Mn distances of 2.9Å characteristic of the cluster constrained in ranciéite must have been modified to a conformation more typical of hollandite where two Mn-Mn distances are 2.7Å and two are 3.3Å (Sauer and Yachandra, 2004; Loll et al., 2005). The manganese cluster, once sequestered in this reaction center and stabilized by the calcium atom, could have oxidized two water molecules, passing four protons and four electrons from the hydrogen to the cell for biosynthesis while the oxygen went to waste (Fig. 23) (Russell and Hall, 2001). Without the calcium atom the structure would be irreversibly reduced to the cubane moiety of the spinel hausmannite (Russell and Hall, 2002, fig. 6). As it is the structure readily reverts to the oxidized form. Thus we conclude that a minimum of genetic control was required for a hydrated [$CaMn_4$] complex

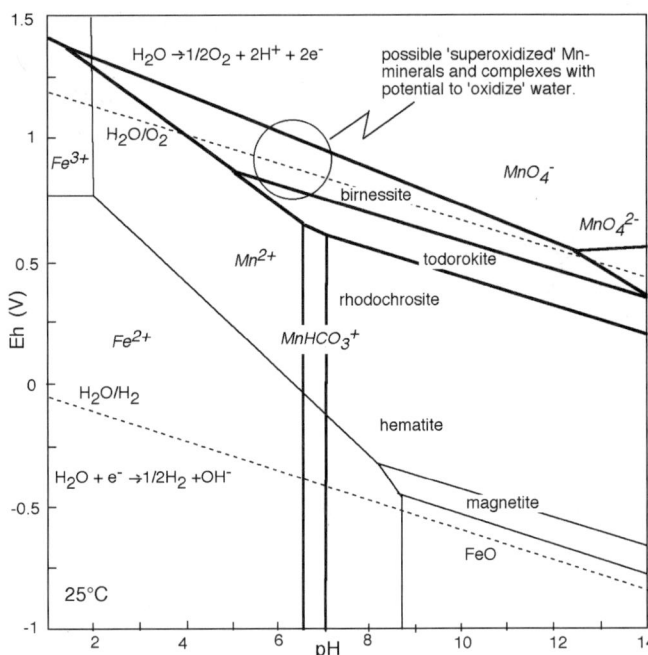

Figure 23. Mn and Fe plots (GWB) of species and phases in similar conditions. Activities of Mn^{2+} and $Fe^{2+} = 10^{-6}$; fugacity of $CO_2 = 1$ (~3000 times the present atmosphere). Pyrolusite is suppressed to favor hydrated and mixed valence oxides and hydroxides such as birnessite [$(Na,K,Ca)(Mg,Mn)Mn_6O_{14} \cdot 5H_2O$] and the thermodynamically uncharacterized ranciéite [$CaMn_4O_9 \cdot 3H_2O$]. In theory, clusters of birnessite and ranciéite have the potential to oxidize water to the peroxide at low pH.

to take up a shape within the membrane that would have facilitated the oxidation of water (Figs. 2D, 23, 24). Invagination and evolution of the protein led to the emergence of the cyanobacteria, organisms that were eventually to change the face of the planet to its present blue-green cast (Dismukes et al., 2001).

Unless one considers the Banded Iron Formations to be a good indicator (Holm, 1987; Dymek and Klein, 1988), the impact of oxygenic photosynthesis is not discernable until the early Proterozoic (Bekker et al., 2004). However, Rosing (1999) and Rosing and Frei (2004), remarking on the apparently high rates of organic production, advance morphological, as well as stable and radiogenic isotopic evidence, to favor an origin of oxygenic photosynthesis prior to 3.7 Ga. And there is independent sulfur isotope evidence for the presence of trace atmospheric oxygen by 3.5 Ga (Ohmoto et al., 1993; Shen and Buick, 2004). Because we can see no reason why, given the appropriate conditions, the emergence of oxygenic photosynthesists from anoxygenic photosynthetic bacteria should have taken any longer than the emergence of life, these inferences, surprising to some (e.g., Blank, 2004), seem entirely reasonable. The long lag time between the photobacterial production of oxygen waste and its appearance as an atmospheric gas in the early Proterozoic (Farquhar et al., 2000) must then be explained by the reductive capacity of the planet, particularly of reduced iron, organic detritus, and biogenic

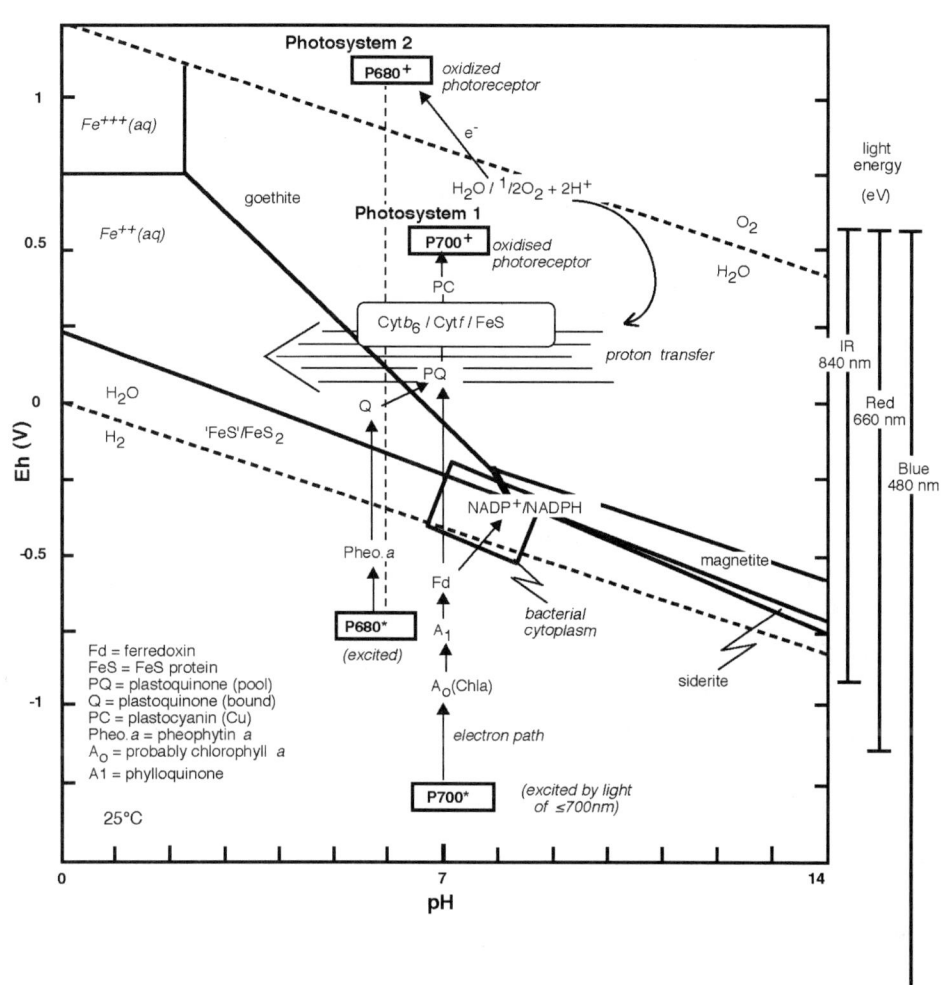

Figure 24. Electron pathways of oxygenic photosynthesis for cyanobacteria for pH = 7 though photosystem 2 (PS2) is displaced for clarity. In non-cyclic photophosphorylation NADP$^+$ is reduced by electrons from PS1 to make NADPH; PS2 provides an electron to replenish PS1 whereas dissociation of water to evolve oxygen provides the electron to replenish PS2. In cyclic photophosphorylation only PS1 is used and NADP$^+$ does not receive an electron; protons are transferred out of the cell, generating a pmf that produces ATP. Photosynthetic production of NADPH, ATP and protons contribute to the synthesis of carbohydrate from CO_2. Light provides an alternative source of redox energy for H_2 provision of chemical energy (Fig. 2) although FeS remains at the core of the energy-transfer system (cf. Blankenship, 2002, Figure 11.7).

methane (Lécuyer and Ricard, 1999; Kasting, 2001; Catling et al., 2001; Bjerrum and Canfield, 2002; Bekker et al., 2004).

CONCLUSIONS

The onset of life was not a haphazard affair but the metastable evolutionary outcome of focused reactions between hydrothermal hydrogen and bicarbonate in the ocean, with inputs from trace metals, phosphate, and ammonia. Considering the fragility of some organic polymers, especially RNA, and the theoretical and experimental support for pyrophosphate and organic synthesis at moderate temperatures, ~40 °C is a likely temperature for life's emergence (Moulton et al., 2000). Life would have emerged as soon as such a temperature was realized—a state probably reached, at least intermittently, by 4.3 Ga (Nisbet and Sleep, 2001). Convection cells operating within the oceanic crust in the relative tranquility of ridge flanks or in the deep ocean floor would have produced hydrothermal mounds—mounds that also would have acted as chemical fractionation reactors. The free energy of reaction between H_2 and CO_2 increases with decreasing temperature though the kinetics become ever more sluggish (Shock, 1992). The main products of the reaction are likely to have been acetate and water (Huber and Wächtershäuser, 1997). Other products such as glycine (amino acetate), other amino acids, and traces of nucleic acids constituted the first organic molecules (Russell and Hall, 1997, 2002). At this moderate temperature the acetate reaction had to be catalyzed (Shock et al., 1998). Mackinawite (FeS) and greigite (NiFe$_5$S$_8$) nanocrysts comprising the first membranes are significant in this respect, and plausible evolutionary steps may be imagined between these and the fully fledged enzymes with [NiFe$_4$S$_5$] centers involved in the generation of acetate to this day (Russell and Martin, 2004). Energy from the acetate reaction, augmented by a natural protonmotive force, the consequence of the pH and redox gradients acting across the semiconducting and semipermeable membrane, was coupled to the formation of pyrophosphate. In turn the energy in the phosphate bond drove polymerization. Small quantities of amino acids, metal-bearing clusters, and eventually RNA precursors, self-organized to become involved in the more efficient generation of peptides and acetate waste, a thermodynamic imperative. RNA genetic regulators eventually evolved to the state where they could be passed on to offspring and be shared with their neighbors (Mellersh, 1993; Hanczyc et al., 2003; Koonin and Martin, 2005). Amyloidal peptides and other polymers eventually took over from iron sulfides as membrane and cell wall constituents.

Pressurized microflow and circulation reactors could be used in parallel and series to test these aspects of the model (Russell et al., 2003; Braun and Libchaber, 2004).

The first gene-swapping cellular cooperative would have emerged at moderate temperature where chemical and electrochemical gradients were high near the surface of the mound. These acetogenic cells were the ancestors of the Bacteria, organisms that evolved while still within the mound to use electron acceptors other than CO_2 and Fe^{III}, perhaps in the order Mn^{IV} and S^0 (Baymann et al., 2001). But free energy was also to be had deeper within the mound where temperatures were higher, though gradients may have been lower. Here a small number of cells may have been able to withstand higher temperatures where the methane reaction was favored. Eventually specializing in generating methane waste, these methanogenic cells differentiated from the parent population (the last common community) to become the forerunners of the Archaea, the second domain of prokaryotic life (Martin and Russell, 2003). Both the Bacteria and the Archaea expanded into environments offering comparable conditions. Although nutrient supply was at a premium, only the oceanic sediments and crust could support these early vulnerable communities. Living syntrophically, the prokaryotes inaugurated the deep biosphere, well out of the way of the early and late bombardments (Parkes et al., 1990). Here they developed the capacity to use SO_4^{2-} and N_2 as electron acceptors.

Convection remained a driving force for evolution. At a large scale, plate tectonics drove portions of the deep biosphere into the photic zone on ocean island chains. Supplied with nutrients (e.g., H_2 and H_2S) and trace metals from subaerial hot springs (Ca, Fe, Mn, Ni, Co) and protected by chemical and detrital sediment, some of the Bacteria adapted the power of photons to drive electron transfer and thereby metabolism. The transition elements could be photo-oxidized, only to be re-reduced by organic molecules. Accumulation of ambient manganese within certain photosynthetic bacteria is likely to have protected them from UV injury (cf. Daly et al., 2004). Once within the cell, a ranciéite cluster $[CaMn_4O_9 \cdot 3H_2O]$ was co-opted to act as the oxygen evolving center $[CaMn_4] \pm 2H_2O)$ which, indirectly excited by photons, extracted the hydrogen (as protons and electrons) from water. The protons and electrons were used in the reduction of CO_2 for biosynthesis while oxygen was emitted as waste. There is no compelling reason to assume life required billions of years to configure the process of oxygenic photosynthesis and several reasons to suppose that this biosynthetic pathway had emerged before our stratigraphic record began ca. 3.75 Ga (Rosing and Frei, 2004). It can therefore be inferred that the core metabolic cycles were in place by that time (Pace, 2002). Reaction with biogenic methane, the use of oxygen as an electron acceptor, and the oxidation of ferrous iron kept oxygen concentrations vanishingly low until the early Proterozoic (Lécuyer and Ricard, 1999; Catling et al., 2001).

Although compartmentalized, the biosphere overall is autotrophic. Bernal (1960) put it well: "Life, geologically speaking, consists of the interference with secondary lithosphere-atmosphere reactions so as to produce a small but ever-renewed stock of organic molecules." Energized mainly by the sun's rays, the two essential contributors to this stock are (i) hydrogen, gained for the most part from water through photosynthesis with a minor component released during the hydrous oxidation of ferrous iron, and (ii) carbon dioxide released during volcanism, metamorphism, subduction, and reoxidation or fermentation of biogenic detritus. Of course there are all kinds of continuing interactions between the biosphere and lithosphere. However, we can conclude that the autogenic view of the emergence of life posed here sits well with the way the biosphere operates, and operated, right from its inception.

ACKNOWLEDGMENTS

We thank Bill Martin, Anthony Mellersh, Everett Shock, Wolfgang Nitschke, James Milner-White, John Allen, Nick Arndt, Jacques Meyer, and Minik Rosing for discussions, although not all of the views expressed here may coincide with theirs.

REFERENCES CITED

Abe, M., and Ooe, M., 2001, Tidal history of the Earth-Moon dynamical system before Cambrian age: Journal of the Geodetic Society of Japan, v. 47, p. 514–520.

Adman, E.T., Sieker, L.C., and Jensen, L.H., 1973, The structure of a bacterial ferredoxin: Journal of Biological Chemistry, v. 248, p. 3987–3996.

Allen, J.F., 2004, Cytochrome b_6f: Structure for signaling and vectorial metabolism: Trends in Plant Science, v. 9, p. 130–137, doi: 10.1016/j.tplants.2004.01.009.

Allen, J.F., 2005, A redox switch hypothesis for the origin of two light reactions in photosynthesis: FEBS Letters, v. 579, p. 963–968, doi: 10.1016/j.febslet.2005.01.015.

Allen, D.A., and Seyfried, W.E., 2003, Compositional controls on vent fluids from ultramafic-hosted hydrothermal systems at mid-ocean ridges: An experimental study at 400°C, 500 bars: Geochimica et Cosmochimica Acta, v. 67, p. 1531–1542, doi: 10.1016/S0016-7037(02)01173-0.

Amend, J.P., and Shock, E.L., 1998, Energetics of amino acid synthesis in hydrothermal ecosystems: Science, v. 281, p. 1659–1662, doi: 10.1126/science.281.5383.1659.

Amend, J.P., and Shock, E.L., 2001, Energetics of overall metabolic reactions of thermophilic and hyperthermophilic Archaea and Bacteria: FEMS Microbiology Reviews, v. 25, p. 175–243, doi: 10.1016/S0168-6445(00)00062-0.

Anbar, A.D., and Holland, H.D., 1992, The photochemistry of manganese and the origin of banded iron formations: Geochimica et Cosmochimica Acta, v. 56, p. 2595–2603, doi: 10.1016/0016-7037(92)90346-K.

Arndt, N.T., 1999, Why was flood volcanism on submerged continental platforms so common in the Precambrian?: Precambrian Research, v. 97, p. 155–164, doi: 10.1016/S0301-9268(99)00030-3.

Arndt, N.T., and Chauvel, C., 1990, Crust of the Hadean Earth: Bulletin of the Geological Society of Denmark, v. 39, p. 145–151.

Bada, J.L., 2004, How life began on Earth: A status report: Earth and Planetary Science Letters, v. 226, p. 1–15, doi: 10.1016/S0012-821X(04)00470-4.

Bahcall, J.N., Pinsonneault, M.H., and Basu, S., 2001, Solar models: Current epoch and time dependences, neutrinos, and helioseismological properties: Astrophysical Journal, v. 555, p. 990–1012, doi: 10.1086/321493.

Baltscheffsky, M., Schultz, A., and Baltscheffsky, H., 1999, H+-PPases: A tightly membrane-bound family: FEBS Letters, v. 457, p. 527–533, doi: 10.1016/S0014-5793(99)90617-8.

Ban, N., Nissen, P., Hansen, J., Moore, P.B., and Steitz, T.A., 2000, The complete atomic structure of the large ribosomal subunit at 2.4 Å resolution: Science, v. 289, p. 905–920, doi: 10.1126/science.289.5481.905.

Banks, D.A., 1985, A fossil hydrothermal worm assemblage from the Tynagh lead-zinc deposit in Ireland: Nature, v. 313, p. 128–131, doi: 10.1038/313128a0.

Banks, D.A., and Russell, M.J., 1992, Fluid mixing during ore deposition at the Tynagh base-metal deposit, Ireland: European Journal of Mineralogy, v. 4, p. 921–931.

Baymann, F., Brugna, M., Mühlenhoff, U., and Nitschke, W., 2001, Daddy, where did (PS)I come from?: Biochimica et Biophysica Acta, v. 1507, p. 291–310.

Baymann, F., Lebrun, E., Brugna, M., Schoepp-Cothenet, B., Giudici-Orticoni, M.T., and Nitschke, W., 2003, The redox protein construction kit: pre-last universal common ancestor evolution of energy-conserving enzymes: Philosophical Transactions of the Royal Society of London, ser. B, v. 358, p. 267–274, doi: 10.1098/rstb.2002.1184.

Beinert, H., Holm, R.H., and Münck, E., 1997, Iron-sulfur clusters: Nature's modular, multipurpose structures: Science, v. 277, p. 653–659, doi: 10.1126/science.277.5326.653.

Bekker, A., Holland, H.D., Wang, P.-L., Stein, H.J., Hannah, J.L., Coetzee, L.L., and Beukes, N.J., 2004, Dating the rise of atmospheric oxygen: Nature, v. 427, p. 117–120, doi: 10.1038/nature02260.

Béland, P., and Allen, T.F.H., 1994, The origin and evolution of the genetic code: Journal of Theoretical Biology, v. 170, p. 359–365, doi: 10.1006/jtbi.1994.1198.

Bernal, J.D., 1960. The problem of stages in biopoesis, in Florkin, M. ed., Aspects of the origin of life: New York, Pergamon Press, p. 30–45.

Bethke, C., 1996, Geochemical reaction modeling: Oxford, UK, Oxford University Press, 397 p.

Beutner, R., 1913, New electric properties of a semipermeable membrane of copper ferrocyanide: Journal of Physical Chemistry, v. 17, p. 344–360, doi: 10.1021/j150139a006.

Bischoff, J.L., and Seyfried, W.E., 1978, Hydrothermal chemistry of seawater from 25° to 350°C: American Journal of Science, v. 278, p. 838–860.

Bjerrum, C.J., and Canfield, D.E., 2002, Ocean productivity before about 1.9 Gyr ago limited by phosphorus adsorption onto iron oxides: Nature, v. 417, p. 159–162, doi: 10.1038/417159a.

Blank, C.E., 2004, Evolutionary timing of the origins of mesophilic sulphate reduction and oxygenic photosynthesis: A phylogenomic dating approach: Geobiology, v. 2, p. 1–20, doi: 10.1111/j.1472-4677.2004.00020.x.

Blankenship, R.E., 2002, Molecular mechanisms of photosynthesis: Oxford, Blackwell Science.

Bonatti, E., Simmons, E.C., Breger, D., Hamlyn, P.R., and Lawrence, J., 1983, Ultramafic rock/seawater interaction in the oceanic crust: Mg-silicate (sepiolite) deposit from the Indian Ocean floor: Earth and Planetary Science Letters, v. 62, p. 229–238, doi: 10.1016/0012-821X(83)90086-9.

Bonomi, F., Werth, M.T., and Kurtz, D.M., 1985, Assembly of $Fe_nS_n(SR)_2$ (n=2,4) in aqueous media from iron salts, thiols and sulfur, sulfide, thiosulfide plus rhodonase: Inorganic Chemistry, v. 24, p. 4331–4335, doi: 10.1021/ic00219a026.

Bounama, C., Franck, S., and von Bloh, W., 2001, The fate of the Earth's ocean: Hydrology and Earth System Sciences, v. 5, p. 569–575.

Boyce, A.J., Coleman, M.L., and Russell, M.J., 1983, Formation of fossil hydrothermal chimneys and mounds from Silvermines, Ireland: Nature, v. 306, p. 545–550, doi: 10.1038/306545a0.

Braun, D., and Libchaber, A., 2004, Thermal force approach to molecular evolution: Physical Biology, v. 1, p. 1–8 (DOI: 10.1088/1478-3967/1/1/P01), 8 p.

Brochier, C., and Philippe, H., 2002, A non-hyperthermophilic ancestor for bacteria: Nature, v. 417, p. 244, doi: 10.1038/417244a.

Burns, R.G., and Burns, V.M., 1979, Manganese oxides, in Burns, R.G., ed., Marine minerals: Washington, D.C., Mineralogical Society of America, Reviews in Mineralogy, v. 6, p. 1–46.

Cairns-Smith, A.G., 1982, Genetic takeover and the mineral origins of life: Cambridge, UK, Cambridge University Press, 477 p.

Cairns-Smith, A.G., Hall, A.J., and Russell, M.J., 1992, Mineral theories of the origin of life and an iron sulphide example: Origins of Life and Evolution of the Biosphere, v. 22, p. 161–180, doi: 10.1007/BF01808023.

Campbell, I.H., Griffiths, R.W., and Hill, R.I., 1989, Melting in an Archaean mantle plume: Heads it's basalts, tails it's komatiites: Nature, v. 339, p. 697–699, doi: 10.1038/339697a0.

Canuto, V.M., Levine, J.S., Augustsson, T.R., and Imhoff, C.L., 1982, UV radiation from the young Sun and oxygen and ozone levels in the prebiological palaeoatmosphere: Nature, v. 296, p. 816–820, doi: 10.1038/296816a0.

Catling, D.C., Zahnle, K.J., and McKay, C.P., 2001, Biogenic methane, hydrogen escape, and the irreversible oxidation of early Earth: Science, v. 293, p. 839–843, doi: 10.1126/science.1061976.

Chafetz, H.S., Akdim, B., Julia, R., and Reid, A., 1998, Mn- and Fe-rich black travertine shrubs: Bacterially (and nanobacterially) induced precipitates: Journal of Sedimentary Research, v. 68, p. 404–412.

Chen, K., Hirst, J., Camba, R., Bonagura, C.A., Stout, C.D., Burgess, B.K., and Armstrong, F.A., 2000, Atomically defined mechanism for proton transfer to a buried redox centre in a protein: Nature, v. 405, p. 814–817, doi: 10.1038/35015610.

Cockell, C.S., and Knowland, J., 1999, UV radiation screening compounds: Biological Reviews, v. 74, p. 311–345, doi: 10.1017/S0006323199005356.

Cody, G.D., 2004, Transition metal sulfides and the origin of metabolism: Annual Review of Earth and Planetary Sciences, v. 32, p. 569–599, doi: 10.1146/annurev.earth.32.101802.120225.

Cody, G.D., Boctor, N.Z., Brandes, J.A., Filley, T.R., Hazen, R.M., and Yoder, H.S., 2004, Assaying the catalytic potential of transition metal sulfides for abiotic carbon fixation: Geochimica et Cosmochimica Acta, v. 68, p. 2185–2196, doi: 10.1016/j.gca.2003.11.020.

Cole, W.J., Kaschke, M., Sherringham, J.A., Curry, G.B., Turner, D., and Russell, M.J., 1994, Can amino acids be synthesised by H_2S in anoxic lakes?: Marine Chemistry, v. 45, p. 243–256, doi: 10.1016/0304-4203(94)90007-8.

Corliss, J.B., Baross, J.A., and Hoffman, S.E., 1981, An hypothesis concerning the relationship between submarine hot springs and the origin of life on Earth: Proceedings, 26th International Geological Congress, Geology of Oceans Symposium, Paris, July 7–17, 1980, Oceanologica Acta, no. SP, p. 59–69.

Crick, F.H.C., 1968, The origin of the genetic code: Journal of Molecular Biology, v. 38, p. 367–379, doi: 10.1016/0022-2836(68)90392-6.

da Silva, J.J.R.F., and Williams, R.J.P., 1991, The biological chemistry of the elements: Oxford, UK, Clarendon Press, 561 p.

Daly, M.J., Gaidamakova, E.K., Matrosova, V.Y., Vasilenko, A., Zhai, M., Venkateswaran, A., Hess, M., Omelchenko, M.V., Kostandarithes, H.M., Makarova, K.S., Wackett, L.P., Fredrickson, J.K., and Ghosall, D., 2004, Accumulation of Mn(II) in Deinococcus radiodurans facilitates gamma-radiation resistance: Science, v. 306, p. 1025–1028, doi: 10.1126/science.1103185.

Daniel, R.M., and Danson, M.J., 1995, Did primitive microorganisms use non-heme iron proteins in place of NAD/P?: Journal of Molecular Evolution, v. 40, p. 559–563, doi: 10.1007/BF00160501.

Darnault, C., Volbeda, A., Kim, E.J. Legrand, P., Vernède, X., Lindahl, P.A., and Fontecilla-Camps, J.C., 2003, Ni-Zn-[Fe_4-S_4] and Ni-Ni-[Fe_4-S_4] clusters in closed and open α subunits of acetyl-CoA synthase/carbon monoxide dehydrogenase: Nature Structural Biology, v. 10, p. 271–279, doi: 10.10.1038/nsb912.

Davies, G.F., 1992, On the emergence of plate tectonics: Geology, v. 20, p. 963–966, doi: 10.1130/0091-7613(1992)020<0963:OTEOPT>2.3.CO;2.

Deamer, D.W., 1985, Boundary structures are formed by organic components of the Murchison carbonaceous chondrite: Nature, v. 317, p. 792–794, doi: 10.1038/317792a0.

de Duve, C., 1991, Blueprint for a cell: The nature and origin of life: Burlington, North Carolina, Neil Patterson Publishers, 275 p.

de Zwart, I.I., Meade, S.J., and Pratt, A.J., 2004, Biomimetic phosphoryl transfer catalysed by iron(II)-mineral precipitates: Geochimica et Cosmochimica Acta, v. 68, p. 4093–4098, doi: 10.1016/j.gca.2004.01.028.

Dismukes, G.C., Klimov, V.V., Baranov, S.V., Kozlov, Yu.N., DasGupta, J., and Tyryshkin, A., 2001, The origin of atmospheric oxygen on Earth: The innovation of oxygenic photosynthesis: Proceedings of the National Academy of Sciences of the United States of America, v. 98, p. 2170–2175, doi: 10.1073/pnas.061514798.

Dobbek, H., Svetlitchnyi, V., Gremer, L., Huber, R., and Meyer, O., 2001, Crystal structure of a carbon monoxide dehydrogenase reveals a [Ni-4Fe-5S] cluster: Science, v. 293, p. 1281–1285, doi: 10.1126/science.1061500.

Doukov, T.I., Iverson, T.M., Seravalli, J., Ragsdale, S.W., and Drennan, C.L., 2002, A Ni-Fe-Cu center in a bifunctional carbon monoxide dehydrogenase/acetyl-CoA synthase: Science, v. 298, p. 567–572, doi: 10.1126/science.1075843.

Douville, E., Charlou, J.L., Oelkers, E.H., Bienvenu, P., Colon, C.F.J., Donval, J.P., Fouquet, Y., Prieur, D., and Appriou, P., 2002, The rainbow vent fluids (36°14′N, MAR): The influence of ultramafic rocks and phase separation on trace metal content in Mid-Atlantic Ridge hydrothermal fluids: Chemical Geology, v. 184, p. 37–48, doi: 10.1016/S0009-2541(01)00351-5.

Drennan, C.L., Heo, J., Sintchak, M.D., Schreiter, E., and Ludden, P.W., 2001, Life on carbon monoxide: X-ray structure of *Rhodospirillium rubrum* Ni-Fe-S carbon monoxide dehydrogenase: Proceedings of the National Academy of Sciences of the United States of America, v. 98, p. 11973–11978, doi: 10.1073/pnas.211429998.

Drobner, E., Huber, H., Wächtershäuser, G., Rose, D., and Stetter, K.O., 1990, Pyrite formation linked with hydrogen evolution under anaerobic conditions: Nature, v. 346, p. 742–744, doi: 10.1038/346742a0.

Dymek, R.F., and Klein, C., 1988, Chemistry, petrology and origin of banded iron-formation lithologies from the 3800 Ma Isua Supracrustal Belt: West Greenland: Precambrian Research, v. 39, p. 247–302, doi: 10.1016/0301-9268(88)90022-8.

Eck, R.V., and Dayhoff, M.O., 1966, Evolution of the structure of ferredoxin based on living relics of primitive amino acid sequences: Science, v. 152, p. 363–366.

Einsle, O., Tezcan, F.A., Andrade, S.L.A., Schmid, B., Yoshida, M., Howard, J.B., and Rees, D.C., 2002, Nitrogenase MoFe-protein at 1.16Å resolution: A central ligand in the FeMo-cofactor: Science, v. 297, p. 1696–1700, doi: 10.1126/science.1073877.

Elston, T., Wang, H., and Oster, G., 1998, Energy transduction in ATP synthase: Nature, v. 391, p. 510–513, doi: 10.1038/35185.

Eschenmoser, A., 1988, Vitamin B12: Experiments concerning the origin of its molecular structure: Angewandte Chemie International Edition in English, v. 27, p. 5–39, doi: 10.1002/anie.198800051.

Farquhar, J., Bao, H., and Thiemens, M.H., 2000, Atmospheric influence of Earth's earliest sulfur cycle: Science, v. 289, p. 756–758, doi: 10.1126/science.289.5480.756.

Ferris, F.G., Jack, T.R., and Bramhill, B.J., 1992, Corrosion products associated with attached bacteria at an oil field water injection plant: Canadian Journal of Microbiology, v. 38, p. 1320–1324.

Ferris, J.P., and Hagan, W.J., 1984, HCN and chemical evolution: The possible role of cyano compounds in prebiotic synthesis: Tetrahedron, v. 40, p. 1093–1120, doi: 10.1016/S0040-4020(01)99315-9.

Ferris, J.P., Hill, A.R., Liu, R., and Orgel, L.E., 1996, Synthesis of long prebiotic oligomers on mineral surfaces: Nature, v. 381, p. 59–61, doi: 10.1038/381059a0.

Ferris, J.P., Joshi, P.C., Edelson, E.H., and Lawless, J.G., 1978, HCN: A plausible source of purines, pyrimidines and amino acids on the primitive Earth: Journal of Molecular Evolution, v. 11, p. 293–311, doi: 10.1007/BF01733839.

Filtness, M.J., Butler, I.B., and Rickard, D., 2003, The origin of life: The properties of iron sulphide membranes: Transactions of the Institution of Mining and Metallurgy, Applied Earth Science, v. 112B, p. 171–172.

Finklea, S., Cathey, S., and Amma, E.L., 1976, Investigation of the bonding mechanism in pyrite using the Mössbauer effect: Acta Crystallographica, v. A32, p. 529–537.

Foley, S.F., Buhre, S., and Jacob, D.E., 2003, Evolution of the Archaean crust by delamination and shallow subduction: Nature, v. 421, p. 249–252.

Fontecilla-Camps, J.C., and Ragsdale, S.W., 1999, Nickel-iron-sulfur active sites: Hydrogenase and CO dehydrogenase: Advances in Inorganic Chemistry, v. 47, p. 283–333.

Forterre, P., 1995, Thermoreduction, a hypothesis for the origin of prokaryotes: C.R: Academy of Science, v. 318, p. 415–422.

Forterre, P., 2002, A hot story from comparative genomics: reverse gyrase is the only hyperthermophile-specific protein: Trends in Genetics, v. 18, p. 236–237, doi: 10.1016/S0168-9525(02)02650-1.

Früh-Green, G.L., Kelley, D.D., Bernascono, S.M., Karson, J.A.. Ludwig, K.A., Butterfield, D.A., Boschi, C., and Proskurowski, G., 2003, 30,000 years of hydrothermal activity at the Lost City vent field: Science, v. 301, p. 495–498, doi: 10.1126/science.1085582.

Gaffey, M.J., 1997, The early solar system: Origins of Life and Evolution of the Biosphere, v. 27, p. 185–203, doi: 10.1023/A:1006578315384.

Garrels, R.M., and Christ, C.L., 1965, Minerals, solutions and equilibria: New York, Harper and Row, 450 p.

Gaucher, E.A., Thomson, J.M., Burgan, M., and Benner, S.A., 2003, Inferring the palaeoenvironment of ancient bacteria on the basis of resurrected proteins: Nature, v. 425, p. 285–288, doi: 10.1038/nature01977.

Geptner, A., Kristmannsdóttir, H., Kristjánsson, J.K., and Marteinsson, V.Th., 2002, Biogenic saponite from an active submarine hot spring, Iceland: Clays and Clay Minerals, v. 50, p. 174–185, doi: 10.1346/000986002760832775.

Glansdorff, N., 1999, On the origin of operons and their possible role in evolution toward thermophily: Journal of Molecular Evolution, v. 49, p. 432–438.

Goldschmidt, V.M., 1937, The principles of distribution of chemical elements in minerals and rocks: Journal of the Chemical Society, v. 1937, p. 655–673.

Goldschmidt, V.M., 1952, Geochemical aspects of the origin of complex organic molecules on Earth, as precursors to organic life: New Biology, v. 12, p. 97–105.

Haeckel, E., 1892, The history of creation, Volume 1 (4th edition), translated by E.R. Lankester: London, Kegan Paul, Trench, Trübner and Co., 422 p.

Haldane, J.B.S., 1929, The origin of life: Rationalist Annual, v. 3, p. 3–10.

Hall, D.O., Cammack, R., and Rao, K.K., 1971, Role for ferredoxins in the origin of life and biological evolution: Nature, v. 233, p. 136–138, doi: 10.1038/233136a0.

Hall, A.J., Boyce, A.J., and Fallick, A.E., 1994, A sulphur isotope study of iron sulphide in the late Precambrian Dalradian Ardrishaig Phyllite Formation, Knapdale, Argyll: Scottish Journal of Geology, v. 30, p. 63–71.

Hanczyc, M.M., Fujikawa, S.M., and Szostak, J.W., 2003, Experimental models of primitive cellular compartments: Encapsulation, growth and division: Science, v. 302, p. 618–622, doi: 10.1126/science.1089904.

Hansson, Ö., and Wydrzynski, T., 1990, Current perceptions of Photosystem II: Photosynthesis Research, v. 23, p. 131–162, doi: 10.1007/BF00035006.

Heinen, W., and Lauwers, A.M., 1996, Organic sulfur compounds resulting from the interaction of iron sulfide, hydrogen sulfide and carbon dioxide in an anaerobic aqueous environment: Origins of Life and Evolution of the Biosphere, v. 26, p. 131–150, doi: 10.1007/BF01809852.

Heinen, W., and Lauwers, A.M., 1997, The iron-sulfur world and the origins of life: Abiotic synthesis from metallic iron, H_2S and CO_2: A comparison of the thiol generating $FeS/HCl(H_2S)/CO_2$-system and its $Fe^0/H_2S/CO_2$-counterpart: Proceedings Koninklijke Nederlandse Akademie van Wetenschappen, Amsterdam, v. 100, p. 11–25.

Helz, G.R., Miller, C.V., Charnock, J.M., Mosselmans, J.F.W., Pattrick, R.A.D., Garner, C.D., and Vaughan, D.J., 1996, Mechanism of molybdenum removal from the sea and its concentration in black shales: EXAFS evidence: Geochimica et Cosmochimica Acta, v. 60, p. 3631–3642.

Hemley, J.J., Cygan, G.L., Fein, J.B., Robinson, G.R., and D'Angelo, W.M., 1992, Hydrothermal ore-forming processes in the light of studies in rock-buffered systems: I. Iron-copper-zinc-lead sulfide solubility relations: Economic Geology and the Bulletin of the Society of Economic Geologists, v. 87, p. 1–22.

Hennet, R.J.-C., Holm, N.G., and Engel, M.H., 1992, Abiotic synthesis of amino acids under hydrothermal conditions and the origin of life: A perpetual phenomenon?: Die Naturwissenschaften, v. 79, p. 361–365, doi: 10.1007/BF01140180.

Holm, N.G., 1987, Possible biological origin of banded iron formations from hydrothermal solutions: Origins of Life and Evolution of the Biosphere, v. 17, p. 229–250.

Huber, C., and Wächtershäuser, G., 1997, Activated acetic acid by carbon fixation on (Fe,Ni)S under primordial conditions: Science, v. 276, p. 245–247, doi: 10.1126/science.276.5310.245.

Huber, C., and Wächtershäuser, G., 1998, Peptides by activation of amino acids on (Fe,Ni)S surfaces: Implications for the origin of life: Science, v. 281, p. 670–672, doi: 10.1126/science.281.5377.670.

Huber, C., and Wächtershäuser, G., 2003, Primordial reductive amination revisited: Tetrahedron Letters, v. 44, p. 1695–1697, doi: 10.1016/S0040-4039(02)02863-6.

Huber, C., Eisenreich, W., Hecht, S., and Wächtershäuser, G., 2003, A possible primordial peptide cycle: Science, v. 301, p. 938–940, doi: 10.1126/science.1086501.

Janecky, D.R., and Seyfried, W.E., 1983, The solubility of magnesium-hydroxide-sulfate-hydrate in seawater at elevated temperatures and pressures: American Journal of Science, v. 283, p. 831–860.

Jernigan, R., Raghunathan, G., and Bahar, I., 1994, Characterization of interactions and metal ion binding sites in proteins: Current Opinions in Structural Biology, v. 4, p. 256–263, doi: 10.1016/S0959-440X(94)90317-4.

Jimenez-Sanchez, A., 1995, On the origin and evolution of the genetic code: Journal of Molecular Evolution, v. 41, p. 712–716.

Joyce, G.F., 1989, RNA evolution and the origins of life: Nature, v. 338, p. 217–224, doi: 10.1038/338217a0.

Joyce, G.F., Visser, G.M., van Boeckel, C.A.A., van Boom, J.H., Orgel, L.E., and Westrenen, J., 1984, Chiral selection in poly(C)-directed synthesis of oligo(G): Nature, v. 310, p. 602–604, doi: 10.1038/310602a0.

Kakegawa, T., Noda, M., and Nannri, H., 2002, Geochemical cycles of bioessential elements on the early Earth and their relationships to the origin of life: Resource Geology, v. 52, p. 83–89.

Karsten, J.L., Klein, E.M., and Sherman, S.B., 1996, Subduction zone geochemical characteristics in ocean ridge basalts from the southern Chile ridge: Implications of modern subduction systems for the Archean: Lithos, v. 37, p. 143–161, doi: 10.1016/0024-4937(95)00034-8.

Kashefi, K., Tor, J.M., Holmes, D.E., Gaw Van Praagh, C.V., Reysenbach, A.-L., and Lovley, D.R., 2002, *Geoglobus ahangari*, gen. nov., sp., nov., a novel hyperthermophile capable of oxidizing organic acids and growing autotrophically on hydrogen with Fe (III) serving as the sole electron acceptor: International Journal of Systematic and Evolutionary Microbiology, v. 52, p. 719–728, doi: 10.1099/ijs.0.01953-0.

Kasting, J.F., 1993, Earth's early atmosphere: Science, v. 259, p. 920–926.

Kasting, J.F., 2001, The rise of atmospheric oxygen: Science, v. 293, p. 819–820, doi: 10.1126/science.1063811.

Kasting, J.F., and Ackerman, T.P., 1986, Climatic consequences of very high carbon dioxide levels in the earth's early atmosphere: Science, v. 234, p. 1383–1385.

Kasting, J.F., and Brown, L.L., 1998, The early atmosphere as a source of biogenic compounds, in Brack, A., ed., The molecular origins of life: Cambridge, UK, Cambridge University Press, p. 35–56.

Kell, D.B., 1988, Protonmotive energy-transducing mechanisms: some physical principles and experimental approaches, in Anthony, C. ed., Bacterial energy transduction: London, Academic Press, p. 429–490.

Kelley, D.S., Karson, J.A., Blackman, D.K., et al., 2001, An off-axis hydrothermal vent field near the Mid-Atlantic Ridge at 30° N: Nature, v. 412, p. 145–149, doi: 10.1038/35084000.

Kelley, D.S., Karson, J.A., Früh-Green, G.L., et al., 2005, A serpentinite-hosted ecosystem: The Lost City hydrothermal field: Science, v. 307, p. 1428–1434, doi: 10.1126/science.1102556.

Koga, Y., Kyuragi, T., Nishihara, M., and Sone, N., 1998, Did archaeal and bacterial cells arise independently from noncellular precursors? A hypothesis stating that the advent of membrane phospholipid with enantiomeric glycerophosphate backbones caused the separation of the two lines of descent: Journal of Molecular Evolution, v. 46, p. 54–63.

Konecny, J., Schöniger, M., and Hofacker, G.L., 1995, Complementary coding conforms to the primeval comma-less code: Journal of Theoretical Biology, v. 173, p. 263–270, doi: 10.1006/jtbi.1995.0061.

Koonin, E.V., and Martin, W., 2005, On the origin of genomes and cells within inorganic compartments: Trends in Genetics, v. 21, p. 647–654.

Krishnarao, J.S.R., 1964, Native nickel-iron alloy, its mode of occurrence, distribution and origin: Economic Geology and the Bulletin of the Society of Economic Geologists, v. 59, p. 443–448.

Krupp, R.E., 1994, Phase relations and phase transformations between low temperature iron sulfides mackinawite, greigite and smythite: European Journal of Mineralogy, v. 6, p. 389–396.

Lalou, C., Reys, J.-L., Brichet, E., Arnold, M., Thompson, G., Fouquet, Y., and Rona, P.A., 1993, New age data for Mid-Atlantic Ridge hydrothermal sites: TAG and Snakepit chronology revisited: Journal of Geophysical Research, v. 98, p. 9705–9713.

Lécuyer, C., and Ricard, Y., 1999, Long-term fluxes and budget of ferric iron for the redox states of the Earth's mantle and atmosphere: Earth and Planetary Science Letters, v. 165, p. 197–211, doi: 10.1016/S0012-821X(98)00267-2.

Leduc, S., 1911, The mechanism of life: London, Rebman, 172 p.

Leja, J., 1982, Surface chemistry of froth flotation: New York, Plenum Press, 640 p.

Leman, L., Orgel, L., and Ghadiri, M.R., 2004, Carbonyl sulfide-mediated prebiotic formation of peptides: Science, v. 306, p. 283–286, doi: 10.1126/science.1102722.

Lindahl, P.A., 2002, The Ni-containing carbon monoxide dehydrogenase family: Light at the end of the tunnel?: Biochemistry, v. 41, p. 2097–2105, doi: 10.1021/bi015932+.

Liu, S.V., Zhou, J., Zhang, C., Cole, D.R., Gajdarziska-Josifovska, M., and Phelps, T.J., 1997, Thermophilic Fe(III)-reducing bacteria from the deep subsurface: The evolutionary implications: Science, v. 277, p. 1106–1109, doi: 10.1126/science.277.5329.1106.

Loll, B, Kern, J., Saenger, W., Zouni, A., and Biesiadka, J., 2005, Towards complete cofactor arrangement in the 3.0Å resolution structure of photosystem II: Nature, v. 438, p. 1040–1044, doi: 10.1038/nature04224.

Luther, G.W., 2004, Kinetics of the reactions of water, hydroxide ion and sulfide species with CO_2, OCS and CS: Frontier molecular orbital considerations: Aqueous Geochemistry, v. 10, p. 81–97.

Luther, G.W., Theberge, S.M., and Rickard, D.T., 1999, Evidence for aqueous clusters as intermediates during zinc sulfide formation: Geochimica et Cosmochimica Acta, v. 63, p. 3159–3169, doi: 10.1016/S0016-7037(99)00243-4.

Macalady, J., and Banfield, J.F., 2003, Molecular geomicrobiology: Genes and geochemical cycling: Earth and Planetary Science Letters, v. 209, p. 1–17, doi: 10.1016/S0012-821X(02)01010-5.

MacDermott, A.J., Tranter, G.E., and Trainor, S.J., 1992, The search for large parity-violating energy differences finds fruit in thiosubstituted DNA analogues: Chemical Physics Letters, v. 194, p. 152–156, doi: 10.1016/0009-2614(92)85525-F.

Macleod, G., McKeown, C., Hall, A.J., and Russell, M.J., 1994, Hydrothermal and oceanic pH conditions of possible relevance to the origin of life: Origins of Life and Evolution of the Biosphere, v. 24, p. 19–41, doi: 10.1007/BF01582037.

Margolis, S.V., Ku, T.L., Glasby, G.P., Fein, C.D., and Audley-Charles, M.G., 1978, Fossil manganese nodules from Timor: Geochemical and radiochemical evidence for deep-sea origin: Chemical Geology, v. 21, p. 185–198, doi: 10.1016/0009-2541(78)90044-X.

Marshall, W.L., 1994, Hydrothermal synthesis of amino acids: Geochimica et Cosmochimica Acta, v. 58, p. 2099–2106, doi: 10.1016/0016-7037(94)90288-7.

Marteinsson, V.Th., Kristjánsson, J.K., Kristmannsdöttir, H., et al., 2001, Discovery of giant submarine smectite cones on the seafloor in Eyjafjordur, Northern Iceland, and a novel thermal microbial habitat: Applied and Environmental Microbiology, v. 67, p. 827–833, doi: 10.1128/AEM.67.2.827-833.2001.

Martin, B., and Fyfe, W.S., 1970, Some experimental and theoretical observations on the kinetics of hydration reactions with particular reference to serpentinization: Chemical Geology, v. 6, p. 185–202.

Martin, W., and Russell, M.J., 2003, On the origin of cells: An hypothesis for the evolutionary transitions from abiotic geochemistry to chemoautotrophic prokaryotes, and from prokaryotes to nucleated cells: Philosophical Transactions of the Royal Society of London, ser. B, v. 358, p. 27–85, doi: 10.1098/rstb.2002.1183.

McCollom, T., and Seewald, J.S., 2003, Experimental constraints on the hydrothermal reactivity of organic acids and acid anions: I. Formic acid and formate: Geochimica et Cosmochimica Acta, v. 67, p. 3625–3644, doi: 10.1016/S0016-7037(03)00136-4.

Mellersh, A.R., 1993, A model for the prebiotic synthesis of peptides which throws light on the origin of the genetic code and the observed chirality of life: Origins of Life and Evolution of the Biosphere, v. 23, p. 261–274, doi: 10.1007/BF01581903.

Mellersh, A.R., and Wilkinson, A.-S., 2000, RNA bound to a solid phase can select an amino acid and facilitate subsequent amide bond formation: Origins of Life and Evolution of the Biosphere, v. 30, p. 3–7, doi: 10.1023/A:1006620421068.

Michel, H., and Deisenhofer, J., 1988, Relevance of the photosynthetic reaction center from purple bacteria to the structure of photosystem II: Biochemistry, v. 27, p. 1–7, doi: 10.1021/bi00401a001.

Miller, S.L., 1992, The prebiotic synthesis of organic compounds as a step toward the origin of life, in Schopf, J.W., ed., Major events in the history of life: Boston, Jones and Bartlett, p. 1–28.

Miller, S.L., and Bada, J.L., 1988, Submarine hot springs and the origin of life: Nature, v. 334, p. 609–611, doi: 10.1038/334609a0.

Milner-White, E.J., and Russell, M.J., 2005, Nests as sites for phosphates and iron-sulfur thiolates in the first membranes: 3 to 6 residue anion-binding motifs: Origins of Life and Evolution of the Biosphere, v. 35, p. 19–27, doi: 10.1007/s11084-005-4582-7.

Mitchell, P., 1967, Proton-translocation phosphorylation in mitochondria, chloroplasts and bacteria: Natural fuel cells and solar cells: FASEB, v. 26, p. 1370–1379.

Morse, J.W., and Arakaki, T., 1993, Adsorption and coprecipitation of divalent metals with mackinawite (FeS): Geochimica et Cosmochimica Acta, v. 57, p. 3635–3640, doi: 10.1016/0016-7037(93)90145-M.

Moulton, V., Gardner, P.P., Pointon, R.F., Creamer, L.K., Jameson, G.B., and Penny, D., 2000, RNA folding argues against a hot origin of life: Journal of Molecular Evolution, v. 51, p. 416–421.

Mulkidjanian, A.Y., and Junge, W., 1997, On the origin of photosynthesis as inferred from sequence analysis—A primordial UV-protector as common ancestor of reaction centers and antenna proteins: Photosynthesis Research, v. 51, p. 27–42, doi: 10.1023/A:1005726809084.

Muller, A.W.J., 1995, Were the first organisms heat engines? A new model for biogenesis and the early evolution of biological energy conservation: Progress in Biophysics and Molecular Biology, v. 63, p. 193–231, doi: 10.1016/0079-6107(95)00004-7.

Müller, V., 2003, Energy conservation in acetogenic bacteria: Applied and Environmental Microbiology, v. 69, p. 6345–6353, doi: 10.1128/AEM.69.11.6345-6353.2003.

Muth, G.W., Orteleva-Donnelly, L., and Strobel, S.A., 2000, A single adenosine with a neutral pK_a in the ribosomal peptidyl transferase center: Science, v. 289, p. 947–950, doi: 10.1126/science.289.5481.947.

Myers, C.R., and Nealson, K.H., 1988, Bacterial manganese reduction and growth with manganese oxide as the sole electron acceptor: Science, v. 240, p. 1319–1321.

Nealson, K.H., and Stahl, D.A., 1997, Microorganisms and biogeochemical cycles: What can we learn from layered microbial communities, in Banfield, J.F., and Nealson, K.H., eds., Geomicrobiology: Interactions between microbes and minerals: Washington, D.C., Mineralogical Society of America, Reviews in Mineralogy, v. 35, p. 5–34.

Nisbet, E.G., and Sleep, N.H., 2001, The habitat and nature of early life: Nature, v. 409, p. 1083–1091, doi: 10.1038/35059210.

Ohmoto, H., Kakegawa, T., and Lowe, D.R., 1993, 3.4-billion-year-old biogenic pyrites from Barberton, South Africa: Sulfur isotope evidence: Science, v. 262, p. 555–557.

Oparin, A.I., 1924, Proiskhozhdenie Zhizny: Moscow, Rabochiĭ.

Oparin, A.I., 1938, Origin of life: New York, Dover.

Oró, L., 1961, Comets and the formation of biochemical compounds on the primitive earth: Nature, v. 190, p. 389–390.

Oró, J., and Kimball, A.P., 1961, Synthesis of purines under possible primitive Earth conditions. I. Adenine from hydrogen cyanide: Archives of Biochemistry and Biophysics, v. 94, p. 217–227, doi: 10.1016/0003-9861(61)90033-9.

Pace, N.R., 2002, The large scale topology of the tree of life [abs.]: Astrobiology, v. 2, p. 484.

Palandri, J.L., and Reed, M.H., 2004, Geochemical models of metasomatism in ultramafic systems: Serpentinization, rodingitization, and sea floor carbonate chimney precipitation: Geochimica et Cosmochimica Acta, v. 68, p. 1115–1133, doi: 10.1016/j.gca.2003.08.006.

Parkes, R.J., Cragg, B.A., Fry, J.C., Herbert, R.A., and Wimpenny, J.W.T., 1990, Bacterial biomass and activity in deep sediment layers from the Peru margin: Philosophical Transactions of the Royal Society of London, ser. A, v. 331, p. 139–153.

Parkes, R.J., Cragg, B.A., Bale, S.J., Getliff, J.M., Goodman, K., Rochelle, P.A., Fry, J.C., Weightman, A.J., and Harvey, S.M., 1994, Deep bacterial biosphere in Pacific Ocean sediments: Nature, v. 371, p. 410–413, doi: 10.1038/371410a0.

Pavlov, A.A., and Kasting, J.F., 2002, Mass-independent fractionation of sulfur isotopes in Archean sediments: Strong evidence for an anoxic Archean atmosphere: Astrobiology, v. 2, p. 27–41, doi: 10.1089/15311070275362 1321.

Pedersen, K., 1993, The deep subterranean biosphere: Earth Science Reviews, v. 34, p. 243–260, doi: 10.1016/0012-8252(93)90058-F.

Peretó, J.G., Velasco, A.M., Becerra, A., and Lazcano, A., 1999, Comparative biochemistry of CO_2 fixation and the evolution of autotrophy: International Microbiology, v. 2, p. 3–10.

Phoenix, V.R., Konhauser, K.O., Adams, D.G., and Bottrell, S.H., 2001, Role of biomineralization as an ultraviolet shield: Implications for Archean life: Geology, v. 29, p. 823–826, doi: 10.1130/0091-7613(2001)029<0823:ROBAAU>2.0.CO;2.

Pinti, D.L., 2002, The isotopic record of Archean nitrogen and the early evolution of the early Earth: Trends in Geochemistry, v. 2, p. 117.

Pontes-Buarques, M., Tessis, A.C., Bonapace, J.A.P., Monte, M.B.M., Cortés-Lopez, G., de Souza-Barros, F., and Vieyra, A., 2001, Modulation of adenosine 5′-monophosphate adsorption onto aqueous resident pyrite: Potential mechanisms for prebiotic reactions: Origins of Life and Evolution of the Biosphere, v. 31, p. 343–362, doi: 10.1023/A:1011805332303.

Poole, A.M., Jeffares, D.C., and Penny, D., 1999, Early evolution: prokaryotes, the new kids on the block: BioEssays, v. 21, p. 880–889, doi: 10.1002/(SICI)1521-1878(199910)21:10<880::AID-BIES11>3.0.CO;2-P.

Pratt, J.M., 1993, Nature's design and use of catalysts based on Co and the macrocyclic corrin ligand: 4×10^9 years of coordination chemistry: Pure and Applied Chemistry, v. 65, p. 1513–1520.

Prigogine, I., 1978, Time, structure, and fluctuations: Science, v. 201, p. 777–785.

Pullman, B., 1972, Electronic factors in biochemical evolution, in Ponnamperuma, C., ed., Exobiology: North Holland Publishing Company, p. 136–169.

Quayle, R.J., and Ferenci, T., 1978, Evolutionary aspects of autotrophy: Microbiological Reviews, v. 42, p. 251–273.

Raymond, J., Siefert, J.L., Staples, C.R., and Blankenship, R.E., 2004, The natural history of nitrogen fixation: Molecular Biology and Evolution, v. 21, p. 541–554.

Reader, J.S., and Joyce, G.F., 2002, A ribozyme composed of only two different nucleotides: Nature, v. 420, p. 841–844, doi: 10.1038/nature01185.

Ricardo, A., Carrigan, M.A., Olcott, A.N., and Benner, S.A., 2004, Borate minerals stabilize ribose: Science, v. 303, p. 196, doi: 10.1126/science.1092464.

Rickard, D., 1997, Kinetics of pyrite formation by the H_2S oxidation of iron(II) monosulfide in aqueous solutions between 25° and 125°C: The rate equation: Geochimica et Cosmochimica Acta, v. 61, p. 115–134, doi: 10.1016/S0016-7037(96)00321-3.

Rickard, D., Butler, I.B., and Olroyd, A., 2001, A novel iron sulphide switch and its implications for earth and planetary science: Earth and Planetary Science Letters, v. 189, p. 85–91, doi: 10.1016/S0012-821X(01)00352-1.

Righter, K., Drake, M.J., and Yaxley, G., 1997, Prediction of siderophile element metal-silicate partition coefficients to 20 GPa and 2,800°C: The effects of pressure, temperature, oxygen fugacity, and silicate and metallic melt compositions: Physics of the Earth and Planetary Interiors, v. 100, p. 115–134, doi: 10.1016/S0031-9201(96)03235-9.

Romero, I., Gómez-Priego, A., and Celis, H., 1991, A membrane-bound pyrophosphatase from respiratory membranes of *Rhodospirillium rubrum*: Journal of General Microbiology, v. 137, p. 2611–2616.

Rosing, M.T., 1999, ^{13}C-depleted carbon microparticles in >3700-Ma sea-floor sedimentary rocks from West Greenland: Science, v. 283, p. 674–676, doi: 10.1126/science.283.5402.674.

Rosing, M.T., and Frei, R., 2004, U-rich Archaean sea-floor sediments from Greenland—Indications of >3700 Ma oxygenic photosynthesis: Earth and Planetary Science Letters, v. 217, p. 237–244, doi: 10.1016/S0012-821X(03)00609-5.

Rouse, R.C., Peacor, D.R., and Freed, R.L., 1988, Pyrophosphate groups in the structure of canaphite, $Ca_2Na_2PO_7 \cdot 4H_2O$: The first occurrence of a condensed phosphate mineral: American Mineralogist, v. 73, p. 168–171.

Russell, M.J., 1974, Manganese halo surrounding the Tynagh ore deposit, Ireland: A preliminary note: Transactions of the Institution of Mining and Metallurgy, section B, Applied Earth Science, v. 83, p. B65–B66.

Russell, M.J., 1988, Chimneys, chemical gardens and feldspar horizons ± pyrrhotine in some SEDEX deposits: aspects of alkaline environments of deposition, in Zachrisson, E., ed., Proceedings of the Seventh IAGOD Symposium, Stuttgart, Schweizerbartsche Verlagsbuchhandlung, p. 183–193.

Russell, M.J., 2003, On the importance of being alkaline: Science, v. 302, p. 580–581, doi: 10.1126/science.1091765.

Russell, M.J., and Arndt, N.T., 2005, Geodynamic and metabolic cycles in the Hadean: Biogeosciences, v. 2, p. 97–111, doi: 1726-4189/bg/2005-2-97.

Russell, M.J., and Hall, A.J., 1997, The emergence of life from iron monosulphide bubbles at a submarine hydrothermal redox and pH front: Journal of the Geological Society of London, v. 154, p. 377–402.

Russell, M.J., and Hall, A.J., 2001, The onset of life and the dawn of oxygenic photosynthesis: Respective roles of cubane core structures $[Fe_4S_4]^{2+}$ and transient $[Mn_4O_4]^{4+}[OCaO]_2$: Sixth International Conference on Carbon Dioxide Utilization, September 9–14, 2001, Breckenridge, Colorado, Abstracts, p. 49.

Russell, M.J., and Hall, A.J., 2002, From geochemistry to biochemistry: chemiosmotic coupling and transition element clusters in the onset of life and photosynthesis: The Geochemical News, no. 113/October, p. 6–12.

Russell, M.J., and Martin, W., 2004, The rocky roots of the acetyl-CoA pathway: Trends in Biochemical Sciences, v. 29, p. 358–363, doi: 10.1016/j.tibs.2004.05.007.

Russell, M.J., Hall, A.J., and Turner, D., 1989, In vitro growth of iron sulphide chimneys: possible culture chambers for origin-of-life experiments: Terra Nova, v. 1, p. 238–241.

Russell, M.J., Daniel, R.M., and Hall, A.J., 1993, On the emergence of life via catalytic iron sulphide membranes: Terra Nova, v. 5, p. 343–347.

Russell, M.J., Daia, D.E., and Hall, A.J., 1998, The emergence of life from FeS bubbles at alkaline hot springs in an acid ocean, in Wiegel, J., and Adams, M.W.W., eds., Thermophiles: The keys to molecular evolution and the origin of life: Washington, Taylor and Francis, p. 77–126.

Russell, M.J., Hall, A.J., and Mellersh, A.R., 2003, On the dissipation of thermal and chemical energies on the early Earth: The onsets of hydrothermal convection, chemiosmosis, genetically regulated metabolism and oxygenic photosynthesis, in Ikan, R. ed., Natural and laboratory-simulated thermal geochemical processes: Dordrecht, Kluwer Academic Publishers, p. 325–388.

Russell, M.J., Hall, A.J., Cairns-Smith, A.G., and Braterman, P.S., 1988, Submarine hot springs and the origin of life: -correspondence- Nature, v. 336, p. 117.

Russell, M.J., Daniel, R.M., Hall, A.J., and Sherringham, J., 1994, A hydrothermally precipitated catalytic iron sulphide membrane as a first step toward life: Journal of Molecular Evolution, v. 39, p. 231–243, doi: 10.1007/BF00160147.

Samson, I.M., and Russell, M.J., 1987, Genesis of the Silvermines zinc-lead-barite deposit, Ireland: fluid inclusion and stable isotope evidence: Economic Geology and the Bulletin of the Society of Economic Geologists, v. 82, p. 371–394.

Sandars, P.G.H., 2003, A toy model for the generation of homochirality during polymerization: Origins of Life and Evolution of the Biosphere, v. 33, p. 575–587.

Sandiford, M., and McLaren, S., 2002, Tectonic feedback and the ordering of heat producing elements within the continental lithosphere: Earth and Planetary Science Letters, v. 204, p. 133–150, doi: 10.1016/S0012-821X(02)00958-5.

Sauer, K., and Yachandra, V.K., 2004, The water-oxidation complex in photosynthesis: Biochimica et Biophysica Acta – Bioenergetics, v. 1655, p. 140–148.

Schäfer, G., Engelhard, M., and Müller, V., 1999, Bioenergetics of the Archaea: Microbiology and Molecular Biology Reviews, v. 63, p. 570–620.

Schink, B., 1997, Energetics of syntrophic cooperation in methanogenic degradation: Microbiology and Molecular Biology Reviews, v. 61, p. 262–280.

Schoofs, S., Trompert, R.A., and Hansen, U., 2000, The formation and evolution of layered structures in porous media: effects of porosity and mechanical dispersion: Physics of the Earth and Planetary Interiors, v. 118, p. 205–225, doi: 10.1016/S0031-9201(99)00148-X.

Schoonen, M.A.A., Xu, Y., and Bebie, J., 1999, Energetics and kinetics of the prebiotic synthesis of simple organic and amino acids with the FeS-H_2/FeS_2 redox couple as reductant: Origins of Life and Evolution of the Biosphere, v. 29, p. 5–32, doi: 10.1023/A:1006558802113.

Schulte, M.D., and Rogers, K.L., 2004, Thiols in hydrothermal solution: Standard partial molar properties and their role in the organic geochemistry of hydrothermal environments: Geochimica et Cosmochimica Acta, v. 68, p. 1087–1097, doi: 10.1016/j.gca.2003.06.001.

Schulte, M.D., and Shock, E.L., 1995, Thermodynamics of Strecker synthesis in hydrothermal systems: Origins of Life and Evolution of the Biosphere, v. 25, p. 161–173, doi: 10.1007/BF01581580.

Seefeldt, L.C., Dance, I.G., and Dennis, R.D., 2004, Substrate interactions with nitrogenase: Fe versus Mo: Biochemistry, v. 43, p. 1401–1409.

Sen, S., Igarashi, R., Smith, A., Johnson, M.K., Seefeldt, L.C., and Peters, J.W., 2004, A conformational mimic of the MgATP-bound "on state" of the nitrogenase iron protein: Biochemistry, v. 43, p. 1787–1797.

Seyfried, W.M., and Bischoff, J.L., 1981, Experimental seawater-basalt interaction at 300°C, 500 bars: Chemical exchange, secondary mineral formation, and implications for transport of heavy metals: Geochimica et Cosmochimica Acta, v. 45, p. 135–147, doi: 10.1016/0016-7037(81)90157-5.

Shen, Y., and Buick, R., 2004, The antiquity of microbial sulfate reduction: Earth-Science Reviews, v. 64, p. 243–272, doi: 10.1016/S0012-8252(03)00054-0.

Shock, E.L., 1990, Geochemical constraints on the origin of organic compounds in hydrothermal systems: Origins of Life and Evolution of the Biosphere, v. 20, p. 331–367, doi: 10.1007/BF01808115.

Shock, E.L., 1992, Chemical environments of submarine hydrothermal systems: Origins of Life and Evolution of the Biosphere, v. 22, p. 67–107, doi: 10.1007/BF01808019.

Shock, E.L., and Schulte, M.D., 1998, Organic synthesis during fluid mixing in hydrothermal systems: Journal of Geophysical Research, v. 103E, p. 28513–28527, doi: 10.1029/98JE02142.

Shock, E.L., McCollom, T., and Schulte, M.D., 1998, The emergence of metabolism from within hydrothermal systems, in Wiegel, J., and Adams, M.W.W., eds., Thermophiles: The keys to molecular evolution and the origin of life: Washington, Taylor and Francis, p. 59–76.

Smith, J.V., 1981, The first 800 million years of the Earth's history: Philosophical Transactions of the Royal Society of London, ser. A, v. 301, p. 401–422.

Smith, B.E., 2002, Nitrogenase reveals its inner secrets: Science, v. 297, p. 1654–1655, doi: 10.1126/science.1076659.

Sowerby, S.J., and Heckl, W.M., 1998, The role of self-assembled monolayers of the purine and pyridine bases in the emergence of life: Origins of Life and Evolution of the Biosphere, v. 28, p. 283–310, doi: 10.1023/A:1006570726326.

Srere, P., 1987, Complexes of sequential metabolic enzymes: Annual Review of Biochemistry, v. 56, p. 89–124, doi: 10.1146/annurev.bi.56.070187.000513.

Steigerwald, V.J., Beckler, G.S., and Reeve, J.N., 1990, Conservation of hydrogenase and polyferredoxin structures in the hyperthermophilic Archaebacterium *Methanothermus fervidus*: Journal of Bacteriology, v. 172, p. 4715–4718.

Stetter, K.O., 1996, Hyperthermophilic prokaryotes: FEMS Microbiology Reviews, v. 18, p. 149–158, doi: 10.1016/0168-6445(96)00008-3.

Stetter, K.O., and Gaag, G., 1983, Reduction of molecular sulphur by methanogenic bacteria: Nature, v. 305, p. 309–311, doi: 10.1038/305309a0.

Stone, D.A., and Goldstein, R.E., 2004, Tubular precipitation and redox gradients on a bubbling template: Proceedings of the National Academy of Sciences of the United States of America, v. 101, p. 11537–11541, doi: 10.1073/pnas.0404544101.

Svetlitchnyi, V., Dobbek, H., Meyer-Klaucke, W., Meins, T., Thiele, B., Römer, P., Huber, R., and Meyer, O., 2004, A functional Ni-Ni-[4Fe4S] cluster in the monomeric acetyl-CoA synthase from *Carboxydothermus hydrogenoformans*: Proceedings of the National Academy of Sciences of the United States of America, v. 101, p. 446–451, doi: 10.1073/pnas.0304262101.

Taylor, P., Rummery, T.E., and Owen, D.G., 1979, Reactions of iron monosulfide solids with aqueous hydrogen sulfide up to 160°C: Journal of Inorganic Nuclear Chemistry, v. 41, p. 1683–1687, doi: 10.1016/0022-1902(79)80106-2.

Thauer, R.K., 1998, Biochemistry of methanogenesis: A tribute to Marjory Stephenson: Microbiology, v. 144, p. 2377–2406.

Thauer, R.K., Jungermann, K., and Decker, K., 1977, Energy conservation in chemotrophic anaerobic bacteria: Bacteriological Reviews, v. 41, p. 100–180.

Trifonov, E.N., 2000, Consensus temporal order of amino acids and evolution of the triplet code: Gene, v. 261, p. 139–151, doi: 10.1016/S0378-1119(00)00476-5.

Trifonov, E.N., Kirzhner, A., Kirzhner, V.M., and Berezovsky, I.N., 2001, Distinct stages of protein evolution as suggested by protein sequence analysis: Journal of Molecular Evolution, v. 53, p. 394–401, doi: 10.1007/s002390010229.

Urey, H.C., 1952, On the early chemical history of the earth and the origin of life: Proceedings of the National Academy of Sciences of the United States of America, v. 38, p. 351–363.

Van Walraven, H.S., Hollander, E.E., Scholts, M.J.C., and Kraayenhof, R., 1997, The H^+/ATP ratio of the ATP synthetase from the cyanobacterium *Synechoccus* 6716 varies with growth temperature and light intensity: Biochimica et Biophysica Acta, v. 1318, p. 217–224.

Vargas, M., Kashefi, K., Blunt-Harris, E.L., and Lovley, D.R., 1998, Microbial evidence for Fe(III) reduction on early Earth: Nature, v. 395, p. 65–67, doi: 10.1038/25720.

Vaughan, D.J., and Ridout, M.S., 1971, Mössbauer studies of some sulfide minerals: Journal Inorganic Nuclear Chemistry, v. 33, p. 741–747, doi: 10.1016/0022-1902(71)80472-4.

Vaughan, D.J., and Craig, J.R., 1978, Mineral chemistry of natural sulfides: Cambridge, UK, Cambridge University Press, 493 p.

Vermaas, W.F.J., 1994, Evolution of heliobacteria: Implications for photosynthetic reaction center complexes: Photosynthesis Research, v. 41, p. 285–294, doi: 10.1007/BF02184169.

Voet, A.B., and Schwartz, A.W., 1982, Uracil synthesis via HCN oligomerization: Origins of Life and Evolution of the Biosphere, v. 12, p. 45–49, doi: 10.1007/BF00926910.

Von Damm, K.L., 1990, Sea floor hydrothermal activity: black smoker chemistry and chimneys: Annual Review of Earth and Planetary Sciences, v. 18, p. 173–204, doi: 10.1146/annurev.ea.18.050190.001133.

Von Damm, K.L., 2000, Chemistry of hydrothermal vent fluids from 9°-10°N, East Pacific Rise: "Time zero," the immediate post eruptive period: Journal of Geophysical Research, 105B, 11,203–11,222.

Wächtershäuser, G., 1988, Pyrite formation, the first energy source for life: A hypothesis: Systematic and Applied Microbiology, v. 10, p. 207–210.

Wächtershäuser, G., 1992, Groundworks for an evolutionary biochemistry: The iron-sulphur world: Progress in Biophysics and Molecular Biology, v. 58, p. 85–201, doi: 10.1016/0079-6107(92)90022-X.

Walker, J.C.G., and Brimblecombe, P., 1985, Iron and sulfur in the pre-biological ocean: Precambrian Research, v. 28, p. 205–222, doi: 10.1016/0301-9268(85)90031-2.

Wenner, D.B., and Taylor, H.P., 1971, Temperatures of serpentinization of ultramafic rocks based on $^{18}O/^{16}O$ fractionation between coexisting serpentine and magnetite: Contributions to Mineralogy and Petrology, v. 32, p. 165–185, doi: 10.1007/BF00643332.

Westall, F., de Witt, M.J., Dann, J., Van der Gaast, S., de Ronde, C., and Gerneke, R., 2001, Early Archean fossil bacteria and biofilms in hydrothermally influenced sediments from the Barberton Greenstone Belt, South Africa: Precambrian Research, v. 106, p. 93–116, doi: 10.1016/S0301-9268(00)00127-3.

Wicken, J.S., 1987, Evolution, information and thermodynamics: Extending the Darwinian program: New York, Oxford University Press, 243 p.

Wilde, S.A., Valley, J.W., Peck, W.H., and Graham, C.M., 2001, Evidence from detrital zircons for the existence of continental crust and oceans on the Earth 4.4 Gyr ago: Nature, v. 409, p. 175–178, doi: 10.1038/35051550.

Woese, C.R., 1967, The genetic code: The molecular basis for genetic expression: New York, Harper and Row, 200 p.

Woese, C.R., 1998, The universal ancestor: Proceedings of the National Academy of Sciences of the United States of America, v. 95, p. 6854–6859, doi: 10.1073/pnas.95.12.6854.

Woese, C.R., Dugre, D.H., Saxinger, W.C., and Dugre, S.A., 1966, The molecular basis for the genetic code: Proceedings of the National Academy of Sciences of the United States of America, v. 55, p. 966–974.

Woese, C.R., Kandler, O., and Wheelis, M.L., 1990, Towards a natural system of organisms: proposal for the domains Archaea, Bacteria, and Eucarya: Proceedings of the National Academy of Sciences of the United States of America, v. 87, p. 4576–4579.

Wolin, M.J., 1982, Hydrogen transfer in microbial communities, in Bull, A.T., and Slater, J.H. eds., Microbial interactions and communities: London, Academic Press, p. 323–356.

Wolthers, M., 2003, Geochemistry and environmental mineralogy of the iron-sulphur-arsenic system: Geologica Ultraiectina, Mededelingen van de Faculteit Aardwetenschappen Universiteit Utrecht, no. 225, 185 p.

Wolthers, M., Van der Gaast, S.J., and Rickard, D., 2003, The structure of disordered mackinawite: American Mineralogist, v. 88, p. 2007–2015.

Yamagata, Y., Wanatabe, H., Saitoh, M., and Namba, T., 1991, Volcanic production of polyphosphates and its relevance to prebiotic evolution: Nature, v. 352, p. 516–519, doi: 10.1038/352516a0.

Zachara, J.M., Kukkadapu, R.K., Frederickson, J.M., Gorby, Y.A., and Smith, S.C., 2002, Biomineralization of poorly crystalline Fe(III) oxides by dissimilatory metal reducing bacteria (DMRB): Geomicrobiology Journal, v. 19, p. 179–207, doi: 10.1080/01490450252864271.

MANUSCRIPT ACCEPTED BY THE SOCIETY 29 OCTOBER 2005

Geological Society of America
Memoir 198
2006

Early life signatures in sulfur and carbon isotopes from Isua, Barberton, Wabigoon (Steep Rock), and Belingwe Greenstone Belts (3.8 to 2.7 Ga)

N.V. Grassineau*
Department of Geology, Royal Holloway, University of London, Egham, Surrey TW20 0EX, UK

P. Abell[†]
Department of Chemistry, University of Rhode Island, Kingston, Rhode Island 02881, USA

P.W.U. Appel
Geological Survey of Denmark and Greenland (GEUS), Oester Voldgade 10, 1350 Copenhagen, Denmark

D. Lowry
E.G. Nisbet
Department of Geology, Royal Holloway, University of London, Egham, Surrey TW20 0EX, UK

ABSTRACT

Carbon and sulfur isotopes have been measured on samples from four Archean greenstone belts dating from 3.8 Ga to 2.7 Ga, in order to trace metabolic changes as life evolved over this one-billion-year period. In the Isua Greenstone Belt (3.8 Ga), Greenland, $\delta^{34}S$ in sulfide minerals from sedimentary sequences range from −3.8‰ to +3.4‰. $\delta^{13}C_{red}$ measured in BIFs, turbidites and conglomerates vary from −29.6‰ to −14.7‰; this range permits us to hypothesize the presence of hyperthermophilic and chemotrophic species in transient settings, or possibly pelagic photoautotrophic microbes, or both. In the Barberton Greenstone Belt, South Africa, sulfide minerals show $\delta^{34}S$ values from +1.5‰ to +5.6‰. Black shales have $\delta^{13}C_{red}$ values from −32.4‰ to −5.7‰, suggesting that oxygenic photosynthetic and sulfate-reducing bacteria were present by ca. 3.24 Ga. The $\delta^{13}C_{red}$ measured in the stromatolites of Steep Rock Lake (3.0 Ga), Ontario, Canada, are from −30.6‰ to −21.6‰, giving clear evidence for occupation of a shallow water environment by cyanobacteria. The wide isotopic ranges for $\delta^{34}S$ in sulfides from −21.1‰ to +16.7‰ and for $\delta^{13}C_{red}$ in carbon-rich cherts and black shales from −43.4‰ to −7.2‰ in the Belingwe Greenstone Belt, Zimbabwe, indicate that photosynthetic microbial mat communities were well established at 2.7 Ga. In these well-preserved Late Archean formations, modern-style biological sulfur and carbon cycles may have been in operation. The $\delta^{34}S$ and $\delta^{13}C_{red}$ ranges, respectively 37‰ and 36‰, indicate a great variety of biological processes interacting with each other.

Keywords: early life, sulfur and carbon isotopes, Archean, Isua, Belingwe, Barberton, Steep Rock Lake.

*nathalie@gl.rhul.ac.uk
[†]Deceased.

Grassineau, N.V., Abell, P., Appel, P.W.U., Lowry, D., and Nisbet, E.G., 2006, Early life signatures in sulfur and carbon isotopes from Isua, Barberton, Wabigoon (Steep Rock), and Belingwe Greenstone Belts (3.8 to 2.7 Ga), *in* Kesler, S.E., and Ohmoto, H., eds., Evolution of Early Earth's Atmosphere, Hydrosphere, and Biosphere—Constraints from Ore Deposits: Geological Society of America Memoir 198, p. 33–52, doi: 10.1130/2006.1198(02). For permission to copy, contact editing@geosociety.org. ©2006 Geological Society of America. All rights reserved.

INTRODUCTION: USE OF STABLE ISOTOPES IN STUDYING EARLY LIFE

The best record of early life is in sedimentary rocks from Archean greenstone belts. This study uses isotopic analyses of carbon- and sulfide-rich deposits from four Archean belts to seek evidence for the onset of major metabolic processes. The Isua Greenstone Belt (ca. 3.8 Ga) in West Greenland and the Belingwe Greenstone Belt (2.7 Ga) in Zimbabwe are the two main areas studied. Additional $\delta^{13}C$ and $\delta^{34}S$ data were obtained from the Barberton Greenstone Belt in South Africa and Steep Rock Lake succession in northwestern Ontario, Canada.

Metabolic processes produce distinctive isotopic fractionations when selecting carbon and sulfur from chemical or organic substrates accessible to the organisms. The recognizable $\delta^{34}S$ and $\delta^{13}C$ signatures in the sediments after burial, both in residual organic matter and in associated minerals, are particularly useful in efforts to understand the nature and extent of microbial activity (e.g., Nisbet and Fowler, 1999; Grassineau et al., 2001a, 2002). These isotopic fingerprints are among the very few biological signals remaining in Archean rocks. Other evidence for early microbial phylogeny comes from molecular studies (Woese, 1987; Pace, 1997; review in Nisbet and Sleep, 2001). In this study, stable isotopes have been used to infer the presence of prokaryotic communities in relics of microbial mats from Archean rocks.

Carbon-bearing compounds in the modern earth system have very different $\delta^{13}C$ values (Table 1). Biological fractionations produce wide $\delta^{13}C$ ranges, with two distinctive signatures dominating modern organic carbon. The first, from –28‰ to –22‰, is due to the fractionation by the Rubisco I enzyme during photosynthesis (Sirevåg et al., 1977; Pierson, 1994), in aerobic and microaerobic environments. This characteristic carbon isotopic fractionation occurs when carbon supply is abundant (Erez et al., 1998) and taken from seawater in exchange with a CO_2-rich atmosphere. The second is for methanogenic archaea that produce metabolic methane cycling, generating more fractionated $\delta^{13}C$, with a range of values mainly between –40‰ and –30‰. Methanotrophs or sulfate reducers that gain organic matter from this methane produce $\delta^{13}C$ down to –80‰ (Coleman et al., 1981). An additional process is anoxygenic photosynthesis with fractionations that are usually smaller (as *Chloroflexus*: –20‰ to –10‰ [e.g., van der Meer et al., 2001]; or Rubisco II [e.g., Robinson et al., 2003; Schidlowski, 2001]). Conversely, the inorganic carbon byproduct of the photosynthesis has a narrow range represented by most Phanerozoic marine carbonates (Table 1).

In the Archean, carbon degassing from the mantle entered the atmosphere and thence the ocean. Carbon extracted by sedimentation from the Late Archean ocean/atmosphere system was fractionated between a relatively small ^{13}C-depleted organic carbon portion ($\delta^{13}C \sim -28$‰), and a larger residual reservoir of ^{13}C-enriched inorganic carbon ($\delta^{13}C \sim 0$‰) (e.g., Schidlowski et al., 1975; Abell et al., 1985a, 1985b). Studies of Archean biology (e.g., Hayes et al., 1983; Schidlowski et al., 1983) have recognized this distinct carbon isotopic fractionation as marking the onset of the global-scale processing of carbon by cyanobacteria, now known to be via Rubisco I. The rapid and successful production of cyanobacteria then induced the start of carbon management by oxygenic photosynthesis that may have been sudden in geological terms. As O_2 is the process by-product, aerobic or at least microaerobic environments of deposition would be expected in rocks that showed the characteristic isotopic signature. Although an aerobic facies, organic $\delta^{13}C$ signature, or $\delta^{13}C_{carb}$ signature of 0‰ may not be evidence of cyanobacteria individually, the presence of all three is diagnostic. The start date of photosynthetic processes in the Archean is very controversial, and more isotopic investigation is needed.

Anoxygenic photosynthesis predates the oxygenic process in standard phylogenetic models (see discussion in Nisbet and Sleep, 2001). Therefore, there may be an interval in the geological record where the $\delta^{13}C$ fractionation between organic matter and carbonate was less than –28‰. However, it is still in debate whether large-scale carbonate deposition occurred before oxygenic photosynthesis. Most geologists concur that methanogenesis dates back to the Archean (Hayes, 1994; Nisbet and Sleep, 2001), and may well predate photosynthesis, though Cavalier-Smith (2002) has argued that methanogens did not appear until the late Proterozoic and hence a very "light methane" signature would not be expected. Thus the "geological" consensus that methanogenesis is ancient needs testing. Primary $\delta^{13}C$ of marine carbonate and organic carbon in the Archean are still not well constrained, but the wide $\delta^{13}C_{red}$ range so far obtained (–52‰ to –13‰; Strauss and Moore, 1992) suggests that biological processes were operational. A key target is to seek evidence for oxygenic photosynthesis ($\delta^{13}C_{carb} \sim 0$‰; $\delta^{13}C_{red} \sim -28$‰) or anaerobic conditions within the Archean.

Two isotopically distinct primary sulfur reservoirs dominate today: homogeneous seawater sulfate, and sulfide derived from the mantle (Table 1). $\delta^{34}S$ of sulfides in modern sediments ranges from –60‰ to +20‰ (Hoefs, 1997). Significant isotopic fractionations can be caused by inorganic processes, especially in hydrothermal fluids, but biological activities produce even greater fractionation (e.g., Schidlowski et al., 1983), particularly

TABLE 1. MAIN ISOTOPIC RESERVOIRS WITHIN THE EARTH SYSTEM

Reservoirs and domains	$\delta^{34}S$‰$_{CDT}$	$\delta^{13}C$‰$_{PDB}$
Seawater	Sulfate + 21.0 ± 0.2‰[†]	Carbonate +0.6 ± 1.6‰
Atmosphere	–30 to +30‰[†]	CO_2 –8‰[†]
		CH_4 –47‰[‡]
Mantle	+ 0.3 ±0.5‰[§]	CO_2 –8 to –5‰[#]
		CH_4 –33 to –15‰[#]
Igneous rocks	–11 to +9‰[†]	Graphite –33 to –6‰[#]
Modern sediments	–60 to +20‰[††]	–30 to –10‰[†]
Petroleum	–8 to +32‰[†]	–34 to –18‰[‡]
Natural gas	–8 to +32‰[†]	CH_4 –80 to –14‰[‡]
Coal	–30 to +24‰[†]	–35 to –15‰[†]

[†] In Faure (1986).
[‡] Stevens (1988).
[§] e.g. Sakai et al. (1984).
[#] Taylor (1986).
[††] Hoefs (1997).

in repetitive recycled reactions (e.g., Canfield and Teske, 1996). The resulting fractionation between seawater sulfate and precipitated biogenic sulfides in sediments has been up to 80‰ since the beginning of the Phanerozoic, generating ^{34}S-enrichment in seawater sulfate, which has averaged between +10‰ to +30‰ at different periods of Earth's history.

δ^{34}S of Archean seawater sulfate remains unknown, as sulfate is very rare in most early sediments, but is estimated to be close to 0‰–2‰ (Ohmoto, 1992). Then oceans contained modest amounts of sulfate (e.g., Veizer et al., 1989; Canfield and Teske, 1996), and Habicht and Canfield (1996) suggested that high SO_4 concentrations in seawater did not occur before the rise of atmospheric oxygen in post-Archean time. On the other hand, Ohmoto et al. (1993) interpreted the 9‰ δ^{34}S range for 3.4–3.2 Ga pyrite in one specimen (South Africa) as evidence for microbial reduction of seawater sulfate, suggesting that seawater already had a high SO_4 content. Shen et al. (2001) and Shen and Buick (2004) reached the same conclusion with values of +3‰ to +9‰ in 3.47 Ga sulfate samples from northwestern Australia. Thus there is currently no consensus.

The evolution of the sulfur cycle is also disputed, as there are few constraints on the timing of the first appearance of key parts of the cycle. Biochemical evidence from the "standard" model of microbial phylogenetic evolution (Woese, 1987; Pace, 1997) indicates strongly, though circumstantially, the great antiquity of both S-oxidation and S-reduction processes. Many biochemical processes rely on a S-containing enzyme, and thus at least some S-utilizing bacteria are Archean (e.g., Londry and Des Marais, 2003). Ohmoto and Felder (1987) suggested that microbial sulfate reduction was established by 3.5 Ga, and further examples have confirmed this (Barberton, South Africa, at 3.4 Ga [Ohmoto et al., 1993; Kakegawa and Ohmoto, 1999]; North Pole, northwestern Australia, at 3.47 Ga [Shen and Buick, 2004]). However, some authors consider that though sulfate reduction may be old, some of the sulfur cycle evolved only in the Proterozoic (Blank, 2004), and that the complete cycle with important S-isotope fractionation only appeared ca. 0.86–1.0 Ga (Canfield and Teske, 1996). Hitherto, the δ^{34}S range of Archean sulfide minerals was considered to be less than 10‰ (Cameron, 1982; Habicht and Canfield, 1996), and up to 13‰ in mineral deposits (Ohmoto, 1992). However, Grassineau et al. (2001a, 2002) reported a wider range of 37‰ at 2.7 Ga, suggesting that the sulfur cycle was already well on its way to full operation in the Late Archean at 2.7 Ga, with evidence of sulfate-reduction, sulfur-oxidation, and possibly disproportionation. Furthermore Shen et al. (2001) found a range of 16‰ at 3.47 Ga. This demonstrates that the sulfur cycle is of great antiquity (Grassineau et al., 2001a).

ANALYTICAL TECHNIQUES

δ^{13}C for reduced carbon and δ^{34}S were analyzed with a VG/Fisons/Micromass "Isochrom-EA" system, consisting of an elemental analyzer (EA1500 Series 2) online to an Optima mass spectrometer operating in He continuous flow mode (Matthews and Hayes, 1978, for carbon; Grassineau et al., 2001b, for sulfur). This high-resolution technique measures local isotopic fractionations in samples at sub-millimeter scale, which is necessary to detect specific activities of the biological communities, otherwise undetectable in larger samples.

A precision of ± 0.1‰ for δ^{34}S was obtained on hand-picked sulfide minerals as small as 0.8 mg for pyrite. The six standards analyzed, including NBS123, NBS127, and IAEA-S3, cover a range from −31.6‰ to +20.3‰. The samples for reduced carbon analysis were first treated in 20% HCl at 120 °C for 12 h. A precision better than ± 0.1‰ in δ^{13}C was obtained on hand-picked samples of 0.07 mg for pure carbon, to 30 mg for whole rock with 0.1 wt% C. The standards measured, including NBS21 and IAEA-CO9, cover a range from −47.1‰ to +3.3‰. Blank contamination from tin capsules is <34 ppm C, as measured in the laboratory, but all samples with less than 200 ppm C, where the blank isotopic effect might be significant for interpreting the results, have been rejected from the data set.

Pure carbonate samples (0.5 mg) were measured using an Isocarb automated carousel connected to a PRISM mass spectrometer. Impure carbonates were analyzed using a modified Micromass Multiflow connected to an Isoprime mass spectrometer. The system requires as little as 200 μg of pure carbonate or up to 50 mg of whole rock powder with 0.5% carbonate. Internal precision is better than ±0.07‰ for δ^{13}C and ±0.10‰ for δ^{18}O for both systems. The standards used are NBS19 limestone and an internal laboratory calcite.

EARLY ARCHEAN: THE ISUA GREENSTONE BELT, WEST GREENLAND

The Isua Greenstone belt (IGB), West Greenland, is a volcano-sedimentary relic exposed in an arcuate belt surrounded and locally intruded by a variety of tonalitic gneisses. The belt is dominated by thick sequences of mafic pillow lavas intercalated with numerous beds of iron formations (Appel et al., 1998; Fedo et al., 2001; Myers, 2001; Polat et al., 2003). Its minimum age constraint of 3.7 Ga comes from U-Pb of zircons in granitic sheets crosscutting the belt (Nutman et al., 1997). Sulfides in the IGB are mainly pyrrhotite and pyrite. They occur mainly as thin stratabound layers and disseminations throughout the rocks. Pyrite is most abundant in BIF and metacherts, often in association with fuchsite. Pyrrhotite dominates in pillow lavas, often associated with chalcopyrite in discordant quartz veins and veinlets (Appel, 1979). The IGB furthermore hosts sedimentary sequences, comprising mica schists or garnet mica schists, with or without staurolite as well as chemical sediments such as chert and iron formation. Rollinson (2002) observed five different metamorphic and structural domains, based on garnet study (Fig. 1). The southwestern zone has undergone two main metamorphic events (Rollinson, 2002) at 3.74 Ga (Frei and Rosing, 2001: Pb/Pb on metabasalts) and 2.84 Ga (Frei et al., 1999: Pb/Pb on magnetite). The latest episode was at high metamorphic grade with widespread metasomatic overprinting (Rose et al., 1996; Frei et

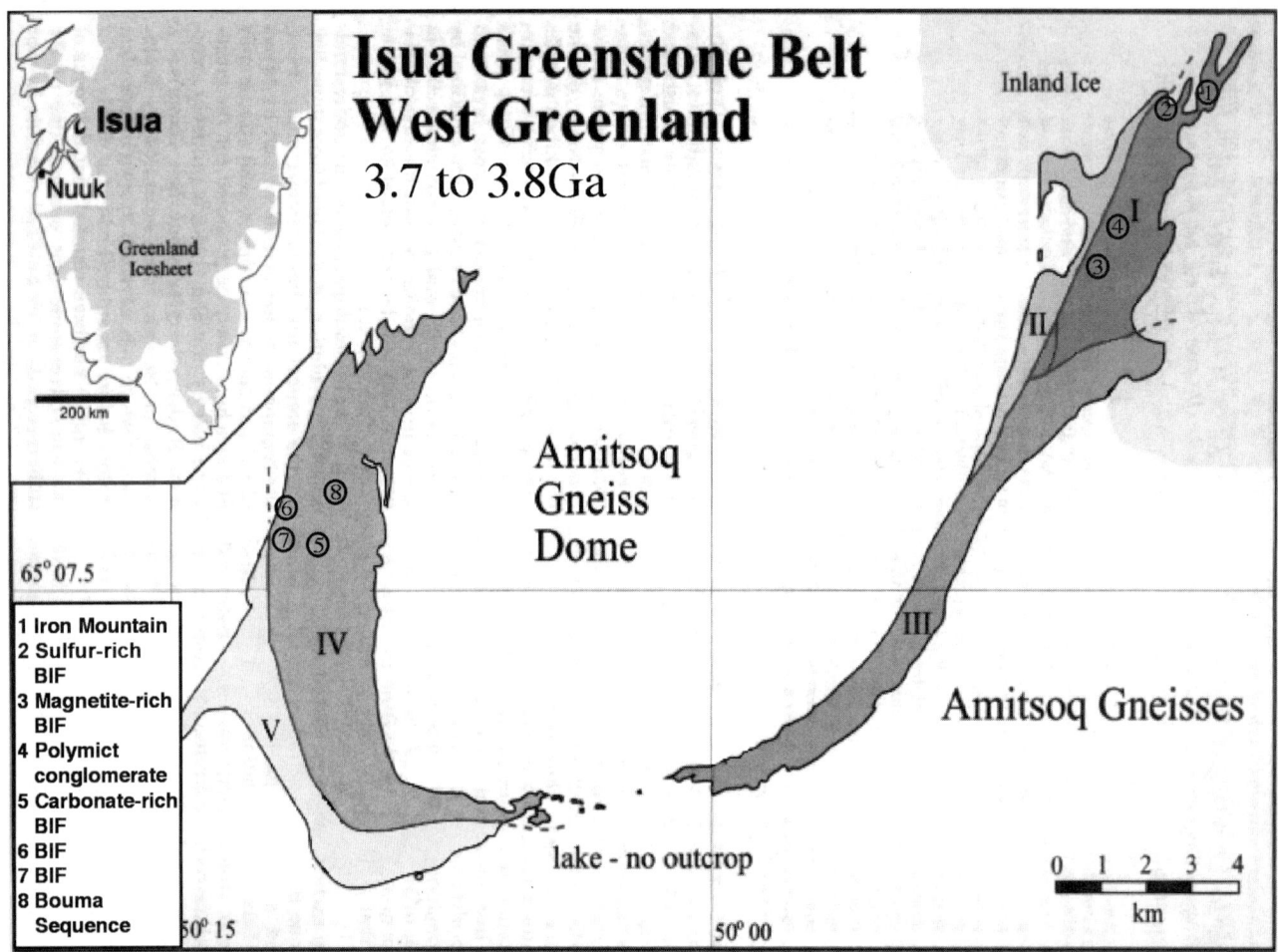

Figure 1. Map of Isua Greenstone Belt (3.7–3.8 Ga) showing the five structural domains defined by various authors and compiled by Rollinson (2002) (modified from Rollinson, 2002). Domain I underwent low-strain metamorphism at 3.69 Ga. Domains II and V recorded two high-grade metamorphic events both at 3.74 Ga. Domains III and IV underwent the same metamorphisms with higher intensity, and a late event at ca. 2.8 Ga. However, Domain IV preserved some low-strain areas. The localities studied are shown.

al., 1999, Rollinson, 2002). Conversely, the northeast zone has experienced only one event of lower intensity, which took place at 3.69 Ga (Frei et al., 1999 [Pb-Pb on magnetite and tremolite]) and no deformation thereafter (Moorbath and Kamber, 1998). The latter authors suggest that this age might represent the depositional period. This low-strain area has well-preserved primary sedimentary and igneous features (Appel et al., 1998; Rollinson, 2002). Several carbonate alteration events took place in the belt. The earliest occurred during the deposition of the sediments and extrusion of lava flows (J.S. Myers, personal commun.). Later carbonate alteration especially affected the western part of the IGB from 3.7 to 2.8 Ga (Moorbath and Whitehouse, 1996; Rose et al., 1996; Frei and Rosing, 2001). The first author carried out fieldwork in the IGB in 1998 and 1999, as part of the Isua Multi-disciplinary Research Project.

Sedimentary Formations Sampled

Sampling of the sedimentary rocks of the Isua Greenstone Belt focused on the regions that had the lowest metamorphic grade and were affected only by the early 3.74–3.69 Ga event, as these are most likely to preserve evidence of biological activity. Such areas are much rarer in the southwest than in the northeast (Rollinson, 2002). Eight formations were sampled in this study (Fig. 1). Six of the sampled units are banded-iron formations (BIF), three from the western and three from the northeastern part of the belt. The northeastern BIF samples are actinolite-bearing metachert and iron formation consisting of magnetite-rich bands alternating with grunerite-rich bands, or magnetite-rich bands alternating with quartz-bands. The largest example is the Iron Mountain Formation, a chert-BIF of alternating layers of quartz and magnetite with local amphibole-rich bands (Frei et al., 1999; Myers, 2001). The BIF samples from the west, including one of carbonate facies, are far more strongly deformed by higher-grade metamorphism than those from the northeast.

Another group of samples are metasedimentary rocks from the northwest that were formed by turbidity currents (Bouma sequence; Nutman, 1986). They consist of a series of medium-grained quartzites grading into fine-grained metapelites containing numerous small carbon particles (Rosing, 1999). In addition, a metaconglomerate was sampled in the northeast. The poorly sorted polymict assemblage consists mainly of quartz pebbles (metachert) with few pebbles of amygdaloidal volcanic rocks and BIF clasts (Fedo, 2000).

Previous Stable Isotope Studies at Isua

Monster et al. (1979) found a narrow range of −1.0‰ to +2.6‰ for sulfides in the IGB. More recently, Strauss (2003) gives a variation from −3.0‰ to +1.0‰, and Mojzsis et al. (2003) reported values of −0.9‰ to +2.2‰ on two pyrite grains (Fig. 2). These values around 0‰ are in the range of magmatic hydrothermal sulfides and provide no positive evidence that sulfur-utilizing bacteria were active during deposition.

Previous studies have obtained a wide range for $\delta^{13}C_{red}$ (reduced carbon) from −28‰ to −6‰ (Fig. 3) (Perry and Ahmad, 1977; Schidlowski et al., 1979; Hayes et al., 1983; Naraoka et al. 1996; Rosing, 1999; Ueno et al., 2002; van Zuilen et al., 2002), suggesting biological fractionation. For example, Rosing (1999) proposed a biogenic origin for the graphite grains in the Bouma sequence, with $\delta^{13}C_{red}$ of −19‰ close to the original depositional isotopic signature.

New Results Obtained for the Isua Belt

Sulfur Isotope Results and Discussion

The sulfide minerals analyzed from the sedimentary sequences (Table 2, Figs. 2A, 2B) are mainly disseminated pyrite, with only a few secondary pyrrhotites from the carbonate facies BIF. The $\delta^{34}S$ obtained for the pyrites are from −3.8‰ to +3.4‰ ($n = 19$) (Grassineau et al., 2000). This 7.2‰ range is larger than previously measured in Isua. The extent of the range found in this study suggests that more diverse processes operated than hitherto thought. This result is particularly significant because the largest range of values is seen in the northeastern sulfide-rich BIF and Iron Mountain, which have the lowest metamorphic grade in the belt (Rollinson, 2002).

The $\delta^{34}S$ has different ranges for each BIF. In the Iron Mountain, the values are only positive (+3.1‰ to +3.4‰; $n = 3$). In contrast, the two other northeastern BIFs show mainly negative values (−3.8‰ to +1.1‰; $n = 9$). The metaconglomerate with an average of +0.4 ± 0.5‰ ($n = 4$) is very homogeneous and close to 0‰, similar to the values of Monster et al. (1979). The only western BIF analyzed for $\delta^{34}S$ gives an average of −0.7 ± 0.1‰ ($n = 3$).

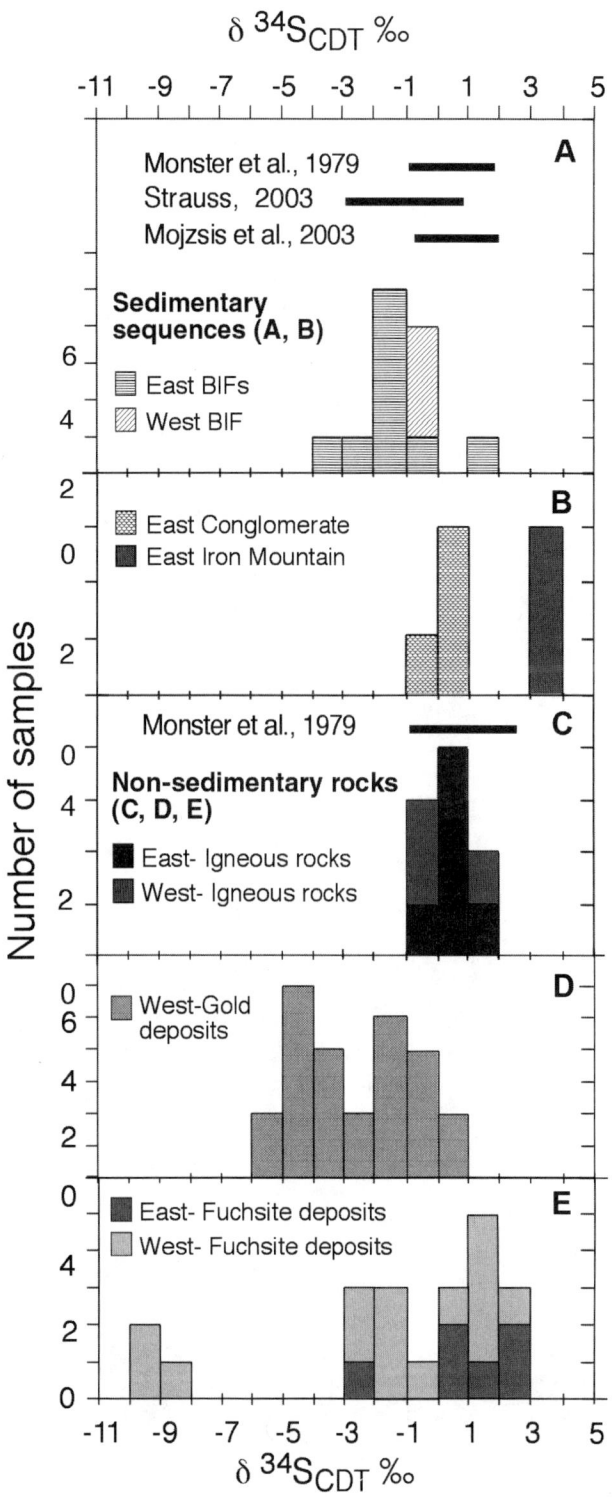

Figure 2. Distributions of $\delta^{34}S$ obtained for sulfide minerals from the IGB. (A, B) $\delta^{34}S$ for the sulfides in the metasediments, with a range of 7.2‰. (C) $\delta^{34}S$ in the sulfides from igneous rocks. (D, E) $\delta^{34}S$ for the gold and fuschite deposits. Results from previous studies are represented by thick horizontal bars.

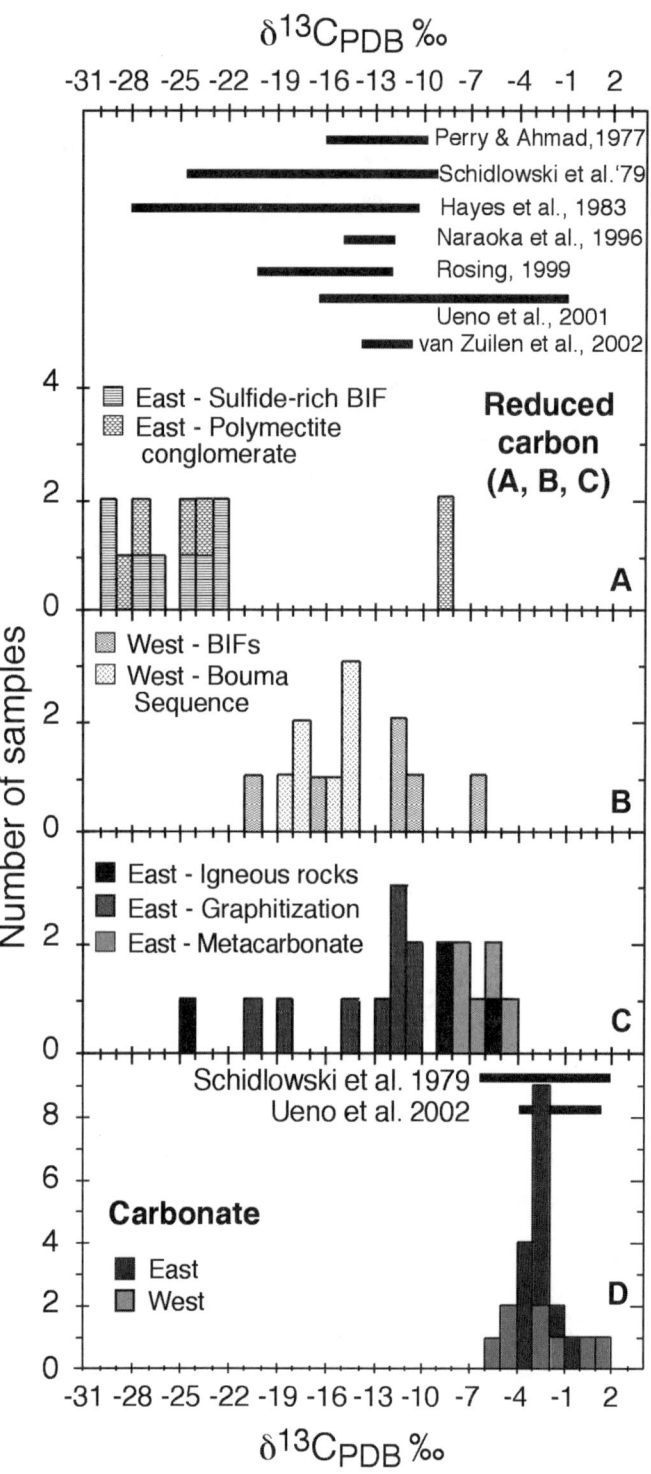

Figure 3. Distributions of $\delta^{13}C$ obtained for the IGB. (A, B) $\delta^{13}C_{red}$ for the metasediments, compared with values in non-sedimentary rocks (C). (D) $\delta^{13}C_{carb}$ from the western carbonate facies BIF and the secondary metacarbonates in the east of the belt. Results from previous studies are represented by thick horizontal bars.

TABLE 2. ISUA GREENSTONE BELT RESULTS (3.8–3.7 Ga)

Lithology	$\delta^{34}S$‰	$\delta^{13}C_{red}$‰	Carbon wt%	$\delta^{13}C_{carb}$‰
East				
Iron Mountain Metachert				
462-401Na	+3.2			
462-401Nb	+3.4			
462-401Nc	+3.1			
Sulfide-rich BIF				
462-439a	−2.5			
462-440a	−1.1	−22.7	0.02	
462-440b	−1.9	−29.6	0.03	
462-440c		−27.4	0.02	
462-441a		−29.4	0.02	
462-441b		−26.6	0.02	
462-442a	−1.6	−23.4	0.02	
462-443a	−0.0	−22.6	0.03	
462-443b		−24.3	0.02	
465-722a	−3.8			
465-723a	−1.2			
465-724a	+1.1			
Magnetite-rich BIF				
462-423a	−1.2			
Polymict conglomerate				
465-726a	+0.3			
465-726b	+0.9			
465-727a	+0.7	−8.6	0.12	
465-727b	−0.4	−8.7	0.09	
465-728a		−27.2	0.03	
465-728b		−23.4	0.02	
465-728c		−24.6	0.02	
465-729a		−28.6	0.03	
West				
Bouma sequence				
466-855b		−14.7	0.12	
466-857a		−16.0	0.52	
466-857b		−14.9	0.44	
466-859a		−14.9	0.10	
466-859b		−18.4	0.09	
466-863a		−17.7	0.12	
466-863b		−17.9	0.14	
Carbonate-rich BIF				
462-497a				−3.1
462-499c				−2.3
462-00a				−3.0
462-01				−3.9
462-08a	−0.8	−16.1	0.07	−0.4
462-08ab	−0.6	−20.1	0.05	−1.0
462-09a	−0.7	−6.5	0.11	−2.2
462-09b				−1.7
Other western BIFs				
466-876a		−12.0	0.02	
466-876c		−11.3	0.03	
466-888a		−10.1	0.18	
Hydrothermal deposits (mainly in SW Isua)				
Gold deposits	−5.4 to +0.6‰			
N = 25	(−2.5 ± 1.8‰)[†]			
Fuchsite deposits	−10.0 to +3.0‰			
N = 21	(−1.1 ± 3.8‰)[†]			

[†]Averages are in parenthesis with 1 s.d.

In order to provide a basis for comparison, pyrites from igneous formations and a slightly later Tarssartôq dyke (3.47 Ga) have been analyzed. The $\delta^{34}S$ ranges obtained are between −0.5 and +1.0‰ ($n = 9$) and would suggest that high-temperature fluids circulating in the belt had a signature of +0.4 ± 0.5‰ (Fig. 2C). This interpretation is supported by pyrite from metabasalts that show a narrow range of +0.5 ± 0.1‰ ($n = 2$).

Further comparison has been made with sulfides from two 3.7–3.8 Ga mineral deposits hosted by the western area of the IGB (Pb-Pb on galena; Frei and Rosing, 2001). These formed from low to moderate temperature hydrothermal fluids (Figs. 2D,

2E). $\delta^{34}S$ from disseminated galena and sphalerite associated with gold mineralization in a tonalite sheet varies from –5.4‰ to +0.6‰ ($n = 25$) (Grassineau and Appel, 2000). More remarkably, the range of –10.0‰ to +3.0‰ ($n = 21$) obtained for disseminated sulfides associated with quartz-fuchsite deposits is the widest yet found in Isua. Such wide isotopic fractionation is most likely created by sulfate-sulfide reactions (Ohmoto, 1992).

Western Samples. The carbonate-facies BIF is dominated by siderite, with secondary pyrrhotite in veins and along fractures. The low to moderate oxygen fugacity and high pH, constrained by the presence of magnetite, kept the isotopic variation relatively small for the hydrothermal fluids generated during metamorphism (Ohmoto, 1986). If there was any $\delta^{34}S$ signature left by organisms then it has been overprinted by the 2.84 Ga metamorphic event and re-homogenized as pyrrhotite. This is shown by the narrow range of values. Two main primary sources for sulfur can be suggested: the meta–pillow lavas and sulfur derived from seawater during the deposition of the BIF on the seafloor. The homogenized $\delta^{34}S$ value of –0.7‰ obtained for the BIF is lighter than the magmatic sources (+0.5‰). Although the shift in values of 1.2‰ could be explained entirely by hydrothermal processes remobilizing magmatic sulfur, it is possible that another source caused this shift during re-homogenization, possibly primary sulfur in the BIF.

Northeastern Samples. In contrast, the better preserved and less deformed metasediments in the northeast are more likely to preserve the original $\delta^{34}S$ signatures. They show a larger $\delta^{34}S$ range of 7.2‰ ($n = 26$), but values are different for each formation. In the polymict metaconglomerate, $\delta^{34}S$ is homogenous (+0.4 ± 0.5‰; n = 4). The pyrites are from the matrix and the clasts and show similar $\delta^{34}S$. This is expected, as the clasts consist of mafic volcanic rocks and metacherts (Fedo, 2000). The detrital pyrites probably originated as magmatic sulfur, at ~0‰. There is no sign of a biological signature, which should have given more isotopic variation.

The three samples analyzed from the Iron Mountain metachert are very homogeneous ($\delta^{34}S$ of +3.2 ± 0.2‰). These $\delta^{34}S$ values are heavier than those of the five other BIFs. The pyrites are idiomorphic and located close to or in veins, indicating that the secondary sulfides were introduced by hydrothermal fluids that may have been produced by the early metamorphism. The sulfur in these fluids could have originated from different sources: (1) sulfur already present in situ in the form of barite (barite traces exist in the metachert) reduced by interaction with hydrothermal fluids (e.g., Ohmoto and Godhaber, 1997); 2) sulfur from volcanic formations in the belt. This view is supported by Lepland et al. (2002) who interpreted REE patterns for apatites in the Iron Mountain as a result of a pervasive fluid from mixed sources.

Sulfide- and magnetite-rich BIFs from the northeast have a relatively wide and mainly negative range of $\delta^{34}S$ (4.9‰; from –3.8‰ to +1.1‰; $n = 9$) (Fig. 2A). This heterogeneity stands out compared to the results of the two previously mentioned northeastern formations. Komiya et al. (1999) analyzed metabasites and suggested that Domain I had undergone retrograde metamorphism to greenschist facies, in which case sulfide would not have survived with a primary isotopic signature. Rollinson (2002), on the other hand, determined a temperature for the metamorphism not higher than 520 °C by analyzing metapelites. So it is likely that this fairly moderate event did not re-homogenize $\delta^{34}S$; hence the isotopic variation remained completely or partially preserved in the sulfides.

The sulfide-rich BIF contains alternating quartz and magnetite bands, but with up to 5% of pyrites in some zones, mostly in quartz but also within magnetite. Some pyrite crystals are agglomerated in clusters, suggesting that they are not re-crystallized by a post-metamorphic event. The wider range of $\delta^{34}S$ and the lower metamorphic grade make these BIFs the most likely to preserve a primary bacterial signature, and it is possible that a much wider range produced by bacterial sulfate reduction has been narrowed by metamorphic homogenization. It cannot be ruled out though that the range was produced by hydrothermal magmatic sulfur, or that there was a mixing between biogenic sulfur and hydrothermal sulfur (+0.5‰) during metamorphism. However, it would be difficult to fractionate inorganically the hydrothermal source to significantly lower $\delta^{34}S$ without involving a second source of lighter sulfur, except at extremely low pH, or at higher f_{O_2} in the presence of sulfates (Rye and Ohmoto, 1974). Sulfate though has so far not been found in these particular samples.

Carbon Isotopic Results and Discussion

$\delta^{13}C$ of reduced carbon from the sedimentary sequences has been measured after acid treatment (Table 2, Figs. 3A, 3B). The $\delta^{13}C_{red}$ range is from –29.6‰ to –6.5‰ ($n = 27$) (Grassineau et al., 2000). The carbon content of the Isua rocks is between 0.02 and 0.52 wt% ($n = 27$). The carbon was not visible in hand specimen and the samples analyzed were whole rock chips. Where known from the literature the carbon is referred to as graphite; otherwise it is called reduced carbon. In addition, carbonates were measured with $\delta^{13}C_{carb}$ of –2.2 ± 1.1‰. The two northeastern formations analyzed are the sulfide-rich BIF and the polymict conglomerate with $\delta^{13}C_{red}$ as light as –29.6‰ (0.03 wt% C) ($n = 8$). With one exception, the values are lighter than –22.6‰.

The $\delta^{13}C_{red}$ range in the western formations is from –20.1‰ to –6.5‰ ($n = 13$). The results obtained from the dark layers of the Bouma sequence are from –18.4‰ to –14.7‰ ($n = 7$) with carbon contents from 0.09 to 0.52 wt%. The values are similar to those of Rosing (1999) (–19.1‰ to –11.4‰). The two small BIFs associated with ultramafic rocks have homogeneous $\delta^{13}C_{red}$ of –11.1‰ ± 0.8‰ ($n = 3$). The carbonate-facies BIF shows the entire isotopic range obtained in the west, indicating that more than one process occurred. Carbonates from this formation, mainly siderite, give $\delta^{13}C_{carb}$ values from –3.9‰ to –0.4‰ ($n = 8$).

To differentiate between carbon in the sediments and the other formations, we measured 18 reduced carbon samples of igneous rocks and graphites from hydrothermally affected units in the northeast part of the belt (Fig. 3C). Carbon contents range from 0.02–3.5 wt%. Nine graphite samples from amphibolite facies give a wide range of $\delta^{13}C_{red}$ from –20.3‰ to –10.2‰ (–14.8 ± 5.4‰). These values fall roughly within the range already reported by previous workers (Fig. 3). $\delta^{13}C_{red}$ was also measured in five metacarbonate samples, with a narrow spread of –6.3 ± 1.3‰. Finally two ultramafic rocks were analyzed with values of –8.5 and –5.8‰, and two volcano-felsic rocks, with –24.2 and –8.6‰. The $\delta^{13}C_{red}$ range (close to 20‰) is large in these formations, but two main sources can be pointed out. The heavier values between –8.6 and –5.8‰ are most likely of magmatic origin, whereas the lighter range of –14.6‰ to –10.2‰ may record the circulation of high-grade metamorphic fluids during the early event (Perry and Ahmad, 1977; Naraoka et al., 1996; van Zuilen et al., 2002, 2003). These authors suggested that the siderite decomposed to graphite at the high temperatures reached by the regional metamorphism. This graphite would be ^{13}C-enriched during isotopic re-equilibration with the siderite (Ueno et al., 2002).

The $\delta^{13}C_{carb}$ range of metacarbonates for the northeastern part is wider (from –5.7‰ to +1.2‰) than for the western samples (from –3.9‰ to –1.0‰) (Fig. 3D). The homogenization of $\delta^{13}C$ in the western carbonate facies BIF is consistent with the higher grade; in fact, most of the carbonates here were most likely remobilized during the 2.84 Ga metasomatic event (Rose et al., 1996). The wider variation shown in the northeastern carbonates reflects the lower grade of the Domain I metamorphism, where the isotopic overprint was less effective.

Western Samples. Comparison between the results shows that $\delta^{13}C_{red}$ is generally lighter in the sedimentary rocks than in the ultramafic and mafic rocks, with a wider range of 23.1‰. This is not the case for the two small western BIFs; the value of –11.1 ± 0.8‰ indicates that the high-grade metamorphism has overprinted the $\delta^{13}C$, with a re-equilibration toward the generalized graphite value of –12‰ obtained in other parts of the belt.

Despite the proximity of the three western BIFs, the metamorphic imprint is less obvious in the carbonate facies BIF. In contrast to $\delta^{34}S$, the $\delta^{13}C_{red}$ range for this BIF is wide, 13.6‰, indicating that the re-homogenization was not as efficient as in the other western BIFs, and that some areas of the formation may have partially retained the primary value. The lightest $\delta^{13}C$, at –20.1‰, is similar to the original value suggested by Rosing (1999) for kerogen in the Bouma sequence nearby. Though metamorphic overprinting is likely, –20.1‰ might be primary too: if so, it could represent the burial of organic matter within the BIF. On the other hand, $\delta^{13}C_{red}$ at –6.5‰ from the same BIF appears to be related to the late (2.84 Ga) high-grade metamorphism. The carbonate in this rock is mainly siderite, which locally may have decomposed to graphite during the high temperatures reached during the 2.84 Ga event.

The last western formation studied is the sedimentary outcrop interpreted as turbidites displaying a Bouma sequence. The $\delta^{13}C_{red}$ range measured in this study (–18.4‰ to –14.7‰; $n = 7$) might represent a primary biological signature, but the possibility of mixing with the –12‰ high-temperature metamorphic carbon cannot be ruled out. Rosing (1999), who obtained the least modified $\delta^{13}C$ at –19.1‰, inferred that this value represents planktonic-like organisms. Though by no means proving the case for biogenicity, the evidence suggests that the most depleted $\delta^{13}C$ (here –18.4‰) might record biological activity at 3.8 Ga.

Northeastern samples. The two northeastern sedimentary formations show lighter values. With the exception of two samples, all $\delta^{13}C_{red}$ are more depleted than –22‰ (Fig. 3A). This is consistent with the lower metamorphic grade in this area, implying that more original signatures might have been preserved. The sulfide-rich BIF values average –25.7 ± 2.7‰, with one $\delta^{13}C_{red}$ at –29.6‰, the lightest found in this study of Isua. These samples consist mainly of alternating quartz and magnetite bands. There is no visible carbon in the quartz-layers in thin sections, so it is likely associated with the magnetite. The range of 6‰ in samples with low reduced carbon content does not suggest a high-temperature fluid source (as with $\delta^{34}S$, a fluid would have left a more homogenized $\delta^{13}C$ range for reduced carbon). Thus the preferred interpretation is that some of this reduced carbon had an organic origin. The $\delta^{13}C_{red}$ distribution possibly records two carbon sources, syn-sedimentary organic carbon and a small component of inorganic CO_2 from post-depositional hydrothermal fluids circulating through the belt. If so, because metamorphism can enrich the graphite in ^{13}C by up to 20‰ at 450 °C (Schidlowski, 2001), the primary signature of the organic component in these BIFs may have been even lighter than –29‰. In this case, the biological environment could have been anaerobic.

Fedo (2000) described the polymict conglomerate as detrital with a mafic origin for some of the clasts. $\delta^{34}S$ in this study (+0.4 ± 0.5‰) supports this interpretation, and the two $\delta^{13}C_{red}$ at –8.6‰ are also similar to the results obtained for metabasalts (Figs. 3A, 3C). However, four values of –28.6‰ to –23.4‰ indicate a second carbon source. They are close to the sulfide-rich BIF values and may record an organic origin. Could traces of life be found in a conglomerate? Preservation is possible and the metamorphic grade that affected the metaconglomerate was low enough not to cause significant shift in primary signatures. The view that the original environment hosted life and preserved an organic signature is debatable. The shallow subaqueous setting for the conglomerate deposition suggested by Fedo (2000) could host life. Therefore the value of –28‰ might be close to the primary signature of the organic activity. It is interesting to compare this with the isotopic signature of oxygenic photosynthesis, though this is no more than speculation in the absence of other supporting evidence.

In summary, metamorphism has caused widespread overprinting of the original $\delta^{13}C$ signatures and especially the $\delta^{34}S$ values of the IGB. However, some areas have better preserved the primary isotopic compositions. This is the case for the western turbiditic rocks, and more particularly in the northeastern part, which was subjected to slightly less strain and lower-grade metamorphism (Rollinson, 2002, 2003). The results presented here are not inconsistent with the hypothesis that there is a record of planktonic life in the western part (Rosing, 1999). From the

TABLE 3. FIG TREE GROUP AND STEEP ROCK LAKE RESULTS

Lithology	$\delta^{34}S$‰	n	$\delta^{13}C_{red}$‰	Carbon wt%	N	$\delta^{13}C_{carb}$‰	N
Fig Tree Group (3.24 Ga) S- and C-rich Black shales	+1.5 to +5.6‰ (+2.3 ± 1.0‰)	15	−32.4 to −5.7‰ (−22.4 ± 9.1‰)	0.02–0.25%	18	−4.9 to −4.4‰ (−4.7 ± 0.2‰)	6
Steep Rock Lake (3.0 Ga) Stromatolites	+5.0‰	1	−30.6 to −21.6‰ (−25.4 ± 2.0‰)	0.22–2.29%	32	+0.1 to +2.9‰ (+2.0 ± 0.6‰)	56[†]

Note: Averages are in parenthesis with 1 s.d.
[†]Including 9 analyses by P. Abell (by conventional line technique).

results obtained in the eastern BIFs and metaconglomerate, biogenic processes, possibly including methanogenesis and photosynthesis (Rosing and Frei, 2004), might have been in operation at 3.7 Ga.

MIDDLE ARCHEAN: THE BARBERTON GREENSTONE BELT

The Barberton Greenstone Belt, located in the Kaapvaal Craton in South Africa, is one of the best-preserved mid-Archean successions. The regional metamorphic grade is low to moderate, much lower than in the IGB, and though there are many shear zones and décollement horizons, the sequences of rocks generally have experienced only low strain (Viljoen and Viljoen, 1969). Among the components of the belt are the Onverwacht Group (mainly mafic and ultramafic formations) and two clastic and chemical sedimentary sequences, the Moodies and Fig Tree Groups (Paris, 1987).

This study is based on the Fig Tree Group, which consists of ferruginous cherts, greywackes, shales, BIFs, and pelites, deposited in a submarine setting (Paris, 1987). The material, collected by J. Kramers and C. Siebert (Bern, Switzerland), comes from a horizontal borehole composed of greywackes and shales, drilled from the Fairview mine in an ESE direction out of the orebody toward the Sheba mine, in the central part of the belt (Siebert, 2003; Kramers et al., 2004). The Fig Tree Group has an age between 3.26 and 3.22 Ga from the measurements of zircons in tuff layers (Lowe and Byerly, 1999).

Previous Stable Isotope Investigations

Many carbon and sulfur isotope measurements have been made on the sedimentary groups. Sulfides in the Fig Tree Group show $\delta^{34}S$ ranges of −0.4‰ to +4.0‰ (Strauss and Moore, 1992), +1.2‰ to +3.9‰ (de Ronde et al., 1992), and −0.9‰ to +4.4‰ (Kakegawa and Ohmoto, 1999). An average of +3.4 ± 0.2‰ for barite (Strauss and Moore, 1992) suggests that mid-Archean seawater sulfate was not highly fractionated. Studies in the Moodies and Onverwacht Groups show similar ranges (e.g., Ohmoto et al. 1993: −3.0‰ to +8.6‰ in pyrites).

Carbon-rich shales in the Fig Tree Group contain up to 14.0 wt% of organic carbon with $\delta^{13}C_{red}$ from −29.5‰ to −9.5‰ (de Ronde and Ebbesen, 1996). Other studies give ranges of −32.8‰ to −24.3‰ (Hayes et al., 1983), or −35.4‰ to −27.0‰ (Strauss and Moore, 1992). The $\delta^{13}C$ in carbonates are from −4.5‰ to −2.0‰ (de Ronde et al., 1992). All these authors came to the conclusion that photosynthesis involving marine organisms was occurring at the time.

New Isotopic Results and Discussion of the Fig Tree Group

The newly analyzed samples consist of three 10–16-cm-long cores of sulfide-rich carbonaceous shales. Similar ranges to those previously obtained have been found on samples analyzed at millimeter scale, with $\delta^{34}S$ in pyrites and pyrrhotites ranging from +1.0‰ to +5.6‰ ($n = 15$) (Table 3). The sulfide minerals are an assemblage of agglomerated fine grains, as seen in Figure 4. $\delta^{13}C_{red}$ varies from −32.4‰ to −5.7‰ ($n = 18$) in samples with carbon contents up to 0.25 wt%, with no direct correlation between the C content and $\delta^{13}C$. Of the samples analyzed, 73%

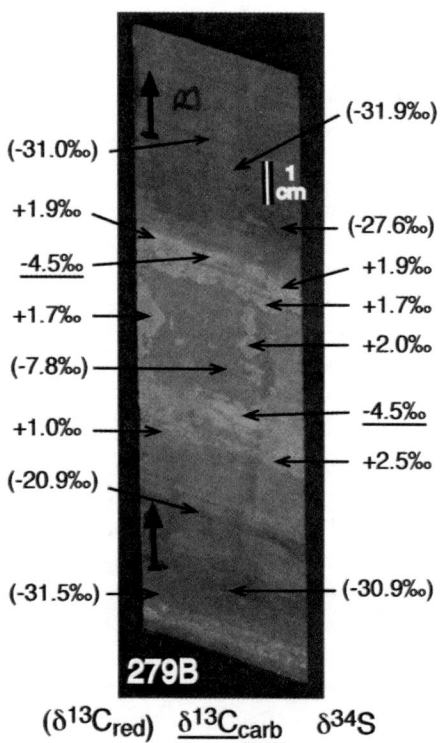

Figure 4. Detailed isotopic analyses of 279B dark shale sample from the Fig Tree Group (3.24 Ga). The core has a central 5-cm-thick layer rich in sulfides and secondary carbonate. At each end, $\delta^{13}C_{red}$ is homogeneous around −31‰.

are lighter than –20‰. Pervasive calcite in bands has a very narrow $\delta^{13}C_{carb}$ range of –4.9‰ to –4.4‰ ($n = 6$).

A detailed section of one of the samples, presented in Figure 4, illustrates the isotopic variations obtained. Core 279B has thin dark laminations, and a central sulfide- and carbonate-rich layer ~5 cm thick. There is a wide range of $\delta^{13}C_{red}$, from –31.9‰ to –7.8‰, but at both ends of the core $\delta^{13}C_{red}$ is homogeneous, with –31.4 ± 0.4‰; the $\delta^{13}C$ tends to become heavier toward the center, reaching –7.8‰ inside the "central" layer. This layer has peripheral 5-mm-thick carbonate-rich bands with very homogeneous $\delta^{13}C_{carb}$ of –4.5‰, similar to the average value found by de Ronde et al. (1992). The sulfides, mainly pyrrhotite on the outside edges of this zone, and pyrites in the center in bleb-like shapes, give a relatively narrow range of $\delta^{34}S$ averaging +1.8 ± 0.4‰. However, a variation of 3.64‰ over 1.5 cm in another sample suggests that re-homogenization of $\delta^{34}S$, if it took place, was not complete, especially considering the low-grade greenschist facies metamorphism that the rocks have experienced (Siebert, 2003).

Kakegawa and Ohmoto (1999), who obtained similar $\delta^{34}S$ variations in pyrite grains at micro-scale within the Fig Tree Group, suggested both organic and inorganic origins for the pyrite crystals. They pointed out that coarse-grained pyrites associated with quartz veins, with $\delta^{34}S$ from +1.1‰ to +3.6‰, have an inorganic origin and are more likely precipitated from high-temperature hydrothermal fluids. Although the values of pyrites in the present study are in a similar range, none are associated with quartz veins and they are agglomerates of fine rounded grains, which parallel the sedimentary bedding, suggesting that they are a sedimentary or diagenetic feature. It is impossible to rule out a partial re-homogenization of the sulfides, but the $\delta^{34}S$ values of +1.5‰ to +5.6‰ found locally seem to indicate at least some microbial reduction of the seawater sulfate. This was suggested by Kakegawa and Ohmoto (1999), who concluded that the Barberton sea at 3.4–3.2 Ga was bearing an appreciable amount of sulfate.

The isotopically homogeneous carbonate ($\delta^{13}C_{carb} \sim –4.5‰$) does not crosscut the structure of the shale. It occurs with pyrite adjacent to a redox interface, outside which is pyrrhotite and reduced carbon with $\delta^{13}C$ around –31‰. It is likely that this is a result of syngenetic or diagenetic hydrothermal processes. Shales enclosed by the carbonate-rich zones show some bleaching (Fig. 4), and significant exchange between the primary organic carbon and carbon from the hydrothermal fluid. The $\delta^{13}C_{red}$ varies from –27.6‰ to –5.7‰ close to the carbonate. This suggests that only the $\delta^{13}C$ values around –31‰ represent a primary organic carbon signature. An alternative explanation for the relatively heavy $\delta^{13}C_{red}$ values is that they are a record of organisms using waste carbon dioxide or bicarbonate from methanotrophs. Either way they are evidence for biological activity.

The view that light $\delta^{13}C_{red}$ values and $\delta^{34}S$ values of sulfides are biogenic is in agreement with Kakegawa and Ohmoto (1999). The light $\delta^{13}C$ of –31‰ suggest that the dark shales may be of deep-water facies, deposited in anoxic zone below the photic zone.

LOWER MIDDLE ARCHEAN: THE STEEP ROCK GROUP

The Steep Rock Group from the Wabigoon Greenstone Belt in northwestern Ontario, Canada, contains an Archean carbonate platform, the Mosher Carbonate Formation, which is 500 m thick and made of laminated carbonate, mainly limestone, in part stromatolitic (Wilks and Nisbet, 1985, 1988; Kusky and Hudleston, 1999). Regional metamorphic grade is broadly lower greenschist. The age of the Steep Rock Group is not fixed. It rests unconformably on a ca. 3.0 Ga gneissic terrane, the Marmion Complex (Davis and Jackson, 1985; Wilks and Nisbet, 1988). The upper unit of the belt includes metavolcanic rocks with an age of 2.93 Ga (Davis, 1993 cited in Kusky and Hudleston, 1999) but it is partly allochthonous. It is thus likely but not certain that the Steep Rock Lake stromatolites are ~2.9–3.0 Ga old. The well-preserved stromatolites clearly show biogenic structures on both small and large-scales (Wilks and Nisbet, 1985; 1988). Cryptozoon structures, branching walled and un-walled columnar forms up to 20 cm, are succeeded at the top of the unit by spectacular domal structures up to 3 m in diameter. The presence of these very well preserved primary structures, ooliths and oncolites, indicates that strain was very low.

The samples studied are three 1-m-long hand drill cores taken at different sites from near the larger domal structures. The carbonate is bluish-white, a mixture of calcite and dolomite, with undulating bands of black material. These laminae are described as organic kerogen (Hayes et al., 1983). Few analyses have been carried out on the formation; Hayes et al. (1983) obtained three $\delta^{13}C_{red}$ values with a range from –26.4‰ to –22.1‰, and Schidlowski et al. (1983), three carbonates with $\delta^{13}C_{carb}$ from +1.1‰ to +2.0‰. The authors inferred that these values were the result of biological activity.

Results from Steep Rock Lake Stromatolites

The $\delta^{13}C_{red}$ results (Table 3) obtained for the stromatolites are from –30.6‰ to –21.6‰ ($n = 32$), and $\delta^{13}C_{carb}$ from +0.1‰ to +2.9‰ ($n = 56$), in agreement with the previously obtained values. The ranges for both reduced and carbonate carbon are narrow, especially the carbonate, with an average of +2.0 ± 0.6‰ (Fig. 5). The reduced carbon isotopic values have a wider spread of 9‰. This could be original variation but might also indicate minor effects of metamorphism. Only one pyrite has been found. It has a $\delta^{34}S$ value of +5.0‰.

The $\delta^{13}C_{red}$ average of –25.4‰ and the $\Delta^{13}C_{red-carb}$ fractionation of 26‰–31‰ are interpreted as clear evidence of biological activity in these stromatolites, more particularly of the carbon fractionation by Rubisco I. Consequently, considering that the biogenic structures were developed in shallow water, on the margin of the Marmion Complex (Kusky and Hudleston, 1999), it is possible to propose that oxygenic photosynthesis was already established in the Steep Rock Group at 3.0 Ga, and that cyanobacteria were fully active. This and the Mushandike

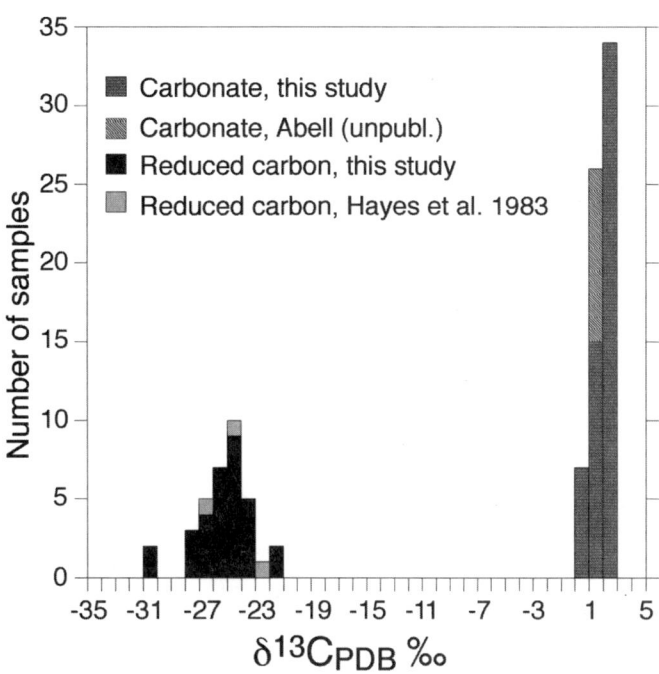

Figure 5. Distributions of $\delta^{13}C_{red}$ and $\delta^{13}C_{carb}$ obtained for Steep Rock Lake stromatolites. $\delta^{13}C_{carb}$ values have a narrow range of +2.0‰ ± 0.6‰, and $\delta^{13}C_{red}$ an average of approximately –26‰.

Formation in Zimbabwe are perhaps the oldest rock units where an unambiguous "modern" carbonate isotopic ratio can be identified (Abell et al., 1985b).

LATE ARCHEAN: THE BELINGWE GREENSTONE BELT

The Belingwe Greenstone Belt in Zimbabwe is one of the best preserved of all Archean successions. The two sequences, the Manjeri and Cheshire Formations, respectively the lowest and highest parts of the Ngezi Group (upper greenstones) contain carbon- and sulfide-rich sediments and stromatolitic limestones that have been affected only by low-grade metamorphism (Martin et al., 1980; Abell et al., 1985a; Bickle and Nisbet, 1993). Bolhar et al. (2002) obtained Pb-Pb ages of 2706 ± 49 Ma for the Manjeri Formation, and 2601 ± 49 Ma for the Cheshire Formation, which is in agreement with Pb-Pb and Sm-Nd ages for the Reliance Formation of 2692 ± 9 Ma from immediately above the Manjeri Formation (Chauvel et al., 1993).

The Manjeri sediments were deposited in a continental basin unconformably upon a 3.5 Ga tonalitic gneiss (Bickle and Nisbet, 1993; Hunter et al., 1998). Four units of this formation in the Nercmar drill core were investigated: at the base, 40 m of organic carbon-rich sediments with intercalated thin sulfide layers (Spring Valley and Shavi Members), 56 m of volcanoclastic rocks (Rubweruchena Member), and at the top, 10 m of organic carbon- and pyrite-rich black shales (Jimmy Member) of which the base is massive pyrite, capped by a thick mafic lava sequence, the Reliance Formation. The study is mainly based on the carbon- and sulfide-rich sedimentary units. The Spring Valley and Shavi Members represent a shallow-water environment (Hunter et al., 1998). At the same stratigraphic level, the discontinuous Rupemba Formation contains well-preserved stromatolites that were deposited on slopes (Martin et al., 1980; Abell et al., 1985a), and that have undergone low-grade metamorphism with a maximum temperature of 200 °C (Abell et al., 1985a). Deposition of the Jimmy Member occurred below wave base but still in the photic zone. It is complexly folded at the contact with the Reliance Formation (Hunter et al., 1998; Grassineau et al., 2002), but with well-preserved sub-rounded and elongate sulfide structures below this.

The Cheshire Formation contains mostly shallow-water sediments, mainly dark and ferruginous shales, ironstones, conglomerates, important limestone reefs, with alternations of lava and tuff. Two units were studied, the first one being the 1500-m-thick dark shales, above the basal volcanic sequence. Their deposition was in shallow water, indicated by ripple mark structures (Martin et al., 1980). The shales are well preserved and show evidence of only low-grade metamorphism. The second unit is the 24-m-thick Macgregor stromatolites, one of the largest reefs of the carbonate formation. They were deposited on a volcanic slope and comprise 22 cycles of environmental change (Martin et al., 1980) that occurred in the basin. Abell et al. (1985a) observed little textural evidence for metamorphism and from $\delta^{18}O_{carb}$ estimated the re-equilibration temperature for these rocks to be 80 °C.

Previous Stable Isotope Investigations

Until recently there were few $\delta^{13}C$ data for the Manjeri black shale units, with ranges of –32.2‰ to –8.8‰ for organic carbon, and –7.2‰ to –5.7‰ for carbonate (from Strauss and Moore, 1992). In contrast, Abell et al. (1985a) made an extensive study on the Cheshire stromatolites (Table 4) with ranges of –0.6‰ to +1.0‰ for the carbonate, and –35.4‰ to –16.9‰ for the reduced carbon. They also obtained a few $\delta^{13}C_{carb}$ values for the Rupemba stromatolites. Some Cheshire black shales analyzed by Yong (1991) show a $\delta^{13}C_{red}$ range of –40.8‰ to –32.0‰, and $\delta^{13}C_{carb}$ values from –13.3‰ to –8.0‰ for the carbonates in veins (Table 4). Some $\delta^{13}C_{red}$ measurements from the Manjeri carbon- and sulfide-rich sediments, with a range of –38‰ to –17‰, have been presented by Grassineau et al. (2001a, 2002). Grassineau et al. (2002) also obtained values of –35.1‰ to –7.3‰ for the Rupemba stromatolites and $\delta^{13}C_{red}$ values of –35.1‰ and 36.6 wt% C for a soft black "ball" ~0.5 mm in diameter found in the carbonate. This sample is bitumen and likely to be close to a pure kerogen composition. Finally a detailed study of $\delta^{34}S$ presented by Grassineau et al. (2001a, 2002) gives a range of –17.6‰ to +16.7‰ in the three sulfide-rich Manjeri units.

TABLE 4. BELINGWE GREENSTONE BELT RESULTS

Lithology	$\delta^{34}S$‰	n	$\delta^{13}C_{red}$‰	Carbon wt%	N	$\delta^{13}C_{carb}$‰	N
Manjeri Formation (2.7 Ga)							
Jimmy Member S- and C-rich sediments	−21.1 to +16.7‰ (−3.4 ± 6.2‰)	206	−38.4 to −20.7‰ (−32.8 ± 4.2‰)	0.04–20.0%[†]	27	−12.5 to −6.7‰ (−9.4 ± 2.1‰)	19
Shavi Member S- and C-rich sediments	−18.4 to +5.4‰ (−3.5 ± 4.5‰)	53	−32.0 to −7.2‰ −25.0 ± 7.6‰	0.13–3.18%	16	–	–
Spring Valley Member S- and C-rich sediments	+1.2 to +2.9‰ (+2.2 ± 0.7‰)	6	−31.9 to −23.3‰ −26.2 ± 4.0‰	1.36–2.66%	3	–	–
Rupemba Member stromatolites	–	–	−35.1 to −7.3‰ (−22.3 ± 4.6‰)	0.40–1.93% 36.65%*	19	−0.7 to +1.4‰ (0.1 ± 0.5‰)	44[§]
Cheshire Formation (2.6 Ga)							
Dark shales	−6.0‰	1	−43.8 to −32.0‰ (−39.5 ± 3.0‰)	0.11–3.19%	39[#]	−13.3 to −4.7‰ (−8.7 ± 2.8‰)	50[#]
McGregor Stromatolites	–	–	−35.4 to −16.9‰ (−28.6 ± 3.3‰)	0.02–1.99%	253[§]	−0.6 to +1.0‰ (+0.2 ± 0.3‰)	463[§]

Note: Averages are in parenthesis with 1 s.d.
[†]High carbon concentrations from samples containing bitumen.
*Pure bitumen ball.
[§]From Abell et al. (1985a), results obtained by McClory (1988).
[#]Compilation of Yong (1991) and this study (McClory and Yong results are from conventional line techniques).

New Results Obtained and Discussion of the Belingwe Greenstone Belt

Clastic Sediments

Further analyses have been made on the carbon- and sulfide-rich shallow-water and subtidal sediments from the Shavi and Jimmy Members and the Cheshire dark shales (Table 4). The results, obtained in pyrites, increase the extensive $\delta^{34}S$ range in the Manjeri units (Figs. 6A, 6B), from −21.1‰ to +16.7‰. This range of nearly 38‰ is the widest range yet recorded in Archean sediments.

In the Jimmy Member alone the $\delta^{34}S$ range equals the total spread found in the belt. There is a bimodal distribution of the $\delta^{34}S$ data (Fig. 6A). The sharp main peak is around 0‰. This value occurs mainly at the top of the Jimmy Member, toward the contact with Reliance volcanics. The proximity of the volcanism, which is taken to be stratigraphically directly above the Manjeri Formation (see discussion by Grassineau et al., 2002), suggests that this value records hydrothermal processes that happened during the deposition of the overlying volcanic unit or soon after, producing local re-homogenization of the primary $\delta^{34}S$. The second and smaller peak in the Jimmy Member histogram is around −5‰, with a skewed distribution on both sides. This value is close to the −6‰ recorded in the only pyrite found in the Cheshire shales. The $\delta^{34}S$ range in the Shavi Member, even if smaller, still has a significant spread of nearly 24‰. The population is distributed around a less pronounced peak of −4‰ (Fig. 6B).

Such large $\delta^{34}S$ ranges suggest microbial sulfate-reduction processes, but locally values have been recorded where $\delta^{34}S$ in the pyrites are above +6‰. This occurs only in one black shale sample of 10 cm length (TR51 at 67.05 m), where the pyrites are in bleb-like shapes with internal concentric structure, which are interpreted to be biogenic features. These ^{34}S-enriched pyrites (up to +17‰) are associated with adjacent blebs of dolomite, which have light $\delta^{13}C_{carb}$ (−12‰ to −10‰).

The reduced carbon for the Manjeri shales has a $\delta^{13}C$ spread of 31‰. The distribution shows a main value around −29‰

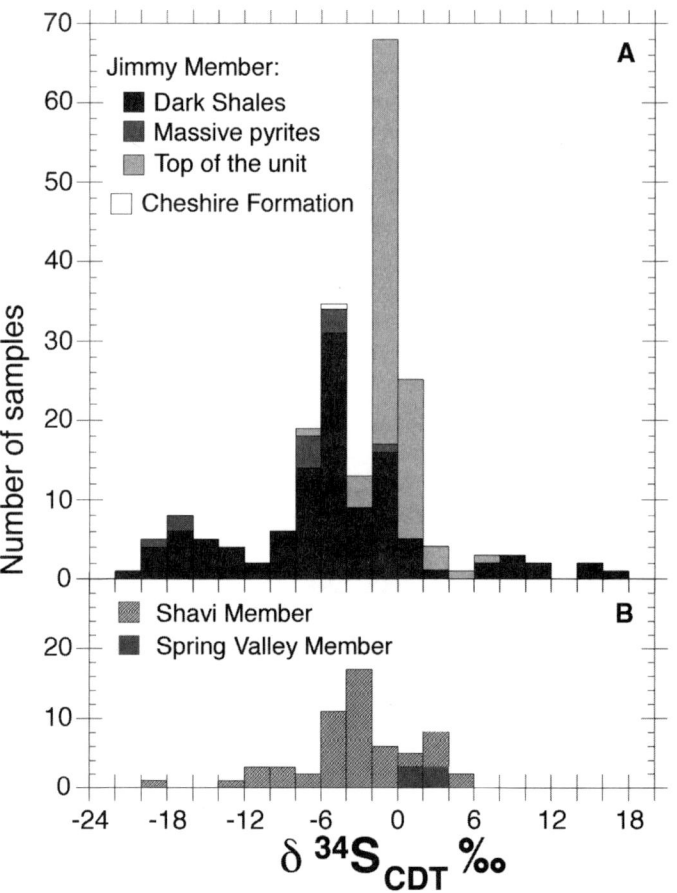

Figure 6. $\delta^{34}S$ distributions for sulfide minerals in three units of carbon- and sulfide-rich sediments of the Manjeri Formation (A, B) (BGB, 2.7 Ga). The distribution is bimodal in the Jimmy Member (A), with a main peak at 0‰, and a second skewed peak at −5‰. For information, the −6‰ value measured for the Cheshire Formation has been added.

(Fig. 7), and carbon concentrations from 0.04 to 20.0 wt‰. A few samples containing "soft" bitumen-like material have a high content of reduced carbon (from 14 to 20 wt%). In detail, the Shavi and Spring Valley Members show an average $\delta^{13}C_{red}$ of −25.0‰ ± 7.6‰ (Fig. 7A), whereas the Jimmy Member is much lighter, with a $\delta^{13}C_{red}$ of −32.8‰ ± 4.2‰ (Fig. 7B).

The wide range found in $\delta^{13}C_{red}$ indicates primary bacterial activity. The main peak value around −30‰ suggests fractionation by Rubisco I (Pierson, 1994). Shavi and Spring Valley units appear to reflect mainly the enzyme signature for oxygenic photosynthesis, but the lighter values in the Jimmy Member more likely record carbon cycling via methanogenesis, in addition to photosynthesis (Grassineau et al., 2001a, 2002). In comparison, data for Cheshire shales are very light, averaging −39.5‰ ± 3.0‰ (Fig. 7C). These occur locally in association with light carbonates (down to −13.3‰; Table 4). These results suggest methanogenic and methanotrophic activities at the time (e.g. Yong, 1991; Grassineau et al., 2002), with methane oxidation by methanotrophs and sulfate reducers to generate the light $\delta^{13}C_{carb}$ values at the oxic/anoxic interface (e.g., Peckmann and Thiel, 2004).

Detailed Study of BES49c

Numerous samples have been analyzed in detail at millimeter scale to search for evidence of microbial consortia, and large isotopic variations have been found for both $\delta^{34}S$ and $\delta^{13}C_{red}$. BES49c, a 5-cm-long core section, roughly at the middle of the Jimmy Member is an example. It is a black shale with a gray chert layer at the bottom, and on the top, a 1.5-cm-thick sulfide layer (Fig. 8). The sedimentary layering in BES49c suggests significant diagenetic compaction.

The $\delta^{34}S$ range is from −19.6‰ to −5.4‰, with large variations of more than 9‰ over distances of 2 or 3 mm. All sulfides are pyrite. In the upper part, they have slightly cataclazed bleb-like shapes, rather rounded, consisting of microcrystalline framboid-like agglomerations, with very small internal cavities, resembling a "sponge" structure. The $\delta^{34}S$ value of −18.6‰ inside the top left large bleb seems to have been sealed from further exchange since the time of its deposition (Fig. 8). The $\delta^{34}S$ variations are evidence of very prolific biological activity. The heaviest value, at −5.4‰, is from a layer of disseminated small pyrite crystals, interbedded in dark chert layers. It is difficult to say if these sulfides are primary, but the −5‰ value is very common in the Jimmy Member and partial homogenization by hydrothermal fluids could have occurred. Most likely the value is a genuine biological signature from isolated activity with a plentiful SO_4 supply, as it is mainly found in scattered interbedded pyrites. The

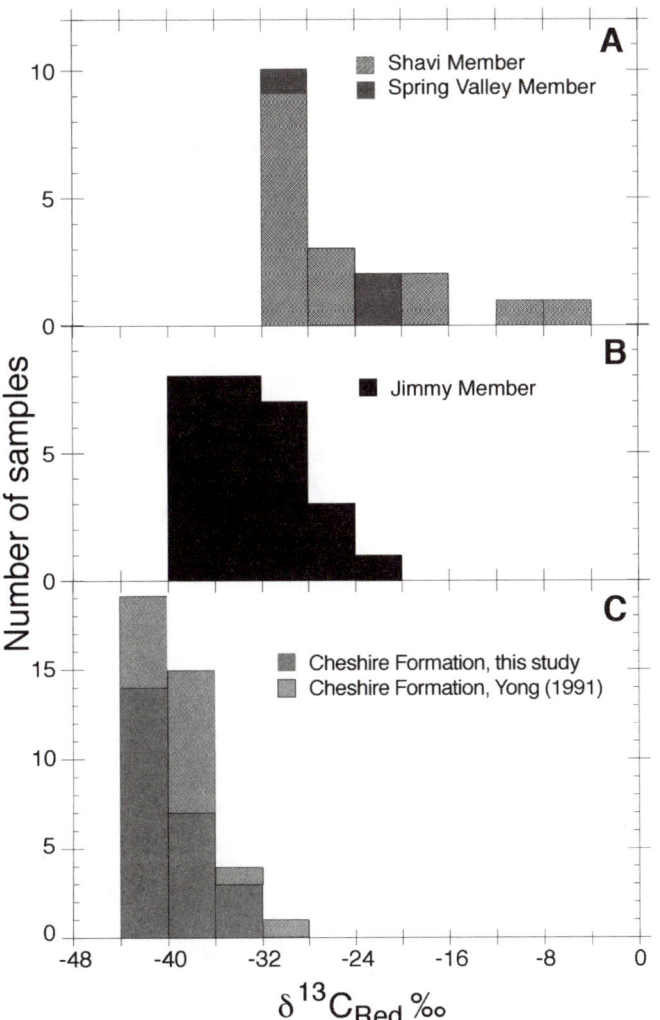

Figure 7. Distribution of $\delta^{13}C_{red}$ from sedimentary sequences in the Belingwe Greenstone Belt. The samples are from various drill cores. The median value is −29‰ for the Shavi Member (A), −33‰ for the Jimmy Member (B), and −40‰ for the Cheshire dark shales (C).

Figure 8. Detailed isotopic study of core sample BES49c, from the Jimmy Member. BES49c is a carbon- and sulfide-rich black shale, with a layer of rounded bleb-like sulfides. The $\delta^{34}S$ values vary between −19.6‰ and −5.4‰, with the heavier values for the dispersed pyrites.

lighter values are typically in ~1–2-cm-size blebs of aggregated microcrystallized sulfides and the isotopic compositions are zoned in these structures. As the variations imply a biogenic origin, these rounded and granulated shapes might be an illustration of what remains of the bacterial communities, with intense biological activity illustrated by large fractionations (Canfield and Teske, 1996). In these multiple micro-ecosystems once hosted by BES49c, the interactions between the bacteria created different fractionation values, depending on the neighboring bacteria and environment. Each value then could represent the identity of a unique bacterial community.

The $\delta^{13}C_{red}$ values in BES49c, –27.5‰ and –31.4‰, are also evidence of biological activity (Fig. 8). They give an idea of the organic processes that had most likely taken place. At the bottom, the signature seems to reflect Rubisco I activity, and consequently the existence of photosynthetic processes, whereas at the top, near the sulfides, the lighter value more likely indicates a combination of photosynthesis and probably methanogenesis. Studying the isotopic variations at the scale of the sample gives an instant snapshot of the interactions between different bacterial activities present at 2.7 Ga.

Carbonate Reefs

Carbon isotopic results obtained for the stromatolites of the Rupemba Member at the base of Manjeri and of the Macgregor Member in the Cheshire are shown in Figure 9. The samples are 1-m-long hand-drilled cores, and they show superbly preserved biological textures. The values of the Macgregor Member are from Abell et al. (1985a) and McClory (1988). Considering the very narrow ranges for $\delta^{13}C_{carb}$ and $\delta^{13}C_{red}$ (Table 4), we concluded that the stromatolites represent nearly unaltered biogenic sediments.

For the Macgregor stromatolites, the average $\delta^{13}C_{carb}$ is +0.2‰ ± 0.3‰ and the $\delta^{13}C_{red}$ values are mostly between –34‰ and –24‰, with a –28.6‰ ± 3.3‰ average (Fig. 9). With these results, a mean value of 28.8‰ for the $\Delta^{13}C_{red-carb}$ fractionation can be inferred. Given the stromatolitic textures, this fractionation is interpreted as the unambiguous signature of Rubisco I activity, indicating that the Macgregor stromatolites were deposited by a cyanobacterial microbial ecology based on oxygenic photosynthesis (Abell et al., 1985a; Grassineau et al., 2002).

Further analyses were made in the Rupemba stromatolites, which underwent more strain and are in a setting where metamorphic recrystallization has been more extensive. The $\delta^{13}C_{carb}$ values range from –0.7‰ to +1.4‰ and average +0.1 ± 0.5‰, which are similar to Macgregor stromatolite values. The samples also analyzed for $\delta^{13}C_{red}$ give values mostly between –23.3‰ and –21.0‰. The <200 °C metamorphism mentioned earlier may have shifted $\delta^{13}C_{red}$ slightly toward heavier values. Compared with the Macgregor stromatolites, the evidence for Rubisco I is less obvious, but it cannot be ruled out because of the probable effect of metamorphism, suggesting that the original values might have been lighter. Alternatively, the 22‰ difference between $\delta^{13}C_{carb}$ and $\delta^{13}C_{red}$ may be a primary fractionation record of Rubisco II

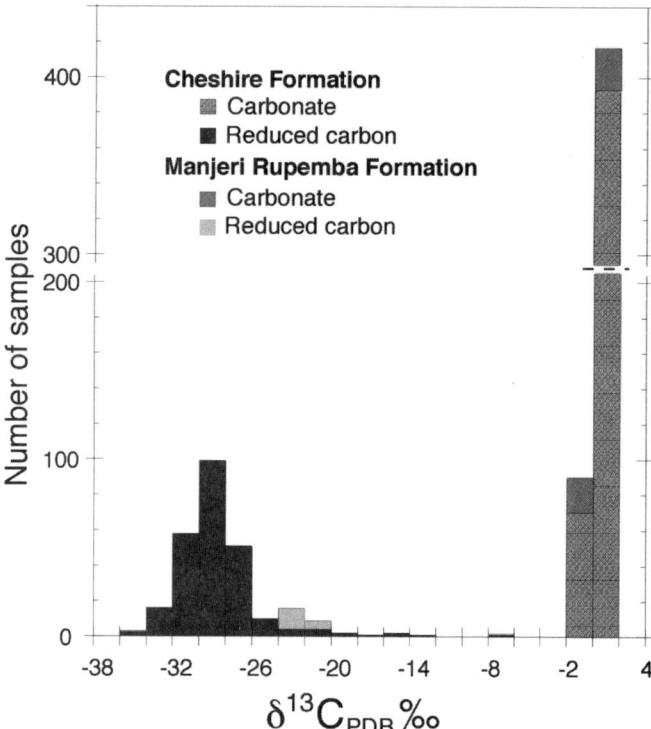

Figure 9. Distributions of $\delta^{13}C_{red}$ and $\delta^{13}C_{carb}$ for the Rupemba and Macgregor (Cheshire Formation) stromatolitic sequences. The McGregor stromatolites have a $\Delta^{13}C_{carb-red}$ around 29‰.

controlling early anoxygenic activity or may be a result of lower pCO_2 in the water column. On the other hand, the single bitumen value of –35.1‰ with 36.6 wt% C (Table 4) from the Rupemba stromatolites indicates that methanogenic processes were occurring either at the base of stromatolitic layers, or post-depositionally if the limestone acted as a hydrocarbon reservoir.

Summary

The wide $\delta^{34}S$ range of nearly 40‰ obtained for pyrite from the Manjeri Formation indicates that the biological sulfur cycle was mostly, if not completely, operational at 2.7 Ga (Grassineau et al., 2001a), well before the post-Archean increase of atmospheric oxygen took place at 2.2 Ga (e.g., Habicht and Canfield, 1996; Blank, 2004). Moreover, the isotopic variations at millimeter scale show that an important diversity of sulfur-dependent organisms, mainly sulfate-reducing bacteria, were present. It is also possible that sulfide-oxidizing bacteria were operational in restricted localities during Jimmy Member deposition, as some anoxygenic photosynthesizers can conduct sulfide oxidation (Cohen et al., 1989). At the interface between anoxygenic and oxygenic conditions, elemental sulfur is liberated and then re-reduced in part to sulfide, by sulfur disproportionation. At the interface, ^{13}C-depleted carbonate can also be formed, as seen in TR51 sample (e.g., Wadham et al., 2004). More recent examples of depleted carbonates associated with sulfide mats and globular sulfides are found near methane seeps (Peckmann et al., 2004). The light carbon is a result of the anaerobic oxidation of the methane in association with sulfate reduction (e.g., Peckmann and Thiel, 2004). An alternative explanation of the positive sulfur values could be Rayleigh fractionation of a finite sulfate reservoir. The data for sulfide below TR51 in the core are on average 15‰ lighter, but they form a separate peak, which does not tail off to ^{34}S-enrichment as expected by Rayleigh fractionation (Fig. 6A). Therefore, a change to an environment typical of more modern methane- and sulfur-rich hydrothermal events seems to be the likely explanation. The relatively smaller range in the Shavi Member (Fig. 6B) suggests that the rate of SO_4 reduction was higher than in the Jimmy Member (e.g., Schidlowski et al., 1983). It is possible that in the Jimmy Member, the SO_4 supply might have temporarily and locally changed, or been limited at the deposition sites, as illustrated by the TR51 sample (δ^{34}S above +6‰).

The wide range of $\delta^{13}C_{red}$ measured in the three Manjeri carbon-rich units reveals the existence of a well-developed carbon cycle. The oxygenic photosynthesis operated by Rubisco I existed in the Shavi and Spring Valley Members. This is consistent with the interpretation of these units as shallow-water sediments (Hunter et al., 1998). At the same stratigraphic level, the Rupemba stromatolites, despite some isotopic adjustment, were most likely generated by the same activity, directed by cyanobacteria. The $\delta^{13}C_{red}$ in the Jimmy Member may represent a complex mixing of signatures from different bacterial processes, recording composite bacterial communities. The Jimmy Member includes finely laminated material probably deposited in quiet water below wave base, deeper than the Shavi Member and with conditions becoming dominantly anaerobic toward the top of the unit, with $\delta^{13}C_{red}$ lighter than −30‰. Further evidence for the introduction of methane into a previously oxic environment comes from the $\delta^{13}C$ of organic carbon. Ten centimeters below TR51, the $\delta^{13}C_{red}$ drops to −37‰ with an associated increase in carbon content from <1 wt% to up to 18 wt% in the overlying 2.4 m. At this stratigraphic level the sulfide isotopic compositions shift from being strongly heterogeneous to having a restricted range typical of hydrothermal sulfur (Fig. 6A). The apparent presence of an "oxygenic photosynthesis" signature, at intervals mainly observed in the lower part of this 2.4 m, may record the infall of dead organisms from the water surface.

Highly evolved prokaryote mat communities operating complex metabolic processes existed at 2.7 Ga (Grassineau et al., 2001a, 2002). They are the result of various microbial ecosystems due to a change of environments, from the subtidal or tidal microbial mats of the Shavi Member to a deeper environment below wave base with the appearance of anoxic processes in the Jimmy Member.

Similar observations are noted in the Cheshire Formation. The dark shales give very light $\delta^{13}C_{red}$ values, down to −44‰, which argue in favor of anaerobic environments. The processes involved might have been occasionally anoxygenic photosynthesis at the oxic/anoxic interface, but more likely methanogenesis

and methanotrophy. The $\delta^{13}C_{red}$ values in the Cheshire stromatolites show the fractionation produced by the Rubisco I enzyme, evidence of oxygenic photosynthesis by cyanobacteria, in a tidal to subtidal environment.

COMPARISON AMONG THE FOUR ARCHEAN GREENSTONE BELTS

Figure 10 (A, B) shows the narrow δ^{34}S ranges obtained for the Barberton and Isua Greenstone Belts. The metamorphic

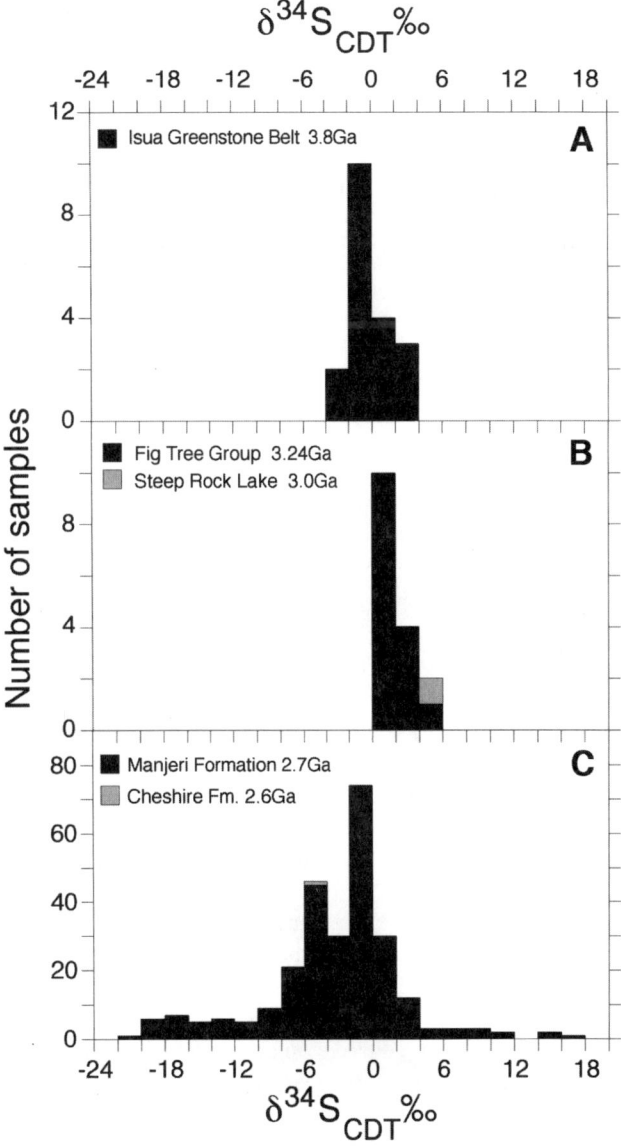

Figure 10. Comparison between δ^{34}S values from this study for sulfide minerals from sedimentary sequences in Early, Middle, and Late Archean. From small ranges, 7.2‰ in the IGB (A) and 4.1‰ in the BGB (B), the range in the Belingwe belt reached 38‰ (C). Considering previous studies in the Barberton belt, it appears that the range slowly expanded through Early and Middle Archean but widened considerably at the dawn of the Late Archean.

events that affected the IGB, including the northeastern part, might have reduced the original range of values, but the average is probably representative of the primary signature. Even if there is no additional evidence, a biological cause for this 4.9‰ range cannot be ruled out. The fractionation is small, perhaps because of a low supply of sulfate, but is still consistent with the presence of sulfate reducers. The presence of high-temperature (>85 °C) hyperthermophiles (Woese, 1987; Nisbet and Fowler, 1996; Nisbet and Sleep, 2001) could result in such a small isotopic fractionation. Evidence for hydrothermal activity at Isua comes from methane- and carbonate-rich inclusions in vesicles from the least deformed pillow lavas (Appel et al., 2001), which are similar in composition to some current off-ridge axis vent fields (Touret, 2003). In such settings inorganic processes may have been significant, with elemental sulfur forming at vents ($\delta^{34}S$ of 0‰ to +1‰). Presumably little sulfate was present in the Isua oceans and around these vents, and sulfur-oxidizing bacteria might have existed (Jannasch, 1989). Under this condition, the fractionation of sulfur due to biological activity will surely have been small, whatever metabolic processes were occurring.

The range of $\delta^{34}S$ values obtained for Barberton samples in this study but also by previous authors (see list of references in the BGB section) is slightly bigger that the one found for Isua. The average is also enriched (+1.4‰) relative to the IGB (–0.2‰). In the detailed study, the growth of replacement carbonate within the formation and preferentially around the sulfide minerals may have been responsible for some re-homogenization. However, the recorded $\delta^{34}S$ range is best explained as evidence of bacterial sulfate reduction, in sediments or in the anoxic water column, at 3.24 Ga (Kakegawa and Ohmoto, 1999).

The impact of metamorphism and secondary processes on the carbon isotopic compositions is not negligible, and perhaps only locally has the $\delta^{13}C$ partially preserved its original signature. This has to be taken into account in assessing claims for biogenicity, especially in Isua, where Schidlowski et al. (1979), Schidlowski (1988, 2001) and Rosing (1999) stressed that results should be corrected for the metamorphism. Only the range of –30‰ to –22‰ in the Isua sedimentary sequences can be discussed (though these values could have been lighter originally), as it indicates the possibility of life. Rosing (1999) suggests the presence of organisms similar to modern plankton in the Isua sea ($\delta^{13}C_{red}$ = –19‰). On the other hand, a process, probably anaerobic, is proposed from the results of –29‰ to –28‰ in the sulfide-bearing BIF and the metaconglomerate. Given that the isotopic range may have been reduced by metamorphism, values as light as –29.6‰ in this BIF suggest that methanotrophic activity took place during deposition, although it is also possible that photosynthetic processes had a wider range of fractionations at the time (Watanabe et al., 1997). Thus a complex community might have been present as early as 3.8 Ga.

In the middle Archean, the $\delta^{13}C_{red}$ range in the Barberton belt is larger and lighter than in the Isua belt, with values down to –32.4‰ (Figs. 11A, 11B). Again, allowance must be made for a small metamorphic shift, and for the possibility that the heavier

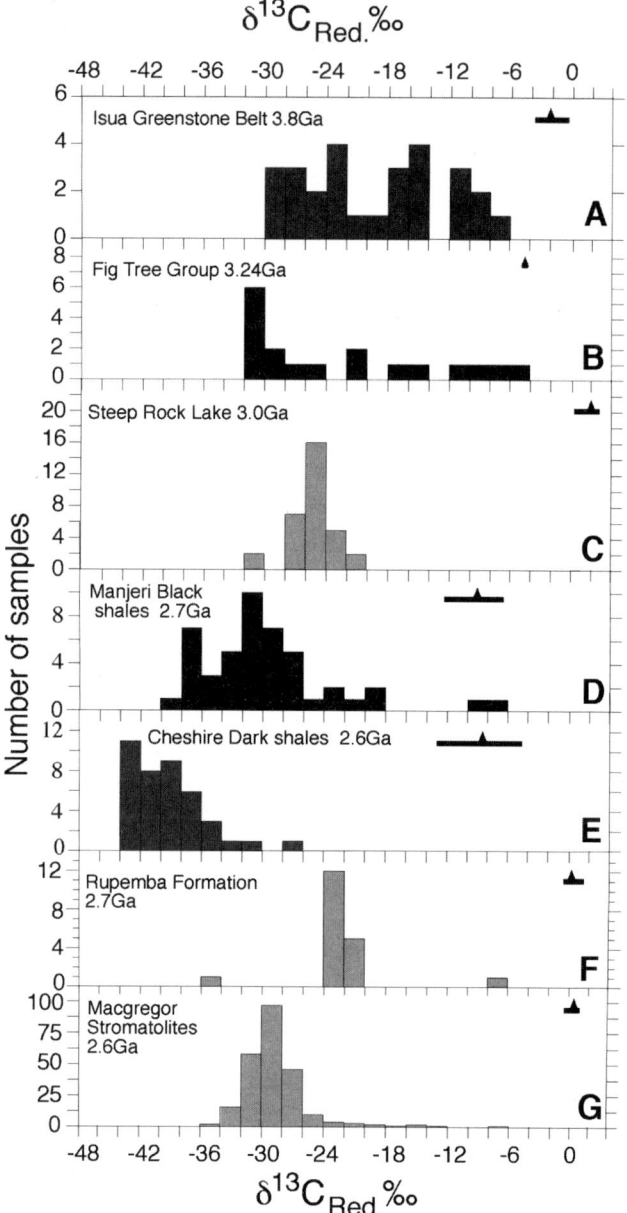

Figure 11. Comparison between $\delta^{13}C_{red}$ values from this study for sedimentary rocks in early (A), middle (B, C), and Late Archean (D–G), with a probable isotopic shift due to metamorphism in the IGB and BGB. The $\delta^{13}C_{red}$ values for Steep Rock Lake and Belingwe are lighter (C, D). The two stromatolites, at 3.0 Ga and 2.6 Ga (C, G), show similar ranges for $\delta^{13}C_{red}$ and $\delta^{13}C_{carb}$. Stromatolites are represented in dotted pattern, black shales in black. The thick black bars for each diagram represent the ranges of $\delta^{13}C_{carb}$ values obtained, with their average value shown by the triangles.

values may be linked to the formation of replacement carbonates. However, the lower part of the range, –32‰ to –26‰, suggests that several bacterial activities are recorded, including photosynthesis and methane-utilizing processes.

By the late Middle Archean, a significant change had occurred. The $\delta^{13}C_{red}$ averaging −25.4‰ and the $\Delta^{13}C_{red\text{-}carb}$ fractionation around 28‰ in Steep Rock Lake (Fig. 11C) are interpreted as clear evidence of biological activity and more particularly of fractionation by Rubisco I. Therefore, as the biogenic structures were developed in shallow water associated with a carbonate platform, it is likely that oxygenic photosynthesis was already established around 3.0 Ga, and that cyanobacteria were fully active at that time.

By the Late Archean, the wide $\delta^{34}S$ range obtained in the Manjeri units strongly indicates an operational biological sulfur cycle at 2.7 Ga (Fig. 10C), with mainly sulfate-reducing bacteria, but also possible sulfur oxidizers. The activities seem to depend on local supplies of sulfate, conditioned by the environment. A fast SO_4 supply will benefit a high rate of sulfate reduction in the Shavi Member, but with a slower rate, additional but restricted sulfur-oxidizing processes could have occurred within the Jimmy Member at the oxic-anoxic interface. The wide $\delta^{13}C_{red}$ range measured in the Manjeri Formation (Fig. 11D) reveals the existence of a well-developed carbon cycle with Rubisco I operational. The Shavi and Spring Valley Members and probably the Rupemba stromatolites indicate oxygenic photosynthesis (Fig. 11F). The lighter $\delta^{13}C_{red}$ values (below −30‰) recorded in the Jimmy Member and Cheshire Formation suggest that processes in addition to photosynthesis were occurring, most likely methanogenesis and methanotrophy (Fig. 11E). Values for $\delta^{13}C_{red}$ in the Macgregor stromatolites (Fig. 11G) indicate oxygenic photosynthesis with cyanobacterial management.

Over this one billion years of Earth's history, the evolution of life appears to have been marked by changes of environmental conditions, as suggested by Nisbet and Sleep (2001). The $\delta^{34}S$ and $\delta^{13}C$ obtained in the three Archean periods suggest that biological processes created wider isotopic fractionations from 3.8 Ga to 2.7 Ga. With better sulfate and oxygen supplies, organisms colonized a wide spectrum of environments on Earth, which increased the biological diversity by the Late Archean.

CONCLUSIONS

In the Early Archean at 3.8 Ga, the $\delta^{34}S$ spread in sedimentary sequences of 7.2‰ around a 0‰ peak suggests that biological activity was focused around high-temperature hydrothermal vents. Sulfur processing could have existed, and possibly methanogenesis ($\delta^{13}C_{red}$ of −29‰ after metamorphism). Photosynthesis might also have occurred.

In the Middle Archean, at 3.24 Ga, microbial activity seems to be more widespread, but mainly under anoxic conditions. Sulfate-reducing processes have been recognized, using seawater sulfate. Methanogenic archaea were probably present, and possibly sulfur oxidizers. For Steep Rock Lake at 3.0 Ga, there is unequivocal evidence of Rubisco I and therefore oxygenic photosynthesis, with a fractionation between organic carbon and residual carbonate of −28‰. Therefore cyanobacteria were well established by the end of the Middle Archean/early Late Archean.

The colonization of the continental margins by the cyanobacterial mats was the beginning of a flourishing of activity that is well illustrated at 2.7 Ga.

In the basal Manjeri Formation and the Cheshire Formation in the 2.7 Ga Belingwe Greenstone Belt, the ranges for $\delta^{13}C_{red}$ and $\delta^{34}S$ are wide, respectively ~37‰ and ~38‰. These values indicate a large diversity of bacterial activities. The important variations found at millimeter scale in samples indicate that these bacterial communities were interacting by various pathways. Oxygenic photosynthesizers and sulfate reducers were operating in microbial mats in shallow water reefs under subtidal and tidal conditions. Organisms below wave base used mainly anoxygenic photosynthesis and sulfate-reducing processes. Methanotrophs and sulfide oxidizers existed at the oxic/anoxic interface, with methanogens below.

ACKNOWLEDGMENTS

Various funding bodies financed this study: Natural Environment Research Council (NERC) and the Leverhulme Trust for the Belingwe work; the Geological Survey of Denmark and Greenland, the Isua Multidisciplinary Research Project, the Danish Research Council, the Bureau of Minerals and Petroleum in Greenland, the Commission of Scientific Research in Greenland, the GEODE project, and the Royal Society for the Isua part. A particular thank is addressed to C. Siebert and J. Kramers for the Fig Tree samples, courtesy of R. LeRoux and C. Rippon (AVGold, ETC division). The authors thank H. Ohmoto, T. Kakegawa and, Y. Shen for their constructive reviews, and S. Kesler for thorough editing of the manuscript.

REFERENCES CITED

Abell, P.I., McClory, J., Martin, A., and Nisbet, E.G., 1985a, Archaean stromatolites from the Ngesi Group, Belingwe Greenstone Belt, Zimbabwe. Preservation and stable isotopes—Preliminary results: Precambrian Research, v. 27, p. 357–383, doi: 10.1016/0301-9268(85)90094-4.

Abell, P.I., McClory, J., Martin, A., Nisbet, E.G., and Kyser, T.K., 1985b, Petrography and stable isotope ratios from Archaean stromatolites, Mushandike Formation, Zimbabwe: Precambrian Research, v. 27, p. 385–398, doi: 10.1016/0301-9268(85)90095-6.

Appel, P.W.U., 1979, Stratabound copper sulfides in a banded iron-formation and in basaltic tuffs in the Early Precambrian Isua supracrustal belt: West Greenland: Economic Geology, v. 74, p. 45–52.

Appel, P.W.U., Fedo, C.M., Moorbath, S., and Myers, J.S., 1998, Recognizable primary volcanic and sedimentary features in a low strain domain of the highly deformed, oldest known (3.7–3.8 Ga) greenstone belt, Isua: West Greenland: Terra Nova, v. 10, p. 57–62, doi: 10.1046/j.1365-3121.1998.00162.x.

Appel, P.W.U., Rollinson, H.R., and Touret, J.L.R., 2001, Remnants of an early Archaean (>3.74 Ga) sea-floor, hydrothermal system in the Isua Greenstone Belt: Precambrian Research, v. 112, p. 27–49, doi: 10.1016/S0301-9268(01)00169-3.

Blank, C.E., 2004, Evolutionary timing of the origins of mesophilic sulphate reduction and oxygenic photosynthesis: A phylogenomic dating approach: Geobiology, v. 2, p. 1–20, doi: 10.1111/j.1472-4677.2004.00020.x.

Bickle, M.J., and Nisbet, E.G., editors, 1993, The geology of the Belingwe Greenstone Belt: A study of the evolution of Archaean continental crust: Geological Society of Zimbabwe Special Publication 2: Rotterdam, A.A. Balkema, 239 p.

Bolhar, R., Hofmann, A., Woodhead, J.D., Hergt, J.M., and Dirks, P., 2002,

Pb- and Nd-isotope systematics of stromatolitic limestones from the 2.7 Ga Ngezi Group of the Belingwe Greenstone Belt: Constraints on timing of deposition and provenance: Precambrian Research, v. 114, p. 277–294.

Cameron, E.M., 1982, Sulphate and sulphate reduction in early Precambrian oceans: Nature, v. 296, p. 145–148, doi: 10.1038/296145a0.

Canfield, D.E., and Teske, A., 1996, Late Proterozoic rise in atmospheric oxygen concentration inferred from phylogenetic and sulphur-isotope studies: Nature, v. 382, p. 127–132, doi: 10.1038/382127a0.

Cavalier-Smith, T., 2002, The neomuran origin of archaebacteria, the negibacterial root of the universal tree and bacterial megaclassification: International Journal of Systematic and Evolutionary Microbiology, v. 52, p. 7–76.

Chauvel, C., Dupre, B., and Arndt, N.T., 1993, Pb and Nd isotopic correlation in Belingwe komatiites and basalts, in Bickle, M.J., and Nisbet, E.G., eds., The geology of the Belingwe Greenstone Belt, Zimbabwe: Harare, Geological Society of Zimbabwe, v. 2, p. 167–174.

Cohen, Y., Gorlenko, V.M., and Bonch-Osmolovskaya, E.A., 1989, Interaction of Sulphur and Carbon Cycles in Microbial Mats, in Brimblecombe, P., and Lein, A. Yu, eds., Evolution of the global biogeochemical sulphur cycle: SCOPE report no. 39: Chichester, UK, John Wiley and Sons, p. 191–238.

Coleman, D.D., Risatti, J.B., and Schoell, M., 1981, Fractionation of carbon and hydrogen isotopes by methane-oxidizing bacteria: Geochimica et Cosmochimica Acta, v. 45, p. 1033–1037, doi: 10.1016/0016-7037(81)90129-0.

Davis, D.W., and Jackson, M.C., 1985, Preliminary U-Pb zircon ages from the Lumby Lake – Marmion Lake area, districts of Kenora and Rainy River, in Wood, J., White, O.L., Barlow, R.B., and Colvine, A.C., eds., Summary of fieldwork and other activities: Ontario Geological Survey Miscellaneous Paper 126, p. 135–137.

de Ronde, C.E.J., and Ebbesen, T.W., 1996, 3.2 b.y. of organic compound formation near sea-floor hot springs: Geology, v. 24, p. 791–794, doi: 10.1130/0091-7613(1996)024<0791:BYOOCF>2.3.CO;2.

de Ronde, C.E.J., Spooner, E.C.T., de Wit, M.J., and Bray, C.J., 1992, Shear zone-related, Au quartz vein deposits in the Barberton greenstone belt, South Africa: Field and petrographic characteristics, fluid properties, and light stable isotope geochemistry: Economic Geology and the Bulletin of the Society of Economic Geologists, v. 87, p. 366–402.

Erez, J., Bouevitch, A., and Kaplan, A., 1998, Carbon isotope fractionation by photosynthetic aquatic microorganisms: Experiments with *Synechococcus* PCC7942, and a simple carbon flux model: Canadian Journal of Botany, v. 76, p. 1109–1118, doi: 10.1139/cjb-76-6-1109.

Faure, G., 1986, Principles of isotope geology (2nd edition): New York, John Wiley and Sons, 589 p.

Fedo, C.M., 2000, Setting and origin for problematic rocks from the >3.7 Ga Isua Greenstone Belt, southern west Greenland: Earth's oldest coarse clastic sediments: Precambrian Research, v. 101, p. 69–78, doi: 10.1016/S0301-9268(99)00100-X.

Fedo, C.M., Myers, J.S., and Appel, P.W.U., 2001, Depositional setting and paleogeographic implications of earth's oldest supracrustal rocks, the > 3.7 Ga Isua Greenstone belt: West Greenland: Sedimentary Geology, v. 141, p. 61–77, doi: 10.1016/S0037-0738(01)00068-9.

Frei, R., and Rosing, M.T., 2001, The least radiogenic terrestrial leads; implications for the early Archean crustal evolution and hydrothermal-metasomatic processes in the Isua supracrustal belt (West Greenland): Chemical Geology, v. 181, p. 47–66, doi: 10.1016/S0009-2541(01)00263-7.

Frei, R., Bridgwater, D., Rosing, M., and Stecher, O., 1999, Controversial Pb-Pb and Sm-Nd isotope results in the early Archean Isua (West Greenland) oxide iron formation: Preservation of primary signatures versus secondary disturbances: Geochimica et Cosmochimica Acta, v. 63, 3/4, p. 473–488.

Grassineau, N.V., and Appel, P.W.U., 2000, Variations in the stable isotope compositions for two gold deposits in the Isua Greenstone Belt, West Greenland Sequences [abs.]: Eos (Transactions, American Geophysical Union), v. 81, no. 48, p. F1267–68.

Grassineau, N.V., Nisbet, E.G., and Fowler, C.M.R., 2000, A stable isotopic search for biological signatures in the Archaean: A comparative study of early (Isua, 3.8 Ga; Greenland) and late Archaean (Belingwe, 2.7 Ga; Zimbabwe) sequences [abs.]: Eos (Transactions, American Geophysical Union), v. 81, no. 48, p. F1257–58.

Grassineau, N.V., Nisbet, E.G., Bickle, M.J., Fowler, C.M.R., Lowry, D., Mattey, D.P., Abell, P., and Martin, A., 2001a, Antiquity of the biological sulphur cycle: Evidence from S and C isotopes in 2.7 Ga rocks of the Belingwe Belt, Zimbabwe: Proceedings of the Royal Society of London, ser. B, v. 268, p. 113–119, doi: 10.1098/rspb.2000.1338.

Grassineau, N.V., Mattey, D.P., and Lowry, D., 2001b, Rapid sulphur isotopic analyses of sulphide and sulphate minerals by continuous flow- isotope ratio mass spectrometry (CF-IRMS): Analytical Chemistry, v. 73, p. 220–225, doi: 10.1021/ac000550f.

Grassineau, N.V., Nisbet, E.G., Fowler, C.M.R., Bickle, M.J., Lowry, D., Chapman, H.J., Mattey, D.P., Abell, P., Yong, J., and Martin, A., 2002, Stable isotopes in the Archaean Belingwe belt, Zimbabwe: Evidence for a diverse microbial mat ecology, in Fowler, C.M.R., Ebinger, C.J., and Hawkesworth, C.J., eds., The early Earth: Physical, chemical and biological development: Geological Society [London] Special Publication 199, p. 309–328.

Habicht, K., and Canfield, D.E., 1996, Sulphur isotope fractionation in modern microbial mats and the evolution of the sulphur cycle: Nature, v. 382, p. 342–343, doi: 10.1038/382342a0.

Hayes, J.M., Kaplan, I.R., and Wedeking, K.W., 1983, Precambrian organic geochemistry, preservation of the record: Chapter 5, in Schopf, J.W., ed., Earth's earliest biosphere: Its origin and evolution: Princeton, New Jersey, Princeton University Press, p. 93–134.

Hayes, J.M., 1994, Global methanotrophy at the Archaean-Proterozoic transition, in Bengtson, S., ed., Early life on Earth: Nobel Symposium 84: New York, Columbia University Press, p. 220–236.

Hoefs, J., 1997, Stable Isotope Geochemistry (4th edition): Berlin, Springer-Verlag, 201 p.

Hunter, M.A.H., Bickle, M.J., Nisbet, E.G., Martin, A., and Chapman, H.J., 1998, A continental extension setting for the Archaean Belingwe Greenstone belt, Zimbabwe: Geology, v. 26, p. 883–886, doi: 10.1130/0091-7613(1998)026<0883:CESFTA>2.3.CO;2.

Jannasch, H.W., 1989, Sulphur emission and transformations at deep sea hydrothermal vents, in Brimblecombe, P., and Lein, A. Yu., eds., Evolution of the global biogeochemical sulphur cycle: SCOPE report no. 39: Chichester, UK, John Wiley and Sons, p. 181–190.

Kakegawa, T., and Ohmoto, H., 1999, Sulfur isotope evidence for the origin of 3.4 to 3.1 Ga pyrite at the Princeton gold mine, Barberton Greenstone Belt, South Africa: Precambrian Research, v. 96, p. 209–224, doi: 10.1016/S0301-9268(99)00006-6.

Komiya, T., Maruyama, S., Masuda, T., Nohda, S., Hayashi, M., and Okamoto, K., 1999, Plate tectonics at 3.8–3.7 Ga: Field evidence from the Isua accretionary complex, Southern West Greenland: Journal of Geology, v. 107, p. 515–554, doi: 10.1086/314371.

Kramers, J.D., Meisel, T., Morel, P., Nägler, T., and Siebert, C., 2004, Clues from PGE and Mo on the early redox history of the atmosphere [abs.]: Geologiska Föreningen, 26th Nordic Geological Winter Meeting, Uppsala, Sweden, v. 126, p. 11.

Kusky, T.M., and Hudleston, P.J., 1999, Growth and demise of an Archean carbonate platform Steep Rock Lake, Ontario, Canada: Canadian Journal of Earth Sciences, v. 36, p. 565–584, doi: 10.1139/cjes-36-4-565.

Lepland, A., Arrhenius, G., and Cornell, D., 2002, Apatite in Early Archean Isua supracrustal rocks, southern West Greenland: Its origin, association with graphite and potential as a biomarker: Precambrian Research, v. 118, p. 221–241, doi: 10.1016/S0301-9268(02)00106-7.

Londry, K.L., and Des Marais, D.J., 2003, Stable carbon isotope fractionation by sulfate-reducing bacteria: Applied and Environmental Microbiology, v. 69, p. 2942–2949, doi: 10.1128/AEM.69.5.2942-2949.2003.

Lowe, D., and Byerly, G.R., 1999, Stratigraphy of the west-central part of the Barberton Greenstone Belt, South Africa, in Lowe, D., and Byerly, G.R., eds., Geologic evolution of the Barberton Greenstone Belt, South Africa. Geological Society of America Special Paper 329, p. 1–36.

Martin, A., Nisbet, E.G., and Bickle, M.J., 1980, Archaean Stromatolites of the Belingwe Greenstone Belt: Precambrian Research, v. 13, p. 337–362, doi: 10.1016/0301-9268(80)90049-2.

Matthews, D.E., and Hayes, J.M., 1978, Isotope-ratio-monitoring gas chromatography-mass spectrometry: Analytical Chemistry, v. 50, p. 1465–1473, doi: 10.1021/ac50033a022.

McClory, J.P., 1988, Carbon and Oxygen Isotopic Study of Archaean Stromatolites from Zimbabwe [Ph.D. thesis]: Kingston, University of Rhode Island, 107 p.

Monster, J., Appel, P.W.U., Thode, H.G., Schidlowski, M., Carmichael, C.M., and Bridgwater, D., 1979, Sulfur isotope studies in Early Archaean sediments from Isua, West Greenland: Implications for the antiquity of bacterial sulfate reduction: Geochimica et Cosmochimica Acta, v. 43, p. 405–413, doi: 10.1016/0016-7037(79)90205-9.

Moorbath, S., and Whitehouse, M.J., 1996, Age of the Isua supracrustal sequence of West Greenland, in Chela-Flores, J., and Raulin, F., eds.,

Chemical evolution: Physics of the origin and evolution of life: Dordrecht, Kluwer Academic Publishers, p. 87–95.

Moorbath, S., and Kamber, B.S., 1998, Re-appraisal of the age of the oldest water-lain sediments, West Greenland, in Chela-Flores, J., and Raulin, F., eds., Exobiology: Matter, energy and information in the origin of evolution of life in the universe. Dordrecht, Kluwer Academic Publishers, p. 81–86.

Mojzsis, S.J., Coath, C.D., Greenwood, J.P., McKeegan, K.D., and Harrison, T.M., 2003, Mass-independent isotope effects in Archean (2.5 to 3.8 Ga) sedimentary sulfides determined by ion microprobe analysis: Geochimica et Cosmochimica Acta, v. 67, p. 1635–1658, doi: 10.1016/S0016-7037(03)00059-0.

Myers, J.S., 2001, Protoliths of the 3.8–3.7 Ga Isua Greenstone Belt: West Greenland: Precambrian Research, v. 105, p. 129–141, doi: 10.1016/S0301-9268(00)00108-X.

Naraoka, H., Ohtake, M., Maruyama, S., and Ohmoto, H., 1996, Non-biogenic graphite in 3.8 Ga metamorphic rocks from the Isua district, Greenland: Chemical Geology, v. 133, p. 251–260, doi: 10.1016/S0009-2541(96)00076-9.

Nisbet, E.G., and Fowler, C.M.R., 1996, Some like it hot: Nature, v. 382, p. 404–406, doi: 10.1038/382404a0.

Nisbet, E.G., and Fowler, C.M.R., 1999, Archaean metabolic evolution of microbial mats: Proceedings of the Royal Society of London, ser. B, v. 266, p. 2375–2382, doi: 10.1098/rspb.1999.0934.

Nisbet, E.G., and Sleep, N.H., 2001, The habitat and nature of early life: Nature, v. 409, p. 1083–1091, doi: 10.1038/35059210.

Nutman, A.P., 1986, The early Archaean to Proterozoic history of the Isukasia area, southern West Greenland: Copenhagen, Grønlands Geologiske Undersøgelse Bulletin, v.154, p. 80.

Nutman, A.P., Bennett, V.C., Friend, C.R.L., and Rosing, M.T., 1997, ~3710 and >3790 Ma volcanic sequences in the Isua (Greenland) supracrustal belt; structural and Nd isotope implications: Chemical Geology, v. 141, p. 271–287, doi: 10.1016/S0009-2541(97)00084-3.

Ohmoto, H., 1986, Stable isotope geochemistry of ore deposits, in Valley, J.W., Taylor, H.P., and O'Neil, J.R., eds., Stable isotopes in high temperature geological processes: Washington, D.C., Mineralogical Society of America, Reviews in Mineralogy, v. 16, p. 491–559.

Ohmoto, H., 1992, Biochemistry of sulfur and the mechanisms of sulfide-sulfate mineralization in Archean oceans, in Schidlowski, M., Golubic, S., Kimberley, M.M., McKirdy, D.M., and Trundinger, P.A., eds., Early organic evolution: Implications for mineral and energy resources. Berlin, Springer-Verlag, p. 378–397.

Ohmoto, H., and Felder, R.P., 1987, Bacterial activity in the warmer, sulphate-bearing, Archaean oceans: Nature, v. 328, p. 244–246, doi: 10.1038/328244a0.

Ohmoto, H., and Godhaber, M., 1997, Sulfur and carbon isotopes, in Barnes, H.L., ed., Geochemistry of hydrothermal ore deposits, Volume 3: New York, John Wiley and Sons, p. 517–612.

Ohmoto, H., Kakegawa, T., and Lowe, D.R., 1993, 3.4-billion-year-old biogenic pyrites from Barberton, South Africa: Sulfur isotope evidence: Science, v. 262, p. 555–557.

Pace, N.R., 1997, A molecular view of microbial diversity and the biosphere: Science, v. 276, p. 734–740, doi: 10.1126/science.276.5313.734.

Paris, I.A., 1987, The Barberton Greenstone succession, South Africa: Implications for modelling the evolution of the Archaean crust: Chichester, UK, John Wiley and Sons Geological Journal (winter thematic issue), v. 22, p. 5–24.

Peckmann, J., and Thiel, V., 2004, Carbon cycling at ancient methane-seeps: Chemical Geology, v. 205, p. 443–467, doi: 10.1016/j.chemgeo.2003.12.025.

Peckmann, J., Thiel, V., Reitner, J., Taviani, M., Aharon, P., and Michalis, W., 2004, A microbial mat of a large sulfur bacterium preserved in a Miocene methane-Seep limestone: Geomicrobiology Journal, v. 21, p. 247–255, doi: 10.1080/01490450490438757.

Perry, E.C., Jr., and Ahmad, S.N., 1977, Carbon isotope composition of graphite and carbonate minerals from 3.8-AE metamorphosed sediments, Isukasia, Greenland: Earth and Planetary Science Letters, v. 36, p. 280–284, doi: 10.1016/0012-821X(77)90210-2.

Pierson, B.K., 1994, The emergence, diversification and role of photosynthetic eubacteria, in Bengtson, S., ed., Early life on Earth: Nobel Symposium 84: New York, Columbia University Press, p. 161–180.

Polat, A., Hofmann, A.W., Münker, C., Regelous, M., and Appel, P.W.U., 2003, Contrasting geochemical patterns in the 3.7–3.8 Ga pillow basalt cores and rims, Isua greenstone belt, Southwest Greenland: Implications for postmagmatic alteration processes: Geochimica et Cosmochimica Acta, v. 67, p. 441–457, doi: 10.1016/S0016-7037(02)01094-3.

Rollinson, H., 2002, The metamorphic history of the Isua Greenstone Belt, West Greenland, in Fowler, C.M.R., Ebinger, C.J., and Hawkesworth, C.J., eds., The early Earth: Physical, chemical and biological development: Geological Society [London] Special Publication 199, p. 329–350.

Rollinson, H., 2003, Metamorphic history suggested by garnet-growth chronologies in the Isua Greenstone Belt, West Greenland: Precambrian Research, v. 126, 3–4, p. 181–196.

Robinson, J.J., Scott, K.M., Swanson, S.T., O'Leary, M.H., Horken, K., Tabita, F.R., and Cavanaugh, C.M., 2003, Kinetic isotope effect and characterization of form II RubisCO from the chemoautotrophic endosymbionts of the hydrothermal vent tubeworm *Riftia pachyptila*: Limnology and Oceanography, v. 48, p. 48–54.

Rose, N.M., Rosing, M.T., and Bridgwater, D., 1996, The origin of metacarbonate rocks in the Archaean Isua supracrustal belt, West Greenland: American Journal of Science, v. 296, p. 1004–1044.

Rosing, M.T., 1999, ^{13}C-depleted carbon microparticles in >3700-Ma sea-floor sedimentary rocks from West Greenland: Science, v. 283, p. 674–676, doi: 10.1126/science.283.5402.674.

Rosing, M.T., and Frei, R., 2004, U-rich Archaean sea-floor sediments from Greenland—Indications of >3700 Ma oxygenic photosynthesis: Earth and Planetary Science Letters, v. 217, p. 237–244, doi: 10.1016/S0012-821X(03)00609-5.

Rye, R.O., and Ohmoto, H., 1974, Sulphur and carbon isotopes and ore genesis: A review: Economic Geology and the Bulletin of the Society of Economic Geologists, v. 69, p. 826–842.

Sakai, H., Des Marais, D.J., Ueda, A., and Moore, J.G., 1984, Concentrations of isotope ratios of carbon, nitrogen and sulfur in ocean-floor basalts: Geochimica et Cosmochimica Acta, v. 48, p. 2433–2441, doi: 10.1016/0016-7037(84)90295-3.

Schidlowski, M., 1988, A 3,800-million-year isotopic record of life from carbon in sedimentary rocks: Nature, v. 333, p. 313–318, doi: 10.1038/333313a0.

Schidlowski, M., 2001, Carbon isotopes as biogeochemical recorders of life over 3.8 Ga of Earth history: Evolution of a concept: Precambrian Research, v. 106, p. 117–134, doi: 10.1016/S0301-9268(00)00128-5.

Schidlowski, M., Eichmann, R., and Junge, C.E., 1975, Precambrian sedimentary carbonates: Carbon and oxygen isotope geochemistry and implications for the terrestrial oxygen budget: Precambrian Research, v. 2, p. 1–69, doi: 10.1016/0301-9268(75)90018-2.

Schidlowski, M., Appel, P.W.U., Eichmann, R., and Junge, C.E., 1979, Carbon isotope geochemistry of the 3.7×10^9 yr old Isua sediments, West Greenland: Implications for the Archaean carbon and oxygen cycles: Geochimica et Cosmochimica Acta, v. 43, p. 189–199, doi: 10.1016/0016-7037(79)90238-2.

Schidlowski, M., Hayes, J.M., and Kaplan, I.R., 1983, Isotopic interferences of ancient biochemistries: Carbon, sulfur, hydrogen, and nitrogen, in Schopf, J.W., ed., Earth's earliest biosphere: Its origin and evolution: Princeton, New Jersey, Princeton University Press, Princeton, v. 7, p. 149–186.

Sirevåg, R., Buchanan, B.B., Berry, J.A., and Troughton, J.H., 1977, Mechanisms of CO_2 fixation in bacterial photosynthesis studied by the carbon isotope fractionation technique: Archives of Microbiology, v. 112, p. 35–38, doi: 10.1007/BF00446651.

Shen, Y., and Buick, R., 2004, The antiquity of microbial sulfate reduction: Earth-Science Reviews, v. 64, p. 243–272.

Shen, Y., Buick, R., and Canfield, D.E., 2001, Isotopic evidence for microbial sulphate reduction in the early Archaean era: Nature, v. 410, p. 77–81, doi: 10.1038/35065071.

Siebert, C., 2003, Molybdenum isotope fractionation and its application to studies of marine environments and surface processes through geological time [Ph.D. thesis]: Bern, Switzerland, University of Bern, 82 p.

Strauss, H., 2003, Sulphur isotope and the early Archaean sulphur cycle: Precambrian Research, v. 126, 3–4, p. 349–361.

Strauss, H., and Moore, T.B., 1992, Abundances and isotopic compositions of carbon and sulfur species in whole rock and kerogen samples, in Schopf, J.W., and Klein, C., eds., The Proterozoic biosphere, Cambridge, UK, Cambridge University Press, Ch. 17, p. 711–798.

Stevens, C.M., 1988, Atmospheric methane: Chemical Geology, v. 71, p. 11–21, doi: 10.1016/0009-2541(88)90102-7.

Taylor, B.E., 1986, Magmatic volatiles: isotopic variation of C, H, and S, in Valley, J.W., Taylor, H.P., and O'Neil, J.R., eds., Stable isotopes in high temperature geological processes: Washington, D.C., Mineralogical Society of America, Reviews in Mineralogy, v. 16, p. 185–226.

Touret, J.L.R., 2003, Remnants of early Archaean hydrothermal methane and brines in pillow-breccia from the Isua Greenstone Belt, West Greenland: Precambrian Research, v. 126, 3–4, p. 219–233.

Ueno, Y., Yurimoto, H., Yoshioka, H., Komiya, T., and Maruyama, S., 2002, Ion microprobe analysis of graphite from ca. 3.8 Ga metasediments, Isua supracrustal belt, West Greenland: Relationship between metamorphism and carbon isotopic Composition: Geochimica et Cosmochimica Acta, v. 66, p. 1257–1268, doi: 10.1016/S0016-7037(01)00840-7.

van der Meer, M.T.J., Schouten, S., van Dongen, B.E., Rijpstra, W.I.C., Fuchs, G., Sinninghe Damsté, J.S., de Leeuw, J.W., and Ward, D.M., 2001, Biosynthetic controls on the ^{13}C contents of organic components in the photoautotrophic bacterium *Chloroflexus aurantiacus*: Journal of Biological Chemistry, v. 276, no. 14, p. 10971–10976, doi: 10.1074/jbc. M009701200.

van Zuilen, M.A., Lepland, A., and Arrhenius, G., 2002, Reassessing the evidence for the earliest traces of life: Nature, v. 418, p. 627–630, doi: 10.1038/nature00934.

van Zuilen, M.A., and Lepland, A., Teranes, J., Fenarelli, J., Wahlen, M., and Arrhenius, G., 2003, Graphite and carbonates in the 3.8 Ga old Isua Supracrustal Belt, southern West Greenland: Precambrian Research, v. 126, p. 331–348.

Veizer, J., Hoefs, J., Ridler, R.H., Jensen, L.S., and Lowe, D.R., 1989, Geochemistry of Precambrian carbonates: I. Archean hydrothermal systems: Geochimica et Cosmochimica Acta, v. 53, p. 845–857, doi: 10.1016/0016-7037(89)90030-6.

Viljoen, M.J., and Viljoen, R.P., 1969, An introduction to the geology of the Barberton granite-greenstone terrain, *in* Haughton, S.H., ed., Upper Mamie Project: Geological Society of South Africa Special Publication 2, p. 9–28.

Wadham, J.L., Bottrell, S., Tranter, M., and Raiswell, R., 2004, Stable isotope evidence for microbial sulphate reduction at the bed of a polythermal high Arctic glacier: Earth and Planetary Science Letters, v. 219, p. 341–355, doi: 10.1016/S0012-821X(03)00683-6.

Watanabe, Y., Naraoka, H., Wronkiewicz, D.J., Condie, K.C., and Ohmoto, H., 1997, Carbon, nitrogen, and sulfur geochemistry of Archean and Proterozoic shales from the Kaapvaal Craton, South Africa: Geochimica et Cosmochimica Acta, v. 61, p. 3441–3459, doi: 10.1016/S0016-7037(97)00164-6.

Wilks, M.E., and Nisbet, E.G., 1985, Archaean stromatolites from the Steep Rock Group, northwestern Ontario, Canada: Canadian Journal of Earth Sciences, v. 22, p. 792–799.

Wilks, M.E., and Nisbet, E.G., 1988, Stratigraphy of the Steep Rock Group, northwestern Ontario: A major Archaean unconformity and Archaean stromatolites: Canadian Journal of Earth Sciences, v. 25, p. 370–391.

Woese, C.R., 1987, Bacterial evolution: Microbiological Reviews, v. 51, p. 221–271.

Yong, J.N., 1991, Stable isotope ratios of shales from the Archaean Cheshire Formation, Zimbabwe [Ph.D. thesis]: Kingston, University of Rhode Island, 127 p.

Manuscript Accepted by the Society 29 October 2005

Fingerprinting the metal endowment of early continental crust to test for secular changes in global mineralization

Christien Thiart*
Africa Earth Observatory Network (AEON) and Department of Statistical Sciences, University of Cape Town, Rondebosch 7701, South Africa

Maarten J. de Wit*
Africa Earth Observatory Network (AEON) and Department of Geological Sciences, University of Cape Town, Rondebosch 7701, South Africa

ABSTRACT

Archean cratons are fragments of old continents that are believed to be more richly endowed with mineral deposits than younger terrains. The mineral deposits of different cratons are also diversely enriched with useful (to humankind) chemical elements. Cratons are therefore mineral diversity hotspots that represent regional geochemical heterogeneities of early Earth, the evidence for which remains encoded on each craton as unique metallogenic "fingerprints." Using six selected elements groups from our extensive in-house GIS database of Gondwana mineral deposits, we derive the metallogenic fingerprints of 11 Archean cratons of the Southern Hemisphere, and compare these against metallogenic fingerprints of the same selected elements in younger crust of three of their host continents (Africa, Australia, and South America). After adjusting the mineral inventory of each craton to account for underexploration of regions lacking infrastructure and other political and economic conditions for mineral investment, we show that mineral deposit density and diversity of Earth's continental lithosphere has decreased with time. We conclude that metallogenic elements were transferred more efficiently from the mantle to the continental lithosphere in the Archean and/or that subsequent (younger than 2.5 Ga) recycling of these elements (mineral deposits) back into the mantle has become more effective. How most of these fragments of old continents inherited their rich and diverse metallogenic characteristics is unresolved, because different cratons are likely to represent only small remnants of once much larger and possibly more varied Archean continents, and part of the total metal inventory of Archean continents must have been recycled back into in the mantle. The latter has implications for understanding the secular change in the redox state of the Archean mantle and fluid envelope.

Keywords: cratons, Gondwana, metallogenesis, mineralization, exploration index, geodynamics.

INTRODUCTION

The lithosphere of Archean cratons (older than 2.5 Ga) is distinct from that of younger continental lithosphere in that they are underlain by relatively thin crust (~30–40 km) and thick mantle lithosphere (up to ~250–300 km; de Wit, 1998; James et al., 2001; Stankiewicz et al., 2002; Fouch et al., 2004). The origin of these cratons is still a matter of intense debate. Part of that debate

---

*thiart@stats.uct.ac.za, maarten@cigces.uct.ac.za

Thiart, C., and de Wit, M.J., 2006, Fingerprinting the metal endowment of early continental crust to test for secular changes in global mineralization, *in* Kesler, S.E., and Ohmoto, H., eds., Evolution of Early Earth's Atmosphere, Hydrosphere, and Biosphere—Constraints from Ore Deposits: Geological Society of America Memoir 198, p. 53–66, doi: 10.1130/2006.1198(03). For permission to copy, contact editing@geosociety.org. ©2006 Geological Society of America. All rights reserved.

centers on the origin of the cratons' mineral deposits and, in turn, how knowledge of these deposits may shed light on the formation of Earth's early continents. However, surprisingly few quantitative data are available in a collective format with which to embark on a comparative study of the riches found in continental crust of different ages. Near surface, the crust of most cratons is well endowed in concentrations of metallic elements useful and economic to humankind (hereafter referred to as mineral deposits), but it is not known with any degree of certainty if the crust of the old continents was mineralized to a greater degree than that of younger continents, as is often assumed intuitively. As a test of such a temporal change, here we compare the mineral endowment of old cratonic crust with that of younger crust, using our in-house mineral deposit database.

Cratons are metallogenically distinct in that their mineral deposits contain different mixtures of metallic elements from craton to craton (Wilsher et al., 1993; Wilsher, 1995; de Wit et al., 1999). For example, the Kaapvaal Craton of South Africa is known to be relatively enriched in gold and platinum group elements (PGE), the Zimbabwe Craton and the Yilgarn Craton of Australia in gold and tungsten, the São Francisco Craton in gold and base metals (Cu/Pb/Zn), and the Amazonian Craton in gold and tin (Groves et al., 1987; Wilsher, 1995; Barley et al., 1998; de Wit et al., 1999; de Wit and Thiart, 2005). We assume that mineral deposits of Archean cratons are contemporaneous with their host rocks and reflect processes of geochemical concentration that operated during the formation of the cratons. In most cases this can be verified if the age of the mineral deposits is constrained by geological relations or isotopic ages. In some cases it is not clear whether these mineral enrichments were inherited from their host Archean craton or were added to the craton at a later stage. For example, the platinum deposits of the Kaapvaal Craton are associated mostly with the igneous rocks of the Bushveld Complex that intruded the center of this craton in the Paleoproterozoic (2.05–2.06 Ga; Cawthorn and Walraven, 1998; Eglington and Armstrong, 2004) and could have been added from the asthenospheric mantle at that time. However, the PGE geochemistry (and that of other siderophile elements like Ni, Cr, Au) of lithospheric mantle xenoliths found in kimberlites across the Kaapvaal Craton (McDonald et al., 1995), as well as that of Archean mid to lower crust (Hart et al., 2004) and of Archean mineral deposits in the greenstone belts of this craton (Tredoux et al., 1989), suggests that both the lithospheric mantle and crust of this craton were already rich in PGE in Archean times. Indeed it has been suggested that the PGE in the parental magmas of the Bushveld Complex may have been inherited from contamination with its underlying PGE-enriched Archean mantle lithosphere (McDonald et al., 1995). Similarly, most of the extraordinary enrichment of the Kaapvaal Craton in gold is hosted by Archean sedimentary rocks of the Witwatersrand Basin and is thought to have been derived largely by erosion of earlier crust, although some gold may have been introduced and/or remobilized at a later time (Frimmel and Minter, 2002; Phillips and Evans, 2004). Gold deposits are also common in older Archean greenstone belts (Herrington et al., 1997). Although not all of these greenstone belt deposits are firmly dated, we assume they reflect inheritance of Archean gold. Thus, unless there is clear evidence to the contrary, some geodynamic process(es) served to concentrate precious elements together into the continental lithosphere of the Kaapvaal Craton during its formation in the Archean (Groves et al., 1987; Tredoux et al., 1989; McDonald et al., 1995).

The mantle lithosphere of most cratons is also invariably enriched in diamonds that range in age from Mesoarchean to Phanerozoic (Hart et al., 1997; Shirey et al., 2002; Jelsma et al., 2004). Because the preservation potential of these minerals in the mantle lithosphere relates mostly to the relatively low heat flow recorded across most cratons, we do not here focus further on diamonds. They do, however, serve to illustrate the resilience of cratons as Archean geologic archives.

We first set out to verify that different Archean cratons have distinct metallogenic patterns, which we refer to as their metallogenic "fingerprints." These fingerprints may reveal something fundamental about the formation of cratons and Earth's earliest continents. Next we address the question of whether a unit of Archean continental crust is more enriched in mineral deposits than that of younger crust. We compare our results from Archean cratons with those from younger crust (younger than 2.5 Ga) at different scales: first at a continental scale (e.g., South America, Australia, Africa), and then at a supercontinental scale using all the known mineral deposits of Gondwana. This reveals important information about the evolution of continental crust.

We are limited in our analyses to the Southern Hemisphere, because our mineral database is confined only to continental fragments of the former supercontinent Gondwana (see below). A shortcoming of our previous work (de Wit and Thiart, 2005) relates to the extent that the distribution of known mineral deposits (as in our database) is skewed because some regions are explored better than others. Less-developed countries with poor infrastructure and political instability are less likely to have had their mineral inventory tested to the same degree as better-developed nations. Here, we attempt to adjust for underexploration by including a socioeconomic measure in our calculations.

Establishing metallogenic fingerprints of different cratons by using selected elements from their mineral deposits allows us to address several controversial tectonic models that contrast early Earth processes with those of the present. For example, modern plate tectonic processes yield specific mineral deposit types in distinct plate tectonic environments (Sawkins, 1990; Windley, 1995). Similar associations, if found in Archean cratons, would argue for plate tectonic processes in the Archean. Conversely, on an early Earth dominated by vertical tectonics (e.g., driven by plume and diapir dynamics, as postulated by some workers [Hamilton, 1998; Zegers and van Keken, 2001; van Kranendonk et al., 2004]), metallogenic provinces would be expected to display distinct metal associations not found in the younger crust of the present continents (e.g., Hutchinson, 1981, 1992). In addition, variations in the total concentration of these elements in continental crust of different ages may be used to test for changes in the rates of related tectonic processes over time. This is one aim of this paper, but there are others: First, can metallogenic elements serve as chemical tracers to establish that the earliest continents were assembled by tectonic processes as diverse as those at present? Second, can the mineral diversity patterns of cratons help us to decipher the recycling history of Earth's continental materials? Third, can mineral diversity patterns (or mineral hotspots) of cratons and younger lithosphere be used to test reconstructions of past supercontinents like Gondwana and Rodinia? Finally, because mineralization in the crust requires concentration processes that often involve large fluid fluxes between the mantle and/or crust and the hydrosphere, it is of interest to ask whether secular change in the total mass of mineral deposits in the crust represents a potential proxy for concomitant changes in chemical fluxes and redox states of Earth's major reservoirs such as its mantle and fluid envelope.

DERIVING METALLOGENIC FINGERPRINTS OF ARCHEAN CRATONS AND YOUNGER CRUST USING SELECTED ELEMENTS FROM THEIR MINERAL DEPOSITS

The Database

The geological and mineral deposit data used for this study are incorporated in a GIS relational database, called GO-GEOID (**GO**ndwana – **GEO**scientific **I**ndexing **D**atabase), housed at our center, AEON. This database is restricted to the major continental fragments of Gondwana; its geological component is based on the geological map of Gondwana (de Wit et al., 1988), whereas the mineral layer of the database was constructed from open-access literature sources (Wilsher, 1995). Originally the mineral layer comprised roughly 10,000 deposits across Gondwana, covering 15 metallogenic elements (commodities). The mineral layer consists of shape files tied to 28 attribute tables, with attributes ranging from commodities, size, type, and age of the mineral deposits, as well as information on the host rocks of the deposits and the references from which the data were obtained. In 1998, in collaboration with the BRGM (France), we embarked on an "added-value" project to create a metallogenic-potential GIS of Gondwana. During this process the original database was updated and revalidated. Roughly 6,000 new deposits were added to the database. In the final stage of the project, this "updated" mineral database was integrated with geostatistical software and a browser (GEORAMA). The end product yielded a metallogenic-potential GIS of Gondwana (available on CD-ROM; for details see: http://gondwana.brgm.fr/index_eng.htm). Thus, currently there are roughly 16,000 deposits in the mineral layer, ranging from active mines to undeveloped occurrences. Figure 1 shows the typical density of these mineral deposits across Gondwana, and from which our data are extracted. The origin and evolution of the database has been described in detail elsewhere (Wilsher et al., 1993; de Wit et al., 1999; Thiart and de Wit, 2000; de Wit et al., 2004).

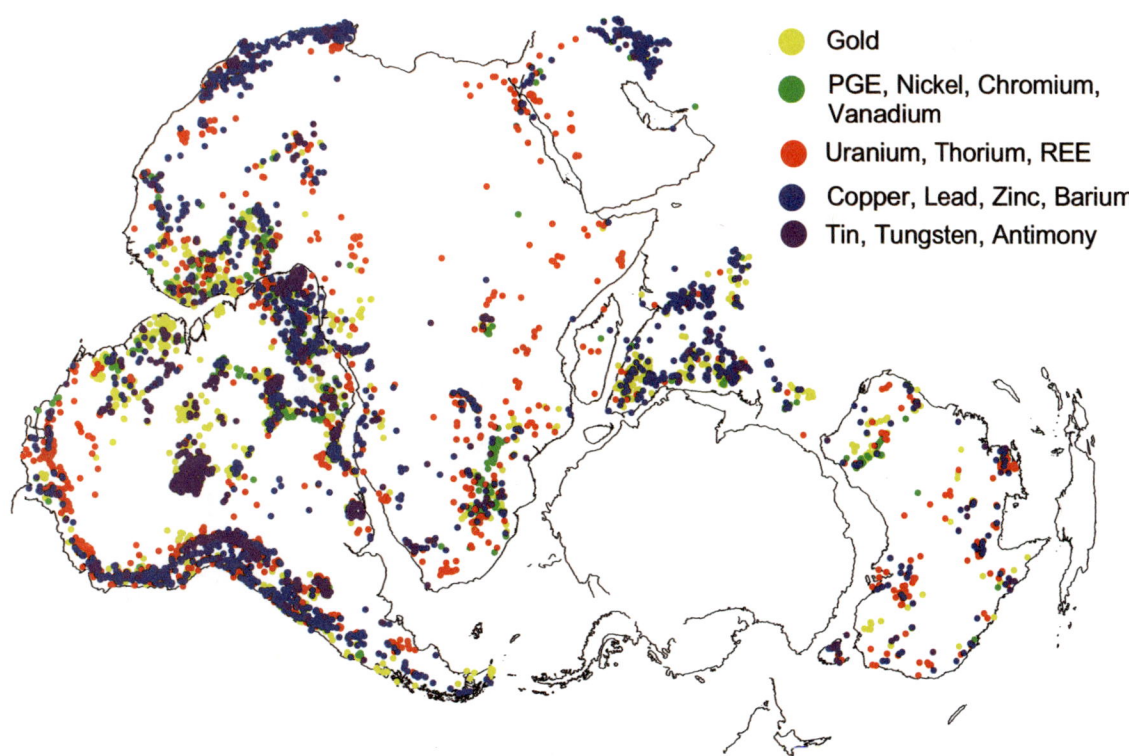

Figure 1. Mineral deposit map of Gondwana. Data from the Gondwana mineral deposit GIS database housed at the Africa Earth Observatory Network (AEON) (de Wit et al., 1999; Thiart and de Wit, 2000; http://www.uct.ac.za/depts/cigces/gondmin.htm).

Figure 2 shows the global distribution of cratons that we selected for this study. There are 11 Gondwana cratons (seven in Africa, two in South America, and two in Australia) for which we have sufficient mineral deposit data in our Gondwana database. For general comparison, we also selected one Canadian craton, the Superior Province. The mineral deposit data for the latter are from the Geological Survey of Canada (Kirkham et al., 1994, 2002; Jenkins et al., 1997; Eckstrand and Good, 2000; Kirkham and Dunne, 2000, 2002; Eckstrand et al., 2002). We emphasize that this Canadian-based data structure is significantly different from our Gondwana database, and because the subset of their data provided to us may not include all known mineral deposits in the Superior Province, the results in the subsequent sections should be evaluated in this light.

Eleven metallogenic elements were selected for our analyses, and these were divided into six element groups according to their geochemical affinities (e.g., lithophile, chalcophile, siderophile), as well as their relative abundances in the database. The six element groups, and the total number of deposits in which these groups occur on each craton, are tabulated in Table 1A. In total there are just over 6000 deposits spread over 12 identified cratons (Table 1A).

Methods

Previously we defined metallogenic fingerprints of fragments of continental crust through their spatial association with a combination of six element groups, and applied this at three scales—cratons, continents, and supercontinents (de Wit and Thiart, 2005). The measure of spatial association (normalized to area) we termed the spatial coefficient, r_{ij}, where

$$r_{ij} = \frac{N(C_i \cap D_j)/N(D_j)}{A(C_i)/A(C_\bullet)}, \qquad (1)$$

in which $A(C_i)$ is the area of the ith craton (C_i), and $A(C_\bullet)$ is the total area of all the cratons in the study $(A(C_\bullet) = \Sigma A(C_i))$. $N(D_j)$ is the total number of deposits in the jth element group (D_j). $N(C_i \cap D_j)$ represents the number of deposits of group D_j in craton C_i.

The spatial coefficient (equation 1) represents the proportion of deposits (say gold, j) of all the jth deposits (in cratons) that occur in the specified craton (i) per unit area of all 12 cratons. Our spatial coefficient is similar to the measure of spatial association used by Mihalasky and Bonham-Carter (2001) to measure the

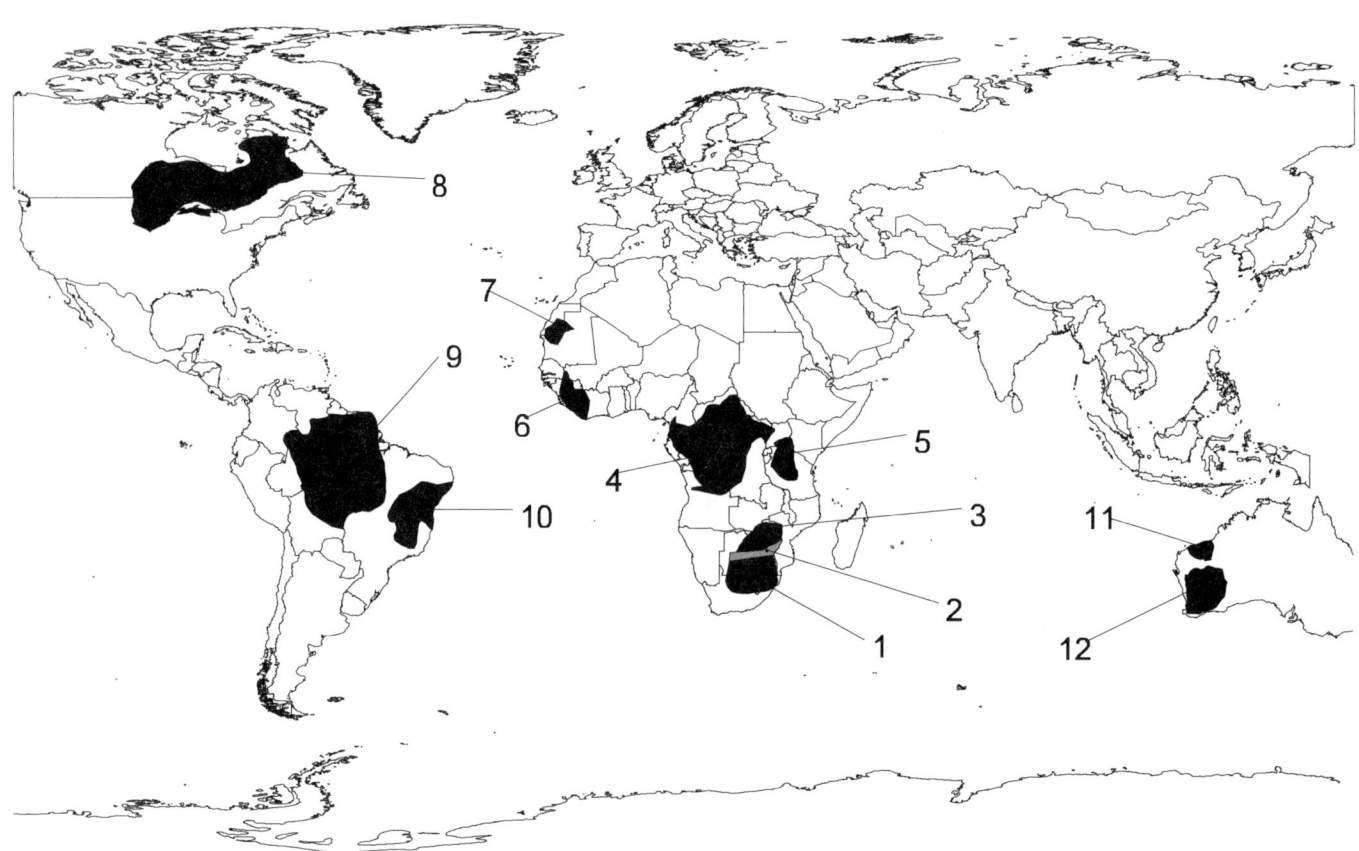

Figure 2. Cratons examined in this study. 1—Kaapvaal; 2—Limpopo; 3—Zimbabwe; 4—Congo; 5—Tanzania; 6—Leo-Man; 7—Requibath; 8—Superior; 9—Amazonia; 10—São Francisco; 11—Pilbara; 12—Yilgarn.

TABLE 1A. NUMBER OF MINERAL DEPOSITS WITH SELECTED ELEMENTS ON GONDWANA'S CRATONS AND THE SUPERIOR PROVINCE OF CANADA

	Au	CrNiPgeTi	CuZnPbBa	SnSb	W	UThREE	Craton total[†]	Area (km^2)
Superior	162	134	52	3	0	0	351	2.73E + 06
Pilbara	7	6	11	1	0	4	29	1.51E + 05
Yilgarn	97	50	17	1	0	11	176	6.25E + 05
Amazonia	434	38	109	645	8	13	1247	3.48E + 06
São Fra.	435	127	462	149	6	66	1245	9.64E + 05
Kaapvaal	894	241	152	96	5	77	1465	6.60E + 05
Limpopo	1	40	33	0	8	9	91	1.59E + 05
Zimbabwe	625	114	106	81	305	15	1246	3.52E + 05
Congo	53	2	24	1	0	17	97	2.31E + 06
Leo-Man	47	22	11	10	2	9	101	3.20E + 05
Tanzania	7	0	3	1	0	5	16	3.33E + 05
Requibath	0	5	21	0	3	4	33	1.68E + 05
Deposits total	2762	779	1001	988	337	230	6097	1.23E + 07

Note: Area is given in km^2 rounded to 2 decimals. The total area of all the cratons is 1.2248155×10^7 km^2.
[†]Craton Total reflects only the total of those deposits in which the elements selected for this study groups occur as its major component(s).

TABLE 1B. NATURAL LOG OF THE SPATIAL COEFFICIENT (r_{ij}) BETWEEN ELEMENT GROUPS AND CRATONS

	Au	CrNiPgeTi	CuZnPbBa	SnSb	W	UThREE	Cra tot
Superior	−1.33	−0.26	−1.46	−4.30			−1.35
Pilbara	−1.58	−0.47	−0.11	−2.50		0.35	−0.95
Yilgarn	−0.37	0.23	−1.10	−3.92		−0.06	−0.57
Amazonia	−0.59	−1.76	−0.96	0.83	−2.48	−1.61	−0.33
São Fra.	0.69	0.73	1.77	0.65	−1.49	1.29	0.95
Kaapvaal	1.79	1.75	1.04	0.59	−1.29	1.83	1.50
Limpopo	−3.58	1.37	0.93		0.60	1.10	0.14
Zimbabwe	2.06	1.63	1.30	1.05	3.45	0.82	1.96
Congo	−2.29	−4.30	−2.06	−5.23		−0.94	−2.47
Leo-Man	−0.43	0.08	−0.87	−0.95	−1.48	0.40	−0.46
Tanzania	−2.37		−2.20	−3.29		−0.22	−2.34
Requibath		−0.76	0.42		−0.43	0.24	−0.93

Note: These values are presented in Figures 4 and 5 as solid bars. No entries—no data.

TABLE 1C. WEIGHTED NATURAL LOG OF THE SPATIAL COEFFICIENT (r_{ij}^w, EQUATION 4) BETWEEN ELEMENT GROUPS AND CRATONS

	Au	CrNiPgeTi	CuZnPbBa	SnSb	W	UThREE	Cra tot
Superior	−2.10	−0.93	−2.21	−5.02			−2.12
Pilbara	−2.31	−1.10	−0.82	−3.18		−0.44	−1.67
Yilgarn	−1.10	−0.40	−1.81	−4.60		−0.86	−1.29
Amazonia	−0.68	−1.75	−1.03	0.79	−2.79	−1.77	−0.41
São Fra.	0.61	0.73	1.70	0.61	−1.80	1.14	0.87
Kaapvaal	1.75	1.80	1.01	0.59	−1.55	1.72	1.46
Limpopo	−3.59	1.46	0.94		0.37	1.03	0.13
Zimbabwe	2.30	1.95	1.55	1.32	3.46	0.99	2.20
Congo	−1.72	−3.64	−1.48	−4.62		−0.44	−1.90
Leo-Man	0.19	0.79	−0.23	−0.28	−1.08	0.96	0.17
Tanzania	−1.71		−1.52	−2.58		0.38	−1.67
Requibath		0.01	1.12		0.02	0.85	−0.25

Note: These values are presented in Figures 4 and 5 as striped bars.

spatial association between lithodiversity and minerals deposits in Nevada. The value of the spatial coefficient ranges from 0 to infinity; it is equal to 1 if there is no spatial association between a craton and an element group (e.g., if the proportion of jth mineral is the same as the proportion of area occupied by the ith craton). For values of $r_{ij} > 1$ (e.g., where the expected number of deposits is greater than by chance), there is a positive association between mineral j and craton i; $r_{ij} < 1$ (e.g., where there are fewer deposits than expected by chance) indicates a negative association. Because all negative associations are compressed in the range from 0 to 1, and all positive associations fall in the range of 1 to infinity, we use the natural log of r_{ij} to eliminate this skewness. Thus, $\ln(r_{ij})$ is a symmetric value around 0: positive associations are greater than 0, and negative associations are less than 0.

The above approach ignores the fact that each study region (e.g., defined cratons) has a different exploration history, a bias that might influence the analyses. Intuitively, we expect, for example, that greater accessibility (e.g., infrastructure) and political stability increases exploration and discovery rates and mining activity. Therefore mineralization in some cratonic areas of less-developed nations may not be adequately represented in our database and analyses. We address this problem by weighting the spatial coefficient (equation 1), with an "exploration index," using socioeconomic data from the World Bank development indicators database (World Bank, 2004). This "weighted" spatial coefficient (defined below in equation 4) can be used with increased confidence to evaluate the metal endowments of different continental fragments.

Rich countries are better explored than poor countries because of their better-developed infrastructure and investment regimes. We tested various world development indicators published by the World Bank (2004) designed to quantify this differential development. Here we use the measure of gross national income per capita (GNI-C, for 2002; Table 2) because it contains the most obtainable data for all the countries of our interest. To derive the GNI-C for a specific craton, we calculate the percentage of area that each country contributes to the total area of that craton. The craton GNI-C is then calculated as a weighted average (by area percentage) of all the GNI-C values of each of the countries involved. For example the Kaapvaal Craton constitutes 14% of Botswana, 4% of Lesotho, 80% of South Africa, and 2% of Swaziland. The GNI-C for the entire Kaapvaal Craton then is as follows: 14%*GNI-C for Botswana + 4%*GNI-C for Lesotho + 80%*GNI-C for South Africa + 0.2%*GNI-C for Swaziland.

To calculate an exploration index, we use the United States (USA) as a benchmark (e.g., the United States has an exploration index of 1). The exploration index for $Craton_i$ is the ratio of the GNI-C for USA to the GNI-C for $Craton_i$

$$\left[\frac{GNI-C\ for\ USA}{GNI-C\ for\ Craton_i} \right]^k \quad (2)$$

where k is a power function. For $k = 1$ our calculated exploration indices range from 1 (for the United States) to 126 (for the poorest craton, the Requibath Craton of northwest Africa). To contain the index to a more manageable range, we apply various power functions (k from ½ to ¼), or use the natural log (ln (equation

TABLE 2. GNI PER CAPITA (GNI-C) FOR COUNTRIES CONTAINING THE SELECTED CRATONS (OR PARTS THEREOFF) ACROSS GONDWANA AND CANADA

Country name	Craton name	Area of craton (%)	GNI-C, Atlas method (current US$)	Craton GNI-C	Continent GNI-C
United States			35400	35400	34120
Canada	Superior	100	22390	22390	
Australia	Pilbara	100	19530	19530	19530
Australia	Yilgarn	100	19530	19530	
Brazil	Amazonia	100	2830	2830	3280
Brazil	SãoFra.	100	2830	2830	
Angola	Congo	9	710		
Cameroon	Congo	4	550		
Central African Republic	Congo	14	250		
Congo	Congo	10	100		
Equatorial Guinea	Congo	1	930		
Gabon	Congo	6	3060		
Sudan	Congo	0	370		
Uganda	Congo	0	240		
Zaire	Congo	56	100	389	
Botswana	Kaapvaal	14	3010		
Lesotho	Kaapvaal	4	550		
South Africa	Kaapvaal	80	2500		
Swaziland	Kaapvaal	2	1240	2470	
Guinea	Leo–Man	40	410		
Ivory Coast	Leo–Man	14	620		
Liberia	Leo–Man	25	140		Africa 650
Mali	Leo–Man	4	240		
Senegal	Leo–Man	3	470		
Sierra Leone	Leo–Man	15	140	330	
Botswana	Limpopo	58	3010		
Mozambique	Limpopo	1	200		
South Africa	Limpopo	13	2500		
Zimbabwe	Limpopo	28	480	2221	
Mauritania	Requibath	53	280		
Saharawi, ADR	Requibath	47	–*	280	
Kenya	Tanzania	8	360		
Tanzania, United Republic of	Tanzania	81	290		
Uganda	Tanzania	11	240	290	
Botswana	Zimbabwe	23	3010		
Mozambique	Zimbabwe	1	200		
Zambia	Zimbabwe	1	340		
Zimbabwe	Zimbabwe	75	480	1069	

Note: Craton GNI-C is the weighted average (by % area). No data is available for the Arab Democratic Republic of Saharawi; GNI-C for Requibath is therefore based on Mauritania.

2), with $k = 1$) of the GNI-C ratios. These exploration indices are given in Figure 3A. For this investigation, we chose to work with the power function $k = 1/3$, as it closely follows the natural log function and has the advantage that it ranges from 1 (United States) to roughly 5 (least explored cratons).

For our analyses at a continental scale we derive the GNI-C ratio for each continent in a similar manner, and use the combined GNI-C of the United States and Canada (North America, top of the last column of Table 2) as our benchmark. The GNI-C for Africa is from the African Development Indicators (2004). The GNI-C for North America and Australia are from the World Bank development indicators database (World Bank, 2004). The GNI-C for South America is from the country data of the World Bank development indicators database (World Bank, 2004). The derived exploration indices for the continents are given in Figure 3B.

From the original mineral deposit data (Table 1A) we can now calculate a proportion (p_{ij}) as the number of deposits in Craton$_i$ and Element Group$_j$ divided by the total number of deposits in the database (6097; Table 1A). This calculated proportion is then multiplied by the exploration index, w_i, where w_i is the ratio defined in equation 2 (with $k = 1/3$) of the ith craton, and divided by the sum of the product of w_i and p_{ij}. Thus

$$p_{ij}^w = \frac{w_i p_{ij}}{\sum_i \sum_j w_i p_{ij}}. \quad (3)$$

This "weighted" proportion (equation 3) is then used in the calculation of the "weighted" spatial coefficient r_{ij}^w:

$$r_{ij}^w = \frac{p_{ij}^w / p_{\bullet j}^w}{A(C_i) / A(C_\bullet)} \quad (4)$$

where p_{ij}^w is the weighted proportion of the ith craton and j^{th} element group; $p_{\bullet j}^w$ is the weighted proportion of the jth element group (total of the jth element group); $A(C_i)$ is the area of the ith craton; and $A(C_\bullet)$ is the total area of all the cratons. This weighted spatial coefficient is treated as before (in this text), in the sense that we again take the natural log of r_{ij}^w. The interpretation will be the same as in that of the unweighted (natural log) spatial coefficient: positive values of natural log (r_{ij}^w) is a measure of the positive spatial association between the ith craton and the jth element group, and negative values of $\ln(r_{ij})$ is a measure of the negative association between the ith craton and j^{th} element group.

RESULTS

Metallogenic Fingerprints of Selected Archean Cratons

The number of mineral deposits of the six selected element groups within each of the 12 cratons is given in Table 1A, and the natural log of the spatial coefficients between the element group and each craton ($\ln(r_{ij})$, r_{ij} defined in equation 1) is given in Table 1B. The natural log of the "weighted" spatial coefficient

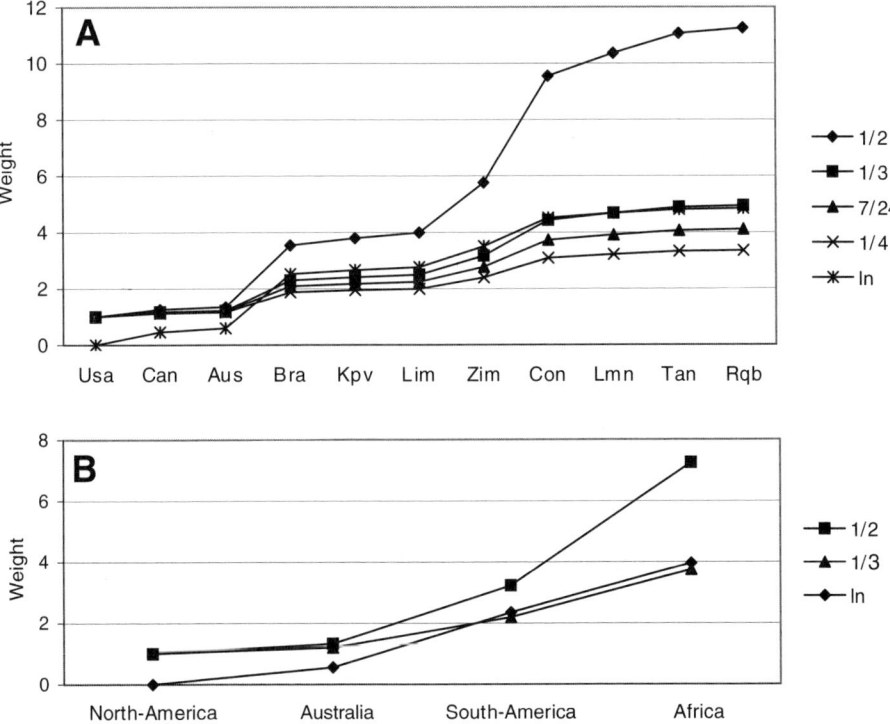

Figure 3. (A) Exploration indices (as a function of the state of development of countries normalized to that of the USA) determined for cratons using data from the World Bank (2004). The Pilbara and Yilgarn Cratons are shown combined under Australia (Aus), and the Amazonia and São Francisco Craton are shown combined as Brazil (Bra). The Superior Province is listed as Canada (Can). The indices are shown at various power functions and as a (natural) log function (see text for explanation). Note the similarity between the natural log function and $k = 1/3$. Con—Congo; Kpv—Kaapvaal; Lim—Limpopo; Lmn—Leo-Man; Rqb—Requibath; Tan—Tanzania; Zim—Zimbabwe. (B) Exploration index determined for the three continents studied, normalized to that of North America. The index is shown at two power functions ($k = 1/2$ and $k = 1/3$), and as a (natural) log function.

(ln (r_{ij}^w), equation 4) for the 12 selected cratons and the six selected element groups is given in Table 1C. Figures 4 and 5 summarize these data in graphical format to illustrate the mineral diversity of each craton. For comparison both the unweighted (ln(r_{ij}), solid bars) and the weighted spatial coefficient (ln(r_{ij}^w), striped bars) are given in the same figure. Figure 4 expresses the diversity among cratons for each set of elements of the six element groups, and provides 12 cratonic (spatial) coefficients for each set of elements. Figure 5 represents the total spatial coefficient combining all sets of elements (e.g., all six element groups) for each craton; these are the "metallogenic fingerprints" of the 12 cratons. Each individual craton has a unique fingerprint even though some may appear very similar (e.g., Congo and Tanzania Cratons). Where values are high (e.g., Zimbabwe) the imprint of total mineralization is high relative to the other cratons: metaphorically a "strong fingerprint." Values between −1 and 1 represent a random association between the total element set and the specific craton; values below 0 (Fig. 4) indicate a negative association for the given element group and the specific craton. The values for the Tanzania and Congo Cratons represent the lowest values, and these values may be unreliable because of the low number of discovered mineral deposits on these cratons (and thus in our database). In turn this corroborates the need to incorporate into the analyses an exploration-index value that reflects the past history of (under)exploration of areas under consideration. Of additional interest here is that the relative differences between the weighted and unweighted results express a degree of underexploration, and therefore provide a signal for potential near-surface mineral deposits still to be found in areas where infrastructure development and exploration investment has lagged behind that of North America. The greatest positive values are those of the Zimbabwe Craton. This reflects a high count of mineral deposits covering all six element groups in Zimbabwe and the relatively small size of this craton. It is the only craton that has no negative spatial coefficient in any of the element groups. The similarity between its weighted and unweighted spatial coefficients indicates a mature degree of exploration throughout this craton, in concert with a long history of intense mineral exploration.

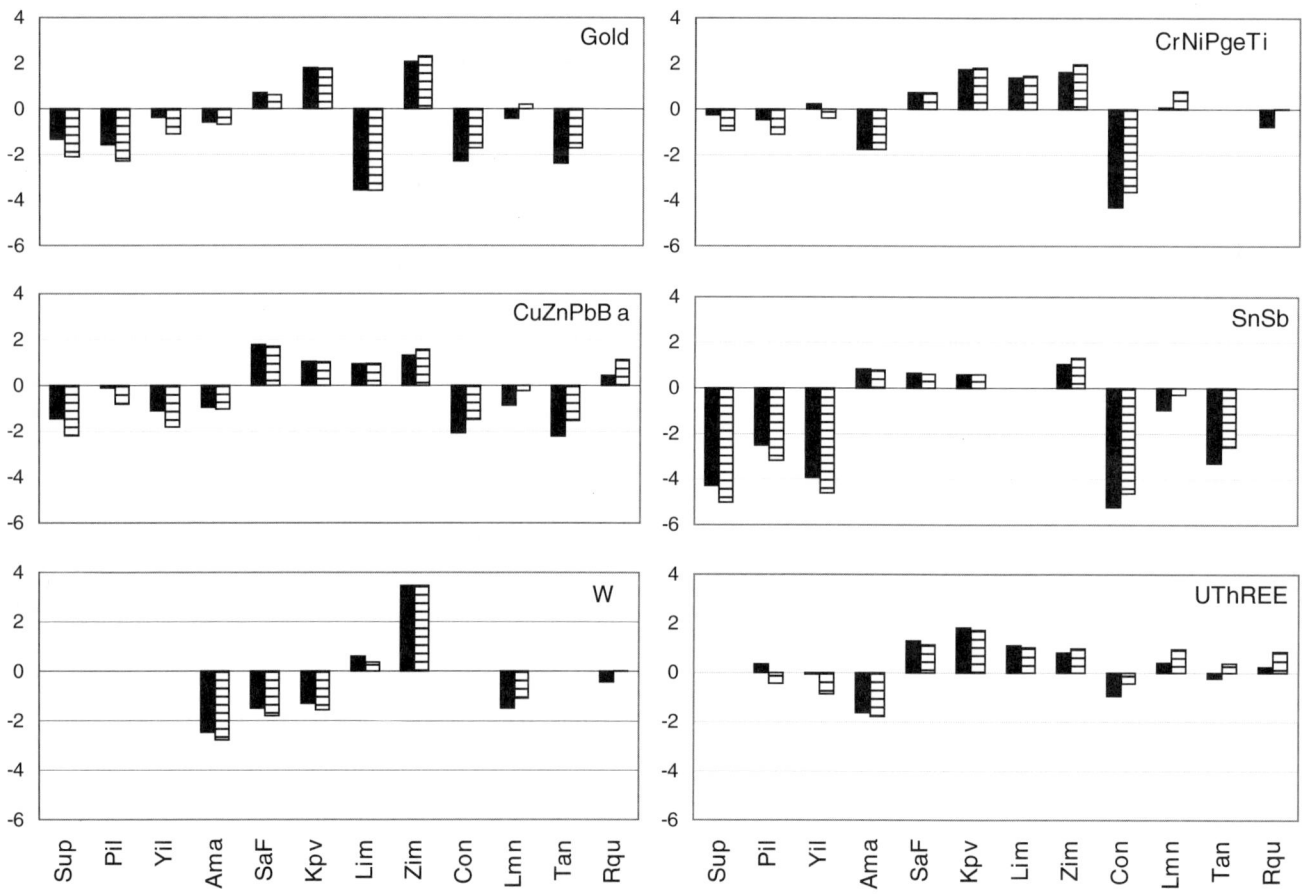

Figure 4. Mineral diversity of Archean cratons. Solid bars represent the "raw" spatial coefficient (ln(r_{ij})); striped bars represent the weighted spatial coefficient (ln(r_{ij}^w)) between specific element(s) and the corresponding crust "weighted" with their exploration indices Note how each craton has a unique diversity of elements that characterize its "metallogenic fingerprint" (see also Fig. 5). The difference between the weighted and unweighted bars represents an exploration potential. See text for further information.

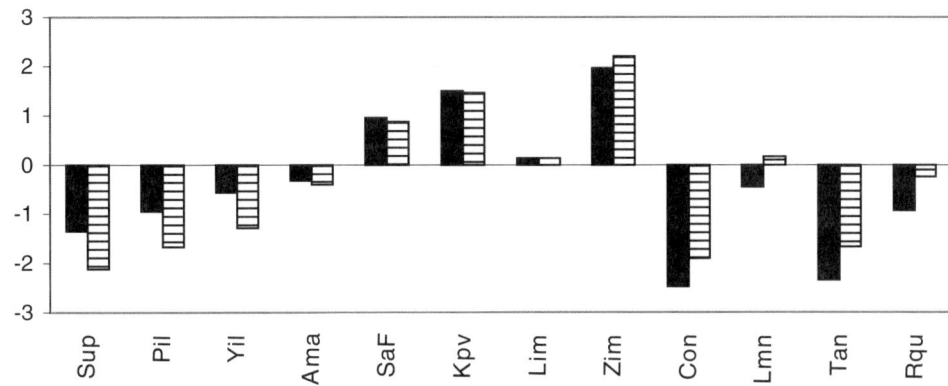

Figure 5. Mineral diversity measure between all combined elements of mineral deposits of the 12 cratons investigated in this study. The solid bars represent the unweighted metallogenic fingerprint (spatial coefficient ($\ln(r_{ij})$)) and the striped bars represent the "weighted" metallogenic fingerprint ($\ln(r_{ij}^w)$) of each craton.

Some negative spatial coefficients might be intuitively surprising (e.g., those for gold in the Superior Province). The spatial coefficient (equation 1) is defined as a ratio between two proportions, that is, the number of deposits (say Au) in a specified craton relative to the number of deposits in all the cratons, divided by the proportion of area that the specific craton occupies relative to that of the area of all the cratons. If the ratio is greater than 1, there is a positive association (more deposits expected than by chance), whereas a value between 0 and 1 results in a negative association (fewer deposits expected than by chance). The natural log of the spatial coefficient for the exceptionally large Superior Province is probably negative because it expresses its mineralization relative to other cratons: in the Superior Province there are only 162 Au deposits in the Canadian database compared to a total of 2762 in all cratons combined; and the Superior province is the largest of all the cratons (2.73×10^6 km^2). We suspect that there are probably a lot more small deposits in the Superior Province that are not incorporated into the Canadian database, because they are in the GO-GEOID database. The results would also change if one could factor in the actual area of outcrop (most of the Superior Province is covered by glacial drift (till, transported overburden). Geophysical and geochemical exploration methods cannot "see" gold deposits through this cover as easily as they can "see" base metal (VMS) deposits. Other cratons with residual overburden would not be affected by this complication. It is probably unwise therefore to compare the two data sets directly, because they were constructed differently and with different specifications. This supports a call to standardize global databases. For the further analysis below we therefore eliminate the Superior Province from our input.

Mineral Inventory on a Continental Scale

The analyses described above were repeated for three large continental fragments (Africa, South America, and Australia) of the former supercontinent Gondwana, to enable direct comparison of the metallogenic inventory of cratons (older than 2.5 Ga; old crust) with mineralization in younger crust (younger than 2.5 Ga) on each individual continent. The mineral deposit data for the old cratons and younger crust are summarized in Table 3A. The natural log of the spatial coefficients is given in Table 3B, and the natural log of the "weighted" spatial coefficients is given in Table 3C. Figure 6 graphically displays the unweighted natural log of the spatial coefficients ($\ln(r_{ij})$, solid bars) against that of the natural log of the weighted coefficient ($\ln(r_{ij}^w)$, striped bars). The total mineral inventory for the continental crust older and younger than 2.5 Ga (O and Y, respectively) can now be directly compared also (Fig. 7). These results clearly depict the mineral deposit diversity between cratons and younger crust; they also depict a real difference (per unit area) in enrichment of specific elements between old and young crust on all three continents. In general the older fragments are more enriched in all element groups, except for tungsten in South America, and for the Sn, Sb, and UThREE groups in Australia. This result is probably independent of the "exposure" problem mentioned above and therefore a strong conclusion.

The strongest spatial coefficient observed is between tin and antimony (Sn and Sb) and the old crust (cratons) of South America. A strong spatial coefficient for Sn and Sb is also observed for Africa, and may be controlled mostly by the West African cratons that were part of the Amazonian Craton until Gondwana breakup. This supports a long (3.0–0.5 Ga) conterminous history for this old crust (Trompette, 1994). We have previously explored this Gondwana-tin association between West Africa and South America to refute a frequently advocated fit between them and eastern North America (where there are no Archean–Mesoproterozoic tin deposits or occurrences), a relation that is important to models of the proposed Mesoproterozoic supercontinent Rodinia (de Wit et al., 1999).

In summary, it seems that the younger crust of the continents in the Southern Hemisphere is relatively enriched in the lithophile element group (UThREE), whereas the older crust of all three continents is affiliated to a greater degree with concentrations of siderophile element group (Cr Ni PGE Ti). In contrast, there is no discernable degree of difference in concentrations of chalcophile element group (base metals) between old and young

TABLE 3A. NUMBER OF MINERAL DEPOSITS OF CRATONS (OLD) AND YOUNGER CRUST (YOUNG) IN THREE CONTINENTS: AFRICA, SOUTH AMERICA, AND AUSTRALIA

	Au	CrNiPgeTi	CuZnPbBa	SnSb	W	UThREE	Total	Area (km^2)
Afr old	1627	424	350	189	323	136	3049	4.31E + 06
Afr young	499	91	768	142	119	491	2110	2.06E + 07
Sam old	869	165	571	794	14	79	2492	4.44E + 06
Sam young	797	119	999	236	828	671	3650	1.07E + 07
Aus old	104	56	28	2	0	15	205	7.75E + 05
Aus young	70	2	15	22	5	157	271	6.01E + 06
Deposit total	3966	857	2731	1385	1289	1549	11777	4.68E + 07

Note: Abbreviations: Afr—Africa, Sam—South America, and Aus—Australia. The total area of younger crust = 3.73×10^7, and that of the cratons = 9.52×10^6 km^2.

TABLE 3B. NATURAL LOG OF THE SPATIAL COEFFICIENT (r_{ij}) OF ELEMENT GROUPS OF OLD CRATONS AND YOUNGER CRUST OF THREE ONTINENTS: AFRICA, SOUTH AMERICA, AND AUSTRALIA

	Au	CrNiPgeTi	CuZnPbBa	SnSb	W	UThREE	Total
Afr old	1.50	1.68	0.33	0.40	1.00	−0.05	1.04
Afr young	−1.25	−1.42	−0.45	−1.46	−1.56	−0.33	−0.90
Sam old	0.84	0.71	0.79	1.80	−2.17	−0.62	0.80
Sam young	−0.13	−0.50	0.47	−0.29	1.03	0.64	0.30
Aus old	0.46	1.37	−0.48	−2.44		−0.54	0.05
Aus young	−1.98	−4.01	−3.15	−2.09	−3.50	−0.24	−1.72

Note: Abbreviations: Afr—Africa, Sam—South America, and Aus—Australia. These values are presented in Figures 6 and 7 as solid bars.

TABLE 3C. WEIGHTED NATURAL LOG OF THE SPATIAL COEFFICIENT (r_{ij}^w) OF ELEMENT GROUPS AND CRATONS (OLD) AS WELL AS YOUNGER CRUST (YOUNG) OF THREE CONTINENTS

	Au	CrNiPgeTi	CuZnPbBa	SnSb	W	UThREE	Total
Afr old	1.73	1.89	0.62	0.78	1.32	0.28	1.32
Afr young	−1.02	−1.22	−0.16	−1.07	−1.24	0.00[†]	−0.62
Sam old	0.53	0.37	0.54	1.65	−2.38	−0.83	0.54
Sam young	−0.44	−0.83	0.22	−0.45	0.82	0.42	0.05
Aus old	−0.44	0.44	−1.32	−3.18		−1.35	−0.80
Aus young	−2.89	−4.94	−4.00	−2.84	−4.31	−1.05	−2.57

Note: Abbreviations: Afr—Africa, Sam—South America, and Aus—Australia. These values are presented in Figures 6 and 7 as striped bars.
[†]Exact value is −0.0026845.

crust. Our integrated results also support general statements that old crust is more richly endowed with mineral deposits than young crust (Fig. 7).

Metallogenesis on a Gondwana Scale

The above analyses were repeated using all the data of the cratons and the three continents combined. This allows us to compare and contrast the mineral inventory of all old (Archean) crust with that of younger crust of the three continents combined as one, as they would have been in Gondwana times (ca. 200–500 Ma). To some degree this corrects a bias in the analyses by concentrations of recent mineral deposits related to one specific present-day plate tectonic environment (such as subduction below South America) that may skew the results and interpretations when using a single continent only. On this subglobal continental scale we combine all old crust (cratons) across all combined continents, but we apply no "weighting" to allow for differential development and exploration; both the relatively "rich" and "poor" of the Southern Hemisphere are dealt with collectively.

The data are summarized in Tables 4A to 4B and plotted in Figure 8. From this it is clear that Archean crust is indeed mineralized to a significantly greater degree than younger crust (except for the strongly lithophile element group UThREE), thus hinting at the possibility that young crust may have inherited at least some metal enrichment during remobilization of its embedded cratons (e.g., tin in South America, gold and PGE in Africa).

DISCUSSION AND CONCLUSIONS

By applying an "exploration index" to our existing mineral deposit data, we have attempted to compare the mineral inventory of different cratons across continents with variable exploration histories. Although our exploration index needs more rigorous testing, incorporating this type of data allows a more informed and robust comparative analysis between mineral riches of different cratons. The results confirm our earlier work (de Wit and Thiart, 2005) that, per unit area of crust, there is a greater concentration of mineral deposits in Archean cratons relative to younger crust. Although the mineral inventory is greater in cratonic crust

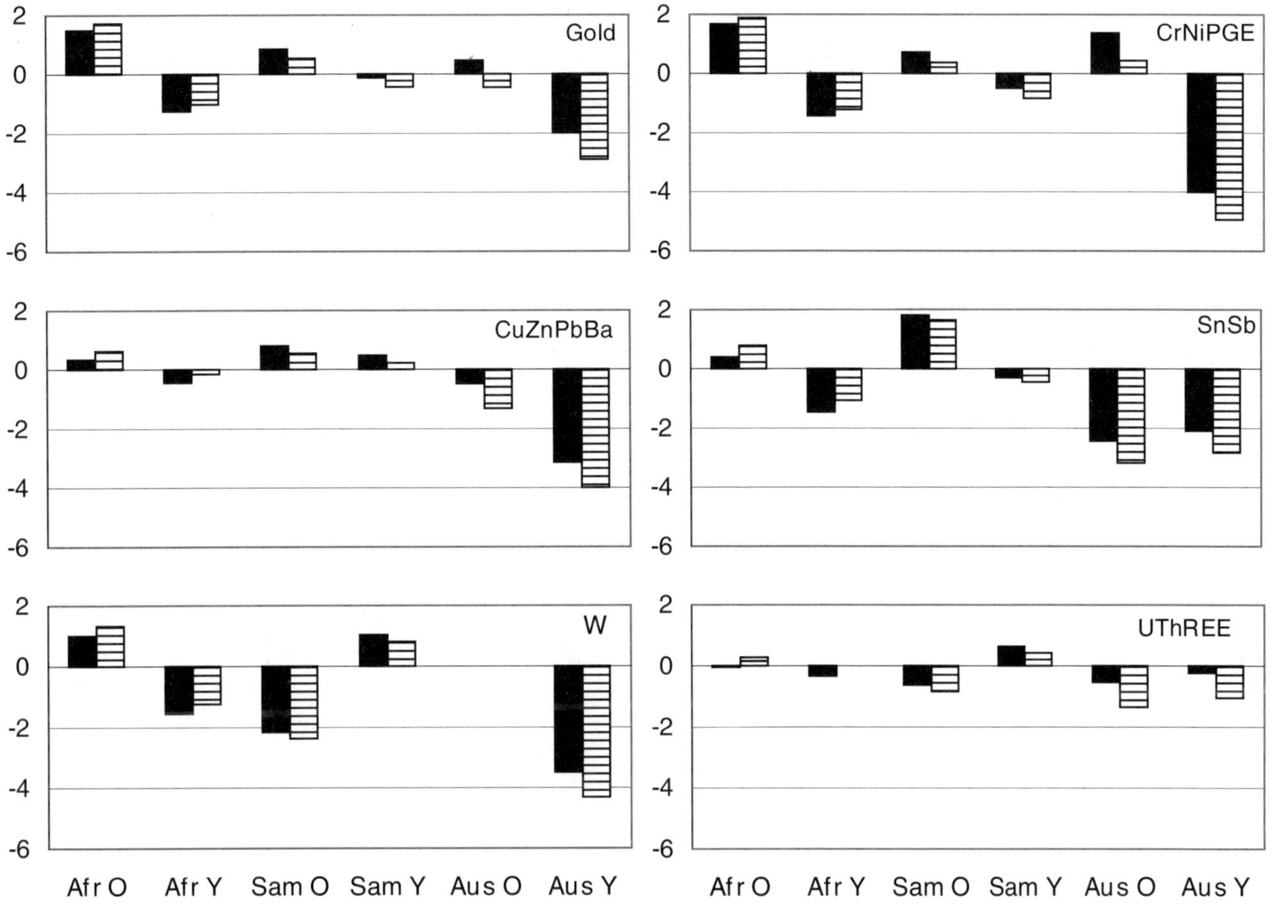

Figure 6. Mineral inventory of three separate continents (Aus—Australia; Afr—Africa; Sam—South America) showing the spatial coefficients ($\ln(r_{ij})$) between their groups of selected elements and two ages of continental crust (older than 2.5 Ga [O] and younger than 2.5 Ga [Y]), The solid bars represent their unweighted spatial coefficient ($\ln(r_{ij})$), and the striped bars represent their "weighted" spatial coefficients ($\ln(r_{ij}^w)$).

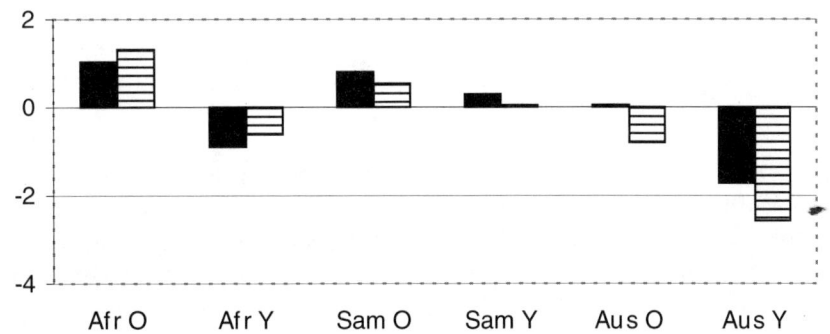

Figure 7. Total mineral diversity between old cratons (O) and younger crust (Y) for three separate continents (Afr—Africa, Sam—South America, and Aus—Australia). The solid bars represent their unweighted metallogenic fingerprint (spatial coefficient ($\ln(r_{ij})$), and the striped bars represent their "weighted" metallogenic fingerprint ($\ln(r_{ij}^w)$). Note that, on each continent, the mineral inventory of its cratons is in general greater than that of its younger surrounding crust. This suggests a decrease in mineral diversity (compare with Fig. 6) and mass per unit area in Earth's crust with time. Exceptions to this are tungsten and the lithophile elements U/Th/REE in South America, and possibly tin and antimony in Australia, but data from cratonic Australia are insufficient for more robust analysis.

TABLE 4A. NUMBER OF MINERAL DEPOSITS OF SELECTED ELEMENT GROUPS ON ALL CRATONS (OLD) AND YOUNGER CRUST (YOUNG) OF THREE CONTINENTS COMBINED AS A SUPER-CONTINENT (GONDWANA)

	Au	CrNiPgeTi	CuZnPbBa	SnSb	W	UThREE	Total	Area (km^2)
Old	2600	645	949	985	337	230	5746	9.52E + 06
Young	1366	212	1782	400	952	1319	6031	3.73E + 07
Total	3966	857	2731	1385	1289	1549	11777	4.68E + 07

TABLE 4B. NATURAL LOG OF THE SPATIAL COEFFICIENT (r_{ij}) BETWEEN ELEMENT GROUPS AND ALL CRATONS (OLD) AS WELL AS YOUNGER CRUST (YOUNG) OF THREE CONTINENTS COMBINED AS A SUPER-CONTINENT (GONDWANA)

	Au	CrNiPgeTi	CuZnPbBa	SnSb	W	UThREE	Total
Old	1.17	1.31	0.54	1.25	0.25	−0.31	0.88
Young	−0.84	−1.17	−0.20	−1.01	−0.08	0.07	−0.44

Note: These values are presented in Figure 8.

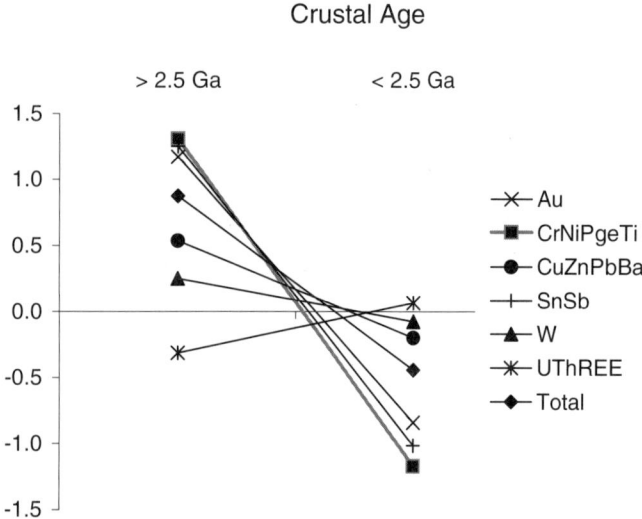

Figure 8. Relative concentration in Archean cratons and younger Gondwana crust of elements in all six element groups analyzed in this study. In all cases (except U/Th/REE) total concentrations are greater in the Archean cratons than in the younger crust of Gondwana. This suggests a decrease in the mass of new mineral deposits per unit area of crust over time.

than younger crust, we also confirm that a significant mineral diversity exists among cratons, and that each craton has a unique metallogenic fingerprint. These differences resemble variations of Phanerozoic mineralization on continents at scales that clearly link mineral deposits to different plate tectonic environments (e.g., oceanic arcs, continental subduction zones; Sawkins, 1990; Windley, 1995). If we assume a similar origin for the processes of mineralization in Archean times, this might provide a strong basis of support for models that advocate that plate tectonics operated on the Archean Earth. For example, many of the Neoarchean cratons have strong Au and base-metal signatures that fit with mineralization of subduction-accretion models as proposed for some of these cratons on the basis of geologic and geophysical evidence (e.g., Herrington et al., 1997; Barley et al., 1998).

Cumulative evidence indicates that cratonic crust is more enriched in mineral deposits than younger crust (younger than 2.5 Ga). The greater concentration of this Archean mineralization may represent more efficient mineralization processes, perhaps related to higher heat and/or volatile loss from the early Earth compared to today (Abbott et al., 1994; de Wit and Hart, 1993; de Wit and Hynes, 1995; Pollack, 1997; de Wit, 1998), in which case our results may be interpreted to reflect greater "partition coefficients" of selected elements between cratonic crust and (now depleted) mantle during the formation of Archean lithosphere (Fig. 9).

However, we cannot rule out the possibility that the higher concentration of Archean mineralization represents a greater preservation potential of cratons (and their mineral deposits) relative to younger continents, as the presence of (Archean) diamonds in cratons might imply. In this case, the greater mineral wealth of cratons may be merely a consequence of greater rates of recycling of young continental crust relative to that of old Archean crust preserved in cratons (Fig. 9). Because the majority of the cratonic crust is preserved at low grades of metamorphism (and is thus representative of the upper crust) this seems an equally valid interpretation.

If this interpretation has merit, it implies significant changes in the efficiency of recycling of young continental crust since the Archean. Because subduction is the principal mechanism by which the hydrosphere recycles into the mantle to generate continental crust enriched in mineral deposits, secular change in crustal mineralization could be used to track secular changes in the chemical composition of Earth's fluid envelope and the redox state of its upper mantle.

Either way, Earth's crust appears to signal a decrease in new mineral deposit mass and diversity through time. But, as a word of caution, our results are based on only one mineral database tied to only one specific exploration index. In addition only two rela-

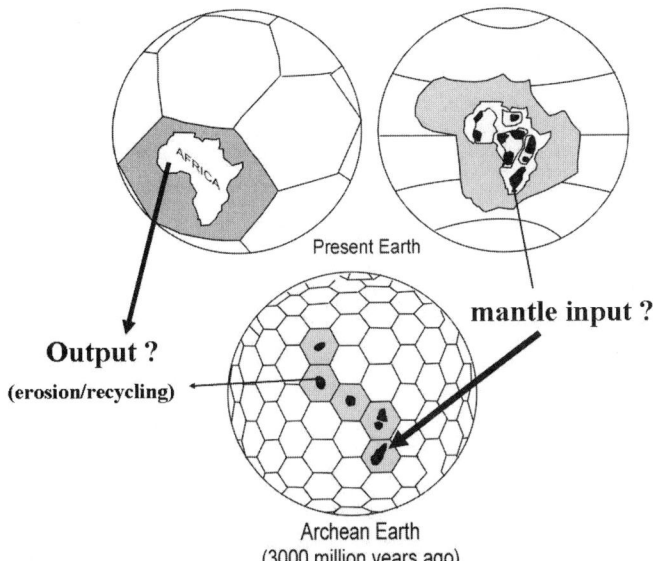

Figure 9. Summary of the main findings of this study as based on the mineral deposit data analyzed in this study. The figures on top schematically illustrate the present Earth with its average-size plates (left), and the African plate with its Archean cratons (black; right); the lower figure shows the possible average size of Archean plates with embedded cratons (black), as based on heat flow arguments (modified from Pollack, 1997, and de Wit and Hart, 1993). Relative to young crust, Archean cratons are mineral diversity hotspots, and these mineral hotspots are diverse, so that each craton retains a unique metallogenic fingerprint. It is not possible at this stage to distinguish with certainty if the mineral wealth and diversity of cratons reflects a decrease in transfer of the selected elements from the mantle to the continental lithosphere from the Archean onward (input), or that the return (output; recycling through erosion and subduction) of continental mineral deposits back into the mantle has become more efficient since Archean times.

tively long time spans are compared (e.g., Archean and younger). It would be wise to explore further, using a different databases and development indices, as well as with greater time resolution, before our general conclusions can be established beyond reasonable doubt.

ACKNOWLEDGMENTS

We are grateful to Graham Bonham-Carter and Fritz Agterberg for their interest and encouragement in our work. Constructive discourse on an early draft by Keith Long, Stephen Kesler, and an anonymous reviewer helped to clarify some of our thoughts and improve this contribution significantly. This work is supported through funds of the South African National Research Foundation (NRF). This is AEON contribution number 008.

REFERENCES CITED

Abbott, D.H., Burgess, L., Linghi, J., and Smith, W.H.F., 1994, An empirical thermal history of the Earth's upper mantle: Journal of Geophysical Research, v. 99, p. 13835–13850, doi: 10.1029/94JB00112.

Barley, M.E., Krapez, B., Groves, D.I., and Kerrich, R., 1998, The Archean bonanza: metallogenic and environmental consequences of the interaction between mantle plumes, lithospheric tectonics and global cyclicity: Precambrian Research, v. 91, p. 65–90, doi: 10.1016/S0301-9268(98)00039-4.

Cawthorn, R.G., and Walraven, F., 1998, Emplacement and crystallization time for the Bushveld Complex: Journal of Petrology, v. 39, p. 1669–1687, doi: 10.1093/petrology/39.9.1669.

de Wit, M.J., 1998, On Archean granites, greenstones, cratons and tectonics: Does the evidence demand a verdict?: Precambrian Research, v. 91, p. 181–226, doi: 10.1016/S0301-9268(98)00043-6.

de Wit, M.J., and Hart, R.A., 1993, Earth's earliest continental lithosphere, hydrothermal flux, and crustal recycling: Lithos, v. 30, p. 309–336, doi: 10.1016/0024-4937(93)90043-C.

de Wit, M.J., and Hynes, A., 1995, The onset of interaction between the hydrosphere and oceanic crust, and the origin of the first continental lithosphere, in Coward, M., and Ries, A., eds., Early Precambrian processes: Geological Society [London] Special Publication 95, p. 1–9.

de Wit, M.J., and Thiart, C., 2005, Metallogenic fingerprints of Archean Cratons, in MacDonald, I., Boyce, A.J., Butler, I.B., and Herrington, R.J., eds., Mineral Deposits and Earth Evolution: Geological Society [London] Special Publication 248, p. 59–70.

de Wit, M.J., Jeffery, M., Bergh, H., and Nicolaysen, L., 1988, Geological map of sectors of Gondwana reconstructed to their disposition ~150 Ma, scale 1:10.000.000: Tulsa, Oklahoma, American Association Petroleum Geologists.

de Wit, M.J., Thiart, C., Doucoure, C.M., and Wilsher, W., 1999, Scent of a supercontinent: Gondwana's ores as chemical tracers—Tin, tungsten and the Neoproterozoic Laurentia- Gondwana connection: Journal of African Earth Sciences, v. 28, p. 35–51, doi: 10.1016/S0899-5362(98)00085-2.

de Wit, M.J., Thiart, C., Doucouré, C.M., Milesi, J.P., Billa, M., Braux, C., and Nicol, N., 2004, The "Gondwana Metal-Potential" GIS, a geological and metallogenic synthesis of the Gondwana supercontinent at 1:10 million scale: Developed and published by CIGCES (South Africa) and BRGM (France). (http://gondwana.brgm.fr/index_eng.htm)

Eckstrand, O.R., and Good, D.J., compilers, 2000, World distribution of nickel deposits: Geological Survey of Canada, Open-File Report 3791a, 3 diskettes.

Eckstrand, O.R., Good, D.J. and Gall, Q., 2002, Ni-PGE-Cr deposits: World Minerals Geoscience Database Project, Mineral Resources Division, Geological Survey of Canada, unpublished database under revision.

Eglington, B., and Armstrong, R.A., 2004, The Kaapvaal craton and adjacent orogens, southern Africa: A geochronology database and overview of the development of the craton: South African Journal of Geology, v. 107, p. 13–32, doi: 10.2113/107.1-2.13.

Frimmel, H.E., and Minter, W.E.L., 2002, Recent developments concerning the geological history and genesis of the Witwatersrand gold deposits, South Africa: Society of Economic Geologists Special Publication 9, p. 17–45.

Fouch M.J., James, D.E., VanDecar, J.C., van der Lee, S., and the Kaapvaal seismic group, 2004, Mantle seismic structure beneath the Kaapvaal and Zimbabwe Cratons: South African: Journal of Geology, v. 107, p. 33–44.

Groves, D.I., Ho, S.E., Rock, N.M.S., Barley, M.E., and Muggeridge, M.T., 1987, Archean cratons, diamonds and platinum: Evidence for long lived crust mantle systems: Geology, v. 15, p. 801–805, doi: 10.1130/0091-7613(1987)15<801:ACDAPE>2.0.CO;2.

Hamilton, W.B., 1998, Archean magmatism and deformation were not products of plate tectonics: Precambrian Research, v. 91, p. 143–181, doi: 10.1016/S0301-9268(98)00042-4.

Hart, J.H., de Wit, M.J., and Tredoux, M., 1997, Refractory trace elements in diamonds: Further clues to the origin of the ancient cratons: Geology, v. 25, p. 1143–1146, doi: 10.1130/0091-7613(1997)025<1143:RTEIDI>2.3.CO;2.

Hart, R.J., McDonald, I., Tredoux, M., de Wit, M.J., Carlson, R.W., Andreoli, M., Moser, D.E., and Ashwal, L.D., 2004, New PGE and Re/Os isotope data from lower crustal sections of the Vredefort dome and a reinterpretation of its "crust on edge" profile: South African Journal of Geology, v. 107, p. 173–184, doi: 10.2113/107.1-2.173.

Herrington, R.J., Evans, D.M., and Buchanan, D.L., 1997, Metallogenic aspects, in de Wit, M.J., and Ashwal L.D., eds., Greenstone belts: Oxford, UK, Oxford University Press, p. 176–220.

Hutchinson, R.W., 1981, Mineral deposits as guides to supracrustal evolution, in O'Connell, R.J., and Fyfe, W.S., eds., Evolution of the earth: Washington D.C, American Geophysical Union, Geodynamics Series, v. 5, p. 120–140.

Hutchinson, R.W., 1992, Mineral deposits and metallogeny: Indicators of Earth evolution, in Schidlowsky, M., Golubic, S., and Kimberley, M.M., eds., Early organic evolution: Implications for mineral and energy resources: Berlin, Springer-Verlag, p. 551–545.

James, D.E., Fouch, M., VanDecar, J., and van der Laan, S., 2001, Tectospheric structure beneath southern Africa: Geophysical Research Letters, v. 28, p. 2485–2488, doi: 10.1029/2000GL012578.

Jelsma, H.A., de Wit, M.J., Thiart, C., Skinner, E.M., Dirks, P.H., Viola, G., Basson, I.J., and Anckar, E., 2004, Preferential distribution along transcontinental corridors of kimberlites and related rocks of southern Africa: South African Journal of Geology, v. 107, p. 301–304, doi: 10.2113/107.1-2.301.

Jenkins, C.L., Vincent, R., Robert, F., Poulsen, K.H., Garson, D.F., and Blondé, J.A., 1997, Index-level database for lode gold deposits of the world: Geological Survey of Canada, Open-File Report 3490, diskette.

Kirkham, R.V., and Dunne, K.P.E., 2000, World distribution of porphyry, porphyry-associated skarn, and bulk-tonnage epithermal deposits and occurrences: Geological Survey of Canada, Open-File Report 3792a, diskette.

Kirkham, R.V., and Dunne, K.P.E., 2002, Sediment-hosted copper deposits: World Minerals Geoscience Database Project, Mineral Resources Division, Geological Survey of Canada, unpublished database under revision.

Kirkham, R.V., Carrière, J.J., Laramée, R.M., and Garson, D.F., compilers, 1994, Global distribution of sediment-hosted stratiform copper deposits and occurrences: Geological Survey of Canada, Open-File Report 2915b, map, report, and diskette.

Kirkham, R.V., Carrière, J.J., Rafer, A., and Born, P., 2002, Sediment-hosted copper deposits: World Minerals Geoscience Database Project, Mineral Resources Division, Geological Survey of Canada, unpublished database under revision.

McDonald, I., de Wit, M.J., Smith, C.B., Bizzi, L., and Viljoen, K.S., 1995, The geochemistry of the platinum-group elements in Brazilian and Southern African kimberlites: Geochimica et Cosmochimica Acta, v. 59, p. 2883–2903, doi: 10.1016/0016-7037(95)00183-2.

Mihalasky, M.J., and Bonham-Carter, G.F., 2001, Lithodiversity and its spatial association with metallic mineral sites, Great Basin of Nevada: Natural Resources Research, v. 10, p. 209–226, doi: 10.1023/A:1012569225111.

Phillips, G.N., and Evans, K.A., 2004, Role of CO_2 in the formation of gold deposits: Nature, v. 429, p. 860–862, doi: 10.1038/nature02644.

Pollack, H.N., 1997, Thermal characteristics of the Archean, in de Wit, M.J., and Ashwal L.D., eds., Greenstone belts: Oxford, UK, Oxford University Press, p. 223–232.

Sawkins, F.J., 1990, Mineral deposits in relation to plate tectonics (2nd and enlarged edition): Berlin, Springer-Verlag, 461 p.

Shirey, S.B., Harris, J.W., Richardson, S.H., Fouch, M.J., James, D.E., Cartigny, P., Deines, P., and Viljoen, F., 2002, Diamond genesis, seismic structure, and the evolution of the Kaapvaal-Zimbabwe Craton: Science, v. 297, p. 1683–1686.

Stankiewicz, J., Chevrot, S., van der Hilst, R.D., and de Wit, M.J., 2002, Crustal thickness, discontinuity depth and upper mantle structure beneath southern Africa: Constraints from body wave conversions: Physics of the Earth and Planetary Interiors, v. 130, p. 235–251, doi: 10.1016/S0031-9201(02)00012-2.

Thiart, C., and de Wit, M.J., 2000, Linking spatial statistics to GIS: Exploring potential gold and tin models of Africa: South African Journal of Geology, v. 103, p. 215–230, doi: 10.2113/1030215.

Tredoux, M., de Wit, M.J., Hart, R.J., Armstrong, R.A., Lindsay, N., and Sellschop, J.P.F., 1989, Platinum group elements in a 3.5 Ga nickel-iron occurrence: Possible evidence of a deep mantle origin: Journal of Geophysical Research, v. 94, B1, p. 795–813.

Trompette, R., 1994, Geology of western Gondwana (2000–500 Ma): Rotterdam, Balkema, 350 p.

van Kranendonk, M.J., Collins, W.J., Hickman, A., and Pawley, M.J., 2004, Critical tests of vertical vs. horizontal tectonic models for the Archean East Pilbara granite-greenstone terrane, Pilbara craton, Western Australia: Precambrian Research, v. 131, p. 173–211, doi: 10.1016/j.precamres.2003.12.015.

Wilsher, W., 1995, The distribution of selected mineral deposits across Gondwana with geodynamic implications [Ph.D. thesis]: Rondebosch, University of Cape Town, 258 p.

Wilsher, W., Herbert, R., Wullschleger, N., Naicker, I., Vitali, E., and de Wit, M.J., 1993, Towards intelligent spatial computing for the earth sciences in South Africa: South African Journal of Science, v. 89, p. 315–323.

Windley, B.F., 1995, The evolving continents (3rd edition): New York, John Wiley and Sons, 385 p.

World Bank, 2004a, World development indicators—Online database: (http://www.worldbank.org/data/onlinedatabases/onlinedatabases.html and http://www.worldbank.org/data/countrydata/countrydata.html).

World Bank, 2004b, African development indicators [drawn from the World Bank Africa Database, published March 2004]: ISBN: 0-8213-5720-4 SKU: 15720, 424 p. (http://www4.worldbank.org/afr/stats/adi2004/default.cfm).

Zegers, T.E., and van Keken, P.E., 2001, Middle Archean continent formation by crustal delamination: Geology, v. 29, p. 1083–1086, doi: 10.1130/0091-7613(2001)029<1083:MACFBC>2.0.CO;2.

MANUSCRIPT ACCEPTED BY THE SOCIETY 29 OCTOBER 2005

Geological Society of America
Memoir 198
2006

Discovery of the oldest oxic granitoids in the Kaapvaal Craton and its implications for the redox evolution of early Earth

Shunso Ishihara
Geological Survey of Japan, Tsukuba, 305-8567, Japan

Hiroshi Ohmoto
Penn State Astrobiology Research Center of the NASA Astrobiology Institute and the Department of Geosciences, Pennsylvania State University, University Park, Pennsylvania 16802, USA

Carl R. Anhaeusser
Economic Geology Research Institute, University of the Witwatersrand, Johannesburg, South Africa

Akira Imai
Kyushu University, Fukuoka, 812-8581, Japan

Laurence J. Robb
Economic Geology Research Institute, University of the Witwatersrand, Johannesburg, South Africa

ABSTRACT

Phanerozoic granitoids have been classified into magnetite and ilmenite series based on the abundance of magnetite, which is related to the Fe_2O_3/FeO ratio of the rock and the oxygen fugacity (f_{O_2}) of its parent magma. We have examined the temporal and spatial distributions of both series in Archean granitoids from the Barberton region and the Johannesburg Dome of the Kaapvaal Craton, South Africa. The oldest syntectonic TTG (tonalite-trondhjemite-granodiorite) granitoids (ca. 3450 Ma in age) were found to be ilmenite series, whereas some intermediate-series granitoids occurred locally. Younger and larger syntectonic TTGs (e.g., the 3230 Ma Kaap Valley plutons) comprise nearly equal quantities of magnetite and ilmenite series. The major 3105 Ma calc-alkaline batholiths (e.g., Nelspruit batholith), emplaced during the late-tectonic stage, comprise mostly magnetite-series granitoids, suggesting that an oxidized continental crust already existed by this time.

The rare earth element ratios and $\delta^{18}O$ values, as well as the Fe_2O_3/FeO ratios, of the Archean magnetite-series granitoids suggest that their magmas were generated from the partial melting of subducted oceanic basalts that had been oxidized by interaction with seawater on mid-oceanic ridges; the processes of magma generation were much like those for Phanerozoic magnetite-series granitoids. This further suggests that the concentrations of oxidants (O_2 and/or SO_4^{2-}) in the Archean oceans were similar to those in Phanerozoic oceans.

Low concentrations of chlorine in the magmas, as well as deep levels of granite erosion, appear to explain the absence of major mineral deposits associated with the Kaapvaal granitoids.

Keywords: Archean, ilmenite series, magnetite series, granitoids, Kaapvaal.

Ishihara, S., Ohmoto, H., Anhaeusser, C.R., Imai, A., and Robb, L.J., 2006, Discovery of the oldest oxic granitoids in the Kaapvaal Craton and its implications for the redox evolution of early Earth, *in* Kesler, S.E., and Ohmoto, H., eds., Evolution of Early Earth's Atmosphere, Hydrosphere, and Biosphere—Constraints from Ore Deposits: Geological Society of America Memoir 198, p. 67–80, doi: 10.1130/2006.1198(04). For permission to copy, contact editing@geosociety.org. ©2006 Geological Society of America. All rights reserved.

INTRODUCTION

Granitoids represent an integrated component of the continental lithosphere, hydrosphere, oceanic crust, and upper mantle, which were amalgamated by dynamic plate motion. Ishihara (1977) has suggested that Phanerozoic granitoids can be classified into two types, magnetite series and ilmenite series, based on the presence or absence of magnetite, which is determined by microscopic observation and modal analyses. However, when the magnetite content of a rock is very low, microscopic observation or modal analysis is difficult; in these situations, magnetite can be easily detected by a magnetic susceptibility measurement, even in the field (Ishihara et al., 2000). Therefore, Ishihara (1977) has proposed a magnetic susceptibility value of 100×10^{-6} emu/g (or 3×10^{-3} SI unit) at $SiO_2 \approx 70\%$ as the boundary between ilmenite and magnetite series. Magnetic susceptibility of ilmenite-series granitoids can be as low as $\sim 0.01 \times 10^{-3}$ SI, and that of magnetite-series granitoids can be as high as $\sim 100 \times 10^{-3}$ SI. Ishihara (1981) has also found that the Fe_2O_3/FeO ratios of granitoids generally increase with increasing values of magnetic susceptibility, from less than 0.5 in weight ratio (or 0.22 in mole ratio) for ilmenite series to greater than 0.5 in magnetite series.

The magnetite- versus ilmenite-series granitoid classification is primarily based on the oxidation state of magma: magnetite series indicate higher f_{O_2} values and ilmenite series denote lower f_{O_2} values at a given temperature and pressure condition. Ishihara (1977) and Czamanske et al. (1981) suggest that the f_{O_2} boundary for the two series probably lies near the NNO (nickel+nickel oxide) buffer line (Fig. 1).

Ishihara (1977) has also recognized that Cu-Au and Cu-Mo porphyry deposits are typically associated with magnetite-series granitoids, whereas W and Sn deposits are found with ilmenite-series granitoids. Such associations of granitoid and ore deposit types occur because the redox state of magma strongly influences the sulfur chemistry (e.g., SO_2, H_2S, and SO_4^{2-}), as well as metal chemistry (e.g., Fe, Cu, Mo, Au, W, and Sn), of magmatic fluids (Burnham and Ohmoto, 1980).

Distinct differences also exist between magnetite- and ilmenite-series granitoids in various geochemical parameters (e.g., O and S isotope ratios; Ishihara et al., 2000; Ishihara and Matsuhisa, 2002; Sasaki and Ishihara, 1979; Ishihara and Sasaki, 2002), as well as in their Fe_2O_3/FeO ratio (Ishihara, 1977, 2004). Such data have been used to suggest that magnetite-series granitoids were generated in subduction zones from the partial melting of hydrated oceanic crust with an Fe_2O_3/FeO ratio that had been increased by reactions with O_2- and SO_4^{2-}-rich seawater at mid-oceanic ridges, whereas ilmenite-series granitoids were generated from the partial melting of normal mantle amphibolites with some contributions of metasediments from the continental crust (e.g., shales, graywackes) (Burnham and Ohmoto, 1980; Ishihara and Matsuhisa, 1999). Note that by seawater-rock interactions, the Fe_2O_3/FeO (weight) ratios of oceanic basalts have typically increased from ~ 0.08 (normal mantle value) to $\sim 0.1 - \sim 1.0$ (average = 0.31) (e.g., Lécuyer and Ricard, 1999).

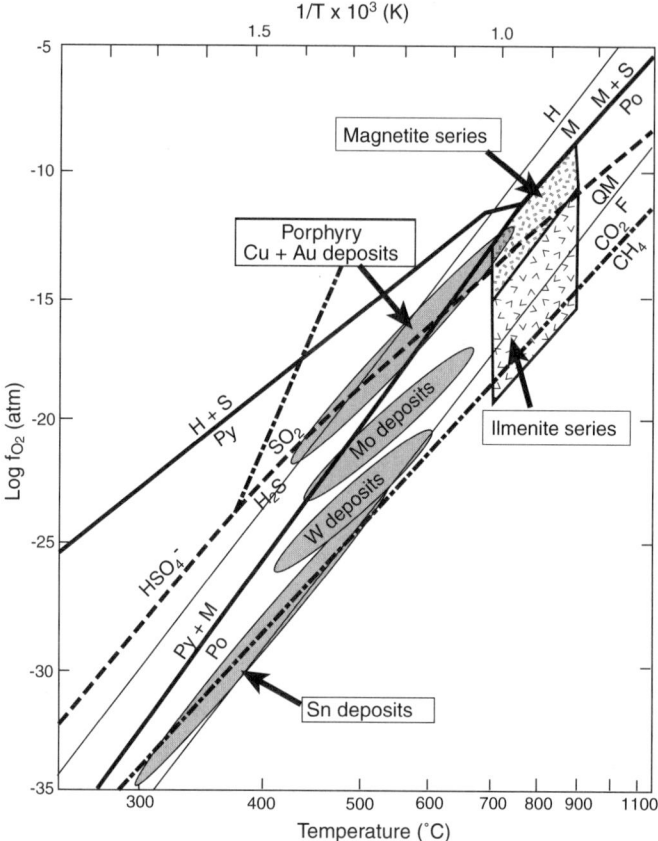

Figure 1. Genetic conditions for magnetite- and ilmenite-series granitoid magmas and depositional conditions for porphyry Cu-Au, porphyry Mo, porphyry Sn, porphyry W, and hydrothermal U deposits (modified after Ohmoto and Goldhaber, 1997). The solid lines show f_{O_2}-T conditions for the following well-known mineral buffers: quartz+fayalite+magnetite (QFM) and magnetite+hematite (MH); the broken lines indicate those for $m_{SO_2}/m_{H_2S} = 1$ and $m_{CO_2}/m_{CH_4} = 1$. Total fluid pressure = 1 kbar. (The nickel+nickel oxide buffer line [NNO] lies approximately halfway between the QFM and HM lines).

Compared with Phanerozoic granitoids, very little is known about the petrochemistry and genesis of Archean granitoids. No study has evaluated the redox state of Archean granitoids. The main objective of this study is, therefore, to determine whether magnetite- and/or ilmenite-series granitoids formed during the Archean. The answer will provide important constraints on the chemical evolution of the mantle, crust, oceans, and atmosphere. We have pursued this objective through magnetic, mineralogical, and geochemical investigations of major granite batholiths and plutons (older than 3150 Ma) in the Kaapvaal Craton, South Africa.

FERRIC/FERROUS RATIOS AND OXIDATION STATE OF GRANITOID MAGMAS (THEORETICAL)

Because our study attempts to relate the magnetic susceptibility and Fe_2O_3/FeO ratios of granitoids to the oxidation state of magmas, it is necessary first to evaluate the relationships among the magnetic susceptibility, relative abundances of Fe-bearing minerals, Fe_2O_3/FeO ratio of granitoids, and the fugacity of oxygen (f_{O_2}) in the magmas.

Magnetic Susceptibility and Fe_2O_3/FeO Ratio of Granitoids

The magnetic susceptibility of granitoids, ranges from 0.01×10^{-3} to 100×10^{-3} SI and is primarily a measure of the amount of magnetite present in the rock, because other magmatic magnetic minerals (e.g., pyrrhotite, ilmenite) are typically much less abundant than magnetite. Magnetic susceptibility of magnetite is $5 \times 10^{-4} m^3 kg^{-1}$ (Thompson and Oldfield, 1986), and that of pyrrhotite and ilmenite is known to be lower by two orders of magnitude. Mafic silicates, common in granitoids such as hornblende and biotite, are lower by three orders of magnitude in magnetic susceptibility, but higher by one order of magnitude in modal abundance, than magnetite. Therefore, the measured magnetic susceptibility represents a combination of these mafic minerals, but heavily depending on the magnetite contents.

Measurement of the magnetic susceptibility indicates that the f_{O_2} conditions of all (or most) granitoid magmas should fall between the quartz+fayalite+magnetite (QFM) and magnetite+hematite (MH) buffer lines (Fig. 1). Oxygen fugacity conditions above the MH buffer line are unlikely because hematite+ferrous silicates are a non-equilibrium assemblage. We may further assume that in granitoids (i) the ferrous component (Fe^{2+} or FeO) mostly resides in ferrous-rich silicates (mostly hornblende and biotite, rarely pyroxene) and magnetite, although some may be found in pyrrhotite and ilmenite; and (ii) the ferric component (Fe^{3+} or Fe_2O_3) resides largely in magnetite, although some may exist in biotite (Czamanske et al., 1981). Certain granitoids contain appreciable amounts of hematite that formed by subsolidus reactions with meteoric water (e.g., Taylor, 1968).

Thus, the mole fractions of Fe in silicates and magnetite (X_{sil} and X_{mt}, respectively) with respect to the total number of moles of Fe in a rock can be expressed as

$$X_{sil} + X_{mt} = 1. \quad (1)$$

Because magnetite (Fe_3O_4) has one FeO and one Fe_2O_3 component, the Fe_2O_3/FeO mole ratio of a rock can be related to X_{mt} as

$$(Fe_2O_3/FeO)_{mol} = X_{mt}/(3 - 2X_{mt}). \quad (2)$$

This positive relationship between the Fe_2O_3/FeO ratio and X_{mt} in rocks is shown in Figure 2. It illustrates that the maximum Fe_2O_3/FeO ratio is 1 in mole ratio (or 2.2 in weight ratio) when all the Fe atoms in rocks are in magnetite. If the Fe_2O_3/FeO ratio exceeds this value, it suggests that a significant amount of secondary hematite crystals occur in the granite.

Because the amount of magnetite in a rock is related to both the X_{mt} and ΣFe content of the rock, the magnetic susceptibility (M) of a rock can be expressed as

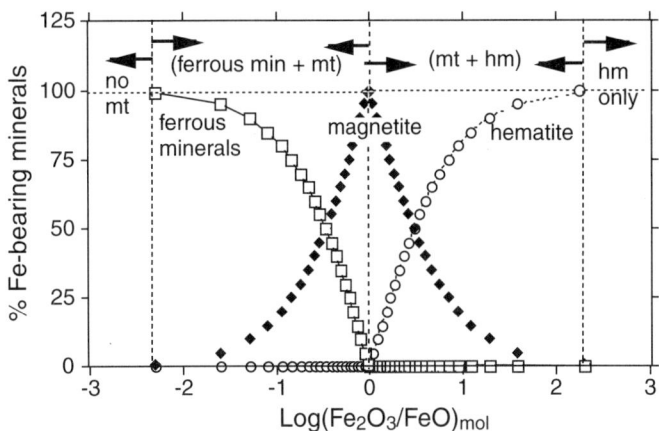

Figure 2. Relationship between the Fe_2O_3/FeO ratios and the relative abundances of magnetite and ferrous-rich silicates in granitoids.

$$M = k\, X_{mt}\, \Sigma Fe \quad (3)$$

where k is a coefficient that relates magnetic susceptibility to the amount of magnetite in a rock. Because the ΣFe content generally decreases with increasing SiO_2 content, the magnetic susceptibility of rocks with a given Fe_2O_3/FeO ratio generally decreases with increasing SiO_2 content.

Relationships between the Fe_2O_3/FeO Ratio and f_{O_2} of Magma

For silicate melts that are undersaturated with $[Fe_2O_3]$ (i.e., $aFe_2O_3 < 1$), the redox equilibria between ferrous and ferric oxides may be written as

$$(Fe^{2+})_{melt} + H^+ + 1/4 O_2 = (Fe^{3+})_{melt} + 1/2 H_2O. \quad (4)$$

For reaction (4), the equilibrium f_{O_2} value is expressed as

$$\log f_{O_2} = 4\log(X_{Fe^{3+}}/X_{Fe^{2+}})_{melt} + 4\log(\gamma_{Fe^{3+}}/\gamma_{Fe^{2+}})_{melt} - 4\log K - 4\log a_{H^+} + 2\log f_{H_2O} \quad (5)$$

where f_i, a_i, X_i, and γ_i are, respectively, the fugacity, activity, mole fraction, and activity coefficient of component i, and K is the equilibrium constant for reaction (4). Equation (5) may be simplified to

$$\log f_{O_2} = 4\log(X_{Fe^{3+}}/X_{Fe^{2+}})_{rock} + C \quad (5')$$

where

$$C = 4\log(\gamma_{Fe^{3+}}/\gamma_{Fe^{2+}})_{rock} - 4\log K - 4\log a_{H^+} + 2\log f_{H_2O} \quad (6)$$

The activity coefficient ratio, $\gamma_{Fe^{3+}}/\gamma_{Fe^{2+}}$, most likely depends on the bulk chemistry of the rock or melt (e.g., SiO_2 content), but here we will assume unity. K depends on temperature (T) and total pressure (P_{total}). Therefore, C has a unique value at a given set of T, pH, f_{H_2O}, and melt (or rock) composition.

The relationship between the $(Fe^{3+}/Fe^{2+})_{melt}$ and mole ratio of $(Fe_2O_3/FeO)_{rock}$ can be expressed as

$$(X_{Fe^{3+}}/X_{Fe^{2+}})_{melt} = k' \times 2(X_{Fe_2O_3}/X_{FeO})_{rock} \quad (7)$$

where k' is a distribution coefficient.
That is,

$$\log(X_{Fe^{3+}}/X_{Fe^{2+}})_{melt} = \log k' + \log 2 + \log(X_{Fe_2O_3}/X_{FeO})_{rock}. \quad (7')$$

Substituting (7') into (5'), we obtain

$$\log f_{O_2} = 4\log(X_{Fe_2O_3}/X_{FeO})_{rock} + 4\log k' + 4\log 2 + C. \quad (8)$$

Equation (8) indicates that, at a given set of T, P_{total}, P_{H_2O}, and bulk rock chemistry, the $\log f_{O_2}$ value proportionally increases to four times the value of $\log(Fe_2O_3/FeO)$mol (Fig. 3). For example, when P_{total} = 1 kb and T = 900 °C, the $\log f_{O_2}$ - $\log(Fe_2O_3/FeO)$mol

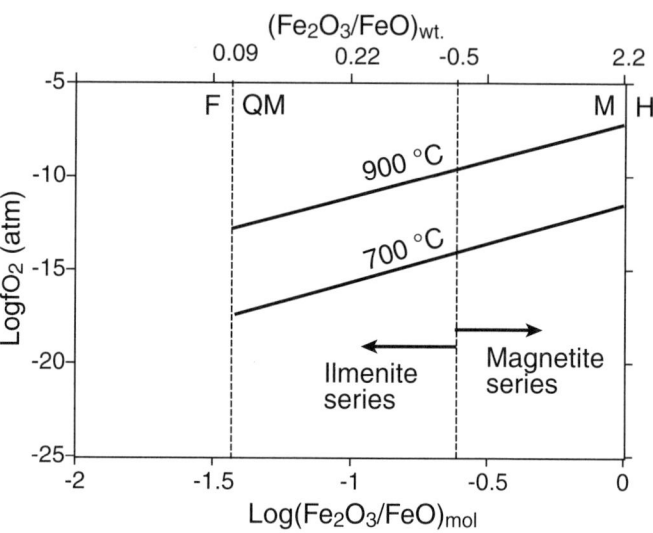

Figure 3. Relationship between the Fe_2O_3/FeO ratios and f_{O_2} values of magmas at P_{total} = 1 kbar and T = 700 °C and 900 °C. Thermodynamic data used in computations are summarized in Ohmoto and Kerrick (1977).

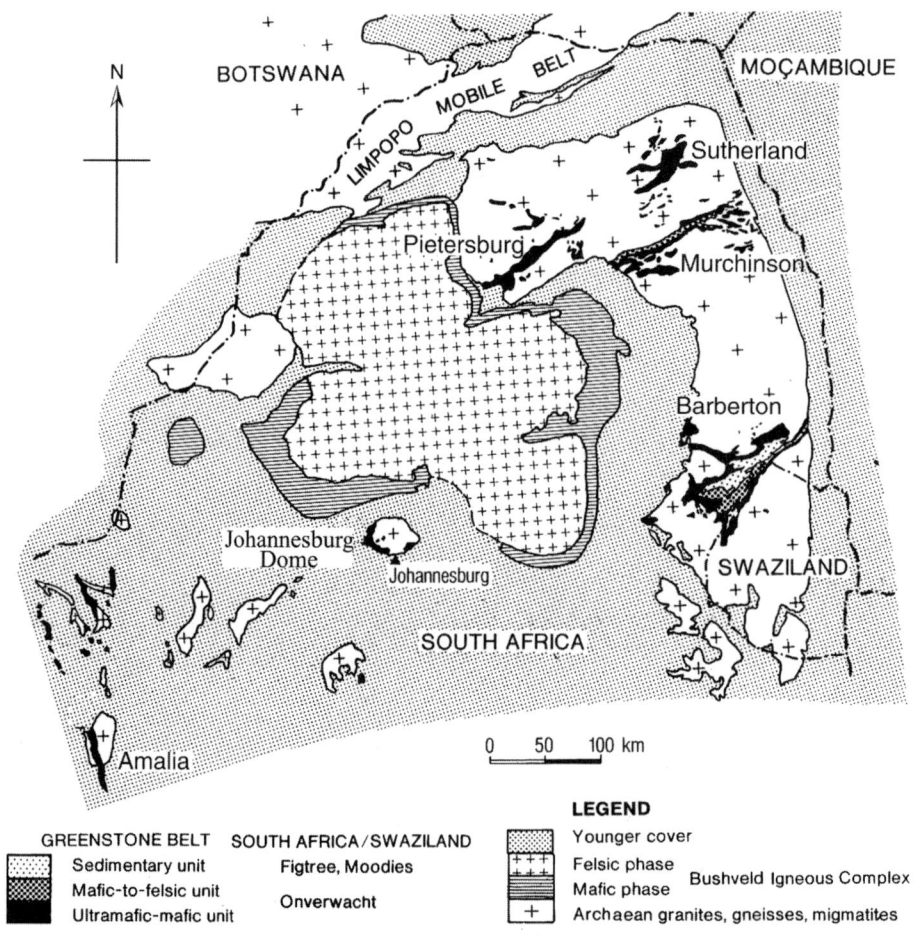

Figure 4. Geologic map of the Kaapvaal Craton, South Africa, showing the locations of the main Archean greenstone belts (Anhaeusser, 1976).

line extends from $\log f_{O_2} = -7.25$ and $\log(Fe_2O_3/FeO)_{mol} = 0$ (i.e., MH buffer) to $\log f_{O_2} = -12.86$ and $\log(Fe_2O_3/FeO)_{mol} = -1.40$ (i.e., QFM buffer). When $P_{total} = 1$ kb and $T = 700$ °C, the line extends from $\log f_{O_2} = -11.70$ and $\log(Fe_2O_3/FeO)_{mol} = 0$ (i.e., MH buffer) to $\log f_{O_2} = -17.35$ and $\log(Fe_2O_3/FeO)_{mol} = -1.41$ (i.e., QFM).

The above examples of the f_{O_2} versus Fe_2O_3/FeO ratio relationship illustrate an important point when comparing the redox state of igneous rocks that crystallized at different T-P conditions; the comparison should be made in terms of the Fe_2O_3/FeO (or Fe^{3+}/Fe^{2+}) ratios of igneous rocks, rather than the f_{O_2} values, because the f_{O_2} value varies depending on temperature and pressure conditions.

GRANITOIDS IN THE KAAPVAAL CRATON

The Kaapvaal Craton (Fig. 4) is one of the oldest and best-preserved Archean continental fragments on Earth. Its assembly during the Archean eon is attributed to a complex combination of processes analogous to modern-day plate tectonics (Poujol et al., 2003). Such processes took place episodically over a 1000 million-year period (ca. 3500–2500 Ma) and involved magmatic arc formation and accretion as well as the tectonic amalgamation of numerous, discrete terranes or blocks (de Wit et al., 1992; Lowe, 1994; Poujol and Robb, 1999).

The oldest rocks so far recognized are located in the Swaziland-Barberton regions on the eastern side of the Kaapvaal Craton (Fig. 4), where ages more than 3600 Ma have been recorded. The early stages of shield development are also best exposed in the Barberton Mountains, where it is now apparent that continent formation took place by magmatic accretion and tectonic amalgamation of small protocontinental blocks. At Barberton, several diachronous blocks that formed between 3600 and 3200 Ma have been identified (Kamo and Davis, 1994), each of which represents a cycle of arc-related magmatism and sedimentation (Lowe, 1999; Poujol et al., 2003).

The Barberton Region

The Archean granitoids in the Barberton region occupy an area of ~20,000 km² (Fig. 5). They are divided into three groups based on their intrusive ages in tectonic history: syntectonic, late-tectonic, and post-tectonic (Anhaeusser and Robb, 1980a; Meyer et al., 1994).

(1) Syntectonic tonalite-trondhjemite-granodiorite (TTG). *S1 substage*: small intrusive bodies with the oldest age (ca. 3450 Ma), e.g., Rooihoogte, Steynsdorp, Stolzburg, Theespruit, and other bodies; and *2 substage*: larger bodies with younger ages (ca. 3230 Ma), e.g., Kaap Valley (KV in Fig. 5) and Nelshoogte plutons.

The syntectonic TTGs are closely associated with ultramafic to mafic volcanic rocks of the greenstone belt and intrude them concordantly. The largest Kaap Valley pluton is circular in form, and intrudes them discordantly. The S1

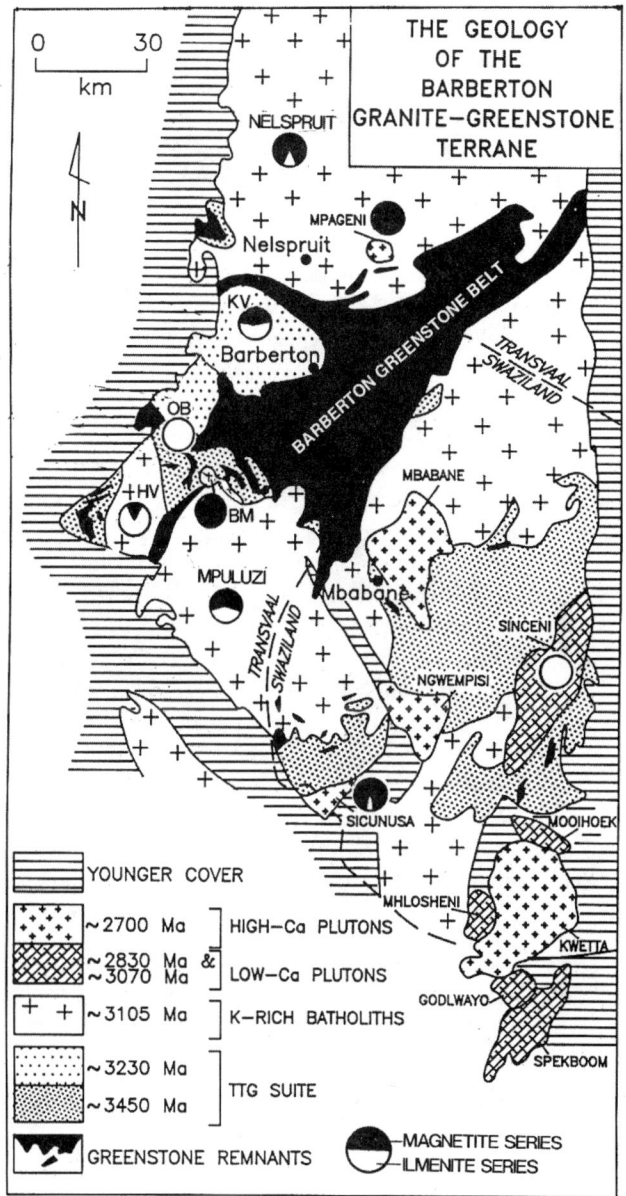

Figure 5. Geologic map of the Barberton region (modified after Ishihara et al., 2002b).

substage granitoids are mostly hornblende-biotite tonalites and are free of magnetite and ilmenite, but contain titanite that occurs as euhedral wedges and mostly subhedral crystals associated with mafic silicates, and which could have been converted from ilmenite. The color of the Z-axis of biotite is generally brown and similar to ilmenite-series biotites of Phanerozoic age. The S2 substage granitoids are also mostly magnetite-free, but the Kaap Valley pluton contains magnetite (up to 1.7% by volume), locally. The magnetite forms polygonal to granular crystals and contains small inclusions of ilmenite and chalcopyrite. In the most mafic intrusive phase, hematite blades occur along the 111 cleavage. However, the hematitization along the cleavage and margin of the crystals is not seen in the magnetite of most magnetite-bearing phases, implying lack of oxidation at the latest stage of their crystallization.

(2) Late-tectonic calc-alkaline granitoids (ca. 3105 Ma); e.g., Nelspruit, Mpuluzi and Heerenveen (HV in Fig. 5). The late-tectonic calc-alkaline granites are batholithic in dimension, and biotite granite in composition, consisting largely of unzoned plagioclase, microcline, and/or orthoclase and quartz. Biotite is the dominant mafic silicate mineral and its Z-axis color is generally greenish brown but rarely green, as is typical for fresh magnetite-series granitoids of Phanerozoic age (Ishihara, 1977). The green biotite, which is found locally, could be an alteration product because it is associated with chloritized biotite. Flaky muscovite that is found in a few samples could also be an alteration product. Magnetite is composed of cubic to polygonal crystals associated with rare hematite blades or stringers. Ilmenite is absent, but titanite is abundant and forms both euhedral wedge-shaped crystals and anhedral aggregates associated with biotite.

(3) Post-tectonic low-Ca (ca. 3070 Ma) and high-Ca (ca. 2700 Ma) granitoids. These rocks include an older, low-Ca biotite granite, and younger, high-Ca alkaline granitoids. The older granites are magnetite free, but the younger granitoids contain magnetite. The younger granitoids of the Boesmanskop and Salisburgkop stocks consist of minor quartz, alkaline amphiboles, and greenish-brown biotite, with more abundant zoned plagioclase and microcline. Euhedral titanite and polygonal to rounded magnetite are common, but ilmenite appears to be absent. No hematitization of the magnetite is observed.

Johannesburg Dome

In the central part of the Kaapvaal Craton, Paleoarchean granitoids similar to those found in the Barberton region form the Johannesburg Dome (~50 km in diameter). Evidence has been presented elsewhere showing that two types of granitoid gneisses exist in this region. The oldest variety (ca. 3340 Ma; Poujol and Anhaeusser, 2001) consists of leuco-biotite trondhjemitic gneisses and associated migmatites that developed on the northern half of the dome (TTG in Fig. 6) (Anhaeusser, 1973, 1999). The younger variety, which is on the southern edge of the dome, consists of hornblende-biotite tonalitic gneisses that yielded a multiple zircon age of ca. 3170 Ma (Anhaeusser and Burger, 1982), and a more recent single zircon emplacement age of ca. 3201 Ma (Poujol and Anhaeusser, 2001).

Following the emplacement of trondhjemite-tonalite gneiss, another episode of magmatism took place on the Johannesburg Dome. This produced intrusions of Mesoarchean potassic granodiorites that occupy an area of batholithic dimension, extending across most of the southern portion of the dome (CA [calc-alkaline], Fig. 6). Two granodiorite phases have been distinguished: one on the southern and southeastern parts of the dome consists mainly of medium-grained, homogeneous, gray granodiorites, whereas a second variety, found mainly on the southwestern part of

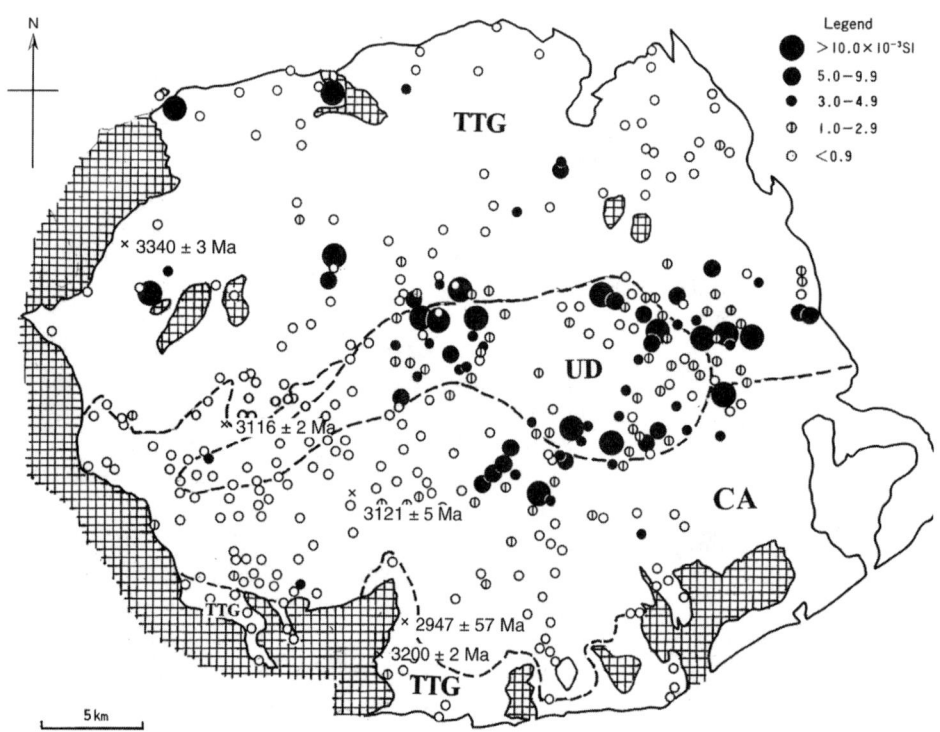

Figure 6. Distribution of the rock facies (TTG, CA, and UD) and magnetic susceptibility values of the Johannesburg Dome.

the dome, consists of porphyritic granodiorites (Anhaeusser, 1973). Zircons extracted from the two granodiorite types yielded ages of ca. 3121 Ma for the homogeneous variety and ca. 3114 Ma for the porphyritic variety (Poujol and Anhaeusser, 2001).

Numerous pegmatite dykes and veins crosscutting the granodiorites are younger than 3114 Ma and may represent the final stages of magmatism associated with a batholith emplacement ca. 3000 Ma.

ANALYTICAL METHODS

Magnetic susceptibilities of 546 hand specimens from the Barberton region and 346 specimens from the Johannesburg Dome were measured using a Kappameter KT-3. Most of the samples used in this study were collected for earlier projects (Robb and Anhaeusser, 1983; Robb et al., 1983; Robb et al., 1986), although ~30 samples were collected especially for this study.

Concentrations of major, trace, and rare earth elements (REE) were determined on seven samples by XRF methods for most elements, titrimetry for FeO, and ICP-MS method for REE. Concentrations of Cl and F, as well as the major element compositions of minerals, were determined using an electron microbe (EPMA) on 452 grains of biotite from eight representative granite samples. Concentrations of Cl, F, and S in apatite crystals (258 spot analyses) from two representative granite samples were also determined using an EPMA.

DISCUSSION OF THE ANALYSES

The magnetic susceptibility of the ~900 samples range from 0.03×10^{-3} to 53×10^{-3} SI, indicating that both ilmenite series ($<3 \times 10^{-3}$ SI) and magnetite series ($>3 \times 10^{-3}$ SI) are present. Approximately 10% of the granitoid samples from the Barberton syntectonic plutons to ~80% in the late-tectonic plutons belong to magnetite series (Fig. 7). Approximately 30% of the samples from the Johannesburg Dome are magnetite series.

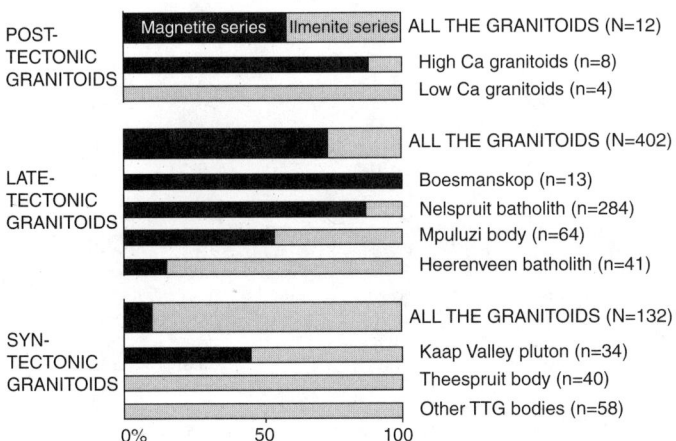

Figure 7. Relative abundances of magnetite- and ilmenite-series granitoids in major batholiths and plutons of the Barberton region

Temporal and Spatial Distributions of Granitoid Types in the Barberton Region

In the Barberton region, all of the earliest S1, as well as S2, substage TTGs of the Nelshoogte pluton possess magnetic susceptibility values less than 3×10^{-3} SI, indicating that they belong to very reduced ilmenite-series granitoids (Fig. 7). Anhaeusser and Robb (1980b) have shown that many of the TTGs interacted with and assimilated greenstone wall rocks that have Fe_2O_3/FeO (weight) ratios much less than 0.3 (Hunter, 1974; Hawkesworth and O'nions, 1977), suggesting that the reduced nature of the TTGs may have been acquired (and/or enhanced) during emplacement, rather than at the magma sources.

The S2 substage TTGs of the Kaap Valley tonalite pluton (Robb et al., 1986) likewise possess generally low magnetic susceptibility values, which suggests they are ilmenite series, but contain sporadic values higher than 3×10^{-3} SI (Ishihara et al., 2002b). The distribution of magnetite-bearing rocks is so erratic that its genetic significance is difficult to evaluate. The post-tectonic, low-Ca biotite granites also appear to be ilmenite series, although the number of samples examined was small.

The late-tectonic calc-alkaline granitoids, which form large batholiths in the Barberton region, intrude the greenstones as well as the older TTGs (Anhaeusser and Robb, 1983). These granitoids, along with the post-tectonic high-Ca granitoids, generally possess magnetic susceptibilities higher than 3.0×10^{-3} SI, indicating they are mostly magnetite series.

In the ca. 3105 Ma Nelspruit batholith, 86% of the 243 samples exhibit magnetic susceptibility values greater than 3×10^{-3} SI (Fig. 7), which are characteristic of magnetite-series granitoids. The Nelspruit batholith is the largest granitic body in the region, and consists dominantly of homogeneous K-rich porphyritic granitic phases, associated potassic gneisses and migmatites that occur together with remnants of older Na-rich gneisses and greenstones, and a homogeneous, medium-grained, granodioritic phase (Robb et al., 1983).

The Mpuluzi batholith (Fig. 5) is the second largest granitoid massif in the region and consists of 53% magnetite series ($n = 64$). In contrast, the smaller Boesmanskop syenite pluton is 100% magnetite series ($n = 13$).

The post-tectonic high-Ca granitoids, found mainly in Swaziland, occur as episodic, discrete intrusions crosscutting all the other Archean rock types. They are hornblende-biotite alkaline granitoids and have high magnetic susceptibilities similar to alkaline granitoids.

Temporal and Spatial Distributions of Granitoid Types in the Johannesburg Dome

Within the Johannesburg Dome area (Fig. 6), the Archean granitoids comprise older TTGs (ca. 3300–3200 Ma) and younger calc-alkaline granitoids (ca. 3100–2900 Ma). Magnetic susceptibility measurements indicate that the TTGs consist mainly of ilmenite series (78% of the measurements) and partly magnetite series (22%); the calc-alkaline granitoids are also largely composed of ilmenite series (83%) with some local (17%) magnetite series (Ishihara et al., 2002a). Fe_2O_3/FeO ratios of bulk rock samples range from 0.05 to 0.72 (Ishihara et al., 2002a) and are generally below 0.5, implying a generally reduced nature for the Johannesburg Dome pluton.

The weakly magnetic granitoids tend to occur in the central part of the Johannesburg Dome around an undifferentiated phase (UD) at the interface between the TTG and calc-alkaline granitoids (Fig. 6).

Chemical Characteristics of Representative Granitoids

Partial chemical analyses were available for ~60 of the samples. Plots of magnetic susceptibility versus SiO_2 content (Fig. 8) show that the Nelspruit granitoids fall within the magnetite-series field (12 of 15 samples). The sporadic occurrence of magnetite-bearing rocks in the Kaap Valley TTGs are also illustrated; one and three samples of the 16 specimens fall within the magnetite- and intermediate-series fields, respectively. Thus, the Kaap Valley TTGs belong essentially to the ilmenite-series granitoids. All other TTGs also have very low magnetic susceptibility values, in contrast to the high values of sodic and/or adakitic granitoids of Phanerozoic ages (e.g., Tanzawa pluton) (Fig. 8).

Chemical compositions of the representative magnetite-series granitoids are presented in Table 1. A sample of magnetite-series Kaap Valley tonalite (LKV7) reveals a much higher

TABLE 1. REPRESENTAIVE CHEMICAL COMPOSITIONS OF TWO ILMENITE-SERIES AND FIVE MAGNETITE-SERIES GRANITOIDS, BARBERTON REGION

	Theespruit-TTG		Kaap Vly	Nelspruit Batholith		Mpuluzi Batholith	
	23-4	23-5	LKV7	C31	D34	22-1	22-2
SiO_2	70.05	75.34	65.47	69.78	71.05	66.49	70.02
TiO_2	0.26	0.09	0.45	0.44	0.35	0.59	0.27
Al_2O_3	15.53	13.73	15.96	14.55	14.47	14.84	14.91
Fe_2O_3	0.49	0.39	2.26	1.88	0.97	2.46	1.24
FeO	1.61	0.80	1.68	1.26	1.57	1.94	1.15
MnO	0.03	0.04	0.05	0.05	0.04	0.08	0.05
MgO	1.22	0.22	2.00	0.69	0.69	0.84	0.35
CaO	2.67	1.49	4.31	1.94	2.05	2.23	1.34
Na_2O	5.54	4.61	5.19	4.67	4.73	5.25	4.53
K_2O	1.78	2.79	0.93	3.56	2.93	3.23	4.79
P_2O_5	0.08	0.03	0.16	0.24	0.13	0.35	0.12
S	<0.01	<0.01	<0.01	<0.01	<0.01	0.05	0.01
H_2O^+	0.43	0.43	1.07	0.61	0.71	0.76	0.43
H_2O^-	0.17	0.09	0.42	0.22	0.17	0.19	0.19
CO_2	0.01	0.08	0.10	0.04	0.05	0.56	0.33
SUM	99.87	100.13	100.05	99.93	99.91	99.86	99.73
Trace elements (ppm)							
Rb	57	176	21	134	100	153	182
Cs	2.7	9.4	<1.5	3.2	1.3	4.0	5.7
Sr	556	126	576	563	467	812	785
Ba	389	227	219	976	470	1276	1993
Zr	110	77	102	262	169	412	293
Hf	3.1	3.4	2.4	6.5	5.6	10.3	7.1
Nb	4.2	13.5	3.0	23.4	14.9	23.2	16.0
Ta	2.3	2.6	2.3	3.2	4.1	2.6	2.2
Y	5	20	7	29	24	34	33
V	20	5	71	15	23	19	<4
Cr	58	41	109	30	77	52	37
Co	11	6	16	7	8	10	6
Ni	27	2	24	4	3	4	1
Cu	3	5	5	13	0	14	3
Zn	50	35	50	76	65	114	53
Pb	10	18	5	28	20	23	27
Ga	17.6	14.4	18.2	19.9	19.5	19.7	16.7
Ge	0.9	1.1	0.9	0.9	0.8	1.0	0.6
Se	0.2	0.2	0.2	0.5	0.4	0.3	0.3
Mo	<0.2	0.2	0.5	1.4	0.7	1.7	1.7
W	1.1	1.9	<1.2	1.9	2.5	<1.4	1.9
Sn	1.5	6.3	1.4	3.3	3.5	6.5	5.0
Cd	0.7	<0.2	0.3	0.3	<0.2	0.3	0.2
Tl	0.6	1.5	0.6	2.8	1.6	1.6	1.9
Bi	0.4	0.3	<0.3	0.9	0.3	<0.3	0.4
Th	4.5	5.8	1.5	15.1	9.9	15.3	20.6
U	0.7	3.0	0.5	1.7	2.7	2.4	1.8
Rare earth elements (ppm)							
La	21.3	8.93	13.7	133	125	109	42.2
Ce	37.4	16.2	29.1	285	244	216	86.2
Pr	3.60	1.65	3.35	29.4	23.5	21.6	9.35
Nd	12.5	5.85	13.7	108	79.3	75.4	34.3
Sm	2.07	1.71	2.71	18.0	13.2	12.3	6.16
Eu	0.673	0.383	0.920	4.15	2.89	2.54	1.28
Gd	1.56	2.21	2.37	12.8	10.0	8.62	5.30
Tb	0.21	0.47	0.32	1.60	1.38	1.17	0.82
Dy	1.10	2.91	1.63	7.43	6.83	5.84	4.41
Ho	0.20	0.60	0.30	1.19	1.19	1.02	0.91
Er	0.58	1.78	0.86	3.09	3.20	2.76	2.65
Tm	0.081	0.295	0.123	0.399	0.427	0.396	0.392
Yb	0.50	1.85	0.75	2.36	2.49	2.44	2.40
Lu	0.076	0.278	0.114	0.331	0.337	0.341	0.359
Y	6.2	19.7	8.8	36.3	36.2	31.7	26.4
Kai (×10^{-3})	0.1	0.1	18.9	19.1	13.9	36.0	14.2
Fe_2O_3/FeO	0.3	0.5	1.4	1.5	0.6	1.3	1.1
Rb/Sr	0.1	1.4	0.0	0.2	0.2	0.2	0.2

Note: Rock type and plutons: 23-4—very fine quartz diorite, foliated, Theespruit; 23-5—Medium, trondjemite, Theespruit; LKV7—Fine, quartz diorite, Kaap Valley; C31—Medium, pink K-feldspar porphyritic biotite granite, Nelspruit; D34—Coarse, pink K-feldspar porphyritic biotite granite, Nelspruit; 22-1—Fine, quartz diorite, Mpuluzi; 22-2—fine, biotite granite, Mpuluzi. Kai, magnetic susceptibility, × 10-3 SI unit.

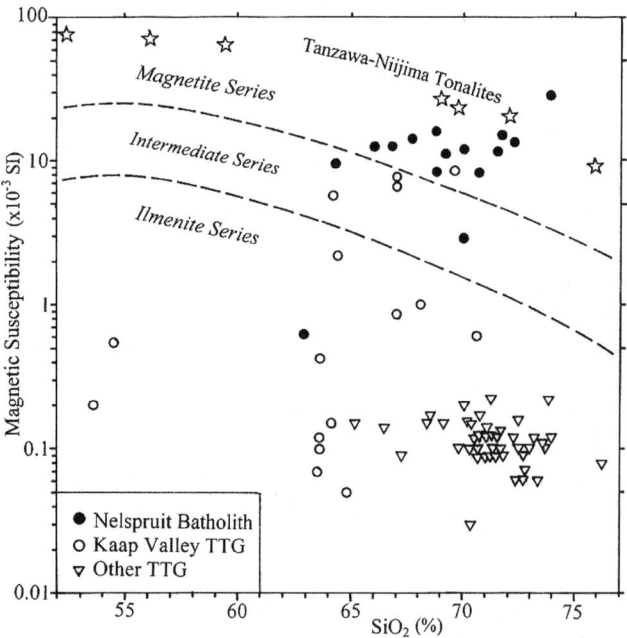

Figure 8. Magnetic susceptibility versus silica content comparison of the TTGs and Nelspruit batholith; SiO_2 data are from Anhaeusser and Robb (1983). The fields for Japanese magnetite- and ilmenite-series granitoids are shown for comparison (separated by dotted lines). Open stars represent data on sodic Miocene gabbroid and granitoids of the Tanzawa-Niijima (Ishihara, unpublished data).

Fe_2O_3/FeO ratio (1.35) than the magnetite-free TTGs of the Theespruit pluton (Fe_2O_3/FeO = 0.30–0.49), indicating a good correlation between magnetic susceptibility and Fe_2O_3/FeO ratio. Trace amounts of V and Cr, which may be substitutes for Fe^{2+} and Fe^{3+} in magnetite, respectively, are higher in the magnetite-bearing rocks than in the magnetite-free rocks.

Other TTG characteristics include high contents of Na_2O and Sr, but low contents of K_2O, Rb, and Ba (Robb and Anhaeusser, 1983) (Table 1). The LKV7 tonalite and 23-4 trondhjemite samples are high in Sr and low in Y, thus exhibiting adakitic characteristics (Drummond et al., 1996). Compared with the late-tectonic granitoids, the TTGs are rich in MgO, Al_2O_3, CaO, Ni, and Co (compare 23-4 and C31 in Table 1).

Magnetite-rich rocks of the Nelspruit and Mpuluzi batholiths have similarly high Fe_2O_3/FeO ratios (0.62–1.49). These batholiths, however, have differing feldspar chemistry; the Nelspruit granites are calcic, whereas the Mpuluzi granites are potassic. The Mpuluzi granites also have higher Rb, Sr, and Ba contents than the Nelspruit granites.

The granitoids generally display enrichments of light rare earth elements (LREE) and depletions of heavy rare earth elements (HREE) with respect to chondrite (Fig. 9). There are almost no Eu anomalies in the studied samples, except for the 23-5 Theespruit-TTG specimen, which, with a higher SiO_2 content and a higher Rb/Sr ratio compared to the other TTGs, represents a highly fractionated phase of the TTGs. The general REE patterns of the studied samples are similar to those of other TTGs (Robb et al., 1986) and garnet-bearing metamorphic protoliths; the HREE-depleted patterns are attributed to garnet and/or hornblende fractionation. The magnetite-series granitoids, particularly of the Mpuluzi potassic granite, are most enriched in REEs, whereas the ilmenite-series TTGs are generally most depleted in REE.

Magnetite crystals in the calc-alkaline magnetite-series granitoids tend to occur together with mafic silicates (mostly biotite). The Mg/Mg+Fe atomic ratios of biotites from the Nelspruit batholith (Table 2) are similar to those of typical magnetite-series biotite granite in the Sanin District, Japan (Fig. 10). The magnetite crystals contain rare hematite blades, which may have formed during the subsolidus stage, corresponding to a high degree of oxidation. Hematitization is only locally and weakly observed in the batholiths.

SIGNIFICANCE OF THE RESULTS

Genesis of Archean Granitoids in the Kaapvaal Craton

The Kaap Valley pluton contains local magnetite-bearing rocks (Figs. 7 and 8), and has, as a whole, higher Fe_2O_3/FeO ratios than the other TTGs. The whole-rock $\delta^{18}O$ values of the Kaap Valley pluton are on average 2‰ higher than those of Quaternary low-K tholeiite (Faure and Harris, 1991). We interpret these data to indicate that the magmas for the Kaap Valley pluton were generated from partial melting of subducted seafloor basalt, which had increased its $\delta^{18}O$ value by low-temperature interaction with seawater.

The Fe_2O_3/FeO ratios of oceanic basalts probably increased from ~0.1 to ~1 by interaction with O_2- and/or SO_4^{2-}-bearing seawater, much like the processes of seawater-rock interaction in modern oceanic crust (e.g., Shanks et al., 1981; Lécuyer and Ricard, 1999). The possibility of SO_4^{2-}-rich Archean seawater is also suggested from the abundance of barite beds associated with felsic volcanic rocks in older greenstones (ca. 3500 Ma Onverwacht Group) and later ferruginous sediments (ca. 3230 Ma Fig Tree Group). The geochemistry of banded iron-formations and associated rocks also led Ohmoto et al. (this volume) to conclude

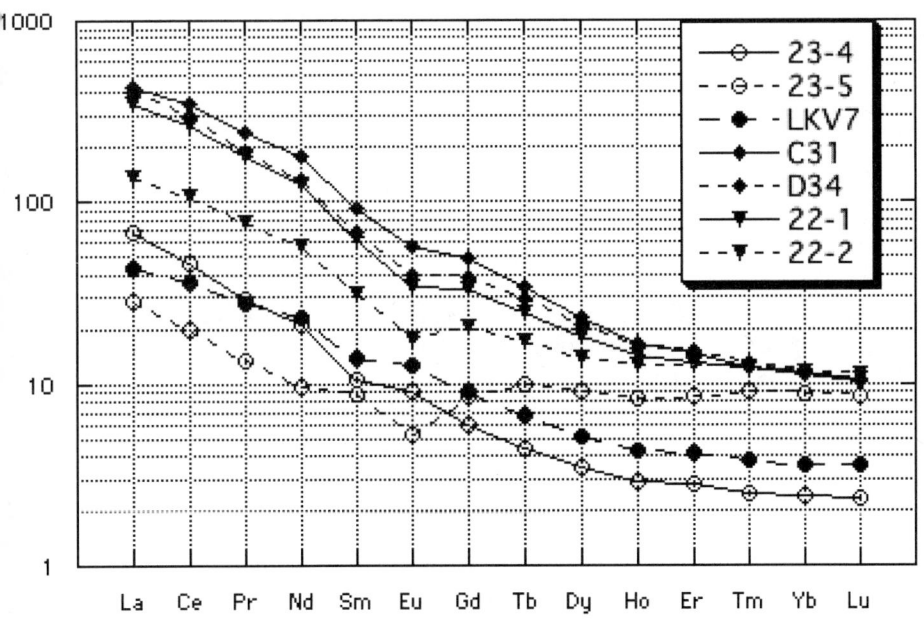

Figure 9. REE patterns of selected magnetite-series (solid symbols) and ilmenite-series (open symbol) granitoids. Sample numbers correspond to those in Table 1.

TABLE 2. CHEMICAL COMPOSITION OF BIOTITES FROM REPRESENTATIVE GRANITOIDS OF THE BARBERTON REGION

	Syn-tectonic granitoids				Late- and post-tectonic granitoids			
	Ilmenite-series		Magnetite-series		Magnetite-series		Magnetite-series	
	Theespruit		Kaap Valley		Boesmankop	Salisburgkop	Nelspruit	Nelspruit
Magnetic susceptibility	0.12×10^{-3}	0.05×10^{-3}	53.0×10^{-3}	19.9×10^{-3}	24.0×10^{-3}	6.5×10^{-3}		
Sample number	23–4	23–5	24–6	LKV–19	22–3	24–7B	C31	D34
	TTG–5	TTG–6	TTG–1	TTG–3				
Number of analysis	85	22	123	16	26	22	94	64
	ave (±1σ)	ave (±1σ)	ave (±1σ)	ave (±1σ)	ave (±1σ)	ave (±1σ)	ave (±1σ)	ave (±1σ)
SiO_2 (wt%)	36.99 (0.36)	35.73 (0.31)	37.16 (0.23)	37.08 (0.54)	39.53 (0.35)	37.66 (0.42)	38.35 (0.26)	37.54 (0.29)
Al_2O_3	14.70 (0.28)	17.26 (0.26)	15.04 (0.23)	14.91 (0.19)	11.25 (0.17)	14.31 (0.41)	14.32 (0.22)	14.47 (0.39)
TiO_2	1.65 (0.29)	1.70 (0.13)	1.74 (0.22)	2.26 (0.60)	0.87 (0.06)	1.41 (0.17)	2.35 (0.29)	2.73 (0.29)
FeO*	19.01 (0.47)	24.07 (0.42)	17.34 (0.44)	17.11 (0.34)	17.79 (0.82)	16.59 (0.37)	16.63 (0.44)	17.12 (0.38)
MnO	0.24 (0.04)	0.44 (0.04)	0.79 (0.06)	0.28 (0.04)	0.55 (0.04)	0.68 (0.04)	0.59 (0.06)	0.59 (0.06)
MgO	11.79 (0.32)	5.92 (0.16)	12.15 (0.37)	12.99 (0.50)	13.91 (0.64)	13.26 (0.41)	11.98 (0.37)	12.04 (0.33)
CaO	0.08 (0.05)	0.04 (0.03)	0.04 (0.03)	0.14 (0.32)	0.01 (0.01)	0.02 (0.01)	0.02 (0.03)	0.02 (0.04)
Na_2O	0.06 (0.03)	0.05 (0.02)	0.08 (0.23)	0.06 (0.03)	0.03 (0.02)	0.06 (0.02)	0.07 (0.02)	0.10 (0.02)
K_2O	9.16 (0.42)	9.28 (0.21)	9.53 (0.27)	9.37 (0.50)	9.76 (0.09)	9.82 (0.11)	9.54 (0.13)	9.48 (0.14)
Cl	0.02 (0.01)	0.01 (0.01)	0.08 (0.01)	0.05 (0.01)	0.22 (0.02)	0.02 (0.01)	0.02 (0.01)	0.04 (0.01)
=O	0.00	0.00	–0.02	–0.01	–0.05	0.00	0.00	–0.01
F	0.25 (0.10)	0.25 (0.13)	0.45 (0.13)	0.13 (0.09)	1.78 (0.21)	1.06 (0.16)	1.06 (0.16)	0.98 (0.13)
=O	–0.06	–0.06	–0.10	–0.03	–0.40	–0.24	–0.24	–0.22
Total	93.88	94.70	94.29	94.33	95.27	94.63	94.68	94.87
Si (atom, O = 22)	5.736 (0.035)	5.630 (0.023)	5.740 (0.029)	5.667 (0.050)	6.222 (0.027)	5.844 (0.062)	5.923 (0.025)	5.798 (0.027)
Al	2.688 (0.045)	3.206 (0.039)	2.739 (0.036)	2.686 (0.030)	2.087 (0.032)	2.617 (0.075)	2.607 (0.034)	2.634 (0.065)
Ti	0.192 (0.034)	0.201 (0.015)	0.203 (0.026)	0.260 (0.070)	0.104 (0.007)	0.164 (0.020)	0.273 (0.034)	0.317 (0.034)
Fe(2+)*	2.465 (0.060)	3.172 (0.068)	2.240 (0.058)	2.187 (0.038)	2.343 (0.118)	2.152 (0.046)	2.148 (0.058)	2.212 (0.051)
Mn	0.032 (0.006)	0.059 (0.006)	0.104 (0.008)	0.036 (0.006)	0.074 (0.006)	0.089 (0.005)	0.077 (0.008)	0.077 (0.009)
Mg	2.725 (0.077)	1.390 (0.036)	2.797 (0.082)	2.960 (0.115)	3.264 (0.139)	3.066 (0.095)	2.758 (0.084)	2.772 (0.077)
Ca	0.013 (0.009)	0.007 (0.004)	0.007 (0.005)	0.023 (0.053)	0.002 (0.002)	0.003 (0.002)	0.003 (0.007)	0.003 (0.007)
Na	0.017 (0.008)	0.016 (0.007)	0.023 (0.007)	0.018 (0.008)	0.010 (0.005)	0.017 (0.007)	0.021 (0.007)	0.030 (0.005)
K	1.812 (0.080)	1.866 (0.036)	1.878 (0.048)	1.826 (0.089)	1.961 (0.020)	1.943 (0.018)	1.879 (0.027)	1.868 (0.028)
Cl	0.006 (0.003)	0.002 (0.002)	0.021 (0.003)	0.012 (0.002)	0.060 (0.006)	0.004 (0.002)	0.005 (0.003)	0.010 (0.003)
F	0.123 (0.051)	0.126 (0.064)	0.222 (0.062)	0.063 (0.045)	0.885 (0.107)	0.520 (0.077)	0.519 (0.081)	0.479 (0.066)
Al (IV)	0.264	0.371	0.260	0.333	–0.222	0.156	0.077	0.202
XMg	0.525 (0.010)	0.305 (0.008)	0.555 (0.013)	0.575 (0.007)	0.582 (0.022)	0.588 (0.012)	0.562 (0.013)	0.556 (0.010)

*Indicates that the total Fe contents are expressed as FeO or Fe(2+).

that the Archean oceans were sulfate-rich and generally oxygenated, except in local basins.

The late-tectonic calc-alkaline granites of the Nelspruit and Mpuluzi batholiths are also composed of typical magnetite-series rocks. They are thought to have been generated by the partial melting of earlier TTGs (Anhaeusser and Robb, 1983; Robb et al., 1983). The Fe_2O_3/FeO weight ratios of earlier TTG magmas may have increased to >0.5 by the same processes as the Kaapvaal plutons (i.e., the partial melting of altered and oxidized subducted mid-oceanic-ridge basalt). The magnetite-series calc-alkaline granites (e.g., Nelspruit batholith) may have been generated from such a TTG lower crust by the heat from the upper mantle and water from dehydration of the altered subducting oceanic crust.

Mineralization Associated with Archean Granitoids

Highly oxidized magnetite-series magmas, rather than ilmenite-series magmas, are more favorable sources for the generation of ore-forming fluids for Cu, Mo, Pb, Zn, Ag, and Au deposits (e.g., Burnham and Ohmoto, 1980). Many such examples are found in the circum-Pacific Rim (Ishihara, 1998). In the Barberton region, orogenic gold deposits that occur in shear zones in the greenstones are the most prominent metallic mineralization type present, but these gold deposits are not considered genetically related to the adjacent TTG and calc-alkaline granitic occurrences.

Along the northern flank of the Barberton greenstone belt, there is a broad temporal overlap between a mineralized, felsic porphyry (ca. 3126 Ma Fairview Mine porphyry; de Ronde et al., 1991) and the ca. 3105 Ma magnetite-series Nelspruit

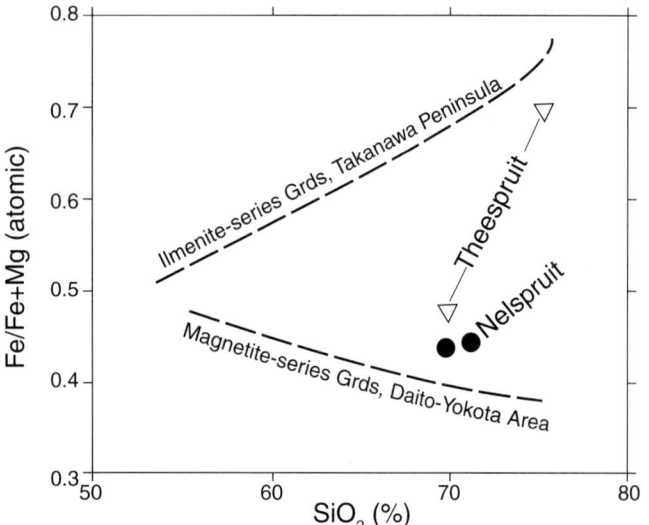

Figure 10. Fe/Mg ratios of the magnetite-series Nelspruit batholith. Japanese data are from Czamanske et al. (1981).

TABLE 3. CHEMICAL COMPOSITION OF APATITES FROM REPRESENTATIVE GRANITOIDS OF THE BARBERTON REGION

	Syn-tectonic granitoids								Late- and post-tectonic granitoids							
	Ilmenite-series				Magnetite-series				Magnetite-series				Magnetite-series			
	Theespruit				Kaap Valley				Boesmankop		Salisburgkop		Nelspruit		Nelspruit	
Magnetic susceptibility	0.12×10^{-3}		0.05×10^{-3}		53.0×10^{-3}		19.9×10^{-3}		24.0×10^{-3}		6.5×10^{-3}					
Sample number	23–4		23–5		24–6		LKV-19		22–3		24–7B		C31		D34	
	TTG–5		TTG–6		TTG–1		TTG–3									
Number of analysis	85		27		104		23		130		42		145		113	
	ave	(±1σ)	ave	(±1σ)	ave	(±1σ)	ave	(±1σ)	ave	(±1σ)	ave	(±1σ)	ave	(±1σ)	ave	(±1σ)
SiO_2 (wt%)	0.11	(0.14)	0.15	(0.06)	0.26	(0.08)	0.16	(0.04)	0.60	(0.17)	0.32	(0.08)	0.27	(0.16)	0.28	(0.18)
Al_2O_3	0.01	(0.05)	0.01	(0.02)	0.01	(0.02)	0.01	(0.01)	0.01	(0.01)	0.01	(0.01)	0.01	(0.01)	0.01	(0.04)
TiO_2	0.01	(0.02)	0.03	(0.03)	0.01	(0.02)	0.01	(0.02)	0.01	(0.02)	0.01	(0.02)	0.01	(0.01)	0.01	(0.02)
FeO*	0.13	(0.10)	0.36	(0.14)	0.13	(0.11)	0.22	(0.08)	0.23	(0.17)	0.16	(0.06)	0.19	(0.09)	0.09	(0.09)
MnO	0.03	(0.03)	0.53	(0.13)	0.10	(0.04)	0.04	(0.02)	0.04	(0.03)	0.10	(0.04)	0.07	(0.03)	0.09	(0.03)
MgO	0.01	(0.03)	0.01	(0.01)	0.01	(0.01)	0.01	(0.01)	0.01	(0.01)	0.01	(0.01)	0.01	(0.01)	0.01	(0.01)
CaO	55.62	(0.38)	54.65	(0.03)	55.44	(0.34)	55.51	(0.26)	54.06	(0.45)	54.77	(0.43)	55.01	(0.56)	55.02	(0.61)
Na_2O	0.00	(0.01)	0.10	(0.02)	0.09	(0.06)	0.03	(0.02)	0.04	(0.04)	0.11	(0.08)	0.06	(0.04)	0.02	(0.02)
K_2O	0.01	(0.02)	0.03	(0.02)	0.02	(0.02)	0.01	(0.01)	0.01	(0.02)	0.02	(0.01)	0.01	(0.01)	0.02	(0.01)
P_2O_5	39.88	(0.53)	39.61	(0.31)	39.04	(0.54)	39.16	(0.34)	38.41	(0.50)	38.80	(0.57)	38.99	(0.65)	38.00	(0.67)
Cl	0.02	(0.01)	0.01	(0.01)	0.03	(0.01)	0.11	(0.05)	0.01	(0.01)	0.01	(0.01)	0.01	(0.01)	0.01	(0.01)
= O	0.00		0.00		-0.01		-0.02		0.00		0.00		0.00		0.00	
F	4.21	(0.31)	4.72	(0.40)	4.47	(0.35)	3.63	(0.33)	4.69	(0.37)	4.86	(0.34)	4.39	(0.45)	4.54	(0.44)
= O	-0.95		-1.06		-1.01		-0.82		-1.06		-1.10		-0.99		-1.02	
SO_3	0.02	(0.02)	0.01	(0.01)	0.40	(0.19)	0.13	(0.04)	0.04	(0.11)	0.36	(0.24)	0.20	(0.07)	0.15	(0.04)
Total	99.11		99.15		99.00		98.19		97.10		98.44		98.23		97.22	

*Indicates that the total Fe contents are expressed as FeO or Fe(2+).

granite batholith (Kamo and Davis, 1994). But the late-tectonic Nelspruit granites may be too large in exposed dimensions, thus exposing deeper parts of the batholiths, to be genetically related to the porphyry that formed at a shallow crustal depth.

Several small cassiterite-bearing pegmatite dikes and veins occur in the Mpuluzi batholith close to the Swaziland border. Although tin mineralization is typically associated with reduced (ilmenite-series) and/or highly fractionated granites (Ishihara, 1981), the redox state of granite related to cassiterite mineralization in the Mpuluzi batholith is unknown.

The TTGs of the Barberton region are adakitic and similar to those in the Au-Cu mineralized regions of the Philippines (Sajona and Maury, 1998), northern Chile (Oyarzun et al., 2001), and Kitakami Mountains (Tsuchiya and Kanisawa, 1994; Ishihara and Murakami, 2004). However, an important difference is their low redox state, which may have limited the concentration of sulfur, and hence S-combined ore metals, to the magmatic fluids (Burnham and Ohmoto, 1980; Ishihara et al., 1988).

Oxidizing conditions of porphyry Cu-related magmas are represented by high SO_3 contents of apatites, such as those in the Philippines (Imai, 2001, 2002, 2004). Chlorine is an important metal carrier in magmatic fluids, and the Cl content of rock-forming apatite is a good indicator of the Cl contents of granitic magmas (Nedachi, 1980). EPMA analyses indicate that the SO_3 contents of apatite crystals in magnetite-series granitoids from the Barberton region approach 0.36 wt.% (Table 3), which is similar to the value of porphyry Cu-related intrusions from the Philippines (Imai, 2001, 2002). However, Cl contents of apatites from the Barberton granitoids are extremely low (below 0.1%) (Fig. 11). These low Cl concentrations in the magnetite-series granitoids were probably an important reason for the scarcity of S-combined metallic mineralizations in the Barberton region.

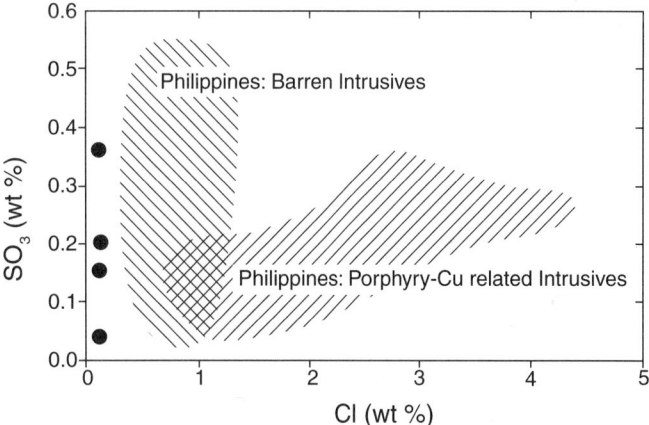

Figure 11. SO_3 and Cl contents of apatite crystals in the magnetite-series Nelspruit and Mpuluzi granitoids (dots). Original data are from Imai (2002), Ishihara et al. (2002b), and this study.

Porphyry Mineralization in Earth's History

Porphyry Cu–Mo and Cu-Au deposits are typically associated with oxidized magnetite-series magmatism, whereas Sn-W deposits generally occur with reduced ilmenite-series granitoids (Fig. 1). Because mineral exploration has been carried out much more extensively than regional granitoid-series studies, ore deposit data may provide a clue to the distributions of the two types of granitoids in Earth's history.

A histogram of the total Cu tonnages in all the discovered porphyry-type deposits, plotted against their geologic age, is shown in Figure 12. It may suggest that the abundance of magnetite-series granitoids, as approximated by porphyry Cu deposits, increased with younger geologic age. The Cenozoic peak in

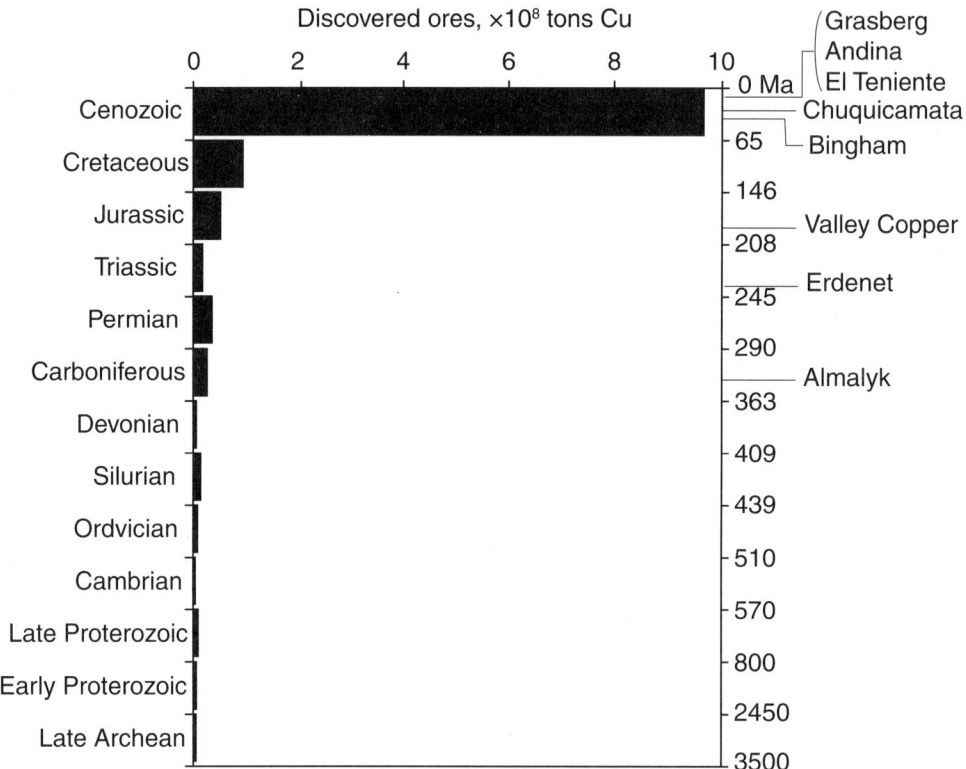

Figure 12. Tonnages of porphyry copper ores versus geologic time. Original data are from Singer et al. (2002).

Figure 12 comes mostly from young and super-large deposits at Grasberg-Ertzberg (3 Ma, 28 million tons [MT] Cu), El Teniente (4 Ma, 56.8 MT Cu), Andina (4 Ma, 37.2 MT Cu), Los Pelambres (10 Ma, 20.5 MT Cu), Chuquicamata (31–34 Ma, 55.2 MT Cu), Escondida (37–38 Ma, 34.3 MT Cu), and Bingham (39–40 Ma, 23.2 MT Cu) (the age and metal tonnage statistics are from Watanabe, 2003). Quaternary volcanoes of the circum-Pacific belt are composed predominantly of magnetite series, and porphyry-type deposits are thought to be forming below these volcanoes (e.g., Mt. Pinatubo; Imai et al., 1993, 1996; Hattori and Keith, 2001).

Most porphyry-related hydrothermal deposits are formed at shallow depths, typically less than ~5 km (Burnham and Ohmoto, 1980). Therefore, the observed trends of decreasing amounts of Cu deposits with older geologic age are likely due to preservation effects. Pillow lavas and high-level granitoids are still locally preserved in the Precambrian terrains. Copper deposits at Malanjkhand, central India (2400 Ma, 6.5 MT Cu), may or may not be porphyry-type (Sarkar et al., 1996; Panigrahi and Mookherjee, 1997). However, the oldest (3300 Ma) Mo-Cu deposit at the Coppin Gap, Pilbara Craton (102 MT; 0.152% Cu, 0.105% Mo; Jones, 1990), is a large porphyry-type deposit. A reconnaissance magnetic survey of the porphyry indicates that it is intermediate between ilmenite and magnetite series. These data, therefore, suggest that the formation of porphyry Cu-Mo and Cu-Au deposits, as well as magnetite-series granitoids, have occurred since at least ~3.3 billion years ago.

Sn deposits, which are typically associated with ilmenite-series granitoids, are found mostly in the Phanerozoic rock (Meyer, 1985). They are concentrated in the late Paleozoic (southwestern Europe) and Triassic–Jurassic (Southeast Asia), reflecting the predominance of ilmenite-series granitoids in these regions. Cenozoic-age Sn-bearing granitoids are limited to the Bolivian Miocene granitic belt and Miocene Outer Zone granitoids of southwestern Japan. Scarcity of the late Cenozoic Sn deposits is likely due to paucity of ilmenite-series granitoids during this period in the circum-Pacific region.

Sizable Sn deposits are known in Precambrian terrains. The late Proterozoic granites (1.0 Ga), of Rondonia, Brazil (Bettencourt and Dall'angol, 1995), and early Proterozoic granites (2.1 Ga) of the Bushveld complex host many cassiterite-pegmatite-greisen-veins. Archean granites also host cassiterite associated with Li and Ta pegmatites. At Greenbush, Western Australia, Sn-Ta-Li pegmatite sheet (2.5 Ga) produced in 1888–1987 the tin concentrate of 26,000 tons (~72% Sn), Ta_2O_5 concentrate of 2300 tons, low iron spodumene concentrate (72% Li_2O) of 39,700 tons and crude kaolin ore of 9500 tons (Hatcher and Clynick, 1990). Ta-pegmatite of the Mount Cassiterite orebody (2.9 Ga) in the Pilbara Craton contains 8785 tons Sn metal (Huston et al., 2001). Archean pegmatite in Zimbabwe is said to contain cassiterite with 114,000 tons Sn metal. Cassiterite mineralization therefore appears to be concentrated in Precambrian time, yet the examples are still too small to draw a definite conclusion.

CONCLUSIONS

Amitsoq-gray tonalite gneisses from West Greenland (ca. 3700 Ma) have Fe_2O_3/FeO ratios that average 0.33 ($n = 23$; Nutman and Bridgwater, 1986), thus being ilmenite series. However, because these tonalites were highly metamorphosed, and metamorphism generally lowers the Fe_2O_3/FeO ratios of rocks by reactions with graphite (Ohmoto and Kerrick, 1977), the original tonalites may have been intermediate or magnetite series. Archean granitoids from the Kaapvaal Craton that are older than ca. 3230 Ma are mostly reduced-type (ilmenite series) with a Fe_2O_3/FeO weight ratio of 0.3–0.5. Because their Fe_2O_3/FeO ratios appear to have been lowered by the assimilation of greenstones, however, their original magmas might also have been more oxidized than normal ilmenite-series magmas. The 3230 Ma Kaap Valley pluton and younger granitoids are mostly magnetite series, suggesting that by ca. 3200 Ma, oxidized protoliths existed in the Kaapvaal continental crust.

Source rocks for the Archean TTGs, as suggested by their REE patterns and geochemical characteristics, appear to have been subducted oceanic crust that was subjected to a garnet-bearing metamorphic grade. The late-tectonic calc-alkaline granitoids may have formed from the partial melting of such older Archean TTGs. Although these rocks form part of the magnetite series, they have low Cl contents. The deep erosion level of the granitoids and low Cl contents of the magmas were probably the primary reasons why no metallic deposits are associated with magnetite-series granitoids in the Barberton region.

Porphyry-type Cu (-Au or -Mo) deposits are the most representative mineralization associated with oxidized, magnetite-series magmatism. The oldest example may be the 3300 Ma Coppin Gap in the Pilbara Craton. The presence of such Cu deposits and the abundance of magnetite-series granitoids in the Kaapvaal Craton suggest that the processes of seawater-rock interaction at mid-oceanic ridges and granite magma generation in subduction zones, as well as the redox chemistry of ocean water, have been basically the same since at least ca. 3200 Ma.

ACKNOWLEDGMENTS

We are grateful to D.A. Singer for providing the world copper tonnage statistics and to S.E. Kesler, B.R. Frost, and E.J. Essene for helpful comments and suggestions on earlier manuscripts. Ohmoto acknowledges support from the National Science Foundation (EAR-9706279, EAR-0229556), the NASA Astrobiology Institute (NCC2-1057, CA#NNA04CC06A), and the NASA Exobiology Program (CA#NNG04GK00G).

REFERENCES CITED

Anhaeusser, C.R., 1973, The geology and geochemistry of the Archaean granites and gneisses of the Johannesburg-Pretoria Dome: Geological Society of South Africa Special Publication 3, p. 361–385.

Anhaeusser, C.R., 1976, Archean metallogeny in southern Africa: Economic Geology, v. 71, p. 16–45.

Anhaeusser, C.R., 1999, Archaean crustal evolution of the central Kaapvaal Craton, South Africa: Evidence from the Johannesburg Dome: South African Journal of Geology, v. 102, p. 303–322.

Anhaeusser, C.R., and Burger, A.J., 1982, An interpretation of U-Pb zircon ages for Archaean tonalitic gneisses from the Johannesburg-Pretoria granite dome: Transactions of the Geological Society of South Africa, v. 85, p. 111–116.

Anhaeusser, C.R., and Robb, L.J., 1980a, Magmatic cycles and the evolution of the Archaean granitic crust in the eastern Transvaal and Swaziland: Geological Society of Australia Special Publication 7, p. 457–467.

Anhaeusser, C.R., and Robb, L.J., 1980b, Regional and detailed field and geochemical studies of Archaean trondhjemitic gneisses, migmatites and greenstone xenoliths in the southern part of the Barberton Mountain Land, South Africa: Precambrian Research, v. 11, p. 373–397, doi: 10.1016/0301-9268(80)90073-X.

Anhaeusser, C.R., and Robb, L.J., 1983, Geological and geochemical characteristics of the Heerenveen and Mpuluzi batholiths south of the Barberton greenstone belt and preliminary thoughts on their petrogenesis: Geological Society of South Africa Special Publication 9, p. 131–151.

Bettencourt, J.S., and Dall'angol, R., eds., 1995, The rapakivi granites of the Rondonia tin province and associated mineralization: Para, Brazil, Federal University, Excursion Guide, 48 p.

Burnham, C.W., and Ohmoto, H., 1980, Late-stage processes of felsic magmatism: Mining Geology Special Issue 8, p. 1–11.

Czamanske, G.K., Ishihara, S., and Atkin, S.A., 1981, Chemistry of rock-forming minerals of the Cretaceous-Paleogene batholith in southwestern Japan and implications for magma genesis: Journal of Geophysical Research, v. 86, p. 10431–10469.

de Ronde, C.E.J., Kamo, S., Davis, D.W., de Wit, M.J., and Spooner, E.T.C., 1991, Field geochemical and U-Pb isotopic constraints from hypabyssal felsic intrusions within the Barberton greenstone belt, South Africa: Implications for tectonics and the timing of gold mineralization: Precambrian Research, v. 49, p. 261–280, doi: 10.1016/0301-9268(91)90037-B.

de Wit, M.J., Roering, C., Hart, R.J., Armstrong, R.A., de Ronde, C.E.J., Green, R.W.E., Tredoux, M., Peberdy, E., and Hart, R.A., 1992, Formation of an Archean continent: Nature, v. 357, p. 553–562, doi: 10.1038/357553a0.

Drummond, M.S., Defant, M.J., and Kepezhinskas, P.K., 1996, Petrogenesis of slab-derived trondhjemite-tonalite-dacite-adakite magmas: Transactions of the Royal Society of Edinburgh, Earth Sciences, v. 87, p. 205–215.

Faure, K., and Harris, C., 1991, Oxygen and carbon isotope geochemistry of the 3.2 Ga Kaap Valley tonalite, Barberton greenstone belt, South Africa: Precambrian Research, v. 52, p. 301–319, doi: 10.1016/0301-9268(91)90085-O.

Hatcher, M.I., and Clynick, G., 1990, Greenbushes tin-tantalum-lithium deposit, in Hughes, F.E., ed., Geology of the mineral deposits of Australia and Papua New Guinea: Melbourne, The Australian Institute of Mining and Metallurgy, p. 599–603.

Hattori, K.H., and Keith, J.D., 2001, Contribution of mafic melt to porphyry copper mineralization: evidence from Mount Pinatubo, Philippines, and Bingham Canyon, Utah, USA: Mineralium Deposita, v. 36, p. 799–806, doi: 10.1007/s001260100209.

Hawkesworth, C.J., and O'nions, R.K., 1977, The petrogenesis of some Archean volcanic rocks from southern Africa: Journal of Petrology, v. 18, p. 487–520.

Hunter, D.R., 1974, Crustal development in the Kaapvaal Craton, I, The Archean: Precambrian Research, v. 1, p. 259–294, doi: 10.1016/0301-9268(74)90002-3.

Huston, D.L., Blewett, R.S., Sweetapple, M., Brauhart, C., Cornelius, H., and Collins, P.L.F., 2001, Metallogenesis of the north Pilbara granite-greenstones, Western Australia—A field guide: Western Australia Geological Survey, Record 2001/11, 87 p.

Imai, A., 2001, Generation and evolution of ore fluids for porphyry Cu-Au mineralization of the Santo Tomas II (Philex) deposit, Philippines: Resource Geology, v. 51, p. 71–96.

Imai, A., 2002, Metallogenesis of porphyry Cu deposits of the western Luzon arc, Philippines: K-Ar ages, SO_3 contents of microphenocrystic apatite and significance of intrusive rocks: Resource Geology, v. 52, p. 147–161.

Imai, A., 2004, Variation of Cl and SO_3 contents of microphenocrystic apatite in intermediate to silicic igneous rocks of Cenozoic Japanese island arcs: Implications for porphyry Cu metallogenesis in the Western Pacific Island Arcs: Resource Geology, v. 54, p. 357–372.

Imai, A., Listanco, E.L., and Fujii, T., 1993, Petrologic and sulfur isotopic significance of highly oxidized and sulfur-rich magma of Mt. Pinatubo, Philippines: Geology, v. 21, p. 699–702.

Imai, A., Listanco, E.L., and Fujii, T., 1996, Highly oxidized and sulfur-rich magma of Mount Pinatubo: Implication for metallogenesis of porphyry copper mineralization in the western Luzon arc, in Newhall, C.G., and Punongbayan, R.S., eds., Fire and mud: Eruptions and lahars of Mount Pinatubo, Philippines: Quezon City, Philippine Institute of Volcanology and Seismology: Seattle, University of Washington Press, p. 865–874.

Ishihara, S., 1977, The magnetite-series and ilmenite series granitic rocks: Mining Geology, v. 27, p. 293–305.

Ishihara, S., 1981, Granitoid series and mineralization: Economic Geology, 75th anniversary volume, p. 458–484.

Ishihara, S., 1998, Granitoid series and mineralization in the Circum-Pacific Phanerozoic granitic belts: Resource Geology, v. 48, p. 219–224.

Ishihara, S., 2004, Redox state of granitoids relative to tectonic setting and Earth history: The magnetite/ilmenite series 30 years later: Transactions of the Royal Society of Edinburgh, Earth Sciences, v. 95, p. 23–33.

Ishihara, S., and Matsuhisa, Y., 1999, Oxygen isotopic constraints on the genesis of the Miocene Outer Zone granitoids in Japan: Lithos, v. 46, p. 523–534.

Ishihara, S., and Matsuhisa, Y., 2002, Oxygen isotopic constraints on the genesis of the Cretaceous-Paleogene granitoids in the Inner Zone of Southwest Japan: Bulletin of the Geological Survey of Japan, v. 53, p. 421–438.

Ishihara, S., and Murakami, H., 2004, Granitoid types related to Cretaceous plutonic Au-quartz vein and Cu-Fe skarn deposits, Kitakami Mountains, Japan: Resource Geology, v. 54, p. 281–298.

Ishihara, S., and Sasaki, A., 2002, Paired sulfur isotopic belts: Late Cretaceous-Paleogene ore deposits in the Inner Zone of Southwest Japan: Bulletin of the Geological Survey of Japan, v. 53, p. 461–477.

Ishihara, S., Hashimoto, M., and Machida, M., 2000, Magnetite/ilmenite-series classification and magnetic susceptibility of the Mesozoic-Cenozoic batholiths in Peru: Resource Geology, v. 50, p. 123–129.

Ishihara, S., Sasaki, A., and Terashima, S., 1988, Sulfur in granitoids and its role for the mineralization, in Proceedings of the 7th Quadrennial IAGOD (International Association of Genesis of Ore Deposits) Symposium: Stuttgart, Schweizerbart'sche Verlag, p. 573–581.

Ishihara, S., Anhaeusser, C.R., and Robb, L.J., 2002a, Granitoid-series evaluation of the Archean Johannesburg Dome granitoids, South Africa: Bulletin of the Geological Survey of Japan, v. 53, p. 1–9.

Ishihara, S., Anhaeusser, C.R., Robb, L.J., and Imai, A., 2002b, Granitoid-series in terms of magnetic susceptibility: A case study from the Barberton region, South Africa: Gondwana Research, v. 5, p. 581–589, doi: 10.1016/S1342-937X(05)70630-4.

Jones, C.B., 1990, Coppin Gap copper-molybdenum prospect, in Hughes, F.E., ed., Geology of the mineral deposits of Australia and Papua New Guinea, Volume I: Melbourne, Australasian Institute of Mining and Metallurgy, Monograph 14, p. 141–144.

Kamo, S.L., and Davis, D.W., 1994, Reassessment of Archean crustal development in the Barberton Mountain Land, South Africa, based on U-Pb dating: Tectonics, v. 13, p. 167–192, doi: 10.1029/93TC02254.

Lécuyer, C., and Ricard, Y., 1999, Long-term fluxes and budget of ferric iron: Implication for the redox states of the Earth's mantle and atmosphere: Earth and Planetary Science Letters, v. 165, p. 197–211, doi: 10.1016/S0012-821X(98)00267-2.

Lowe, D.R., 1994, Accretionary history of the Archean Barberton Greenstone Belt (3.55–3.22 Ga), southern Africa: Geology, v. 22, p. 1099–1102, doi: 10.1130/0091-7613(1994)022<1099:AHOTAB>2.3.CO;2.

Lowe, D.R., 1999, Geologic evolution of the Barberton greenstone belt and vicinity, in Lowe, D.R., and Byerly, G.R., eds., Geologic evolution of the Barberton Greenstone Belt, South Africa: Geological Society of America Special Paper 329, p. 287–312.

Meyer, C., 1985, Ore metals through geologic history: Science, v. 227, p. 1421–1427.

Meyer, F.M., Robb, L.J., Reimold, W.U., and de Bruiyn, H., 1994, Contrasting low- and high-Ca granites in the Archean Barberton Mountain Land, southern Africa: Lithos, v. 32, p. 63–76, doi: 10.1016/0024-4937(94)90021-3.

Nedachi, M., 1980, Chlorine and fluorine contents of rock-forming minerals of the Neogene granitic rocks in Kyushu, Japan: Mining Geology, Special Issue 8, p. 39–48.

Nutman, A.P., and Bridgwater, D., 1986, Early Archaean Amitsoq tonalites and granites of the Isukasia area, southern West Greenland: Development of the oldest-known sial: Contributions to Mineralogy and Petrology, v. 94, p. 137–148, doi: 10.1007/BF00592931.

Ohmoto, H., and Goldhaber, M.B., 1997, Sulfur and carbon isotopes, in Barnes, H.L., ed., Geochemistry of hydrothermal ore deposits (3rd edition): New York, John Wiley and Sons, p. 517–612.

Ohmoto, H., and Kerrick, D.M., 1977, Devolatilization equilibria in graphitic systems: American Journal of Science, v. 277, p. 1013–1044.

Oyarzun, R., Marquez, A., Lillo, J., Lopez, I., and Rivera, S., 2001, Giant versus small porphyry copper deposits of Cenozoic age in northern Chile: Adakitic versus normal calc-alkaline magmatism: Mineralium Deposita, v. 36, p. 794–798, doi: 10.1007/s001260100205.

Panigrahi, M.K., and Mookherjee, A., 1997, The Malanjkhand copper (+molybdenum) deposit, India: Mineralization from a low-temperature ore-fluid of granitoid affinity: Mineralium Deposita, v. 32, p. 133–149, doi: 10.1007/s001260050080.

Poujol, M., and Anhaeusser, C.R., 2001, The Johannesburg Dome, South Africa: New single zircon U-Pb isotopic evidence for early Archaean granite-greenstone development within the central Kaapvaal Craton: Precambrian Research, v. 108, p. 139–157, doi: 10.1016/S0301-9268(00)00161-3.

Poujol, M., and Robb, L.J., 1999, New U-Pb zircon ages on gneisses and pegmatite from south of the Maurchison greenstone belts, South Africa: South African Journal of Geology, v. 102, p. 93–97.

Poujol, M., Robb, L.J., Anhaeusser, C.R., and Gericke, B., 2003, A review of the geochronological constraints on the evolution of the Kaapvaal Craton, South Africa: Precambrian Research, v. 127, p. 181–213, doi: 10.1016/S0301-9268(03)00187-6.

Robb, L.J., and Anhaeusser, C.R., 1983, Chemical and petrogenetic characteristics of Archaean tonalite-trondhjemite gneiss plutons in the Barberton Mountain Land: Geological Society of South Africa Special Publication 9, p. 103–116.

Robb, L.J., Anhaeusser, C.R., and van Nierop, D.A., 1983, The recognition of the Nelspruit batholith north of the Barberton greenstone belt and its significance in terms of Archaean crustal evolution: Geological Society of South Africa Special Publication 9, p. 117–130.

Robb, L.J., Barton, J.M., Kable, E.J.D., and Wallace, R.C., 1986, Geology, geochemistry and isotopic characteristics of the Archean Kaap Valley pluton, Barberton Mountain Land, South Africa: Precambrian Research, v. 31, p. 1–36, doi: 10.1016/0301-9268(86)90063-X.

Sajona, F.G., and Maury, R.C., 1998, Association of adakites with gold and copper mineralization in the Philippines: Comptes Rendus de l'Académie des Sciences de Paris, Sciences de la Terre et des Planètes, v. 326, p. 27–34.

Sarkar, S.C., Kabiraj, S., Bhattacharya, S., and Pal, A.B., 1996, Nature, origin and evolution of the granitoid-hosted early Proterozoic copper-molybdenum mineralization at Malanjkhand: Mineralium Deposita, v. 31, p. 419–431, doi: 10.1007/s001260050049.

Sasaki, A., and Ishihara, S., 1979, Sulfur isotopic composition of the magnetite-series and ilmenite-series granitoids in Japan: Contributions to Mineralogy and Petrology, v. 68, p. 107–115, doi: 10.1007/BF00371893.

Shanks, W.C., Bischoff, J.L., and Rosenbauer, R.J., 1981, Seawater sulfate reduction and sulfur isotope fractionation in basaltic systems: Interaction of seawater with fayalite and magnetite at 200–350°C: Geochimica et Cosmochimica Acta, v. 45, p. 1977–1995, doi: 10.1016/0016-7037(81)90054-5.

Singer, D.A., Berger, V.I., Moring, B.C., 2002, Porphyry copper deposits of the world: Database, maps, and preliminary analysis: U.S. Geological Survey Open-File Report 02-268, 61 p.

Taylor, H.P., Jr., 1968, The oxygen isotope geochemistry of igneous rocks: Contributions to Mineralogy and Petrology, v. 19, p. 1–71, doi: 10.1007/BF00371729.

Thompson, R., and Oldfield, F., 1986, Environmental magnetism: London, Allen and Unwin, 227 p.

Tsuchiya, N., and Kanisawa, S., 1994, Early Cretaceous Sr-rich silicic magmatism by slab melting in the Kitakami Mountains, northeast Japan: Journal of Geophysical Research, v. 99, p. 22205–22220, doi: 10.1029/94JB00458.

Watanabe, Y., 2003, Porphyry copper deposits, in, Shikazono, N., Nakano, T., and Hayashi, K., eds., Resource and environmental geology: Tokyo, Society of Resource Geology, p. 35–44 [in Japanese].

Manuscript Accepted by the Society 29 October 2005

Geological Society of America
Memoir 198
2006

Secular variations of N-isotopes in terrestrial reservoirs and ore deposits

R. Kerrich
Department of Geological Sciences, University of Saskatchewan, 114 Science Place, Saskatoon S7N 5E2, Canada

Y. Jia*
CSIRO Exploration and Mining, School of Geosciences, Monash University, P.O. Box 28, Victoria 3800, Australia

C. Manikyamba
S.M. Naqvi
National Geophysical Research Institute, Hyderabad 500 007, India

ABSTRACT

New $\delta^{15}N$ analyses combined with a literature compilation reveal that shale kerogen, VMS-micas, and late-metamorphic vein micas show a secular trend from enriched values in the Archean, through intermediate values in Proterozoic terranes, to the Phanerozoic mode of 3‰–4‰. Kerogen in metashales from the 2.7 Ga Sandur Greenstone Belt, eastern Dharwar Craton, India, is characterized by $\delta^{15}N$ 13.1‰ ± 1.3‰, and C/N 303 ± 93. A second population has $\delta^{15}N$ 3.5‰ ± 0.9‰, and C/N 8 ± 0.4, close to the Redfield ratio of modern microorganisms, and is interpreted as precipitates of Proterozoic or Phanerozoic oilfield brines that penetrated the Archean basement. Kerogen from 1.7 Ga carbonaceous shales of the Cuddapah Basin average 5.0‰ ± 1.2‰, close to the mode at 3‰–4‰ for kerogen and bulk rock of Phanerozoic sediments. Biotites from late-metamorphic quartz-vein systems of the 2.6 Ga Kolar gold province, E. Dharwar Craton, that proxy for average crust, are also enriched at 14‰–21‰ for three samples, confirming that the N-budget of the hydrothermal fluids is dominated by sedimentary rocks. Muscovites from altered volcanic rocks in 2.7 Ga Abitibi belt VMS deposits have $\delta^{15}N$ 12‰–20‰, in keeping with published data for shale kerogen from the same terrane, whereas equivalents in the 1.8 Ga Jerome VMS span 11.7‰–14.1‰.

^{15}N-enriched values in Precambrian rocks cannot be caused by N-isotopic shifts due to metamorphism or Rayleigh fractionation because (1) pre-, and post-metamorphic samples from the same terrane are both enriched in ^{15}N; (2) there is no covariation of $\delta^{15}N$ with N, C/N ratios, or metamorphic grade; and (3) the magnitude of fractionations of 1‰ (greenschist) to 3‰ (amphibolite facies) during progressive metamorphism of sedimentary rocks, as constrained from empirical observations and experimental studies, is very small. Nor can ^{15}N-enriched values stem from long-term preferential diffusional loss of ^{14}N, as samples were selected from terranes where $^{40}Ar/^{39}Ar$ ages are within a few million years of concordant U-Pb ages; nitrogen is structurally bound in micas, whereas Ar is not.

It is possible that the ^{15}N-enriched values stem from a different N-cycle in the Archean, with large biologically mediated fractionations, yet the magnitude of the

---

*Corresponding author, present address: Silvercorp Metal Inc., 1588-609 Granville Street, Vancouver, B.C., V7Y 1G5 Canada; yefei_jia@yahoo.com.

Kerrich, R., Jia, Y., Manikyamba, C., and Naqvi, S.M., 2006, Secular variations of N-isotopes in terrestrial reservoirs and ore deposits *in* Kesler, S.E., and Ohmoto, H., eds., Evolution of Early Earth's Atmosphere, Hydrosphere, and Biosphere—Constraints from Ore Deposits: Geological Society of America Memoir 198, p. 81–104, doi: 10.1130/2006.1198(05). For permission to copy, contact editing@geosociety.org. ©2006 Geological Society of America. All rights reserved.

fractionations between atmospheric N_2 and organic nitrogen observed exceeds any presently known, and chemoautotrophic communities tend to depleted values. Earlier results on Archean cherts show a range of $\delta^{15}N$ from –6‰ to 30‰. Given the temporal association of chert–banded iron formation (BIF) with mantle plumes, the range is consistent with mixing between mantle N_2 of –5‰ and the ^{15}N-enriched marine reservoir identified in this study. The ^{15}N-enriched Archean atmosphere-hydrosphere reservoir does not robustly constrain Archean redox-state. We attribute the ^{15}N-enriched reservoir to a secondary atmosphere derived from CI-chondrite-like material and comets with $\delta^{15}N$ of +30‰ to +42‰. Shifts of $\delta^{15}N$ to its present atmospheric value of 0‰ can be accounted for by a combination of early growth of the continents with sequestration of atmospheric N_2 into crustal rocks, and degassing of mantle N ~–5‰. If Earth's surface environment became oxygenated ca. 2 Ga, then there were no associated large N-isotope excursions.

Keywords: nitrogen, isotopes, kerogen, hydrothermal-systems, gold, VMS, Precambrian, Paleozoic, secular change.

INTRODUCTION AND SCOPE

The present nitrogen cycle is quite well documented, including N-isotope fractionations accompanying organic and inorganic transfers of N between terrestrial reservoirs (Delwiche and Steyn, 1970; Macko et al., 1987; Rau et al., 1987; Williams et al., 1995; Kao and Liu, 2000). There are abundant data on modern organic compounds in the biosphere and sediments (Peters et al., 1978; Sweeney et al., 1978; Mazuka et al., 1991; Sadofsky and Bebout, 2004, and references therein), and Phanerozoic rocks (Haendel et al., 1986; Bebout and Fogel, 1992; Williams et al., 1995; Mingram and Bräuer, 2001, and references therein). Holloway and Dahlgren (2002) give a recent review.

The isotopic compositions of N in the mantle (–5‰), organic compounds in sediments (0‰ to –6‰ with a mean of –4‰), atmosphere (0‰), Phanerozoic black shales (kerogen, 3‰–4‰), and granites (range 0 to10‰) are well constrained (Fig. 1). Isotopic differences in terrestrial reservoirs stem from (1) nitrification, denitrification, N-limitation, and other metabolic or inorganic reactions of the near-surface N-cycle (Delwiche and Steyn, 1970; Wada et al., 1975; Sweeney et al., 1978; Saino and Hattori, 1980; Macko et al., 1987; Rau et al., 1987; Hoch et al., 1994; Pinti and Hashizume, 2001; Lehmann et al., 2002); (2) equilibrium fractionation in the geosphere, for example fluid-rock interaction (Hanschmann, 1981); and (3) kinetic effects associated with mantle degassing (Marty and Zimmermann, 1999; Cartigny and Ader, 2003, and references therein), or Rayleigh fractionation during metamorphic devolatilization (Bebout and Fogel, 1992; Mingram and Bräuer, 2001) (Fig. 1).

However, little is known about the early evolution of the N-cycle because records in Archean and Proterozoic rocks are sparse and the oxidation state of the Archean atmosphere-hydrosphere system is uncertain (Ohmoto, 1997, 2004; Holland, 1999; Phillips et al., 2001). Only a few data for N contents and N-isotopes have been published on Archean shales (Zhang, 1988; Jia and Kerrich, 2000, 2004a, 2004b), cherts and stromatolite-bearing sediments (Gibson et al., 1985, 1986), and chert–iron forma-

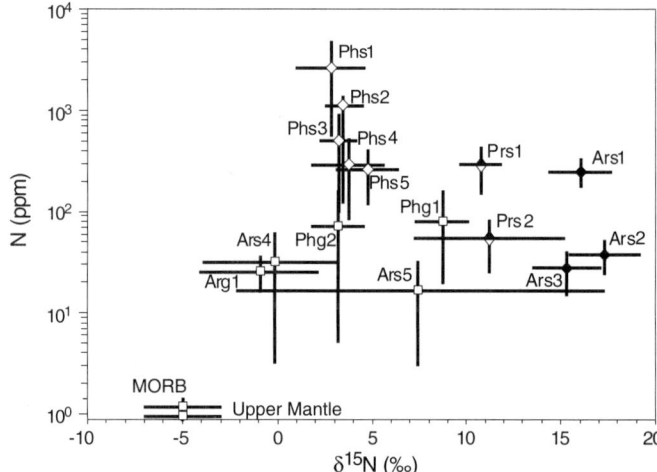

Figure 1. Nitrogen isotope compositions and concentrations in various geological reservoirs. Data represent mean value plus one standard deviation from the following sources: Archean (Ars): shale kerogen, Ars1 and Ars2 (Jia and Kerrich, 2004b) and Ars3 (Jia and Kerrich, 2000); Meso- and Neoarchean chert kerogen, Ars4 and Ars5 (Beaumont and Robert, 1999). Proterozoic shale kerogen (Prs): Prs1 (Jia and Kerrich, 2004b) and Prs2 (Boyd and Philippot, 1998; Haendel et al., 1986). Phanerozoic sediments and sedimentary rocks (Phs): Phs1 (Sephton et al., 2002), Phs2 (Williams et al., 1995; Kao and Liu, 2000), Phs3 (Bebout and Fogel, 1992; Busigny et al., 2003), Phs4 (Haendel et al., 1986; Mingram and Bräuer, 2001), and Phs5 (Peters et al., 1978). Archean granitoids (Arg): Arg1 (Jia and Kerrich, 1999, 2000); Phanerozoic granite (Phg): Phg1 (Boyd et al., 1993), Phg2 (Bebout et al., 1999). Mid-oceanic ridge basalt (MORB) source N (1–2 ppm and –5 ± 2‰) and upper mantle N (0.27 ± 0.16 ppm and –5 ± 2‰; Marty and Dauphas, 2003, and references therein).

tion (Hayes et al., 1983; Sano and Pillinger, 1990; Beaumont and Robert, 1999; Pinti et al., 2001a). N-isotope data on hydrothermal ore deposits, Precambrian or Phanerozoic, are limited (Jia and Kerrich, 1999, 2000; Jia et al., 2001, 2003a).

Nitrogen-isotopes have been used to address a variety of questions: (1) the isotopic composition of the Archean atmosphere (Sano and Pillinger, 1990; Beaumont and Robert, 1999); (2) secular variation of redox state of the atmosphere and oceans (Beaumont and Robert, 1999); (3) bacterial metabolic pathways (Pinti and Hashizume, 2001; Pinti et al., 2001a); (4) origin of Earth's atmosphere-hydrosphere (Javoy, 1998; Tolstikhin and Marty, 1998; Jia and Kerrich, 2004a, 2004b); (5) the N-isotope characteristics of Archean sedimentary rocks (Hayes et al., 1983; Zhang, 1988; Jia and Kerrich, 2000, 2004a, 2004b); (6) sources of hydrothermal fluids involved in orogenic, or shear zone-hosted mesothermal, gold deposits (Jia and Kerrich, 1999, 2000; Jia et al., 2001, 2003a); (7) nitrogen budgets in convergent margins (Bebout and Fogel, 1992; Sadofsky and Bebout, 2004); and (8) recycling of sedimentary rocks into the mantle (Marty and Dauphas, 2003).

Several observations emerge from the limited database: (1) kerogen in Mesoarchean (3.4–2.9 Ga) cherts are ^{15}N-depleted relative to Neoarchean (2.9–2.5 Ga) counterparts; (2) kerogen in Archean black shales is ^{15}N-enriched compared to Mesoarchean cherts and Phanerozoic equivalents; (3) the δ^{15}N of hydrothermal K-micas in orogenic gold deposits of Phanerozoic accretionary terranes is comparable to that of contemporaneous shales; and (4) K-micas in Archean gold deposits are as enriched as contemporaneous shale kerogen in ^{15}N. Consequently, Archean cherts and shales may sample different N-reservoirs, and shales record a secular variation of δ^{15}N (Fig. 1).

In this paper, new N-isotope data are reported for 2.7 Ga and 1.8 Ga carbonaceous shales from India to further test for secular variations. The first data for 2.7 and 1.8 Ga volcanic hosted massive base metal (Cu-Zn-Pb) sulfide (VMS) deposits are also presented, together with new data for hydrothermal K-micas from the ca. 2.6 Ga Kolar gold province, India. We compile existing data together with the new results to evaluate the δ^{15}N of various lithologies through time. From this database we consider the origin of N in orogenic gold, rare metal pegmatite, and VMS deposits, examine the implications of secular variations of δ^{15}N, and address the question of whether N-isotopes record "the great oxygenation event" ca. 2.3 Ga.

GEOLOGICAL SETTING

Carbonaceous Shales

Carbonaceous metashales were sampled from the 2.7 Ga Sandur Greenstone Belt (SGB), and the Paleoproterozoic Cuddapah Basin, India (Fig. 2). The Sandur belt is one of a series of composite tectonostratigraphic supracrustal terranes, separated by granitoids, in the eastern Dharwar Craton (Manikyamba et al., 1997). The eastern SGB is dominated by arc-related tholeiitic to calc-alkaline flows. Minor sedimentary units include polymictic conglomerate, graywacke, and carbonaceous shales. Rhyolites have zircon U-Pb ages of 2658 ± 14 Ma using the SHRIMP technique (Nutman et al., 1996). The western SGB is prevalently 2.7 Ga high-Mg basalts and komatiites. Shales were sampled from Vibhutigudda and Bhimangundi (Fig. 2). Fossil cyanobacteria have been described from carbonaceous cherts at the former locality (Naqvi et al., 1987; Venkatachala et al., 1990). The eastern arc and western plateau terranes accreted post 2.6 Ga (Manikyamba et al., 1997; Naqvi et al., 2002). Metamorphic grade varies from greenschist facies in the west to mid-amphibolite facies in the east (Manikyamba et al., 1997). Deformation is of low intensity except proximal to faults or shear zones.

The intracratonic Cuddapah Basin developed over 1.9–1.7 Ga (Fig. 2). Siliciclastic rocks are prevalent in this Proterozoic sequence, including conglomerates, current-bedded and rippled arenites, and diverse shales (Nagaraja-Rao et al., 1987). Well-preserved stromatolitic units and microfossils in cherts associated with stromatolites are present (Schopf and Prasad, 1978; Nagaraja-Rao et al., 1987). Samples of carbonaceous shales were obtained from near Mangampeta and Marcapur. Metashales are at prehnite-pumpellyite facies in the former locality, and at greenschist facies in the latter. The basin's western margin lies unconformably on Archean craton, whereas the eastern margin tectonically underlies the Eastern Ghat Mobile Belt (EGMB), which was thrust over the basin. From dating of anorthosite, alkali plutons, and granitoids emplaced along the accretionary zone, accretion of the EGMB, which was part of Antarctica before the breakup of Gondwanaland, to the Dharwar Craton is estimated to have occurred ca. 1600 Ma (Dasgupta and Sengupta, 2003). Carbonaceous shales were sampled from fresh rock in road or rail cuttings for both the Sandur belt and Cuddapah Group.

Kolar Gold Province

The Kolar terrane is one of several supracrustal greenstone sequences in the eastern Dharwar Craton. Tholeiitic basalts and komatiites dated at 2.7 Ga are prevalent, at amphibolite facies. Gold mineralization was coeval with brittle-ductile deformation (Hamilton and Hodgson, 1986) associated with accretionary tectonics (Balakrishnan et al., 1999). Biotite-rich alteration selvedges bounding gold-bearing quartz-calcite veins were obtained from the Oriental reef (see Siddaiah and Rajamani, 1989, and references therein). Nitrogen, as NH_4^+, may substitute for K given similar valence and ionic radius (Honma and Itihara, 1981; Bos et al., 1988). Accordingly, K-silicates are preferred minerals for N-isotope studies of hydrothermal ore systems.

VMS Deposits

VMS deposits reflect the conjunction of magmatic, hydrothermal, and sedimentary processes proximal to the seafloor. The Abitibi greenstone, composite, supracrustal terrane, dated to ca. 2.7 Ga, is the largest and best-preserved greenstone belt, with several VMS districts (for a review see Jackson and Fyon, 1991). Metamorphic grade is prehnite-pumpellyite to greenschist facies, and deformation intensity is low except proximal to regional shear zones. The VMS deposits formed on the seafloor during a hiatus of bimodal volcanism in several areas, notably Kidd Creek, Ontario, and the Noranda and Matagami districts of Que-

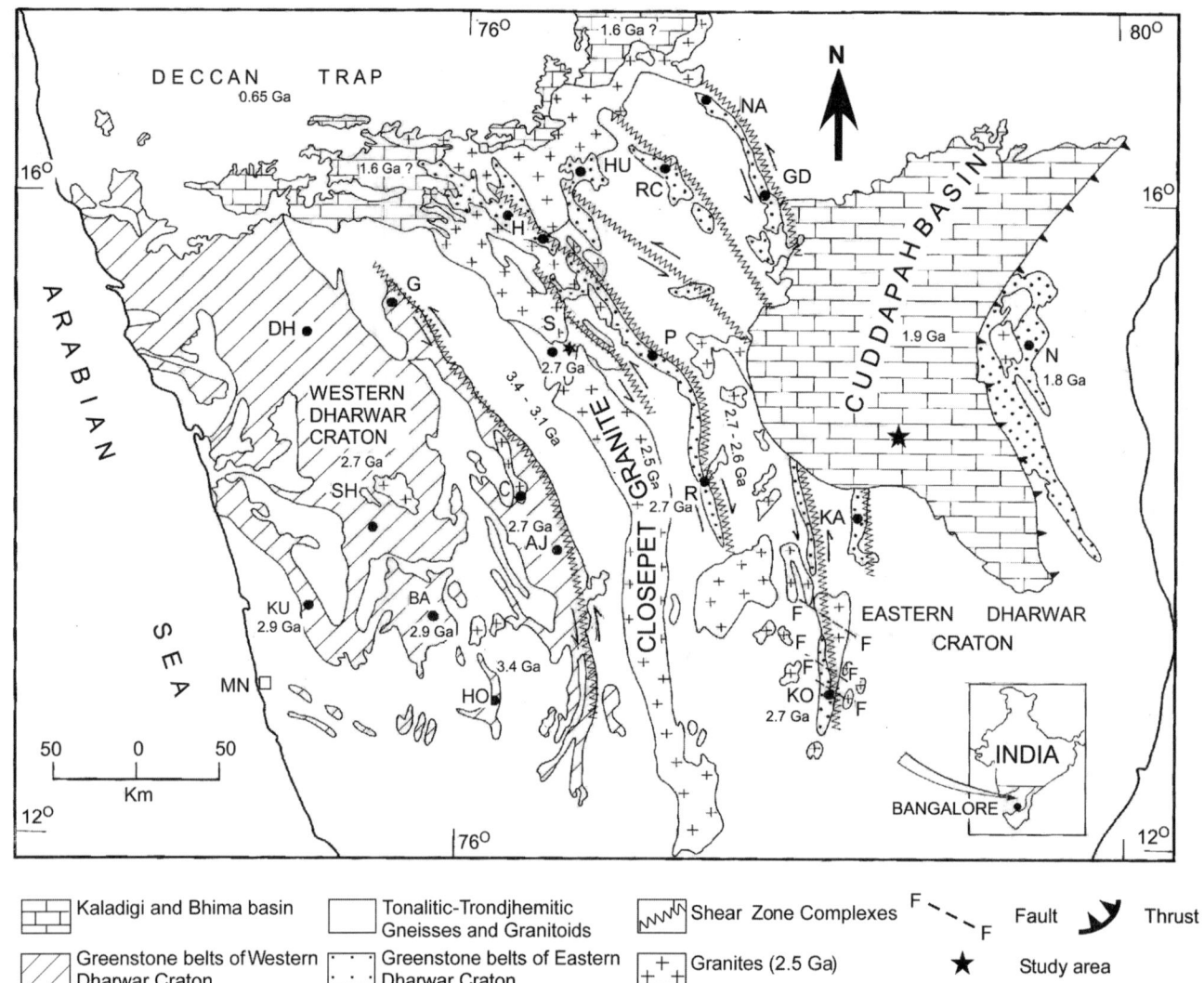

Figure 2. Simplified geological map of the Dharwar Craton showing the distribution of greenstone belts and shear zone complexes from the western and eastern Dharwar Craton. (AJ) Ajjanahalli, (BA) Bababudan, (C) Chitradurga, (DH) Dharwar, (G) Gadag, (GD) Gadwal, (H) Hungund, (HO) Holenarsipur, (HU) Hutti, (KA) Kadri, (KO) Kolar, (KU) Kudremukh, (MN) Mangalore, (N) Nellore, (NA) Narayanpet, (P) Penakacherla, (R) Ramagiri, (RC) Raichur, (S) Sandur, (SH) Shimoga. Inset shows the location of the main map (modified after Sreeramachandra Rao, 2001).

bec. Alteration of footwall volcanic rocks, by modified seawater-derived hydrothermal fluids, generated domains of muscovite and chlorite bearing alteration (Franklin et al., 1981). Muscovite-rich samples were obtained from mafic volcanic flows in the footwall of the 2.7 Ga Kidd Creek VMS deposit, Ontario, and from felsic pyroclastic units subjacent to the Amulet deposit, Noranda, and Mattagami Lake deposit, Matagami, Quebec. Geological relationships of Abitibi VMS deposits have been reviewed by Franklin et al. (1981), Bleeker et al. (1999), and Hannington et al. (1999). Seawater-rock ratios in the footwall are known to be large compared to possible mantle or magmatic contributions (Beaty and Taylor, 1982; Costa et al., 1983).

The Jerome VMS deposit is located in the upper of two cycles of bimodal magmatism of the 1.8 Ga Ash Creek Group, Arizona. Seafloor massive sulfides and mineralized breccias are present, and "black smoker" chimneys are preserved. Muscovite-rich, hydrothermally altered felsic volcanic rocks were sampled from the footwall. Metamorphic grade is greenschist facies (Anderson et al., 1971; Sangster and Scott, 1976; Lindberg and Gustin, 1987).

SAMPLE DESIGN

Studies of progressively metamorphosed sedimentary rocks show shifts of ~1‰ in $\delta^{15}N$ from protoliths to greenschist facies counterparts, and ≤3‰ to amphibolite facies (Haendel et al., 1986; Bebout and Fogel, 1992; Mingram and Bräuer, 2001; Busigny et al., 2003; Jia, 2004). Alternatively, some authors attributed variations of 27‰–36‰ in their data sets to shifts due to Rayleigh fractionation accompanying metamorphism (Beaumont and Robert, 1999; Pinti et al., 2001a). The former studies are all of siliciclastic sedimentary rocks whereas the latter are of kerogen from chert–banded iron formation (BIF). Consequently, differences between the studies could be sample or environment dependent, rather than due to metamorphism, or alternatively they could reflect assumptions in Rayleigh modeling.

Given the time-series and spatial association of chert–iron formation with volcanic sequences erupted from mantle plumes (Isley and Abbott, 1999; Condie et al., 2001), mantle N may be incorporated into kerogen in chert–iron formation, but probably not in distal carbonaceous shales. Accordingly, we selected carbonaceous shales distal from chert-BIF to obtain a marine biogenic signature. The magnitude of metamorphic shifts of $\delta^{15}N$ values was estimated by analysis of pre-metamorphic and late-metamorphic materials from the same terrane. Analyses of fine-grained hydrothermal K-micas from VMS deposits allow comparison with published data for kerogen in carbonaceous shales from the same 2.7 Ga terrane. In turn, data from these two types of pre-metamorphic samples are compared with late-metamorphic hydrothermal micas from the Kolar gold deposit, and with published data for micas from 2.7 Ga late-metamorphic gold deposits. Alteration associated with gold deposits overprints regional metamorphic fabrics and the deposits retain primary fluid inclusions, equilibrium quartz-muscovite oxygen isotope fractionations, and upper plateau $^{40}Ar/^{39}Ar$ ages. Accordingly the deposits have not been overprinted by a later metamorphic event (Kerrich and Cassidy, 1994; McCuaig and Kerrich, 1998).

ANALYTICAL METHODS

Separation and Analysis

Previous studies of Precambrian carbonaceous sedimentary rocks revealed low bulk rock N contents. Accordingly, kerogen was separated for analysis of N and $\delta^{15}N$, revealing that kerogen N dominated the whole rock N-budget. Minor N is likely to be in K-silicates (Hayes et al., 1983; Beaumont and Robert, 1999; Jia and Kerrich, 2004b). Kerogen was separated using the following procedure. All selected sedimentary rock samples were washed in dichloromethane-ethanol to remove modern organic contamination, ground in a steel mortar to <50 μm, and then treated with HCl and HF to remove carbonates and silicates using the technique of Durand and Nicaise (1980). All kerogens in this study have been screened by X-ray diffraction (XRD); graphite peaks were at or below detection, in keeping with greenschist facies shale and the results of Landis (1971). Pure muscovite separates, where N as NH_4^+ substitutes for K, from gold and VMS deposits were obtained by standard mineral separation procedures.

The analytical techniques used in this study involved a high-precision continuous flow-isotope ratio mass spectrometer (CF-IRMS) at the Soil Science Laboratory, University of Saskatchewan, and followed the techniques used by Jia and Kerrich (2000). Analytical precision (reproducibility, 1σ; $n \geq 3$) is typically ~0.3‰ for kerogen $\delta^{15}N$, ~0.5‰ for muscovite $\delta^{15}N$, and ~0.3‰ for kerogen $\delta^{13}C$. The long-term reproducibilities for international nitrogen isotope standard materials are as follows: IAEA-N1 = 0.54 ± 0.07‰ ($n = 15$, accepted value 0.53‰); IAEA-N2 = 20.35 ± 0.08‰ ($n = 10$, accepted value 20.41‰); and for the internal laboratory standard material BLN.SOIL, 5.20 ± 0.21‰ ($n = 20$, accepted value 5.15‰). Ten replicate analyses of "in house" muscovite separates of samples KAII-1 and CD11-21-1 yielded mean $\delta^{15}N$ values of 19.4‰ ± 0.09‰ and 3.4‰ ± 0.06‰, respectively (Jia et al., 2003b). Nitrogen and carbon concentrations were obtained from each sample based on system calibration using known standards. Blanks for this technique are <0.075 μg N_2 for routine runs of the types of samples analyzed in this study. Isotope data are reported in standard δ-notation relative to atmospheric N_2 for nitrogen and to the Peedee Belemnite limestone (PDB) standard for carbon.

Isobaric Interferences and Modern Organic Contamination

Studies show that isobaric interference of carbon monoxide (CO) may produce erroneous nitrogen isotope ratios. One percent CO in an analyte would cause ~7‰ errors in $\delta^{15}N$ (Beaumont et al., 1994). The analytical approaches used in this study have carefully eliminated any such isobaric interferences.

Contamination by organic nitrogen compounds during the preparation of the kerogen residue could change the initial isotopic composition. However, samples were treated with dichloromethane-ethanol, as in other studies of Precambrian rocks (Beaumont and Robert, 1999). Consequently, the very positive $\delta^{15}N$ values of Archean samples cannot be attributed to contamination because modern organic nitrogen compounds have $\delta^{15}N$ of close to 0‰, or negative down to −6‰ (Nadelhoffer and Fry, 1988; Sachs and Repeta, 1999; Kao and Liu, 2000), and modern kerogen ranges from 1‰ to 6‰ (Williams et al., 1995; Ader et al., 1998; Kao and Liu, 2000; Sephton et al., 2002). Also, several samples have the extremely ^{13}C-depleted values of −33‰ to −48‰, characteristic of some Archean kerogen (Wellmer et al., 1999); accordingly, significant contamination by modern hydrocarbons can be ruled out.

Total Inorganic Carbon (TIC)/Total Organic Carbon (TOC)

Total inorganic carbon, and total organic carbon were determined on carbonaceous shales using a Leco CR-12 carbon analyzer, following the procedure of Wang and Anderson (1994).

RESULTS

Data are reported in Table 1 and Figure 3. Also shown in Tables 2 and 3, for comparison, are compilations of recent data on nitrogen contents and $\delta^{15}N$ values of Archean to Phanerozoic siliciclastic sediments, chert-BIF, granitoids, and ore deposits.

TABLE 1. NITROGEN AND CARBON ISOTOPIC COMPOSITIONS AND C/N ATOMIC RATIOS OF KEROGEN FROM INDIAN PRECAMBRIAN CARBONACEOUS SHALES, AND FROM HYDROTHERMAL K-MICAS VMS AND GOLD DEPOSITS

Era	Sample no.	N (ppm)	$\delta^{15}N$ (‰)	C (ppm)	$\delta^{13}C$ (‰)	C/N (atomic ratio)	TOC/TIC
Carbonaceous shales[†]							
Proterozoic							
Mangambeta Locality (P3)							
	HCLCB1	454	5.6	15868	−30.3	41	
	HCLCB2	506	5.9	15322	−30.7	35	
	HCLCB3	419	5.5	14809	−29.9	41	
	HCLCB4	464	6.4	8756	−30.6	22	
	HCLCB6	370	6.8	11173	−31.0	35	
	Mean ± 1standard derivation	442 ± 51	6.0 ± 0.6	13,185 ± 3085	−30.5 ± 0.4	35 ± 8	
Marcapur locality (P3)							
	HCLCB8	289	3.8	26297	−28.6	106	
	HCLCB9	334	4.4	28111	−28.4	98	
	HCLCB10	415	4.1	46828	−28.5	132	
	HCLCB11	385	3.1	37070	−28.6	112	
	HCLCB12	313	4.0	8857	−27.5	33	
	Mean ± 1standard derivation	347 ± 52	3.9 ± 0.5	297,430 ± 14,105	−28.3 ± 0.5	96 ± 37	
Archean							
Bhimangundi Locality (P2)							
	HCLC1	838	3.4	5690	−32.0	8	0.1505
	HCLC4	847	3.1	5810	−32.0	8	0.1405
	HCLC9	811	2.6	4980	−31.9	7	0.1598
	HCLC10	867	4.7	5522	−32.0	7	0.1462
	Mean ± 1standard derivation	841 ± 24	3.5 ± 0.9	5500 ± 365	−32.0 ± 0.1	8 ± 0.4	0.149 ± 0.008
Vibutigudda Locality (P1)							
	HCLC11	88	14.6	19233	−28.3	256	0.0241
	HCLC12	73	14.1	11197	−26.0	179	0.0309
	HCLC14	80	13.0	17729	−26.3	259	0.0575
	HCLC17	57	11.8	21165	−26.7	435	0.0591
	HCLC19	64	13.4	20807	−28.1	380	0.0516
	HCLC22	76	11.4	19952	−28.3	305	0.0451
	Mean ± 1standard derivation	73 ± 11	13.1 ± 1.3	18,345 ± 3710	−27.3 ± 1.1	303 ± 93	0.045 ± 0.014
VMS deposit							
Archean							
Kidd Creek deposit							
	K1441	15	18.6				
	K9943	10	16.3				
Ansil deposit							
	91GIA	25	16.4				
	30EA	16	12.4				
Matagami							
	M80-7	14	20.1				
	M80-8	35	18.4				
	M80-9	28	12.9				
Proterozoic							
Jerome							
	J92-A	47	12.7				
	J92-B	33	14.1				
	J92-C	28	11.7				
	J92-D	39	12.4				
Gold deposits							
Archean							
Kolar							
	K432	18	19.4				
	K577	32	21.0				
	K616	47	13.7				

[†]N and C contents referenced to whole rock powders.

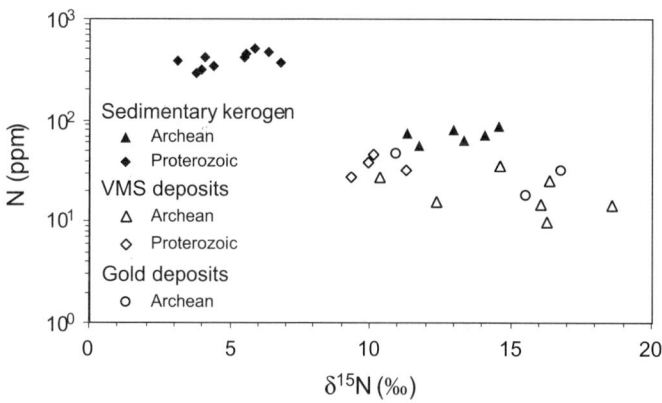

Figure 3. Variations of N contents and nitrogen isotopic compositions of Indian Precambrian carbonaceous shales and the Kolar gold deposit, and Precambrian VMS deposits. Uncertainties for $\delta^{15}N$ values and N content are all smaller than the plot symbols. Data sources are in Table 1.

Precambrian Kerogen

Carbonaceous shales from the Archean Dharwar Craton have two compositional-isotopic populations. The six kerogens from Vibutigudda (P1) are characterized by relatively uniform N (73 ± 11 ppm) and C (18,345 ± 3710 ppm) contents, with C/N ratios of 179–435. The $\delta^{15}N$ values (12‰ to 15‰) are much higher than those of most Phanerozoic sedimentary rocks (1‰–6‰; Table 3). Four kerogens from Bhimangundi define a second population (P2); they feature greater N (841 ± 24 ppm) but lower C (5500 ± 365 ppm) contents, giving uniform C/N ratios of 7–8. Both the $\delta^{15}N$ values (3.5‰ ± 0.9‰) and $\delta^{13}C$ (−32.0‰ ± 0.1‰) are lower than counterparts from Vibutigudda (Table 1). Proterozoic carbonaceous shales constitute a third population (P3). Their average N (395 ppm) and C (21,309 ppm) contents are greater than in P1, but average C/N ratios (73) and $\delta^{15}N$ are lower (Table 1, Fig. 3). Shale kerogens at the Mangampeta locality are enriched by ~2‰ relative to greenschist facies equivalents at Macapur. Given shifts of ~1‰ from sedimentary protoliths to greenschist facies in several studies (see Discussion section), the 2‰ difference may reflect primary variations of $\delta^{15}N$.

P2 "Archean" samples are characterized by C/N values close to the Redfield ratio, averaging 6.6 (C/N = 106:16) of modern organic compounds (Chen et al., 1996; Fraga et al., 1998). In contrast, P1 and most Archean kerogens are characterized by C/N spanning 35–600 (Hayes et al., 1983; Sano and Pillinger, 1990; Beaumont and Robert, 1999; Jia and Kerrich, 2004b). P2 samples also feature the conjunction of less-dispersed compositional and isotopic values than the other two Precambrian populations (P1 and P3), with total organic carbon (TOC) to total inorganic (TIC) ratios greater than P1 (Table 1). The data plot to the lower end member of the colinear array, in $\delta^{15}N$ vs. $\delta^{13}C$ coordinates, of organic compounds in modern marine sediments of Peters et al. (1978). Considering these lines of evidence together, we interpret P2 data in terms of infiltration of Proterozoic or Phanerozoic hydrocarbon-bearing formation brines locally into Archean basement, forming a secondary hydrocarbon signature. Penetration of Proterozoic and Phanerozoic formation brines, and Quaternary groundwaters, into Archean crust has been documented for several cratons (for a review see Kerrich and Ludden, 2000).

Hydrothermal K-micas from Volcanogenic Massive Sulfide (VMS) and Gold Deposits

The $\delta^{15}N$ values of hydrothermal biotite from 2.6 Ga quartz vein systems at Kolar are 14‰–21‰ and overlap the range for P1 kerogen in the eastern Dharwar Craton; both sets of samples are enriched in ^{15}N compared to Proterozoic or Phanerozoic N-reservoirs (Tables 1, 3; Figs. 1, 3). K-micas generated by modified seawater-derived hydrothermal fluids in 2.7 Ga Abitibi VMS deposits are enriched in ^{15}N at Kidd Creek (16‰ to 19‰), Amulet (12‰–16‰), and Matagami (13‰–20‰), overlapping the range of $\delta^{15}N$ values of kerogen in metashales from this terrane. All micas from the Archean and Proterozoic VMS deposits have low N contents of 10–25 ppm (Tables 1, 3; Fig. 3). In the 1.8 Ga Jerome VMS deposit, four $\delta^{15}N$ values of muscovite are 12‰–14‰, intermediate between Phanerozoic kerogen or bulk sedimentary rocks, and Archean VMS counterparts or kerogen (Tables 2, 3).

COMPARISON WITH THE LITERATURE

Mantle and Diamonds

From worldwide sampling of diamonds, Boyd et al. (1987, 1992) reported negative $\delta^{15}N$ values (−8.7‰ to −1.7‰) with a mode of −6‰ to −5‰. Independently, Cartigny et al. (1997, 1998) showed that mantle $\delta^{15}N$ values of diamonds were between −8‰ and −5‰ (see also Marty and Humbert, 1997; Marty and Zimmermann, 1999; Javoy and Pineau, 1991). The diamonds are mostly Archean, as old as 3.2 Ga (Richardson et al., 1984, 2001). Accordingly, upper mantle $\delta^{15}N$ may have been uniform at about −6 ± 1‰ since ~3.2 Ga (Fig. 1). However, lower $\delta^{15}N$ values of down to −20‰ were found in some rare diamonds, signifying another N reservoir (Cartigny et al., 1997).

Marty (1995) estimated the N content of undegassed mid-oceanic ridge basalt (MORB) to be 1–2 ppm; these values are inferred from (1) observed covariations of $N_2/^{36}Ar$, $^{40}Ar/^{39}Ar$, and $^3He/^4He$ in MORB; (2) an assumption that nitrogen, like He and CO_2, behaves as an incompatible element during partial melting of rocks; and (3) the mantle carbon content, which is ~400 ppm. The upper mantle would then have an N content of ~0.16 ppm (Porcelli and Turekian, 2004). According to Cartigny and Ader (2003), uncertainties in these estimates for MORB may arise from isotopic fractionation during partial degassing of basaltic rocks. Alternatively, the mantle N content is estimated at ~40 ppm from $\delta^{13}C$-N systematics for diamond, assuming that N is not highly incompatible (Cartigny et al., 2001).

TABLE 2. SUMMARY OF N-ISOTOPIC COMPOSITIONS OF SEDIMENTARY ROCKS

Era / Sample location	Age (Ma)	N (ppm) range	N (ppm) mean ± 1σ	$\delta^{15}N_{air}$ (‰)† range	$\delta^{15}N_{air}$ (‰)† mean ± 1σ	C_{org} (ppm) range	C_{org} (ppm) mean ± 1σ	C/N ratio range	C/N ratio mean ± 1σ	Refs.
Archean										
S. Africa, Canada and W. Australia‡	3500–2600	3–33	27 ± 27 (22)	–4.8–29.9	3.6 ± 6.9 (22)	23–1438	457 ± 493 (22)	53–454	185 ± 117 (22)	1
Cherts or dolomitic cherts, S. Africa and W. Australia‡	3500–2500	1.04–1.89 (μmole)	1.31 ± 0.33 (5)	0.8–3.2	2.3 ± 1.0 (5)	114–260 (μmole)	187 ± 54 (5)	90–250	150 ± 64 (4)	2
Shales, Western Australia‡	2500	1.18 (μmole)		5.7		125 (μmole)		148		2
West Greenland and Western Australia	3800–2800	0.7–13.4	2.7 ± 3.7 (21)	–3.8–12.2	6.7 ± 4.4 (12)					3
South Africa and Western Australia	3300–2600	0.7–13.4		–0.1–1.5						4
Vibutigudda locality, India‡	2700	57–88	75 ± 11 (6)	11.4–14.6	13.1 ± 1.3 (6)	11,200–21,165	18,345 ± 3,710 (6)	179–435	305 ± 95 (6)	5
Abitibi greenstone belt, Canada‡	2700	15–51	28 ± 13 (8)	11.8–17.3	15.3 ± 1.8 (8)	14,610–242,500	119,920 ± 81,550 (6)	51–928	390 ± 385 (6)	6
Abitibi greenstone belt, Canada‡	2700	97–310	245 ± 71 (10)	14.8–17.9	16.0 ± 1.7 (10)					7
Penhalonga Formation, Botswana‡	2700	26–60	39 ± 14 (5)	15.7–19.9	17.3 ± 1.9 (5)	311–1,811	807 ± 671 (5)	134–685	285 ± 230 (5)	7
Proterozoic										
Ashanti belt, Ghana, West Africa‡	2200–2100	140–575	300 ± 147 (11)	9.3–12.6	10.8 ± 1.1 (11)	76,790–107,330	90,245 ± 11,470 (8)	208–698	450 ± 230 (8)	7
Cherts or dolomitic cherts, S. Africa and W. Australia‡	2000–800	0.90–2.85 (μmole)	1.80 ± 0.66 (11)	2.8–9.9	5.7 ± 2.1 (11)	57–203 (μmole)	133 ± 45 (11)	40–200	83 ± 45 (11)	2
Shales, Australia‡	2000–800	1.86–3.79 (μmole)	2.62 ± 0.77 (6)	3.0–8.5	4.8 ± 2.4 (6)	90–155 (μmole)	134 ± 24 (6)	40–83	54 ± 17 (6)	2
Canada and Australia	2100–700	3–106	13 ± 11 (11)	2.1–7.7	3.8 ± 1.8 (11)	7–348	91 ± 95 (11)	31–148	93 ± 38 (11)	1
Mangambeta and Marcapur localities, India‡	1900–1700	289–506	395 ± 70 (10)	3.1–6.8	5.0 ± 1.2 (10)	8,755–46,830	21,310 ± 12,880 (10)	22–132	65 ± 40 (10)	5
Moine succession, Scotland	1500–1025	140–422	307 ± 99 (7)	8.4–16.6	13.7 ± 2.6 (7)					8
Tetsa Formation, British Columbia, Canada‡	1600	570–590	580 ± 20 (5)	3.6–4.5	4.0 ± 0.4 (5)	16,000–18,200	17,220 ± 880 (5)	32–38	35 ± 3 (5)	7
Sachsisches Erzgebirge, Germany	Neoproterozoic	35–115	55 ± 30 (9)	6.8–17.0	11.3 ± 4.0 (9)					9
Phanerozoic										
Sachsisches Erzgebirge, Germany	Ordovician	14–650	297 ± 207 (32)	3.2–10.8	3.7 ± 1.9 (32)					9
Erzgebirge schists, Germany	Ordovician	76–896	282 ± 230 (30)	1.2–10.5	3.9 ± 2.2 (30)					10
Anthracites, Germany and Pennsylvania‡	Carboniferous			2.7–5.1	4.0 ± 0.8 (18)					11
Black shales, British Columbia, Canada‡	Jurassic	400–9300		–0.6–5.0	2.8 ± 1.9 (17)	1,600–40,700	16,170 ± 11,245 (17)	1–14	7 ± 4 (17)	12
Catalina Schist, California, U.S.A.	Cretaceous	30–1075	390 ± 293 (46)	1.0–5.9	3.0 ± 1.2 (46)					13
Louisiana sandstone, U.S.A.	Cretaceous		1600 ± 100 (64)		3.1 ± 1.4 (51)					14
Louisiana sandstone, U.S.A.‡	Cretaceous	100–3100		2.6–3.6	3.2 ± 0.3 (26)					14
Sediments, North America	Tertiary			2.4–9.9	6.1 ± 1.8 (55)					15
Sediments, Izu-Bonin-Mariana margin	Tertiary	18–661	277 ± 154 (36)	–0.9–8.2	4.7 ± 1.7 (36)					16
Sediments, western Alps, Europe	Tertiary	169–1721	586 ± 490 (16)	2.6–4.8	3.6 ± 0.7 (16)					17
Arctic Ocean sediments	Tertiary	2,100–17,800	1,105 ± 1080 (37)	4.5–7.7	6.1 ± 0.8 (37)	2,600–18,000	9,705 ± 4,695 (37)	4–10	7 ± 2 (37)	18
Taiwan watershed, China‡	Tertiary	700–800	800 ± 130 (5)	3.9–4.0‡	3.9 ± 0.1 (5)	3,700–4,500	3,980 ± 340 (5)	5–7	6 ± 1 (5)	19

Note: Refs: 1—Beaumont and Robert (1999); 2—Hayes et al. (1983); 3—Sano and Pillinger (1990); 4—Pinti et al. (2001); 5—this study; 6—Jia and Kerrich (2000); 7—Jia and Kerrich (2004a,b); 8—Boyd and Philippot (1998); 9—Haendel et al. (1986); 10—Mingram and Bräuer (2001); 11—Ader et al. (1998); 12—Sephton et al. (2002); 13—Bebout and Fogel (1992); 14—Williams et al. (1995); 15—Peters et al. (1978); 16—Sadofsky and Bebout (2004); 17—Busigny et al. (2003); 18—Schubert and Calvert (2001); 19—Kao and Liu (2000).
†Numbers in the parentheses are the numbers of analyzed samples.
‡N-isotopic compositions were determined on kerogens isolated from sedimentary rocks.

TABLE 3. SUMMARY OF N-ISOTOPIC COMPOSITIONS OF HYDROTHERMAL MICAS FROM OROGENIC GOLD DEPOSITS AND VMS DEPOSITS

Era / Sample location	Age (Ma)	N (ppm)[†] range	N (ppm)[†] mean ± 1σ	$\delta^{15}N_{air}$ (‰)[†] range	$\delta^{15}N_{air}$ (‰)[†] mean ± 1σ	Refs.
Archean						
Norseman terrane, Western Australia[‡]	2700	20–70	32 ± 20 (12)	10.9 to 23.7	17.3 ± 4.3 (12)	1
Abitibi greenstone belt, Canada[§]	2700	10–35	20 ± 9 (7)	12.4 to 20.1	16.5 ± 2.9 (7)	2
Abitibi greenstone belt, Canada[‡]	2700	20–205	95 ± 72 (40)	10.1 to 21.0	16.3 ± 2.9 (40)	1, 3
Harare greenstone belt, Zimbabwe[‡]	2700	22–25	23 ± 1.5 (4)	17.7 to 23.4	20.7 ± 3.0 (4)	4
Dharwar, South India[‡]	2500	18–47	32 ± 15 (3)	13.7 to 21.0	18.0 ± 3.8 (3)	2
Proterozoic						
Ashanti belt, Ghana, West Africa[‡]	2200–2100	1160–1900	1655 ± 250 (9)	9.1 to 12.0	10.2 ± 1.5 (9)	4
Trans-Hudson orogen, Canada[‡]	1800	105–120	112 ± 6 (6)	6.6 to 7.4	7.2 ± 0.4 (6)	4
Jerome, USA[§]	1800	1028–47	37 ± 8 (4)	11.7 to 14.1	12.7 ± 1.0 (4)	2
Moine succession, Scotland*	1500–1025	435–1739	1048 ± 520 (21)	7.4 to 16.2	13.6 ± 2.3 (21)	5
Phanerozoic						
Lachlan fold belt, SE Australia[‡]	Ordovician	861–895	733 ± 103 (20)	2.8 to 4.5	3.5 ± 0.4 (20)	6
Western Qilian orogen, North China[‡]	210 to 225	650–2510	260 ± 760 (8)	1.7 to 6.6	4.0 ± 2.0 (8)	4
North American Cordillera[‡]	190 to 50	130–3500	1535 ± 1080 (100)	1.6 to 6.1	3.0 ± 1.2 (100)	7
Western Alps, Italy[‡]	Tertiary	280–290	288 ± 5 (4)	4.5 to 5.5	4.9 ± 0.6 (4)	4

Note: Refs: 1—Jia and Kerrich (1999); 2—this study; 3—Jia and Kerrich (2000, 2004b,c); 4—Jia and Kerrich (2002); 5—Boyd and Philippot (1998); 6—Jia et al. (2001); 7—Jia et al. (2003a).
[†]Numbers in the parentheses are the numbers of analyzed samples.
[‡]Hydrothermal micas from orogenic gold deposits.
[§]Hydrothermal micas from VMS deposits.
*Micas from amphibolite facies sedimentary rocks.

Siliciclastic Sediments

Phanerozoic and Modern Sediments

Sedimentary rocks have similar bulk nitrogen isotopic compositions as indigenous kerogen, consistent with the bulk rock N budget being dominated by kerogen N. The total range of $\delta^{15}N$ in Phanerozoic bulk sedimentary rocks is 1.0‰–10.8‰ (Fig. 1, Table 2). However, the averages of most rock data sets cluster between 3‰ and 5‰, within one standard deviation, the global average being 3.65‰ ± 0.55‰ (Table 2). Similarly, the total range of kerogen is –2.0‰ to 6.0‰, but averages of each data set cluster between 1.5‰ and 4.6‰, and the global average is 3.7‰ (Table 2).

Recent marine sediments have a $\delta^{15}N$ range of –3‰–9‰, with an average of 3.9‰ ± 1.1‰ (Table 2; Peters et al., 1978; Sweeney et al., 1978; Rau et al., 1987; Williams et al., 1995; Schubert and Calvert, 2001; Sadofsky and Bebout, 2004). Variations in $\delta^{15}N$ have been attributed to varying mixtures of terrestrially derived (^{13}C and ^{15}N depleted) and enriched marine-derived organic compounds, given a linear correlation between $\delta^{15}N$ and $\delta^{13}C$ (Peters et al., 1978; Sweeney et al., 1978; Minoura et al., 1997). Alternatively, according to Sadofsky and Bebout (2004) variations may stem from some combination of changes in bioproductivity, complex diagenetic processes, and differing proportions of marine and terrestrial organic matter in continental margin versus intraoceanic arc settings.

A larger range of N-isotope compositions has been recorded from specific niches. $\delta^{15}N$ values of –12‰ to 4‰ are documented from chemoautotrophic bacteria in warm seeps on the seafloor (Conway et al., 1994). Reduced nitrate availability may generate sedimentary organic compounds (SOC) with $\delta^{15}N$ as low as –2.7‰. In contrast, intense denitrification, characterized by a kinetic isotope effect, may generate $\delta^{15}N$ values as high as 19‰ (Cline and Kaplan, 1975; Sweeney et al., 1978; Rau et al., 1987). For example, Mazuka et al. (1991) report $\delta^{15}N$ values of 0.9‰–18.9‰, with a mode of 8‰, for Pliocene and younger marine sediments at ODP site 724.

The principal process that controls changes in isotope composition of organic nitrogen is thermal decomposition of SOC during diagenesis. Preferential release of ^{14}N, given lower energy to break ^{14}N-^{12}C than ^{15}N-^{12}C bonds, shifts the residual kerogen to ~3‰ (Delwiche and Steyn, 1970; Wada et al., 1975; Macko et al., 1987), implying a flux of ^{15}N depleted nitrogen back to the surface. Wada et al. (1975) and Sweeney et al. (1978) also considered that denitrification may be responsible for nitrogen isotope fractionation between atmospheric and organic nitrogen in kerogens, because denitrification, a bacterial process, involves preferential loss of ^{14}N to the atmosphere, leaving nitrate enriched in ^{15}N; nitrate is the source of marine organic matter, which retains its isotopic signature in marine sedimentary rocks. Williams et al. (1995) report two stages of N production during progressive burial: from microbiological activity during diagenesis, and then release of N in the "oil window" when NH_4^+ becomes incorporated into illite during the smectite-illite transition. The remaining organic matter as kerogen in sedimentary rocks increases in $\delta^{15}N$ relative to precursors. Freudenthal et al. (2001) report a decrease from 1000 to 250 ppm N during diagenesis of recent sediments accompanied by a shift of 1.9‰ by Rayleigh fractionation.

Fractionations between kerogen N (3.2‰ ± 0.3‰) and fixed N in mudstones (3.0‰ ± 1.4‰) are small (Williams et al., 1995). Given that Phanerozoic upper crust is dominated by recycled sedimentary rocks (Taylor and McLennan, 1985), and the kerogen-sediment fractionation is insignificant, bulk upper crust has a nitrogen isotope composition of 3‰–4‰, indistinguishable from kerogen (see Fig. 1 for references).

In summary, notwithstanding extreme $\delta^{15}N$ values in special niches, sedimentation, maturation, and diagenesis clearly generate kerogen with a restricted range of values (Figs. 1, 4E, 4G; Table 2). The most robust comparison for this study is of Precambrian carbonaceous shales with modern and Phanerozoic counterparts (Tables 1 and 2).

Proterozoic

Carbonaceous shales sampled in the 2.2 Ga Paleoproterozoic lower greenschist facies Birimian volcanic-sedimentary terrane, Ghana, have $\delta^{15}N$ values ranging from 9.3 to 12.6‰ (Table 2, Fig. 4C) (Jia and Kerrich, 2004b). Boyd and Philippot (1998) reported bulk rock values of 8.4–16.6 ‰ for the mid-amphibolite facies Moine succession, Scotland, deposited between 1.5 and 1.0 Ga (Table 2). Micas separated from these samples yield similar isotopic compositions of 7.4‰–16.2‰. Assuming an oxidized atmosphere-hydrosphere system by this time in the Mesoproterozoic, and therefore contemporaneous organic matter similar to $\delta^{15}N$ in modern marine sediments of –3‰ to 9‰ (Peters et al., 1978; Sweeney et al., 1978; Rau et al., 1987; Williams et al., 1995; Schubert and Calvert, 2001; Sadofsky and Bebout, 2004), they concluded that metamorphism to amphibolite facies had induced a shift of ≥ 8‰. However, if shifts to amphibolite facies are ≤3‰ (see Discussion, below), primary values may have been 3‰–12‰. Haendel et al. (1986) documented ^{15}N-enriched values for the Proterozoic Saschsisches Erzebirge, Germany (11.3‰; Table 2). The 1.6 Ga Tetsa samples, and Cuddapah kerogen data of this study have $\delta^{15}N$ values that are indistinguishable from the Phanerozoic data set. In summary, three of the five Proterozoic data sets are enriched in ^{15}N compared to Phanerozoic counterparts (Table 2; Figs. 1, 4C).

Archean

Hayes et al. (1983) reported an extensive database on Precambrian sedimentary rocks spanning 3.8–0.8 Ga, including X-ray characteristics, H and C contents, and δD and $\delta^{13}C$ values. They reported N contents and $\delta^{15}N$ on a subset of ten samples 3.4 to 1.6 Ga in age, of which five are cherts mostly associated with BIF, two are shales, and three are stromatolitic dolomites. $\delta^{15}N$ values range from 0.8‰–5.7‰. Chert kerogen has a similar range of $\delta^{15}N$ values as in the larger database for Precambrian cherts of Beaumont and Robert (1999). Kerogen N in chert could be mixtures of mantle N having $\delta^{15}N$ of ~–5‰ with enriched oceanic N.

Zhang (1988) documented $\delta^{15}N$ ranging from 2‰ to 39‰ in Archean carbonaceous metashales, the most enriched values being in greenschist facies shales from the 2.6 Ga Ventersdorp Group, South Africa (Hayes et al., 1983). Greenschist facies carbonaceous shales in the 2.7 Ga Archean Abitibi Greenstone Belt are characterized by variably enriched $\delta^{15}N$ values from 12‰, with an average of 15.3‰ ± 1.8‰ (Table 2, Fig. 4A) (Jia and Kerrich, 2000). Two other ca. 2.7 Ga Archean data sets on unweathered rocks in drill core from the greenschist facies Penhalonga Formation, Botswana, and western Abitibi Greenstone Belt also have overlapping enriched averages $\delta^{15}N$ of 17.3‰ ± 1.9‰, and 16.0‰ ± 1.7‰, respectively (Table 2; Figs. 1, and 2A) (Jia and Kerrich, 2004b, 2004c). Contemporaneous samples of this study from the Dharwar Craton endorse systematically ^{15}N-enriched sedimentary kerogen in Neoarchean carbonaceous shales (Table 1, Fig. 1).

Precambrian Chert–Iron Formation

Beaumont and Robert (1999) analyzed nitrogen concentrations and isotopic compositions of kerogen from Precambrian cherts, ranging in age from 3.5 to 0.7 Ga. They found a very large range of $\delta^{15}N$ values, from –4.7‰ to 30.0‰, with generally low bulk rock N contents of 2–106 ppm. There is a general increase of $\delta^{15}N$ values from –4.7‰ to 5.9‰ in the Mesoarchean (3.5 to 2.9 Ga), through –2.5‰ to 30.0‰ in the Neoarchean (2.8 to 2.5 Ga), to 2.1‰ to 7.7‰ in the Proterozoic (<2.5 to 0.6 Ga). They concluded that Archean atmospheric N_2 had a similar isotopic composition to the present, and that the secular change of $\delta^{15}N$ records the "great oxygenation event" (Fig. 4A).

Using a step-heating procedure, Pinti et al. (2001a) documented $\delta^{15}N$ ranging from –7‰ to 20‰ in Archean chemical sedimentary rocks (3.8–2.8 Ga). Nitrogen released at low-temperature steps, in all samples, had negative or near present-day atmospheric $\delta^{15}N$, which was interpreted to reflect modern atmospheric or/and organic compound contamination; on the other hand, for all but one sample the nitrogen extracted from high-temperature steps showed positive $\delta^{15}N$, with most between 2‰ and 20‰; enriched values were modeled as secondary shifts by Rayleigh volatilization during metamorphism (Pinti et al., 2001a).

Granitoids

Phanerozoic

In a study of the Cornubian batholith of southwest England, Boyd et al. (1993) reported N concentrations of 8–187 ppm and $\delta^{15}N$ values in the range of 5.2‰–10.2‰. According to Bebout et al. (1999), the early Devonian Skiddaw peraluminous granite in the English Lake District has nitrogen contents of 49 ± 27 ppm and $\delta^{15}N$ values of 3.5‰ ± 1.1‰ ($n = 7$). Given that Phanerozoic peraluminous granites have metasedimentary precursors (Hawkesworth and Kemp, 2004), these results are consistent with melting of sedimentary rocks having initial $\delta^{15}N$ of 2‰–6‰, which is shifted by ~3‰ during metamorphism prior to partial melting (Table 2).

Archean

Sparse data for 2.7 Ga tonalitic rocks in the Uchi subprovince, Superior Province, range from –5.3 to +5.2‰, averaging –0.9‰ (Jia and Kerrich, 1999, 2000). The Archean tonalite-

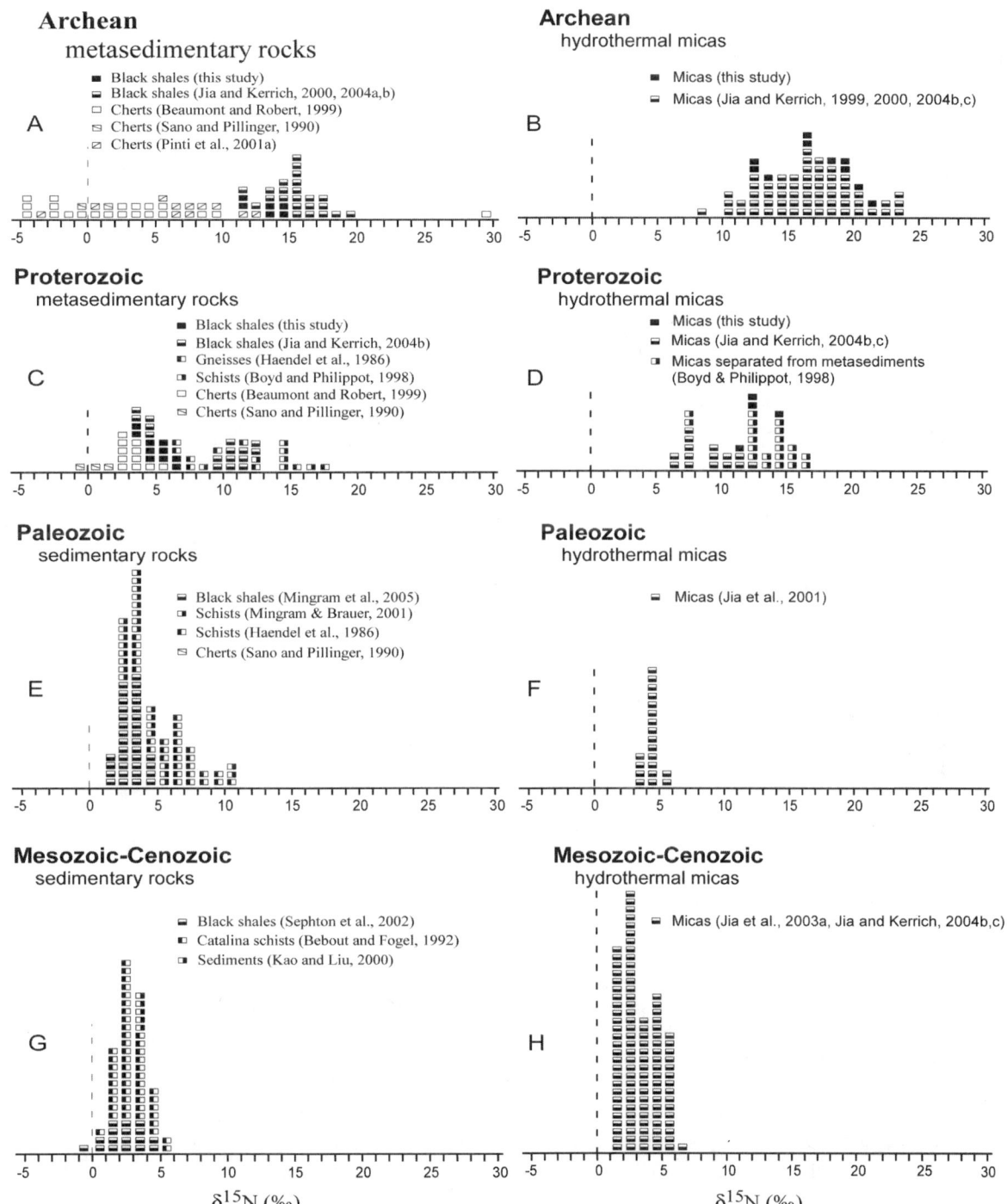

Figure 4. Histograms showing variations in $\delta^{15}N$ of sedimentary and/or metasedimentary rocks (left side: A, C, E, and G) and hydrothermal micas (right side: B, D, F, and H) of Archean to Phanerozoic age. The figure displays, except for cherts, common secular evolution of both sedimentary rocks and hydrothermal micas. Data sources are in Tables 1–4, and published data.

trondhjemite-granodiorite (TTG) suite is considered to have formed by partial melting of basaltic oceanic crust on a subducting slab (Drummond and Defant, 1990; Martin, 1999; Smithies, 2000). Archean diamond data indicate that Archean oceanic crust may have had a $\delta^{15}N$ similar to that of modern MORB (averaging −5‰). Secondary isotopic shifts could arise from some combination of seawater alteration and addition of carbonaceous sediments, which are unconstrained, followed by a shift of ≤3‰ during subduction metamorphism prior to partial melting. Nitrogen contents of the tonalites at 5–27 ppm are comparable with Phanerozoic metaluminous granitoids (Hall, 1999), implying similar concentrations of N in the source.

Ore Deposits

Orogenic Gold

Orogenic, or mesothermal, lode gold deposits constitute a distinct class of Au- and Ag-rich structurally hosted vein systems. They developed post peak-metamorphism, late in the development of accretionary orogenic belts. This deposit type is common in Neoarchean greenstone belts and Paleoproterozoic terranes, and formed continuously through the Phanerozoic, including large metallogenic provinces in the lower Paleozoic Lachlan orogen and Jurassic–Tertiary of the North American Cordillera. The vein systems extend tens of kilometers laterally, and up to 6 km vertically (McCuaig and Kerrich, 1998; Goldfarb et al., 2001).

Muscovite, or more rarely biotite, is abundant as an alteration phase, and therefore a candidate for determining the N-isotope composition of the hydrothermal fluids and source reservoir. According to H, C, and O-isotope data, and K-Cs-Rb-Ba systematics, the veins precipitated from metamorphic fluids. Given a "deep later" P-T-t path where peak-metamorphism migrated down through the crust, metamorphic fluids were probably generated syn peak-metamorphism at mid-crustal levels, and advected to shallow crustal levels where vein minerals precipitated post peak-metamorphism (McCuaig and Kerrich, 1998). In the 2.7 Ga Abitibi Greenstone Belt, the deposition of sedimentary-volcanic host rock sequences was temporally separated from precipitation of the post-metamorphic hydrothermal micas by only 30 m.y. (Corfu et al., 1989; Kerrich and Cassidy, 1994).

Stable (H, C, O, and S) and radiogenic (Sr, Nd, and Pb) isotope studies of these vein systems of Archean to Phanerozoic age suggest that the fluids acquired a signature of bulk crust, as expected for fluids evolved during regional metamorphism (for reviews see McCuaig and Kerrich, 1998; Hagemann and Cassidy, 2000; Kerrich et al., 2000). The dilute, aqueous dominated, but carbonic- and N-bearing fluids, were generated by dehydration of hydroxyl-silicates for H_2O and NH_4^+, and by decarbonation of carbonates and/or oxidation or hydrolysis of organic compounds for aqueous carbonic species and N_2 (Jia and Kerrich, 2004b).

Hydrothermal micas in Phanerozoic gold quartz veins have $\delta^{15}N$ values similar to those in contemporaneous sedimentary rocks (Tables 1–3, Fig. 4). This result is in keeping with the bulk mid to upper crustal N budget being dominated by sedimentary N. Crustal sedimentary rocks contain $(0.22 \pm 0.10) \times 10^{20}$ mol N_2, mostly in shales where N abundances are hundreds to thousands ppm. The mass of N is comparable in lower crustal igneous rocks (Zhang and Zindler, 1993, and references therein) (Table 5)

Averages of $\delta^{15}N$ values for deposits in accretionary orogenic belts are as follows: 3.5‰ ± 0.4‰ for the Paleozoic Lachlan fold belt in southeastern Australia; 3.0‰ ± 1.2‰ for the western North American Cordillera, which hosts many Jurassic–early Tertiary gold-bearing quartz vein systems from the Mother Lode of southern California to Alaska; 4.0‰ ± 2.0‰ for Mesozoic quartz veins in the western Qilian orogen, North China; and 4.9‰ ± 0.6‰ for the middle to late Tertiary Monte Rosa quartz veins in the Alpine orogen (Table 2; Figs. 4F, 4H) (Jia and Kerrich, 2004b).

$\delta^{15}N$ values for hydrothermal micas from gold-bearing quartz veins in the 2.1 Ga Ashanti belt of Ghana, and the 1.8 Ga Trans-Hudson orogen, Canada, are 7‰–10‰, intermediate between Archean and Phanerozoic counterparts (Table 3, Fig. 4D) (Jia and Kerrich, 2004b).

For the Archean Superior Province, Jia and Kerrich (1999, 2000) reported data for hydrothermal micas from nine 2.7 Ga deposits (Abitibi belt: Kerr-Addison, Dome, Hollinger, Goldhawk, Beaumont; Wawa belt: Hemlo; Geraldton-Beardmore belt: Geraldton, Pickle Crow; Red Lake belt: Red Lake); the total range of $\delta^{15}N$ values is 11.8‰–21.0‰, with a mean of 16.3‰ ± 3.0‰ (Table 3, Fig. 2B). Hydrothermal micas from the Norseman terrane, Western Australia, and Harare Greenstone Belt, Zimbabwe, are also characterized by ^{15}N-enriched values, averaging 17.3‰ ± 4.3‰ and 20.7‰ ± 3.0‰, respectively. Including the ^{15}N-enriched data for the Kolar deposit of this study, high $\delta^{15}N$ values of late-metamorphic hydrothermal micas have been recorded on four Neoarchean cratons (Tables 1, 3; Fig. 4B).

VMS Deposits

$\delta^{15}N$ values of micas in the 1.8 Ga Jerome deposit are within the range for Proterozoic siliciclastic rocks and kerogen. Hall (1989) documented an increase of N content from 1 ppm to 182 ppm (average 53 ppm), correlated with secondary K addition, in basalts altered by seawater. High N contents of 144–238 ppm have been reported from submarine hydrothermal fluids in the Sea of Cortez (Von Damm et al., 1985). Nitrogen in zones of potassic submarine hydrothermal alteration could be sourced in seawater, organic-rich sediments, hydrolysis of K-silicates in the host volcanic rocks, or some combination. Given bottom seawater N contents of 0.5 ppm (Létolle, 1980), we tentatively interpret this variably enriched signature in the VMS alteration zone as reflecting N-bearing organic compounds in ambient sediments. Micas in the Archean deposits have the same high $\delta^{15}N$ values as contemporaneous shale kerogen, consistent with a sedimentary N source.

Rare Metal Pegmatites

Rare metal pegmatites containing Mo and W occur in peraluminous, S-type domains of the post-metamorphic 2.6 Ga Preissac-Lacorne post-tectonic batholith, Abitibi Greenstone Belt, Canada (Feng and Kerrich, 1992). Muscovites are characterized by $\delta^{15}N$ values of 1.6‰–5.3‰, with an average +3.2‰ (Jia and Kerrich, 1999, 2000). Peraluminous granites are considered to result from melting of average mid-crust (Sylvester et al., 1997; Hawkesworth and Kemp, 2004). Neoarchean greenstone belts are dominated by syn-tectonic TTG batholiths, with supracrustal volcanic-sedimentary sequences. Notwithstanding uncertainties in melt-residue fractionations, these results are consistent with fusion of a mix of relatively ^{15}N-depleted tonalites and ^{15}N-enriched clastic sedimentary rocks (Table 1, Fig. 1).

Other Deposits

Nitrate deposits formed in arid climates possess up to 163,000 ppm N, and $\delta^{15}N$ ~0‰ consistent with atmospheric deposition of N (Böhlke et al., 1997). Sparse data on NH_4^+-bearing K-feldspars from a variety of epithermal springs and deposits in the western United States range from 2700 to 19,000 ppm N, with $\delta^{15}N$ of −0.6‰–12.3‰ interpreted as N mobilized from sedimentary rocks (Krohn et al. 1993).

DISCUSSION

Nitrogen isotopic compositions of Archean shale kerogen, hydrothermal gold quartz vein systems that proxy for average crust, and seawater-altered micas in the VMS deposits all show ^{15}N-enriched values. The three types of samples show parallel secular trends from 15‰–24‰ at 2.7 Ga, through intermediate values in the Proterozoic, to 3‰–4‰ in the Phanerozoic (Tables 1–3, Fig. 4). We first constrain the magnitude of possible isotope fractionation during metamorphism or Rayleigh volatilization, then address possible long-term (chronic) diffusional loss of N.

Effects of Metamorphism

The sample design of this study involves comparing pre-metamorphic kerogens from carbonaceous shales with late-metamorphic hydrothermal micas from the eastern Dharwar Craton. Also, published data for pre-, and late-metamorphic samples from other Precambrian terranes are compared. These sample types permit testing for large shifts of $\delta^{15}N$ during metamorphism, as proposed by Boyd and Philippot (1998), Beaumont and Robert (1999), and Pinti et al. (2001a).

From inspection of Tables 1–3 and Figure 4, it is clear that the $\delta^{15}N$ values of sedimentary kerogen are consistent with those of post peak-metamorphic hydrothermal micas through geological time. Jia et al. (2001, 2003a) and Jia and Kerrich (2004c) showed that hydrothermal micas in the Paleozoic Lachlan accretionary orogen and the Mesozoic–Cenozoic western North American Cordillera have $\delta^{15}N$ of 3.5‰ ± 0.4‰ ($n = 20$) and 3.0‰ ± 1.2‰ ($n = 100$) respectively, comparable to Phanerozoic bulk metasedimentary rocks and average crustal values (Haendel et al., 1986; Bebout and Fogel, 1992; Mingram and Bräuer, 2001; Busigny et al., 2003), and Phanerozoic kerogens (Williams et al., 1995; Ader et al., 1998; Kao and Liu, 2000; Sephton et al., 2002). This result endorses the use of hydrothermal micas to proxy for crust in the Precambrian.

Mean $\delta^{15}N$ values of the 2.2–2.1 Ga pre-metamorphic carbonaceous shales and post peak-metamorphic hydrothermal micas hosted in the Birimian sediments overlap at 10.8‰ ± 1.1‰ and 10.2‰ ± 1.5‰, respectively (Tables 2, 3; Figs. 4C, 4D). Kerogen and hydrothermal biotites from the eastern Dharwar Craton are both enriched in ^{15}N. Similarly, $\delta^{15}N$ values of the 2.7 Ga Three Nations carbonaceous shales (16.0‰ ± 1.7‰) and VMS micas (16.5‰ ± 2.9‰) from the Abitibi belt are both enriched, and comparable to the isotopic composition of hydrothermal micas (16.3‰ ± 2.9‰) in the same terrane (Figs. 4A, 4B) (Jia and Kerrich, 1999, 2000, 2004c). The hydrothermal quartz-mica veins precipitated post peak-metamorphism (Kerrich and Cassidy, 1994), and accordingly their $\delta^{15}N$ values cannot have been influenced by peak-metamorphic conditions.

These results demonstrate minimal isotope fractionation of nitrogen between metasedimentary rocks, hydrothermal fluids, and minerals precipitated from the fluids. They also rule out any significant shifts in $\delta^{15}N$ during metamorphism to greenschist facies, or mid amphibolite for the Sandur shales, consistent with the results of previous studies as indicated below.

Progressively Metamorphosed Sediments

Bebout and Fogel (1992) reported data for progressively metamorphosed sedimentary rocks of the Catalina Schist complex, California (Tables 2, 4; Fig. 5A). They obtained $\delta^{15}N$ of 2.2‰ ± 0.6‰ in low-grade rocks (350 °C) and 4.3‰ ± 0.8‰ in amphibolite facies equivalents (600 °C). They calculated fluid-rock (N_2-NH_4^+) N-isotope fractionations of −1.5‰ ± 1‰ with the Rayleigh distillation equation at temperatures ranging from 350 to 600 °C.

Mingram and Bräuer (2001) also found shifts of <2‰ in $\delta^{15}N$ from low-grade carbonaceous shales (300 °C) at 2.2‰ ± 0.6‰, through greenschist facies equivalent (470 °C) at 3.5‰ ± 0.9‰, to amphibolite facies mica schists (550 °C) at 3.9‰ ± 0.8‰ (Table 4, Fig. 5B). In a more recent report on N-isotope fractionation due to metamorphism, Busigny et al. (2003) found that the $\delta^{15}N$ values of metasedimentary rocks from the Schistes Lustrés complex (Western Alps), which were subducted from shallow level to depth of 90 km, are between 3.1‰ and 4.8‰ and show no systematic isotopic shifts with increasing metamorphic grade (Tables 2, 4; Fig. 5C). A more closed system behavior may explain why shifts were smaller than in the Catalina.

TABLE 4. EMPIRICAL AND EXPERIMENTAL STUDIES ON NITROGEN ISOTOPE FRACTIONATIONS DURING METAMORPHISMS

Author/location	Geological setting	isotopic composition	Isotope fractionations
Empirical observations			
Bebout and Fogel (1992) Catalina complex, California, USA	The complex contains lawsonite to amphibolite facies metasedimentary rocks that represent metamorphism at temperatures of 350°–750 °C and pressures corresponding to 15–45 km depth.	2.2–4.3‰	Calculated bulk fluid-rock fractionations = –1.5 ± 1‰
Haendel et al. (1986) Sächsisches Erzgebirge, Germany	The metasedimentary rocks consist of Phanerozoic phyllites and schists to Proterozoic gneisses, with corresponding metamorphic grades from low greenschist to amphibolite facies.	(a) 5.1‰ in phyllites, (b) 7.4‰ for schists, (c) 11.8‰ for gneisses	Range of $\delta^{15}N$ reflects age, not metamorphic grade
Jia and Kerrich (2000) Quartz veins, Canada and western Australia	Veins are hosted in metamorphic terranes. Metamorphism ranges from low greenschist facies (270°–300 °C/1–2 kb) to lower amphibolite facies (420°–500 °C/2–4 kb).	16.3–17.2‰	Do not show significant isotope fractionations
Mingram and Bräuer (2001) Erzgebirge Schists, Germany	The Paleozoic schists contain progressively metamorphosed sedimentary rocks from low-grade unit (300 °C/2kb), garnet-phyllite(470 °C/9kb), to mica schist unit (550 °C/12kb).	(a) 2.2‰ in low-grade, (b) 3.5‰ for phyllites, (c) 3.9‰ in schists	Only 1.7‰ shift from low-grade to amphibolite facies rocks
Busigny et al. (2003) Schistes Lustres nappe western Alps, Europe	The complex is an homogeneous sequence of pelagic sediments subducted to depths of 0–90 km. Metamorphic grade increases from lower greenschist (300 °C/0.8 GPa) to amphibolite (650 °C/3.0 GPa) facies.	2.6–4.8‰	No specific trend with increasing metamorphic grade
Jia (2004) Cooma complex, Australia	The Paleozoic complex contains low-grade greenschist (300 °C/2–4kb) to high-grade granulite facies (730 °C/ 2–4kb) metasedimentary rocks.	(a) 2.4–3.2 ‰ in greenschist facies rocks, (b) 3.8–4.3‰ for amphibolite facies zone; (c) 12.3–12.9‰ in granulite facies rocks	< 2‰ shift from greenschist to amphibolite facies rocks. Large fractionations of 8–10‰ at granulite facies
Experimental studies			
Ader et al. (1998) Anthracites, USA and Germany	The coals represent a wide range of rank from anthracite to metaanthracite. Experimental simulation of the denitrogenation process was conducted at 600 °C and 2 kb conditions.	4.40–4.45‰	No shift due to change of facies or hydrothermal disturbance

Haendel et al. (1986) documented shifts of ~1‰ from sediments to greenschist facies counterparts, and ≤3‰ to amphibolite facies in progressively metamorphosed siliciclastic sequences of the Sachsisches Erzgebirge, Germany (Table 4, Fig. 5D). However, Haendel et al. (1986) emphasized that the total range of $\delta^{15}N$ may not be attributable to progressive metamorphism alone because (1) the different sedimentary lithologies may have had intrinsically different primary N contents and $\delta^{15}N$ values; (2) there were multiple metamorphic events in some sequences, but not others; (3) there is a range of age of the lithologies from Ordovician to Late Precambrian; and (4) the Precambrian gneisses alone are characterized by the most ^{15}N-enriched values.

Hence, there is systematic ^{15}N-enrichment in the three Proterozoic data sets (see Table 2), the Erzegebirge ($\delta^{15}N$ = 11.3‰ ± 4.0‰), Moine (13.7‰ ± 2.6‰), and Ashanti belt (10.8‰ ± 1.1‰), notwithstanding the fact that the former two are at amphibolite facies, but the Ashanti belt is prevalently lower greenschist facies (Tables 2 and 4, Jia and Kerrich, 2004b). Accordingly, the "metamorphic trend" of Haendel et al. (1986) could be reinterpreted in terms of a secular evolution of crustal N and $\delta^{15}N$, in keeping with this study

Experimental and Theoretical Studies

Ader et al. (1998) observed a decrease of nitrogen content but uniformity of $\delta^{15}N$ in anthracite at temperatures of up to 600 °C, in keeping with the empirical studies of N in progressively metamorphosed sedimentary rocks as indicated above (Table 4). Hanschmann (1981) calculated nitrogen isotope fractionations between NH_4^+ in solid and fluid (N_2) phases; interpolation of the data yields fractionation of ~–2.25‰ at temperatures of 350–600 °C, in accord with the results of Bebout and Fogel (1992) and Haendel et al. (1986). Collectively, studies of progressively metamorphosed terranes, theoretical and empirical studies, as well as data for hydrothermal micas, show that the isotope composition

Figure 5. $\delta^{15}N$ versus N content for progressively metamorphosed sedimentary rocks. Values on the right-hand side are mean (and median) $\delta^{15}N$ plus one standard deviation, and numbers in parentheses are the sample sizes. Data sources: A, Bebout and Fogel (1992); B, Mingram and Bräuer (2001); C, Busigny et al. (2003); and D, Haendel et al. (1986).

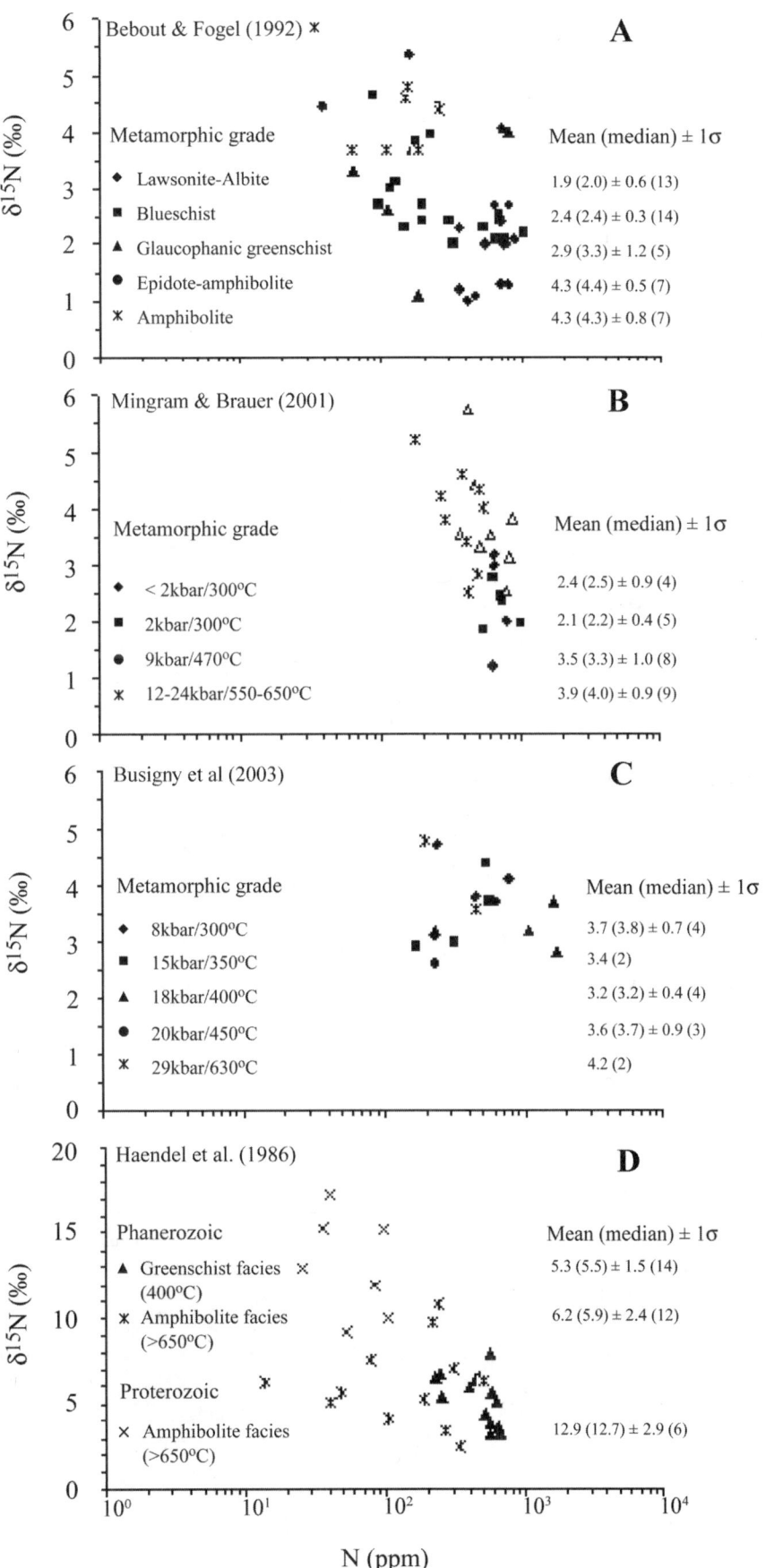

of N_2 in metamorphic fluids is close to that of the source rock reservoir (Hanschmann, 1981; Bebout and Fogel, 1992; Jia et al., 2001, 2003a).

Rayleigh Processes

Rayleigh fractionation generates large isotopic shifts where volatilization nears completion (Valley, 1986). Pinti et al. (2001a) model their range of data as a shift of ≥27‰ from primary values of −7‰, stemming from Rayleigh devolatilization using fractionation factors of Hanschmann (1981) and no constrained f (the fraction of residual nitrogen in the rocks) values. Modeling was on a mix of data from 3.8 Ga amphibolite facies chert-BIF and 3.4 Ga greenschist facies cherts. Two ^{15}N-enriched samples of Pinti et al. (2001a) are amphibolite facies metasedimentary rocks having C/N ratios of ~2000 and 5000 that have experienced multiple Archean metamorphic events and a Proterozoic disturbance (e.g., Rollinson. 2002). Of the remaining four samples, two metamorphosed to high-grade have δ^{15}N intermediate between two low-grade samples. Accordingly, there is a data cluster, rather than a correlation of δ^{15}N values with either C/N or metamorphic grade, and hence no progressive Rayleigh fractionation with increasing grade (Fig. 6 of Pinti et al., 2001a).

Within the seven Precambrian data sets there are only small trends of increasing δ^{15}N to lower N contents as expected for Rayleigh processes (Table 2; Figs. 3, 4). As a corollary, in the Onverwacht data of Beaumont and Robert (1999) two samples with the most positive δ^{15}N have greater N contents than two of the more negative.

Rayleigh effects can also be addressed via the quartz-mica vein systems, which are precipitated from fluids generated by loss of volatile species during progressive metamorphism. If Rayleigh effects were significant during metamorphic devolatilization, then H, C, O, N, and S isotope values would be depleted in the veins relative to ambient crust, but this is not the case (McCuaig and Kerrich, 1998).

Constraints from C/N and δ^{13}C

Precambrian kerogens collected in this study do not show correlations between C/N ratios and δ^{13}C or δ^{15}N values (Fig. 6). Both δ^{15}N and δ^{13}C values of samples, for each given age, shift ≤3‰ over the span of C/N ratios, consistent with previous estimates (Haendel et al., 1986; Watanabe et al., 1997).

A range of δ^{13}C values (−18‰ to −48‰) has been reported from greenschist facies sedimentary units in 2.7 Ga Superior Province supracrustal terranes (Schoell and Wellmer, 1981; Wellmer et al., 1999). According to Wellmer et al. (1999), no correlations exist between δ^{13}C and H/C or C/N. Rather, they interpreted variations in δ^{13}C values of kerogens as resulting from organic matter synthesized by different metabolic pathways. The two Paleoproterozoic sample sets from the Cuddapah Basin (−27‰ and −31‰) are interpreted in this way (Table 1). Two Archean kerogen sample sets from the Abitibi belt and

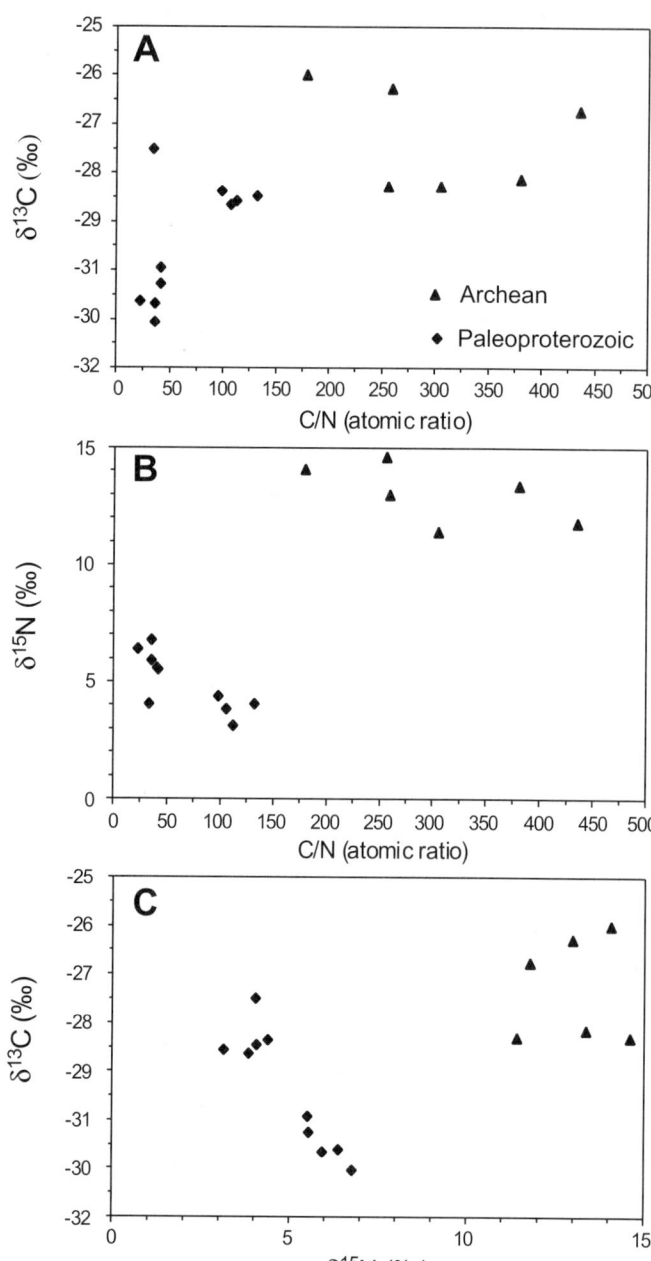

Figure 6. δ^{13}C versus C/N (A), δ^{15}N versus C/N (B), and δ^{13}C versus δ^{15}N (C) of kerogen in Precambrian carbonaceous shales of India.

Botswana, each from greenschist facies carbonaceous shales, possess distinct δ^{13}C populations averaging −20‰ and −26‰, respectively (Table 1 of Jia and Kerrich, 2004b), consistent with different processes during deposition, as suggested by Wellmer et al. (1999). These results complicate the evaluation of metamorphic effects on kerogen samples compiled in Table 2, including new data from India, but endorse preservation of primary signatures (Table 1, Fig. 4).

Long-Term Diffusional Effects

Possible long-term kinetically induced diffusional loss of ^{14}N from kerogen or hydrothermal micas has been evaluated with respect to the K-Ar system. Whereas N, as NH_4^+, substitutes for K in K-silicates, radiogenic Ar gas is not structurally sited in silicates. Diffusion-dependent blocking temperatures are known for amphibole (~500 °C), biotite (300 °C), and muscovite (350 °C) (McDougall and Harrison, 1988). The sample design for kerogen and N-isotope analysis in four terranes 2.7 Ga in age and three Proterozoic terranes was predicated on U-Pb zircon and $^{40}Ar/^{39}Ar$ ages on granitoids within a few tens of million years. Given undetectable Ar loss in these terranes, N-isotope values should also be primary (Jia and Kerrich, 2004b).

Given that Ar-loss is grain-size and temperature dependent (McDougall and Harrison, 1988), the compliance of ^{15}N-enriched data between fine-grained pre-metamorphic micas in VMS deposits and kerogen, with coarse-grained late-metamorphic micas from the same terrane provides further constraints against preferential diffusional loss of ^{14}N (Table 3).

The effects on preferential loss of ^{14}N of (1) structural sitting of N between kerogen and silicates (Jia and Kerrich, 2004b), (2) metamorphism, and (3) age can also be evaluated from data sets for Phanerozoic rocks. There are no obvious differences in $\delta^{15}N$ between these three classes of samples from the Tertiary to Cambrian (Table 5, Fig. 7).

Hydrothermal Alteration

Kerogen for this study, as well as other Precambrian kerogens analyzed by Jia and coworkers, was selected for minimal secondary disturbance on the following criteria: (1) preservation of sedimentary structures; (2) large distance from VMS or gold deposits; (3) large distance from faults; (4) coherent REE patterns and Eu anomalies; and (5) coherent LILE/REE/HFSE systematics (see Jia and Kerrich, 2004b) (Table 2).

IMPLICATIONS AND CONCLUSIONS

Precambrian Oxidation State

Two conflicting models based on various lines of evidence have been developed for the evolution of oxygen in the atmosphere in the Precambrian: low pO_2 in the Archean, with a rapid increase in oxygen ca. 2.3 Ga (Holland, 1999, 2002 and references therein), or alternatively, an oxygenated atmosphere from the early Archean (Ohmoto, 1997, 2004; and references therein). Beaumont and Robert (1999) interpreted their secular isotopic "trend" as a record of changes in the redox potential of Earth. Negative $\delta^{15}N$ values reflect an unspecified metabolic isotopic fractionation under anoxic conditions, with microorganisms using reduced forms of nitrogen, whereas positive $\delta^{15}N$ values reflect an increase in pO_2 after the Paleoproterozoic, which promoted the biologic production of nitrate species (Beaumont and Robert, 1999).

Beaumont and Robert reported data for 12 Mesoarchean cherts (33 analyses), 9 Neoarchean cherts, of which one is split into bedded and homogeneous domains (19 analyses), and 12 Proterozoic cherts (21 analyses). Some samples were analyzed once, others had duplicate or triplicate analytes prepared, and measurements of some analyses were made more than once. Statistically, for a given population multiple analyses of a sample reveal "within sample" variance, whereas differences between samples represent "between sample" variance (Searle, 1971).

TABLE 5. A COMPILATION OF DATA FOR N-ISOTOPES IN PHANEROZOIC SEDIMENTARY KEROGENS AND ROCKS, AND POST PEAK-METAMORPHIC HYDROTHERMAL K-MICAS

Metamorphic facies / Sample location	Age	N content (ppm) Range	N content (ppm) mean ± 1σ	$\delta^{15}N$ (permil) Range	$\delta^{15}N$ (permil) mean ± 1σ	Refs.
Unmetamorphosed:						
(1) Kerogen, Taiwan	Tertiary	700–800	800 ± 100 (5)	3.9–4.0	3.9 ± 0.1 (5)	1
(1) Lavagna, western Alps, Europe	Tertiary	226–762		3.1–4.7		2
(4) Kerogen, Louisiana	Cretaceous	100–3100		2.6–3.6	3.2 ± 0.3 (26)	4
(5) Sandstone, Louisiana	Cretaceous	400–3500	1601 ± 100 (64)	0.2–3.2	3.1 ± 1.4 (51)	4
(9) Kerogen, British Columbia	Late Jurassic	400–9300		-0.6–5.0	2.8 ± 1.9 (17)	7
(13) Mudstone, English Lake District	Ordovician	730–910	476 ± 147 (10)	3.1–4.2	3.7 ± 0.3 (10)	9
Subgreenschist - greenschist facies:						
(6) Catalina complex, California	Cretaceous	100–1100	530 ± 260 (27)	1.1–3.1	2.2 ± 0.6 (27)	5
(8) K-micas, North American Cordillera	190 to 50 Ma	130–3500	1535 ± 1080 (100)	1.6–6.2	3.0 ± 1.2 (100)	6
(10) K-micas, W. Qinlin orogen, N. China	Mesozoic	650–2510	1260 ± 760 (8)	1.7–5.6	4.0 ± 2.0 (8)	3
(11) Kerogen, Pennsylvania	Carboniferous			4.1–5.1	4.6 ± 0.4 (10)	8
(12) Kerogen, Germany	Carboniferous			2.7–3.7	3.2 ± 0.4 (8)	8
(14) Slate, north Wales	Cambrian	129–168	148 (2)	3.9–5.2	4.5 (2)	9
(15) K-micas, Lachlan fold belt, SE. Australia	Paleozoic	650–895	733 ± 103 (20)	2.8–4.5	3.5 ± 0.4 (20)	10
Greenschist - amphibolite facies:						
(2) Schistes Lustrés, western Alps, Europe	Tertiary	169–1721	586 ± 490 (16)	2.6–4.8	3.6 ± 0.7 (16)	2
(3) K-micas, W. Alps	Tertiary	280–290	288 ± 5 (4)	4.4–5.1	4.9 ± 0.6 (4)	3
(7) Catalina complex, California	Cretaceous	35–810	190 ± 150 (14)	3.6–5.9	4.3 ± 0.8 (14)	5

Note: Refs.: 1—Kao and Liu (2000); 2—Busigny et al. (2003); 3—Jia and Kerrich (2004b,c); 4—Williams et al. (1995); 5—Bebout and Fogel (1992); 6—Jia et al. (2003a); 7—Sephton et al. (2002); 8—Ader et al. (1998); 9—Bottrell et al. (1988); 10—Jia et al. (2001).

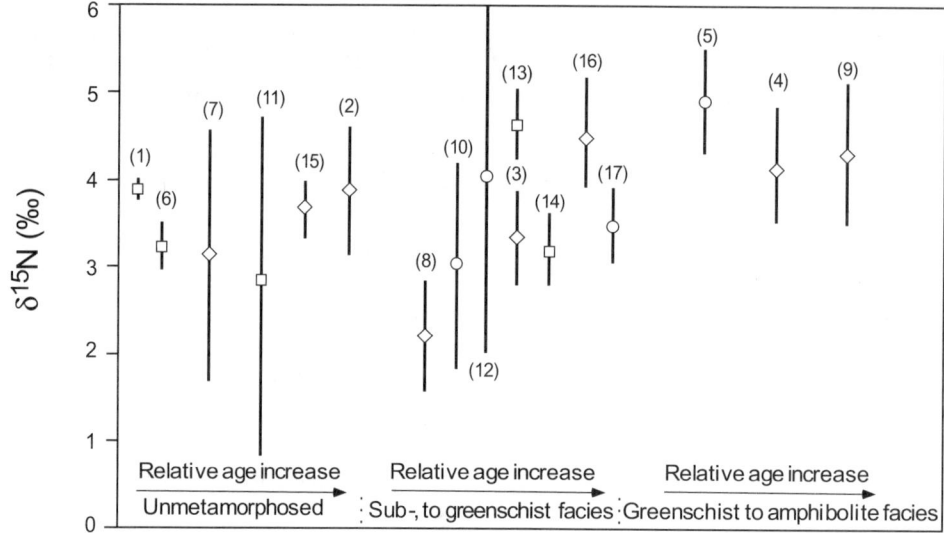

Figure 7. Variation in N-isotopic composition of Phanerozoic kerogen (square), sedimentary rocks (diamond), and hydrothermal K-micas (circle). Open symbols represent mean values of $\delta^{15}N$. Vertical bars represent ± 1 standard deviation of N-isotope data. See Table 5 for data sources and locations (numbers in parentheses).

Consequently, samples with multiple analyses are over-represented as plotted by Beaumont and Robert (1999), Holland (2002), and Marty and Dauphas (2003).

When the data are replotted to take into account "between sample" variance alone, the major shift is from $\delta^{15}N$ averages of –0.5‰ in Mesoarchean chert kerogen, through 7.2‰ in Neoarchean kerogen to Paleoproterozoic and Neoproterozoic averages of 3.9 ‰ and 2.7‰ respectively (Figs. 8, 9). If the model of an anoxic environment until the "great oxidation event" ca. 2.3 Ga is correct, then a rapid rise of pO_2 is not reflected in the Archean to Proterozoic N-isotope record, with the implication that the N cycle was not profoundly changed ca. 2.3 Ga. The redox state of the early atmosphere, and the timing and mechanism(s) of redox transitions, remain controversial (Kasting, 1993; Ohmoto, 1997; Holland, 1999; Phillips et al., 2001; Ohmoto, 2004) and beyond the scope of this paper. New approaches are required as suggested by Anbar and Knoll (2002).

Chemoautotrophs

Pinti et al. (2001a) reported N and C contents and $\delta^{15}N$ from a variety of Archean chert-BIF, where $\delta^{15}N$ spans from –7‰– 20‰. They identify the most depleted values of –7‰, obtained in high-temperature heating steps, as primary and related to metabolic isotopic fractionation of NH_4^+ by chemoautotrophic bacteria at Archean hydrothermal vents. In a related discussion paper, Pinti and Hashizume (2001) also suggested that ^{15}N-depleted compositions in Archean cherts could stem from input of mantle N, but their preferred explanation for values of –6‰ obtained in low-temperature steps by Sano and Pillinger (1990), and ^{15}N-depleted values in the data set of Beaumont and Robert (1999), is for chemoautotrophic processes.

Chert-BIF

All data sets from Precambrian chert-BIF, metamorphosed from greenschist to amphibolite facies, are characterized by (1) some depleted $\delta^{15}N$ values (–5‰); (2) a large range of $\delta^{15}N$ values, up to 30‰, in greenschist facies cherts; and (3) an absence of correlation of $\delta^{15}N$ with either N content or C/N ratios. The third observation rules out Rayleigh fractionation as a cause of enriched values.

Precambrian chert and BIF are often interbedded; their spatial and temporal association with volcanic sequences erupted from mantle plumes has been used as evidence for a hydrothermal origin of these sediments (Barley et al., 1998, and references therein). Isley and Abbott (1999) and Condie et al. (2001) established a statistical correlation of the time-series for BIF and volcanic sequences erupted from mantle plumes, corroborating the empirical association (Fig. 9).

Consequently, the large spread of values reported by Beaumont and Robert (1999) and Pinti et al. (2001a) may be primary. Pinti et al. (2001a) measured $\delta^{15}N$ of –7.4‰ in sample Pano D-136 from 3.5 Ga cherts of the Pilbara Craton, Western Australia, for which $^{40}Ar/^{36}Ar$ at 58,500 and $N_2/^{36}Ar$ ratios are close to those estimated for the upper mantle of $\delta^{15}N = -5 \pm 2$‰, $^{40}Ar/^{36}Ar = 42,000$, and $N_2/^{36}Ar = (5 \pm 2) \times 10^6$ (Marty and Zimmermann, 1999). Independently, Pinti et al. (2001b) reported Xe-isotope evidence for input of mantle volatiles to the Pilbara cherts.

In light of these observations and constraints on the magnitude of metamorphic $\delta^{15}N$ shifts to ≤3‰, we reinterpret the range

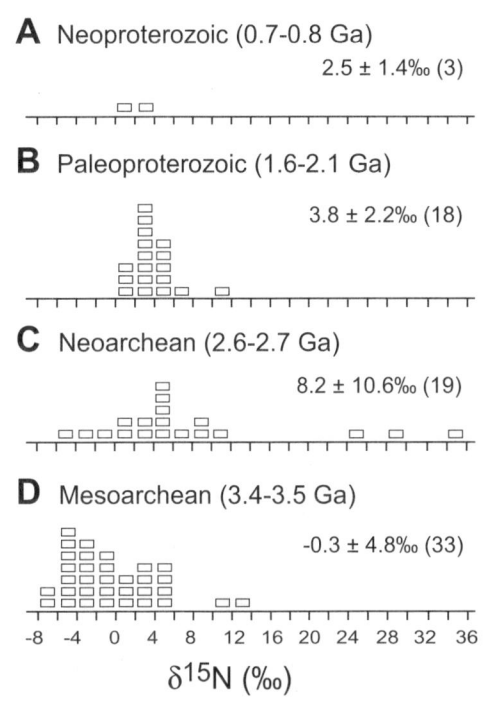
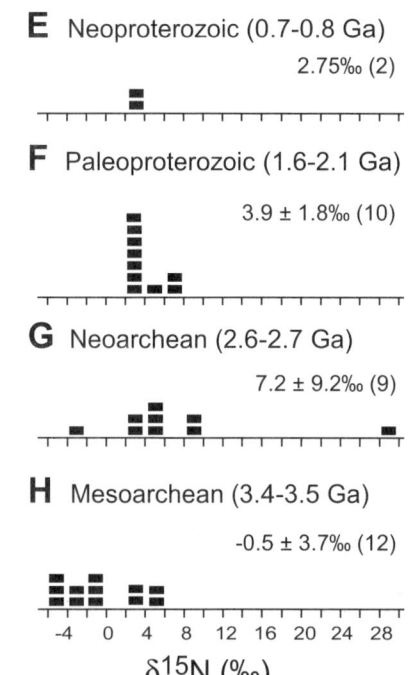

Figure 8. N-isotope compositions of Precambrian cherts and/or iron formation (Data are from Beaumont and Robert, 1999). (A–D) Histograms reproduced from Beaumont and Robert (1999) who plot all their analyses. (E–G) Histograms plotted in this study based on averages for replicate analyses of the same sample. Average and one standard deviation are listed above each of the eight histograms, with numbers in parentheses on left and right sides representing numbers of analyses and of actual samples, respectively.

of $\delta^{15}N$ values from depleted to enriched in all of the data sets for Precambrian chert-BIF as mixing between a ^{15}N-depleted mantle source emitted from plumes and a ^{15}N-enriched marine sedimentary kerogen component, identified by Jia and Kerrich (2004a, 2004b) and this study (Fig. 3A). Given a mantle plume association, BIFs are not a record of near-surface redox conditions.

Secular Trends

Nitrogen contents of micas from orogenic gold deposits mirror the secular variation in N content of carbonaceous shales (Table 2, Fig. 4B). This result is consistent with the breakdown of sedimentary kerogen as the primary source of N in the hydrothermal fluids from which the micas were deposited, with dehydration of K-micas as a secondary source of N. This interpretation is also in keeping with the second-order observation that Archean micas are characterized by a larger range of $\delta^{15}N$ than Phanerozoic equivalents (Tables 2, 3, 5). Jia et al. (2001) accounted for this distribution in that Archean terranes are volcanic dominated, with subordinate sedimentary units, whereas Phanerozoic accretionary belts possess a higher proportion of turbidites. Consequently, metamorphism of Archean terranes would generate metamorphic fluids having a variable N content and a $\delta^{15}N$ depending on the proportions of metaigneous (low N content, depleted $\delta^{15}N$) and metasedimentary (high N content, enriched $\delta^{15}N$) rocks. Metamorphism of Phanerozoic terranes would generate fluids in which the N-budget was dominated by metasedimentary rocks. This interpretation is also consistent with fluid inclusions, characterized by high N_2 content (0.6–99.0 mol) from quartz-gold veins in organic-rich slate belts (Bottrell et al., 1988; Ortega et al., 1991; de Ronde et al., 1992).

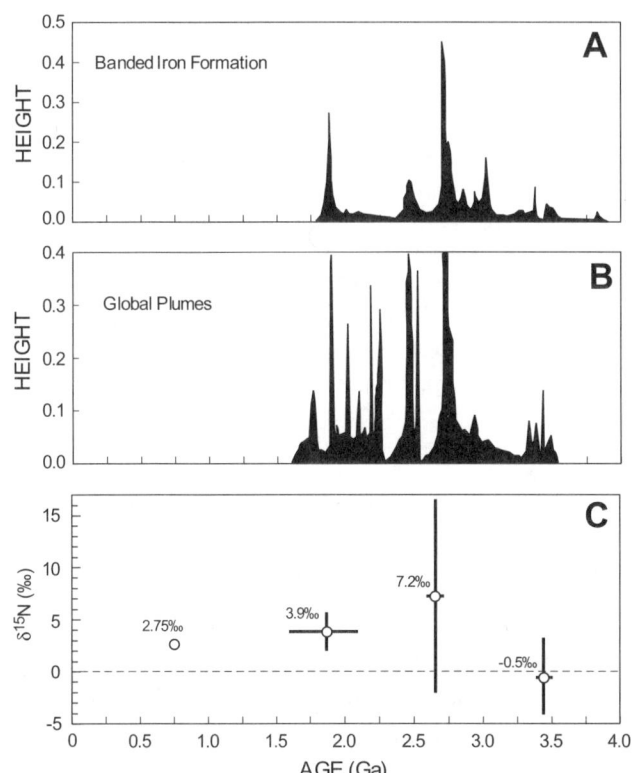

Figure 9. Time-series of occurrences of banded iron formation (A) and global plumes (B) from Isley and Abbott (1999). N-isotopic compositions of chert–iron formation (C) from Beaumont and Robert (1999) as plotted in Figure 8 D–G. These time-series are generated by summing Gaussian distributions of unit area using mean ages and standard deviations.

Hayes et al. (1983) documented a general trend of increasing $\delta^{13}C$ with decreasing organic C content for Precambrian sedimentary rocks. They modeled the trend as a Rayleigh fractionation where $\delta^{13}C$ of residual organic C shifted by ~6‰ for each factor-of-ten loss of C. However, both $\delta^{13}C$ and TOC overlap for their Archean and Proterozoic data sets (Figure 8 of Hayes et al., 1983). In addition, differences of $\delta^{13}C$ and $\delta^{15}N$ between populations of samples within given Archean and Proterozoic terranes are preserved (Table 1) (Wellmer et al., 1999; Jia and Kerrich, 2004b). We interpret increasing bulk N contents of sedimentary rocks as follows: Most Archean siliciclastic sequences are first-cycle volcanogenic turbidites shed off of bimodal arcs into tectonically active basins. Stable passive margins developed in the Proterozoic, receiving cratonic detritus at relatively slow sedimentation rates (Taylor and McLennan, 1985). Accordingly, organic compounds were diluted by detritus in the former setting relative to the latter (Jia et al., 2001).

The implication from the shale and mica record is that N-fixing microorganisms have progressively drawn down atmospheric N_2, to sequester it as NH_4^+ in crustal K-silicates, increasing the crustal N-inventory from an original basaltic value of 1–2 ppm (Table 6). These processes concurrently shift atmospheric and crustal $\delta^{15}N$ values down. An analogous process is sequestration of atmospheric CO_2, with transfer to the crustal carbonate and kerogen budgets. According to Delsemme (1998; 2001), the crustal carbonate budget translates into a Hadean atmosphere with 20 times the present CO_2.

Origin of the Archean Atmosphere-Hydrosphere

Archean sedimentary kerogens and crustal hydrothermal systems as recorded in this study, together with compilations from the literature, signify systematically ^{15}N-enriched values of 15‰–24‰ (Tables 2, 3; Fig. 4). If the isotopic fractionation between kerogens and atmosphere in the Precambrian was approximately the same as at present, then it implies a ^{15}N-enriched atmosphere in the Archean (13‰–21‰ ca. 2.7 Ga). It is possible that the ^{15}N-enriched values stem from a different N-cycle in the Archean, with large biologically mediated fractionations, yet the magnitude of the fractionations exceeds any presently known.

Kramers (2003) conducted an evaluation of Earth's volatile budget. He compared volatile element abundances in the outer Earth reservoirs (OER: atmosphere-hydrosphere, continental crust, MORB-source mantle) with average carbonaceous chondrites (CC) on an Al-normalized basis. Normalized abundances of I, Br, N, and C decrease relative to CC in a manner consistent with mass-dependent hydrodynamic loss during energetic impacts and extreme UV (EUV) solar radiation. Kramers accounted for overabundance of H and Cl by their residence dominantly in an ocean, and for Ne-isotope characteristics of mantle plumes by incorporation of ~10^{-7} of Earth's volatile budget from solar atmosphere. The results of this study are consistent with the model of Kramers (2003). An implication is that if isotopic fractionation of N accompanied its hydrodynamic loss, then Earth's secondary atmosphere could have acquired an initial $\delta^{15}N$ >CI chondrite.

Several authors have proposed an E-chondrite model for Earth based on rare ^{15}N depleted diamonds (Javoy, 1998; Tolstikhin and Marty, 1998). Those diamonds at ~–25‰ (Cartigny et al., 1997, 1998) are close to the range of $\delta^{15}N$ in E-chondrites of –15‰ to –43‰ (Kung and Clayton, 1978; Grady et al., 1986). However, a largely pure E-chondritic mantle can be ruled out from chemical and isotopic compositional data of Earth (e.g., Allègre et al., 2001; Drake and Righter, 2002), Cr-isotope data of the mantle and various classes of meteorites (Shukolyukov and Lugmair, 1998), and ^{15}N-enriched Archean carbonaceous shales (this study).

The best material to fit the isotope characteristics for ^{15}N-enriched atmosphere and carbonaceous shales is CI-chondrite-like material having $\delta^{15}N$ of 30‰–42‰ (Kerridge, 1985) at the end of Earth accretion ca. 4.5 Ga, and/or comets formed in the vicinity of Jupiter, which likely had the same N-isotopic composition as CI-chondrites, because they originated in the same zone of the early solar system. Volatiles in CI-chondrites and comets may be similar (Delsemme, 2001).

After considering the $\delta^{18}O$ and other isotopes, and the major element abundances (e.g., Mg/Si vs. Al/Si) in Earth and chondritic meteorites, Drake and Righter (2002) ruled out either E- or CI-chondrites as source materials for Earth. Rather, they considered Earth to have formed by accretion in a narrow feeding annulus at ~1 astronomical unit (AU), close to its present orbital radius. They argued that Earth's budget of water cannot have been acquired primarily from a late veneer of CI chondrites in the asteroid belt, and that comets originating in the Oort cloud at 2.6 AU can deliver no more than 50% of Earth's minimum water budget; consequently, Earth's water is indigenous. However, Robert (2001) argued that the source of Earth's water is consistent with current understanding of the water content of the asteroid belt, as inferred from the chondritic meteorite record and from the mean of D/H ratios of clay minerals in carbonaceous chondrites, which is close to standard mean ocean water. According to Morbidelli et al. (2000), from early stages of accretion to late stage gas-free sweep-up of planetesimals, water was delivered to Earth from a mix of objects in the asteroid belt, in the vicinity of the giant planets, and in the Kuiper Belt.

If the ^{15}N-enriched values of Archean samples reflect a commensurately enriched atmosphere, then it is possible that N is

TABLE 6. GLOBAL NITROGEN INVENTORIES

	Mass × 10^{24} (g)[†]	Abundance[‡]	Mass N (g)	%
Atmosphere	5.1×10^{-3}	78%[1]	3.9×10^{21}	30.1
Continental crust	2.2×10^{1}	50 ppm[2]	1.1×10^{21}	8.5
Upper mantle	6.2×10^{2}	0.27 ppm[3]	1.7×10^{20}	1.3
Lower mantle	2.9×10^{3}	2.7 ppm[3]	7.8×10^{21}	61.1

[†]The mass of the atmosphere is from Jacobson et al. (2000). The Continental crust, upper mantle, and lower mantle are calculated based on their respective (percentage) 0.374%, 10.3%, and 49.2% of the earth's mass at 5.98×10^{27} (g).
[‡]Abundances are from: 1—Kramers (2003); 2—Wedepohl (1995); 3—Marty And Dauphas (2003).

not dominantly indigenous, but was acquired from ^{15}N- enriched solar system materials including CI-chondrite type material and comets. The secular trend of ^{15}N in shale kerogen documented in this study reflects a corresponding atmospheric trend from >21‰ ca. 2.7 Ga to 0‰ now: the shift can be accounted for by a combination of (1) mantle degassing (continuous addition of N_2 with $\delta^{15}N$ −5‰); (2) progressive sequestration of atmospheric N_2 into crustal rocks by nitrogen fixing organisms, with a return flux of ^{15}N-depleted N_2 stemming from diagenetic fractionation; and (3) recycling of ^{15}N-enriched Archean sediments into the mantle. Such nitrogen recycling has been proposed by Zhang and Zindler (1993). Given the limited database for Archean rocks this interpretation is of necessity speculative, and other explanations may emerge. Isotopic mass balance for N is difficult given uncertainties in mantle N-content, mass of subducted material, and the conservative or volatile behavior of N at convergent margins. Kerrich and Jia (2004) attempted a simplified model, which indicated that subduction has caused a shift of <0.1‰ in the mantle $\delta^{15}N$ using the "chert"-like kerogen recycling assumptions of Marty and Dauphas (2003).

ACKNOWLEDGMENTS

We thank Steve Kesler and Hirishi Ohmoto for the invitation to submit this manuscript. We are grateful to M. Stocki for assistance with the nitrogen analysis in the Department of Soil Sciences, University of Saskatchewan; P. Lindgren for orchestrating a field trip to Jerome; and K.M. Ansdell, T. Oberthür, B. Pratt, A. Still, and S. Vearncombe for providing some of the samples. Y. Jia acknowledges receipt of a CSIRO postdoctoral fellowship, an honorary position at Monash University, and a research grant from SEG Foundation, Inc., USA. R. Kerrich acknowledges a Natural Sciences and Engineering Research Council (NSERC) Discovery Grant, an NSERC MFA grant, and the George McLeod endowment to the Department of Geological Sciences, University of Saskatchewan. The critiques of H. Ohmoto, S. Kesler, D. L. Pinti and an anonymous journal reviewer significantly improved an earlier version of the manuscript.

REFERENCES CITED

Ader, M., Boudou, J., Javoy, M., Goffe, B., and Daniels, E., 1998, Isotope study on organic nitrogen of Westphalian anthracites from the Western Middle field of Pennsylvanian (U.S.A) and from the Bramsche Massif (Germany): Organic Geochemistry, v. 29, p. 315–323, doi: 10.1016/S0146-6380(98)00072-2.

Allègre, C., Manhès, G., and Lewin, È., 2001, Chemical composition of the Earth and the volatility control on planetary genetics: Earth and Planetary Science Letters, v. 185, p. 49–69, doi: 10.1016/S0012-821X(00)00359-9.

Anbar, A.D., and Knoll, A.H., 2002, Proterozoic ocean chemistry and evolution: A bioinorganic bridge?: Science, v. 297, p. 1137–1142, doi: 10.1126/science.1069651.

Anderson, A., Jr., Clayton, R.N., and Mayeda, T.K., 1971, Oxygen isotope thermometry of mafic igneous rocks: Journal of Geology, v. 79, p. 715–729.

Balakrishnan, S., Rajamani, V., and Hansen, G.N., 1999, U-Pb ages for zircon and titanite from the Ramagiri area, southern India: Evidence for accretionary origin of the eastern Dharwar craton during the Late Archean: Journal of Geology, v. 107, p. 69–86, doi: 10.1086/314331.

Barley, M.E., Krapez, B., Groves, D.I., and Kerrich, R., 1998, The Late Archaean bonanza: Metallogenic and environmental consequences of the interaction between mantle plumes, lithosphere tectonics and global cyclicity: Precambrian Research, v. 91, p. 65–90, doi: 10.1016/S0301-9268(98)00039-4.

Beaty, D.W., and Taylor, H.P., Jr., 1982, Some petrologic and oxygen isotopic relationships in the Amulet Mine, Noranda, Quebec, and their bearing on the origin of Archean massive sulfide deposits: Economic Geology and the Bulletin of the Society of Economic Geologists, v. 77, p. 95–108.

Beaumont, V., and Robert, F., 1999, Nitrogen isotope ratios of kerogens in Precambrian cherts: A record of the evolution of atmosphere chemistry: Precambrian Research, v. 96, p. 63–82, doi: 10.1016/S0301-9268(99)00005-4.

Beaumont, V., Agrinier, P., Javoy, M., and Robert, F., 1994, Determination of the CO contribution to the $^{15}N/^{14}N$ ratio measured by mass-spectrometry: Analytical Chemistry, v. 66, p. 2187–2189, doi: 10.1021/ac00085a039.

Bebout, G.E., and Fogel, M.L., 1992, Nitrogen-isotope compositions of metasedimentary rocks in the Catalina Schist, California: Implications for metamorphic devolatilization history: Geochimica et Cosmochimica Acta, v. 56, p. 2839–2849, doi: 10.1016/0016-7037(92)90363-N.

Bebout, G.E., Ryan, J.G., Leeman, W.P., and Bebout, A.E., 1999, Fractionation of trace elements by subduction-zone metamorphism – effect of convergent-margin thermal evolution: Earth and Planetary Science Letters, v. 171, p. 63–81, doi: 10.1016/S0012-821X(99)00135-1.

Bleeker, W., Parrish, R.R., and Sager, K.A., 1999, High-precision U-Pb geochronology of the late Archean Kidd Creek Deposit and Kidd volcanic complex: Economic Geology Monograph 10, p. 43–69.

Böhlke, J.K., Eriksen, G.E., and Revesz, K., 1997, Stable isotope evidence for an atmospheric origin of desert nitrate deposits in northern Chile and southern California, USA: Chemical Geology, v. 136, p. 135–152, doi: 10.1016/S0009-2541(96)00124-6.

Bos, A., Duit, W., Eerden, M.J., and Jansen, J.B.H., 1988, Nitrogen storage in biotite: An experimental study of the ammonium and potassium partitioning between IM-phlogopite and vapour at 2 kb: Geochimica et Cosmochimica Acta, v. 52, p. 1275–1283, doi: 10.1016/0016-7037(88)90281-5.

Bottrell, S.H., Carr, L.P., and Dubessy, J., 1988, A nitrogen-rich metamorphic fluid and coexisting minerals in slates from North Wales: Mineralogical Magazine, v. 52, p. 451–457.

Boyd, S.R., and Philippot, P., 1998, Precambrian ammonium biogeochemistry: A study of the Moine metasediments, Scotland: Chemical Geology, v. 144, p. 257–268, doi: 10.1016/S0009-2541(97)00135-6.

Boyd, S.R., Mattey, D.P., Pillinger, C.T., Milledge, H.J., Mendelssohn, M., and Seal, M., 1987, Multiple growth events during diamond genesis: An integrated study of carbon and nitrogen isotopes and nitrogen aggregation state in coated stone: Earth and Planetary Science Letters, v. 86, p. 341–353, doi: 10.1016/0012-821X(87)90231-7.

Boyd, S.R., Pillinger, C.T., Milledge, H.J., Mendelssohn, M., and Seal, M., 1992, C and N isotopic composition and the infrared absorption spectra of coated diamonds: Evidence for the regional uniformity of CO_2-H_2O rich fluids in lithospheric mantle: Earth and Planetary Science Letters, v. 109, p. 633–644, doi: 10.1016/0012-821X(92)90121-B.

Boyd, S.R., Hall, A., and Pillinger, C.T., 1993, The measurement of $\delta^{15}N$ in crustal rocks by static vacuum mass spectrometry: Application to the origin of the ammonium in the Cornubian batholith, southwest England: Geochimica et Cosmochimica Acta, v. 57, p. 1339–1347, doi: 10.1016/0016-7037(93)90070-D.

Busigny, V., Cartigny, P., Philippot, P., Ader, M., and Javoy, M., 2003, Massive recycling of nitrogen and other fluid-mobile elements (K, Rb, Cs, H) in a cold slab environment: Evidence from HP to UHP oceanic metasediments of the Schistes Lustrés nappe (western Alps, Europe): Earth and Planetary Science Letters, v. 215, p. 27–42, doi: 10.1016/S0012-821X(03)00453-9.

Cartigny, P., and Ader, M., 2003, A comment on "The nitrogen record of crust-mantle interaction and mantle convection from Archean to Present" by B. Marty and N. Dauphas: Earth and Planetary Science Letters, v. 216, p. 425–432, doi: 10.1016/S0012-821X(03)00505-3.

Cartigny, P., Boyd, S.R., Harris, W.J., and Javoy, M., 1997, Nitrogen isotopes in peridotitic diamonds from Fuxian, China: The mantle signature: Terra Nova, v. 9, p. 175–179, doi: 10.1046/j.1365-3121.1997.d01-26.x.

Cartigny, P., Harris, J.W., Phillips, D., Girard, M., and Javoy, M., 1998, Subduction-related diamonds?: The evidence for a mantle-derived origin from coupled $\delta^{13}C$–$\delta^{15}N$ determinations: Chemical Geology, v. 147, p. 147–159, doi: 10.1016/S0009-2541(97)00178-2.

Cartigny, P., Jendrzejewski, J.N., Pineau, F., Petit, E., and Javoy, M., 2001, Volatiles (C, N, Ar) variability in MORB and the respective roles of mantle source heterogeneity and degassing: The South West Indian Ridge's case: Earth and Planetary Science Letters, v. 194, p. 241–247, doi: 10.1016/S0012-821X(01)00540-4.

Chen, C.T., Gong, A.G.C., Wang, S.L., and Bychkov, A.S., 1996, Redfield ratios and regeneration rates of particulate matter in the Sea of Japan as a model of closed system: Geophysical Research Letters, v. 23, p. 1785–1788, doi: 10.1029/96GL01676.

Cline, J.D., and Kaplan, I.R., 1975, Isotope fractionation of dissolved nitrate during denitrification in the Eastern Tropical North Pacific: Marine Chemistry, v. 3, p. 271–299, doi: 10.1016/0304-4203(75)90009-2.

Condie, K.C., Des Marais, D.J., and Abbott, D., 2001, Precambrian superplumes and supercontinents: A record in carbonaceous shales, carbon isotopes, and paleoclimates?: Precambrian Research, v. 106, p. 239–260, doi: 10.1016/S0301-9268(00)00097-8.

Conway, N.M., Kenicutt, M.C., and Van Dover, C.L., 1994, Stable isotopes in the study of marine chemosynthetic-based ecosystems, in Lajtha, K., and Michener, R.H., eds., Stable isotopes in ecology and environmental science: Oxford, UK, Blackwell Scientific, p. 158–186.

Corfu, F., Krogh, T.E., Kwok, Y.Y., and Jensen, L.S., 1989, U-Pb zircon geochronology in the southwestern Abitibi greenstone belt, Superior Province: Canadian Journal of Earth Sciences, v. 26, no. 9, p. 1747–1763.

Costa, U.R., Barnett, R.L., and Kerrich, R., 1983, The Mattagami Lake Mine Archean Zn-Cu sulfide deposit, Quebec: Hydrothermal coprecipitation of talc and sulfides in a sea-floor brine pool—Evidence from geochemistry, $^{18}O/^{16}O$, and mineral chemistry: Economic Geology and the Bulletin of the Society of Economic Geologists, v. 78, p. 1144–1203.

Dasgupta, S., and Sengupta, P., 2003, Indo-Antarctic correlation: A perspective from the Eastern Ghats granulite belt, India, in Yosida, M., Windley, F., and Dasgupta, S., eds., Proterozoic East Gondwana: Supercontinent assembly and breakup: Geological Society [London] Special Publication 206, p.131–143.

Delsemme, A., 1998, Our Cosmic Origin from the Big Bang to the Emergence of Life and Intelligence: Cambridge, UK, Cambridge University Press, 269 p.

Delsemme, A., 2001, An argument for the cometary origin of the biosphere: American Scientist, v. 89, p. 432–442, doi: 10.1511/2001.5.432.

Delwiche, C., and Steyn, P.L., 1970, Nitrogen isotope fractionations in soils and microbial reactions: Environmental Science and Technology, v. 4, p. 929–935, doi: 10.1021/es60046a004.

de Ronde, C.E.J., Spooner, E.T.C., De Wit, M.J., and Bray, C.J., 1992, Shear zone-related Au quartz vein deposits in the Barberton greenstone belt, South Africa: Field and petrological characteristics, fluid properties, and light stable isotope geochemistry: Economic Geology and the Bulletin of the Society of Economic Geologists, v. 88, p. 366–402.

Drake, M.J., and Righter, K., 2002, Determining the composition of the Earth: Nature, v. 416, p. 39–44, doi: 10.1038/416039a.

Drummond, M.S., and Defant, M.J., 1990, A model for trondhjemite-tonalite-dacite genesis and crustal growth via slab melting: Archean to modern comparisons: Journal of Geophysical Research, v. B95, p. 21503–21521.

Durand, B., and Nicaise, G., 1980, Procedures for kerogen isolation, in Burand, B., ed., Kerogen: Paris, Technip, p. 35–53.

Feng, R., and Kerrich, R., 1992, Geodynamic evolution of the Southern Abitibi and Pontiac terranes: Evidence from the geochemistry of granitoid magma series (2700 Ma–2630 Ma): Canadian Journal of Earth Sciences, v. 29, p. 2266–2286.

Fraga, F., Ríos, A.F., Pérez, F.F., and Figueiras, F.G., 1998, Theoretical limits of oxygen/carbon and oxygen/nitrogen ratios during photosynthesis and mineralization of organic matter in the sea: Scientia Marina, v. 62, p. 161–168.

Franklin, J.M., Lydon, J.W., and Sangster, D.F., 1981, Volcanic-associated massive sulfide deposits: Economic Geology, 75th Anniversary Volume, p. 485–627.

Freudenthal, T.T., Wagner, T., Wezhöfer, F., Zabel, M., and Wefer, T., 2001, Early diagenesis of organic matter from sediments of the eastern subtropical Atlantic: Evidence from stable nitrogen and carbon isotopes: Geochimica et Cosmochimica Acta, v. 65, p. 1795–1808, doi: 10.1016/S0016-7037(01)00554-3.

Grady, M.M., Wright, I.P., Carr, L.P., and Pillinger, C.T., 1986, Compositional differences in enstatite chondrites based on carbon and nitrogen stable isotope measurements: Geochimica et Cosmochimica Acta, v. 50, p. 2799–2813, doi: 10.1016/0016-7037(86)90228-0.

Gibson, E.K., Carr, L.P., and Pillinger, C.T., 1985, Nitrogen isotopic composition of Archean samples: Evidence of the earth's early atmosphere? [abs.]: Lunar and Planetary Science, v. XVI, p. 270–272.

Gibson, E.K., Carr, L.P., Gilmour, I., and Pillinger, C.T., 1986, Earth's atmosphere during the Archean as seen from carbon and nitrogen isotopic analysis of sediments [abs.]: Lunar and Planetary Science, v. XVII, p. 258–259.

Goldfarb, R.J., Groves, D.I., Gardoll, S., 2001, Orogenic gold and geologic time; a global synthesis: Ore Geology Reviews, v. 18, p. 1–75.

Haendel, D., Mühle, K., Nitzsche, H., Stiehl, G., and Wand, U., 1986, Isotopic variations of the fixed nitrogen in metamorphic rocks: Geochimica et Cosmochimica Acta, v. 50, p. 749–758, doi: 10.1016/0016-7037(86)90351-0.

Hagemann, S.G., and Cassidy, K.F., 2000, Archean orogenic lode gold deposits, in Hagemann, S.G., and Brown, P.E., eds., Gold in 2000: Reviews in Economic Geology, v. 13, p. 9–68.

Hamilton, J.V., and Hodgson, C.J., 1986, Mineralization and structure of the Kolar gold field, India, in MacDonald, A.J., ed., Gold '86: Willowdale, Ontario, Konsult International, p. 270–283.

Hannington, M.D., Barrie, C.T., and Bleeker, W., 1999, The giant Kidd Creek volcanogenic massive sulfide deposit, western Abitibi Subprovince, Canada; summary and synthesis: Economic Geology Monograph 10, p. 661–672.

Hanschmann, G., 1981, Berechnung von isotopieeffekten auf quantenchemischer grundlage am beispiel stickstoffhaltiger molecule: Zfl-Mitteilungen, v. 41, p. 19–31.

Hall, A., 1989, Ammonium in spilitized basalts of southwest England and its implications for the recycling of nitrogen: Geochemical Journal, v. 23, p. 19–23.

Hall, A., 1999, Ammonium in granites and its petrogenetic significance: Earth-Science Review, v. 45, p. 145–165, doi: 10.1016/S0012-8252(99)00006-9.

Hayes, J.M., Kaplan, I.R., and Wedeking, K.W., 1983, Precambrian organic geochemistry, preservation of the record, in Schopf, W.J., ed., Earth's Earliest Biosphere: Cambridge University Press, p. 92–134.

Hoch, M.P., Fogel, M.L., and Kirchman, D.L., 1994, Isotope fractionation during ammonium uptake by marine microbial assemblages: Geomicrobiological Journal, v. 12, p. 113–127.

Holland, H.D., 1999, When did the Earth's atmosphere become oxic? A Reply: Geochemical News, v. 100, p. 20–23.

Holland, H.D., 2002, Volcanic gases, black smokers, and the great oxidation event: Geochimica et Cosmochimica Acta, v. 66, p. 3811–3826, doi: 10.1016/S0016-7037(02)00950-X.

Holloway, J.M., and Dahlgren, R.A., 2002, Nitrogen in rock: Occurrences and biogeochemical implications: Global Biogeochemical Cycles, v. 16, doi:10.1029/2002GB001862.

Honma, H., and Itihara, Y., 1981, Distribution of ammonium in minerals of metamorphic and granite rocks: Geochimica et Cosmochimica Acta, v. 45, p. 983–988, doi: 10.1016/0016-7037(81)90122-8.

Isley, A.E., and Abbott, D.H., 1999, Plume-related mafic volcanism and the deposition of banded iron formation: Journal of Geophysical Research, v. 104, p. 15461–15477, doi: 10.1029/1999JB900066.

Jackson, S.L., and Fyon, J.A., 1991, The western Abitibi subprovince in Ontario: Ontario Geological Survey, Special Volume 4, pt. 1, p. 405–482.

Jacobson, M.C., Charlson, R.J., Rodhe, H., and Orians, G.H., 2000, Earth system science from biogeochemical cycles to global change: San Diego, Academic Press, 527 p.

Javoy, M., 1998, The birth of the Earth's atmosphere: The behavior and fate of its major elements: Chemical Geology, v. 147, p. 11–25, doi: 10.1016/S0009-2541(97)00169-1.

Javoy, M., and Pineau, F., 1991, The volatiles record of a "popping" rock from the Mid Atlantic Ridge at 14° N: Chemical and isotopic composition of gas trapped in the vesicles: Earth and Planetary Science Letters, v. 107, p. 598–611, doi: 10.1016/0012-821X(91)90104-P.

Jia, Y., 2004, Nitrogen isotope compositions of the Cooma metamorphic complex, Australia: Implications for ^{15}N-enriched reservoirs: Geochimica et Cosmochimica Acta, v. 68, Suppl. 1, p. A42–A42.

Jia, Y., and Kerrich, R., 1999, Nitrogen isotope systematics of mesothermal lode gold deposits: Metamorphic, granitic, meteoric water, or mantle origin?: Geology, v. 27, p. 1051–1054, doi: 10.1130/0091-7613(1999)027<1051: NISOML>2.3.CO;2.

Jia, Y., and Kerrich, R., 2000, Giant quartz vein systems in accretionary orogenic belts: The evidence for a metamorphic fluid origin from $\delta^{15}N$ and $\delta^{13}C$ studies: Earth and Planetary Science Letters, v. 184, p. 211–224, doi: 10.1016/S0012-821X(00)00320-4.

Jia, Y., and Kerrich, R., 2004a, A reinterpretation of the crustal N-isotope record: Evidence for a ^{15}N-enriched Archean atmosphere?: Terra Nova, v. 16, p. 102–108, doi: 10.1111/j.1365-3121.2004.00535.x.

Jia, Y., and Kerrich, R., 2004b, Nitrogen 15-enriched Precambrian kerogen and hydrothermal systems: Geochemistry, Geophysics, and Geosystems, v. 5, doi:10.1029/2004GC000716.

Jia, Y., and Kerrich, R., 2004c, Secular evolution of δ^{15}N in crustal fluids: Implications for the origin of Earth's early atmosphere and hydrosphere, in Wanty, R.B., and Seal, R.R., eds., Proceedings of the 11th International Symposium on Water-Rock Interaction: New York, A.A. Balkema, p. 945–949.

Jia, Y., Li, X., and Kerrich, R., 2001, Stable isotope (O, H, S, C, and N) systematics of quartz vein systems in the turbidite-hosted Central and North Deborah gold deposits of the Bendigo gold field, central Victoria, Australia: Constraints on the origin of ore-forming fluids: Economic Geology and the Bulletin of the Society of Economic Geologists, v. 96, p. 705–721.

Jia, Y., Kerrich, R., and Goldfarb, R., 2003a, Genetic constraints of orogenic gold-bearing quartz vein systems in the western North American Cordillera: Constraints from a reconnaissance study of δ^{15}N, δD, and δ^{18}O: Economic Geology and the Bulletin of the Society of Economic Geologists, v. 98, p. 109–213.

Jia, Y., Kerrich, R., Gupta, A.K., and Fyfe, W.S., 2003b, ^{15}N enriched Gondwana lamproites, eastern India: Crustal N in the mantle source: Earth and Planetary Science Letters, v. 215, p. 43–56, doi: 10.1016/S0012-821X(03)00426-6.

Kao, S.J., and Liu, K.K., 2000, Stable carbon and nitrogen isotope systematics in a human-disturbed watershed (Lanyang-His) in Taiwan and the estimation of biogenic particular organic carbon and nitrogen fluxes: Global Biogeochemical Cycles, v. 14, p. 189–198, doi: 10.1029/1999GB900079.

Kasting, J.F., 1993, Earth's early atmosphere: Science, v. 259, p. 920–926.

Kemp, A.I.S., and Hawkesworth, C.J., 2004, 3.11. Granitic perspectives on the generation and secular evolution of the continental crust, in Rudnick, R., ed., The crust: Amsterdam, Elsevier, Treatise on Geochemistry, v. 3,p. 349–410.

Kerrich, R., and Cassidy, K.F., 1994, Temporal relationships of lode-gold mineralization to accretion, magmatism, metamorphism and deformation, Archean to present: A review: Ore Geology Reviews, v. 9, p. 263–310, doi: 10.1016/0169-1368(94)90001-9.

Kerrich, R., and Jia, Y., 2004, A comment on "The nitrogen record of crust-mantle interaction and mantle convection from Archean to Present" by B. Marty and N. Dauphas: Earth and Planetary Science Letters, v. 225, p. 435–440, doi: 10.1016/j.epsl.2004.07.004.

Kerrich, R., and Ludden, J., 2000, The role of fluids during formation and evolution of the Southern Superior Province lithosphere: A review: Canadian Journal of Earth Sciences, v. 37, p. 135–165, doi: 10.1139/cjes-37-2-3.135.

Kerrich, R., Goldfarb, R., Groves, D., Garvin, S., 2000, The geodynamics of world-class gold deposits: Characteristics, space-time distribution, and origins: Reviews of Economic Geology, v. 13, p. 501.544.

Kerridge, J.F., 1985, Carbon, hydrogen and nitrogen in carbonaceous chondrites: Abundances and isotopic compositions in bulk samples: Geochimica et Cosmochimica Acta, v. 49, p. 1707–1714, doi: 10.1016/0016-7037(85)90141-3.

Kramers, J.D., 2003, Volatile element abundance patterns and an early liquid water ocean on Earth: Precambrian Research, v. 126, p. 379–394, doi: 10.1016/S0301-9268(03)00106-2.

Krohn, M.D., Kendall, C., Evans, R.J., and Fries, T.L., 1993, Relations of ammonium minerals at several hydrothermal systems in the western U.S.: Journal of Volcanology and Geothermal Research, v. 56, p. 401–413, doi: 10.1016/0377-0273(93)90005-C.

Kung, C.C., and Clayton, R.N., 1978, Nitrogen abundances and isotopic compositions in stony meteorites: Earth and Planetary Science Letters, v. 38, p. 421–435, doi: 10.1016/0012-821X(78)90117-6.

Landis, C.A., 1971, Graphitization of dispersed carbonaceous material in metamorphic rocks: Contributions to Mineralogy and Petrology, v. 30, p. 34–45, doi: 10.1007/BF00373366.

Lehmann, M.F., Bernasconi, S.M., Barbieri, A., and McKenzie, J.A., 2002, Preservation of organic matter and alteration of its carbon and nitrogen isotope composition during simulated and in situ early sedimentary diagenesis: Geochimica et Cosmochimica Acta, v. 66, p. 3573–3584, doi: 10.1016/S0016-7037(02)00968-7.

Létolle, R., 1980, Nitrogen-15 in the natural environment, in Fritz, P. and Fontes, J.C., eds., Handbook of environmental isotope geochemistry: Amsterdam, Elsevier, p. 407–433.

Lindberg, P., and Gustin, M.S., 1987, Field-trip guide to the geology, structure, and alteration of the Jerome, Arizona ore deposits, in Davis, G.H., and VandenDolder, E.M., eds., Geologic diversity of Arizona and its margins: Excursions to choice areas: Geological Society of America, 100th Annual Meeting, Field-trip Guidebook: Arizona Bureau of Mines and Mineral Technology Special Paper 5, p. 176–187.

Macko, E.S., Fogel, M.L., Hare, P.E., and Hoering, T.C., 1987, Isotopic fractionation of nitrogen and carbon in the synthesis of amino acids by microorganisms: Chemical Geology, v. 65, p. 79–92, doi: 10.1016/0009-2541(87)90196-3.

Manikyamba, C., Naqvi, S.M., Gnaneshwar-Rao, T., Balaram, V., Ramesh, S.L., and Reddy, G.L.N., 1997, Geochemical heterogeneities of metagraywackes from the Sandur schist belt; implications for active plate margin processes: Precambrian Research, v. 84, p. 117–138, doi: 10.1016/S0301-9268(97)00022-3.

Martin, H., 1999, Adakitic magmas: Modern analogues of Archean granitoids: Lithos, v. 46, p. 411–429, doi: 10.1016/S0024-4937(98)00076-0.

Marty, B., 1995, Nitrogen content of the mantle inferred from N_2-Ar correlation in oceanic basalts: Nature, v. 377, p. 326–329, doi: 10.1038/377326a0.

Marty, B., and Dauphas, N., 2003, The nitrogen record of crust-mantle interaction and mantle convection from Archean to present: Earth and Planetary Science Letters, v. 206, p. 397–410, doi: 10.1016/S0012-821X(02)01108-1.

Marty, B., and Humbert, F., 1997, Nitrogen and argon isotopes in oceanic basalts: Earth and Planetary Science Letters, v. 152, p. 101–112, doi: 10.1016/S0012-821X(97)00153-2.

Marty, B., and Zimmermann, L., 1999, Volatiles (He, C, N, Ar) in mid-ocean ridge basalts: Assessment of shallow-level fractionation and characterization of source composition: Geochimica et Cosmochimica Acta, v. 63, p. 3619–3633, doi: 10.1016/S0016-7037(99)00169-6.

Mazuka, A.N.N., Macko, S.A., and Pedersen, T.F., 1991, Stable carbon and nitrogen isotope compositions of organic matter from sites 734 and 725, Oman Margin: Proceedings of the Ocean Drilling Program, Scientific Results, v. 117, p. 561–586.

McCuaig, T.C., and Kerrich, R., 1998, P-T-t-deformation-fluid characteristics of lode gold deposits: Evidence from alteration systematics: Ore Geology Reviews, v. 12, p. 381–453, doi: 10.1016/S0169-1368(98)00010-9.

McDougall, I., and Harrison, T.M., 1988, Geochronology and thermochronology by the ^{40}Ar/^{39}Ar Method: New York, Oxford University Press, 212 p.

Mingram, B., and Bräuer, K., 2001, Ammonium concentration and nitrogen isotope composition in metasedimentary rocks from different tectonometamorphic units of the European Variscan Belt: Geochimica et Cosmochimica Acta, v. 65, p. 273–287, doi: 10.1016/S0016-7037(00)00517-2.

Mingram, B., Hoth, P., Lüders, V., and Harlov, D., 2005, The significance of fixed ammonium in Palaeozoic sediments for the generation of nitrogen-rich natural gases in the North German Basin: International Journal of Earth Sciences, v. 94, p. 1010–1022.

Minoura, K., Hoshino, K., Nakamura, T., and Wada, E., 1997, Late Pleistocene-Holocene paleoproductivity circulation in Japan Sea: Sea-level control on δ^{13}C and δ^{15}N records of sediment organic material: Paleogeography, Paleoclimatology, Paleoecology, v. 135, p. 41–50.

Morbidelli, A., Chambers, J., Lunine, J.I., Petit, J.M., Robert, F., Valsecchi, G.B., and Cyr, K.E., 2000, Source regions and timescales for the delivery of water to the Earth: Meteorite and Planetary Science, v. 35, p. 1309–1320.

Nadelhoffer, K.J., and Fry, B., 1988, Controls on natural ^{15}N and ^{13}C abundances in forest soil organic matter: Soil Science Society of America Journal, v. 52, p. 1633–1640.

Nagaraja-Rao, B.K., Rajurkar, S.T., Ramalingaswamy, G., and Ravindra, B.B., 1987, Stratigraphy, structure, and evolution of the Cuddapah Basin: Geological Society of India Memoir 6, p. 33–86.

Naqvi, S.M., Venkatachala, B.S., Shukla, M., Kumar, B., Natarajan, R., and Sharma, M., 1987, Silicified cyanobacteria from the cherts of Archaean Sandur schist belt; Karnataka, India: Journal of the Geological Society of India, v. 29, p. 535–539.

Naqvi, S.M., Uday, R.B., Subba, R.D.V., Manikyamba, C., Nirmal, C.S., Balarm, V., and Srinivasa, S.D., 2002, Geology and geochemistry of arenite-quartzwacke from the late Archaean Sandur schist belt; implications for provenance and accretion processes: Precambrian Research, v. 114, p. 177–197, doi: 10.1016/S0301-9268(01)00227-3.

Nutman, A.P., McGregor, V.R., Friend, C.R.L., Bennett, V.C., and Kinny, P.D., 1996, The Itsaq Gneiss Complex of southern West Greenland; the world's most extensive record of early crustal evolution (3900–3600 Ma): Precambrian Research, v. 78, p. 1–39, doi: 10.1016/0301-9268(95)00066-6.

Ohmoto, H., 1997, When did the Earth's atmosphere become oxic? Geochemical News, v. 93, p. 12–13, 26–27.

Ohmoto, H., 2004. Archean atmosphere, hydrosphere and biosphere, in Eriksson, P.G., Alterman, W., Nelson, D.R., Mueller, W.U., and Catuneanu, O., eds., The Precambrian Earth: Tempos and events: Amsterdam, Elsevier, Developments in Precambrian Geology, v. 12, p. 361–388.

Ortega, L., Vindel, E.M., and Beny, C., 1991, C-O-H-N fluid inclusions associated with gold-stibnite mineralization in low-grade metamorphic rocks, Mari Rosa mine, Cacers, Spain: Mineralogical Magazine, v. 55, p. 235–247.

Peters, K.E., Sweeney, R.E., and Kaplan, I.R., 1978, Correlation of carbon and nitrogen stable isotope ratios in sedimentary organic matter: Limnology and Oceanography, v. 23, p. 598–604.

Phillips, G.N., Law, J.D., and Myers, R.E., 2001, Is the redox-state of the Archean atmosphere constrained?: Society of Economic Geologists Newsletter, v. 47, p. 9–18.

Pinti, D.L., and Hashizume, K., 2001, ^{15}N-depleted nitrogen in early Archean kerogens: Clues on ancient marine chemosynthetic-based ecosystems? A comment to Beaumont, V., and F. Robert, F. (1999): Precambrian Research, v. 96, p. 62–82: Precambrian Research, v. 105, p. 85–88, doi: 10.1016/S0301-9268(00)00100-5.

Pinti, D.L., Hashizume, K., and Matsuda, J., 2001a, Nitrogen and argon signatures in 3.8 to 2.8 Ga metasediments: Clues on the chemical state of the Archean ocean and the deep biosphere: Geochimica et Cosmochimica Acta, v. 65, p. 2301–2315, doi: 10.1016/S0016-7037(01)00590-7.

Pinti, D.Z., Matsuda, J., and Maruyama, S., 2001b, Anomalous xenon in Archean cherts from Pilbara craton, Western Australia: Chemical Geology, v. 175, p. 387–395, doi: 10.1016/S0009-2541(00)00331-4.

Porcelli, D., and Turekian, K.K., 2004, The history of planetary degassing as recorded by noble gases, in Keeling, R, ed., The atmosphere. Amsterdam, Elsevier, Treatise on Geochemistry, v. 4, p. 281–318.

Rau, G.H., Arthur, M.A., and Dean, W.E., 1987, ^{15}N/^{14}N variations in Cretaceous Atlantic sedimentary sequence: Implication for past changes in marine nitrogen biogeochemistry: Earth and Planetary Science Letters, v. 82, p. 269–279, doi: 10.1016/0012-821X(87)90201-9.

Richardson, S.H., Gurney, J.J., Erlank, A.J., and Harris, J.W., 1984, Origin of diamonds in old enriched mantle: Nature, v. 310, p. 198–202, doi: 10.1038/310198a0.

Richardson, S.H., Shirey, S.B., Harris, J.W., and Carlson, R.W., 2001, Archean subduction recorded by Re-Os isotopes in eclogitic sulfide inclusions in Kimberley diamonds: Earth and Planetary Science Letters, v. 191, p. 257–266.

Robert, F., 2001, The origin of water on earth: Science, v. 293, p. 1056–1058, doi: 10.1126/science.1064051.

Rollinson, H., 2002, The metamorphic history of the Isua Greenstone Belt, West Greenland, in Fowler, C.M.R., Ebinger, C.J., and Hawkesworth, C.J., eds., The early Earth: Physical, chemical and biological development: Geological Society [London] Special Publication 199, p. 328–350.

Sachs, J.P., and Repeta, S., 1999, Oligotrophy and nitrogen fixation during eastern Mediterranean sapropel events: Science, v. 286, p. 2485–2488, doi: 10.1126/science.286.5449.2485.

Sadofsky, S.J., and Bebout, G.B., 2004, Nitrogen geochemistry of subducting sediments: New results from the Izu-Bonin-Mariana margin and insights regarding global nitrogen subduction: Geochemistry, Geophysics, and Geosystems, v. 5, Q03I15, doi: 10.1029/2003GC000543.

Saino, T., and Hattori, A., 1980, ^{15}N natural abundance in oceanic suspended particulate matter: Nature, v. 283, p. 752–754, doi: 10.1038/283752a0.

Sangster, D.F., and Scott, S.D., 1976, Precambrian, strata-bound, massive Cu-Zn-Pb sulfide ores of North America, in Wolf, K.H., ed., Cu, Zn, Pb, and Ag deposits: New York, Elsevier, Handbook of Strata-Bound and Stratiform Ore Deposits, v. 6, p. 129–222.

Sano, Y., and Pillinger, C.T., 1990, Nitrogen isotopes and N_2/Ar ratios in cherts: An attempt to measure time evolution of atmospheric δ^{15}N value: Geochemical Journal, v. 24, p. 315–325.

Schoell, M., and Wellmer, F.W., 1981, Anomalous δ^{13}C depletion in early Precambrian graphites from Superior Province, Canada: Nature, v. 290, p. 696–699, doi: 10.1038/290696a0.

Schopf, J.W., and Prasad, K.N., 1978, Microfossils in Collenia-like stromatolites from the Proterozoic Vempalli Formation of the Cuddapah basin, India: Precambrian Research, v. 6, p. 347–366, doi: 10.1016/0301-9268(78)90022-0.

Schubert, C.J., and Calvert, S.E., 2001, Nitrogen and carbon isotopic composition of marine and terrestrial organic matter in Arctic Ocean sediments: Implications for nutrient utilization and organic matter composition: Deep-Sea Research, v. 48, p. 789–810, doi: 10.1016/S0967-0637(00)00069-8.

Searle, S.R., 1971, Topics in variance component estimation: Biometrics, v. 27, p. 1–76.

Sephton, M.A., Amor, K., Franchi, I.A., Wignall, P.B., Newton, R., and Zonneveld, L.P., 2002, Carbon and nitrogen isotope disturbances and an end-Norian (Late Triassic) extinction event: Geology, v. 30, p. 1119–1122, doi: 10.1130/0091-7613(2002)030<1119:CANIDA>2.0.CO;2.

Shukolyukov, A., and Lugmair, G.W., 1998, Isotopic evidence for the Cretaceous-Tertiary impactor and its types: Science, v. 282, p. 927–929, doi: 10.1126/science.282.5390.927.

Siddaiah, N.S., and Rajamani, V., 1989, The geologic setting, mineralogy, geochemistry, and genesis of gold deposits of the Archean Kolar schist belt, India: Economic Geology and the Bulletin of the Society of Economic Geologists, v. 84, p. 2155–2172.

Smithies, R.H., 2000, The Archaean tonalite-trondhjemite-granodiorite (TTG) series is not an analogue of Cenozoic adakite: Earth and Planetary Science Letters, v. 182, p. 115–125, doi: 10.1016/S0012-821X(00)00236-3.

Sreeramachandra Rao, K., 2001, Regional surveys and exploration for gold in the greenstone-granite terranes of Andhra Pradesh: Geological Survey of India Special Publication 58, p. 11–27.

Sweeney, R.E., Liu, K.K., and Kaplan, I.R., 1978, Oceanic nitrogen isotopes and their uses in determining the source of sedimentary nitrogen, in Robinson, B.W., ed., Stable isotopes in the earth sciences: DSIR Bulletin, v. 220, p. 9–26.

Sylvester, P.J., Campbell, I.H., and Bowyer, D.A., 1997, Niobium/uranium evidence for early formation of the continental crust: Science, v. 275, p. 521–523, doi: 10.1126/science.275.5299.521.

Taylor, S.R., and McLennan, S.M., 1985, The continental crust: Its composition and evolution: Blackwell Scientific Publications, p.312.

Tolstikhin, I.N., and Marty, B., 1998, The evolution of terrestrial volatiles: A review from helium, neon, argon, and nitrogen isotope modeling: Chemical Geology, v. 147, p. 27–52, doi: 10.1016/S0009-2541(97)00170-8.

Valley, J.W., 1986, Stable isotope geochemistry of metamorphic rocks: Reviews in Mineralogy, v. 16, p. 445–489.

Venkatachala, B.S., Manoj, S., Mukund, S., and Naqvi, S.M., 1990, Archaean microbiota from the Donimalai Formation, Dharwar Supergroup, India: Precambrian Research, v. 47, p. 27–34, doi: 10.1016/0301-9268(90)90028-O.

Von Damm, K.L., Edmond, J.M., Measures, C.I., and Grant, B., 1985, Chemistry of submarine hydrothermal solutions at Guaymas Basin, Gulf of California: Geochimica et Cosmochimica Acta, v. 49, p. 2221–2237, doi: 10.1016/0016-7037(85)90223-6.

Wada, E., Kadonaga, T., and Natsuo, S., 1975, ^{15}N abundance in naturally occurring substances and global assessment of denitrification from isotopic viewpoint: Geochemical Journal, v. 9, p. 139–148.

Wang, D., and Anderson, D.W., 1994, Determination of soil pedogenic carbonates by stable carbon isotope ratios [abs.]: Canadian Society of Soil Science Annual Conference, v. 74, p. 357.

Watanabe, Y., Naraoka, H., Wronkiewicz, D., Condie, K.C., and Ohmoto, H., 1997, Carbon, nitrogen, and sulfur geochemistry of Archean and Proterozoic shales from the Kaapvaal Craton, South Africa: Geochimica et Cosmochimica Acta, v. 61, p. 3441–3459, doi: 10.1016/S0016-7037(97)00164-6.

Wellmer, F.W., Berner, U., Hufnagel, H., and Wehner, H., 1999, Carbon isotope geochemistry of Archean carbonaceous horizons in the Timmins area: Economic Geology Monograph 10, p. 441–456.

Williams, L.B., Ferrell, R.E., Jr., Hutcheon, I., Bakel, A.J., Walsh, M.M., and Krouse, H.R., 1995, Nitrogen isotope geochemistry of organic matter and minerals during diagenesis and hydrocarbon migration: Geochimica et Cosmochimica Acta, v. 59, p. 765–779, doi: 10.1016/0016-7037(95)00005-K.

Zhang, Y., and Zindler, A., 1993, Distribution and evolution of argon and nitrogen in Earth: Earth and Planetary Science Letters, v. 117, p. 331–345, doi: 10.1016/0012-821X(93)90088-Q.

Zhang, D., 1988, Nitrogen concentrations and isotopic compositions of some terrestrial rocks [Ph.D. thesis]: Chicago, University of Chicago, 157 pp.

MANUSCRIPT ACCEPTED BY THE SOCIETY 29 OCTOBER 2005

The sedimentary setting of Witwatersrand placer mineral deposits in an Archean atmosphere

Wyatt Ernest Lawrence Minter
Department of Geosciences, University of Cape Town, Rondebosch 7700, South Africa

ABSTRACT

The 3.05 Ga U-Pb dating of uraninite grains in Dominion and Witwatersrand conglomerates has established that they were older than the onset of Witwatersrand sedimentation at 2.97 Ga. and therefore that they are detrital in origin. A precise Re-Os isochron age of 2.99 Ga obtained for rounded pyrite grains associated with the uraninite indicates that the pyrite is also detrital. Evidence of detrital forms in the gold has confirmed its placer origin prior to modification during metamorphism. Furthermore, rhenium-depletion ages ranging from 3.5 to 2.9 Ga for Witwatersrand gold support numerous other lines of evidence that have been used in the past to interpret the gold, uranium and pyrite concentrates as Archean placers.

The sedimentary and stratigraphic history of the Witwatersrand succession indicates that uraninite, pyrite, and gold were part of a sub-aerial sediment load over a period of 180 m.y. Net sedimentation rate during the accumulation of the Central Rand Group is estimated at less than 14 m per million years, reflecting stratigraphic losses due to repeated reworking. Individual paleoplacers would have been exposed over areas of up to 400 km^2 but collectively they covered a region exceeding 2000 km^2 and contained more than 243 million tons of pyrite, 1.5 million tons of uraninite, and 80,000 tons of gold.

Because there is no record of either detrital uraninite or pyrite in Proterozoic red-bed sediments on the Kaapvaal Craton, it is concluded that a change in the composition of the atmosphere took place there after 2.64 Ga.

Keywords: paleoplacer, Witwatersrand, Central Rand Group, Dominion Group, West Rand Group, Klipriviersberg Group.

INTRODUCTION

Early concepts that the Witwatersrand ore deposits represented a giant composite of coalescing sub-aerial placers (Pretorius, 1975) led to the opinion that for the detrital uraninite and pyrite mineralization to have been preserved it must have been in equilibrium with the atmosphere. Therefore, because uraninite and pyrite are unstable in Earth's present atmosphere it has been proposed that the atmosphere during the late Archean had a lower oxygen content, and that by the time Proterozoic red beds were being deposited it had changed to an oxygen-rich atmosphere in which detrital uraninite and pyrite could not survive.

Converse concepts that the Witwatersrand ore deposits were the result of hydrothermal infiltration into the pebbly sandstone host rocks have been raised repeatedly (Pretorius, 1975; Barnicoat, et al., 1997; Phillips and Law, 2000) and acceptance would of course negate using Witwatersrand mineralogy to indicate a time of low atmospheric oxygen availability.

A recent review, however, of the status of the debate concerning these two genetic concepts (Frimmel and Minter, 2002a),

E-mail: lminter@absamail.co.za

has highlighted a precise rhenium-osmiridium isochron age of 2.99 Ga for Witwatersrand pyrite, and rhenium-depletion ages ranging from 3.5 to 2.9 Ga for Witwatersrand gold (Kirk et al., 2001). These findings confirm that the pyrite and gold are older than the onset of Witwatersrand sedimentation and support the detrital origin of the mineralization previously indicated by a 3.05 Ga date for the uraninite (Rundle and Snelling, 1977).

The purpose of this account then is to describe the sedimentary environments, within which the placer minerals survived transport and deposition, by referring to interpretations based on numerous systematic sedimentological studies conducted since 1964.

STRUCTURAL AND STRATIGRAPHIC SETTING

The "Witwatersrand basin" may be thought of as a set of cratonic successor basins overlapping each other and resting on middle Archean granite-greenstone terrane (Fig. 1) that is dated at 3086 ± 3 Ma (Robb et al., 1992). The first stage of basin development is recorded in the Dominion Group rocks deposited in a continental rift. The basal unit comprises a thin sequence of fluvial sediments including placers that contain uraninite, which has been dated at 3050 ± 50 Ma (Rundle and Snelling, 1977). These placers, which are extensive and represent the largest unmined inferred uranium resource in the world, were followed and buried by extensive bimodal volcanism dated at 3074 ± 6 Ma (Armstrong et al., 1991). The standard deviation of ± 50 Ma for the Rundle and Snelling date for the detrital uraninite allows one to place its age at between 3086 and 3074 Ma.

The second stage of basin development is recorded in the rocks of the West Rand Group. They comprise a 5150 m thickness of sandstones and siltstones deposited after 2970 Ma (Barton et al., 1989) in a shallow marine setting, which transgressed the Dominion Group to the north and west. A small number of fluvial intervals containing minor paleoplacers (referred to as reefs) occur, particularly in the Government Subgroup (Fig. 2). A laterally extensive lava formation near the top of the West Rand Group has been dated at 2914 ± 8 Ma (Armstrong et al., 1991), providing a minimum age for the West Rand Group. The rate of sedimentation was at least 92 m/m.y.

The third stage of basin development is recorded in the Central Rand Group of rocks that unconformably overlie the West Rand Group (Fig. 1). It comprises a thickness of 2880 m that is dominated by fluvial cycles of coarse siliciclastic sedimentary rocks, which are separated by erosional unconformities. The upper age limit of this sequence is set by the age of overlying lavas of the Klipriviersberg Group at 2714 ± 8 Ma (Armstrong et al., 1991).

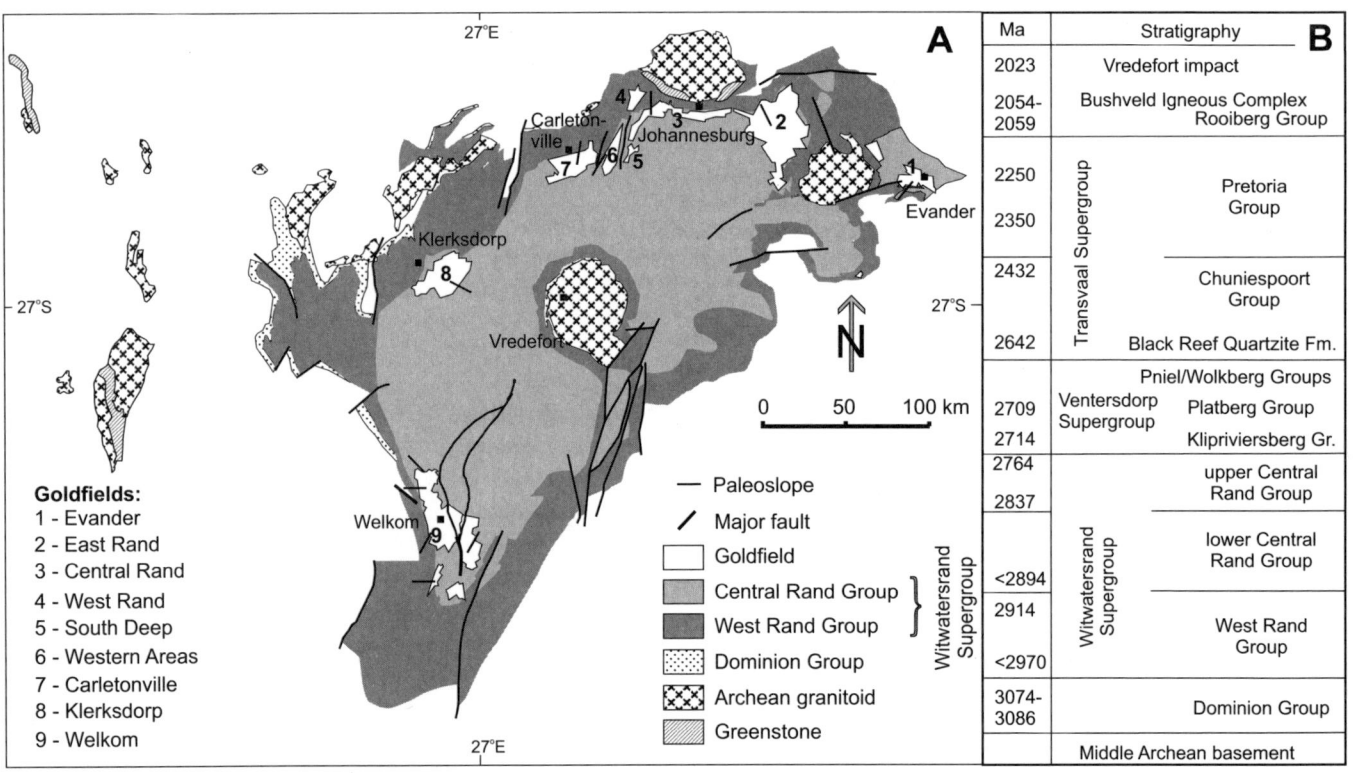

Figure 1. (A) Simplified surface and subsurface geological map of the Witwatersrand Basin (Frimmel and Minter, 2002a), also showing the distribution of Archean granitoid domes, the location of the principal goldfields, major faults, and paleocurrent directions of Witwatersrand reefs (from Minter and Loen, 1991, and Frimmel and Minter, 2002b). (B) Main stratigraphic units and ages (for sources see text).

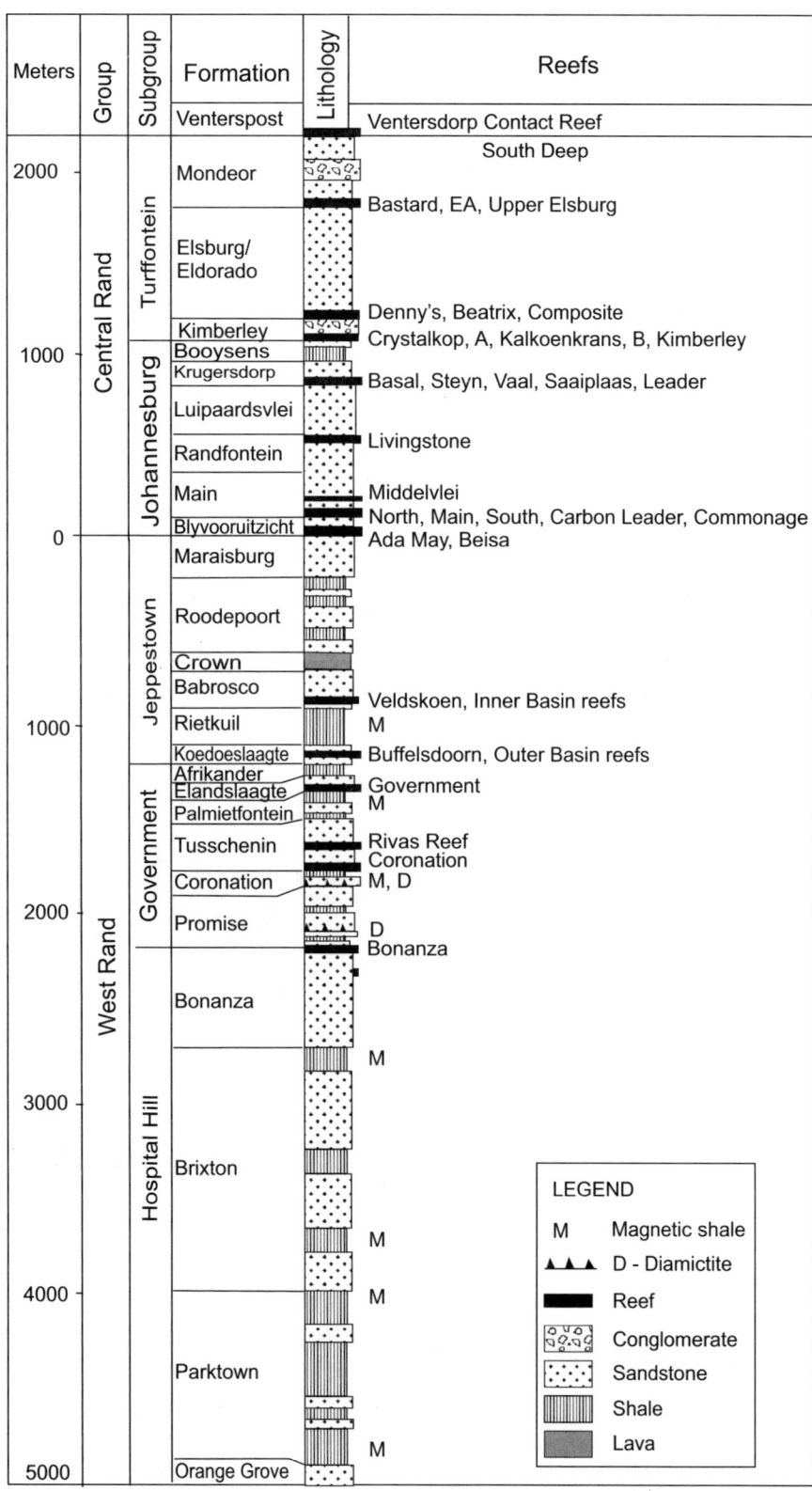

Figure 2. Generalized stratigraphic column for the Witwatersrand Supergroup, exemplified by a section through the Klerksdorp goldfield, as proposed by the South African Committee for Stratigraphy (unpubl.); also listed are the stratigraphic positions of the main auriferous conglomerate (reef) horizons in the various goldfields (Frimmel and Minter, 2002b).

The net rate of sedimentation of the Central Rand Group was only 14 m/m.y., reflecting the cumulative amount of degradation associated with the numerous unconformities (Sadler, 1981). This truncation and reworking of the stratigraphy was apparently up to seven times more prevalent than in the underlying West Rand Group succession and might account for the greater incidence in the Central Rand Group of detrital heavy mineral concentrations represented by the paleoplacer orebodies that have been exploited during the past 118 yr of mining.

SEDIMENTOLOGY

Stratigraphic and paleocurrent data indicate that the Central Rand Group was deposited in a shrinking continental basin. The paleoplacers were deposited at locations peripheral to the basin in a variety of sedimentary environments that ranged from alluvial fans to braid plains and braid deltas (Minter, 1978; Minter and Loen 1991). In general, the paleoplacers are hosted by gray-colored, medium-grained to coarse-grained sandstones, composed of quartz-arenite, that occupied fluvial channels in braided drainage systems and contained a generally oligomictic gravel component as lags, bars, and sheets (Fig. 3). Primary sedimentary structures within the host sediments indicate unidirectional paleoflow directions into the basin (Fig. 1). The top surfaces of the placer sediment packages were leveled off either by wind deflation in temporarily inactive parts of the depositional environments, or by wave action after inundation by a body of water (Fig. 4). Lag deposits, especially on basal unconformities, degradation scour surfaces, clast-supported conglomerates, and winnowed reworked top-surfaces are preferentially mineralized (Fig. 5).

Mineral Concentration

Although the clastic sediment loads that produced paleoplacers would have traveled in suspension during flood conditions, and been initially sorted by settling processes as the flow velocity decreased, it has become evident from detailed studies that selective entrainment of the less dense particles during bedload transport (Slingerland and Smith, 1986) was the most effective concentrating mechanism of smaller, more dense particles. Figure 6 illustrates a number of consecutive beds of heavy-mineral lags, dominated by rounded detrital pyrite, which demonstrate the process.

Measurements of the areal size frequency distribution of rounded pyrite in both the Basal and Steyn paleoplacers of the Welkom goldfield (Minter, 1978) demonstrate further how selec-

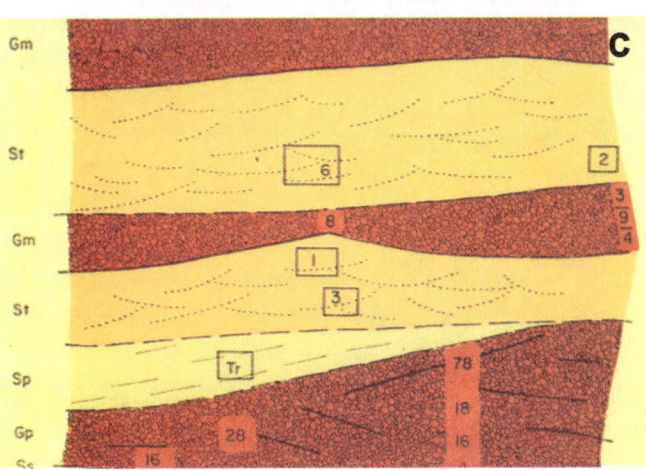

Figure 3. (A) Emerging longitudinal gravel bar in the Kicking Horse River, at Field, British Columbia, with lateral accretion of sand facies draping the left-hand side, and sand dunes occupying the channel below water level. Shovel for scale. (B) A remnant pillar of the Elsburg No. 5 Reef at Western Reefs Gold Mine in the Klerksdorp Goldfield. The face of the pillar is 2 m high and exposes a transverse section through a mineralized longitudinal conglomerate bar with laterally accreted planar cross-bedded sandstone to the left and overlying trough cross-bedded sandstone filling the channel (Smith and Minter, 1980). (C) A diagram illustrating sample sites that indicate the distribution of gold in the paleoplacer lithofacies described above. The blocks numbered from Tr (Trace) to 78 represent gold content measured in grams per ton and illustrate association between gold tenor and particular lithofacies.

Figure 4. An underground exposure of the Leader Reef at Free State Saaiplaas Gold Mine in the Welkom Goldfield. The stope face represents a section across a channel filled with clean gray-colored paleoplacer sediment. Note the higher bed relief to the left and full channel depth to the right. The arrangement of inclined conglomerate facies indicates that gravel shed off the high relief and lined the channel, which was then filled by a lens-shaped sheet of trough cross-bedded sand. The sequence was topped by a planar layer of gravel that reflects reworking of the surface by either eolian deflation or wave action. Fine-grained rounded pyrite, gold, and an assemblage of other heavy detrital minerals are located in the gravel facies, particularly in the bottom and top units where gold concentrations of up to 100 g/t occur, and distributed as cross-bedded laminations in the sand facies. The sediment overlying and underlying the gray paleoplacer sediment has a yellowish color due to its argillaceous content. Hammer handle (30 cm) for scale.

Figure 6. In this sample of the Basal Reef from Free State Geduld Gold Mine in the Welkom Goldfield, imbricated pebbles lying on pyrite layers indicate that the current flowed from the left, winnowing the underlying layer of pyrite grains and removing them in the turbulent zone upstream of pebbles, which then tilted into the depression. The small ellipsoidal pyrite pebbles have also been imbricated in the same sense as the quartz pebbles have been. The open framework of the dense pyrite bed has, in this instance, provided spaces in which finer-grained gold and other small heavy mineral particles were trapped. This resembles the bed of a Malaysian tin jig. Width of sample is 15 cm.

Figure 5. A bed of conglomerate in the Elsburg No. 5 Reef at Western Reefs Gold Mine in the Klerksdorp Goldfield, which illustrates selective accumulation of rounded pyrite and other heavy detrital minerals in the upper, clast-supported, portion of the conglomerate. This is interpreted to be the result of selective separation of fine-grained dense minerals from passing sand facies and their protection by the gravel framework from removal. Width of image is 20 cm.

tive entrainment produces a regional measure of sorting insofar as the size of rounded pyrite decreases from a mean diameter of 7 mm to 1 mm over a transport distance of 25 km.

Calculations of the threshold shear stress required to entrain the detrital mineral grains concentrated together in a set of cross-bedded laminations (Figs. 7, 8) indicated that at the velocities associated with dunes (trough cross-bedding), which is the most common bedform preserved in the quartz arenite lithofacies of paleoplacers, the quartz, zircon, pyrite, chromite and gold particles could be entrained (Table 1). The size frequency distribution of the disc-shaped gold grains is lognormal, as for natural sediments, with a mean nominal diameter of 136 μ and a standard deviation of 0.48, which indicates good sorting.

Inclined degradation surfaces sampled in a section across a distal Steyn Reef channel on Free State Saaiplaas Gold Mine in the Welkom Goldfield (Minter, 1989) illustrates how selective entrainment reworks placer sediment to leave a concentrated assemblage of denser finer-grained minerals. The correlation between zircon, uraninite, and gold demonstrates this (Figs. 9A–C). The greater the reworking becomes, the pebblier the lag concentrate becomes, and the greater the detrital mineral concentration is. The highest concentrations collecting in this way occur

Figure 7. A rare specimen of cross-bedded heavy minerals from the Basal Reef at Free State Geduld Gold Mine, containing abundant gold particles that have retained their original detrital forms (Minter et al., 1993b). Measurement of these grains has provided an opportunity to compare the relative sizes of associated heavy minerals in the same foreset laminations. The four tangential foreset planes visible converge toward the right where the toeset meets the bottomset. Width of specimen is 8 cm.

on major distal disconformities or unconformities like those associated with the Vaal Reef, the Basal and Steyn Reefs, the Carbon Leader Reef, and the Crystalkop Reef. The duration of these longer-term surfaces of degradation is also displayed by the Chemical Index of Alteration of the strata immediately underlying the unconformity (Frimmel and Minter, 2002a). Values of up to 90% have been recorded (Fig. 10).

The spatial and temporal association of paleoplacer mineralization with the surface of degradation is amply demonstrated by contemporaneous loadcasting of channel sediments in the Saaiplaas Reef, into the underlying thixotropic sediments of the Harmony Formation (Fig. 11). This soft-sediment deformation of paleoplacer deposits in erosion channels above the Main Reef, the Kimberley Reef, and the Black Reef deposits has been objectively described by Papenfus (1957). Attempted mining of such disrupted paleoplacer beds has been frustrated and generally unsuccessful.

Sedimentary Settings

Most of the goldfields shown in Figure 1 represent multiple composites of any one of these sub-environments or combinations of a number of different sub-environments (Minter, 1991). Each deposit might range from 100 to 400 km² in area and composites might exceed 600 km².

Alluvial fans containing paleoplacers are limited to the most proximal preserved remnants of the basin margin. Examples include the Livingstone Reefs on the West Rand, the Massives at Western Areas on the Far West Rand, and the EA Reefs in the Eldorado Formation of the northern part of the Welkom goldfield (Minter et al., 1986). The EA sediments, for example, form

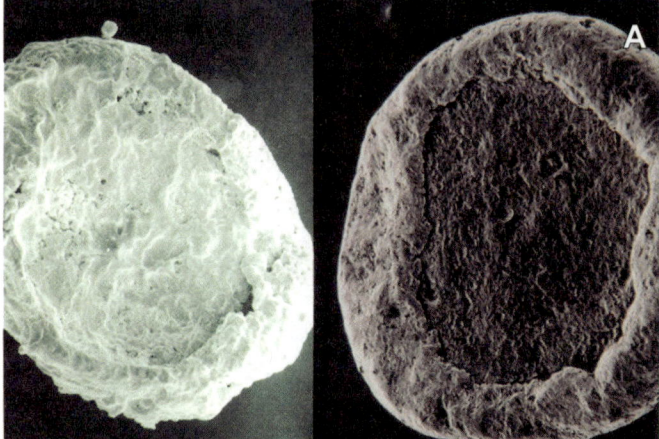

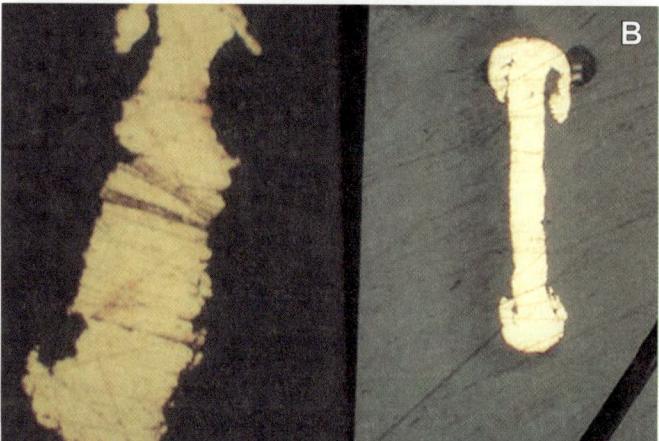

Figure 8. (A) Left: preserved disc-shaped 300-μ-diameter toroidal gold grains, representative of thousands extracted (Minter et al., 1993) from the specimen illustrated in Figure 7, which demonstrate their detrital origin. Right: Toroid from Baaga stream in Yakutia which is reworking an ancient wind deflation surface (Minter, 1999). (B) Sections through toroids from the Basal Reef (left) and the Baaga stream (right) indicating a unique form believed to have been formed by the impact (peening) of sand grains on gold flakes in an eolian environment (Minter, 1999). (C) A concentrate of sand-blasted platinoid and gold grains from a modern raised beach placer deposit in New Zealand, South Island. An evolution of 200-μ-diameter toroidal gold shapes from primary rims to almost hollow spherical forms is evident (Mitchell, 1996).

TABLE 1. RELATIONSHIP BETWEEN MEASURED
AND THEORETICAL PLACER GRAIN DIAMETERS

Mineral	Specific gravity	Observed size (mm)	Theoretical size (mm) Quartz basis	Threshold shear stress (dynes/cm^2)
Quartz	2.65	0.384		2.41
Zircon	4.68	0.133	0.225	16.7
Pyrite	5.00	0.158	0.213	11.2
Gold	19.32	0.136	0.072	76.3

Note: Bed shear stresses in dune regimen for median grain size, D50 = 0.4mm, range from 20 to 80 dynes/cm^2 (James and Minter, 1999).

a bajada of coalescing alluvial fan lobes that seem to be spaced at 1500 m intervals along the north-striking folded structure that defines the western border of the Welkom goldfield. These fans extend downdip for 750 m eastward into the depository (Gray, et al., 1994). They were shed off the upturned western margin of the Central Rand Group as a syn-sedimentary response to active tectonic compression during the final stages of Central Rand Group deposition. At the Target mine, individual fans comprise up to 15 repeated sequences of oligomictic mineralized conglomerates, quartz-arenites, and lithic arenites in a 200-m-thick succession. They evidently derived placer concentrates by reworking subcropping paleoplacers in the upturned footwall.

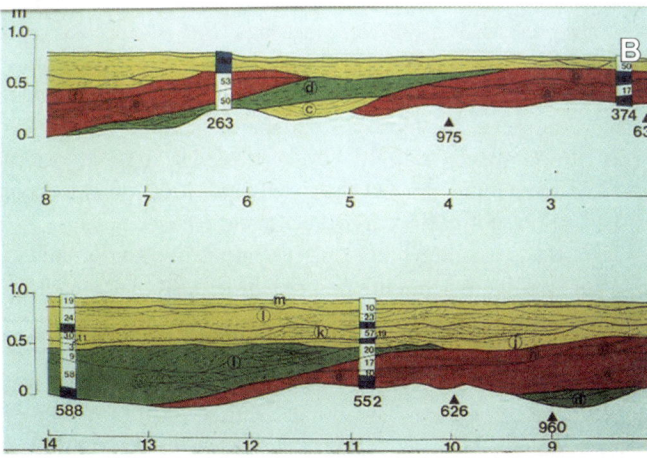

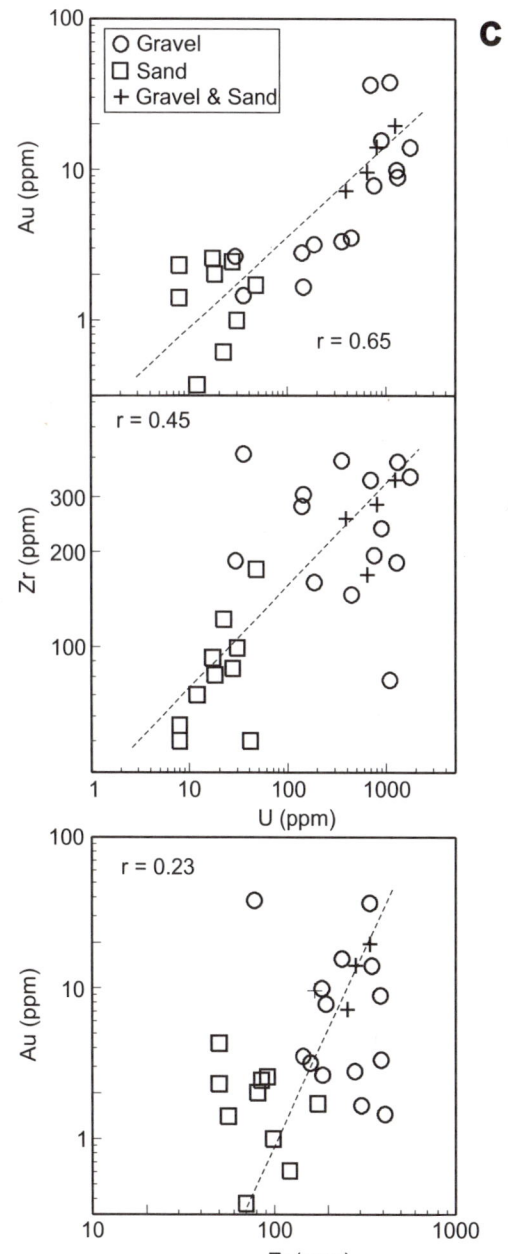

Figure 9. (A) A transverse underground exposure of the inner bend of a channel edge in the Steyn Reef at Free State Saaiplaas Gold Mine in the Welkom Goldfield. Inclined degradation surfaces on which reworking has taken place have permitted selective entrainment of the sand fraction to leave behind pebbles and fine-grained dense detrital minerals like gold, uraninite, and zircon. Clinorule (1 m) for scale. Photo located at meter mark 7 shown in Section B. (B) Section illustrating with color packages of sediment wedges separated by master bedding plane surfaces of degradation. Vertical and horizontal scale in meters. (C) The results of chemical analysis indicate correlations between these minerals that support their collective hydraulic placer origin (Frimmel and Minter, 2002a).

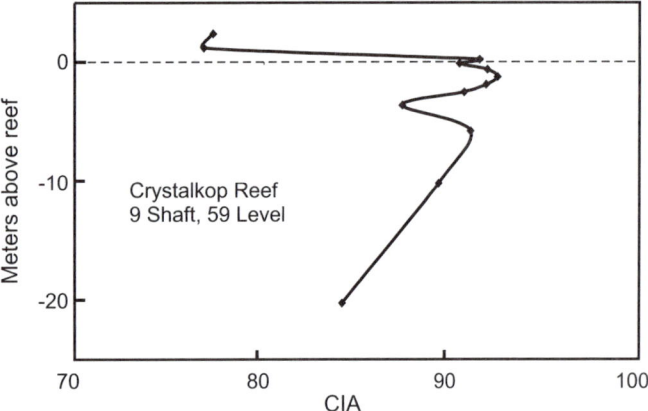

Figure 10. Variation in Chemical Index of Alteration (CIA) with distance from the Crystalkop Reef as measured in a profile at 9 Shaft, 59 Level, Vaal Reefs Gold Mine, in the Klerksdorp Goldfield (Frimmel and Minter, 2002a).

Figure 11. An exposure of the Saaiplaas Reef in Crosscut 100 on 74 Level, 4 Shaft at President Steyn Gold Mine in the Welkom Goldfield. The paleoplacer channel, which originally comprised of light gray-colored quartz arenite containing trough cross-bedded laminations of rounded pyrite, and overlay a basal pebbly lag concentrate, now displays contemporaneous loadcasting into the thixotropic underlying (waxy yellow) argillaceous sandstone of the Harmony Formation in such a way that the pebble lag surrounds the oval-shaped loadcast and the cross-bedded quartz arenite interior is deformed. Mineralization is associated with the pebble lag surface. Geological pick head for scale.

Another paleoplacer associated with alluvial fan development is the Beatrix Reef, which occurs at the southern end of the Welkom goldfield. It lies on the pediment of an early Eldorado Formation alluvial fan where Aandenk paleoplacers outcropped. The recycled detrital minerals eroded from the Aandenk were evidently spread out as a concentrated deflation surface downdip from the outcrop strike. These were distributed by shallow channels and preserved in arenite lithofacies as laminae, in the matrix of clast-supported conglomerates, and on eolian deflation surfaces associated with ventifacts. These were then buried beneath a prograding Eldorado Formation fan (Minter et al., 1988). In the distal black mudstone lithofacies of the prograding Eldorado fan, heavy minerals were entrained from the paleosurface by debris flows and suspended in an unsegregated dispersion quite atypical of the average paleoplacer. The Beatrix Reef environment covered an area of ~78 km^2 in which mixed lithofacies are recorded, indicating fluvial, eolian, and semi-arid mud-cracked playa lake sediments.

Many of the Witwatersrand paleoplacers were deposited in gravelly braid plains resembling the outwash sheets in southern Iceland (Minter, 1982a, p. 133). Examples are found in the Leader Reef in the Welkom goldfield (Smith and Minter, 1980; Minter et al., 1986) and in the Main Reef and Kimberley Reefs of the West Rand and Central Rand goldfields. A classic example of a braid-plain channel-fill paleoplacer is illustrated in an exposure of the Kalkoenkrans paleoplacer at the Oryx Goldmine (Fig. 12).

The most extensive paleoplacers in the Witwatersrand were braid deltas associated with shorelines (McPherson et al., 1987). They include the Basal and Steyn Reefs (Minter, 1978), the Vaal Reef (Minter, 1976, Antrobus et al., 1986), the Carbon Leader Reef (Buck and Minter, 1985), and the Kimberley UK9 Reefs (Armstrong, 1966). Active flowing fluvial drainage systems that deposited the placers were on average 50 cm deep and covered only a small part of the deposits at any one time. Therefore the remaining subaerial regions were exposed to eolian deflation. Ventifacts recovered from many of the paleoplacers attest to this (Fig. 13), and it is likely that wind deflation played an important role in condensing the stratigraphy and thereby concentrating mineral accumulations (Minter, 1999). These deflation surfaces are preserved as thin ventifact lags (Fig. 14) on planar to gently undulating unconformity surfaces, possibly representing groundwater-table control on deflation, and as planar winnowed surfaces on gravel-bar tops. Measurements of facet orientations on in situ ventifacts in the Crystalkop Reef in the Klerksdorp goldfield have indicated that the paleowind direction was from 12° east of present-day north (Minter et al., 1993a).

The Basal and Steyn paleoplacers in the Welkom goldfield were adjacent braid deltas that entered the depositional area from sources to the west and southwest, respectively, and terminated in a standing body of water (Fig. 15) (Minter, 1982b). The Steyn paleoplacer is slightly younger than the Basal paleoplacer in that it has truncated and overlapped the southern edge of the Basal paleoplacer fan. Both paleoplacers were eventually inundated by water and buried beneath siltstone deposited from suspension (khaki shale). They have been preserved over an area of ~600 km^2. In the case of the Steyn paleoplacer, not only were the pebbles sorted by size downstream, from 70 mm in diameter in the truncated proximal zone to 11 mm in the most distal zone, but so was the pyrite, from a mean diameter of 0.33 mm to 0.13 mm (Minter, 1978, 1991), thus supporting its detrital origin (Fig. 16). The detrital nature of accompanying uraninite is illustrated by its correlation with zircon and gold (Frimmel and Minter, 2002a)

Figure 12. This Kalkoenkrans Reef exposure at the Oryx Gold Mine in the Welkom Goldfield records the tapered edge of a heavily mineralized longitudinal gravel bar located at the base of the channel. It has been buried beneath white trough cross-bedded quartz arenite (dunes constituting the face of the exposure), which filled the channel, and then during abandonment was covered by black mud (on the roof of the exposure) that dried and cracked in the subaerial environment. The mud polygon boundaries have filled with light-colored sandstone.

Figure 13. (A) Ventifact from the Crystalkop Reef at Vaal Reefs Gold Mine. Scale in mm divisions; (B) Ventifact 20 mm in diameter from the Vaal Reef at Hartebeestfontein Gold Mine, in the Klerksdorp Goldfield.

and by a steadily increasing uranium/gold ratio down the paleoslope, which is a result of mineral sorting. Uraninite is therefore generally enriched in the more distal fluvial facies of Witwatersrand paleoplacers (Minter, 1978).

The Witwatersrand alluvial-fan and braid-plain models can be extended to other ancient pyrite- and uraninite-bearing paleoplacers. These include the Ventersdorp Contact Reef and the Black Reef, which overlie the Witwatersrand (Minter, 1991), paleoplacers in the Pongola Supergroup (Minter, 1991), the Bababudan paleoplacers in India (Srinivasan and Ojakangas, 1986), paleoplacers on the Pilbara Craton in Australia (England et al., 2002), paleoplacers at Elliot Lake (Theis, 1976) on the Canadian Shield and elsewhere in North America (Roscoe and Minter, 1993), and the Mocda and Jacobina paleoplacers on the São Francisco Craton in Brazil (Minter, et al., 1990, Garayp et al., 1991, and Minter, 1991).

DETRITAL MINERALOGY

From the position of an observer, standing at the southern edge of the Basal and Steyn Reef deposit (Fig. 15), the view to the north for 30 km and to the east for 25 km would have been a gray, pebbly sand flat mineralized with bronze-colored pyrite, black ilmenite and chromite, pink zircon, and black uraninite, in that order of abundance, with lesser amounts of gold. The gold content across such surfaces as seen in the Carbon Leader Reef, the Vaal Reef, and the Steyn and Basal Reefs averages about a million ounces per square kilometer.

The pyrite is heterogeneous in origin, up to six species occurring, chemically distinguished by different sulfur isotopic signatures (England et al., 2002), and has been abraded to coarse- to fine-grained rounded forms, the outer margins truncating internal structures and textures (Fig. 17). Concentrations of pyrite range from 50,000–200,000 ppm.

Figure 14. A polished slab of the Vaal Reef from Stilfontein Gold Mine, which represents a thin pebbly sandstone lag containing ventifacts and fine-grained rounded detrital pyrite. Note thin carbon seam on basal contact. Diameter of large quartz pebble is 20 mm.

Figure 15. An artist's impression of the distribution of the Steyn (left) and Basal (right) paleoplacers in the Welkom goldfield, facing west. The drainage patterns, and distal depositional limits, are based on underground data and borehole intersections. The upslope front, which was truncated by unconformable onlapping formations (Minter, 1982b) extends for 30 km to the north. Black dots mark the sites of discovery boreholes.

Figure 16. An exposure illustrating brassy-colored granular pyrite concentrated along the upper parts of cross-bedded foresets (inclined to the right) in the Steyn Reef at President Steyn Gold Mine in the Welkom Goldfield.

The ilmenite is fine-grained and occurs throughout the Witwatersrand sequence, concentrated on all scour surfaces. In Witwatersrand ores it has been altered during syn-depositional weathering to leucoxene, a composite of rutile needles and authigenic quartz. In the Composite Reef on the West Rand, the TiO_2 content ranges between 420 and 4000 ppm (Tucker, 1980), which is similar to contents of 500–11,700 ppm at Elliot Lake (Theis, 1976) and 1310 ppm in the Moeda in Brazil (Minter et al., 1990).

Chromite is common throughout, and averages ~3000 ppm in the Steyn Reef (Frimmel and Minter, 2002a). Zircon is also abundant throughout. In the Composite Reef the ZrO_2 content ranges between 83 and 1455 ppm (Tucker, 1980) and in the Steyn Reef ~410 ppm, which are similar to values between 10 and 600 ppm at Elliot Lake (Theis, 1976), and 54 ppm in the Moeda (Minter et al., 1990). There are significant correlations between all these minerals and uraninite, illustrating an hydraulic control over detrital mineral concentration (Slingerland and Smith, 1986).

Uraninite was first identified in Witwatersrand ores by Cooper (1923) and then quantified by Weston Bourret in 1944–45 during the Manhattan Project (Bourret, 1975). In the original survey in 1944, uranium was also recorded from Black Reef on the East Rand at Vogelstruisbult where a gold and pyrite rich specimen assayed 3350 ppm U_3O_8. This is an exceptionally high value because average ore grades at Modderfontein East were between 26 and 37 ppm. Recent borehole intersections of Black Reef from the East Rand (Camden-Smith, 2004, personal commun.) range from 2 to 37 ppm over the full width.

Figure 17. An enlargement of part of Figure 6 to illustrate how the margins of coarse-grained rounded pyrites display an abraded truncation of internal layers and crystal structures in the pyrite. The large pyrite grain in the center of the image is 5 mm in diameter.

The first "yellow cake" was produced from a pilot plant at Blyvooruitzicht in the Carletonville goldfield in 1951. Commercial production began in 1952 at West Rand Consolidated in the West Rand goldfield where the Monarch Reef, represented by a small-pebble distal placer, was mined at a mean grade of 2860 ppm U_3O_8, ten times richer than the average Witwatersrand ore.

Uranium in the Witwatersrand ores also occurs in the form of brannerite, an alteration product of uraninite, which is attached to leucoxene. It is interesting to note that the ratio of uraninite to brannerite in Witwatersrand ores varies. In the Welkom goldfield this variation appears to be systematic insofar that it changes up the stratigraphic column from 8.7 in the Steyn Reef to zero in the Beatrix Reef (Minter et al., 1988). All the uranium in the Black Reef, which is the uppermost reef in the sequence, reports to brannerite (England, 2004, personal commun.). Could this be related to repeated reworking in an increasingly oxygenated evolving atmosphere?

Up to 15% of the uranium in the Vaal Reef is attached to leucoxene and requires a high-pressure leach to extract. According to Camisani-Calzolari et al. (1985), the average metallurgical recovery of uranium has been between 20% and 80% at an average grade of 271 ppm. The Elliot Lake paleoplacers produced an average grade of 1000 ppm and the grade of a typical Moeda intersection is 157 ppm (F.E. Renger, 2004, personal commun.).

It was estimated in 1974 (Armstrong, 1975) that 35% of known world uranium resources were contained in quartz pebble conglomerates in South Africa and Canada and that the slimes dumps alone contained 500 million pounds of U_3O_8. The Arab oil embargo in 1973 apparently stimulated the installation of nuclear power reactors, increased the demand for uranium and therefore its market price, and generated worldwide exploration for further resources. Processing the pyrite, uranium, and gold contents of the Witwatersrand slimes dumps became a lucrative exercise, as demonstrated by the Ergo operation, and most of these deposits were exploited. However, Chernobyl and other disasters resulted in lobbies against nuclear power and the price fell to the point where closure of many of these operations took place.

DISCUSSION

Today there is general agreement that the majority of Witwatersrand gold particles, which appear late in the paragenetic sequence, are indeed hydrothermal precipitates. Thus, the debate has shifted to the question of whether the source of the gold in the auriferous hydrothermal fluid(s) was proximal, fluvially deposited, detrital gold within the conglomerate beds, subsequently mobilized by post-depositional fluids over short distances (modified placer model), or whether the source of the gold was external to their host rocks. The latter case would imply long-distance chemical transport by hydrothermal fluids (hydrothermal model).

The modified paleoplacer model has achieved strong support from the discovery of two distinct morphological types of gold that occur together on a millimeter scale; one type displays morphological features typical of detrital gold, whereas the other occurs as irregularly shaped aggregates or well-crystallized, euhedral overgrowths (Minter et al., 1993b). In this model, transport of detrital gold particles into the host sediments is believed to have taken place by fluvial processes. Physical deformation of these particles as a result of post-burial compaction, which pressed sand or heavy-mineral grains into the surfaces of discs, is evident (Minter et al. 1993b). Chemical mobilization of the gold as a result of short-range (micrometer- to meter-scale) transport, took place largely within the host rock, by infiltrating hydrothermal fluids and/or degradation of in situ hydrocarbon or hydrous phases.

In Brazil, comparable Paleoproterozoic sedimentary cover rocks above greenstone belts also contain auriferous quartz pebble conglomerates with a few large gold deposits at the base of fluvial and beach sequences (e.g., Jacobina deposit in Bahia and Moeda deposit, Quadrilatero Ferrifero, in the northern and southern São Francisco Craton, respectively). An estimated 50 tons (t) of Au have been recovered from conglomerates in the Serra de Jacobina region (Teixeira et al., 2001). A foreland setting for the Jacobina Basin has been proposed (Ledru et al., 1997), with U-Pb data for detrital zircon and a Rb-Sr whole-rock age for a post-collisional granitoid in the Jacobina-Contendas-Mirante belt providing maximum and minimum age constraints for the Basin fill of 2086 and 1883 Ma, respectively. The bulk of the gold extracted from the Jacobina deposits came from pyritic quartz pebble conglomerate and minor quartz arenite intercalations of the lowest formation, interpreted as representing alluvial fan and braided stream deposits. The gold occurs as fibrous to oval-shaped particles, typically associated with euhedral pyrite. Detrital phases include zircon and chromite, whereas metamorphic/hydrothermal phases comprise pyrrhotite and pyrite, fuchsite, white mica, minor andalusite, rutile, and tourmaline (Ledru et al., 1997).

By analogy with the Witwatersrand, the likely hydrothermal gold source was detrital gold in the local conglomerate beds. A possibly more intense metamorphic or hydrothermal overprint would have led to a complete obliteration of the original detrital microtextures and morphological forms of the gold, as is the case in some Witwatersrand reefs (e.g., the Ventersdorp Contact Reef). The S isotopic composition of the pyrite points to a magmatic origin (Teixeira et al., 2001), but it remains unclear whether the hydrothermal pyrite is related to sulfidation of originally detrital Fe- oxides or reflects mobilized detrital pyrite as in the Witwatersrand.

In contrast to the Witwatersrand, the Moeda deposits (Minter et al., 1990) are not in a thick siliciclastic succession but rather at the base of only a thin (120 m) siliciclastic, predominantly arenitic succession, whose age is loosely constrained between 2.8 and 2.2 Ga. The gold occurs together with heavy mineral concentrates, predominantly coarse-grained rounded pyrite, in the matrix of cobble conglomerates that form discrete bars and sheets above an angular unconformity. Sediment transport was from the northwest into individual basins that were separated by basement ridges of Archean chert and iron formation. Mineralogy, geochemistry, and gold concentrations are similar to those of the Witwatersrand deposits (Minter et al., 1990).

Apart from gold, associated mineral phases that are key to any plausible interpretation of the Witwatersrand deposits are pyrite and uraninite. Between 1952 and 1975, up to 1.5 million tons (Mt) of U_3O_8 were produced from Witwatersrand quartz pebble conglomerates at an average grade of 271 ppm U_3O_8. Iron oxides are conspicuously lacking, with the principal Fe-bearing phase being pyrite. Both pyrite and uraninite occur predominantly in rounded form. In the modified placer model, these rounded pyrite and uraninite grains are considered detrital, as is the gold (Frimmel, 1997). In contrast, in the hydrothermal models, they are interpreted as pseudomorphic replacements of detrital Fe-oxides, Fe-pisolite, ferricrete, banded iron formation, and Fe-rich shale (Phillips and Law, 2000) and/or products of post-depositional dissolution and re-precipitation mechanisms (Phillips and Myers, 1989; Barnicoat et al., 1997). As both pyrite and uraninite are sensitive to redox conditions, the genetic interpretations of rounded pyrite and uraninite grains in the Witwatersrand metasedimentary rocks constrain models of gold genesis as well as understanding of the evolution of the Archean atmosphere (for contrasting views see Rasmussen and Buick, 1999, and Ohmoto et al., 1999).

There are other uraninite and pyrite placer deposits that bear many similarities to those of the Witwatersrand, although none of these contain significant amounts of gold. These include the 2.9–2.6 Ga Bababudan Group in India (Srinivasan and Ojakangas, 1986) and the 2.45 Ga Elliot Lake Group, Huronian Supergroup, in Canada (Krogh et al., 1984, Sutton and Maynard, 1993). Interestingly, in contrast to the Witwatersrand, none of the quartz pebbles in the Elliot Lake Group carry an O isotopic signature comparable with orogenic vein quartz (Vennemann et al., 1995). The lack of gold in these deposits might therefore simply reflect a lack of suitable source rocks in the eroded hinterland.

Uraninite and pyrite occur throughout the Witwatersrand succession from Dominion at 3086 Ma to the upper Central Rand at 2837 Ma and then into the Ventersdorp Contact Reef at 2714 Ma. They occur again in the Black Reef dated at 2642 Ma (Eriksson, et al., 1995). This is the last known placer deposit before the onset of red beds on the Kaapvaal Craton.

Although the oxidation state of the contemporaneous atmosphere plays a crucial role in the mineralogy of the detrital minerals associated with the gold, it does not control the distribution of detrital gold particles as evidenced by the Fe-oxide bearing auriferous conglomerates of the Tarkwa Basin. The second largest known paleoplacer gold deposits outside the Kaapvaal Craton are hosted by the Tarkwaian System of Ghana, whose age is constrained between 2133 and 2097 Ma (Oberthür et al., 1998; Pigois et al., 2003). Past production was ~310 t Au (Pigois et al., 2003). Reserves in 2002 totaled 131 Mt at 1.6 g/t for 215 t Au within a resource of 340 Mt at 1.6 g/t for 534 t Au (Gold Fields Ltd., unpublished annual report, 2003).

The known amount of gold in that depository is thus two orders of magnitude less than in the Witwatersrand. Similar to the Witwatersrand, the gold is hosted by quartz pebble conglomerate beds, interbedded with sandstone units, all of which experienced greenschist facies metamorphism that led to recrystallization and redistribution of many of the minerals, including gold. In contrast to the Witwatersrand, however, heavy minerals associated with the gold are hematite and magnetite rather than pyrite, in addition to rutile and zircon. Uranium-bearing minerals are rare to absent and there is no significant bitumen. The individual gold grains are located in the matrix, concentrated along heavy-mineral foresets and small fractures, on quartz-pebble boundaries, in pressure-solved quartz between quartz pebbles, as inclusions in hematite derived from the metamorphic oxidation of detrital magnetite, and associated with metamorphic chlorite and white mica (Hirdes and Nunoo, 1994; Pigois et al., 2003).

Another analogue to Witwatersrand-type deposits is sited in the Roraima Supergroup, in northern South America (Santos et al., 2003). It consists of largely undeformed sandstones, minor conglomerates, shale, and felsic ash-fall tuffs, all of which rest on a 2.25–2.00 Ga granitoid-greenstone terrane (Trans-Amazon province) of the Guyana Shield in the northwest of the craton. That terrane contains lode-gold deposits and probably represents a continuation of the lode-gold bearing, Eburnean granitoid-greenstone belts of West Africa. The Roraima sedimentary rocks represent braided fluvial sediments deposited in alluvial plain to subaerial braided delta, and possibly shallow marine environments, for which a foreland setting has been suggested (Santos et al., 2003). The fluvial sediments appear enriched in Au with a background concentration in the arenaceous fraction of 10 ppb (i.e., more than the average 6 ppb for Witwatersrand arenites) and maximum gold grades in quartz pebble conglomerates on basal degradation surfaces of as much as 26 g/t over 5 cm thickness. In contrast to the other examples above, no significant metamorphic or hydrothermal overprint is indicated for the Roraima host rocks. Similar to the Witwatersrand deposits there is an apparent sedimentological control on Au grade and, combined with the occurrence of gold as well rounded nuggets, the evidence supports a paleoplacer model. The age of the Roraima auriferous conglomerates is constrained by a U-Pb single zircon age of 1901 ± 1 Ma for a tuff bed overlying a conglomerate bed (H.E. Frimmel, 2001, personal commun.). The O isotopic composition of the quartz pebbles indicates derivation of the vast majority of the pebbles from orogenic vein quartz (Minter et al., 2002). This, together with paleocurrent directions and detrital zircon ages (Santos et al., 2003), points to a source for the sediment, and by implication the gold, in the 2.0 Ga Trans-Amazon greenstone belt to the north and northeast of the Roraima Basin.

It is evident from this comparison that the Witwatersrand deposits are not unique in terms of style of mineralization. Furthermore, it has to be recognized that Witwatersrand-style gold mineralization occurred not only in a number of cratons, but also at different times. It cannot be ascribed to a single, outstanding, gold-forming event.

The distribution of these deposits is global and over vast areas, 31,000 km^2 in the Witwatersrand, 15,000 km^2 in the Elliot Lake, and 17,000 km^2 in the Moeda. Because there is no evidence for any correlation between organic carbon in the reefs and the pyrite and uraninite content, it is unlikely that localized redox conditions could have preserved these detrital minerals. If the 2200 Ma age of the Moeda is correct (Kirk, et al., 2002, and 2004, personal commun.), the rate of change in the oxygen content of the atmosphere might have been very sudden (Maarten de Wit, 2004, personal commun.). One may well consider this time of change in the atmospheric composition to be the boundary between the Archean and the Proterozoic Eras.

ACKNOWLEDGMENTS

The invitation and subvention received from Hiroshi Ohmoto and Steve Kesler to participate in the Pardee symposium is gratefully acknowledged.

REFERENCES CITED

Antrobus, E.S.A., Brink, W.C.J., Brink, M.C., Caulkin, J., Hutchinson, R.I., Thomas, D.E., van Graan, J.A., and Viljoen, J.J., 1986, The Klerksdorp goldfield, *in* Anhaeusser, C.R., and Maske, S., eds, Mineral deposits of Southern Africa, Volume 1: Johannesburg, Geological Society of South Africa, p. 549–598.

Armstrong, F.C, 1975, Genesis of uranium- and gold-bearing Precambrian quartz-pebble conglomerates: U.S. Geological Survey Professional Paper 1161-A.

Armstrong, G.C., 1966, A sedimentological study of the U.K.9 Kimberley reefs in part of the East Rand [M.S. thesis]: Johannesburg, University of the Witwatersrand, 65 p.

Armstrong, R.A., Compston, W., Retief, E.A., William, L.S., and Welke, H.J., 1991, Zircon ion microprobe studies bearing on the age and evolution of the Witwatersrand triad: Precambrian Research, v. 53, p. 243–266, doi: 10.1016/0301-9268(91)90074-K.

Barnicoat, A.C., Henderson, I.H.C, Knipe, R.J., Yardley, B.W.D., Napier, R.W., Fox, N.P.C., Kenyon, A.K., Muntingh, D.J., Strydom, D., Winkler, K.S., Lawrence, S.R., and Cornford, C., 1997, Hydrothermal gold mineralization in the Witwatersrand basin: Nature, v. 386, p. 820–824, doi: 10.1038/386820a0.

Barton, E.S., Compston, W., Williams, I.S., Bristow, J.W., Hallbauer, D.K., and Smith, C., 1989, Provenance ages for the Witwatersrand Supergroup and the Ventersdorp Contact Reef: Constraints from ion-microprobe U-Pb ages of detrital zircons: Economic Geology and the Bulletin of the Society of Economic Geologists, v. 84, p. 2012–2019.

Bourret, W., 1975, Investigation of Witwatersrand uranium-bearing quartz-pebble conglomerates in 1944–45, *in* Armstrong, F., ed., Genesis of uranium- and gold-bearing Precambrian quartz-pebble conglomerates: U.S. Geological Survey Professional Paper 1161-A, p. A1–A6

Buck, S.G., and Minter, W.E.L., 1985, Placer formation by fluvial degradation of an alluvial fan sequence: The Proterozoic Carbon Leader placer, Witwatersrand Supergroup, South Africa: Journal of the Geological Society [London], v. 142, p. 757–764.

Camisani-Calzolari, F.A.G.M., de Klerk, W.J., and van der Merwe, P.J., 1985, Assessment of South African uranium resources: Methods and results: Transactions of the Geological Society of South Africa, v. 88, p. 83–97.

Cooper, R.A., 1923, Mineral constituents of Rand concentrates: Journal of the Chemical, Metallurgical and Mining Society of South Africa, v. 24, no. 4, p. 90–95.

England, G.L., Rasmussen, B., Krapez, B., and Groves, D., 2002, Palaeoenvironmental significance of rounded pyrite in siliciclastic sequences of the Late Archaean Witwatersrand Basin: Oxygen-deficient atmosphere or hydrothermal alteration?: Sedimentology, v. 49, p. 1133–1156, doi: 10.1046/j.1365-3091.2002.00479.x.

Eriksson, P.G., Hatting, P.J., and Altermann, W., 1995, An overview of the geology of the Transvaal Sequence and Bushveld Complex, South Africa: Mineralium Deposita, v. 30, p. 98–111.

Frimmel, H.E., and Minter, W.E.L., 2002a, Recent developments concerning the geological history and genesis of the Witwatersrand gold deposits, South Africa: Society of Economic Geologists Special Publication, v. 9, p. 17–45.

Frimmel, H.E., and Minter, W.E.L., 2002b, An overview of geological processes that controlled distribution of gold in the Witwatersrand deposits, *in* Cooke, D.R., and Pongratz, J., eds., Giant ore deposits, characteristics, genesis and exploration: Hobart, Tasmania, Centre for Ore Deposit and Exploration Studies: CODES Special Publication 4, p. 221–242.

Frimmel, H.E., 1997, Detrital origin of hydrothermal Witwatersrand gold- a review: Terra Nova, v. 9, p. 192–197, doi: 10.1046/j.1365-3121.1997. d01-23.x.

Garayp, E., Minter, W.E.L., Renger, F.E., and Siegers, A., 1991, Moeda placer gold deposits in the Ouro Fino Syncline, Quadrilatero Ferrifero, Brazil, p. 601–608, *in* Ladeira, F.A., ed., Brazil Gold '91: Rotterdam, A.A. Balkema, 823 p.

Gray, N.K., Tucker, R.F., and Kershaw, D.J., 1994, The Sun project. 1. Discovery of a major new Witwatersrand goldfield, *in* XVth Council of Mining and Metallurgical Institutions Congress: Johannesburg, South African Institute of Mining and Metallurgy, v. 3, p. 95–102.

Hirdes, W., and Nunoo, B., 1994, The Proterozoic paleoplacers at Tarkwa Gold Mine, SW Ghana: Sedimentology, mineralogy, and precise age dating of the Main Reef and West Reef, and bearing on the investigations on the source area aspects, *in* Oberthür, T., ed., Metallogenesis of selected gold deposits in Africa: Geologisches Jahrbuch, v. D100, p. 247–311.

James, C.S., and Minter, W.E.L., 1999, Experimental flume study of the deposition of heavy minerals in simulated Witwatersrand sandstone unconformity: Economic Geology and the Bulletin of the Society of Economic Geologists, v. 94, p. 671–688.

Kirk, J., Ruiz, J., Chesley, J., Titley, S., and Walshe, J., 2001, A detrital model for the origin of gold and sulfides in the Witwatersrand basin based on Re-Os isotopes: Geochimica et Cosmochimica Acta, v. 65, p. 2149–2159, doi: 10.1016/S0016-7037(01)00588-9.

Kirk, J., Ruiz, J., Chesley, J., Walshe, J., and England, G., 2002, A major Archean, gold- and crust-forming event in the Kaapvaal craton, South Africa: Science, v. 297, p. 1856–1858, doi: 10.1126/science.1075270.

Krogh, T.E., Davis, D.W., and Corfu, F., 1984, Precise U-Pb zircon and baddeleyite ages for the Sudbury area, in Pye, E., ed., The geology and ore deposits of the Sudbury Structure: Ontario Geological Survey Special Volume 1, p. 431–446.

Ledru, P., Milesi, J.P., Johan, V., Sabate, P., and Maluski, H., 1997, Foreland basins and gold-bearing conglomerates: A new model for the Jacobina Basin (São Francisco Province, Brazil): Precambrian Research, v. 86, p. 155–176, doi: 10.1016/S0301-9268(97)00048-X.

McPherson, J.G., Shanmugan, G., and Moiola, R.J., 1987, Fan-deltas and braid deltas: Varieties of coarse-grained deltas: Geological Society of America Bulletin, v. 99, p. 331–340, doi: 10.1130/0016-7606(1987)99<331: FABDVO>2.0.CO;2.

Minter, W.E.L., 1976, Detrital gold, uranium and pyrite concentrations related to sedimentology in the Precambrian Vaal Reef placer, Witwatersrand, South Africa: Economic Geology and the Bulletin of the Society of Economic Geologists, v. 71, p. 157–176.

Minter, W.E.L., 1978, A sedimentological synthesis of placer gold, uranium and pyrite concentrations in Proterozoic Witwatersrand sediments, in Miall, A.D., ed, Fluvial sedimentology: Canadian Society of Petroleum Geologists Memoir 5, p. 801–829.

Minter, W.E.L., 1982a, The golden Proterozoic, in Tankard, A.J., Jackson, M.P.A., Eriksson, K.A., Hobday, D.K., Hunter, D.R., and Minter, W.E.L., eds., Crustal evolution of southern Africa: 3.8 billion years of earth history: Berlin, Springer-Verlag, p. 115–150.

Minter, W.E.L., 1982b, The Witwatersrand goldfield: Optima, v. 30, no. 4.

Minter, W.E.L., 1989, Witwatersrand placer concentrates associated with inclined surfaces [abs.], in Programme and Abstracts, 4th International Conference on Fluvial Sedimentology, Sitges, Spain: International Association of Sedimentologists, p. 185.

Minter, W.E.L., 1991, Ancient placer gold deposits, in Foster, R.P., ed., Gold metallogeny and exploration: London, Blackie and Sons, p. 283–308.

Minter, W.E.L., 1999, Irrefutable detrital origin of some Witwatersrand gold and evidence of eolian signatures: Economic Geology and the Bulletin of the Society of Economic Geologists, v. 94, p. 665–670.

Minter, W.E.L., and Loen, J.S., 1991, Palaeocurrent dispersal patterns of Witwatersrand gold placers: South African Journal of Geology, v. 94, p. 70–85.

Minter, W.E.L., Hill, W.C.N., Kidger, R.D., Kingsley, C.S., and Snowden, P.A., 1986, The Welkom goldfield, in Anhaeusser, C.R., and Maske, S., eds., Mineral deposits of southern Africa, Volume 1: Johannesburg, Geological Society of South Africa, p. 497–535.

Minter, W.E.L., Feather, C.E., and Glathaar, C.W., 1988, Sedimentological and mineralogical aspects of the newly discovered Witwatersrand placer deposit that reflect Proterozoic weathering, Welkom gold field, South Africa: Economic Geology and the Bulletin of the Society of Economic Geologists, v. 83, p. 481–491.

Minter, W.E.L., Renger, F.E., and Siegers, A., 1990, Early Proterozoic gold placers of the Moeda Formation within the Gandarela Syncline, Minas Gerais, Brazil: Economic Geology and the Bulletin of the Society of Economic Geologists, v. 85, p. 943–951.

Minter, W.E.L., Cheatle, A., and Hartnady, C.J.H., 1993a, Signatures left by Archean paleowinds in the Witwatersrand gold deposits and their geotectonic significance [abs.], in Abstracts and Program, Stratigraphic record of global change, Pennsylvania State University, University Park, August 8-12: Society of Economic Paleontologists and Mineralogists, p. 59.

Minter, W.E.L., Goedhart, M.L., Knight, J., and Frimmel, H.E., 1993b, Morphology of Witwatersrand gold grains from the Basal Reef: Evidence for their detrital origin: Economic Geology and the Bulletin of the Society of Economic Geologists, v. 88, p. 237–248.

Minter, W.E.L., Frimmel, H.E., Kirk, J., and Vennemann, T., 2002, Paleoplacer gold potential in the Early Proterozoic Roraima Group: Geological Society of America Abstracts with Programs, v. 34, no. 6, p. 517.

Mitchell, M., 1996, Alluvial platinum group minerals from southern New Zealand [Ph.D. thesis]: Dunedin, New Zealand, University of Otago, 277 p.

Oberthür, T., Vetter, U., Davis, D.W., and Amanor, J.A., 1998, Age constraints on gold mineralization and Paleoproterozoic crustal evolution in the Ashanti belt of southern Ghana: Precambrian Research, v. 89, p. 129–143, doi: 10.1016/S0301-9268(97)00075-2.

Ohmoto, H., Rasmussen, B., Buick, R., and Holland, H.D., 1999, Redox state of the Archean atmosphere: Evidence from detrital heavy minerals in ca. 3250–2750 Ma sandstones from the Pilbara Craton, Australia: Comment and Reply: Geology, v. 27, p. 1151–1152, doi: 10.1130/0091-7613(1999)027<1151:RSOTAA>2.3.CO;2.

Papenfus, J.A., 1957, The geology of the Government gold mining areas and new state areas, with particular reference to channel sediments and the origin of minerals of economic importance [Ph.D. thesis]: Pretoria, University of South Africa,.

Phillips, G.N., and Law, J.D.M., 2000, Witwatersrand gold fields: Geology, genesis and exploration: Society of Economic Geology Reviews, v. 13, p. 439–500.

Phillips, G.N., and Myers, R.E., 1989, Witwatersrand gold fields, Part II: An origin for Witwatersrand gold during metamorphism and associated alteration, in Keays, R.R., Ramsay, W.R.H., and Groves, D.I., eds., The geology of gold deposits: The perspective in 1988: Economic Geology Monograph 6, p. 598–608.

Pigois, J.-P., Groves, D.L., Fletcher, I.R., McNaughton, N.J., and Snee, L.W., 2003, Age constraints on Tarkwaian palaeoplacer and lode-gold formation in the Tarkwa-Damang district: SW Ghana: Mineralium Deposita, v. 38, p. 695–714, doi: 10.1007/s00126-003-0360-5.

Pretorius, D.A., 1975, The depositional environment of the Witwatersrand goldfields: A chronological review of speculations and observations: Minerals, Science and Engineering, v. 7, p. 18–47.

Rasmussen, B., and Buick, R., 1999, Redox state of the Archean atmosphere: Evidence from detrital heavy minerals in ca. 3250–2750 Ma sandstones from the Pilbara Craton, Australia: Geology, v. 27, p. 115–118, doi: 10.1130/0091-7613(1999)027<0115:RSOTAA>2.3.CO;2.

Robb, L.J., Davis, D., Kamo, S.L., and Meyer, F.M., 1992, Ages of altered granites adjoining the Witwatersrand Basin with implications for the origin of gold and uranium: Nature, v. 357, p. 677–680, doi: 10.1038/357677a0.

Roscoe, S.M., and Minter, W.E.L., 1993, Pyritic paleoplacer gold and uranium deposits, in Kirkham, R.V., Sinclair, W.D., Thorpe, R.I., and Duke, J.M., eds., Mineral deposit modelling: Geological Association of Canada Special Paper 40, p. 103–124.

Rundle, C.C., and Snelling, N.J., 1977, The geochronology of uraniferous minerals in the Witwatersrand Triad: An interpretation of new and existing U-Pb data on rocks and minerals from the Dominion Reef, Witwatersrand and Ventersdorp Supergroups: Philosophical Transactions of the Royal Society of South Africa, v. 286, p. 567–583.

Sadler, P.M., 1981, Sediment accumulation rates and the completeness of stratigraphic sections: Journal of Geology, v. 89, p. 569–584.

Santos, J.O.S., Potter, P.E., Reis, N.J., Hartmann, L.A., Fletcher, I.R., and McNaughton, N.J., 2003, Age, source, and regional stratigraphy of the Roraima Supergroup and Roraima-like outliers in northern South America based on U-Pb geochronology: Geological Society of America Bulletin, v. 115, p. 331–348, doi: 10.1130/0016-7606(2003)115<0331:ASARSO>2.0.CO;2.

Slingerland, R., and Smith, N.D., 1986, Occurrence and formation of water-laid placers: Annual Review of Earth and Planetary Sciences, v. 14, p. 113–147, doi: 10.1146/annurev.ea.14.050186.000553.

Smith, N.D., and Minter, W.E.L., 1980, Sedimentological controls of gold and uranium in two Witwatersrand paleoplacers: Economic Geology and the Bulletin of the Society of Economic Geologists, v. 75, p. 1–14.

Srinivasan, R., and Ojakangas, R.W., 1986, Sedimentology of quartz-pebble conglomerates and quartzites of the Archean Bababudan Group, Dharwar craton, South India—Evidence for early crustal stability: Journal of Geology, v. 94, p. 199–214.

Sutton, S.J., and Maynard, J.B., 1993, Petrology, mineralogy, and geochemistry of sandstones of the lower Huronian Matinenda Formation: Resemblance to underlying basement rocks: Canadian Journal of Earth Sciences, v. 30, p. 1209–1223.

Teixeira, J.B.G., de Souza, J.A.B., da Silva, M.d.G., Leite, C.M.M., Barbosa, J.F., Coelho, C.E.S., Abram, M.B., Filho, V.M.C., and Iyer, S.S.S., 2001, Gold mineralization in the Serra de Jacobina region, Bahia Brazil: Tectonic framework and metallogenesis: Mineralium Deposita, v. 36, p. 332–344.

Theis, N.J., 1976, Uranium-bearing and associated minerals in their geochemical and sedimentological context, Elliot Lake, Ontario [Ph.D. thesis]: Kingston, Ontario, Queen's University, 158 p.

Tucker, R.F., 1980, The sedimentology and mineralogy of the Composite Reef on Cooke Section, Randfontein Estates Gold Mine, Witwatersrand, South Africa [M.S. thesis], Johannesburg, University of the Witwatersrand, 355 p.

Vennemann, T.W., Kesler, S.E., Frederickson, G.C., Minter, W.E.L., and Heine, R.R., 1995, Oxygen isotope sedimentology of gold- and uranium-bearing Witwatersrand and Huronian Supergroup quartz-pebble conglomerates: Economic Geology and the Bulletin of the Society of Economic Geologists, v. 91, p. 322–342.

MANUSCRIPT ACCEPTED BY THE SOCIETY 29 OCTOBER 2005

Witwatersrand gold-pyrite-uraninite deposits do not support a reducing Archean atmosphere

Jonathan Law
CSIRO c/- 6 Genoa Court, Mount Waverley, 3149, Victoria, Australia

Neil Phillips
School of Earth Sciences, University of Melbourne c/- P.O. Box 3, Central Park, 3145, Victoria, Australia

ABSTRACT

The first serious suggestion that the Archean atmosphere was reducing was based on the interpretation of round uraninite and pyrite grains in the Witwatersrand Basin in the early 1950s. It was then inferred that these minerals were detrital and that they reflected equilibrium with a reducing Archean atmosphere.

Over the past 20 years the understanding of the Witwatersrand Basin has changed dramatically with more integrated studies of the basin and the recognition of widespread alteration in close spatial association with the mineralization in every goldfield. Post-depositional mobility of gold, sulfur, and uranium during alteration is widespread and supports hydrothermal ore genesis, or at least substantial modification of the original mineral assemblage. Pseudomorphic replacements of pre-existing detrital minerals (e.g., pyrite after titano-magnetite), and precipitation and/or chemical rounding to generate round mineral shapes (e.g., uraninite in carbon seams) have all been documented.

The recognition that the carbon seams formed by the post-depositional introduction and maturation of migrated hydrocarbons is a dramatic departure from earlier models of coalified algal material deposited with the sediments. The enrichment of both gold and uraninite in carbon seams implies that these minerals are hydrothermal and that their shapes do not reflect detrital processes. Uranium mobility in basinal waters may in fact require a relatively oxidizing atmosphere.

None of the existing arguments for the Witwatersrand mineralization unambiguously support a placer or modified placer model for the mineralization. Consequently, round uraninite and pyrite of the Witwatersrand Basin do not provide support for a reducing Archean atmosphere.

Keywords: atmosphere, Archean, Witwatersrand, pyrite, hydrothermal.

INTRODUCTION

Since the early 1950s, it has been suggested that the Archean atmosphere was "reducing," and that it may not have contained significant amounts of free oxygen until ca. 2 Ga. However, the geological evidence for a reducing Archean atmosphere is both ambiguous and controversial (see reviews by Dimroth and Kimberley, 1976; Clemmey and Badham, 1982; Palmer et al., 1987; Ohmoto, 1996; Phillips et al., 2001).

In fact it was the discovery of widespread uraninite in the Witwatersrand Basin in the 1950s that led to the first serious suggestion of a reducing Archean atmosphere. This hypothesis arose

---

*E-mail: law_jonathan@hotmail.com; neil.phillips@bigpond.com

Law, J., and Phillips, N., 2006, Witwatersrand gold-pyrite-uraninite deposits do not support a reducing Archean atmosphere, *in* Kesler, S.E., and Ohmoto, H., eds., Evolution of Early Earth's Atmosphere, Hydrosphere, and Biosphere—Constraints from Ore Deposits: Geological Society of America Memoir 198, p. 121–141, doi: 10.1130/2006.1198(07). For permission to copy, contact editing@geosociety.org. ©2006 Geological Society of America. All rights reserved.

from combining the widely held beliefs that the gold grains in the Witwatersrand conglomerates were detrital in origin, and that round uraninite and pyrite grains spatially associated with the gold probably had a similar origin. As the uraninite grains could not have survived detrital transportation under oxidizing atmospheric conditions, a reducing Archean atmosphere was proposed. Subsequently, paleosols were inferred in and around the Witwatersrand Basin (and elsewhere), and their chemical composition was inferred to support the reducing atmosphere hypothesis (Button and Tyler, 1981).

These models for a reducing Archean atmosphere were developed in an era of overwhelming dominance of the placer and modified placer models for the Witwatersrand gold mineralization. As a result, research focused on the sedimentology of the host rocks, and there was a widespread belief that Witwatersrand rocks were "unaltered" and "unmetamorphosed," effectively precluding the possibility of a hydrothermal origin for the mineralization. Over the past 20 years this picture has changed dramatically with more integrated studies of the basin and the recognition of widespread alteration in close spatial association with the mineralization in every goldfield. In the light of these developments, it is appropriate to critically review new and existing data for the Witwatersrand to assess their relevance as constraints on the Archean atmosphere. In particular we review the arguments linking mineralization to the Archean atmosphere, i.e., whether or not the gold, pyrite, and uraninite are primary detrital grains.

The possible link between mineralization and the Archean atmosphere has important implications for ore genesis beyond the Witwatersrand. Although a reducing Archean atmosphere is "built in" to many existing genetic models for a variety of orebodies, the alternative of an oxidizing atmosphere would require critical re-evaluation of the origins of many Archean orebodies that assume reducing atmospheric conditions. Also, new exploration opportunities for Witwatersrand-style exploration targets may exist in younger sedimentary basins.

DEFINITION OF ATMOSPHERIC REDOX STATE

Most models for the early atmosphere assume that it was reducing when Earth formed, and subsequently changed to be oxidizing as it is today. A reducing atmosphere in the very earliest Archean is common to all models and, in the absence of geological constraints, is based exclusively on conceptual studies of the early evolution of Earth, Sun, and the solar system. Two basic models have been proposed:

1. **An oxidizing Archean atmosphere:** an oxidizing atmosphere evolved very early in the evolution of Earth and was already well developed in the early Archean (i.e., by 3.5 Ga; Palmer et al., 1987, 1989; Ohmoto, 1996).
2. **A reducing Archean atmosphere:** Earth's atmosphere was reducing until ca. 2.4–2.0 Ga, when a proliferation of photosynthetic organisms resulted in gradually increasing levels of oxygen in the atmosphere (Krupp et al., 1994; Holland et al., 1994; Kasting, 1993; 2001; Rye and Holland, 1998; Farquhar et al., 2000).

There are several different understandings of what constitute "reducing" and "oxidizing" conditions in the early atmosphere. Holland (1984) has used the composition of inferred "paleosols" to express O_2 levels relative to present atmospheric compositions, whereas others have argued that such calculations are premature before conclusive evidence is presented demonstrating that the paleosols are unaltered, or indeed are paleosols at all (Palmer et al., 1987; 1989).

We base our working definition of atmospheric redox state on the geochemistry of iron, and this definition is implicit in many other studies. For example, supporters of the "reducing" model assume that the level of atmospheric O_2 is too low to stabilize significant Fe^{3+} (e.g., when drawing conclusions on iron mobility in inferred "paleosols"; Holland, 1984; 1994). Iron is one of the most important components in Earth's crust, and knowledge of its behavior and its oxidation states is possibly the most important factor related to the early atmosphere affecting our understanding of early geological processes. The chemical behavior of iron is strongly influenced by whether near-surface environments, including aqueous solutions, had abundant Fe^{2+} and Fe^{3+}, or simply Fe^{2+}. If the former (i.e., oxidizing atmosphere), then many near-surface processes would be similar to today; if the latter (i.e., reducing atmosphere), then none of the near-surface processes that involve and require abundant iron in two oxidation states (e.g., laterite formation and iron pisolith formation) would be operative. An "oxidizing" environment in terms of this definition does not necessarily imply an atmospheric composition exactly the same as today: *it means oxidizing enough to facilitate geological processes that require the presence of two oxidation states of iron.* Calculations of the amount of free oxygen necessary to stabilize ferric iron minerals suggest that free oxygen may have been several orders of magnitude less than the present atmospheric level (e.g., Krupp et al., 1994; Holland, 1994). Any more precise definition is premature given uncertainties in interpreting basic geological relationships and then inferring the associated chemical environment. For example, absolute thermodynamic calculations require many assumptions on the origin and evolution of the mineral assemblages, chemical boundary conditions, and kinetic controls on reaction progress.

The stability of detrital pyrite and uraninite are directly linked to atmospheric redox conditions. In a reducing atmosphere, detrital uraninite and pyrite would be meta-stable along with a suite of other sulfide minerals, and iron might be leached from detrital iron oxides. In an oxidizing atmosphere, many iron oxide minerals would be stable and uranium would be highly soluble, destabilizing detrital uraninite. Oxidation of U^{4+} minerals such as uraninite and brannerite occurs readily under modern atmospheric conditions, making their persistence during detrital transport highly unlikely in an oxidizing atmosphere (Davidson, 1953, 1957; Davidson and Cosgrove, 1955; Holland, 1994), though remotely possible given localized modern occurrences (Robinson and Spooner, 1984; Maynard et al., 1991).

OVERVIEW OF PYRITE, GOLD AND URANIUM IN CONGLOMERATES

Geological Overview and Distribution

Gold, pyrite, and uraninite in fluvial sediments of continental basins have been variably described as Witwatersrand-type, conglomerate-hosted, quartz pebble-associated, and quartz arenite-associated mineralization. Each of these descriptors focuses on the nature of the host rocks and each implicitly relies on a placer origin of the mineralization to provide useful discrimination from other deposit styles. In fact, the genesis of conglomerate-hosted pyrite and uraninite ore bodies was once thought to be so well understood that the most commonly used name for the orebodies included the genetic descriptor "paleoplacer."

Numerous occurrences of gold (with or without pyrite and uraninite) in conglomerates are known from around the world but only a few have been significant economic producers (Fig. 1). Of all these, the Witwatersrand Basin in South Africa is by far the largest producer of both gold and uranium. All-time Witwatersrand production now exceeds 50,000 t of gold from seven goldfields. In each goldfield there is typically one reef horizon that has produced the majority of the gold (e.g., Kimberley Reef at Evander, Main Reef Leader in the Rand goldfields, Carbon Leader Reef at Carletonville, Vaal Reef at Klerksdorp, and Basal Reef at Welkom). With the exception of Evander, all goldfields also contain one or more subordinate reef horizons that are economic over smaller areas, and in the case of the West Rand goldfield there are at least ten such reefs (Phillips and Law, 2000).

Both placer and hydrothermal models have been proposed for Witwatersrand-type pyrite, uraninite, and gold mineralization. Each of these models acknowledges the relationship between gold distribution and rock type but provides differing explanations. According to the *unmodified placer model* gold and uranium were introduced as placer concentrations during sedimentation. The model assumes that detrital gold, pyrite, and uraninite grains have the same shape, composition, and location now as they had at their time of burial and hence can be used to constrain source terranes and their compositions. The *modified placer model* is similar but assumes that although gold and related minerals were detrital in origin they have been locally remobilized and recrystallized after burial to account for secondary grain shapes and compositions. A scale of movement in the millimeter to centimeter range is generally inferred. The *hydrothermal replacement model* proposes that the gold and uranium mineralization is not detrital and was introduced by the circulation of hydrothermal fluids in the basin after burial (see Robb and Meyer, 1995 and Phillips and Law, 2000 for opposing views on the merits of these models).

The placer and modified placer models have dominated thinking throughout much of the twentieth century and have been used to support the reducing Archean atmosphere hypothesis.

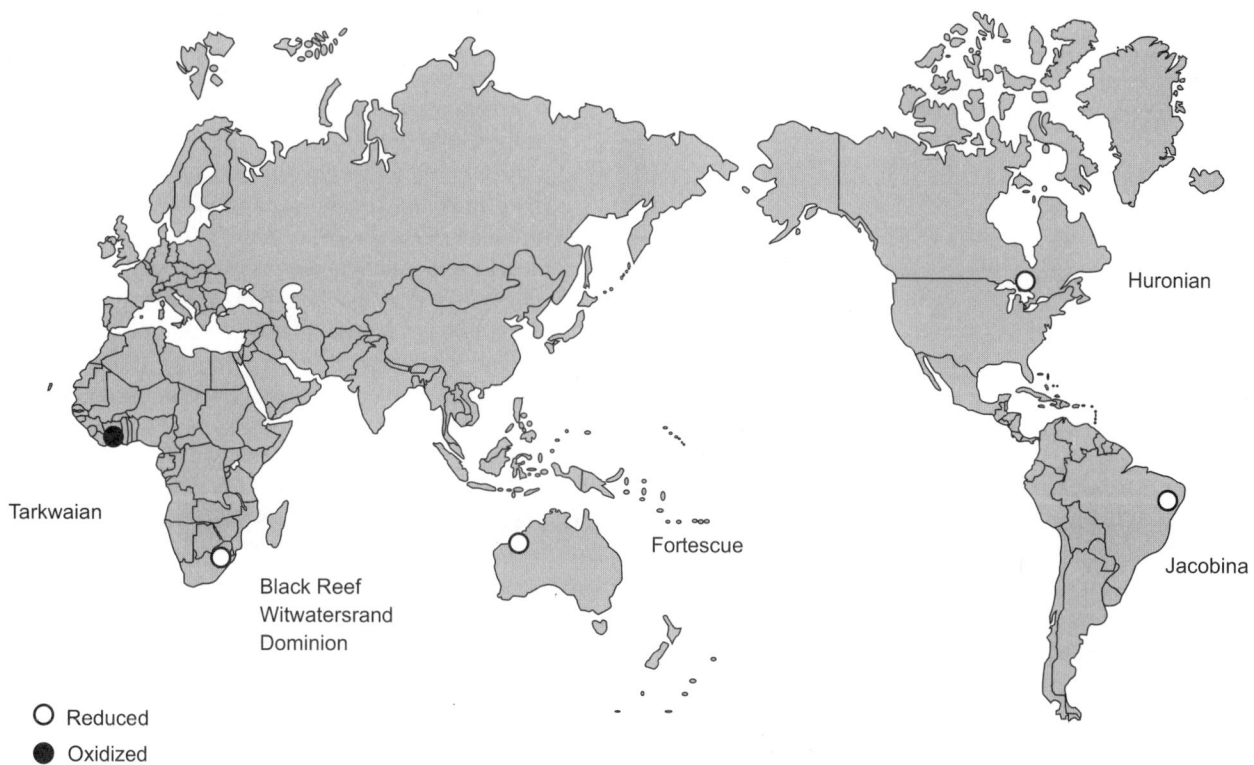

Figure 1. Location of major reduced gold-pyrite-uraninite conglomerates and the oxidized Tarkwaian gold-magnetite-hematite conglomerates.

Time Distribution of "Oxidized" and "Reduced" Witwatersrand-type Deposits

All pyrite-, gold-, and uraninite-bearing conglomerates are older than 1900 Ma and together span the transition from a reducing to oxidizing atmosphere proposed by the reducing Archean atmosphere hypothesis (i.e., between 2.4 and 2.0 Ga; Fig. 2). In South Africa, Witwatersrand-type gold, pyrite, and uranium mineralization is sporadically represented over a considerable period of time from local occurrences in the Archean greenstone basement, through the Dominion Reef, Witwatersrand Supergroup, and Black Reef Formation of the Transvaal Sequence (Fig. 3).

The Jacobina deposits are the youngest economically significant "reduced" conglomerate-hosted gold-pyrite-uranium deposits. Sedimentation at Jacobina is bracketed between 2086 Ma (age of the youngest detrital zircon) and 1883 Ma (age of intrusive post-tectonic granite) (Milesi et al., 2002). In contrast, the oxidized conglomerate-hosted gold ores such as those at Tarkwa have been inferred to postdate the oxygenation of the atmosphere and have been dated between 2124 ± 9 Ma (the age of the youngest detrital zircon) and 1991 ± 12 Ma (the age of intruding granites) by Bossière et al. (1996). Uraninite and pyrite are absent from the Tarkwaian ores and iron oxides dominate the heavy mineral suite. Available geochronological constraints for oxidized mineralization at Tarkwa and reduced mineralization at Jacobina indicate that they may overlap in time. The transition between reduced and oxidized styles of conglomerate-hosted gold mineralization has been used to support a transition from a reduced to oxidizing atmosphere; however, the available age data indicate that they are broadly coeval and cannot both reflect global atmospheric composition.

Common Features of Reduced Deposits

In spite of the variety in terms of age and geological setting, some aspects of the gold-pyrite conglomerates listed in Table 1 (from South Africa and around the world) are remarkably consistent and imply that similar processes have operated in each:

- All deposits are located in clastic continental sediments closely associated with active uplift during sedimentation and associated unconformity development.
- All contain elevated gold and uranium, although the relative abundance of each varies dramatically both within and among deposits.
- All contain round pyrite grains.
- Detrital magnetite and ilmenite are virtually absent in all deposits.
- All contain uraninite and a variety of U-Ti minerals, including brannerite and leucoxene, that at least in part reflect post-depositional modification of pre-existing uraniferous minerals.
- All deposits have an overprint of diagenetic and hydrothermal alteration that modifies the primary detrital mineralogy.
- All contain migrated hydrocarbons intimately associated with uranium minerals and inferred to be syn- or post-uranium mineralization in age.
- All contain gold that is mostly paragenetically late in the mineralization sequence and is variably interpreted either as hydrothermal in origin or as remobilized detrital gold.

Although the Witwatersrand gold production is more than 100 times larger than any of the other reduced deposits known from elsewhere in the world, the genetic models proposed for each are similar.

THE WITWATERSRAND BASIN

Structural and Sedimentological Setting

The Witwatersrand orebodies are confined to a tectonically preserved remnant of a more extensive sedimentary basin with a complex syn- and post-depositional history (Phillips and Law, 2000). The Witwatersrand sedimentary sequence (Witwatersrand Supergroup) was deposited on the stable granite-greenstone crust of the Kaapvaal Craton. The preserved structural basin is elon-

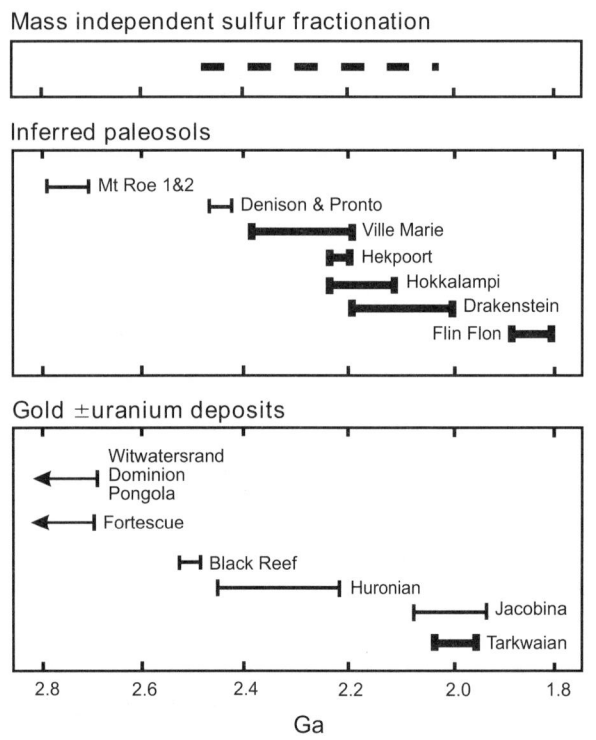

Figure 2. Age constraints on the deposition of the major global gold-pyrite-uraninite occurrences relative to inferred atmospheric composition based on altered zones interpreted as "paleosols" (from Holland and Rye, 1997) and the mass independent fractionation of sulfur (from Farquhar et al., 2000). The oxidation state inferred by these authors is indicated by line thickness (narrow—reducing; bold—oxidizing). The transition from mass dependent to mass independent styles of sulfur fractionation is indicated by a dotted line.

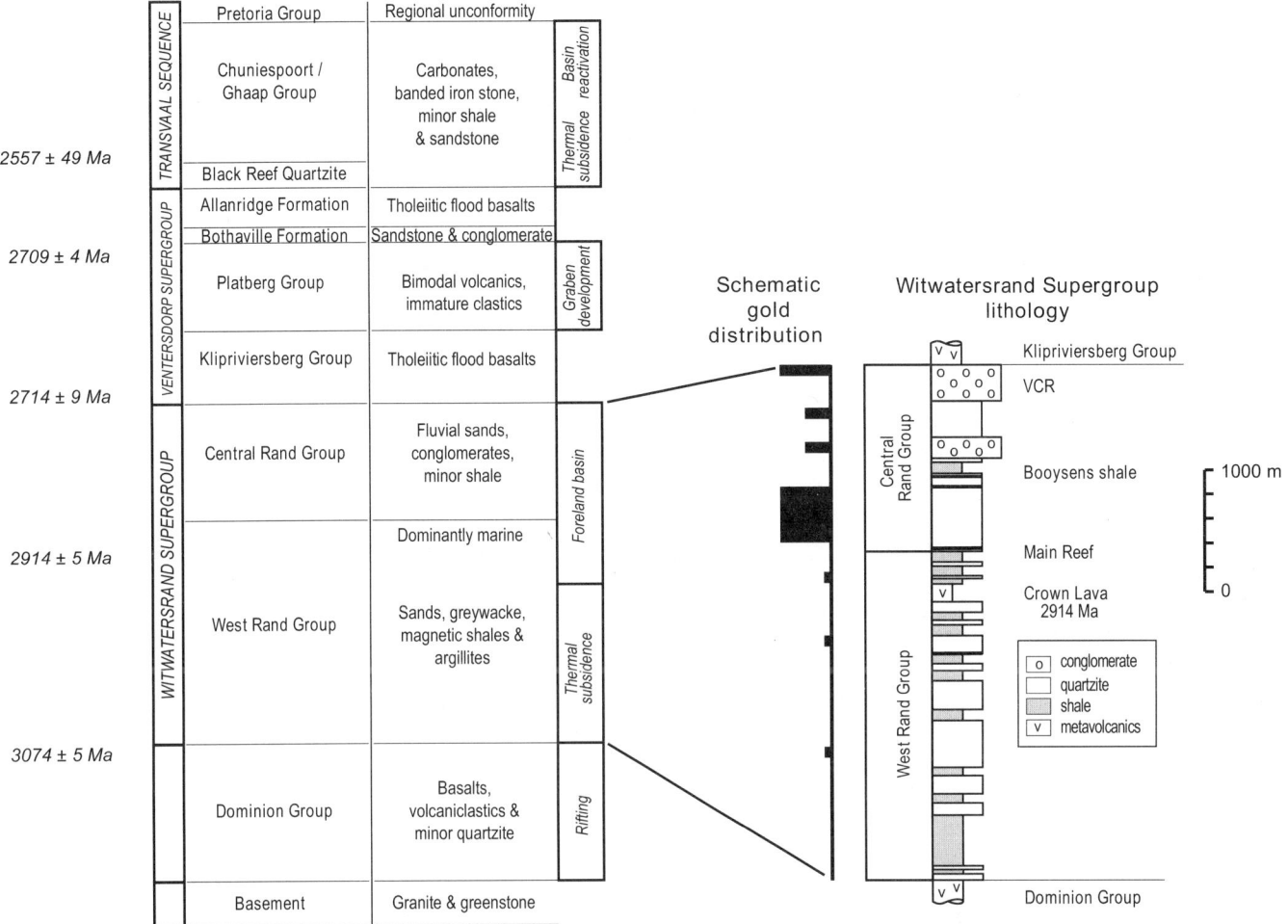

Figure 3. Generalized stratigraphy of the Kaapvaal Craton in the vicinity of the Witwatersrand Basin showing the positions of the main mineralized horizons. Ages refer to volcanic horizons in the stratigraphy at the positions shown.

TABLE 1. SUMMARY OF SELECTED BASINS WITH CONGLOMERATE-HOSTED GOLD AND/OR URANIUM MINERALIZATION AND DOMINANT MINERALOGICAL ASSOCIATIONS

Basin, location	Redox state	Mineralogical associations				
		Gold	Pyrite	Uraninite	Hydrocarbon	Iron oxides
Jacobina, Brazil	Reduced	XX	XX	XX	XX	-
Tarkwaian, Ghana	Oxidized	XX	-	-	-	XX
Huronian, Canada	Reduced	X	XX	XX	XX	-
Transvaal (Black Reef Formation only), South Africa	Reduced	XX	XX	XX	XX	-
Witwatersrand, South Africa	Reduced	XX	XX	XX	XX	-
Fortescue, Australia	Reduced	X	XX	XX	XX	-
Dominion, South Africa	Reduced	X	XX	XX	XX	-

Note: XX—common component; X—minor component; —absent to very rare. Note that absolute abundances of minerals vary dramatically within and between ore bodies. "Redox state" refers to the presence of either pyrite (reduced) or iron oxide (oxidized) in the ore assemblage (Krupp et al., 1994).

gate to the NE and is ~350 km long, 150 km wide, and up to 8 km thick. In a general sense, the stratigraphic record preserves a progressively upward-coarsening depositional sequence dominated by marine sedimentary rocks toward the base and clastic continental sedimentary rocks toward the top (Fig. 3). The lower marine sequence was considerably more extensive than the currently preserved structural basin. The upper parts of the stratigraphic sequence were deposited in response to local syn-depositional deformation near the currently preserved margin (McCarthy, 1994; Coward et al., 1995). Most of the gold mineralization is in the upper sequence of continental sedimentary rocks.

The Witwatersrand Supergroup unconformably overlies the Dominion Group and surrounding Archean granite-greenstone basement and is itself overlain unconformably by the Ventersdorp

Supergroup. All three units are Archean in age. The Ventersdorp is overlain by variable thicknesses of the Archean to Proterozoic Transvaal Sequence and generally flat-lying Mesozoic sedimentary rocks of the Karoo Sequence. Rapid lateral variations in the thickness of the cover sequence around the Witwatersrand Basin result from regional unconformities and local structural complexity. Dikes and sills of Ventersdorp, Bushveld, and Karoo age are common throughout the Basin.

The basement granite-greenstone terranes are mostly from 3.3 Ga to 3.1 Ga in age (Robb et al., 1990a, 1990b), and volcanic rocks from the Dominion Group have been dated at 3.07 Ga (Armstrong et al., 1991). By synthesizing available age constraints, Robb et al. (1990a) inferred that that Witwatersrand sedimentation commenced ca. 3.0 Ga and was terminated by extrusion of the Klipriviersberg volcanic rocks ca. 2714 Ma. Within the Witwatersrand Supergroup, the only reliable age is 2914 Ma from the Crown metabasalt toward the top of the West Rand Group (Armstrong et al., 1991).

The structural evolution of the basin has been reviewed by Coward et al. (1995), who have linked basin formation to the progressive tectonic evolution of the surrounding region during the Archean:

- Pre-Witwatersrand rifting and deposition of the Dominion Group comprising dominantly basaltic and felsic volcanic rocks and minor sedimentary rocks.
- Post-Dominion thermal subsidence and deposition of the dominantly clastic marine lower Witwatersrand succession.
- A progressive change to a compressional tectonic regime and the development of an emergent fold-thrust belt along the northern and western basin margins. The upper Witwatersrand succession thus comprises a generally upward-coarsening succession of dominantly fluvial sedimentary rocks forming an asymmetrical foredeep in the north and west, thinning progressively toward the south and east.
- Extrusion of the Klipriviersberg flood basalts during the waning stages of compression that effectively terminated Witwatersrand sedimentation at 2714 Ma.
- Post-Klipriviersberg extension and deposition of the Platberg Group sedimentary and bimodal volcanic rocks in asymmetric grabens related to northwest-southeast extension ca. 2709 Ma.

The overlying Transvaal Sequence covers an area much greater than the Witwatersrand Basin and comprises extensive clastic and chemical sedimentary rocks and minor volcanic rocks. At the base of the Transvaal, the Black Reef Formation unconformably overlies the Ventersdorp Supergroup and a range of older stratigraphic units, and is developed over a wide area of the Kaapvaal Craton both within and around the Witwatersrand Basin. It comprises a basal conglomerate overlain by quartzite and black shale of fluvial and marine origin, respectively (Tankard et al., 1982; Els et al., 1995). Platform carbonates of the Chuniespoort or Ghaap Groups and fine-grained clastic rocks of the Pretoria Group successively overlie the Black Reef. A Pb-Pb whole rock age from the Chuniespoort Group of the Transvaal sequence of 2557±49 Ma suggests that the base of the Transvaal may be much earlier and that the lower Transvaal rocks are of late Archean age (Jahn et al., 1990). The stratigraphically highest gold mineralization in the Witwatersrand goldfields has been attributed to the mineralized conglomerate at the base of the Black Reef Formation.

Metamorphism and Alteration

Metamorphic assemblages indicate that greenschist facies conditions of 300–400 °C and 2–3 kbars were reached in each goldfield (Phillips and Law, 1994). Metamorphism was accompanied by widespread alteration that affected virtually the entire upper Witwatersrand succession in every goldfield and resulted in the progressive conversion of pre-existing detrital and diagenetic assemblages to muscovite, chlorite, pyrophyllite, and chloritoid. Pyrite is very widespread in samples in the Central Rand Group. Fluid flow has been channeled along bedding-subparallel brittle-ductile faults particularly in heterogeneous sedimentary units overlying unconformity surfaces (Phillips, 1988; Barnicoat et al., 1997). Alteration crosscuts stratigraphy, is locally focused by thick, regionally persistent shales, and is spatially related to mineralization (Phillips and Law, 2000). Isotopic ages from alteration assemblages typically reflect younger resetting events, but a minimum age on alteration is inferred from large-scale extensional faults of Platberg age that displace both alteration and mineralization on all scales (i.e., alteration and mineralization predate 2709 ± 4 Ma; Phillips and Law, 2000). Mineralization in the Black Reef and specifically pyrophyllite alteration and related deformation in mineralized Black Reef at South Deep Mine are inconsistent with a pre-Platberg age. This evidence has been used to infer hydrothermal gold introduction with at least some of this gold post–Black Reef in age (Wall et al., 2004).

In addition to the regional-scale alteration, mesoscopic chlorite veinlets, retrogression around shear zones, and late micas and calcite are all common small-scale features in the Witwatersrand and not unlike what is found in other sedimentary basins. These small-scale alteration systems typically overprint the regional alteration and are not related to mineralization.

Alteration has resulted in the loss of mobile cations, including Na, Ca, and locally K in quartzites and conglomerates, to stabilize the highly aluminous alteration assemblages. Given the similarity between cation loss during alteration and that during weathering, several authors have argued that the unusual bulk-rock compositions reflect weathering at the source, during transport, or after deposition of the sediment (Sutton et al., 1990; Reimer, 1985). However, recent studies have shown that alteration zones cut stratigraphy on a local and regional scale and are spatially related to bedding-subparallel shear zones and brittle fractures that channeled fluid flow (Barnicoat et al., 1997; Phillips and Law, 1997; 2000). The large-scale addition of K is a feature of alteration, but less likely during weathering.

Alteration of Witwatersrand sedimentary rocks has even greater implications for the interpretation of the Archean atmosphere in that many of the early examples of reduced "paleosols" from the Witwatersrand are now recognized as alteration zones (Palmer et al., 1987; 1989). Similarly, studies of arenite textures and unaltered sedimentary rocks show extensive evidence of hydrothermal alteration superimposed on preexisting weathering and/or diagenetic assemblages (Phillips et al., 1990; Law et al., 1990; Law, 1991; Phillips and Law, 2000)

Mineralization and Paragenesis

Gold, uranium, pyrite, and hydrocarbons are important components of all "Witwatersrand-style" deposits worldwide although their relative abundance varies dramatically both within and among deposits (Table 1). In the Witwatersrand, paragenetic studies on the timing of gold grains and related minerals have yielded diametrically opposing conclusions: that some (but not all) of the gold is texturally early (i.e., detrital; Minter, 1999) or that the gold is late (i.e., hydrothermal or remobilized; Davidson, 1960; Ramdohr, 1958). Nevertheless, there is now a broad consensus in the literature that the bulk of the gold is texturally late with (Frimmel et al., 1993; Minter, 1999) or without (Phillips and Law, 1997; Barnicoat et al., 1997) a significant placer gold contribution. Differences still exist as to the timing and paragenetic relationships among the most closely related ore components, i.e., the gold, carbon, uraninite and pyrite; in spite of local complexities, however, most studies indicate a similar temporal sequence of early uraninite followed by hydrocarbons followed by gold (see review in Phillips and Law, 2000; see also Barnicoat et al., 1997; England, 1999; England et al., 2001).

Phillips and Law (2000) have suggested that this paragenetic sequence is compatible with the inferred evolution of the basin with uranium introduction by meteoric water during syn–Central Rand Group tectonic uplift, followed by diagenetic maturation and migration of hydrocarbons, and final introduction of gold by hydrothermal processes. In contrast, Barnicoat et al. (1997) and Jolley et al. (1999) also support a hydrothermal model but suggest that the gold, uranium, and hydrocarbon relate to the one fluid event.

On a larger scale, detailed structural studies of alteration and deformation show a network of bedding-subparallel shears, faults, quartz veins, and fractures that control alteration (Law and Spencer, 1992; Jolley et al., 1999; 2004). Individual fractures within these deformation zones host carbon seams that contain a substantial proportion of the mineralization in many orebodies. In the case of the Carbon Leader Reef, the mineralization is locally hosted entirely within the carbon seam. More commonly, mineralization is hosted by a variety of rock types but hydrocarbon nodules and seams are commonly important. A complete progression from discrete individual nodules to aggregated nodules and finally continuous seams along the margins of shear zones suggests that the hydrocarbons were emplaced during deformation (Law and Spencer, 1992).

The recognition that the carbon seams formed by the post-depositional introduction and maturation of migrated hydrocarbons is a dramatic departure from earlier models that interpreted the seams as coalified material derived from the burial of algal material interbedded with the sediments. The enrichment of both gold and uraninite in carbon seams implies that these minerals are hydrothermal in origin and that their round shapes do not reflect detrital rounding. U mobility in basinal waters may in fact require a relatively oxidizing atmosphere.

THE DILEMMA OF WITWATERSRAND MINERALOGY

All of the economically mineralized Witwatersrand conglomerates accumulated on long-lived regional unconformities and some are moderately enriched in durable minerals such as zircon and chromite. On the basis of their grain shape, chemical and mechanical character, and textural setting, these two minerals almost certainly represent detrital placer accumulations. By inference, many other heavy minerals should have been accumulated at the same time, including magnetite, garnet, ilmenite, and possibly hematite. Most of these minerals exist in significant quantities in potential source rocks in southern Africa (e.g., magnetite in banded iron formations) and they are well represented in the overlying and underlying stratigraphic units spatially removed from mineralization (Fig. 4) (Fuller, 1958; Phillips and Law, 2000; Meyer et al., 1990). However, any visual examination shows the reefs to be devoid of iron oxides (except as magnetite preserved within BIF clasts; e.g., Hirdes and Saager, 1983) whereas U and Ti oxides including rutile, brannerite, leucoxene, and uraninite are common. Iron-bearing minerals are common, but Fe-oxides are virtually absent.

Another dilemma comes from the implications for the source area if the sulfides are inferred to be detrital. Some very common sulfides (especially of the base metals) are under-represented, and some unusual ones are over-represented (e.g., gersdorffite and cobaltite), despite no evidence to suggest this relationship in any reasonable Archean source area. This discrepancy between the predicted placer assemblage and the observed Witwatersrand Reef assemblage must be examined in light of the post-depositional history of the sedimentary sequence but even with local remobilization, the base metal sulfides should be better represented given likely source areas.

Despite the emphasis on sedimentological aspects of the reefs in the modern literature, there are relatively few published whole rock geochemical data for the orebodies and especially for the surrounding lithologies. One consequence is that marked chemical differences between reefs and their enclosing lithologies are not widely reported. A comparison of compositional differences between the Leader Reef in the Welkom goldfield and the surrounding upper Witwatersrand arenites (Fig. 5) illustrates several key relationships with implications for ore genesis:

1. There is a dramatic enrichment in the orebody of gold and uranium relative to all other elements.

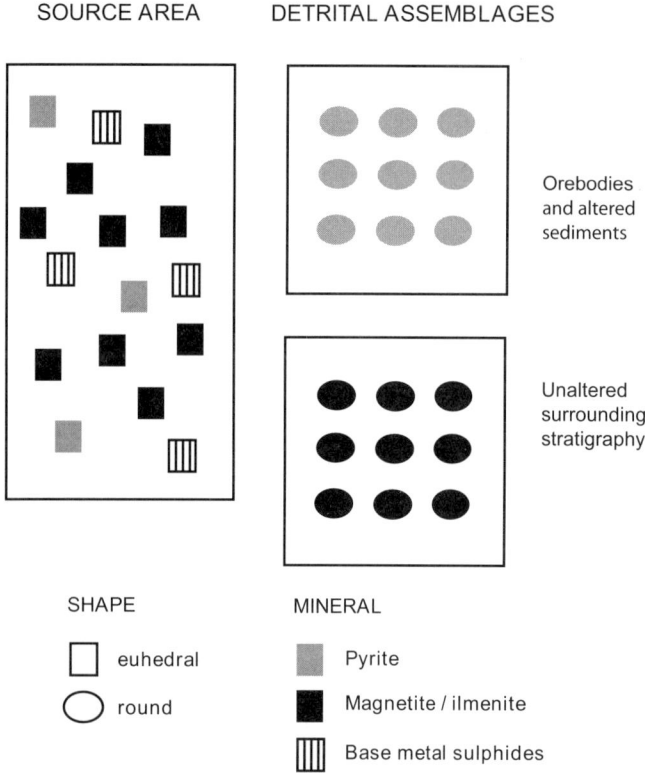

Figure 4. Schematic illustration of the "Witwatersrand dilemma" with respect to the mineralogy of likely source rocks, the orebodies, and other Archean stratigraphic units. The inferred granite-greenstone terrane is likely to contain iron oxides including magnetite, ilmenite, and a complex suite of sulfides including pyrite and base metals. Given a reducing atmosphere, each of these components should be well represented in placer mineralization. In contrast, the orebodies contain round pyrite to the virtual exclusion of base metals (excluding Ni and Co, which are moderately enriched) and the "black sand" or iron oxide component typical of modern placers. Away from the orebodies and associated alteration the overlying and underlying units (in the West Rand Group) contain iron oxides but commonly lack any round sulfides. This distribution pattern favors post-depositional sulfidation as the most likely control on the observed abundances of heavy minerals.

2. Major elements constituting the silicate assemblage are similar in both units with the exception of Mg and Fe, reflecting the increased abundance of chlorite and iron sulfides in the ore assemblage.
3. Cr and Zr, which reflect detrital concentrations of chromite and zircon, respectively, are slightly enriched in the Leader Reef. However, the concentrations do not support substantial enrichment of the heavy mineral suite, especially not enough to explain the enormous enrichment of gold and uranium. Fe and Ti also reflect a substantial detrital contribution either as detrital pyrite or as the missing "black sands" component of the reef, including magnetite and ilmenite (prior to sulfidation of the reefs).
4. Chalcophile metals including Ni, Co, As, and Sb are considerably enriched in the Leader Reef. Although As and Sb are common in many hydrothermal gold deposits, Ni and Co are rare, although they are well represented in hydrothermal uranium deposits, notably the unconformity-associated deposits in Australia and Canada. Cu and Zn are moderately enriched in the Leader Reef but their absolute abundances are relatively low in both the reef and enclosing quartzites.

Enrichments in iron sulfides and associated chalcophile metals similar to those in the Leader Reef in Welkom have been reported from many other Witwatersrand reefs (e.g., Tucker, 1980; Pretorius, 1976a, 1976b; Minter 1978).

Ore Mineralogy: Pyrite, Gold and Uraninite

Pyrite

One of the most striking features of the Witwatersrand conglomerates is the concentration of pyrite. The relationship between pyrite and gold is clear in a stratigraphic sense; most reef horizons have significantly more pyrite than average for the upper Witwatersrand although not all pyrite rich zones are auriferous. In a lateral sense, there is also a general correlation

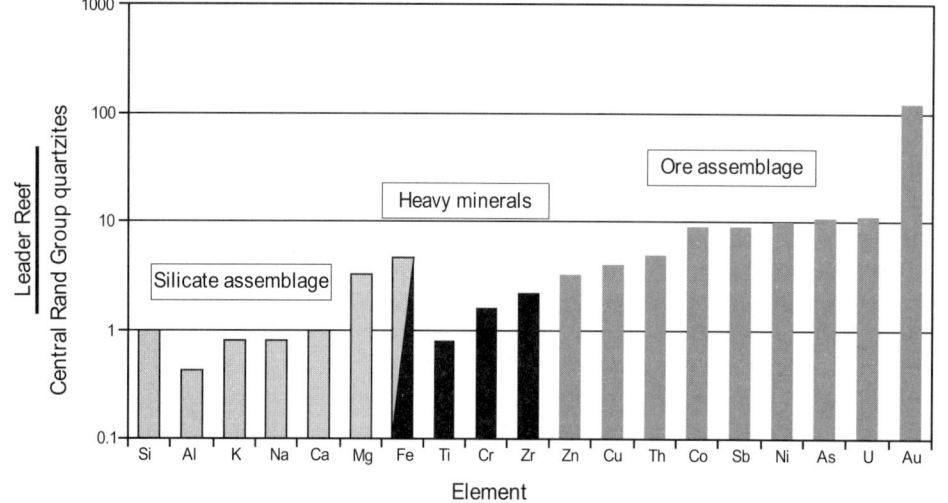

Figure 5. Ratios of element concentrations (geometric means) for the Leader Reef and surrounding unmineralized quartzite from the Central Rand Group (quartzite data from Law, 1991; Leader Reef geochemistry from Callow, 1989, personal commun.). Gold is enriched in a suite of chalcophile metals including Ni, Co, As, and Sb. Cu and Zn are moderately enriched but absolute abundances in all lithologies are low. Fe, Ti, Zr, and Cr form part of the heavy mineral suite reflecting chromite, zircon, and iron/titanium oxides, respectively.

between modal pyrite abundance and gold grades within some reefs but this is less clear when data from many reefs are plotted together (Phillips and Law, 2000). Pyrite takes many forms and has been extensively studied and morphologically classified (e.g., Saager, 1970; see summary in Hallbauer, 1986).

An overwhelming proportion of round pyrite grains are spatially related to scours and other sedimentological features (Fig. 6A) and to heavy minerals that almost certainly form part of the detrital suite (e.g., zircon, chromite). There is thus broad agreement that round pyrite grains are located in positions controlled by the sedimentary dynamics during deposition of the sediments. The atmosphere debate has thus centered on whether the grains were deposited as pyrite or as detrital iron oxide minerals that were pseudomorphously replaced by pyrite.

Three main types of round ("buckshot") pyrite are common although other textural varieties are also present (see summary in Hallbauer, 1986):

1. Compact round pyrites (Fig. 6B) comprising single pyrite grains with round shapes and diameters of less than 1 mm that have been widely interpreted as detrital. Alternatively they could be rounded magnetite and titano-magnetite grains that were sulfidized after burial. Round pyrite grains commonly host gold in crosscutting, post-depositional fractures (Fig. 6C).

2. Porous round or "mudball" pyrite comprising round aggregates of numerous smaller pyrite grains that are 1 mm to 1 cm in diameter and commonly display radial, concentric, or oolitic textures (Figs. 7A [lighter colored nodules to the right of scale bar], 7C) (Ramdohr, 1958; Hallbauer, 1986). An interesting feature of these pyrite grains is that they are considerably larger than other heavy minerals, such as zircon, chromite, and normal (compact round) pyrite found in the conglomerates (e.g., Coetzee, 1965). Porous round pyrites could represent syn-sedimentary growth of sulfides at or near the depositional site (e.g., Hallbauer, 1986) or post-depositional sulfidation of iron oxide pisoliths (Phillips and Law, 2000).

3. Laminated round pyrites (Fig. 6D) comprising banded layers of fine-grained pyrite giving the appearance of rounded fragments of pyrite. These may represent banded sulfide clasts or replacement of other iron-bearing minerals in a banded precursor (e.g., ferruginous chert).

Euhedral pyrite grains (Fig. 6E) and euhedral overgrowths on round grains (Fig. 6F) are very common in samples from Witwatersrand mines. There is also abundant evidence for sulfidation of a variety of other minerals throughout the Witwatersrand goldfields (Ramdohr, 1958; Phillips and Dong, 1993; Myers et al., 1993) (Figs. 8A–C). In some round pyrite grains, remnants of

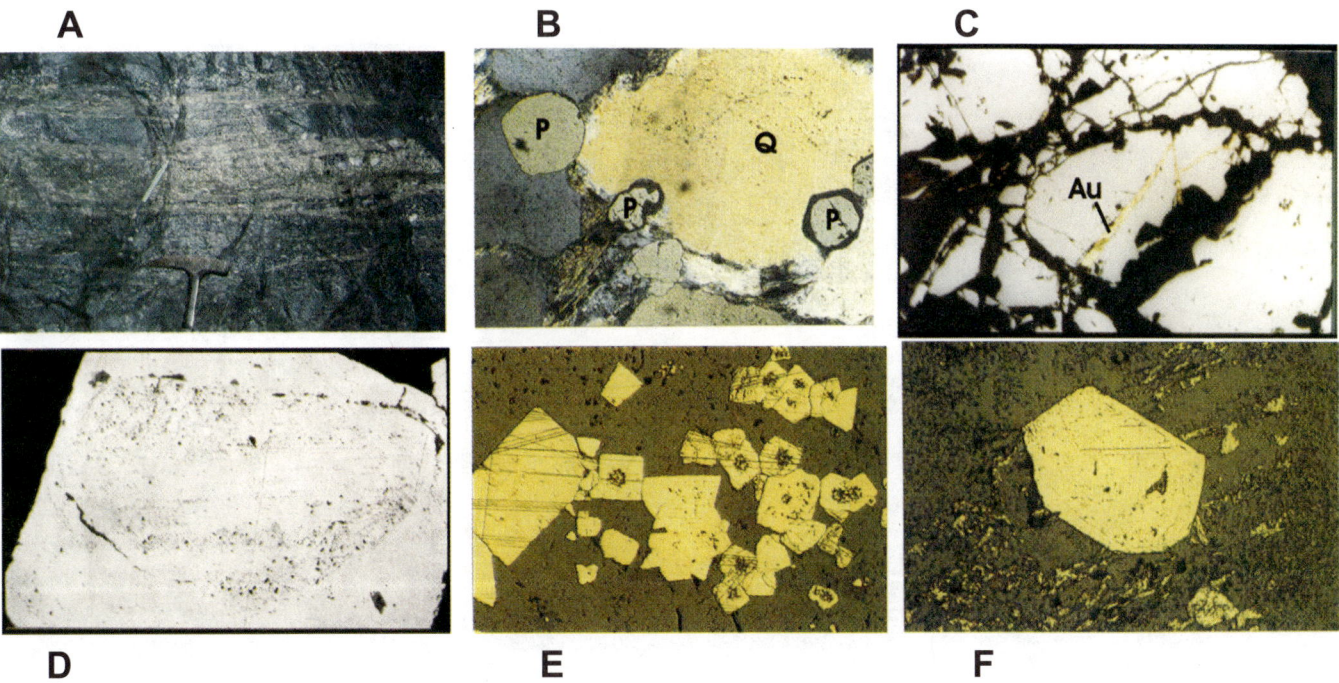

Figure 6. Photomicrographs. (A) Pyrite on subhorizontal sedimentary bedding surfaces, Black Reef, Consolidated Modderfontein mine, showing the strong sedimentological control on the distribution of iron typical of most orebodies. (B) Compact round pyrite (P), detrital quartz (Q), and phyllosilicate matrix (part reflected and part cross-polarized). Field of view is 1.3 mm. From Phillips and Dong, 1993. Pressure solution has resulted in pyrite indenting detrital quartz grains. (C) Brecciated, compact, round pyrite with gold filling fractures, Ventersdorp Contact Reef. Reflected light. Field of view is 0.5 mm. (D) Banded, porous, round pyrite with secondary overgrowth. Reflected light. Field of view is 2 mm. (E) Cubic pyrite overgrowths on spheroidal cores (etched by HNO3). Field of view is 1.3 mm. From Phillips and Dong, 1993. (F) Pyrite overgrowth on a round pebble with a skeletal texture indicated by differences in relief (reflected light). Field of view is 0.66 mm. From Phillips and Dong, 1993.

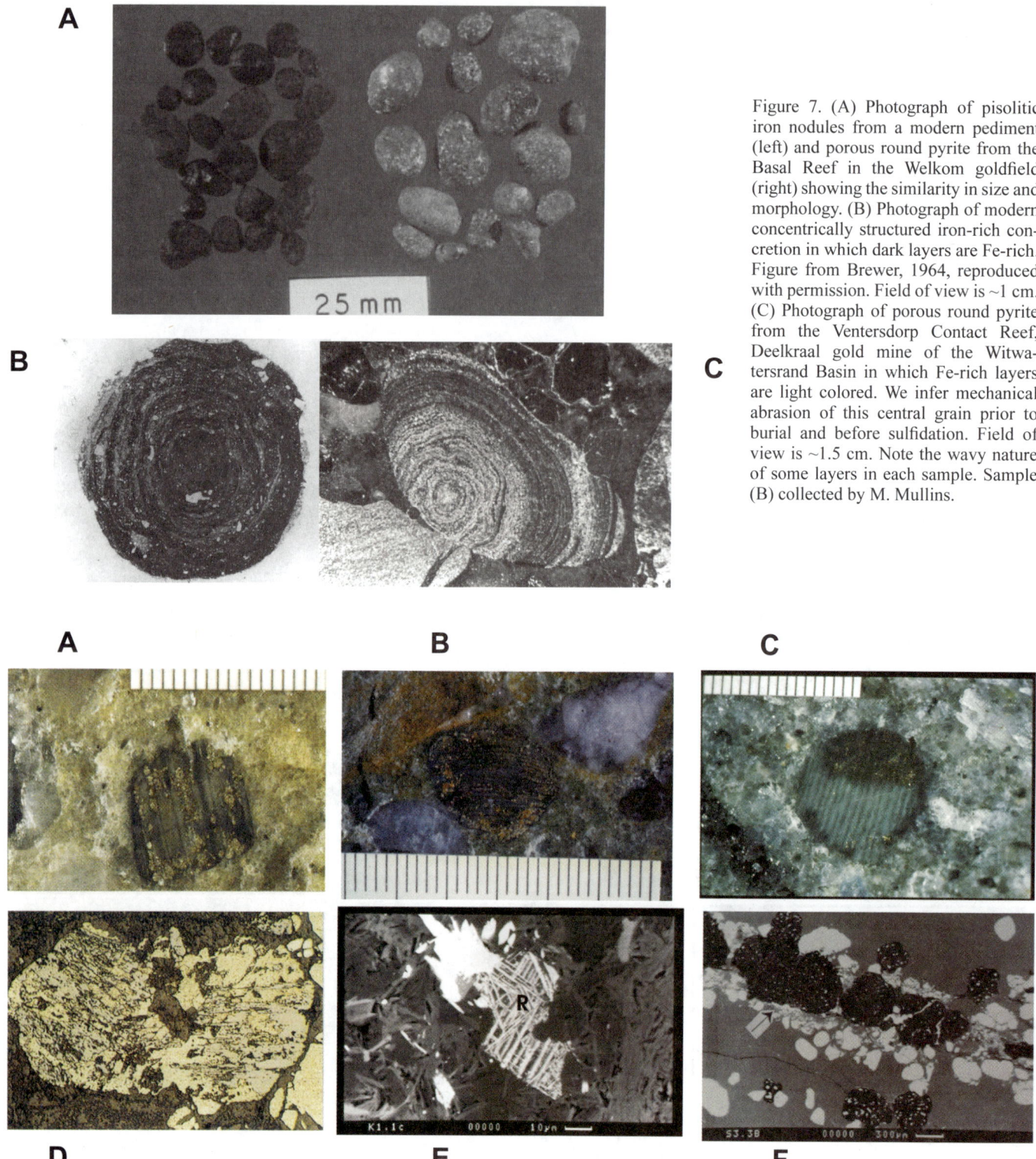

Figure 7. (A) Photograph of pisolitic iron nodules from a modern pediment (left) and porous round pyrite from the Basal Reef in the Welkom goldfield (right) showing the similarity in size and morphology. (B) Photograph of modern concentrically structured iron-rich concretion in which dark layers are Fe-rich. Figure from Brewer, 1964, reproduced with permission. Field of view is ~1 cm. (C) Photograph of porous round pyrite from the Ventersdorp Contact Reef, Deelkraal gold mine of the Witwatersrand Basin in which Fe-rich layers are light colored. We infer mechanical abrasion of this central grain prior to burial and before sulfidation. Field of view is ~1.5 cm. Note the wavy nature of some layers in each sample. Sample (B) collected by M. Mullins.

Figure 8. Photos of replacement textures in Witwatersrand orebodies. (A–C) Partially sulfidized chert pebbles with increased sulfidation around the pebble margin, reflecting sulfidation after burial. Scale bars in millimeters. (D) Inferred former titano-magnetite grain with lamellae of rutile (light gray) formed during Ti-exsolution. The grain matrix of magnetite is replaced by pyrite (light yellow) reflected light. Field of view is 1.3 mm. From Phillips and Dong, 1993. (E) SEM photomicrograph of rutile needles (R) with a texture indicating dissolution of titano-magnetite with the preservation of ilmenite exsolution lamellae, Kimberley Reef, Evander Goldfield. Photo by Andy Barnicoat. (F) SEM photomicrograph of nodular hydrocarbon (black) with inclusion of uraninite and galena (white). Nodules are located along a fracture zone in association with round pyrite (light gray), quartz, and chlorite. Pyrite is locally corroded by hydrocarbon, and gold is indicated by the arrow. Photo by Andy Barnicoat.

unsulfidized ilmenite are preserved (Fig. 8D) (Phillips and Dong, 1993). In other cases, exsolution lamellae of rutile are preserved within round pyrite grains indicating their origin by sulfidation of titano-magnetite (Fig. 9D). Rare examples of leached titano-magnetite and ilmenite grains have also been described (Fig. 8E). MacLean and Fleet (1989 have described some compact round pyrite grains with growth zones truncated by round grain boundaries (Figs. 9A, 9B).

Sulfur isotope studies of pyrite grains from the Witwatersrand provide conflicting evidence partly because of conflicting data sets depending on the scale of observation (Fig. 10). Conventional bulk grain analyses show little departure from 0‰ (see review in Phillips and Law, 2000, p. 478). In contrast, SHRIMP (sensitive high-resolution ion microprobe) ion microprobe analysis of micron-sized spots on individual grains show variations of 6‰ within single pyrite grains, 9‰ between adjacent touching grains, and 11‰ in single samples (Eldridge et al., 1993; Phillips and Law, 2000), and 21‰ in round porous pyrite (England, 1999).

Gold

There is now widespread agreement that the majority of Witwatersrand gold grains have secondary grain shapes and/or occur in textural sites that postdate deposition of the sediments (e.g., Frimmel, 1997; Frimmel and Gartz, 1997; Barnicoat et al., 1997; Phillips and Law, 2000). Some possible exceptions have been noted by Minter (1999) but the interpretation of these grain shapes remains controversial (Barnicoat et al., 2001).

Individual Witwatersrand gold grains are compositionally homogenous with respect to Ag and Hg (Utter, 1979; Hirdes and Saager, 1983; von Gehlen, 1983; Hallbauer, 1986; Oberthür and Saager, 1986; Reid et al., 1986; Frimmel et al., 1993). On the scale of a single hand specimen, greater variability has been described. Some samples show significant between-grain variability whereas all grains in other samples are compositionally homogeneous. In a study of the Basal Reef, Frimmel and Gartz (1997) demonstrated substantial within-sample variability and inferred a detrital origin for the gold. However, there is a strong mineralogical control on gold grain composition (Fig. 11) with relatively restricted compositional variability in grains spatially associated with chlorite and pyrite and far greater variability within grains associated with quartz grains. This pattern implies an in situ control on gold compositions.

A recent study on within-sample gold homogeneity has described grain shapes from a single sample of the Basal Reef in the Welkom goldfield (Minter et al., 1993; Frimmel et al., 1993).

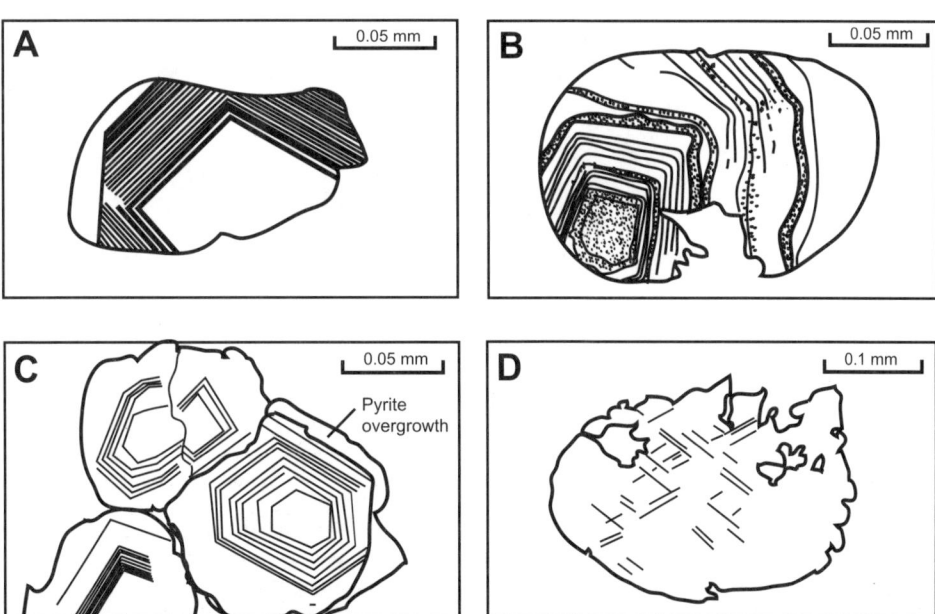

Figure 9. Sketches illustrating important processes responsible for round pyrite grains in the Witwatersrand. (A, B) Compact round pyrite grains with oscillatory polygonal to colloform growth banding defined by As-rich and As-poor bands, Basal Reef, Welkom Goldfield. KMnO$_4$ stained. Redrawn from MacLean and Fleet (1989). MacLean and Fleet have argued that the truncation of growth banding by round grain margins indicates that the grain was pyrite at the time of rounding and that these textures preclude sulfidation of pre-existing round minerals. (C) Oscillatory-zoned pyrite from the hydrothermal ores of the Agnico-Eagle gold mine showing sector zoned core, polyhedral growth banding and irregular margin. KMnO$_4$ stained. Redrawn from Fleet et al. (1989). Original zoned grains have been partly dissolved and overgrown by later low-As pyrite. Dissolution of this type provides an alternative rounding mechanism for pyrite grains in the Witwatersrand such as (A) and (B). (D) Round pyrite grain from the Witwatersrand showing characteristic titano-magnetite exsolution lamellae preserved as rutile. The original titano-magnetite grain has been sulfidized to form pyrite (redrawn from Ramdohr, 1958). Sulfidation may have occurred pre- or post-burial.

132   J. Law and N. Phillips

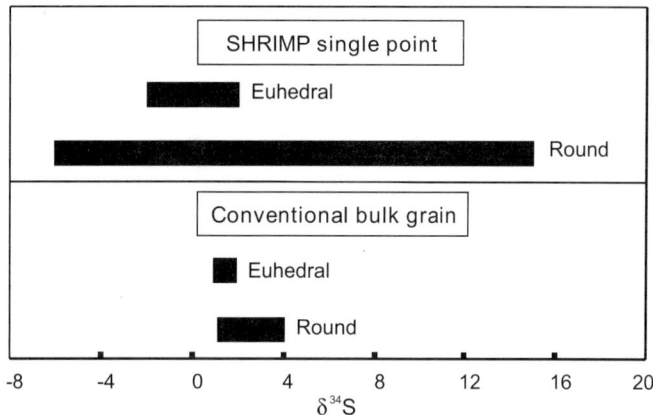

Figure 10. Summary of traditional "bulk-grain" and SHRIMP microprobe analyses of sulfur isotopes for Witwatersrand pyrites (data from Hoefs et al., 1968; Palmer, 1986; Eldridge et al., 1993; England, 1999; Phillips and Law, 2000).

These authors have inferred that 75% of the gold preserves primary detrital gold shapes and that the remaining 25% has typical hydrothermal characteristics. Other workers studying exactly the same sample inferred that all of the gold grains are hydrothermal in origin (Barnicoat et al., 2001), that locations of the grains are controlled by fractures, and that their shapes reflect intergrowths with other secondary minerals. Irrespective of the origin of the gold, the compositions and compositional ranges of all grains are similar, requiring either a remarkably homogeneous source terrane (if detrital) or a post-depositional control on gold compositions (Fig. 12).

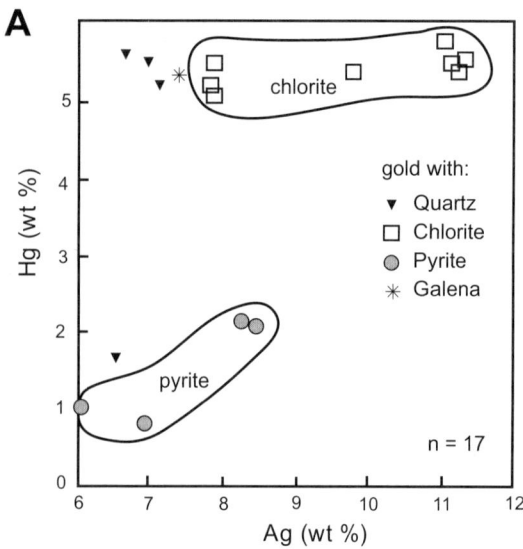

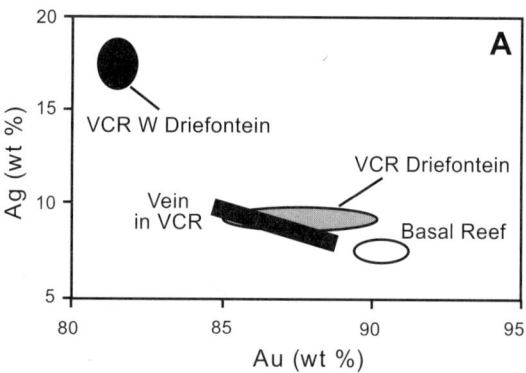

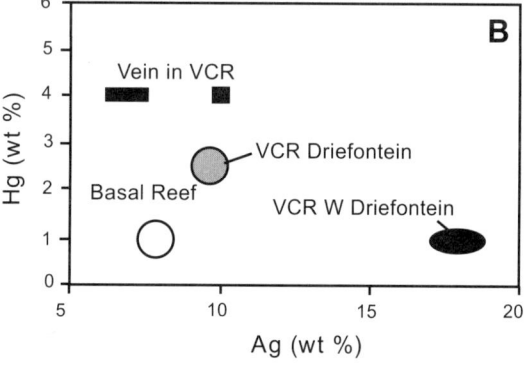

Figure 12. Compositional variability in Witwatersrand gold grains from the Ventersdorp Contact Reef (VCR) and the Basal Reef. (A) Ag versus Au, and (B) Hg versus Ag. Each field represents the variability in gold grain compositions for one or more samples from the localities indicated and discussed in the text. Compositional variability for any likely source terrane is likely to span the entire range illustrated and should be reflected in random detrital samples. Gold grain compositional variability between sites is limited and implies post-depositional controls on gold grain composition. Data compiled from Frimmel et al., 1993 and Frimmel and Gartz, 1997.

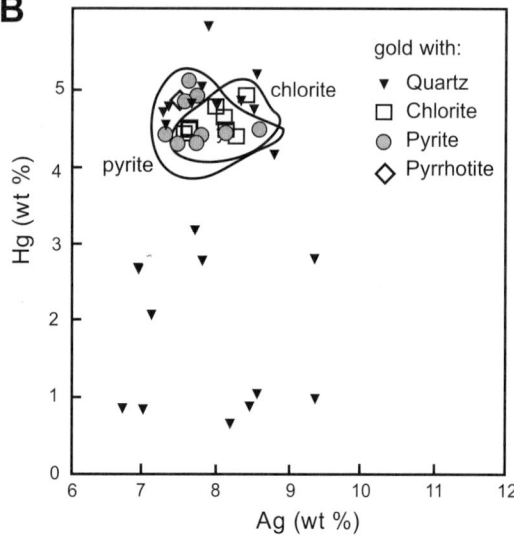

Figure 11. Within-sample compositional variation in Hg and Ag for gold particles in two samples from the Ventersdorp Contact Reef, Klerksdorp goldfield (reproduced from Frimmel and Gartz, 1997). Compositional variability is reduced in spatial association with pyrite and chlorite implying a post-depositional control on gold grain compositions.

On a regional scale, there are substantial variations in the compositions of gold grains among samples. For example, gold grains in two samples from the Ventersdorp Contact Reef (VCR) at East Driefontein mine are compositionally distinct from those at West Driefontein mine some 10 km away on the same stratigraphic horizon (Fig. 12). The degree of variability at each of these locations is similar to that reported by Frimmel et al. (1993) and discussed previously.

Uraninite

Round to "muffin" shaped uraninite grains are common in the Witwatersrand reefs (see summaries in Liebenberg, 1955, Saager, 1968, and England et al., 2001). They typically occur as rounded to sub-rounded grains although rare euhedral grains have been reported. Grains are typically less than 250 μm in diameter and occur in two dominant associations: (1) as isolated grains often spatially associated with, and partially disaggregated and replaced by, hydrocarbons (Fig. 13); or (2) as concentrated clusters within bedding-subparallel fractures known as carbon seams (e.g., Fig. 8F).

IMPLICATIONS OF WITWATERSRAND MINERALOGY FOR THE ARCHEAN ATMOSPHERE

The placer model for Witwatersrand gold has been the mainstay of the reducing Archean atmosphere hypothesis since the 1950s. Over the past 20 years the geological framework for the Witwatersrand has changed significantly and many of the assumptions underpinning the placer model are no longer valid. In particular, the recognition of post-depositional alteration in and around each of the orebodies requires that the placer model

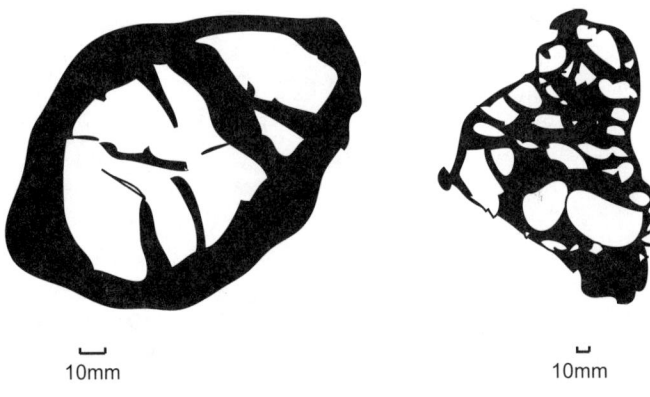

Figure 13. Sketches of an SEM photomicrograph showing the progressive fragmentation and dissolution of uraninite grains (white) by hydrocarbon (black). Drawn from photographs by England et al., 2001. The nodule in left (A) is coated with brannerite and florencite (too small to illustrate). The nodule in right (B) is highly fragmented and rounding of individual fragments reflects the partial dissolution uraninite by hydrocarbons. The nodule is surrounded by chlorite, pyrite and gold (not illustrated).

and its implications for the Archean atmosphere be carefully reassessed. Three questions are critical: (1) Are the round pyrite grains detrital? (2) Do gold compositions reflect atmospheric composition? (3) Are round uraninite grains detrital?

Are the Round Pyrite Grains Detrital?

In addition to weathering and mechanical abrasion, post-depositional mechanisms that have modified Witwatersrand detrital assemblages include chemical rounding, grain dissolution/replacement and sulfidation. The significance of these processes for interpretation of round pyrite grains is discussed below.

Chemical Rounding of Iron Sulfides

Many of the textural varieties described above display complex internal structure that is locally truncated by round grain margins (e.g., Ramdohr, 1958). Several recent studies of the Witwatersrand have highlighted the aggressive nature of hydrothermal alteration in the reefs, especially in association with hydrocarbons, suggesting the potential for widespread dissolution and modification of original grain shapes (Gray et al., 1998; Jolley et al., 1999; Phillips and Law, 2000; England et al., 2001). Some compact round pyrite grains have euhedral growth zones truncated by round grain boundaries (Figs. 9A, 9B; MacLean and Fleet, 1989). Similar grains, with textures similar to the Witwatersrand, have been described by the same authors (Fleet et al., 1989) from the greenstone hosted hydrothermal ores at the Agnico-Eagle Mine of the Abitibi greenstone belt of Canada (Fig. 9C) where hydrothermal pyrite grains have been modified by later hydrothermal processes. The Witwatersrand pyrites may thus have formed by similar hydrothermal processes, but MacLean and Fleet have argued that these textures demonstrate that the pyrites were rounded by detrital processes and deposited as a sulfide phase. In the case of the hydrothermal Agnico-Eagle mineralization, this conclusion is demonstrably wrong and similar textures cannot be used to constrain the origin of Witwatersrand sulfides.

Dissolution of Iron Oxides

Iron is highly mobile in reduced, low-sulfur fluids near Earth's surface. It is thus possible that iron could be leached from detrital iron oxides during transport under reducing atmospheric conditions. If such a process occurred, it would be difficult to constrain the extent of the dissolution. However, in some cases textural evidence suggests that some titano-magnetite and ilmenite grains did survive transport to the basin and were subsequently leached in situ to leave skeletal titanium oxides that are probably too delicate to have survived detrital transport (Fig. 8E).

The dissolution of heavy minerals by reduced basinal fluids is a common process in sedimentary basins (Morton and Hallsworth, 1999) and has been important in the Witwatersrand. However, the relative importance of this process and the location of the dissolution process (i.e., in transport versus in situ) remain effectively unconstrained.

Sulfidation of Detrital Iron Oxides

Reduced Witwatersrand-type pyrite and uraninite mineralization is invariably associated with round pyrite and the virtual absence of the "black sand" components typical of modern placers, including magnetite, ilmenite, and hematite, all of which are common components of any likely Witwatersrand source area. Several authors have suggested that these minerals were sulfidized, dissolved, or replaced prior to deposition (e.g., Reimer and Mossman, 1990). Others have argued that because detrital magnetite and ilmenite are common detrital phases in stratigraphy above and below the Witwatersrand orebodies, post-depositional processes must be responsible for their removal (Dimroth and Kimberley, 1976; Clemmey and Badham, 1982).

There is abundant supporting evidence of in situ sulfidation of a variety of other minerals throughout the Witwatersrand goldfields (Figs. 8A–C). Furthermore, euhedral pyrite grains indicate the presence of a fluid capable of mobilizing gold and sulfur and the potential to dissolve and/or re-precipitate pyrite during alteration. This process is widely accepted, but the location and timing of the sulfidation are unconstrained and could have occurred at source, during transport into the basin, or after deposition. Diagnostic textural relationships such as replacement parallel to clast margins are the exception rather than the rule, and the extent of the sulfidation is thus unconstrained.

Perhaps the most telling argument is the regional distribution of the round sulfide assemblages. In each example of reduced Witwatersrand-type pyrite and uraninite mineralization worldwide (Table 1), round pyrites are common, black sand components are virtually absent, and the evidence for sulfidation, in the form of secondary euhedral pyrites and partially sulfidized mineral grains, is widespread. If sulfidation occurred outside the basin, the detrital assemblage (moving up the sequence) must have changed from oxide-dominated to sulfide-only, and then back again, to account for the stratigraphic distribution of assemblages noted above. Furthermore, this process must have been repeated in several localities throughout the world. In situ sulfidation by hydrothermal processes appears to be more likely.

If the current isotopic distribution has not been reset during retrogression, the wide range of sulfur isotopic values for pyrite (Fig. 10) reflects either a source area that supplied heterogeneous detrital pyrite or sulfidation by solutions with variable sulfide-sulfate ratios. The detrital pyrite model (with the variable $\delta^{34}S$ thus reflecting processes in the source terrane) has the conflicting requirement of a reducing atmosphere to stabilize pyrite during transport, and a quite oxidizing atmosphere to provide the wide range of $\delta^{34}S$ values. If such grains were not the result of an oxidizing atmosphere, a suitable mechanism to generate the observed variability in the absence of an oxygenated atmosphere must be defined.

On the basis of the abundance of pyrite, the absence of iron and iron-titanium oxides, and the textural evidence for pseudomorphous replacement, the most reasonable conclusion is that the origin of any specific round pyrite grain in the Witwatersrand and other altered sedimentary sequences is likely to be ambiguous on the basis of shape alone. This ambiguity currently prevents any definite conclusions to be made about the original proportions of pyrite in the conglomerates and precludes the use of pyrite to constrain the composition of the Archean atmosphere.

This does not necessarily preclude the presence of some detrital pyrite grains—it is possible that uplift of older basin sediments exposed local sources of detrital pyrite that was formed during diagenesis, to erosion and redeposition. The extent of this process and its implications for the composition of the Archean atmosphere cannot be determined in the absence of data on other key variables such as the duration of transport and the physical and chemical nature of the depositional environment.

Redox gradients in the near-surface are commonly extreme, and many modern near-surface waters are not in equilibrium with the atmosphere. Even within modern weathering profiles, the redox potential changes significantly with both time and position in the profile, often in proximity to the water table. Inferences of atmospheric composition based on mineral assemblages and paleosols thus require a clear demonstration of equilibrium with the atmosphere, and that remains a challenge for most studies.

Do Gold Compositions Reflect Atmospheric Composition?

Detrital gold grains in equilibrium with the modern atmosphere are commonly (but not universally) depleted in silver relative to their inferred source, reflecting the preferential leaching of silver in an oxygenated atmosphere (Morrison et al., 1991). In contrast, Witwatersrand gold grains typically contain significant silver (around 10 wt%; range 0–30 wt%). It has thus been suggested that compositional differences between gold in modern placers and in the Witwatersrand may reflect differences in atmospheric composition. This hypothesis is based on two key assumptions: (1) Witwatersrand gold is detrital, and (2) there has been no secondary remobilization and/or post-depositional alteration of gold chemistry (i.e., assuming and unmodified placer model).

There is now a broad consensus in the literature that compositions of gold grains in the Witwatersrand reflect post-depositional processes either by local remobilization and/or re-equilibration of detrital gold (e.g., Frimmel et al., 1993; Frimmel and Gartz, 1997) or by a hydrothermal origin for the gold (e.g., Barnicoat et al., 1997; Phillips and Law, 2000). Gold grain compositions (e.g., Figs. 10, 11) show limited variability, with a range less than that observed in gold grains from auriferous quartz veins cutting the orebodies. As a result, within-sample variability cannot be used to suggest that the grains are derived from a variety of different source rocks and is compatible with a hydrothermal genesis.

In summary, textural studies indicate that the overwhelming majority of gold grains are in structural sites that were not present at the time of deposition and cannot be detrital (unmodified) in origin. Local remobilization of any detrital grains may be possible but gold compositions now reflect post-depositional processes and cannot be used to constrain atmospheric compositions at the time of sedimentation or to date the age of inferred source rocks for detrital gold.

Are Round Uraninite Grains Detrital?

The interpretation of round uraninite grain shapes is controversial. Davidson (1960, p. 155) drew attention to round pitchblende grains in vein deposits from Freiberg, Germany. More recently Phillipe et al. (1993) showed that epigenetic uraninite grains in unconformity uranium deposits of the Athabasca Basin are round and have compositions similar to those in the Witwatersrand. These grains demonstrate that roundness is not necessarily a function of detrital transport.

The textural interpretation of isolated uraninite grains in the Witwatersrand is particularly difficult because of the intimate association of uraninite and carbon in the reefs. This association has variably been ascribed to sedimentological (Minter, 1976; Hallbauer, 1986) or chemical processes (Phillips and Law, 1997; 2000), and it is now widely accepted that the carbon is derived from the thermal maturation of migrated hydrocarbons (e.g., Phillips et al., 1990; Gray et al., 1998; Jolley et al., 2004). Round nodules of carbon, known as "flyspeck carbon," are not volumetrically abundant in the Witwatersrand, but they are found in virtually every reef and are intimately related to the distribution of uranium. Although these blebs have been interpreted as detrital grains (e.g., Hallbauer, 1986), it is now believed that they are genetically related to the carbon seams and were precipitated around uraniferous minerals by radiolytic polymerization (McCready and Parnell, 1998). Most Witwatersrand carbon nodules and seams contain high uranium concentrations disseminated throughout and imply substantial post-depositional mobility of uranium.

There are at least two post-depositional processes that could account for the intimate association of uraninite with migrated hydrocarbon material. If the uraninite pre-dates the hydrocarbons, then it could precipitate migrating hydrocarbons by radiolytic polymerization (McCready and Parnell, 1998; England et al., 2001). Importantly, this process requires that hydrocarbons postdate the uranium minerals, but does not differentiate between detrital and early hydrothermal uranium mineralization. Alternatively, if the uraninite postdates or is synchronous with the hydrocarbon-bearing fluid, the carbon may represent a localized site of reduction capable of precipitating uraninite from solution.

Criteria to differentiate these processes are not clear-cut. In some cases, the textural evidence suggests that carbon replaced and progressively dismembered pre-existing uraninite grains (e.g., Liebenberg, 1955; Smits, 1984; England et al., 2001) (Figs. 13A, 13B). However, mesoscopic relationships indicate that carbon seams in fractures subparallel to stratigraphy host substantial uraninite and are thus incompatible with a simple detrital origin for at least this part of the mineralization (cf. England et al., 2001). In other cases, uraninite and other uraniferous minerals are finely disseminated within the hydrocarbons and are probably epigenetic precipitates (Fig. 8F). The process of nodule formation results in round, discrete uraninite grains that reflect either precipitation of uraninite or disaggregation of pre-existing grains. In the latter case, rounding of pre-existing grains may be an inevitable consequence of the disaggregation process. England et al. (2001) have argued that the hydrocarbons are precipitated around detrital uraninite grains. Other workers argue that uraninite and hydrocarbon are both localized along structurally controlled fractures (e.g., Law and Spencer, 1992; Barnicoat et al., 1997; Jolley et al., 1999; 2004; Phillips and Law, 2000).

In addition to uraninite, brannerite and uraniferous leucoxene are important uranium-bearing minerals in many Witwatersrand reefs (Hallbauer, 1986), but the concentrations and proportions of these minerals vary greatly on local and regional scales. Textural studies show that interaction of Ti-bearing detrital minerals on unconformity surfaces with uranium in solution has formed widespread secondary brannerite and uraniferous leucoxene (Davidson, 1953; 1957; 1960; Liebenberg, 1955; Ramdohr, 1958; Hallbauer, 1986). The migration of uranium after deposition of the Witwatersrand sediments provides a clear indication of dissolution from surrounding grains or the introduction of uranium by hydrothermal solutions. Similarly, the widespread distribution of round uraninite grains in hydrocarbon seams that postdate the deposition of the sediments demonstrates that round shapes are not related to sedimentary rounding and that uranium has been mobile after deposition of the sediments.

In summary, round uraninite grains in hydrocarbon seams in brittle fractures and along shear zones cannot reflect detrital processes and imply substantial uranium mobility after deposition. Aggressive dissolution of uraninite by hydrocarbons together with widespread uranium mobility casts further doubt on the significance of any grain shapes. Round uraniferous hydrocarbon nodules in granitic rocks surrounding the basin probably reflect the same alteration and have been dated between 2.7 Ga and 2.0 Ga and thus postdate Witwatersrand sedimentation (Klemd, 1999).

Even if it is accepted that some or all of the round uraninite grains in the Witwatersrand, in Elliot Lake, and in the Pilbara Craton are primary detrital minerals, the presence of uraninite as a detrital mineral in the Indus River (Maynard et al., 1991) begs the question of how definitive this criterion can really be in constraining the Archean atmosphere. If the Archean atmosphere truly stabilized uraninite, it is surprising how rare uraninite is in the Archean geological record.

Given the evidence for post-depositional alteration in the Witwatersrand, Rasmussen and Buick (1999) completely discard the Witwatersrand ores as a convincing site of detrital pyrite and uraninite, and focus on heavy mineral assemblages in other largely unmineralized Archean sediments such as those of the Pilbara Craton of Western Australia. These authors have identified pyrite, uraninite and siderite in Archean sedimentary rocks of inferred detrital origin and argue that because the sediments are believed to be unaltered, they are more likely to reflect the composition of the Archean atmosphere. However, the presence of sericite in many of these samples (typically a hydrothermal phase in the Witwatersrand; Law, 1991) and migrated hydrocarbons suggests that they have not escaped post-depositional alteration processes, and this study suffers from the same limitations as those outlined above for the Witwatersrand.

Possible Evidence for an Oxidizing Archean Atmosphere

The atmosphere debate on the Witwatersrand has focused largely on the current heavy mineral assemblages in the belief that they are detrital. Given the evidence for post-depositional changes after burial, inferences of minerals existing prior to alteration may be more relevant to constraining atmospheric composition. For example, there is good reason to believe that vital information may be contained within the sediments immediately overlying the unconformity. These sediments may contain material derived from truncated soil profiles and specifically from the B-horizon in which Fe is concentrated in modern profiles (Anand, 1995). Fe-oxyhydroxide concretions (e.g., pisoliths) formed in areas of seasonal rainfall, for example, are durable, coarse-grained, and have a distinctive physical appearance (Brewer, 1964). Therefore, they should be identifiable even after considerable transport, metamorphism, and alteration (Fig. 7).

The genesis of the porous, round pyrite can be subdivided into models postulating original pyrite and those postulating sulfidation of another mineral. The sulfidation origin for Witwatersrand porous round pyrites was initially proposed by Clemmey (1981) and developed by Phillips and Myers (1989). There is a striking similarity in both morphology and size between the porous round pyrite and modern Fe-concretions (Fig. 7). Furthermore, the distribution of the concretions in residual conglomerates on unconformities is exactly as predicted from modern analogues (Smith and Perdrix, 1982). The mechanical migration and accumulation of these pisoliths can be predicted from their density by analogy with modern sedimentological sorting processes (Anand, 1995). Oxyhydroxide minerals in the pisoliths would be highly susceptible to sulfidation after burial because of their high Fe content. In contrast, ferruginous chert has a lower Fe content and generally undergoes only partial alteration to pyrite (Phillips and Dong, 1993). Textural evidence for sulfidation to form these pyrite grains is equivocal. The distribution of porous round pyrite grains is entirely restricted to unconformity surfaces near the proximal basin margin, and thus they lie within the phyllosilicate-plus-pyrite alteration envelope associated with Witwatersrand mineralization (Phillips and Law, 2000). Unconformity surfaces outside the alteration envelope would be required to determine the pre-alteration mineralogy. Comparable sulfide nodules are uncommon elsewhere in the Archean except where associated with similar mineralization and are thus unlikely to reflect a general process related to the Archean atmosphere.

Traditionally, the origin of the porous round pyrite grains has been linked to sulfidic muds accumulated in the intertidal channel areas of fluvial systems, hence the colloquial name "mud ball pyrite." These sulfidic muds have been attributed to a reducing atmosphere (Hallbauer, 1986) or to exhalative activity around the margins of the Witwatersrand Basin (Hutchinson and Viljoen, 1988). The former explanation requires an atmosphere different from the present to preserve the pyrite during transport, and fails to explain the association of the "mud balls" with unconformities. The hydrothermal activity required for the latter model appears unlikely given the prevalence of the porous round pyrites in the Witwatersrand Supergroup, the Ventersdorp Contact Reef, and the Black Reef at the base of the Transvaal Sequence. Both ideas are built on the premise that the Witwatersrand pyrite is detrital. As shown above, this premise may not be sound and is not independently verified.

This concept of original Fe-rich pisoliths requires a Precambrian atmosphere oxidizing enough to stabilize two forms of iron in different parts of the Archean soil horizon with at least some zone where ferric iron is abundant. No simple relationship between the observed mineralogy and the composition of the atmosphere is predicted for two reasons. First, atmospheric redox state is only one of several thermodynamic and kinetic constraints on the stability of pyrite and uraninite, and simple assumptions regarding other variables and the whole chain of weathering, erosion, transport, and depositional effects on individual grains are unlikely to reflect any single process reliably (Robinson and Spooner, 1984). For example, detrital uraninite and pyrite grains are known from modern sediments, albeit in minor quantities, and round uraninite and pyrite grains are also known from epigenetic deposits. Second, all sedimentary basins are subject to complex post-depositional alteration and diagenesis of the original detrital suite by basinal fluids that are commonly, but not universally, reducing (Phillips et al., 1990; Morton and Hallsworth, 1999). The impact of these fluids must be recognized and understood before the redox state of the atmosphere at the time of deposition can be interpreted.

Given the abundant evidence for post-depositional modification of the Witwatersrand detrital assemblage and the evidence for post-depositional alteration, sulfidation, and widespread mobility of gold and uranium, we conclude that the mineral assemblage of the Witwatersrand is not a reliable indicator of a reducing atmosphere. In fact, reinterpretation of some porous round pyrites as sulfidized iron pisoliths may require a relatively oxidizing environment to stabilize iron in both its ferric and ferrous forms. The widespread inferred mobility of uranium in the Witwatersrand may also require an oxygenated fluid to stabilize U^{6+} in solution. Similar sedimentary environments in the younger rock record are frequently altered by inflowing oxygenated meteoric waters that reflect prevailing atmospheric conditions.

Significance of "Old" Isotopic Ages

A large range of mineral and whole rock ages have been published for the Witwatersrand that reflect pre-, syn-, and post-depositional events in the basin (see summaries in Robb and Meyer, 1995 and Robb et al., 1990a). Of particular interest to the atmosphere debate are "old" ages for gold, pyrite, and uraninite that purportedly pre-date sedimentation and thus imply a detrital origin for the mineralization.

Re-Os Isochron "Ages" for Gold and Pyrite

Kirk et al. (2001; 2002) have recently used the Re-Os isochron technique in an attempt to directly date Witwatersrand gold and pyrite. These authors assumed that pyrite and gold form part of the heavy mineral assemblage, and used randomly selected grains from a single sample of the Vaal Reef to construct an isochron with an implied age of 3010 ± 20 Ma.

The Vaal Reef sample is described as "fine-grained quartzitic conglomerate with angular to sub-rounded horizontally fractured quartz clasts 5-10 mm in size. The conglomerate is clast supported and is approximately 10% matrix and 90% clasts. The matrix consists of sericite, fine-grain quartz and carbon seams/patches. The majority of the pyrite/arsenopyrite, and ~10% of the visible gold are within the quartz and sericite matrix, while ~90% of the gold, the uraninite and some of the pyrite are confined to the carbonaceous material" (Kirk et al., 2002).

We interpret the sericitic matrix as part of the phyllosilicate alteration assemblage typical of Witwatersrand orebodies. Much of the gold is located in the carbon seam, whereas the sulfides are from the quartzite matrix. Recent studies by several different authors (Parnell, 1996; Gray et al., 1998; Jolley et al., 1999; 2004; Phillips and Law, 2000; England et al., 2001) concluded that the carbon seams postdate deposition of the sediments and reflect migration of hydrocarbons during post-depositional alteration. Although the precise paragenetic setting of grains used to construct the isochron remains unknown, it is likely that the grains are from different host lithologies, with gold from the carbon seam and pyrite from the host conglomerate. In the case of hydrocarbon-hosted gold, textural evidence invariably shows that the gold was introduced after deposition of the hydrocarbon (Ramdohr, 1958; Saager, 1968; Barnicoat et al., 1997; Phillips and Law, 1997) and is thus not co-genetic with the round pyrite grains as inferred by Kirk et al. (2002).

An important challenge for the Re-Os studies is thus to demonstrate that all grains were part of an isotopically homogeneous initial Os^{187}-Os^{188} reservoir, which is the foundation of isochron-based dating techniques. In practice this generally means that all grains need to be sourced from a grain population from a single sample all sharing a common genesis. Re-Os dating of round pyrite grains from Steyn Reef in the Welkom goldfield by the same authors clearly reflects this problem and yields imprecise "isochron" ages of 3490 ± 900 Ma. There is no reason to believe that the individual pyrite grains analyzed are derived from an isotopically homogeneous source and therefore we assign no age significance to the "isochron." In any event, the 3490 Ma age fails to discriminate between detrital and hydrothermal mineralization given the 1800 Ma error.

Post-depositional changes in the composition of gold and pyrite grains are now well established for the Witwatersrand and make the use of gold and pyrite geochemistry for dating of dubious value. In particular, homogenization and/or secondary introduction of gold grains, and crosscutting pyrite veins and overgrowths on round pyrites, are likely to affect isotopic systematics.

U-Pb "Ages" for Uraninite

The pioneering work of Rundle and Snelling (1977) on the geochronology of uraniferous minerals in the Witwatersrand has been widely quoted in the literature in support of the placer model. In their 1977 review of their own and available existing data, these authors infer two discrete age groups: (1) an older population at 3050 ± 50 Ma that predates deposition of the Witwatersrand Basin; and (2) a younger population at 2040 ± 100 Ma that postdates the deposition of the Witwatersrand.

Rundle and Snelling argue that the data reflect detrital uraninite derived from a 3050 Ma source followed by a resetting of the U-Pb system at 2040 Ma within a closed system with limited secondary introduction of uranium. However, they also note that "it would be virtually impossible to distinguish between disturbed detrital systems and disturbed systems with both detrital and authigenic components." These ages have been widely reported as "uraninite" ages (e.g., Robb and Meyer, 1995); however, a review of the original paper highlights some fundamental shortcomings in the approach used by Rundle and Snelling (1977) and especially the interpretation of the data by other authors.

First, the "ages" were obtained on "portions of the matrix, avoiding as far as possible the pebbles, cut in the form of cubes ~2cm in size" and thus effectively reflect "bulk rock" analyses rather than individual mineral grains. Second, each age group reported by Rundle and Snelling has been estimated from a series of samples based on their distribution on the Concordia diagram, with many samples showing evidence for both lead and uranium loss. As a result, individual age estimates are highly variable. Third, no paragenetic information on uranium minerals is reported to provide independent geological constraints on the likely validity of the ages. For example, it has now been well documented that many uraninite grains, particularly those hosted by carbon seams filling structural dilation zones, cannot be of detrital origin and must reflect substantial mobility of uranium after burial. Galena is also a common secondary mineral in the reefs and associated quartz veins, suggesting an open system on the scale of the samples discussed above. Similarly, many secondary uranium minerals are now well documented in the reefs and form an unknown proportion of the material sampled. Many of these minerals are poor hosts for lead that was presumably mobilized during alteration. It seems improbable in the light of this petrographic evidence that a closed U-Pb system has operated on either a mineral or whole rock scale.

Rundle and Snelling also report data for "uraninite" and "thucolite" that are apparently from individual mineral separates. Thucolite refers to "thorium-uranium-carbon-oxygen"-bearing material or "carbon nodules" to use the terminology in this paper. These ages range from 2760 Ma for a sample from the Dominion Group to 2540 Ma through 2000 Ma for

Witwatersrand samples and are consistently younger than the host sedimentary rocks (although they were thought to be older at the time of the analyses).

In light of these uncertainties and limitations, it is unlikely that the reported ages accurately reflect the age of uraninite or other detrital uranium-bearing minerals. In fact, the younger population reported by Rundle and Snelling could equally have been used to support a hydrothermal origin for the mineralization (although these ages also suffer from the limitations outlined above).

Lithological and Chemical Controls on Witwatersrand Mineralization

Given the close spatial association between gold and uranium in Witwatersrand orebodies, it is possible that there is a genetic link between the two metals. A common assumption has been that both minerals are concentrated in response to sedimentary sorting processes (e.g., Smith and Minter, 1980; Hallbauer, 1986; Smits, 1984; Roscoe and Minter, 1993). These authors point out that concentrations of gold and uranium are commonly correlated over several orders of magnitude, and that metal concentrations are also broadly correlated with sedimentary facies (Figs. 14A,

14B). For example, Smith and Minter (1980) describe better grades in quartzites and conglomerates with pyritic foresets and a general increase in grade from sandstone through conglomerate to carbon seams. Similar associations are widely reported from the Witwatersrand and the implied link between sedimentary facies and gold grade has been used to support a placer model for the mineralization.

From a chemical standpoint these data can also be interpreted to reflect post-depositional mineralizing processes (Phillips and Law, 2000):

- an association of better gold grades with round pyrite, reflecting sulfidation of detrital Fe-oxide grains and precipitation of gold transported as sulfur complexes;
- an association of better uranium grades with detrital titanium minerals, reflecting precipitation of uranium by reaction with Ti to form brannerite.

The increases in grade could thus reflect either differences in depositional concentration processes between sandstones and conglomerates during sedimentation (placer model) or elevated iron and titanium concentrations as detrital phases during conglomerate deposition (hydrothermal model). However, given the secondary origin of the carbon, the elevated grades in carbon-rich lithologies are more likely to reflect reduction by organic

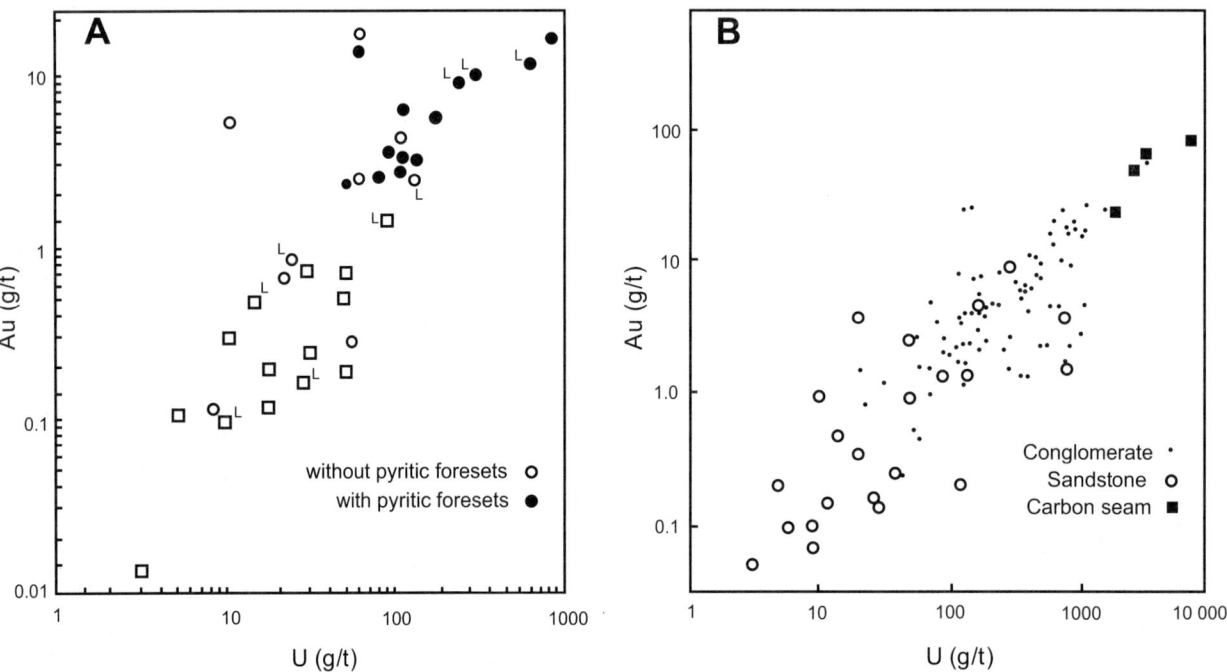

Figure 14. Plots of Au versus uranium for selected Witwatersrand orebodies illustrating the link between lithology, mineralogical associations, and increasing metal content. (A) Elsburg number 5 Reef and Leader Reef (symbols marked L), Klerksdorp and Welkom goldfields, respectively. Redrawn from Smith and Minter, 1980. Smith and Minter have interpreted the lithological control on gold grade in terms of a detrital control on mineralization. Alternatively, the strong association of increased metal concentrations with sulfide-rich facies may reflect a chemical control on the mineralization (i.e., elevated pyrite in gold-rich samples via sulfidation of detrital iron-bearing minerals). (B) Leader Reef, Welkom goldfield. Redrawn from Smith and Minter, 1980. Carbon seams are not present at the time of sedimentation, and consistent metal ratios in seams and associated sedimentary facies may imply a post-depositional control on the mineralization.

material and hence hydrothermal ore genesis. These mineralogical associations provide strong chemical reasons for the association of mineralization with specific host lithologies.

A major criticism of the placer model has been its inability to explain the variety of mineralized sedimentary rock types either on theoretical grounds or by analogy with young placer deposits (Phillips and Law, 2000). This limitation is most obvious in the case of carbon seams that are locally major host lithologies for gold and uranium. In contrast, the chemical associations of gold with iron (pyrite) and carbon, and uranium with carbon and titanium, are clear in all the major orebodies and crosscut sedimentary facies.

ACKNOWLEDGMENTS

We are grateful to Steve Kesler, Hiroshi Ohmoto, and Michael Kimberley for the invitation to attend the 2002 symposium on the "Evolution of the early atmosphere, hydrosphere, and biosphere." We would also like to thank coworkers in this field who have been instrumental in shaping our ideas, especially Russell Myers, Judy Palmer, Guoyi Dong, and Martin Hughes. Richard Spencer was a key person in recognizing the structural controls on hydrocarbon seams during mapping at Kinross Mine. Bill Fyfe has provided insightful comments on the early Earth and its atmosphere. We also thank Rob Hough, Katy Evans, and Martin Hughes for their constructive comments on an early draft of this manuscript. Holly Stein and Judy Hannah are thanked for discussions on Re-Os and their early review of this section of the manuscript. We are grateful to Steve Kesler and two anonymous reviewers for insightful and constructive reviews that greatly improved the paper. JL acknowledges an Honorary Research Fellowship at the School of Geosciences, Monash University. NP acknowledges the support of the School of Geosciences, Monash University, and the School of Earth Sciences, University of Melbourne.

REFERENCES CITED

Anand, R.R., 1995, Genesis and classification of ferruginous regolith materials in the Yilgarn Craton: Implications for mineral exploration: Townsville, Australia, Economic Geology Research Unit, James Cook University, Contribution 54, p. 1–4.

Armstrong, R.A., Compston, W., Retief, E.A., Williams, I.S., and Welke, H.J., 1991, Zircon ion microprobe studies bearing on the age and evolution of the Witwatersrand triad: Precambrian Research, v. 53, p. 243–266, doi: 10.1016/0301-9268(91)90074-K.

Barnicoat, A.C., Henderson, I.H.C., Knipe, R.J., Yardley, B.W.D., Napier, R.W., Fox, N.P.C., Kenyon, A.K., Muntingh, D.J., Strydom, D., Winkler, K.S., Lawrence, S.R., and Cornford, C., 1997, Hydrothermal gold mineralization in the Witwatersrand Basin: Nature, v. 386, p. 820–824, doi: 10.1038/386820a0.

Barnicoat, A.C., Phillips, G.M., Law, J.D.M., Walshe, J.L., Phillips, G.N., and Fox, N.P.C., 2001, Refuting the irrefutable: A new look at a well-known sample of Witwatersrand gold mineralisation: Townsville, Australia, Economic Geology Research Unit, James Cook University, Contribution 59, p. 16–17.

Bossière, G., Bonkoungou, I., Peucat, J.-J., and Pupin, J.-J., 1996, Origin and age of the Paleoproterozoic conglomerates and sandstones of the Tarkwaian Group in Burkino Faso: West Africa: Precambrian Research, v. 80, p. 153–172, doi: 10.1016/S0301-9268(96)00014-9.

Brewer, R., 1964, Fabric and mineral analysis of soils: New York, John Wiley and Sons, 420 p.

Button, A., and Tyler, N., 1981, The character and economic significance of paleoweathering and erosion surfaces in southern Africa: Economic Geology, 75th Anniversary Volume, p. 686–709.

Clemmey, H., 1981, Some aspects of the genesis of heavy mineral assemblages in Lower Proterozoic uranium-gold conglomerates: Mineralogical Magazine, v. 44, no. 366, p. 399–408.

Clemmey, H., and Badham, N., 1982, Oxygen in the Precambrian atmosphere: An evaluation of the geological evidence: Geology, v. 10, p. 141–146, doi: 10.1130/0091-7613(1982)10<141:OITPAA>2.0.CO;2.

Coetzee, F., 1965, Distribution and grain size in gold, uraninite, pyrite and certain other heavy minerals in gold-bearing reefs of the Witwatersrand Basin: Transactions of the Geological Society of South Africa, v. 68, p. 61–88.

Coward, M.P., Spencer, R.M., and Spencer, C.E., 1995, Development of the Witwatersrand Basin, South Africa, in Coward, M.P., and Ries, A.C., eds., Early Precambrian processes: Geological Society [London] Special Publication 95, p. 243–269.

Davidson, C.F., 1953, The gold-uranium ores of the Witwatersrand: Mining Magazine (London), v. 88, p. 73–85.

Davidson, C.F., 1957, On the occurrence of uranium in ancient conglomerates: Economic Geology and the Bulletin of the Society of Economic Geologists, v. 52, p. 668–693.

Davidson, C.F., 1960, The present state of the Witwatersrand controversy: Mining Magazine (London), v. 102, p. 84–95, 149–159, 222–229.

Davidson, C.F., and Cosgrove, M.E., 1955, On the importance of uraninite as a detrital mineral: Geological Survey of Great Britain Bulletin, v. 10, p. 74–80.

Dimroth, E., and Kimberley, M.M., 1976, Precambrian atmospheric oxygen: Evidence in the sedimentary distribution of carbon, sulfur, uranium and iron: Canadian Journal of Earth Sciences, v. 13, p. 1161–1185.

Eldridge, C.S., Phillips, G.N., and Myers, R.E., 1993, Sulfides in the Witwatersrand gold fields: New perspectives on old sediments via SHRIMP: Geological Society of America Abstracts with Programs, v. 25, no. 6, p. 278–279.

Els, B.G., Van den Berg, W.A., and Mayer, J.J., 1995, The Black Reef Quartzite Formation in the Western Transvaal: Sedimentological and economic aspects, and significance for basin evolution: Mineralium Deposita, v. 30, p. 112–123, doi: 10.1007/BF00189340.

England, G., 1999, Constraints on the origin of the Witwatersrand Au-U deposit, South Africa: Applying new techniques to an old problem, in Dunphy, J.M., and Hagemann, S.G., eds., Seminar on recent advances in ore genesis, December 1999: Perth, Centre for Strategic Mineral Deposits, University of Western Australia, 3 p.

England, G.E., Rasmussen, B., Krapez, B., and Groves, D.I., 2001, The origin of uraninite, bitumen nodules, and carbon seams in Witwatersrand gold-uranium-pyrite ore deposits, based on a Permo-Triassic analogue: Economic Geology and the Bulletin of the Society of Economic Geologists, v. 96, p. 1907–1920.

Farquhar, J., Bao, H., and Thiemens, M., 2000, Atmospheric influence of Earth's earliest sulfur cycle: Science, v. 289, p. 756–758, doi: 10.1126/science.289.5480.756.

Fleet, M.E., MacLean, P.J., and Barbier, J., 1989, Oscillatory-zoned As-bearing pyrite from strata-bound and stratiform gold deposits: An indicator of ore fluid evolution: Economic Geology Monograph, v. 6, p. 356–362.

Frimmel, H.E., 1997, Detrital origin of hydrothermal Witwatersrand gold—A review: Terra Nova, v. 9, p. 192–197, doi: 10.1046/j.1365-3121.1997.d01-23.x.

Frimmel, H.E., and Gartz, V.H., 1997, Witwatersrand gold particle chemistry matches model of metamorphosed, hydrothermally altered placer deposits: Mineralium Deposita, v. 32, p. 523–530, doi: 10.1007/s001260050119.

Frimmel, H.E., Le Roux, A.P., Knight, J., and Minter, W.E.L., 1993, A case study of the postdepositional alteration of the Witwatersrand Basal reef gold placer: Economic Geology and the Bulletin of the Society of Economic Geologists, v. 88, p. 249–265.

Fuller, A.O., 1958, A contribution to the petrology of the Witwatersrand System: Transactions of the Geological Society of South Africa, v. 61, p. 10–50.

Gray, G.J., Lawrence, S.R., Kenyon, A.K., and Cornford, C., 1998, Nature and origin of 'carbon' in the Archaean Witwatersrand Basin, South Africa: Journal of the Geological Society [London], v. 155, p. 39–59.

Hallbauer, D.K., 1986, The mineralogy and geochemistry of Witwatersrand pyrite, gold and uranium, and carbonaceous matter, in Anhaeusser, C.R., and Maske, S., eds, Mineral deposits of Southern Africa: Johannesburg, Geological Society of South Africa, p. 731–752.

Hirdes, W., and Saager, R., 1983, The Proterozoic Kimberley reef placer in the Evander gold field, Witwatersrand, South Africa: Gebruder Borntrager, Monograph Series on Mineral Deposits, Berlin, no. 20, 101 p.

Hoefs, J., Nielsen, H., and Schidlowski, M., 1968, Sulfur isotope abundances in pyrite from the Witwatersrand conglomerates: Economic Geology and the Bulletin of the Society of Economic Geologists, v. 63, p. 975–977.

Holland, H.D., 1984, Chemical evolution of the atmosphere and oceans: Princeton, New Jersey, Princeton University Press, 582 p.

Holland, H.D., 1994, Early Proterozoic atmospheric change, in Bengtson, S., ed., Early life on Earth: Nobel Symposium 84: New York, Columbia University Press, p. 237–244.

Holland, H.D., and Rye, R.O., 1997, Evidence in pre-2.2 Ga paleosols for the early evolution of atmospheric oxygen and terrestrial biota: Comment and reply: Geology, v. 25, p. 857–859, doi: 10.1130/0091-7613(1997)025<0857:EIPGPF>2.3.CO;2.

Holland, H.D., Kuo, P.H., and Rye, R.O., 1994, O_2 and CO_2 in the Late Archaean and Early Proterozoic atmosphere: Mineralogical Magazine, v. 58A, p. 424.

Hutchinson, R.W., and Viljoen, R.P., 1988, Re-evaluation of the gold source in Witwatersrand ores: South African Journal of Geology, v. 91, p. 157–173.

Jahn, B.M., Bertrand-Satarti, J., Morin, N., and Mace, J., 1990, Direct dating of stromatolitic carbonates from the Schmidtsdrift Formation (Transvaal Dolomite), South Africa, with implications for the age of the Ventersdorp Supergroup: Geology, v. 18, p. 1211–1214, doi: 10.1130/0091-7613(1990)018<1211:DDOSCF>2.3.CO;2.

Jolley, S.J., Henderson, I.H.C., Barnicoat, A.C., and Fox, N.P.C., 1999, Thrust-fracture network and hydrothermal gold mineralization: Witwatersrand Basin, South Africa, in McCaffrey, K.J.W., Longergan, L., and Wilkinson, J.J., eds, Fractures, fluid flow and mineralization: Geological Society [London] Special Publication 155, p. 153–165.

Jolley, S.J., Freeman, S.R., Barnicoat, A.C., Phillips, G.M., Knipe, R.J., Pather, A., Fox, N.P.C., Strydom, D., Birch, M.T.G., Henderson, I.H.C., and Rowland, T.W., 2004, Structural controls on Witwatersrand gold mineralization: Journal of Structural Geology, v. 26, p. 1067–1086, doi: 10.1016/j.jsg.2003.11.011.

Kasting, J.F., 1993, Earth's early atmosphere: Science, v. 259, p. 920–926.

Kasting, J.F., 2001, The rise of atmospheric oxygen: Science, v. 293, p. 819–820, doi: 10.1126/science.1063811.

Kirk, J., Ruiz, J., Chesley, J., Titley, S., and Walshe, J., 2001, A detrital model for the origin of gold and sulfides in the Witwatersrand Basin based on Re-Os isotopes: Geochimica et Cosmochimica Acta, v. 65, p. 2149–2159, doi: 10.1016/S0016-7037(01)00588-9.

Kirk, J., Ruiz, J., Chesley, J., Walshe, J., and England, G., 2002, A major Archean, gold- and crust-forming event in the Kaapvaal Craton: Science, v. 297, p. 1856–1858, doi: 10.1126/science.1075270.

Klemd, R., 1999, A comparison of fluids causing post-depositional hydrothermal alteration in Archaean basement granitoids and the Witwatersrand Basin: Mineralogy and Petrology, v. 66, p. 111–122, doi: 10.1007/BF01161724.

Krupp, R.K., Oberthür, T., and Hirdes, W., 1994, The early Precambrian atmosphere and hydrosphere: Thermodynamic constraints from mineral deposits: Economic Geology and the Bulletin of the Society of Economic Geologists, v. 89, p. 1581–1598.

Law, J.D.M., 1991, Alteration and geochemistry of Witwatersrand metasediments in the Welkom area: Implications for mineralization [M.Sc. thesis]: Johannesburg, University of Witwatersrand, 198 p.

Law, J.D.M., and Spencer, R.M., 1992, A detailed study of the Kimberley Reef on portions of Kinross and Winkelhaak Mines: Johannesburg, Gencor, Mineral Resources Division, Unpublished Geological Research Report, 94 p.

Law, J.D.M., Bailey, A.C., Cadle, A.B., Phillips, G.N., and Stanistreet, I.G., 1990, Reconstructive approach to the classification of Witwatersrand 'quartzites': South African Journal of Geology, v. 93, p. 83–92.

Liebenberg, W.R., 1955, The occurrence and origin of gold and radioactive minerals in the Witwatersrand System, the Dominion Reef, the Ventersdorp Contact Reef, and the Black Reef: Transactions of the Geological Society of South Africa, v. 58, p. 101–227.

MacLean, P.J., and Fleet, M.E., 1989, Detrital pyrite in the Witwatersrand gold fields of South Africa: Evidence from truncated growth banding: Economic Geology and the Bulletin of the Society of Economic Geologists, v. 84, p. 2008–2011.

Maynard, J.B., Ritger, S.D., and Sutton, S.J., 1991, Chemistry of sands from the modern Indus River and the Archean Witwatersrand basin: Implications for the composition of the Archean atmosphere: Geology, v. 19, p. 265–268, doi: 10.1130/0091-7613(1991)019<0265:COSFTM>2.3.CO;2.

McCarthy, T.S., 1994, The tectono-sedimentary evolution of the Witwatersrand Basin with special reference to its influence on the occurrence and character of the Ventersdorp Contact Reef—A review: South African Journal of Geology, v. 97, no. 3, p. 247–259.

McCready, A.J., and Parnell, J., 1998, A Phanerozoic analogue for Witwatersrand-type uranium mineralization: Uranium-titanium-bitumen nodules in Devonian conglomerate/sandstone, Orkney, Scotland: Transactions of the Institute of Mining and Metallurgy, v. 107, p. B89–B96.

Meyer, F.M., Tainton, S., and Saager, R., 1990, The mineralogy and geochemistry of small-pebble conglomerates from the Promise Formation in the West Rand and Klerksdorp areas: South African Journal of Geology, v. 93, p. 118–134.

Milesi, J.P., Ledru, P., Marcoux, E., Johan, V., Lerouge, C., Sabate, P., Bailly, L., Respaut, J.P., and Skipwith, P., 2002, The Jacobina Paleoproterozoic gold-bearing conglomerates, Bahia, Brazil: A 'hydrothermal shear-reservoir' model: Ore Geology Reviews, v. 19, p. 95–136, doi: 10.1016/S0169-1368(01)00038-5.

Minter, W.E.L., 1976, Detrital gold, uranium, and pyrite concentrations related to sedimentology in the Precambrian Vaal Reef placer, Witwatersrand, South Africa: Economic Geology and the Bulletin of the Society of Economic Geologists, v. 71, p. 157–176.

Minter, W.E.L., 1978, A sedimentological synthesis of placer gold, uranium and pyrite concentrations in Proterozoic Witwatersrand sediments, in Miall, A.D., ed., Fluvial sedimentology: Canadian Society of Petroleum Geology Memoir 5, p. 801–829.

Minter, W.E.L., 1999, Irrefutable detrital origin of the Witwatersrand gold and evidence of eolian signatures: Economic Geology and the Bulletin of the Society of Economic Geologists, v. 94, p. 665–670.

Minter, W.E.L., Goedhart, M., Knight, J., and Frimmel, H.E., 1993, Morphology of Witwatersrand gold grains from the Basal Reef: Evidence for their detrital origin: Economic Geology and the Bulletin of the Society of Economic Geologists, v. 88, p. 237–248.

Morrison, G.W., Rose, W.J., and Jaireth, S., 1991, Geological and geochemical controls on the silver content (fineness) of gold in gold-silver deposits: Ore Geology Reviews, v. 6, p. 333–364, doi: 10.1016/0169-1368(91)90009-V.

Morton, A.C., and Hallsworth, C.R., 1999, Processes controlling the composition of heavy mineral assemblages in sandstones: Sedimentary Geology, v. 124, p. 3–29, doi: 10.1016/S0037-0738(98)00118-3.

Myers, R.E., Zhou, T., and Phillips, G.N., 1993, Sulphidation in the Witwatersrand goldfields: Evidence from the Middelvlei Reef: Mineralogical Magazine, v. 57, p. 395–405.

Oberthür, T., and Saager, R., 1986, Delineation of different mineral facies in the Carbon Leader Reef Placer, Carletonville Goldfield, Witwatersrand, and their relation to sedimentology and gold distribution: Geological Society of South Africa, 21st Congress, Extended abstracts, p. 161–165.

Ohmoto, H., 1996, Evidence in pre-2.2 Ga paleosols for the early evolution of atmospheric oxygen and terrestrial biota: Geology, v. 24, p. 1135–1138, doi: 10.1130/0091-7613(1996)024<1135:EIPGPF>2.3.CO;2.

Palmer, J.A., 1986, Palaeoweathering in the Witwatersrand and Ventersdorp Supergroups [M.Sc. thesis]: Johannesburg, University of the Witwatersrand, 166 p.

Palmer, J.A., Phillips, G.N., and McCarthy, T.S., 1987, The nature of the Precambrian atmosphere and its relevance to Archaean gold mineralization, in Ho, S.E., and Groves, D.I., eds, Recent advances in understanding Precambrian gold deposits: Perth, Geology Department and Extension, University of Western Australia, Publication 11, p. 327–340.

Palmer, J.A., Phillips, G.N., and McCarthy, T.S., 1989, Paleosols and their relevance to Precambrian atmospheric composition: Journal of Geology, v. 97, p. 77–92.

Parnell, J., 1996, Phanerozoic analogues for carbonaceous matter in Witwatersrand ore deposits: Economic Geology and the Bulletin of the Society of Economic Geologists, v. 91, p. 55–62.

Phillipe, S., Lancelot, J.R., Clauer, N., and Pacquet, A., 1993, Formation and evolution of the Cigar Lake uranium deposit based on U-Pb and K-Ar isotope systematics: Canadian Journal of Earth Sciences, v. 30, p. 720–730.

Phillips, G.N., 1988, Widespread fluid infiltration during metamorphism of the Witwatersrand goldfields: Generation of chloritoid and pyrophyllite: Journal of Metamorphic Geology, v. 6, p. 311–332.

Phillips, G.N., and Dong, G., 1993, Chert-plus-pyrite pebbles in Witwatersrand goldfields: Townsville, Australia, Economic Geology Research Unit, James Cooke University, Contribution 47, 15 p.

Phillips, G.N., and Law, J.D.M., 1994, Metamorphism of the Witwatersrand goldfields: A review: Ore Geology Reviews, v. 9, p. 1–31, doi: 10.1016/0169-1368(94)90017-5.

Phillips, G.N., and Law, J.D.M., 1997, Hydrothermal origin for Witwatersrand gold: Society of Economic Geologists Newsletter, v. 31, p. 26–33.

Phillips, G.N., and Law, J.D.M., 2000, Witwatersrand gold fields: Geology, genesis and exploration: Society of Economic Geologists Reviews, v. 13, p. 439–500.

Phillips, G.N., and Myers, R.E., 1989, The Witwatersrand goldfields, II: An origin for Witwatersrand gold during metamorphism and associated alteration: Economic Geology Monograph 6, p. 598–608.

Phillips, G.N., Law, J.D.M., and Myers, R.E., 1990, The role of fluids in the evolution of the Witwatersrand Basin: South African Journal of Geology, v. 93, p. 54–69.

Phillips, G.N., Law, J.D.M., and Myers, R.E., 2001, Is the redox state of the Archaean atmosphere constrained?: Society of Economic Geologists Newsletter, v. 47, no. 1, p. 9–18.

Pretorius, D.A., 1976a, Gold in the Proterozoic sediments of South Africa: Systems, paradigms, and models, in Wolf, K.H., ed., Au, U, Fe, Mn, Hg, Sb, W and P deposits: Amsterdam, Elsevier, Handbook of Strata-Bound and Stratiform Ore Deposits, v. 7, p. 1–27.

Pretorius, D.A., 1976b, The nature of the Witwatersrand gold-uranium deposits, in Wolf, K.H., ed., Au, U, Fe, Mn, Hg, Sb, W and P deposits: Amsterdam, Elsevier, Handbook of Strata-Bound and Stratiform Ore Deposits, v. 7, p. 29–88.

Pretorius, D.A., 1981, Gold and uranium in quartz-pebble conglomerates: Economic Geology, 75th Anniversary Volume, p. 117–138.

Rasmussen, B., and Buick, R., 1999, Redox state of the Archean atmosphere: Evidence from detrital heavy minerals in ca. 3250–2750 Ma sandstones from the Pilbara Craton, Australia: Geology, v. 27, p. 115–118, doi: 10.1130/0091-7613(1999)027<0115:RSOTAA>2.3.CO;2.

Ramdohr, P., 1958, New observations on the ores of the Witwatersrand in South Africa and their genetic significance: Transactions of the Geological Society of South Africa, v. 61, Annexure, 50 p.

Reid, A.M., le Roux, A.P., and Minter, W.E.L., 1986, Electron-microprobe analyses of Witwatersrand gold particles from the Vaal placer: Geocongress '86, Geological Society of South Africa, Johannesburg, Abstracts, p. 179–181.

Reimer, T.O., 1985, Volcanic rocks and weathering in the Early Proterozoic Witwatersrand Supergroup, South Africa: Geological Survey of Finland Bulletin, v. 331, p. 33–49.

Reimer, T.O., and Mossman, D.J., 1990, Sulfidization of Witwatersrand black sands: From enigma to myth: Geology, v. 18, p. 426–429, doi: 10.1130/0091-7613(1990)018<0426:SOWBSF>2.3.CO;2.

Robb, L.J., and Meyer, F.M., 1995, The Witwatersrand Basin, South Africa: Geological framework and mineralisation processes: Ore Geology Reviews, v. 10, p. 67–94, doi: 10.1016/0169-1368(95)00011-9.

Robb, L.J., Davis, D.W., and Kamo, S.L., 1990a, U-Pb ages on single detrital zircon grains from the Witwatersrand basin, South Africa: Constraints on the age of sedimentation and on the evolution of granites adjacent to the basin: Journal of Geology, v. 98, p. 311–328.

Robb, L.J., Meyer, F.M., Ferraz, M.F., and Drennan, G.R., 1990b, The distribution of radioelements in Archaean granites of the Kaapvaal Craton, with implications for the source of uranium in the Witwatersrand Basin: South African Journal of Geology, v. 93, p. 5–40.

Robinson, A., and Spooner, E.T.C., 1984, Can the Elliot Lake uraninite-bearing quartz pebble conglomerates be used to place limits on the oxygen content of the early Proterozoic atmosphere?: Journal of the Geological Society [London], v. 141, p. 221–228.

Roscoe, S.M., and Minter, W.E.L., 1993, Pyritic paleoplacer gold and uranium deposits, in Kirkham, R.V., Sinclair, W.D., Thorpe, R.I. and Duke, J.M., eds., Mineral deposit modelling: Geological Association of Canada Special Paper 40, p. 103–124.

Rundle, C.C., and Snelling, N.J., 1977, The geochronology of uraniferous minerals in the Witwatersrand Triad; an interpretation of new and existing U-Pd age data on rocks and minerals from the Dominion Reef, Witwatersrand and Ventersdorp Supergroups: Philosophical Transactions of the Royal Society of London, v. 286, p. 567–583.

Rye, R.O., and Holland, H.D., 1998, Paleosols and the evolution of atmospheric oxygen: A critical review: American Journal of Science, v. 298, p. 621–672.

Saager, R., 1968, Newly observed ore-minerals from the Basal Reef in the Orange Free State goldfield in South Africa: Economic Geology and the Bulletin of the Society of Economic Geologists, v. 63, p. 116–123.

Saager, R., 1970, Structures in pyrite from the Basal Reef in the Orange Free State gold-field: Transactions of the Geology Society of South Africa, v. 73, p. 29–46.

Smith, N.D., and Minter, W.E.L., 1980, Sedimentological controls of gold and uranium in two Witwatersrand paleoplacers: Economic Geology and the Bulletin of the Society of Economic Geologists, v. 75, p. 1–14.

Smith, R.E., and Perdrix, J.L., 1982, Pisolitic laterite geochemistry for detecting massive sulphide deposits, in Geochemical exploration in deeply weathered terrains: Floreat, Western Australia, CSIRO Report, p. 128–139.

Smits, G., 1984, Some aspects of the uranium minerals in the Witwatersrand sediments of the early Proterozoic: Precambrian Research, v. 25, p. 37–59, doi: 10.1016/0301-9268(84)90023-8.

Sutton, S.J., Ritger, S.D., and Maynard, J.B., 1990, Stratigraphic control of chemistry and mineralogy in metamorphosed Witwatersrand quartzites: Journal of Geology, v. 98, p. 329–341.

Tankard, A.J., Jackson, M.P.A., Eriksson, K.A., Hobday, D.K., Hunter, D.R., and Minter, W.E.L., 1982, Crustal evolution of Southern Africa: New York, Springer-Verlag, 523 p.

Tucker, R.F., 1980, The sedimentology and mineralogy of the Composite Reef on Cooke Section, Randfontein Estates Gold Mine, Witwatersrand, South Africa [M.Sc. thesis]: Johannesburg, University of the Witwatersrand, 355 p.

Utter, T., 1979, The morphology and silver content of gold from the Upper Witwatersrand and Ventersdorp systems of the Klerksdorp gold field, South Africa: Economic Geology and the Bulletin of the Society of Economic Geologists, v. 74, p. 27–44.

von Gehlen, K., 1983, Silver and mercury in single gold grains from the Witwatersrand and Barberton, South Africa: Mineralium Deposita, v. 18, p. 529–534, doi: 10.1007/BF00204496.

Wall, V.J., Mason, R.N., and Hall, G.C., 2004, Hydrothermal alteration and gold mineralization in the Witwatersrand—When, where and why?: Geoscience Africa 2004, University of the Witwatersrand, Johannesburg, Abstracts, p. 685 - 686.

MANUSCRIPT ACCEPTED BY THE SOCIETY 29 OCTOBER 2005

Geological Society of America
Memoir 198
2006

Evidence from sulfur isotope and trace elements in pyrites for their multiple post-depositional processes in uranium ores at the Stanleigh Mine, Elliot Lake, Ontario, Canada

Kosei E. Yamaguchi

Institute for Research on Earth Evolution (IFREE), Japan Agency for Marine-Earth Science and Technology (JAMSTEC), 2-15 Natsushima, Yokosuka, Kanagawa 237-0061, Japan, and NASA Astrobiology Institute

Hiroshi Ohmoto

Penn State Astrobiology Research Center of the NASA Astrobiology Institute and the Department of Geosciences, The Pennsylvania State University, University Park, Pennsylvania 16802, USA

ABSTRACT

The ca. 2.45 Ga pyritic uraniferous quartz-pebble conglomerate (UQC) of the Matinenda Formation of the Elliot Lake Group, Huronian Supergroup, was used in this study to investigate the origin of pyrite. A laser-microprobe was used for analysis of the sulfur isotopic compositions of individual pyrite grains, and an electron-probe microanalyzer was used for analysis of the trace element compositions of pyrite grains with overgrowth texture. We found a variation in $\delta^{34}S$ values among pyrite crystals (73 analyses) of various size and morphologies that occur in a small (~1 cm³) rock chip: the total range in $\delta^{34}S$ is −9.0‰ to +5.5‰ with respect to CDT (Cañon Diablo Troilite) with a mean value of +0.6‰ ±2.1‰ (1σ). The widest range of ~15‰ is found among euhedral pyrite grains whereas variations of ~4‰ to ~6‰ are common in anhedral, subhedral, and rounded grains of pyrite. These values are in marked contrast to the $\delta^{34}S$ values of pyrite from the Matinenda Formation that were obtained by previous investigators using bulk-rock sulfur isotope analyses. We found variable concentrations of Co (below detection to 4700 ppm), Ni (to 1900 ppm), and As (to 3400 ppm) among individual pyrite crystals and within single grains with overgrowth textures. These elemental concentrations are markedly different between core and overgrowth parts of pyrite. We demonstrate that the pyrite grains in the Paleoproterozoic UQC have been isotopically, chemically, and morphologically modified by post-depositional processes, suggesting that the pyrite grains have undergone multiple generations. The results of the present study cannot be explained solely by a detrital process. Therefore, one cannot use the preserved morphology and chemistry of pyrite (and possibly uraninite) to represent the original features at the time of deposition to support the hypothesis of an anoxic atmosphere prior to 2.2 Ga.

Keywords: sulfur isotope, pyrite, laser ablation, trace element, Elliot Lake, and sulfate reduction.

*kosei@jamstec.go.jp

Yamaguchi, K.E., and Ohmoto, H., 2006, Evidence from sulfur isotope and trace elements in pyrites for their multiple post-depositional processes in uranium ores at the Stanleigh Mine, Elliot Lake, Ontario, Canada, *in* Kesler, S.E., and Ohmoto, H., eds., Evolution of Early Earth's Atmosphere, Hydrosphere, and Biosphere—Constraints from Ore Deposits: Geological Society of America Memoir 198, p. 143–156, doi: 10.1130/2006.1198(08). For permission to copy, contact editing@geosociety.org. ©2006 Geological Society of America. All rights reserved.

INTRODUCTION

Origin of Uraninite and Pyrite and the Rise of Atmospheric Oxygen

The rise of atmospheric O_2 is of great importance; it directly relates to the chemical evolution of the atmosphere and oceans, and the origin and evolution of life on Earth. The current view of the rise of atmospheric O_2 on Earth ca. 2.2–2.0 Ga (e.g., Cloud, 1968; Holland, 1984, 1994; Kasting, 1993; references therein) is based on several lines of geological and geochemical observations. These observations include the mode of occurrence and genesis of uranium ore (e.g., Roscoe, 1969; Dimroth and Kimberley, 1976; Minter, 1976; Phillips and Myers, 1981; Maynard et al., 1991; Robb and Meyer, 1995; Barnicoat et al., 1997). However, a consensus among scientists on the timing and mechanism for the rise of atmospheric oxygen has not been formed, resulting in vigorous controversy.

Enrichment of uranium and pyrite in uraniferous quartz-pebble conglomerate (UQC) has been found at many locations in the world; Witwatersrand, South Africa, and Elliot Lake–Blind River, Canada (Figs. 1, 2) are two of the best known. The occurrence of apparently detrital grains of uraninite and pyrite in UQC that deposited before 2.2 Ga has been cited as evidence for a reducing atmosphere, because those minerals are unstable in oxidizing environments (e.g., Grandstaff, 1980). However, controversy continues as to whether such uraninite and pyrite are detrital (Minter, 1976), epigenetic / hydrothermal (Davidson, 1957; Dimroth and Kimberley, 1976; Phillips et al., 1987; Phillips and Myers, 1981; Barnicoat et al., 1997), or a hybrid of these two types (modified placer: Holmes, 1957; Pienaar, 1963; Robertson, 1968, 1986; Roscoe, 1969, 1973, 1981; Theis, 1979; Myers, 1981; Ruzicka, 1981, 1988, 1989; Robinson and Spooner, 1982, 1984a, 1984b; Mossman and Farrow, 1992; Frimmel et al., 1993; Robb and Meyer, 1991, 1995). Some researchers (e.g., Holland, 1984, 1994) have suggested that, after the rise of atmospheric O_2, groundwater-type U deposits began to form by the mixing of oxic groundwater containing dissolved U and reduced fluid containing petroleum-like organic matter (OM) and/or reduced minerals such as pyrite. The oldest unequivocally groundwater-type U deposit that has been recognized so far is the 2.0 Ga Oklo U deposit in Gabon (e.g., Gauthier-Lafaye and Weber, 1989). The key question that arises here is whether the mode of U mineralization is the same throughout geologic history or different before and after 2.2–2.0 Ga.

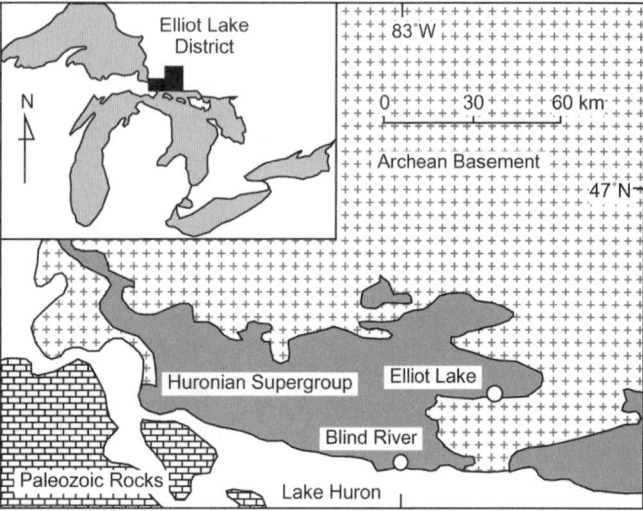

Figure 1. Arial distribution of the Huronian Supergroup in east Ontario. Modified after Bennett et al. (1991).

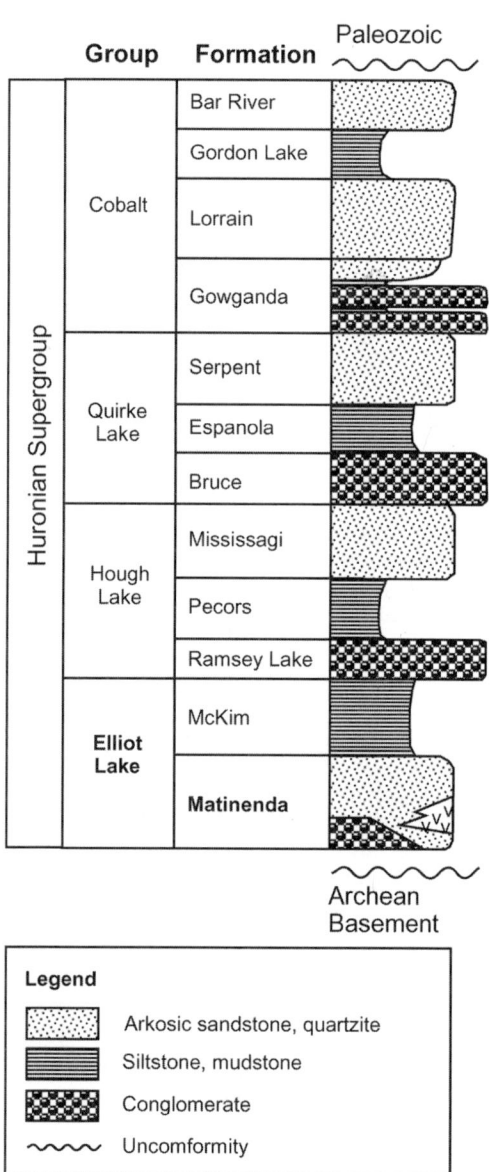

Figure 2. Generalized stratigraphy of the Huronian Supergroup. Modified after Young (1991).

A detrital origin for the uraninite and pyrite in UQC has been based on observed textural features, such as grain size and shape (Ramdohr, 1958, 1969), and related sedimentological criteria (e.g., Robertson and Steenland, 1960; Roscoe, 1969; Minter, 1976; Theis, 1979), intergrain chemical heterogeneity (Grandstaff, 1974), and the high Th and rare earth element content of the uraninite. Hydraulic sorting, an important sedimentological criterion to identify the "detrital" hypothesis, depends on the size, shape, and specific gravity of the sediment grains/particles. When applying such criteria, it is necessary to assume that the observed characteristics of the present (preserved) state of the grains represent their original state at the time of deposition, or at least represent a modified state that is reasonably traceable to their original state. Otherwise, one would be just looking at grains that were modified in shape and size by post-depositional processes, which could lead to an erroneous conclusion for their origin.

Preliminary Observations and Purpose of Study

During reconnaissance observation of approximately ten samples from the pyritiferous UQC from the Matinenda Formation of the Paleoproterozoic Huronian Supergroup, we noticed that pyrite grains are not uniform in shape and size at the scale of a thin section. Additionally, there are generally two textures of pyrite: porous and non-porous. These preliminary observations suggest that pyrite grains in the UQC were modified by post-depositional processes and thus could not be regarded as detrital grains showing original features at the time of deposition. Therefore, it is necessary to investigate each pyrite grain for its origin. To detect microscale variation in chemical and isotopic compositions among individual pyrite grains in the UQC, we utilized micro-analytical methods: S isotope analysis by in situ laser ablation and trace element analysis by electron-probe microanalyzer (EPMA). For trace element analyses, we focused on three elements: As, Ni, and Co. Arsenic can substitute for S; Ni and Co can substitute for Fe. Assuming the degree of substitution of these three trace elements for FeS_2 depends on the environments of mineralization of pyrite, it is expected that pyrite of different origins will show different trace element characteristics.

GEOLOGICAL SETTING

The geological setting of the Huronian Supergroup has been described by Young (1991) and Bennett et al. (1991). The Paleoproterozoic Huronian Supergroup is exposed in an area ~340 km E-W and ~300 km N-S, on the northern margin of Lake Huron as part of the Canadian Shield (Fig. 1). The Huronian Supergroup, which has been regarded as one of the best-exposed Paleoproterozoic successions in the world, consists of sedimentary and subordinate volcanic rocks unconformably overlying the Archean basement of the Superior Province.

A generalized stratigraphic column for the Huronian Supergroup is given in Figure 2. The supergroup is subdivided into four stratigraphic groups. The Elliot Lake Group, the oldest, consists of volcanic rocks and clastic sedimentary rocks, and contains economically significant Witwatersrand-type U deposits. The three overlying groups, in ascending stratigraphic order, are the Hough Lake Group, Quirke Lake Group, and Cobalt Group. Each group has a sedimentary cycle of conglomerate, mudstone, siltstone, and coarse arenite. The Huronian Supergroup contains the world's oldest well-documented and widespread glaciogenic rocks (Nesbitt and Young, 1982). Paleosols are preserved at the base of the Huronian succession. Studies of the Huronian Supergroup have provided various lines of evidence interpreted as reflecting an increase in oxygen content of Earth's atmosphere during deposition of the Huronian succession, which occurred ca. 2.5–2.2 Ga (e.g., Hattori et al., 1983a). The tectonic setting of the Huronian Supergroup has recently been interpreted in terms of a Wilson cycle of ocean opening and closing. Krough et al. (1984) and Corfu and Andrews (1986) gave geochronological constraints on the deposition of the Huronian Supergroup using the U-Pb zircon method. They bracketed the deposition of the Huronian Supergroup between 2450 +25/-10 Ma and 2219.4 ±3.5 Ma.

The Matinenda Formation, the lowermost formation of the Elliot Lake Group, contains the world-famous pyritiferous UQC (Figs. 2–4). The Formation is up to 180 m thick in the Elliot Lake area (Robertson, 1981) and is mainly composed of cross-bedded

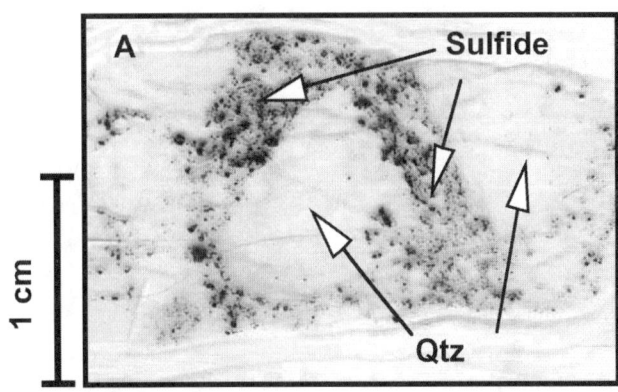

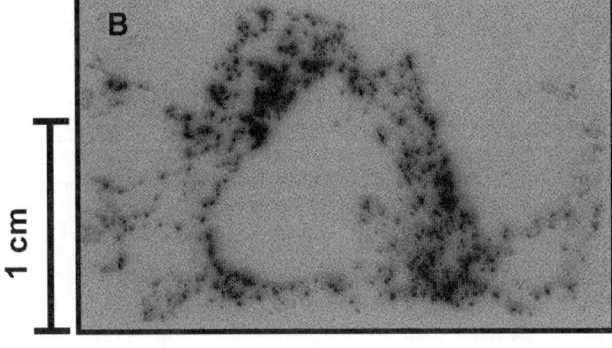

Figure 3. (A) Scanned digital image of a UQC thin section. (B) X-ray autoradiography; ten-day exposure. Radiogenic parts are indicated in black. Note close association of sulfide with U-bearing minerals in the matrix of the sample.

146 K.E. Yamaguchi and H. Ohmoto

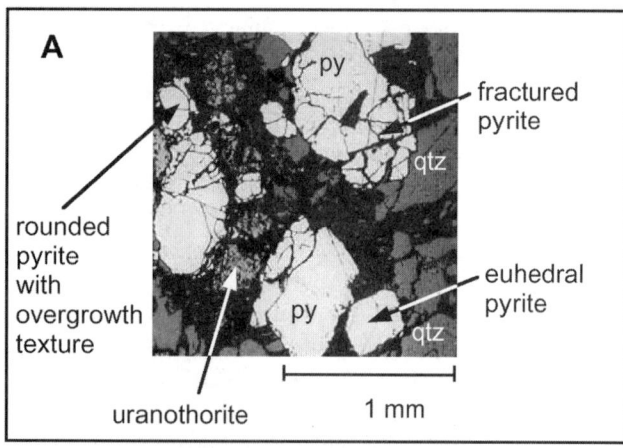
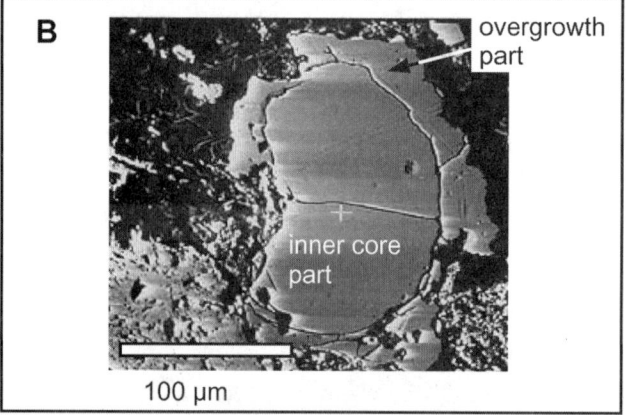

Figure 4. Examples of the sample. (A) Photomicrograph showing fractured pyrite, rounded pyrite with overgrowth texture, and euhedral pyrite. Uranothorite is sandwiched between pyrite grains. (B) Backscattered electron (BSE) image of pyrite with overgrowth texture (zoom of the pyrite grain in the upper left corner of A). (C) Photomicrograph of porous (micro-fractured?) pyrite surrounded by non-porous pyrite. The thin section was dipped in dilute nitric acid for ~1 min to enhance the mineral grain boundaries. Scale bars are different among the photos.

arkosic sandstones, poorly sorted quartz-pebble conglomerate, with uraniferous parts near the base (Young, 1991). The arkosic sandstones consist of poorly sorted quartz and feldspar grains in a sericite matrix. The Matinenda Formation has three units of a ~3-m-thick ore horizon, which has several cycles of ~40 cm conglomerate and sandstone beds separated by ~10-m-thick quartzite beds. The conglomerate samples are oligomictic and contain poorly sorted, well-rounded, and tightly packed quartz pebbles, chert clasts, feldspar, sericite, U-bearing minerals, and various morphologies of pyrite (Figs. 3, 4). Pyrite and U-bearing minerals are concentrated in the matrix (Fig. 3), and the content of pyrite, the commonest sulfide mineral in the UQC, is 6–10% of the ore (Bennett et al., 1991). There is no major difference in the mode of occurrence for these minerals among different units of UQC horizons in the Matinenda Formation. The Matinenda Formation's detrital material was produced during intense weathering in a warm humid environment (Nesbitt and Young, 1982), and its UQC units were deposited in a braided fluvial environment that developed on Archean basement, with a dominantly southeasterly paleocurrent direction (Fralick and Miall, 1989).

During the Penokean orogeny (ca. 1.7–1.9 Ga), rocks of the southern Huronian Supergroup were subjected to low-pressure, dynamothermal metamorphism, with a maximum grade equivalent to the staurolite zone of the amphibolite facies (Bennett et al., 1991; Hu et al., 1998). In the Sault Ste. Marie–Elliot Lake area, the Huronian succession is relatively thin and the metamorphic grade is low (sub-greenschist facies; Robertson, 1981; Farrow and Mossman, 1988; Bennett et al., 1991).

SAMPLES

Samples of pyritiferous U-ore for this study were taken in 1995 at the Stanleigh Mine (permanently closed since 1996) in the Elliot Lake district, Ontario, Canada. The samples belong to the ca. 2.45 Ga Matinenda Formation (Krough et al., 1984) and have suffered from minor burial metamorphism of sub-greenschist facies and little or no modern weathering (Robertson, 1981; Farrow and Mossman, 1988; Bennett et al., 1991). Therefore, they are suitable for isotope and trace element analysis. Sulfide (mainly pyrite) and U-bearing minerals are in a matrix of quartz-pebble conglomerate. Figure 3 shows a high-resolution digital image of a sample together with its X-ray autoradiography (radioactive parts appear in black). Various morphologies of pyrite (i.e., pyrite with overgrowth texture, euhedral pyrite, subhedral pyrite, anhedral pyrite, porous pyrite, non-porous pyrite, fractured pyrite, and non-fractured pyrite; an estimate of their relative abundance is given later) are observed and a few examples are shown in Figure 4. Pyrite commonly occurs intergrown with various other minerals. Classification of pyrite morphologies as shown above is based on shapes of pyrite grains in a polished specimen of the examined UQC sample. Although rounded, non-porous, and non-fractured pyrite grains were not found in the examined specimen, the scheme of pyrite classification used in this study is basically comparable to that used in

Saager (1981), in which pyrite grains in the UQC of the Witwatersrand Basin in South Africa were used.

There is no major difference in the mode of occurrence of pyrite (and uraninite) among different units of UQC layers in the Matinenda Formation, and on that basis we have assumed that the examined 1 cm^3 sample is representative of the UQC units of the Matinenda Formation. This assumption may be supported from the observation by Meyer et al. (1990), who use similar UQC samples in the Witwatersrand Supergroup in South Africa (e.g., Pretorius, 1981), that there is no fundamental chemical difference between pyrites from the various UQC horizons. However, there is a limitation that we do not know the exact depositional setting of the examined 1 cm^3 specimen in the braided stream environment where quartz-pebble conglomerates were deposited.

ANALYTICAL METHOD

Laser Microprobe Analysis

Sulfur isotopic compositions were determined using a stable isotope mass spectrometer (VG Prism II) with a Nd-YAG laser ablation system at the Pennsylvania State University (Kakegawa, 1997). Notation for S isotopic values is with respect to CDT (Cañon Diablo Troilite): $\delta^{34}S = (R_{sample}/R_{standard} - 1) \times 1000$, where $R = {^{34}S}/{^{32}S}$. The precision and reproducibility is ±0.2‰, and laboratory standard SO$_2$ gas, calibrated with an international standard, was used for data acquisition. The experimental scheme and conditions are described elsewhere in detail (Crowe et al., 1990; Kelley and Fallick, 1990; Kakegawa, 1997) and therefore are only briefly described here. The polished 1 cm^3 sample was placed in a sample chamber and viewed through a CCD camera. A Nd-YAG laser, whose beam radius is ~50 µm, vaporized a pyrite grain, and in some cases its surrounding material (organic matter) in O$_2$ atmosphere, resulting in a crater with a radius ≤200 µm. The generated gas mixture (SO$_2$, CO$_2$, H$_2$O, various hydrocarbons, and excess O$_2$) was then sent to a graphite-furnace operating at ~950 °C to oxidize completely. After cryogenic removal of impurities, SO$_2$ was transferred to the mass spectrometer for isotope ratio measurement. Correction of isotope fractionation (1‰) induced during laser ablation was made according to Kakegawa (1997). To monitor the instrumental fractionation of S isotopes, the standard SO$_2$ gas was measured frequently for its S isotope composition between samples.

Electron Microprobe Analysis

Concentrations of As, Ni, and Co in pyrites were measured by electron-probe microanalyzer (EPMA; Cameca SX50) at the Material Research Institute of the Pennsylvania State University. The EPMA is equipped with four wavelength spectrometers and one energy dispersive spectrometer. The operating conditions for quantitative analysis of the above elements were as follows: the accelerating voltage was 25 kV; the beam current was 15 nA; Kα line was used for As, Ni, and Co; the normal beam diameter was 2 µm; and the counting time was kept 120 seconds. The detection limit was 100–200 ppm for all elements examined. Pyrite grains with overgrowth texture were chosen for the trace element study by EPMA to examine the differences in their chemical composition between the outer overgrowth part and inner core part.

RESULTS

Sulfur Isotope Ratios

A large variation in $\delta^{34}S$ values was found among individual pyrite grains of various size (~0.1 to ~2.7 mm diameter) and morphology (euhedral, subhedral, anhedral, or rounded) from a single ~1 cm^3 rock specimen (Fig. 5). The S isotope compositions are summarized in Table 1. Sizes of pyrite crystals were estimated on the basis of the longest and shortest diameter / edge measured for each pyrite grain on the polished section (data not shown). We found no clear statistical correlation between the measured crystal size and the $\delta^{34}S$ values of pyrite examined. However, negative $\delta^{34}S$ values were obtained from relatively small (<200 µm; in many cases euhedral) pyrite crystals. The $\delta^{34}S$ values of 73 data points range from –9.0‰ to +5.5‰ with a mean $\delta^{34}S$ value of +0.6‰. Hattori et al. (1983b) analyzed $\delta^{34}S$ values of pyrite in the Matinenda Formation using the conventional bulk-sample method and reported a range from –1.5‰ to +1.2‰ with a mean of +0.2‰ for 75 data points. The range observed in the present study (14.5‰) is much greater than that for previously published data (2.7‰) reported by Hattori et al. (1983b), whereas the mean values are very similar (only 0.4‰ difference). Comparison of histograms for $\delta^{34}S$ values of pyrite in the UQC of the Matinenda Formation between this study and those of Hattori et al. (1983b, 1986) is not readily possible, because Hattori et al. (1983b) presented only the range, mean, and number of data points they obtained and Hattori et al. (1986) (see Table 2 of Hattori et al., 1986) presented 19 data points instead of 75 data points used in Hattori et al. (1983b). Conventional bulk-rock S isotope analysis requires relatively large number of samples, and inevitably homogenizes the potentially variable S isotope compositions of different pyrite grains. Microanalytical methods utilizing laser and ion beams often reveal true variations in chemical and isotopic compositions of geological materials (e.g., Eldridge et al., 1993). We believe that the range of $\delta^{34}S$ values determined by the laser microprobe method in this study represents the true range of $\delta^{34}S$ values of pyrite in the sample.

The S isotopic compositions of individual grains of pyrite are separated according to their morphologies, and are shown in Figure 5B–E. Classification of pyrite morphologies are based on the cross sections of pyrite crystals that appeared on the polished surface of the examined rock specimen. Twenty-four pyrite grains are observed as euhedral. Euhedral pyrite has the

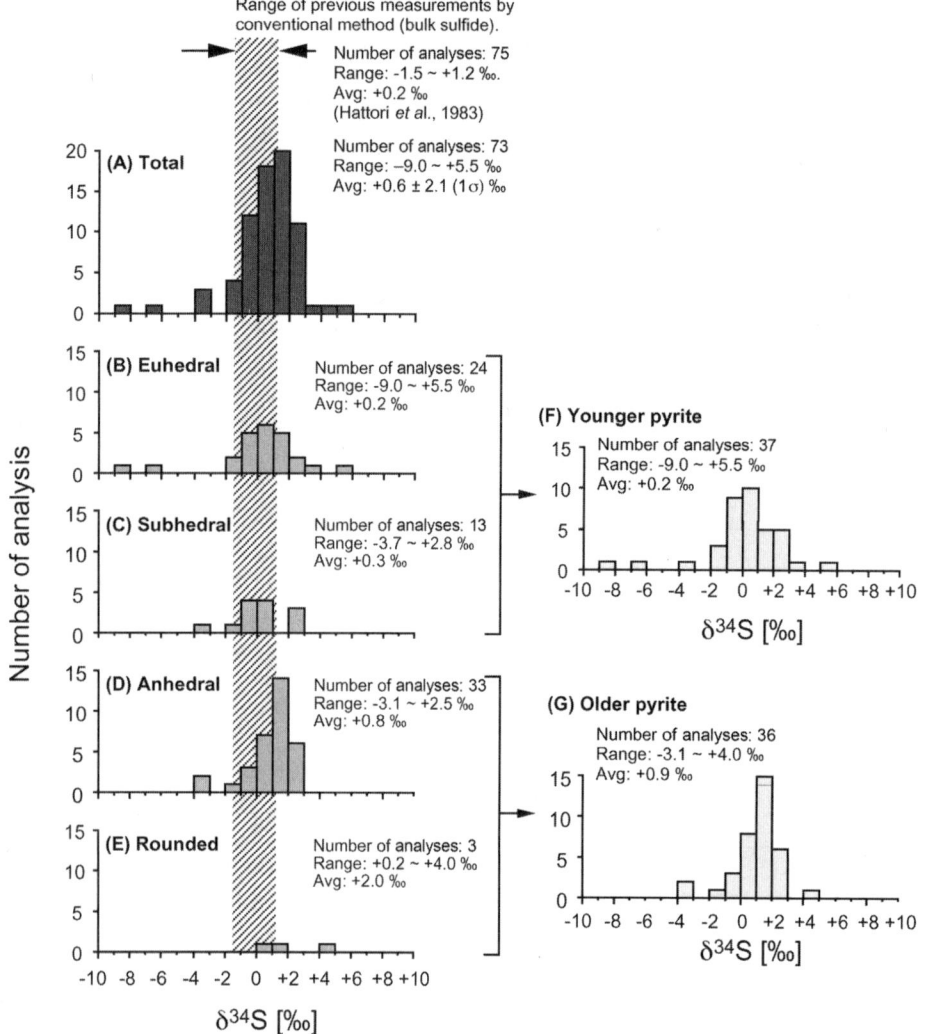

Figure 5. Histogram of S isotopic compositions of individual grains of pyrite in the ~1 cm³ rock specimen of the Paleoproterozoic UQC, Matinenda Formation, Elliot Lake Group, Ontario, Canada. Sulfur isotopic compositions are expressed by delta notation from the standard CDT. Total number of analyses is 73. Overall distribution shows a wide range from –9.0‰ to +5.5‰, with a mean value of +0.6 ± 2.1 (1σ)‰. (A) Total analysis. (B) Euhedral pyrite. (C) Subhedral pyrite. (D) Anhedral pyrite. (E) Rounded pyrite. (F) Younger pyrite: combination of (B) and (C). (G) Older pyrite: combination of (D) and (E). Hatched column indicates the range in S isotopic compositions of previous bulk measurements by Hattori et al. (1983a, 1983b). Note the (previously undetected) larger spread of S isotopic compositions revealed by the microanalytical method those by conventional bulk method.

widest variation in $\delta^{34}S$ value, ranging from –9.0‰ to +5.5‰ with a mean of +0.2‰. The most positive and negative $\delta^{34}S$ values of total data points are from euhedral pyrite. The mean $\delta^{34}S$ value for euhedral pyrite is less than the mean of the total data points. Thirteen subhedral pyrite grains show a range in $\delta^{34}S$ value from –3.7‰ to +2.8‰ with a mean of +0.3‰. The mean value is similar to that of euhedral pyrite. We use the terms "anhedral" for pyrite that we cannot group into euhedral, subhedral, or rounded pyrite. The range in $\delta^{34}S$ values of anhedral pyrite grains is –3.1‰ to +2.5‰ and the mean value is +0.8‰. This group of pyrite is dominant (data points $n = 33$) in the examined sample of pyritiferous UQC. We found only three grains of rounded pyrite in the examined sample. Their $\delta^{34}S$ values are +0.2‰, +1.8‰, and +4.0‰, with a mean value of +2.0‰. It should be noted that the mean and median $\delta^{34}S$ values for each group of pyrite grains (see Table 2) are all different from one another, and that the number of euhedral pyrite grains ($n = 24$ in total) that fall outside the range observed by Hattori et al. (1983b) (i.e., –1.5‰ to +1.2‰) is comparable ($n = 11$) to the number that fall within that range ($n = 13$).

Euhedral pyrite and subhedral pyrite grains are combined into one group termed "younger pyrite," whereas anhedral pyrite is considered "older pyrite." The histograms of their $\delta^{34}S$ values are shown in Figure 5F and 5G. This grouping is based on interpretation of the $\delta^{34}S$ data and trace element data of pyrite grains, which is discussed later in detail.

Ratios of the numbers of each type of pyrite grain were used as an approximate estimate of the relative abundance of each type of pyrite grain. According to this rough estimate, euhedral pyrite is ~33%, subhedral pyrite is ~18%, anhedral pyrite is ~45%, and rounded pyrite is ~4%. The older pyrite and the younger pyrite are found to be nearly equal in abundance (49% and 51%, respectively). From microscopic observation, such estimates for the relative abundance of different types of pyrite grains in the examined sample are found to be not much different from those in the other samples obtained from differ-

TABLE 1. SUMMARY OF SULFUR ISOTOPE ANALYSIS BY LASER MICROPROBE METHOD

Analysis number	Morphology of pyrite crystal	$\delta^{34}S$ (‰)	Analysis number	Morphology of pyrite crystal	$\delta^{34}S$ (‰)
1	Euhedral	+1.82	38	Euhedral	+1.83
2	Anhedral	+1.09	39	Anhedral	−0.75
3	Anhedral	+1.85	40	Euhedral	+0.39
4	Euhedral	−1.11	41	Euhedral	+0.95
5	Anhedral	+0.98	42	Anhedral	+0.59
6	Anhedral	+0.94	43	Euhedral	−0.31
7	Euhedral	+0.61	44	Subhedral	−3.72
8	Euhedral	−0.61	45	Anhedral	+1.26
9	Subhedral	−0.02	46	Euhedral	+1.75
10	Subhedral	+0.89	47	Euhedral	+1.03
11	Euhedral	−0.28	48	Anhedral	+0.91
12	Anhedral	+1.74	49	Anhedral	+1.03
13	Anhedral	−3.05	50	Euhedral	+3.05
14	Anhedral	+1.76	51	Euhedral	+1.63
15	Euhedral	+0.99	52	Anhedral	+0.49
16	Euhedral	+0.94	53	Euhedral	−1.82
17	Anhedral	+1.09	54	Anhedral	+2.29
18	Anhedral	0.00	55	Anhedral	+2.46
19	Euhedral	−0.34	56	Subhedral	−0.18
20	Rounded	+0.24	57	Anhedral	−1.80
21	Anhedral	+1.78	58	Anhedral	+1.50
22	Euhedral	+0.51	59	Anhedral	+1.24
23	Anhedral	+2.11	60	Anhedral	+1.65
24	Subhedral	+0.66	61	Anhedral	+1.13
25	Euhedral	−6.36	62	Anhedral	+2.22
26	Subhedral	+2.32	63	Subhedral	+0.45
27	Anhedral	−0.78	64	Subhedral	−0.15
28	Anhedral	+1.26	65	Anhedral	+2.16
29	Anhedral	−3.09	66	Euhedral	−8.98
30	Anhedral	+0.19	67	rounded	+1.78
31	Subhedral	+0.29	68	Euhedral	+2.53
32	Subhedral	−0.80	69	Anhedral	−0.11
33	Euhedral	v2.13	70	Subhedral	+2.59
34	Subhedral	−1.64	71	Rounded	+4.04
35	Anhedral	+1.18	72	Euhedral	−0.12
36	Subhedral	+2.76	73	Anhedral	+2.42
37	Euhedral	+5.45			

TABLE 2. STATISTICAL ANALYSIS (INCLUDING T-TEST) OF THE SULFUR ISOTOPE COMPOSITIONS FOR EACH GROUP OF PYRITE TYPES

Sample group	All	Pyrite type				Hattori et al. (1983b)
		Anhedral	Euhedral	Subhedral	Rounded	
Number of data points (n)	73	33	24	13	3	75
Degree of freedom (d.f. = n−1)	72	32	23	12	2	74
Mean	0.59	0.84	0.24	0.27	2.02	0.2
Median	0.95	1.13	0.78	0.29	1.78	-
Maximum	+5.45	+2.46	+5.45	+2.76	+4.04	+1.2
Minimum	−8.98	−3.09	−8.98	−3.72	+0.24	−1.5
Variance	4.34	1.97	8.34	3.13	3.65	-
Standard deviation (1σ)	2.08	1.40	2.89	1.77	1.91	-
T_{cal} (α = 0.05)	5.48	3.24	4.35	3.05	-	-
T_{table} (α = 0.05)	1.99	2.04	2.07	2.18	-	-

Note: t_{cal}—t-value calculated from the obtained data sets; t_{table}—t-value found in statistical textbook.

ent units of UQC in the Matinenda Formation. Such similarity supports the idea that the examined samples of this study may be representative of UQC in the Matinenda Formation.

Because of the apparent similarities between the shape of frequency histograms for the obtained data sets and statistical "normal" distribution we statistically analyzed the significance of variations in the obtained $\delta^{34}S$ values for each group of pyrite grains. We used the Student's *t*-test, and the results of are summarized in Table 2.

When all samples are considered, the upper 0.05 point (i.e., α = 0.05) of the *t*-distribution ($t_{0.05}$) with 72 (= n−1; *n* is number of samples) degrees of freedom (d.f.) is found from the *t*-test table to be 1.99 (calculated *t*-value is 5.48). For anhedral pyrite grains, $t_{0.05}$ with d.f. = 32 (*n* = 33) is 2.04 (calculated *t*-value is 3.24). For euhedral pyrite grains, $t_{0.05}$ with d.f. = 23 (*n* = 24) is 2.07 (calculated *t*-value is 4.35). And for subhedral pyrite grains, $t_{0.05}$ with d.f. = 12 (*n* = 13) is 2.18 (calculated *t*-value is 3.05). All the *t*-values found from the *t*-test table for all, anhedral, euhedral, and subhedral pyrite grains (1.99, 2.04, 2.07, and 2.18, respectively) are smaller than the corresponding calculated values (5.48, 3.24, 4.35, and 3.05, respectively). Therefore, the null hypothesis (i.e., the histogram of the obtained data sets can be approximated by the normal distribution) is rejected at the level of significance α = 0.05. We suggest that the claim "the histograms of the obtained data sets are statistically different from and not approximated by the normal distribution" is strongly supported by the data.

Trace Element Ratios

Trace element data are summarized in Figure 6 and Table 3. Arsenic concentrations range from 200 to 3400 ppm for the outer overgrowth part of pyrite grains, whereas the inner cores range from below detection limit (~100 ppm) to 1000 ppm and clearly contain less As. Nickel concentrations are <2000 ppm for both types of pyrite. Distribution patterns look similar for both types of pyrite and it is impossible to distinguish between the overgrowth part and core part based on their Ni contents. Co concentrations range from 400 to 4700 ppm for overgrowth pyrite, and 100–2300 ppm for core pyrite. Higher Co contents are found in the overgrowth part. More than half of the data points (eight points) for Co content of the overgrowth part are more than 1200 ppm and appear evenly distributed, whereas the remaining five data points are between 400 and 600 ppm. With one exception, Co contents are <900 ppm for the core part, and the mode of these analyses is between 400 and 600 ppm.

In addition to the absolute content of trace elements, their As/Ni and Ni/Co ratios are useful for distinguishing between overgrowth and core pyrite (Table 3). For example, the As/Ni ratio of grain A is <1 for the core part, whereas the overgrowth part has an As/Ni ratio >1 (grain A). With one exception, the Ni/Co ratios of grain A also show a clear distinction between the two parts (Ni/Co < 1 for the overgrowth and <1 for the core). As/Co ratios do not indicate a clear distinction between the two parts. These elemental ratios differ for grains B and C, where overgrowth pyrite has generally higher As/Ni ratios and lower Ni/Co ratios than core pyrite, with some exceptions.

The concentrations of trace elements are different between various morphologies of pyrite grains. Euhedral pyrite grains typically have higher As and Ni concentrations than subhedral, anhedral, and rounded pyrite. Rounded pyrite grains have lower As, Ni, and Co concentrations than the other types of pyrite (the data are not presented in this paper).

DISCUSSION

Origins of S in Pyrite Crystals

To account for the observed variations in the $\delta^{34}S$ values and trace element contents of pyrite in UQC, one may attribute them to detrital transport of pyrite grains from source rocks that already had such variations. The source rocks are likely to have been the Archean basement, granitic and basaltic rocks underlying the Huronian Supergroup. The $\delta^{34}S$ values of detrital pyrite, if any, are expected to be close to 0‰, because the parental pyrite crystals in the source rocks typically have $\delta^{34}S$ values close to 0‰ (Ohmoto and Goldhaber, 1997). Furthermore, extensive physical breakdown and mixing of detrital pyrite grains, if any, during transport in ancient rivers would probably have erased any potential original variability in the $\delta^{34}S$ values and trace element contents of pyrite in the source rocks. Thus, variation in the $\delta^{34}S$ values of pyrite as large as 14.5‰—observed in this study—is probably not caused by the inherited variation of those in the source rocks.

The similarity of the overall average $\delta^{34}S$ values of pyrite in the UQC between individual analyses (+0.6‰; this study) and bulk-rock analyses (+0.2‰; Hattori et al., 1983b) may suggest that the majority of pyrite grains in the UQC are indeed detrital in origin. However, simple detrital process of source-rock pyrite with $\delta^{34}S$ values of 0‰ cannot explain the asymmetrical distribution of pyrite $\delta^{34}S$ values (i.e., skewed toward heavier values) shown in Figure 5A. Furthermore, observation of (1) overgrowth textures for pyrite grains, (2) irregularly shaped anhedral pyrite,

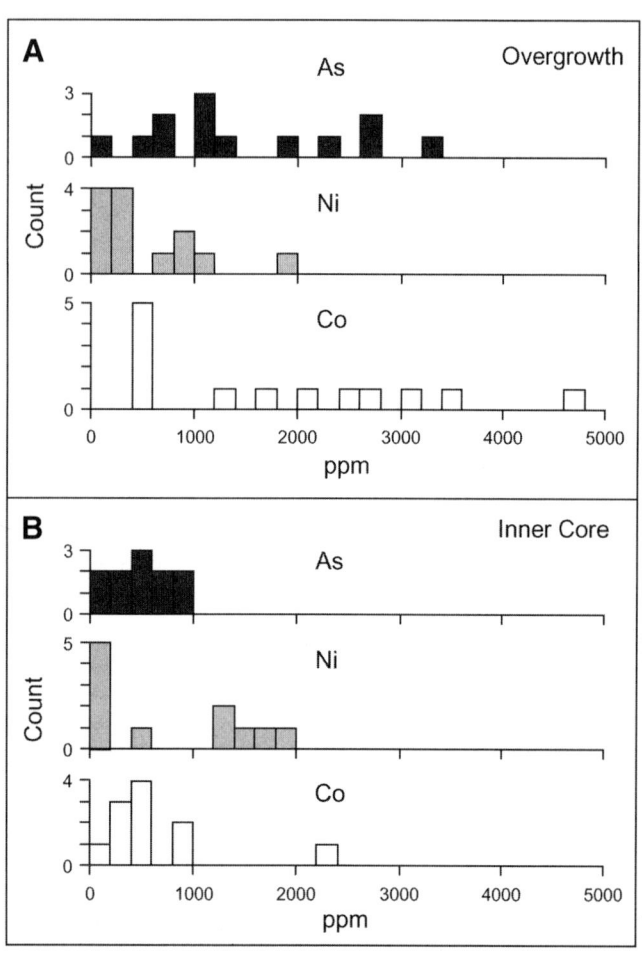

Figure 6. Histogram of trace element (As, Ni, and Co) concentrations of pyrite grains with overgrowth texture in the Paleoproterozoic UQC, Matinenda Formation, Elliot Lake Group, Ontario, Canada. Elemental concentration was measured by EPMA. (A) Overgrowth part of pyrite grains. (B) Inner core part of pyrite grains.

TABLE 3. CONCENTRATION OF TRACE ELEMENTS (As, Ni, AND Co) IN PYRITE GRAINS WITH OVERGROWTH TEXTURES

Sample ID	Overgrowth or core	Concentration (ppm)			Wt. ratio (ppm/ppm)		
		As	Ni	Co	As/Ni	As/Co	Ni/Co
Grain A							
A-1	Overgrowth	1210	730	1230	1.7	1.0	0.6
A-2	Overgrowth	1080	400	1630	2.7	0.7	0.2
A-3	Overgrowth	3390	810	2790	4.2	1.2	0.3
A-4	Overgrowth	2670	930	4660	2.9	0.6	0.2
A-5	Overgrowth	1030	130	500	7.9	2.1	0.3
A-6	Overgrowth	2660	1930	2420	1.4	1.1	0.8
A-7	Overgrowth	780	540	2150	1.4	0.4	0.3
A-8	Overgrowth	540	310	490	1.7	1.1	0.6
A-9	Overgrowth	680	310	410	2.2	1.7	0.8
A-10	Core	340	1540	580	0.2	0.6	2.7
A-11	Core	680	1280	2290	0.5	0.3	0.6
A-12	Core	440	1330	890	0.3	0.5	1.5
A-13	Core	730	1650	490	0.4	1.5	3.4
A-14	Core	600	1190	830	0.5	0.7	1.4
A-15	Core	600	1820	390	0.3	1.5	4.7
Grain B							
B-1	Overgrowth	2250	100	3030	22.5	0.7	0.0
B-2	Core	980	110	520	8.9	1.9	0.2
B-3	Core	340	180	360	1.9	0.9	0.5
Grain C							
C-1	Overgrowth	180	330	3450	0.5	0.1	0.1
C-2	Overgrowth	1120	b.d.	550	-	2.0	-
C-3	Overgrowth	1980	160	520	12.4	3.8	0.3
C-4	Core	110	110	410	1.0	0.3	0.3
C-5	Core	910	120	140	7.6	6.5	0.9
C-6	Core	b.d.	b.d.	270	-	-	-

Note: b.d.—Below detection (<100 ppm).

and (3) variable grain sizes strongly supports the theory of pyrite formation after deposition of the host conglomerate. Similarly, some authors suggest that textural evidence exists showing some pyrite grains formed by replacement of rounded pebbles of cherts and shales during diagenesis and/or later hydrothermal events (Myers et al., 1993; Phillips and Dong, 1993). These observations are fully consistent with our suggestion that non-detrital pyrite is present in the UQC (Feather, 1981; Simpson and Bowles, 1981; Roscoe, 1969, 1973, 1981).

Activity of sulfate-reducing bacteria (SRB) in anaerobic environments during transportation of sediments in rivers and/or early diagenetic stage of host sediments may be responsible for the observed spread of $\delta^{34}S$ values. During bacterial sulfate reduction (BSR), the lighter S isotope ^{32}S in sulfate is preferentially reduced using OM to form dissolved sulfide, which then reacts with reactive Fe to eventually precipitate pyrite (e.g., Berner, 1984). The source of sulfate could have been oxidative weathering of sulfide minerals in the source rock region, hydration of volcanic SO_2, or oxidation of volcanic H_2S followed by hydration. The $\delta^{34}S$ values of sulfate that formed by oxidation of igneous sulfides would have been ~0‰ because of the small kinetic isotopic effect ($\Delta_{SO_4\text{-}FeS_2} = \delta^{34}S_{FeS_2} - \delta^{34}S_{SO_4}$) (Fry et al., 1988; Ohmoto and Goldhaber, 1997). BSR in closed systems with $\Delta_{SO_4\text{-}FeS_2}$ of –9‰ may explain the result obtained in this study. This magnitude is in good agreement with the proposed $\Delta_{SO_4\text{-}FeS_2}$ value of ~–15‰ to ~–5‰ ca. 2.5 Ga (Kakegawa et al., 1994, 1998; Kakegawa, 1997). However, such $\Delta_{SO_4\text{-}FeS_2}$ is rather small when compared to that for typical SRB in the modern environments (tens of permil; e.g., Kaplan and Rittenburg, 1964). This discrepancy may be attributable to environmental conditions for SRB such as (1) elevated rate of BSR induced by generally warm environments of early Earth (e.g., Ohmoto et al., 1993); (2) abundance of readily available OM for SRB (e.g., Canfield, 2001); and/or (3) low sulfate concentration (e.g., Harrison and Thode, 1958; Boudreau and Westrich, 1984). Remnant OM (algal mat or pyrobitumen) has been found in the matrix of UQC in the Elliot Lake area (e.g., Nagy and Mossman, 1988, 1992; Mossman et al., 1993; Nagy, 1993).

Thermochemical sulfate reduction (TSR) caused by mixing of hydrothermal (reduced and petroleum-bearing) fluids derived from a deeper part of the sedimentary basins and sulfate-bearing oxic groundwater from the surface could have been responsible for the observed spread in the $\delta^{34}S$ values (Kiyosu, 1980; Ohmoto and Lasaga, 1982; Ohmoto and Goldhaber, 1997). Upon reaction with OM or petroleum, which migrated as deep fluids at elevated temperature, dissolved sulfate formed by dissolution of evaporite minerals or oxidation of sulfide minerals (or connate seawater) was thermochemically reduced to form pyrite. This process is virtually the same as for groundwater-type U ores of younger ages (e.g., Gauthier-Lafaye and Weber, 1989). The petroleum may have been ubiquitous in Archean and Paleoproterozoic time (Dutkiewicz et al., 1998; Buick et al., 1998). A close association of pyrite with sericite may suggest that both formed by high-temperature hydrothermal fluids, which may or may not be related to the mild metamorphism that occurred during the Penokean orogeny ca. 1.7 Ga (Roscoe, 1969; Willingham et al., 1985; Bennett et al., 1991) in the Elliot Lake district or to impact-induced metamorphism that occurred during the Sudbury event ca. 1.8 Ga (e.g., Krough et al., 1984). Post-depositional fluids passed through the Matinenda Formation during diagenesis and sub-greenschist facies metamorphism at estimated temperatures of 80–200 °C and 280–350 °C, respectively (Dutkiewicz et al., 2003). These temperature windows overlap with those that are permissive of TSR (e.g., Ohmoto and Goldhaber, 1997; Machel, 2001). TSR may explain the formational mechanism of pyrite, especially pyrite with heavier $\delta^{34}S$ values (Ohmoto and Goldhaber, 1997).

Origins of Ni, Co, and As in Pyrite Crystals

Cobalt and nickel have been the most widely applied chemical indicators for the environment of pyrite formation (Fleischer, 1955; Loftus-Hills and Solomon, 1967). Volcanogenic-hydrothermal pyrite is generally characterized by a high Co content with a low Ni/Co (<1) ratio (Hawley and Nichol, 1961; Roscoe, 1965; Kimberley et al., 1980). Nickel is more abundant in magmatic pyrite (Hawley and Nichol, 1961) and in sedimentary (syngenetic/diagenetic) pyrite than in volcanogenic-hydrothermal pyrite (Loftus-Hills and Solomon, 1967). Kimberley et al. (1980) found that Co is substantially more abundant than Ni in all UQC samples. The Ni/Co ratios of Roscoe (1969) range from 0.3 to 1.0, whereas the Ni/Co ratios of Kimberley et al. (1980) range from 0.0 to 0.2. These low Ni/Co ratios are comparable to those for overgrowth pyrite of the present study, suggesting a volcanogenic-hydrothermal origin. However, some core pyrite also has a low Ni/Co ratio (<1), which is comparable to overgrowth pyrite. Pyrite in the Archean Samreid Lake VMS (volcanogenic massive sulfides) deposit near Elliot Lake displays Ni/Co ratios similar to those in the Paleoproterozoic UQC at Elliot Lake (Friedman, 1959). This suggests that pyrite in the UQC was derived (i.e., detrital) from such Archean volcanic pyrite and/or that the process of pyrite formation in such Archean volcanogenic-hydrothermal pyrite and in the UQC were the same.

Compared with Co and Ni, As is less often used as a chemical indicator of the environment of pyrite formation in the UQC in the Elliot Lake area. However, the As content of pyrite in this study shows interesting characteristics. Arsenic is relatively enriched in the overgrown part of pyrite (generally >1000 ppm for overgrowth and <1000 ppm for core). Previous studies (Gendron et al., 1986; Sullivan and Aller, 1996) have proposed that in the early diagenesis of modern marine sediments, As and Co were mobilized in association with the (reductive) dissolution of Fe- and Mn-hydroxide in a reducing environment. An enrichment of As in the overgrowth part of pyrite grains can be explained by the involvement of such an As-bearing reducing fluid, possibly derived from a deeper part of the sedimentary basin. Such process of fluid migration, originally proposed to explain the observed characteristics of the S isotopic compositions of pyrite in the UQC, may also explain the observed characteristics of their trace element compositions.

Models for the Formation of Pyrite in the Uranium Ore at Elliot Lake

Mechanical grinding during physical weathering and transport in rivers must have been responsible for cracking / rounding pyrite grains under an O_2-poor atmosphere (Fig. 7B) (e.g., Holland, 1984), where most pyrite and uraninite are resistant to chemical weathering and therefore survive during weathering and transportation in rivers (Fig. 7C1). Under an O_2-rich atmosphere, the pyrite grains would have been completely decomposed to form sulfate in solution and Fe-oxides, such as they are today (Fig. 7C2). In a special environment where uplift, erosion, transport, and burial are rapid, pyrite may survive as detrital grains even under an O_2-rich atmosphere (Robertson, 1968, 1981, 1986; Maynard et al., 1991). However, considering the existing sedimentological data on the lower Huronian Supergroup, which favor rather low-energy fluvial environments (e.g., Fralick and Miall, 1989) and intense weathering (e.g., Maynard et al. 1991), detrital pyrite, if any, would not have survived if the sediments were overlain by an O_2-rich atmosphere.

In subsequent early diagenetic stages (Figs. 7D1, 7D2), diagenetic pyrite could have formed by BSR as overgrowths on pre-existing pyrite grains or as newly formed euhedral pyrite. Framboidal pyrite may form and, through later aggregation, result in the formation of porous pyrite. Some of the overgrown parts of pyrite, small isolated non-porous (compact) euhedral/subhedral pyrite crystals, and minor pyrrhotite crystals could have formed by TSR (Fig. 7F). Anhedral pyrite grains also could have formed by connecting pre-existing pyrite grains with newly precipitated pyrite by a thermochemical process. Sulfur isotopic evidence supports the above scenario; small isolated euhedral pyrite crystals, which presumably formed by TSR, tend to have $\delta^{34}S$ values different from 0‰. Trace element characteristics of the overgrowth part of pyrite and of small isolated euhedral pyrite crystals are very similar in that As and Co are enriched. However, the core part of pyrite with overgrowth textures has relatively depleted concentrations of As and Co compared to the overgrowth part.

After the Penokean orogeny and/or Sudbury impact event, the dissolution of pyrite grains may have occurred again (Fig. 7G). Dissolution features of the rim of the overgrown part of pyrite suggest dissolution after the formation of pyrite overgrowth (Figs. 7E, 7G). Some small isolated euhedral pyrite grains also have this dissolution feature at their rim. This could probably have been caused by the infiltration of O_2-bearing groundwater into the conglomerate unit.

CONCLUSIONS AND IMPLICATIONS

The ca. 2.45 Ga pyritic UQC of the Matinenda Formation of the Elliot Lake Group, Huronian Supergroup, was used in this study to investigate the origin of pyrite. A laser-microprobe was used for analysis of the S isotopic compositions of individual pyrite grains and EPMA was used to analyze the trace element compositions of pyrite grains with overgrowth texture. Together

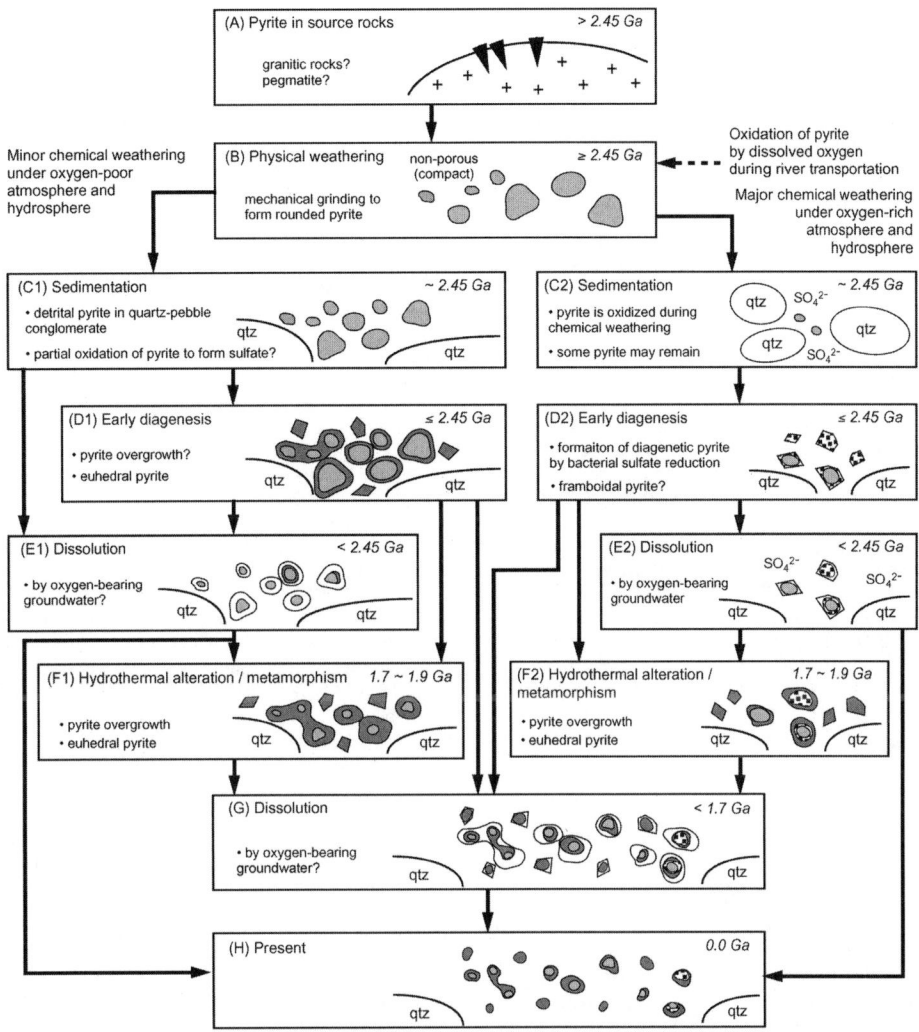

Figure 7. A model for pyrite mineralization in the Paleoproterozoic UQC in the Elliot Lake district, Ontario, Canada. Sizes of the quartz pebbles and pyrite grains are adjusted for the sake of clarity in the figure (in reality, pyrite grains are by far smaller than quartz pebbles). See text for explanation of the stages.

with microscopic examinations of the samples, it is demonstrated that (1) various morphological types of pyrite (euhedral, subhedral, anhedral, rounded, overgrowth) are present; (2) S isotopic compositions of the pyrite range from –9.0‰ to +5.5‰; (3) euhedral pyrite, which formed after the sedimentation of the host rocks, tends to have large negative or positive $\delta^{34}S$ values compared to other pyrites; and (4) trace element (As, Ni, and Co) compositions of overgrowths and cores of pyrite grains are different. These results suggest that the preserved pyrite grains in the Paleoproterozoic UQC samples had been isotopically, chemically, and morphologically overprinted by post-depositional processes, such as early diagenesis and/or hydrothermal alteration.

Preservation of detrital minerals that are labile under oxidative weathering conditions, such as pyrite and uraninite in UQC, has been widely used by many researchers to argue for a reducing atmosphere prior to 2.2 Ga. Recognition of such minerals mostly depends on their morphological features such as shape and size. However, such sedimentological observation has been based on the preserved state of the pyrite and uraninite in the UQC, which is very unlikely to retain the original characteristics at the time of sedimentation of the host rocks. Our results cast doubt on the use of evidence of "detrital" pyrite (and uraninite by inference) in the UQC of the Matinenda Formation for a reducing atmosphere ca. 2.45 Ga. Such an interpretation requires a clear demonstration of morphological, isotopic, and chemical features for truly detrital grains of uraninite and pyrite in the UQC samples that have not been altered by any later process.

ACKNOWLEDGMENTS

We appreciate G. Bennett for his field assistance, D. Walizer and M. Angelone for their technical assistance, and Y. Watanabe and K. Hattori for discussion. Detailed and constructive reviews by M. Goldhaber, B. Maynard, and S. Kesler were very helpful for improving the early manuscript, and therefore greatly appreciated. We thank the Stanleigh Mine for access to the underground mine samples. This study was financially supported by the NASA Astrobiology Institute (NCC2-1057), the NASA Exobiology Program (NAG5-9089), the National Science Foundation (EAR-9706279), and the Japanese Ministry of Education, Culture, Sports, Science, and Technology (MEXT; #07041081) to H.O.

REFERENCES CITED

Barnicoat, A.C., Henderson, I.H.C., Knipe, R.J., Yardley, B.W.D., Napler, R.W., Fox, N.P.C., Kenyon, A.K., Muntingh, D.J., Strydom, D., Winskler, K.S., Lawrence, S.R., and Cornford, C., 1997, Hydrothermal gold mineralization in the Witwatersrand Basin: Nature, v. 386, p. 820–824, doi: 10.1038/386820a0.

Bennett, G., Dressler, B.O., Robertson, J.A., 1991, The Huronian Supergroup and associated intrusive rocks: Geological Survey of Ontario Special Volume 4-I.

Berner, R.A., 1984, Sedimentary pyrite formation: An update: Geochimica et Cosmochimica Acta, v. 48, p. 605–615, doi: 10.1016/0016-7037(84)90089-9.

Boudreau, B.P., and Westrich, J.T., 1984, The dependence of bacterial sulfate reduction on sulfate concentration in marine sediments: Geochimica et Cosmochimica Acta, v. 48, p. 2503–2516, doi: 10.1016/0016-7037(84)90301-6.

Buick, R., Rasmussen, B., and Krapez, B., 1998, Archean oil: Evidence for extensive hydrocarbon generation and migration 2.5–3.6 Ga: American Association of Petroleum Geologists Bulletin, v. 82, p. 50–69.

Canfield, D.E., 2001, Isotope fractionation by natural populations of sulfate-reducing bacteria: Geochimica et Cosmochimica Acta, v. 65, p. 1117–1124, doi: 10.1016/S0016-7037(00)00584-6.

Cloud, P., 1968, Atmospheric and hydrospheric evolution on the primitive Earth: Science, v. 160, p. 729–736.

Corfu, F., and Andrews, A., 1986, A U-Pb age for mineralized Nipissing diabase, Gowganda, Ontario: Canadian Journal of Earth Sciences, v. 23, p. 107–112.

Crowe, D.E., Valley, J.W., and Baker, K.L., 1990, Micro-analysis of sulfur-isotope ratios and zonation by laser microprobe: Geochimica et Cosmochimica Acta, v. 54, p. 2075–2092, doi: 10.1016/0016-7037(90)90272-M.

Davidson, C.F., 1957, On the occurrence of uranium in ancient conglomerates: Economic Geology and the Bulletin of the Society of Economic Geologists, v. 52, p. 668–693.

Dimroth, E., and Kimberley, M.M., 1976, Precambrian atmospheric oxygen: Evidence in the sedimentary distributions of carbon, sulfur, uranium, and iron: Canadian Journal of Earth Sciences, v. 13, p. 1161–1185.

Dutkiewicz, A., Rasmussen, B., and Buick, R., 1998, Oil preserved in fluid inclusions in Archean sandstones: Nature, v. 395, p. 885–888, doi: 10.1038/27644.

Dutkiewicz, A., Ridley, J., and Buick, R., 2003, Oil-bearing CO_2-CH_4-H_2O fluid inclusions; oil survival since the Paleoproterozoic after high temperature entrapment: Chemical Geology, v. 194, p. 51–79, doi: 10.1016/S0009-2541(02)00271-1.

Eldridge, C.S., Phillips, G.N., and Myers, R.E., 1993, Sulfides in the Witwatersrand Goldfields: New perspectives on gold sediments via SHRIMP: Geological Society of America Abstracts with Programs, v. 25, p. A278–279.

Farrow, C.E., and Mossman, D.J., 1988, Geology of Precambrian paleosols at the base of the Huronian Supergroup, Elliot Lake, Ontario, Canada: Precambrian Research, v. 42, p. 107–139, doi: 10.1016/0301-9268(88)90013-7.

Feather, C.E., 1981, Some aspects of Witwatersrand mineralization, with special reference to uranium minerals: U.S. Geological Survey Professional Paper 1161, p. Q1–Q23.

Fleischer, M., 1955, Minor elements in some sulfide minerals, in Bateman, A.M., ed., Economic Geology, 50th Anniversary Volume, Part 2, p. 970–1024.

Fralick, P.W., and Miall, A.D., 1989, Sedimentology of the Lower Huronian Supergroup (Early Proterozoic), Elliot Lake area, Ontario, Canada: Sedimentary Geology, v. 63, p. 127–153, doi: 10.1016/0037-0738(89)90075-4.

Friedman, G.M., 1959, The Samreid Lake sulfide deposit, Ontario, an example of a pyrrhotite-pyrite iron formation: Economic Geology and the Bulletin of the Society of Economic Geologists, v. 54, p. 265–284.

Frimmel, H.E., Le Roex, A.P., Knight, J., and Minter, W.E.L., 1993, A case study of the post-depositional alteration of the Witwatersrand Basal Reef gold placer: Economic Geology and the Bulletin of the Society of Economic Geologists, v. 88, p. 249–265.

Fry, B., Ruf, W., Gest, H., and Hayes, J.M., 1988, Sulfur isotope effects associated with oxidation of sulfides by O_2 in aqueous solution: Chemical Geology, v. 73, p. 205–210.

Gauthier-Lafaye, F., and Weber, F., 1989, The Francevillian (Lower Proterozoic) uranium ore deposits of Gabon: Economic Geology and the Bulletin of the Society of Economic Geologists, v. 84, p. 2267–2285.

Gendron, A., Silverberg, N., Sundby, B., and Lebel, J., 1986, Early diagenesis of cadmium and cobalt in sediments of the Laurentian Trough: Geochimica et Cosmochimica Acta, v. 50, p. 741–747, doi: 10.1016/0016-7037(86)90350-9.

Grandstaff, D.E., 1974, Microprobe analyses of uranium and thorium in uraninite from the Witwatersrand, South Africa, and Blind River, Ontario, Canada: Transactions of the Geological Society of South Africa, v. 77, p. 291–296.

Grandstaff, D.E., 1980, Origin of uraniferous conglomerate at Elliot Lake, Canada, and Witwatersrand, South Africa: Implications for oxygen in the Precambrian atmosphere: Precambrian Research, v. 13, p. 1–26, doi: 10.1016/0301-9268(80)90056-X.

Harrison, A.G., and Thode, H.G., 1958, Mechanism of the bacterial reduction of sulfate from isotope fractionation studies: Transactions of the Faraday Society, v. 54, p. 84–92, doi: 10.1039/tf9585400084.

Hattori, K., Campbell, F.A., and Krouse, H.R., 1983a, Sulphur isotope abundances in Aphebian clastic rocks: Implications for the coeval atmosphere: Nature, v. 302, p. 323–326, doi: 10.1038/302323a0.

Hattori, K., Krouse, H.R., and Campbell, F.A., 1983b, The start of sulfur oxidation in continental environments: About 2.2×10^9 years ago: Science, v. 221, p. 549–551.

Hattori, K., Campbell, F.A., and Krouse, H.R., 1986, Sulphur isotope abundances in sedimentary rocks; relevance to the evolution of the Precambrian atmosphere: Geochemistry International, v. 22, p. 97–114.

Hawley, J.E., and Nichol, I., 1961, Trace elements in pyrite, pyrrhotite, and chalcopyrite of different ores: Economic Geology and the Bulletin of the Society of Economic Geologists, v. 56, p. 467–487.

Holland, H.D., 1984, Chemical evolution of the atmosphere and ocean: Princeton, New Jersey, Princeton University Press, 582 p.

Holland, H.D., 1994, Proterozoic atmospheric change, in Bengtson, S., ed., Early life on Earth: Nobel Symposium 84: New York, Columbia University Press, p. 237–244.

Holmes, S.W., 1957, Pronto Mine, in Gilbert, G., ed., Structural geology of Canadian ore deposits, Volume 2: Montreal, Canadian Institute of Mining and Metallurgy, p. 324–339.

Hu, Q., Evensen, N.M., Smith, P.E., and York, D., 1998, A world in a grain of sand; regional metamorphic history from $^{40}Ar/^{39}Ar$ laser probe analyses of Proterozoic sediments from the Canadian Shield: Precambrian Research, v. 91, p. 287–294, doi: 10.1016/S0301-9268(98)00054-0.

Kakegawa, T., 1997, Sulfur isotope geochemistry of Archean sedimentary rocks [Ph.D. dissertation]: University Park, Pennsylvania State University, 163 p.

Kakegawa, T., Kawai, H., and Ohmoto, H., 1998, Origins of pyrites in the ~2.5 Ga Mt. McRae Shale, the Hamersley District, Western Australia: Geochimica et Cosmochimica Acta, v. 62, p. 3205–3220, doi: 10.1016/S0016-7037(98)00229-4.

Kakegawa, T., Ohmoto, H., Kasahara, Y., Kawai, H., and Hayashi, K., 1994, Biogenic and hydrothermal pyrites in Archean shales from the Pilbara-Hamersley district, Western Australia: Geological Society of America, Abstracts with Programs, v. 26, p. A353–A354.

Kaplan, I.R., and Rittenburg, S.C., 1964, Microbiological fractionation of sulphur isotopes: Journal of General Microbiology, v. 34, p. 195–212.

Kasting, J.F., 1993, Earth's early atmosphere: Science, v. 259, p. 920–926.

Kelley, S.P., and Fallick, A.E., 1990, High precision spatially resolved analysis of $\delta^{34}S$ in sulfides using a laser extraction technique: Geochimica et Cosmochimica Acta, v. 54, p. 883–888, doi: 10.1016/0016-7037(90)90381-T.

Kimberley, M.M., Tanaka, R.T., and Farr, M.R., 1980, Composition of middle Precambrian uraniferous conglomerate in the Elliot Lake–Agnes Lake area of Canada: Precambrian Research, v. 12, p. 375–392, doi: 10.1016/0301-9268(80)90036-4.

Kiyosu, Y., 1980, Chemical reduction and sulfur-isotope effects of sulfate by organic matter under hydrothermal conditions: Chemical Geology, v. 30, p. 47–56, doi: 10.1016/0009-2541(80)90115-1.

Krough, T.E., Davis, D.W., and Corfu, F., 1984, Precise U-Pb zircon and baddeleyite ages for the Sudbury area, in Pye, E.G., Naldrett, A.J., and Gilbin, P.E., eds. The geology and ore deposits of the Sudbury structure: Geological Survey of Ontario Special Volume 1, p. 431–446.

Loftus-Hills, G., and Solomon, M., 1967, Cobalt, nickel and selenium in sulphides as indicators of ore genesis: Mineralium Deposita, v. 2, p. 228–242, doi: 10.1007/BF00201918.

Machel, H.G., 2001, Bacterial and thermochemical sulfate reduction in diagenetic settings—Old and new insights: Sedimentary Geology, v. 140, p. 143–175, doi: 10.1016/S0037-0738(00)00176-7.

Maynard, J.B., Rigger, S.D., and Sutton, S.J., 1991, Chemistry of sands from the modern Indus River and the Archean Witwatersrand basin: Implications for the composition of the Archean atmosphere: Geology, v. 19, p. 265–268, doi: 10.1130/0091-7613(1991)019<0265:COSFTM>2.3.CO;2.

Meyer, F.M., Robb, L.J., Oberthür, T., Saager, R., and Stupp, H.D., 1990, Cobalt, nickel, and gold in pyrite from primary gold deposits and Witwatersrand reefs: South African Journal of Geology, v. 93, p. 70–82.

Minter, W.E.L., 1976, Detrital gold, uranium, and pyrite concentrations related to sedimentology in the Precambrian Vaal Reef Placer, Witwatersrand, South Africa: Economic Geology and the Bulletin of the Society of Economic Geologists, v. 71, p. 157–176.

Mossman, D.J., and Farrow, C.E.G., 1992, Paleosol and ore-forming processes in the Elliot Lake district of Canada, in Schidlowski, M., Golubic, S., Kimberley, M.M., McKirdy, D.M., and Trudinger, P.A., eds. Early organic evolution: Implications for mineral and energy resources: Berlin, Springer-Verlag, p. 67–75.

Mossman, D.J., Goodarzi, F., and Gentzis, T., 1993, Characterization of insoluble organic matter from the lower Proterozoic Huronian Supergroup, Elliot Lake, Ontario: Precambrian Research, v. 61, p. 279–293, doi: 10.1016/0301-9268(93)90117-K.

Myers, B., 1981, Genesis of uranium-gold pyritic conglomerates: U.S. Geological Survey Special Paper 1161 p. AA1–AA26.

Myers, R.E., Zhou, T., and Phillips, G.N., 1993, Sulphidation in the Witwatersrand Goldfields: Evidence from the Middelvlei Reef: Mineralogical Magazine, v. 57, p. 395–405.

Nagy, B., 1993, Kerogens and bitumens in Precambrian uraniferous ore deposits: Witwatersrand, South Africa, Elliot Lake, Canada, and the natural fission reactors, Oklo, Gabon, in Parnell, J., Kucha, H., and Landis, P., eds., Bitumens in ore deposits: Berlin, Springer-Verlag, p. 286–333.

Nagy, B., and Mossman, D.J., 1988, The nature and origin of kerogens in the lower Huronian Supergroup, Elliot Lake region, Ontario, Canada: Terra Cognita, v. 8, p. 219.

Nagy, B., and Mossman, D.J., 1992, Stratiform and globular organic matter in the Lower Proterozoic metasediments at Elliot Lake, Ontario, Canada, in Schidlowski, M., Golubic, S., Kimberley, M.M., McKirdy, D.M., and Trudinger, P.A., eds., Early organic evolution: Implications for mineral and energy resources: Berlin, Springer-Verlag, p. 224–231.

Nesbitt, H.W., and Young, G.M., 1982, Early Proterozoic climates and plate motions inferred from major element chemistry of lutites: Nature, v. 299, p. 715–717, doi: 10.1038/299715a0.

Ohmoto, H., and Lasaga, A.C., 1982, Kinetics of reactions between aqueous sulfates and sulfides in hydrothermal systems: Geochimica et Cosmochimica Acta, v. 46, p. 1727–1745, doi: 10.1016/0016-7037(82)90113-2.

Ohmoto, H., Kakegawa, T., and Lowe, D.R., 1993, 3.4-billion-year-old biogenic pyrites from Barberton, South Africa: Sulfur isotope evidence: Science, v. 262, p. 555–557.

Ohmoto, H., and Goldhaber, M.B., 1997, Sulfur and carbon isotopes, in Barnes, H.L., ed., Geochemistry of hydrothermal ore deposits (3rd edition): New York, John Wiley and Sons, p. 517–611.

Phillips, G.N., and Dong, G., 1993, Chert-plus-pyrite pebbles in the Witwatersrand Goldfield: International Geology Review, v. 36, p. 65–71.

Phillips, G.N., and Myers, R.E., 1981, The Witwatersrand gold field, Part II: An origin for Witwatersrand gold during metamorphism and associated alteration: Economic Geology Monograph 6, p. 311–332.

Phillips, G.N., Myers, R.E., and Palmer, J.A., 1987, Problems with the placer model for Witwatersrand gold: Geology, v. 15, p. 1027–1030, doi: 10.1130/0091-7613(1987)15<1027:PWTPMF>2.0.CO;2.

Pienaar, P.J., 1963, Stratigraphy, petrography, and genesis of the Elliot Lake Group, Blind River, Ontario, including the uraniferous conglomerate: Geological Survey of Canada Bulletin, v. 83, p. 140.

Pretorius, D.A., 1981, Gold and uranium in quartz-pebble conglomerate: Economic Geology, 75th Anniversary Volume, p. 117–138.

Ramdohr, P., 1958, New observations on the ores of the Witwatersrand in South Africa and its genetic significance: Transactions of the Geological Society of South Africa, Annex, v. 61, p. 50.

Ramdohr, P., 1969, The ore minerals and their intergrowths. Oxford, UK, Pergamon Press, 1174 p.

Robb, L.J., and Meyer, F.M., 1991, A contribution to recent debate concerning epigenetic versus syngenetic mineralization processes in the Witwatersrand Basin: Economic Geology and the Bulletin of the Society of Economic Geologists, v. 86, p. 396–401.

Robb, L.J., and Meyer, F.M., 1995, The Witwatersrand Basin, South Africa: Geological framework and mineralization processes: Ore Geology Reviews, v. 10, p. 67–94, doi: 10.1016/0169-1368(95)00011-9.

Robertson, D.S., and Steenland, N.C., 1960, On the Blind River uranium ores and their origin: Economic Geology and the Bulletin of the Society of Economic Geologists, v. 55, p. 659–694.

Robertson, J.A., 1968, Geology of Township 149 and Township 150, District of Algoma: Ontario Department of Mines, Geological Report, v. 57, p. 162.

Robertson, J.A., 1981, The Blind River uranium deposits: The ores and their setting: U.S. Geological Survey Professional Paper 1161-A-BB, p. U1–U23.

Robertson, J.A., 1986, Huronian geology and the Blind River (Elliot Lake) uranium deposits, 7–31, the Pronto Mine, in Evans, E.L., ed., Uranium deposits of Canada: Canadian Institute of Mining and Metallurgy Special Paper 33, p. 36–43.

Robinson, A., and Spooner, T.C., 1982, Source of the detrital components of uraniferous conglomerates, Quirke ore zone, Elliot Lake Ontario, Canada: Nature, v. 299, p. 622–624, doi: 10.1038/299622a0.

Robinson, A., and Spooner, T.C., 1984a, Postdepositional modification of uraninite-bearing quartz-pebble conglomerate from the Quirke ore zone, Elliot Lake, Ontario, Canada: Economic Geology and the Bulletin of the Society of Economic Geologists, v. 79, p. 297–321.

Robinson, A., and Spooner, T.C., 1984b, Can the Elliot Lake uranium-bearing quartz-pebble conglomerate be used to place limits on the oxygen content of the early Proterozoic atmosphere?: Journal of Geology, v. 141, p. 221–228.

Roscoe, S.M., 1965, Geochemical and isotopic studies, Noranda and Mattagami area: Canadian Institute of Mining and Metallurgy Bulletin, v. 58, p. 965–971.

Roscoe, S.M., 1969, Huronian rocks and uraniferous conglomerates: Geological Survey of Canada Special Paper 68-40, 205 p.

Roscoe, S.M., 1973, The Huronian Supergroup, a Paleoaphebian succession showing evidence of atmosphere evolution, in Young, G.M., ed., Huronian stratigraphy and sedimentation: Geological Association of Canada Special Paper 12, p. 31–38.

Roscoe, S.M., 1981, Temporal and other factors affecting deposition of uraniferous conglomerate: U.S. Geological Survey Professional Paper 1161-A-BB, p. W1–W17.

Ruzicka, V., 1981, Some metallogenic features of the Huronian and post-Huronian uraniferous conglomerates: U.S. Geological Survey Professional Paper 1161-A-BB, p. V1–V8.

Ruzicka, V., 1988, Geology and genesis of uranium deposits in the early Proterozoic: Blind River-Elliot Lake basin, Ontario, Canada, in Recognition of uranium deposits: Proceedings of a Technical Committee Meeting, London, 18–20 September 1985: Vienna, Austria, International Atomic Energy Agency, p. 107–130.

Ruzicka, V., 1989, Conceptual genetic models for important types of uranium deposits and areas favorable for their occurrences in Canada, in Uranium resources and geology of North America: Vienna, Austria, International Atomic Energy Agency, TECDOX 500, p. 49–79.

Saager, R., 1981, Geochemical studies on the origin of the detrital pyrites in the conglomerates of the Witwatersrand Goldfields, South Africa: U.S. Geological Survey Professional Paper 1161, p. L1–L17.

Simpson, P.R., and Bowles, J.F.W., 1981, Detrital uraninite and pyrite: Are they evidence for a reducing atmosphere?: U.S. Geological Survey Professional Paper 1161, p. S1–S12.

Sullivan, K.A., and Aller, R.C., 1996, Diagenetic cycling of arsenic in Amazon shelf sediments: Geochimica et Cosmochimica Acta, v. 60, p. 1465–1477, doi: 10.1016/0016-7037(96)00040-3.

Theis, N.J., 1979, Uranium-bearing and associate minerals and their geochemical and sedimentological context, Elliot Lake, Ontario: Geological Survey of Canada Bulletin, v. 304, p. 50.

Willingham, T.O., Nagy, B., Nagy, L.A., Krinsley, D.H., and Mossman, D.J., 1985, Uranium-bearing stratiform organic matter in paleoplacers of the lower Huronian Supergroup, Elliot Lake–Blind River region, Canada: Canadian Journal of Earth Sciences, v. 22, p. 1930–1944.

Young, G.M., 1991, Stratigraphy, sedimentology and tectonic setting of the Huronian Supergroup: Geological Association of Canada, Mineralogical Association of Canada, Society of Economic Geologists, Joint Annual Meeting, Toronto, 1991, Field Trip Guidebook B5: Toronto, Geological Association of Canada, 34 p.

Manuscript Accepted by the Society 29 October 2005

Geological Society of America
Memoir 198
2006

Time constraint for the occurrence of uranium deposits and natural nuclear fission reactors in the Paleoproterozoic Franceville Basin (Gabon)

F. Gauthier-Lafaye*

Centre de géochimie de la surface (CGS), UMR 7517, 1 rue Blessig, 67084 Strasbourg Cedex, France

ABSTRACT

Natural fission reactors at the Oklo uranium deposits in Gabon appear to have formed in a short interval of geologic time during which uranium could migrate to form deposits and the $^{235}U/^{238}U$ ratio was still high enough to trigger fission reactions. At the time of sediment deposition in the ore-hosting Franceville Basin ~2100 m.y. ago, the oxygen deficient atmosphere would have inhibited uranium dissolution and therefore its migration to form deposits. Dissolution and migration of uranium probably began only during later diagenesis after ca. 2050 Ma, and local reduction reactions in the presence of hydrocarbons allowed formation of high-grade uranium deposits. At this time the $^{235}U/^{238}U$ ratio was still significantly higher than it is today, thus triggering nuclear fission reactions. Before 2.0 Ga, the $^{235}U/^{238}U$ ratio was also high enough to allow fission reactions but no mechanisms were able to produce high-grade uranium ores. Thus, oxygen in the atmosphere was probably the main factor controlling the occurrence of natural nuclear fission reactions. This conclusion is in agreement with earlier suggestions that oxygen contents in atmosphere increased during a "transition phase" some 2450–2100 m.y. ago.

Keywords: natural fission reactors, Paleoproterozoic, uranium deposit, atmosphere.

INTRODUCTION

Oklo and other high-grade uranium deposits in the Paleoproterozoic Franceville sedimentary basin in Gabon (Fig. 1) are the only deposits known to have acted as natural nuclear fission reactors. The Franceville Basin also contains the Moanda Mn deposit, the second largest world reserve of manganese. This paper demonstrates that critical behavior of the uranium deposits in the Franceville Basin could have been achieved only during a very specific period of time corresponding to the initial oxygenation of Earth's atmosphere. It is generally admitted that during early Archean, oxygen produced by photosynthetic organisms induced Fe^{2+} oxidation and precipitation in iron deposits (BIFs). At this time oxidation/precipitation of Mn^{2+} was inhibited by the presence of dissolved Fe^{2+} (Roy, 2000) and therefore Mn accumulated in seawater. During late Archean and Paleoproterozoic, the balance of Fe supplied by oceanic hot springs and O_2 from photosynthetic production was altered, and the Fe^{2+} concentration in seawater drastically decreased. Then, oxygen was consumed by oxidation and precipitation of Mn^{2+} and, finally, the concentration of O_2 in the atmosphere started to increase when all Mn in seawater precipitated. Afterward pO_2 in the atmosphere allowed the oxidation of other elements such as uranium, its remobilization and precipitation, and its concentration in high-grade deposits by redox reactions.

---

*fgl@illite.u-strasbg.fr

Gauthier-Lafaye, F., 2006, Time constraint for the occurrence of uranium deposits and natural nuclear fission reactors in the Paleoproterozoic Franceville Basin (Gabon), *in* Kesler, S.E., and Ohmoto, H., eds., Evolution of Early Earth's Atmosphere, Hydrosphere, and Biosphere—Constraints from Ore Deposits: Geological Society of America Memoir 198, p. 157–167, doi: 10.1130/2006.1198(09). For permission to copy, contact editing@geosociety.org. ©2006 Geological Society of America. All rights reserved.

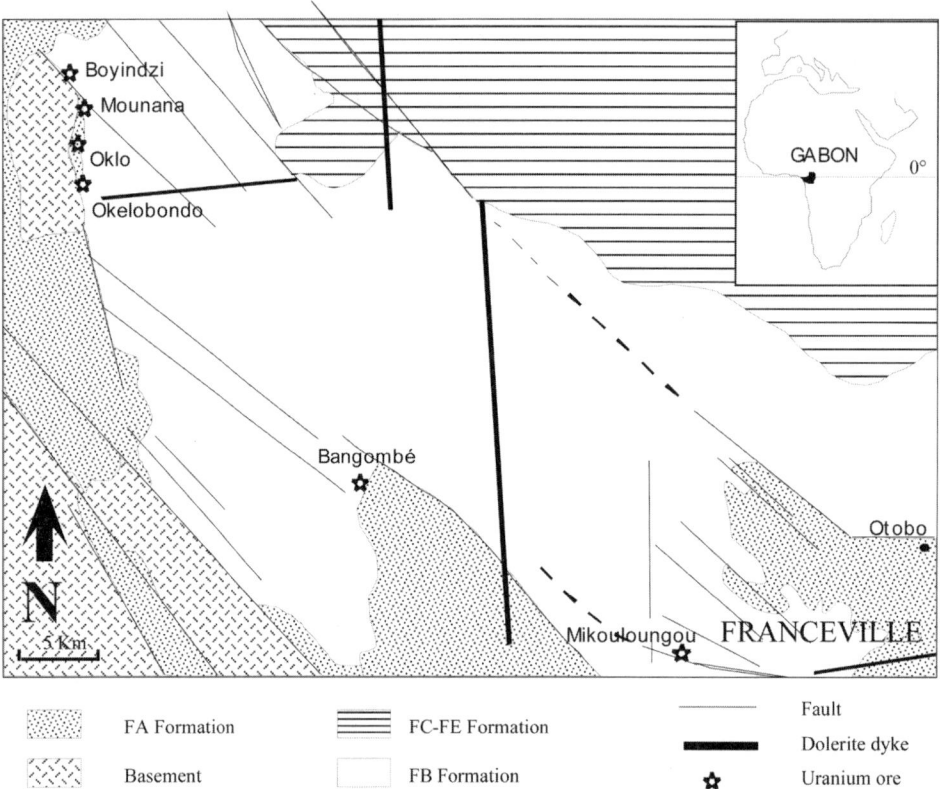

Figure 1. Location of the Oklo and related deposits in Africa (inset) showing their relation to geological units in the Franceville Basin.

Our data agree with a model suggesting an increase of oxygen in the atmosphere between 2150 Ma, the time of deposition of the Francevillian black shales devoid of uranium, and 2050 Ma, which corresponds to the period when uranium was mobilized in its oxidized state (Gauthier-Lafaye and Weber, 2003). Holland (1994) estimates that during this period, the pO_2 level in the atmosphere increased rapidly from <1% to ≥15% present atmospheric level.

The possibility that nuclear fission reactions might have occurred in the past was first suggested by Kuroda in 1956 but we had to wait until 1972 to discover that such reactions did indeed occur 1950 m.y. ago in the Oklo uranium deposits of the Francevillian Series, Gabon (Neuilly et al., 1972). Other uranium deposits of Proterozoic age in the world (Maas and McCulloch, 1990) do not manifest the characteristic uranium and rare earth element (REE) isotopic anomalies of natural nuclear fission reactions, suggesting that fission reactions occurred only in the Paleoproterozoic uranium ore deposits of Gabon. This paper points out the main physical and chemical conditions that were necessary to start and to sustain fission reactions in a natural environment and tries to show that some of crucial conditions are related to the age of the uranium deposits.

REGIONAL GEOLOGICAL SETTING AND STRATIGRAPHY

The unmetamorphosed Francevillian Series is exposed in the Plateau des Abeilles, Franceville, and Lastoursville intracratonic basins. These basins represent the continental platform of the more distal Okondja Basin (Weber, 1968; Ledru et al., 1989). The Gabonese uranium deposits are all located in the Franceville Basin and only this basin will be described here. The stratigraphic column of the Franceville Basin has been subdivided by Weber (1968) into five formations, which are from bottom to top the FA through the FE Formations (Fig. 2)

FA Formation

This formation, which contains all the Francevillian uranium deposits, increases from 100 to 1000 m in thickness from the edge to the central part of the basin. It consists of fluviatile conglomerates and coarse to medium-grained, poorly sorted sandstones overlain by marine well-sorted sandstones that were deposited in tidal to supratidal environments. The FA sandstones of the Franceville Basin are very mature, suggesting that they were affected by many cycles of erosion and sedimentation. Such

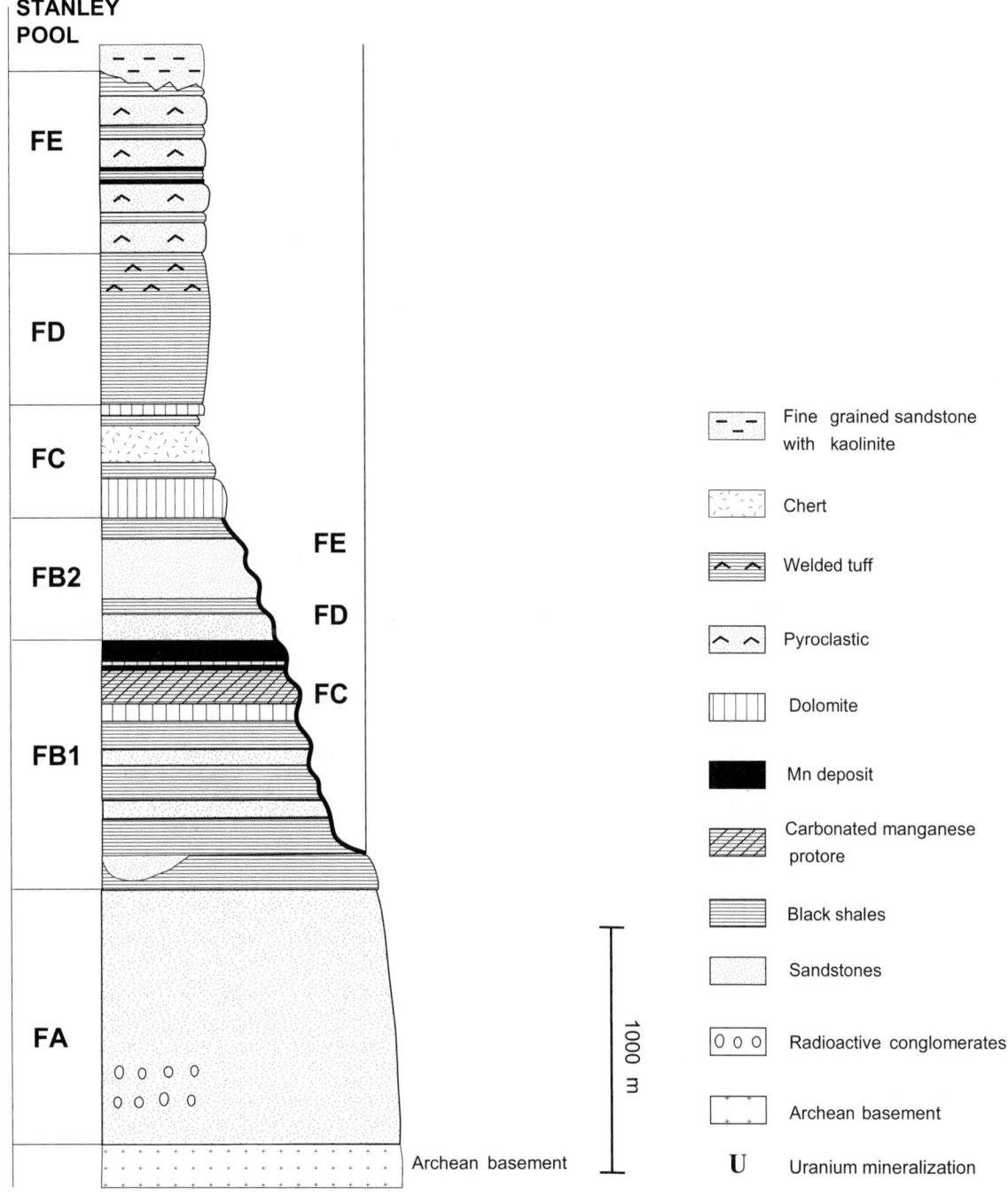

Figure 2. Stratigraphic section for the Franceville Basin showing position of uranium mineralization.

multiple cycles allow the formation of heavy mineral deposits in sedimentological settings such as fluviatile placers. At Oklo, the mineralized sandstones (the uppermost 10 m of the FA Formation) were deposited in a tidal environment of beaches and very wide bars in which coarse-grained, well-sorted sandstones alternate with micaceous, fine-grained sandstones (Gauthier-Lafaye 1986; Gauthier-Lafaye and Weber 1989).

The FA sandstones consist mainly of quartz and muscovite. Variably altered biotite and feldspar (mostly microcline) are abundant in the lower fluviatile sandstones but rare in the overlying marine sandstones. Heavy minerals such as zircon, thorite, and monazite are abundant in conglomerates and coarse-grained sandstones. Well-sorted marine sandstones at the top of the FA Formation have been cemented extensively by an intense silica.

Three sedimentary facies that are distinguished as red-, green-, and black-colored sediments, have been recognized in FA sandstones (Gauthier-Lafaye and Weber, 1989). The contact between sediments of different color is clearly discordant to the bedding, indicating that the coloration is not of sedimentary origin but was acquired during the diagenesis.

The red color is due to hematite impregnation of the argillaceous matrix; hematite also surrounds the quartz grains between their detrital boundaries and later silica overgrowths. The argillaceous matrix consists mainly of illite, and detrital biotites and heavy minerals are highly altered. Detrital monazites are altered to a Th-OH silicate with very low concentrations of U and REE, whereas altered zircons show several growth zones enriched in REE, P, Th, and U (with Th/U = 5–10) (Mathieu et al., 2001). The red-colored sediments also contain dolomite and sulfates (anhydrite, gypsum, and accessory barite) in the fluviatile and tidal environments.

Black sandstones are restricted to the upper parts of the FA Formation. Their color is mainly due to organic matter consisting of solidified petroleum (pyrobitumen) that fills primary and secondary porosity in the sandstone, including fractures (Gauthier-Lafaye and Weber, 1981; Gauthier-Lafaye 1986; Cortial et al., 1990; Nagy et al., 1991, 1993). In the primary porosity, organic matter surrounds detrital quartz grains or is trapped in quartz overgrowths (Fig. 3). Two types of black sandstones can be recognized by the type of alteration of the detrital biotites (Gauthier-Lafaye et al., 1989). The first type contains weakly altered biotites with a composition intermediate between biotite and chlorite. The matrix consists mainly of iron-rich chlorite. In the second type of black sandstone, detrital biotites have been altered to a pale color and resemble muscovite with titanium oxides on their surface. The matrix consists of illite without chlorite and the black color is due to the presence of organic matter in secondary porosity. The first type of black sandstone is interpreted as sediments that were never oxidized, whereas the second type corresponds to sediments that were first oxidized after deposition and then reduced by fluids associated with oil migration.

Green sediments are usually located between the red and the black sandstones. They contain highly altered biotites, which look like "sandwiches" made of flakes of muscovite and green iron-rich chlorite (Gauthier-Lafaye and Weber, 1989). The matrix contains both illite and green, iron-rich chlorite in varying proportions. Pyrite accumulations are common in these green sandstones. Pyrite corrodes both the detrital quartz grains and their overgrowths, and is usually associated with green chlorite. These pyrite accumulations are interpreted as reduced fronts resulting from the flow of reducing fluids through oxidized sediments.

FB Formation

The FB Formation is composed chiefly of black shales that reach a thickness of 400–1000 m. On structural highs, mainly at the edges of the Franceville Basin, these black shales are only few meters thick and grade into dolomites and stromatolite-rich cherts of the FB-C Formation (Fig. 2). The black shales are characterized by concentrations of organic matter (C_{org}) that reach a maximum of 15% (Gauthier-Lafaye and Weber 1989; Cortial 1990). The deposition of these black shales may have occurred during a worldwide black shale event (Shunga event of Melezhik et al., 1999) between ca 1.9 and 2.05 Ga. This event is related to the stromatolite explosion (Semikhatov and Raaben, 1994; Melezhik et al., 1997), and stromatolites are well preserved in the FC cherts. Mn-carbonates are a major component near the top of the formation, where they form the protore of the important manganese deposits of Moanda (Weber, 1968, 1997).

The uranium content of FB black shale samples located in various places in the Franceville Basin near and far from uranium deposits ranges between 3.5 ppm (average of 48 samples *in* Gauthier-Lafaye, 1986) and 10.8 (average of six samples in Mossman et al., 1998). These values are particularly low when compared to those for other black shales compiled by Vine and

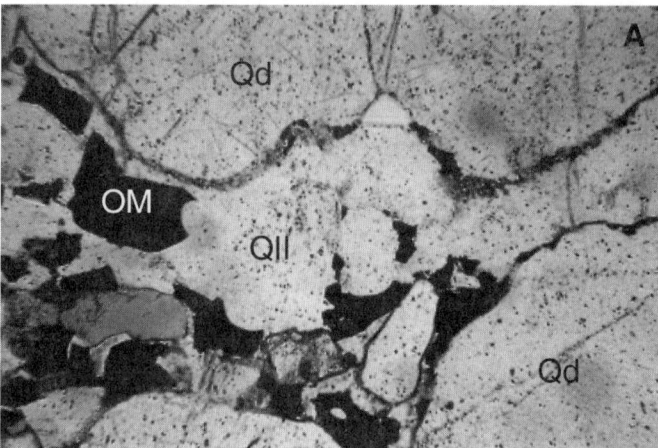

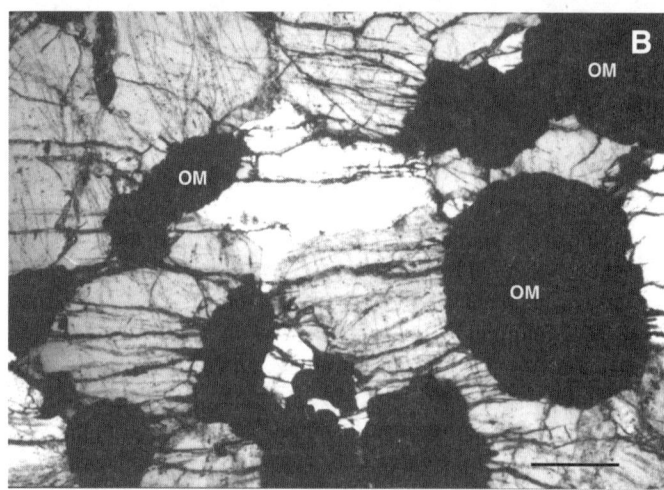

Figure 3. (A) Hydrocarbon trapped in the primary porosity of FA sandstone. Qd—detrital quartz grain; QII—quartz overgrowth; OM—organic matter. (B) Hydrofracturing of a mineralized sandstone. Black nodules are uranium-bearing hydrocarbon.

Tourtelot (1977), which have an average uranium content of 30 ppm, and to the USGS Standard Devonian Oil Shale, which has a uranium content of 48.8 ppm.

FC-FD-FE Formations

The FC Formation is 10–40 m thick and consists chiefly of massive dolomite and thick-banded cherts. Stromatolites, together with organic mats, are well developed in cherts and are described in detail by Bertrand-Sarfati and Potin (1994). The FD Formation consists of black shales, with ignimbrite tuff at the top of the formation. The FE Formation comprises epiclastic sandstones and interlayered shales. The combined FD and FE Formations are more than 1000 m thick in the Franceville Basin, and are probably thicker in the northern part of the Okondja Basin, although this forested area is poorly known.

INTRUSIONS AND DEFORMATION

Syenites and pegmatites of the N'Goutou volcanic complex are interlayered with sediments at the base of FB Formation near the edge of the Okandja Basin. These rocks have an Rb-Sr age of 2143 ± 143 Ma (Bonhomme et al., 1982).

Later dykes of dolerite intrude the Franceville Basin, forming two orthogonal systems oriented NNW-SSE and ENE-WSW. These dolerites were dated by the K-Ar method on feldspathic fractions, giving an age of 970 ± 30Ma (Bonhomme et al., 1982). More recent U-Pb isotopic measurements on zircons from the Oklo dyke yield an age of 860 ± 30 Ma (Evins, 2002).

The Franceville sedimentary succession is only slightly deformed. Maximum deformation occurs in rocks affected by synsedimentary faults activated during a later extensional tectonic episode. This tectonic event was also responsible for hydrofracturing of sandstones (Fig. 3) in the upper part of the FA formation (Gauthier-Lafaye 1986). In the Okondja Basin, deformation was more intense, and the succession is weakly folded. Thereafter, the basin remained remarkably stable until recent uplift resulting in its erosion under the present weathering regime.

DIAGENESIS

The highest diagenetic level reached by the sediments corresponds to the level defined by Winkler (1976) as very low grade metamorphism characterized by crystallization of chlorite and illite, mainly of 1M polytype (Gauthier-Lafaye 1986). 2-M illite is found locally only at the bottom of the FA Formation. In the iron formation, greenalite and stilpnomelane are present; the persistence of greenalite without minnesotaite or grunerite also indicates a minimal metamorphic grade.

Fluid inclusion studies of quartz overgrowths and carbonate cements indicate that the maximum temperature reached in the Franceville Basin during diagenesis was between 180 and 200 °C at a maximum pressure of 1000 ± 200 bars (Openshaw et al., 1978; Gauthier-Lafaye and Weber 1981; Gauthier-Lafaye and Weber 1989; Savary and Pagel 1997; Mathieu 1999). The inclusions contain highly saline brines (28.7 wt.% NaCl eq. to 30 wt.% $CaCl_2$ eq.) that are rich in Li, Br, and SO_4 (Mathieu et al., 2001). Optical and electron microscopic properties and chemical compositions of organic matter have been described in detail by Cortial et al. (1990), Nagy et al. (1993), and Mossman et al. (1993, 1998).

Organic matter of the Francevillian Series has reached high levels of thermal maturity; H/C and O/C ratios are low and the "coke" stage was reached. The true anthracite stage, as defined by using Transmitted Electron Microscopy (TEM), has not been detected, thus suggesting the absence of higher-grade metamorphism (Cortial et al. 1990). The main source for the migrated bitumens has been identified as the black shales of the FB Formation, which were placed below the FA reservoir locally by normal faults. In the upper part of the FA Formation, interlayered shales with very high organic carbon contents might also have served as source rocks. Hydrocarbons were generated when sediments of the FB Formation passed through the oil window (Gauthier-Lafaye, 1986; Mossman, 2001; Mossman et al., 2001).

ISOTOPIC AGES AND EVOLUTION OF THE FRANCEVILLE BASIN

Isotopic Ages

The Francevillian series rests on Archean basement, which is composed dominantly of the granitic Chaillu and North Gabon blocks that were emplaced between 3.9 and 3.2 Ga (Caen-Vachette et al., 1988; Ledru et al., 1989). The Francevillian Series and the geological events that affected the sediments have been dated by several geochronometers (Fig. 4). Deposition of black shales of the FB Formation has been dated by Rb-Sr isotopic analysis of interlayered syenites of the N'Goutou volcanic complex (2143 ± 143 Ma) by Bonhomme et al. (1982). More recently, new U-Pb isotopic analyses on zircons in ignimbrite tuffs from the FE formation at Bidoudouma have been performed (Horie et al., 2004). Results give concordant ages at 2083 ± 6 Ma, which represents the age of deposition of the FD Formation. The time of petroleum generation in the Franceville Basin has been dated by two Sm-Nd isochrons on small authigenic clay fractions of the FB Formation. These give ages of 2099 ± 115 Ma and 2036 ± 79 Ma (Bros et al., 1992; Stille et al., 1993), which are in good agreement with Pb-Pb dating of authigenic clay fractions of the FB black shales (Gauthier-Lafaye et al., 1996a).

The age of the uranium deposit is 2050 ± 30 Ma as determined by classical U-Pb analysis on uraninites (Gancarz, 1978). The date of the fission reactions has been determined with great precision by comparing the fluence of the fission reaction to the amount of fission elements (mainly REE) produced. The result gives an age of 1950 ± 40 Ma (Ruffenach, 1978, 1979; Holliger, 1988; Naudet 1991). Late diagenesis has been dated from clay fractions in the FB Formation using the Rb-Sr method, yielding

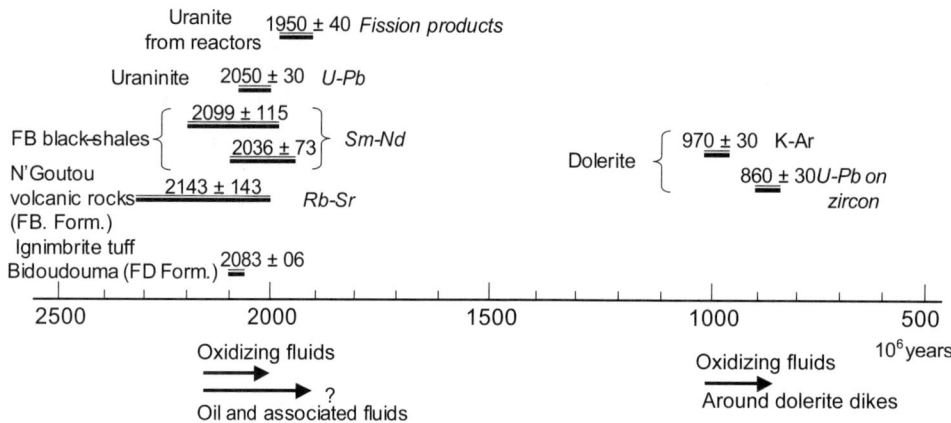

Figure 4. Summary of the main isotopic ages for rocks of the Franceville Basin, the uranium ore deposits, and the natural nuclear fission reactors. All data are referenced in the text.

an age of 1870 ± 50 Ma (Bonhomme et al., 1982). The dolerite dykes were dated by the K-Ar method at 970 Ma ± 30 Ma (Bonhomme et al., 1982). More recently, in situ U-Pb dating of zircons from the Oklo dike yields an age of 860 ± 30 Ma (Evins, 2002). The time of basin uplift is unknown. The only isotopic measurement that is available is a U-Th age of 76,500 ± 6800 yr obtained on secondary torbernite related to the weathering of the Bangombé uranium deposit (Bros et al., 2000).

Evolution of the Franceville Basin

Between the time of deposition of the sediments and the uplift of the Franceville Basin, two main periods can be recognized (Fig. 5). The first period started after deposition of the sediments, continued during early diagenesis, and ended with the extensional tectonic event noted above. During this time, fluids in the FA sandstones and FB black shales escaped from the reservoirs due to compaction of sediments. The chemical composition and oxidation state of the fluids that first migrated into the FA sandstones are not known, but later on, as burial progressed, fluids were probably oxidizing as recorded by oxidation affecting most of the FA sediments and the basal FB black shales. This oxidation, which occurred at the scale of the overall basin, implies recharge by meteoric waters. As burial of the series increased, organic matter in the FB black shales (and black shales interlayered in the tidal sediments of the FA Formation) underwent maturation and reached the "oil window," producing gas and liquid hydrocarbons. Petroleum then migrated

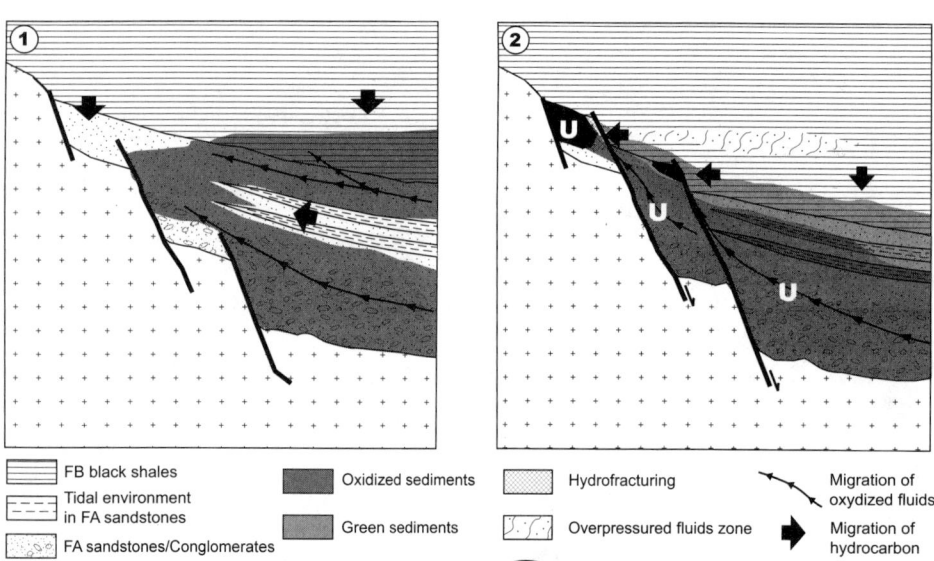

Figure 5. Schematic evolution of the Franceville Basin during the two main periods related to formation of the uranium deposits (see text).

into the FA sandstone reservoir, where it was trapped in the primary porosity (Fig. 3). This organic matter does not contain any uranium, indicating that fluids migrating into the FA reservoir at this time were uranium-free.

The extensional tectonic phase reactivated ancient faults, and accompanying hydraulic fracturing in the FA sandstone caused a change of fluid circulation in the basin (Gauthier-Lafaye 1986; Gauthier-Lafaye and Weber 1989). Fluids responsible for hydrofracturing came from the FB black shales, which have characteristics typical of overpressured shales (Osborne and Swarbrick 1997). Because uranium minerals are associated with the hydraulic fractures, it is believed that this tectonic event occurred during the uranium mineralization episode ca. 2050 Ma. Liquid hydrocarbons concentrated mainly in tectonic traps at the top of anticlines and beside normal faults where hydrofracturing was intense (Fig. 3). In these areas, petroleum moved into the secondary porosity of the sandstones, and the organic matter has a high uranium content. The migration of hydrocarbon and associated reduced fluids coming from the FB black shales was responsible for reduction of previously oxidized sediments, forming reduced fronts that are now represented by the green-sulfide-rich sediments located between the red and black sandstones and shales (Fig. 5).

THE URANIUM DEPOSITS

The Franceville Basin contains five uranium deposits. From north to south (Fig. 1) these are Mounana, the first to be discovered in 1956, Boyindzi, Oklo-Okelobondo, Bangombé, and Mikouloungou. The Oklo-Okelobondo and Bangombé deposits contain the famous natural fission reactors. These deposits have been mined out with the exception of the Bangombé deposit. Combined production from these deposits amounted to 28,000 tons of uranium from 1961 to 1999.

The uranium deposits of the Franceville Basin are located in the upper part of the FA Formation and are associated with tectonic structures that also formed traps for hydrocarbons. In the non-weathered ore most of the uranium is in the form of uraninite and coffinite, which are closely associated with hydrocarbons occupying secondary porosity of the sandstones. The mineralized organic matter consists of solid pyrobitumen with an atomic hydrogen/carbon ratio lower than 0.5 (Cortial et al., 1990).

In the deposits, uranium is associated with calcium, vanadium, iron, lead, barium, zinc, copper, and molybdenum (Gauthier-Lafaye 1986). Up to 6 wt% calcium is disseminated in uranium-bearing minerals (uraninite). Vanadium contents were very high (0.01%–1%) at Mounana, where it was produced, and also at Boyindzi and Mikouloungou. At Mounana, vanadium oxides (karelianite, montroseite, duttonite, corvusite) and silicate (roscoelite) were found in the primary (unweathered) ores, whereas uranium, lead, and barium vanadates occurred in the weathered zones (Geffroy et al., 1964 ; Weber 1968). In the ores with a high uranium content, hematite and sulfides such as pyrite, marcasite, and melnikovite are important locally. Lead is of radiogenic origin and is mainly concentrated in galena. Minium and even native lead have been found only in the fission reactors (Gauthier-Lafaye et al., 1996b). Zinc, copper, and molybdenum are present in minor concentrations. They occur mainly as sulfides (sphalerite, chalcopyrite, chalcocite) in the primary zones and as sulfates in the weathered zones, although molybdenite is also found in wulfenite. Barite is a common accessory mineral except in the vanadium-rich zones (Gauthier-Lafaye et al., 1996b).

In all uranium deposits, two types of ores with different uranium contents may be distinguished. In the first type, uranium is located inside organic matter and its uranium content ranges from 0.1% to 1%. The uranium is then associated with sulfides, mainly pyrite, chlorite, and calcite. $\delta^{13}C$ of this calcite is low (−10 to −15‰) suggesting that part of its carbon has an organic origin (Gauthier-Lafaye et al., 1989). Barren calcite outside the deposits has a $\delta^{13}C$ value of 0% to −5‰. In the second type of uranium ore, the uranium content is high (1%–10%) and uraninite is associated with hematite and illite but not with organic matter. This type of ore is located in highly fractured rocks. Ore transitional between these two types exists. This suggests that both types of uranium ores are products of the same oxidation-reduction process and that the highly fractured zones have been more permeated by the oxidized uranium-bearing fluids.

FORMATION OF THE URANIUM DEPOSITS

During deposition of the FA sandstones and conglomerates, reducing conditions prevailed as suggested by their mineralogical assemblage (fresh biotite in fine non-oxidized sandstones). Uranium-bearing minerals (monazite, thorite, and possibly detrital uraninite) were therefore stable and could have formed deposits similar to the Witwatersrand and Elliot Lake deposits. However, as the burial of the Franceville Series progressed, oxygen in the atmosphere evidently increased. This oxygenation of the atmosphere and hydrosphere is recorded by the primary and secondary oxidation of the FA sandstones that occurred during diagenesis and during the oil migration into the FA reservoir. This oxidation event was also responsible for alteration of the uranium-bearing minerals (monazite and possibly uraninite) and migration of uranium into the petroleum traps where it met reducing conditions favorable for its precipitation.

In the field, the relationship between uranium deposits and oxidation fronts is not conspicuous because the fronts were obscured by the later reduction process. The relation can be seen, however, at Mikouloungou and Otobo and in several places at Oklo where high-grade uranium ores are developed in oxidized fractures. The association of uranium with copper, molybdenum, and vanadium, which are typical of uranium deposits associated with an oxidized front, supports this interpretation (Dahlkamp, 1993). The oxidized uranium-bearing solutions migrated upward through the lower part of the FA sandstones, which have been oxidized. Uraninite precipitated in tectonic traps when these solutions met hydrocarbons sourced from the FB Formation.

The uranium-bearing fluids were very saline and were trapped at 100–170 °C, at a pressure of 800–1200 bars (Mathieu, 1999; Mathieu et al., 2001). Some fluids are a mixture between brines from the FA sandstone reservoir and low Na/Ca fluids typical of formation fluids associated with hydrocarbons in petroleum fields (Carpenter et al., 1974). If we assume a concentration of 10 ppb uranium in the solutions, and given that the FA sandstones are 1 km thick and have 10% porosity, one can calculate that 40,000 tons of uranium, the total estimated reserve of the Franceville Basin, corresponds to ten times the water volume contained within the FA sandstone pore space. This suggests that significant water circulation through the FA reservoir was needed to form the uranium deposits and that this water needed to be recharged with O_2 in order to maintain its oxidizing potential.

The source of the uranium is believed to be detrital heavy minerals including uraninite, which accumulated in several 100-m-thick conglomerates in the lower FA Formation (Gauthier-Lafaye, 1986; Gauthier-Lafaye and Weber, 1989). These conglomerates contain monazite and thorite, but despite a thorough search of the FA conglomerates for detrital uraninites, this mineral has never been found. This may be because the FA sandstones passed through the oxidation event that would have dissolved any uraninite that was present. That oxidation event allowed the formation of the uranium deposits and the occurrence of fission reactions when a critical mass was reached. Therefore, taking into account the chronology of these phenomena, a considerable increase of the oxygen in the atmosphere evidently occurred between deposition of sediments and formation of uranium ore deposits. This happened between 2.14 ± 14 Ga and 2.05 ± 30 Ga. Unfortunately, the large uncertainty in the radiometric data does not allow us to fix the age of the transition stage more precisely.

Following formation of the uranium deposits and the fission reactors, diagenesis of the Franceville Series continued until the FA sandstones reached a depth of 4000 m where temperatures reached 180–200 °C (Gauthier-Lafaye et al., 1989; Mathieu, 1999). This allowed major silicification of the FA sandstones, which reduced the porosity of the rocks and protected the uranium deposits and the reactors from alteration.

THE NATURAL NUCLEAR FISSION REACTORS

Description of the Reactors

Fifteen natural nuclear fission reactors (hereafter reactors) were discovered in two uranium deposits of the Franceville Basin. Fourteen are located in the Oklo-Okelobondo deposit and one in the small deposit of Bangombé which is 30 km from Oklo-Okelobondo (Fig. 1). Geological, mineralogical, and geochemical descriptions of the reactors are given by Gauthier-Lafaye (1986), Gauthier-Lafaye et al. (1989), and Gauthier-Lafaye et al. (1996b), as well as in a report that synthesized a recent European study of fission products in the various reactors (Gauthier-Lafaye et al., 2000). Other major contributions are reported in the proceedings of two IAEA international conferences held in Libreville (1975) and Paris (1978) and in the final report of the European program "Oklo-Natural Analogue: phase I" (Blanc, 1996).

Reactor 9, which is typical, is shown in Figure 6. The size of the reactors is variable, reflecting both differences in original size and later weathering. The biggest reactor (reactor 2) is a lens 12 m long, 18 m deep, and 20–50 cm thick. The reactor at Bangombé is only 8–10 m long, 1 m wide, and a few centimeters thick, although recent weathering has dissolved a large part of this reactor (Gauthier-Lafaye et al., 2000).

The reactor cores consist of a 5–20 cm thick layer of uraninite embedded in clays (illite and chlorite). The uranium content of the core ranges between 40% and 60%. Accessory minerals are mainly sulfides (pyrite and galena), hematite, and phosphates (mainly hydroxyapatite). The clay minerals reflect the thermal gradient during the operation of the reactors (Gauthier-Lafaye et al., 1989; Pourcelot and Gauthier-Lafaye, 1999), with Mg-Al-chlorite and 2M1-illite close to the core and 1M-illite and Fe-chlorite at the edge. In some cases, quartz grains, more or less altered and dissolved, remain in the clays.

Fission Reactions

Fission reactions started in high-grade uranium ore when the uranium content of the sandstones reached 10% (Naudet, 1991). Once fission reactions began, the temperature in the reactor core increased, starting a convective hydrothermal system around the reactor. Core temperatures reached 400 °C (Gauthier-Lafaye et al. 1989; Mathieu et al., 2001) and decreased rapidly toward the edges (thermal gradient around 100 °C/m) with heat transferred largely by conduction (Gerard, 1997; Gerard et al. 1997). Hydrothermal circulation caused the dissolution and migration of 80% of the silica from the mineralized sandstones. This decreased the volume of the sandstones and increased the uranium content of the residual layer, which then consisted mainly of clays and uranium dioxides. Furthermore, it has been shown that in reactor 2,

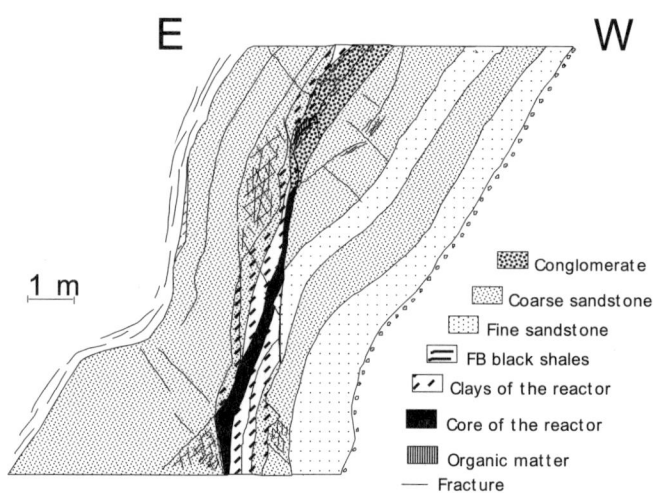

Figure 6. Cross section of reactor 9 in the Oklo deposit.

up to 50% of the present uranium was introduced into the mineralized layer during the fission reaction, probably by the fluids migrating through the reactors. This process allowed a fivefold increase in the amount of uranium per unit volume, forming the reactor core (Gauthier-Lafaye et al., 1989).

Fluid inclusions show that the hydrothermal fluids produced by the fission reactions were similar to the diagenetic fluids, but their temperatures were higher, ranging from 280 to 420 °C (Mathieu, 1999; Mathieu et al., 2001). They also contained traces of CH_4, CO_2, O_2, H_2, and H_2O_2, with the last three species resulting from the radiolysis of water in the core of the reactor (Dubessy et al., 1988; Mathieu et al., 2001). Because H_2 diffuses faster than O_2, very oxidized conditions could have occurred in the reactors, accounting for the precipitation of minium (Pb_3O_4), which needs a high oxidation potential and very low reduced sulfur content in the hydrothermal fluids (Savary and Pagel, 1997). During the fission reaction, CH_4 and CO_2 were released from organic matter at high temperatures (Mathieu, 1999). Uraninite depleted in ^{235}U occurs in bitumen nodules, suggesting that organic matter was still mobile during the fission reactions (Nagy et al., 1993)

To start and sustain a fission reaction, the concentration of neutron "poisons" in the system must be low (Naudet, 1991). In geological systems boron, vanadium, and REE are the most important poisons. Differences in the vanadium concentration of the various deposits in the Franceville Basin may explain the distribution of the reactors.

Conditions Necessary for Fission Reactions

The conditions for the occurrence of nuclear fission reactions in a geological system have been discussed by Naudet (1991). Fission reactions occur spontaneously in ^{238}U, resulting in the production of fast neutrons. If these neutrons are slowed down, they may induce fission in ^{235}U or ^{239}Pu. Nuclear reactions can then be sustained. This can happen in a geological system if at least three conditions are met:

1. The uranium ore must have a high uranium content. This increases the number and density of spontaneous fission events in ^{238}U and therefore increases the probability of inducing fission in ^{235}U, which sustains the fission reaction.

2. The fast neutrons produced by the spontaneous fission of ^{238}U must be slowed down by migration through water or graphite. Naudet (1991) has shown that water should have been the main moderator for neutrons in the Oklo reactors, as it must have been present in the pore space of the sandstones and in clays that were important mineralogical phases in the reactors. Carbon may have played an important role in some reactors with abundant organic matter (Naudet, 1991).

3. The ^{235}U content of the uranium should be high, so that the probability of a neutron hitting a fissile ^{235}U atom and thus sustaining the nuclear reaction is also high. This last condition is satisfied in commercial nuclear plants by increasing the $^{235}U/^{238}U$ ratio of the nuclear fuel. At Oklo this condition was met because ^{235}U was more abundant at that time.

Because the radioactive decay constant for ^{235}U (99.8485 * 10^{-10}) is smaller than that for ^{238}U (1.55125*10^{-10}), the $^{235}U/^{238}U$ ratio changed with time, as shown in Figure 7. Two b.y. ago the ratio was similar to that of fuel in PWR (Pressured Water Reactor) nuclear plants.

There is little chance of finding natural fission reactors younger than 1.9 Ma because the amount of fissile ^{235}U by then was too small to sustain fission reactions in a natural environment. At ages greater than 2.0 Ga, the $^{235}U/^{238}U$ was sufficient to sustain fission in natural reactors and it was much higher than this minimum before ca. 2.5 Ga. This is the main argument for predicting that fission reactors should have been common prior to 2.0 Ga. However, no traces of nuclear fission reactors have been found in pre-2.0 Ga uranium deposits such as those in the Witwatersrand or at Elliot Lake. This means that other conditions must be important as well. Clearly, a high concentration of uranium is of particular importance and this may be achieved only in certain geochemical conditions.

Fulfillment of conditions 1 and 3 allows a uranium deposit to reach the critical mass required to sustain a fission reaction. If the physical (porosity) and chemical (content of various minerals and neutron poisons such as REE, boron, and vanadium) properties of the Oklo sandstones are taken into account, a sustained fission reaction would have occurred at Oklo where the uranium content was 10% throughout one cubic meter of rock (Naudet, 1991). In actuality, the uranium content in the reactors was up to 60%. Such high concentrations in a sedimentary rock indicate that the uranium was enriched during one or more dissolution-precipitation episodes and implies that enough oxygen was available in the water to allow the required oxidation-reduction reactions of U to occur.

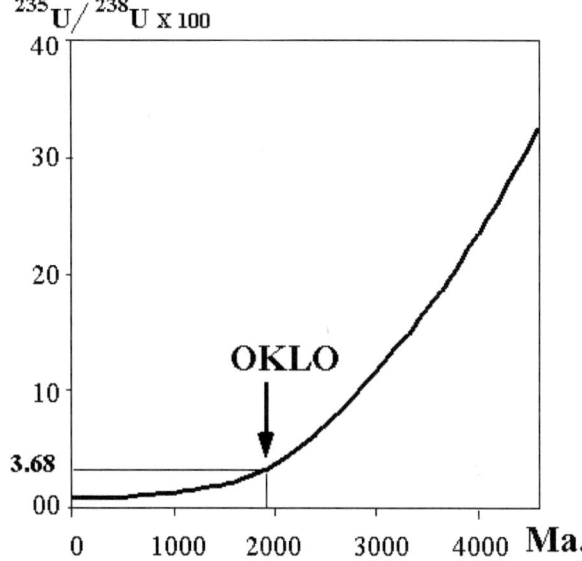

Figure 7. Evolution of $^{235}U/^{238}U$ ratio versus time in Earth.

CONCLUSIONS

The mobilization of uranium in the Franceville Basin occurred during diagenesis ca. 1950 Ma, as the burial of the FA and FB Formations reached depths of 3000–4000 m. Migration of uranium was related to oxidation of the whole FA reservoir, alteration of the detrital uranium-bearing minerals, and precipitation of iron oxides. This required a complete recharge of oxygenated waters circulating in the FA reservoir, which would have required a high level of oxygen in the atmosphere. The oxidized uranium precipitated when it came into contact with hydrocarbons concentrated in tectonic traps.

The occurrence of oxidation-reduction reactions allowed formation of very high grade uranium ore deposits for the first time in Earth history, and this led in turn to nuclear fission reactions. Younger uranium deposits that formed by similar processes did not undergo nuclear fission reactions because the abundance of U^{235} was too low after 1950 Ma. Older deposits did not develop the high grades necessary to induce spontaneous fission because oxygenated fluids necessary to form such deposits were not available. Thus, the natural nuclear reactors in the Franceville Basin record a unique interval in Earth's history, of high ^{235}U and transitional redox conditions, and this accounts for the apparent absence of such remarkable phenomena in older or younger uranium deposits.

ACKNOWLEDGMENTS

The author wishes to thank the Compagnie des Mines d'Uranium de Franceville (COMUF), the Compagnie Générale des Matériaux (COGEMA) and the Commissariat à l'Energie Atomique (CEA) for their financial support when working on the uranium deposits of Gabon and on natural nuclear fission reactors. Two reviewers, A. Wallace and an anonymous one, are also thanked for constructive reviews. This is EOST/CGS/CNRS contribution n° 2004.602-UMR7517.

REFERENCES CITED

Bertrand-Sarfati, J., and Potin, B., 1994, Microfossiliferous cherty stromatolites in the 2000 Ma Franceville Group, Gabon: Precambrian Research, v. 65, p. 341–356, doi: 10.1016/0301-9268(94)90112-0.

Blanc, P.-L., 1996, Oklo—Natural analogue for a radioactive waste repository (phase 1), Volume 1: Acquirement of the project: European Commission, Nuclear Science and Technology Series, VI, 123 p. ISBN 92-827-7448-1.

Bonhomme, M., Gauthier-Lafaye, F., and Weber, F., 1982, An example of Lower Proterozoic sediments: The Francevillian in Gabon: Precambrian Research, v. 18, p. 87–102, doi: 10.1016/0301-9268(82)90038-9.

Bros, R., Stille, P., Gauthier-Lafaye, F., Weber, F., and Clauer, N., 1992, Sm-Nd isotopic dating of Proterozoic clay material: An example from Francevillian sedimentary series, Gabon: Earth and Planetary Science Letters, v. 113, p. 207–218, doi: 10.1016/0012-821X(92)90220-P.

Bros, R., Roos, P., Andersson, P., Holm, E., Griffault, L., and Smellie, J., 2000, Mobilization of fission products and ^{238}U-series radionuclides in the Bangombé reactor zone: Migration trends and time-scale evaluation, in Louvat, D., Michaud, V., and von Maravic, H., eds., OKLO working group proceedings of the third and final EC-CEA workshop on Oklo-phase II, Cadarache, France, 20–21 May 1999: European Commission, Nuclear Science and Technology Series, EUR 19137 EN, p. 227–241.

Caen-Vachette, M., Vialette, Y., Bassot, J.P., and Vidal, P., 1988, Apport de la géochronologie isotopique à la connaissance de la géologie gabonaise: Chronique Recherche Minière, v. 491, p. 35–54.

Carpenter, A.B., Trout, M.L., and Pickett, E.E., 1974, Preliminary report on the origin and chemical evolution of lead- and zinc-rich oil field brines in Central Mississippi: Economic Geology and the Bulletin of the Society of Economic Geologists, v. 69, p. 1191–1206.

Cortial, F., Gauthier-Lafaye, F., Oberlin, A., Lacrampe-Couloume, G., and Weber, F., 1990, Characterization of organic matter associated with uranium deposits in the Francevillian Formation of Gabon (Lower Proterozoic): Organic Geochemistry, v. 15, p. 73–85, doi: 10.1016/0146-6380(90)90185-3.

Dahlkamp, F.J., 1993, Uranium ore deposits: Berlin, Springer-Verlag, 460 p.

Dubessy, J., Pagel, M., Beny, J.M., Christensen, H., Bernard, H., Kosztolanyi, C., and Poty, B., 1988, Radiolysis evidence by H2–O2 and H2-bearing fluid inclusions in three uranium deposits: Geochimica et Cosmochimica Acta, v. 52, p. 1155–1167, doi: 10.1016/0016-7037(88)90269-4.

Evins, L.Z., 2002, Geochronology of the Oklo and Bangombé fossil natural reactors: Tracing the effects of geological events [Ph.D. thesis]: Stockholm, Department of Geology and Geochemistry, Stockholm University, Avhandling för Filosofie Doktorexamen no. 313.

Gancarz, A.J., 1978, U-Pb age (2.05 x 10^9 years) of the Oklo uranium deposit, in Les réacteurs de fission naturels: Vienna, IAEA, p.513–520.

Gauthier-Lafaye, F., 1986, Les gisements d'uranium du Gabon et les réacteurs d'Oklo: Modèle métallogénique de gites à fortes teneurs du Protérozoïque inférieur: Mémoire des Sciences Géologiques, 78: 206 p.

Gauthier-Lafaye, F., and Weber, F., 1981, Les concentration uranifères du Francevillien du Gabon: Leur association avec des gîtes à hydrocarbures fossiles du Protérozoïque inférieur: Comptes Rendus de l'Académie des Sciences, series D, v. 292, p. 69–74.

Gauthier-Lafaye, F., and Weber, F., 1989, The Francevillian (Lower Proterozoic) uranium ore deposits of Gabon: Economic Geology and the Bulletin of the Society of Economic Geologists, v. 84, p. 2267–2285.

Gauthier-Lafaye, F., Weber, F., and Ohmoto, H., 1989, Natural fission reactors of Oklo: Economic Geology and the Bulletin of the Society of Economic Geologists, v. 84, p. 2286–2295.

Gauthier-Lafaye, F., Bros, R., and Stille, P., 1996a, Pb-Pb isotope systematics of diagenetic clays: An example from Proterozoic black shales of the Franceville basin (Gabon): Chemical Geology, v. 133, p. 243–250, doi: 10.1016/S0009-2541(96)00032-0.

Gauthier-Lafaye, F., Holliger, P., and Blanc, P.-L., 1996b, Natural fission reactors in the Franceville basin, Gabon: A review of the conditions and results of a critical event in a geologic system: Geochimica et Cosmochimica Acta, v. 60, p. 4831–4852, doi: 10.1016/S0016-7037(96)00245-1.

Gauthier-Lafaye, F., Ledoux, E., Smellie, J., Louvat, D., Michaud, V., Pérez del Villar, L., Oversby, V., and Bruno, J., 2000, Oklo—Natural analogue (phase II): Behaviour of nuclear reaction products in a natural environment: European Commission, Nuclear Science and Technology Series, Contract FI4W-CT96-0020, Final report EUR 19139 EN, 116 p.

Gauthier-Lafaye, F., and Weber, F., 2003, Natural nuclear fission reactors: time constraints for occurrence and their relation to uranium and manganese deposits and to the evolution of the atmosphere: Precambrian Research, v. 120, p. 81–101, doi: 10.1016/S0301-9268(02)00163-8.

Geffroy, J., Cesbron, F., and Lafforgue, P., 1964, Données préliminaires sur les constituants profonds des minerais uranifères et vanadifères de Mounana (Gabon): Comptes Rendus de l'Académie des Sciences, series D, v. 259, p. 601–603.

Gerard, B., 1997, Modélisation 3D des transferts de chaleur et de fluide dans les formations sédimentaires: Application aux réacteurs d'Oklo (Gabon) [Ph.D. thesis]: Nancy, Institut National Polytechnique de Lorraine, 1997, 251 p.

Gerard, B., Royer, J.J., le Carlier, C., Pagel, M., Scius, H., and Gauthier-Lafaye, F., 1997, 3D modelling of heat and mass transfers around a natural fission reactor, in Louvat, D., and von Maravic, H., eds., Proceedings of the first joint EC-CEA workshop on the Oklo-natural analogue Phase II project (EUR 18314), Sitges, Spain, June 18-20, 1997: Oklo Phase II workshop, EUR Report Series no. 18314, p. 301–308.

Holland, H.D., 1994, Paleoproterozoic atmospheric change, in Bengtson, S., ed., Early life on Earth: Nobel Symposium 84: New York, Columbia University Press, p. 237–244.

Holliger, P., 1988, Ages U-Pb définis in-situ sur oxydes d'uranium à l'analyseur ionique: Méthodologie et conséquences géochimiques: Comptes Rendus de l'Académie des Sciences, series D, v. 307, p. 367–373.

Horie, K., Hidaka, H., and Gauthier-Lafaye, F., 2004, U-Pb Zircon geochronology of the Franceville series at Bidoudouma, Gabon: Abstracts, Proceedings, Goldschmidt Conference, 5–11 June 2004, Copenhagen, p. 511

Kuroda, P.F., 1956, On the nuclear physical stability of the uranium minerals: Journal of Chemical Physics, v. 25, p. 781–782, doi: 10.1063/1.1743058.

Ledru, P., Eko N'Dong, J., Johan, V., Prian, J.P., Coste, B., and Haccard, D., 1989, Structural and metamorphic evolution of the Gabon orogenic belt: Collision tectonics in the Lower Proterozoic: Precambrian Research, 44: 227–241.

Maas, R., and McCulloch, M.T., 1990, A search of fossil nuclear reactors in the Alligator River uranium field, Australia: Constraints from Sm, Gd and Nd isotopic studies: Chemical Geology, v. 88, p. 301–315, doi: 10.1016/0009-2541(90)90095-O.

Mathieu, R., 1999, Reconstitution des paléocirculations fluids et des migrations élémentaires dans l'environnement des réacteurs nucléaires naturels d'Oklo (Gabon) et des argilites de Tournemire France) [Ph.D. thesis]: Nancy, Institut National Polytechnique de Lorraine, 518 p.

Mathieu, R., Zetterström, L., Cuney, M., Gauthier-Lafaye, F., and Hidaka, H., 2001, Alteration of monazite and zircon and lead migration as geochemical tracers of fluid paleocirculations around the Oklo-Okélobondo and Bangombé natural nuclear reaction zones (Franceville basin, Gabon): Chemical Geology, v. 171, p. 147–171, doi: 10.1016/S0009-2541(00)00245-X.

Melezhik, V.A., Fallick, A.E., Makarikhin, V.V., and Lubtsov, V.V., 1997, Links between Palaeoproterozoic palaeogeography and rise and decline of stromatolites: Fennoscandian Shield. Precambrian Research, v. 82, p. 311–348, doi: 10.1016/S0301-9268(96)00061-7.

Melezhik, V.A., Fallick, A.E., Medvedev, P.V., and Makarikhin, V.V., 1999, Extreme 13Ccarb enrichment in ca. 2.0 Ga magnesite-stromatolite-dolomite-red beds association in a global context: A case for the world-wide signal enhanced by a local environment: Earth-Science Reviews, v. 47, p. 1–40, doi: 10.1016/S0012-8252(99)00027-6.

Mossman, D.J., Nagy, B., Rigali, M.J., Gauthier-Lafaye, F., and Holliger, P., 1993, Petrography and paragenesis of organic matter associated with the natural fission reactors at Oklo, Republic of Gabon: A preliminary report: International Journal of Coal Geology, v. 24, p. 179–194, doi: 10.1016/0166-5162(93)90009-Y.

Mossman, D.J., Gauthier-Lafaye, F., Nagy, B., and Rigali, M., 1998, Geochemistry of organic-rich black shales overlying the natural nuclear fission reactors of Oklo, Republic of Gabon: Energy Sources, v. 20, p. 521–539.

Mossman, D.J., Gauthier-Lafaye, F., and Jackson, S., 2001, Carbonaceous substances associated with the Paleoproterozoic natural nuclear fission reactors of Oklo, Gabon: Paragenesis, thermal maturation and carbon isotopic and trace element composition: Precambrian Research, v. 106, p. 135–148, doi: 10.1016/S0301-9268(00)00129-7.

Mossman, D., 2001, Hydrocarbon habitat of the Paleoproterozoic Franceville Series, Republic of Gabon: Energy Sources, v. 23, p. 45–53, doi: 10.1080/00908310151092137.

Nagy, B., Gauthier-Lafaye, F., Holliger, P., Davis, D.W., Mossman, D.J., Leventhal, J.S., Rigali, M., and Parnell, J., 1991, Role of organic matter in containment of uranium and fissiogenic isotopes at the Oklo natural reactors: Nature, v. 354, p. 472–475, doi: 10.1038/354472a0.

Nagy, B., Gauthier-Lafaye, F., Holliger, P., Mossman, D., Leventhal, J., and Rigali, M., 1993, Role of organic matter in the Proterozoic Oklo natural fission reactors, Gabon, Africa: Geology, v. 21, p. 655–658, doi: 10.1130/0091-7613(1993)021<0655:ROOMIT>2.3.CO;2.

Naudet, R., 1991, Oklo: Des réacteurs nucléaires fossiles: Paris, Collection du Commissariat à l'Energie Atomique, 695 p.

Neuilly, M., Bussac, J., Frejacques, C., Nief, G., Vendryes, G., and Yvon, J., 1972, Sur l'existence dans un passé reculé d'une réaction en chaîne naturelle de fission, dans le gisement d'uranium d'Oklo (Gabon): Comptes Rendus de l'Académie des Sciences, series D, v. 275, p. 1847–1849.

Openshaw, R., Pagel, M., and Poty, B., 1978, Phases fluides contemporaines de la diagenèse des grès, des mouvements tectoniques et du fonctionnement des réacteurs nucléaires d'Oklo (Gabon), in Les réacteurs de fission naturels: Vienna, IAEA, TC-119/9, p. 265–296.

Osborne, M.J., and Swarbrick, R.E., 1997, Mechanisms for generating overpressure in sedimentary basins: A re-evaluation: American Association of Petroleum Geologists Bulletin, v. 81, p. 1023–1041.

Pourcelot, L., and Gauthier-Lafaye, F., 1999, Hydrothermal and supergen clays of the Oklo natural reactors: Conditions of radionuclide release, migration and retention: Chemical Geology, v. 157, p. 155–174, doi: 10.1016/S0009-2541(98)00194-6.

Roy, S., 2000, Late Archean initiation of manganese metallogenesis: Its significance and environmental controls: Ore Geology Reviews, v. 17, p. 179–198, doi: 10.1016/S0169-1368(00)00013-5.

Ruffenach, J.C., 1978, Etude des migrations de l'uranium et des terres rares sur une carotte de sondage et application à la détermination de la date des réactions nucléaires, in Les réacteurs de fission naturels: Vienna, IAEA, TC-119/9, p. 441–471.

Ruffenach J.C, 1979, Les réacteurs nucléaires naturels d'Oklo. Paramètres neutroniques, date et durée de fonctionnement, migrations de l'uranium et des produits de fission [Ph.D. thesis Sciences]: Université Paris VII, 351 p.

Savary, V., and Pagel, M., 1997, The effects of water radiolysis on local redox conditions in the Oklo, Gabon, natural fission reactors 10 and 16: Geochimica et Cosmochimica Acta, v. 61, p. 4479–4494, doi: 10.1016/S0016-7037(97)00261-5.

Semikhatov, M.A., and Raaben, M.A., 1994, Dynamics of global diversity of Proterozoic stromatolites: 1. Northern Eurasia, China and India: Stratigraphy and Geological Correlation, v. 2, p. 10–32.

Stille, P., Gauthier-Lafaye, F., and Bros, R., 1993, The neodymium isotope system as a tool for petroleum exploration: Geochimica et Cosmochimica Acta, v. 57, p. 4521–4525, doi: 10.1016/0016-7037(93)90502-N.

Vine, J.W., and Tourtelot, E.B., 1977, Geochemistry of black shale deposits—A summary report: Economic Geology and the Bulletin of the Society of Economic Geologists, v. 65, p. 253–272.

Weber, F., 1968, Une série précambrienne du Gabon: Le Francevillien, sédimentologie, géochimie et relation avec les gîtes minéraux associés. Mémoires Service Cartes Géologiques d'Alsace-Lorraine, 28: 328 p.

Weber, F., 1997, Evolution of lateritic manganese deposits, in Paquet, H., and Clauer, N., eds., Soils and sediments: Mineralogy and geochemistry: Berlin, Springer-Verlag, 369 p.

Winkler, H.G.F., 1976, Petrogenesis of metamorphic rocks: New York, Springer-Verlag, 334 p.

MANUSCRIPT ACCEPTED BY THE SOCIETY 29 OCTOBER 2005

Geological Society of America
Memoir 198
2006

Proterozoic sedimentary exhalative (SEDEX) deposits and links to evolving global ocean chemistry

Timothy W. Lyons*
Anne M. Gellatly
Department of Earth Sciences, University of California, Riverside, California 92521, USA

Peter J. McGoldrick
CODES, ARC Centre of Excellence in Ore Deposits, University of Tasmania, Hobart, TAS 7001, Australia

Linda C. Kah
Department of Earth & Planetary Sciences, University of Tennessee, Knoxville, Tennessee 37996, USA

ABSTRACT

Sedimentary exhalative (SEDEX) Zn-Pb-sulfide mineralization first occurred on a large scale during the late Paleoproterozoic. Metal sulfides in most Proterozoic deposits have yielded broad ranges of predominantly positive $\delta^{34}S$ values traditionally attributed to bacterial sulfate reduction. Heavy isotopic signatures are often ascribed to fractionation within closed or partly closed local reservoirs isolated from the global ocean by rifting before, during, and after the formation of Rodinia. Although such conditions likely played a central role, we argue here that the first appearance of significant SEDEX mineralization during the Proterozoic and the isotopic properties of those deposits are also strongly coupled to temporal evolution of the amount of sulfate in seawater.

The ubiquity of ^{34}S-enriched sulfide in ore bodies and shales and the widespread stratigraphic patterns of rapid $\delta^{34}S$ variability expressed in both sulfate and sulfide data are among the principal evidence for global seawater sulfate that was increasing during the Proterozoic but remained substantially lower than today. Because sulfate is produced mostly through weathering of the continents in the presence of oxygen, low Proterozoic concentrations imply that levels of atmospheric oxygen fell between the abundances of the Phanerozoic and the deficiencies of the Archean, which are also indicated by the Precambrian sulfur isotope record. Given the limited availability of atmospheric oxygen, deep-water anoxia may have persisted well into the Proterozoic in the presence of a growing sulfate reservoir, which promoted prevalent euxinia. Collectively, these observations suggest that the mid-Proterozoic maximum in SEDEX mineralization and the absence of Archean deposits reflect a critical threshold in the accumulation of oceanic sulfate and thus sulfide within anoxic bottom waters and pore fluids—conditions that favored both the production and preservation of sulfide mineralization at or just below the seafloor. Consistent with these evolving global conditions, the appearance of voluminous SEDEX mineralization ca. 1800 Ma coincides generally with the disappearance of banded iron formations—marking the transition from an early iron-dominated ocean to one more strongly influenced by sulfide availability.

*timothy.lyons@ucr.edu

Lyons, T.W., Gellatly, A.M., McGoldrick, P.J., and Kah, L.C., 2006, Proterozoic sedimentary exhalative (SEDEX) deposits and links to evolving global ocean chemistry, *in* Kesler, S.E., and Ohmoto, H., eds., Evolution of Early Earth's Atmosphere, Hydrosphere, and Biosphere—Constraints from Ore Deposits: Geological Society of America Memoir 198, p. 169–184, doi: 10.1130/2006.1198(10). For permission to copy, contact editing@geosociety.org. ©2006 Geological Society of America. All rights reserved.

In further agreement with this conceptual model, Proterozoic SEDEX deposits in northern Australian formed from relatively oxidized fluids that required reduced conditions at the site of mineralization. By contrast, the generally more oxygenated Phanerozoic ocean may have only locally and intermittently favored the formation and preservation of exhalative mineralization, and most Phanerozoic deposits formed from reduced fluids that carried some sulfide to the site of ore precipitation.

Keywords: Proterozoic, SEDEX deposits, sulfur isotopes, ocean chemistry, atmospheric oxygenation.

INTRODUCTION

Sedimentary exhalative (SEDEX) deposits are sulfide-dominated Pb-Zn ore bodies with sphalerite and galena as the principal ore minerals. Although the specifics of this deposit type are complex (Goodfellow et al., 1993; Lydon, 1996; Large et al., 2004), formation through submarine venting of hydrothermal fluids in structurally controlled (rifted) sedimentary basins are universal characteristics. The compilations of Goodfellow et al. (1993) and Lydon (1996) show us that SEDEX mineralization spans a broad portion of geologic time—ranging from ca. 1800 Ma to possible analogues forming today—however, peak abundances are well expressed during discrete intervals of the mid-Proterozoic (1800–1600 Ma) and late Proterozoic to Paleozoic (600–300 Ma). A clear maximum in ore tonnage is observed between 1700 and ca. 1600 Ma.

Despite this broad temporal distribution, Lydon (1996, p. 141) found the lack of early Proterozoic and Archean SEDEX deposits to be enigmatic:

There does not appear to be any compelling reason, based on current understanding, why Sedex deposits should not occur in rocks older than Middle Proterozoic [late Paleoproterozoic]. The onset of the Middle Proterozoic does not seem to coincide with any major permanent change in global climate, ocean water composition, atmosphere composition, or geotectonic processes.

The purpose of this report is to revisit this assertion within the context of the most recent studies of Precambrian ocean-atmosphere chemistry and, in doing so, suggest why such a gap is precisely what the models for early Earth predict. It is not our intent, however, to provide a comprehensive survey of SEDEX mineralization or to develop a model that precludes the possibility of exceptions. Instead, we will suggest that the first-order patterns, and in particular the absence of Archean deposits and the abundance of SEDEX mineralization of Proterozoic age, are completely consistent with a now widely held view of evolving Precambrian ocean chemistry. The essential Proterozoic controls include the first appearance of quantitatively significant sulfate in the ocean, the persistence of deep water oxygen deficiency, the widespread distribution of bacterial hydrogen sulfide within the ocean and underlying sediments, and possibly abundant methane within the Proterozoic ocean-atmosphere system. All of these parameters will be viewed in light of the pronounced ^{34}S enrichments that dominate the metal sulfides of most mid-Proterozoic SEDEX mineralization. More challenging is an explanation of the distribution of major SEDEX mineralization in younger sediments. Tectonic influences, specifically the temporally varying extent of continental rifting, are always the essential backdrop to SEDEX hydrothermal processes. However, in the Phanerozoic, as for the Precambrian, depositional redox may have played a central role in controlling the amount and preservation of metal sulfide accumulation on the seafloor.

BACKGROUND

SEDEX Deposits

The many details of sedimentary exhalative sulfide deposits, including genetic models, are ably reviewed in Goodfellow et al. (1993), Lydon (1996), and Large et al. (2004) and are only briefly summarized here. SEDEX deposits are marked by stratiform, sediment-hosted Zn-Pb-sulfide mineralization, with sphalerite and galena as the principal ore minerals and pyrite typically as the most abundant sulfide phase. These deposits form in sedimentary basins through submarine venting of hydrothermal fluids. SEDEX mineralization is essentially synsedimentary—forming at or just below the seafloor within fault-controlled basins or troughs that are most often related to major intracratonic or continental margin rift zones—although mineralization often postdates the most active phase of rifting (Lydon, 1996),

The deposits range in age from ca. 1800 Ma to possible modern analogues but cluster in two groups: 1800–1600 and 600–300 Ma (Fig. 1). Most Proterozoic deposits formed between ca. 1700 and ca. 1600 Ma, which, from the standpoint of ore mass, is the dominant interval of all SEDEX-related Zn and Pb mineralization. Ore fluids are generally linked to metalliferous formational waters that were heated within the sedimentary basin under the elevated geothermal conditions of the typically extensional tectonic settings. The redox conditions of the basin lithologies exert an important control on the redox state of the metal-transporting fluids (Cooke et al., 2000). These workers distinguish two distinct classes of SEDEX Zn-Pb deposits termed McArthur-type and Selwyn-type. Metal-transporting fluids for the former appear to have been relatively oxidized

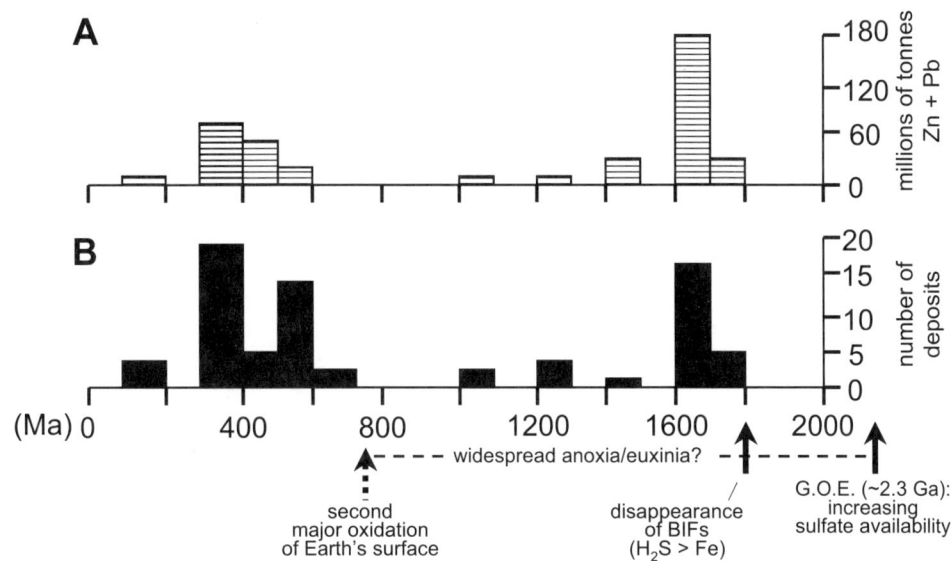

Figure 1. Age distribution of SEDEX mineralization as (A) tons of Zn and Pb per 100 m.y. interval and (B) number of deposits per 100 m.y. interval (after Lydon, 1996). Background for the interpretative details and related references are available in the text—including the G.O.E. (Great Oxidation Event).

(sulfate » sulfide), whereas the latter type formed from more reduced fluids (sulfide ≥ sulfate).

SEDEX deposits are well known for the large amounts of sulfur they contain, which is present in Proterozoic examples almost exclusively as pyrite, base metal sulfides, and pyrrhotite. Barite is much more common in the Phanerozoic deposits. Because the metal-rich ore fluids would by necessity be comparatively sulfide deficient, a substantial portion of the sulfide sulfur must be supplied independently at the site of mineralization. Bacterial reduction of seawater sulfate within anoxic pore fluids or a euxinic (anoxic and H_2S-containing) water column is generally considered an essential source of that sulfide. When present, the $\delta^{34}S$ values of barite are generally consistent with a seawater sulfate source, suggesting barite formation by reaction between hydrothermal barium and sulfate in ambient seawater.

Sulfur Isotopes

Dissimilatory bacterial sulfate reduction (BSR) typically yields hydrogen sulfide that is strongly depleted in ^{34}S relative to the $^{34}S/^{32}S$ ratio in the residual parent sulfate. Under pure-culture laboratory conditions, this sulfide can be depleted by up to 40‰–45‰ relative to the sulfate (Chambers et al., 1975; Canfield, 2001; Detmers et al., 2001). Instantaneous fractionations of this magnitude are observed during BSR even at very low initial sulfate concentrations ranging down to 200 μM, or less than 1% of the concentration in modern seawater. Below ~200 μM, rates of sulfate reduction are limited by sulfate transport across the bacterial cell membrane to the internal sites of enzymatically catalyzed fractionation. Under these conditions, most of the sulfate entering the cell is reduced, and the net fractionation is minimized (Canfield, 2001; Habicht et al., 2002).

The ^{34}S depletions observed in Phanerozoic sedimentary sulfides can exceed 60‰, which may be larger than what is possible via BSR alone. A popular hypothesis for the extreme ^{34}S depletions observed in low-temperature sulfide minerals invokes additional fractionation during bacterial disproportionation of elemental S and other S intermediates, which form through partial oxidation of H_2S (Canfield and Thamdrup, 1994; Habicht and Canfield, 2001; cf. Brunner and Bernasconi, 2005).

The net isotopic fractionation observed between sulfate and sulfide reflects both the collective effects of bacterial S reduction and disproportionation as well as the properties of the sulfate reservoir (Zaback et al., 1993). For example, even in the presence of large fractionations during BSR with coupled disproportionation reactions and diffusional inputs of sulfate, high $\delta^{34}S$ values for sulfate and sulfide can occur in pore-water and water-column systems with weak sulfate renewal relative to the rate of bacterial consumption. By contrast, low $\delta^{34}S$ values—i.e., high net fractionations—are characteristic of systems where sulfate is less limiting, such as in the water column of the Black Sea today (Lyons, 1997). Anomalously high (strongly positive) $\delta^{34}S$ data, as described in this report, require strong limitations in sulfate availability on either a local or global scale.

Sulfur Geochemical Records of Precambrian Ocean-Atmosphere Conditions—An Overview

Relevant details of the Precambrian sulfur cycle are provided in a recent review paper by Lyons et al. (2004) and in the references cited within that synthesis. These details are also summarized schematically in Figure 2. Briefly, recent estimates by Shen et al. (2001) place the onset of BSR at ca. 3.47 Ga (billion years ago). Despite this possibility, extremely low seawater sulfate concentrations (likely less than 200 μM) throughout the Archean beneath an O_2-deficient atmosphere yielded only local examples of the large S isotope fractionations that are often diagnostic of BSR (cf. Ohmoto et al., 1993, Kakegawa et al., 1998, and

Kakegawa and Ohmoto, 1999). These exceptions might record local, evaporative enrichments in seawater sulfate. By ca. 2.3 Ga, sulfate concentrations within seawater increased to levels where fractionations of a few tens of per mil are commonly observed in bacteriogenic pyrite. This spread in the $\delta^{34}S$ data is coincident with the so-called Great Oxidation Event (Holland, 2002; Farquhar and Wing, 2003; Bekker et al., 2004), when redox conditions at Earth's surface are thought by most workers to have shifted fundamentally from reducing to oxidizing. At this time, weathering of sulfide minerals exposed on the continents beneath an O_2-containing atmosphere would have led to appreciable increases in the flux of sulfate to the ocean (Fig. 2)—although oxygen concentrations were still only a small fraction of present atmospheric levels.

Although seawater sulfate concentrations rose in the early Proterozoic, its availability relative to the Phanerozoic ocean remained low throughout most if not all of the Proterozoic. Evidence for low sulfate concentrations—on the order of <5% to 15% of the present-day value of 28–29 mM for the mid-Proterozoic—is seen in the abundance of ^{34}S-enriched pyrite deposited during this interval (Shen et al., 2002, 2003; this report) and in the rapid and large-magnitude S isotope variability observed stratigraphically in many Proterozoic sulfide- and sulfate-bearing sedimentary sequences (Luepke and Lyons, 2001; Hurtgen et al., 2002, 2004; Kah et al., 2004; Gellatly and Lyons, 2005). During the Phanerozoic, by contrast, $\delta^{34}S$ fluctuations of similar magnitude are described by Claypool et al. (1980) for time periods of 10^7 to 10^8 years. By the late Proterozoic, atmospheric oxygen levels rose sufficiently—with a corresponding rise in oceanic sulfate—to support increases in S isotope fractionations between sulfate and sulfide to values typical of the Phanerozoic (i.e., up to and exceeding 60‰). Canfield and Teske (1996) attributed this increased fractionation to evolution among non-photosynthetic sulfide-oxidizing bacteria, which accelerated the production of intermediate S species and thus the disproportionation reactions that can magnify the net isotope effect. However, overall increases in sulfate concentration would also have supported larger net fractionations, and a persistence of rapid $\delta^{34}S$ variability in the marine sulfate reservoir would make fractionations between coeval sulfate and sulfide difficult to evaluate (Lyons et al., 2004; Hurtgen et al., 2005).

The Archean is generally regarded as a time of low oxygen and sulfate availability in the ocean. Under these conditions, high concentrations of dissolved iron supported extensive deposition of banded iron formations (BIFs). Across the Great Oxidation

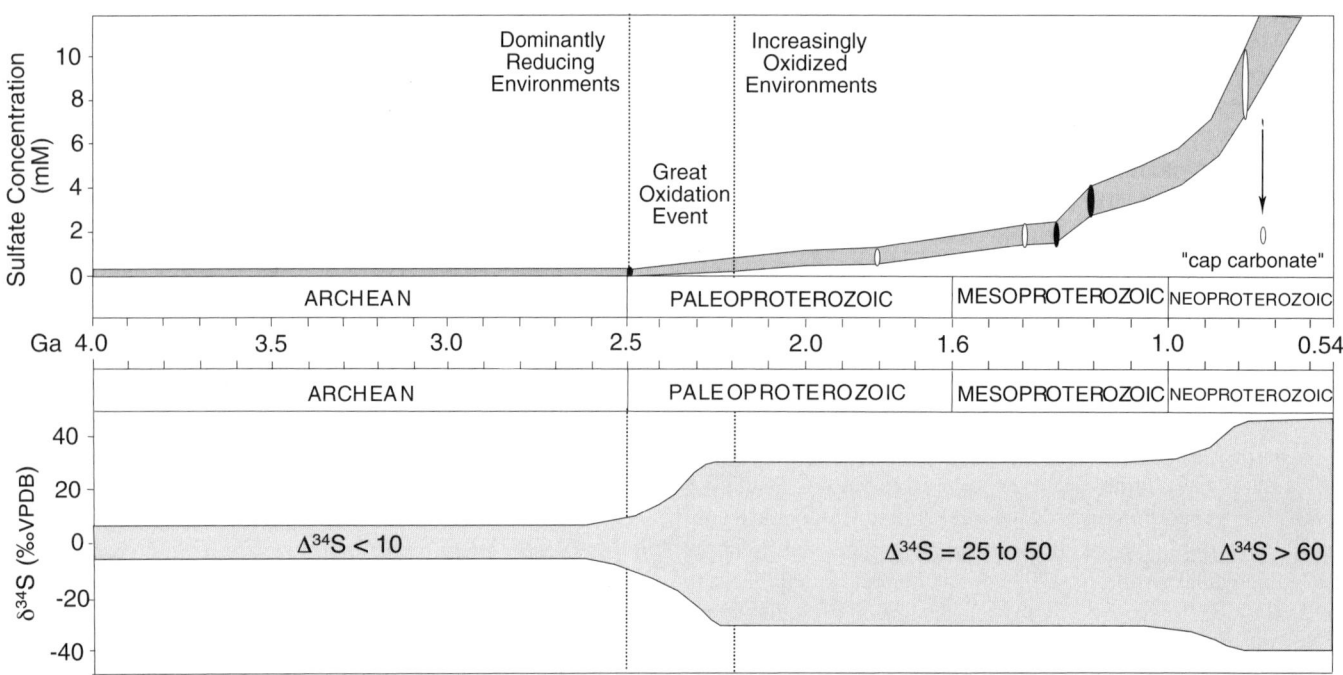

Figure 2. Estimates of sulfate concentration and a schematic of $\delta^{34}S$ values for bacteriogenic pyrite from the Archean and Proterozoic. Proterozoic sulfate concentrations are based on the Kah et al. (2004) model for S isotope variability as recorded in CAS. Archean estimates reflect the low sulfate values required to explain the predominance of small bacterial S isotope fractionations observed during this interval (Habicht et al., 2002). The pyrite $\delta^{34}S$ schematic (lower figure) is based on the compilation of Shen et al. (2001). Increasing fractionations between sulfate and sulfide ($\Delta^{34}S$) during BSR following the G.O.E. are inferred to record sulfate concentrations in the ocean that increased to values in excess of ~200 μM. Canfield and Teske (1996) attributed the increase in $\Delta^{34}S$ in the Neoproterozoic to an increasing role by disproportionating bacteria in the sulfur cycle, although this interpretation has recently been challenged (Hurtgen et al., 2005). Because most of the sulfate in the ocean derives from continental weathering of sulfide minerals beneath an O_2-containing atmosphere, increasing sulfate concentrations in seawater track the oxygenation of the Proterozoic atmosphere.

Event ca. 2.3 Ga, oxygen in the atmosphere and thus sulfate in the ocean increased to levels that supported widespread marine BSR. A recent model argues, however, that under the comparatively low oxygen conditions of the Proterozoic atmosphere, the deep ocean may have remained free of oxygen for more than a billion years beyond the oxidation event (Canfield, 1998). Canfield argued that within this anoxic environment, and in the presence of a growing oceanic sulfate reservoir, sulfate-reducing bacteria would have thrived—giving rise to euxinic conditions throughout much of the deep Proterozoic water column (also Anbar and Knoll, 2002). Recently, Mo isotope data from Proterozoic black shales have confirmed the possibility of widespread oxygen deficiency and perhaps euxinia in the Proterozoic ocean (Arnold et al., 2004). Such conditions may have persisted until the Neoproterozoic when a second major oxygenation of Earth's surface resulted in a widely oxic deep ocean and ultimately the appearance of multicellular organisms (Des Marais et al., 1992; Canfield, 1998). Despite this fundamental shift, the redox transition in the marine realm was complicated by the possibility of persistent or renewed oxygen deficiency within the isolated, ice-covered glacial ocean of the Neoproterozoic snowball Earth episodes (Hoffman et al., 1998; Hurtgen et al., 2002, 2004).

SULFUR ISOTOPE TRENDS

In an earlier study, Lyons et al. (2000) showed a preponderance of high $\delta^{34}S$ values in disseminated bacteriogenic pyrite extracted from organic-rich shales hosting the Sheep Creek and Soap Gulch SEDEX mineralization found within the ca. 1.47 Ga lower Belt Supergroup. The massive pyrite within the mineralized intervals showed similar enrichments. On further examination, strong stratigraphic trends were observed within the host shales, with upsection $\delta^{34}S$ decreases of ~45‰ and 12‰ over thicknesses of 45 m and 134 m, respectively (Fig. 3). Subsequently,

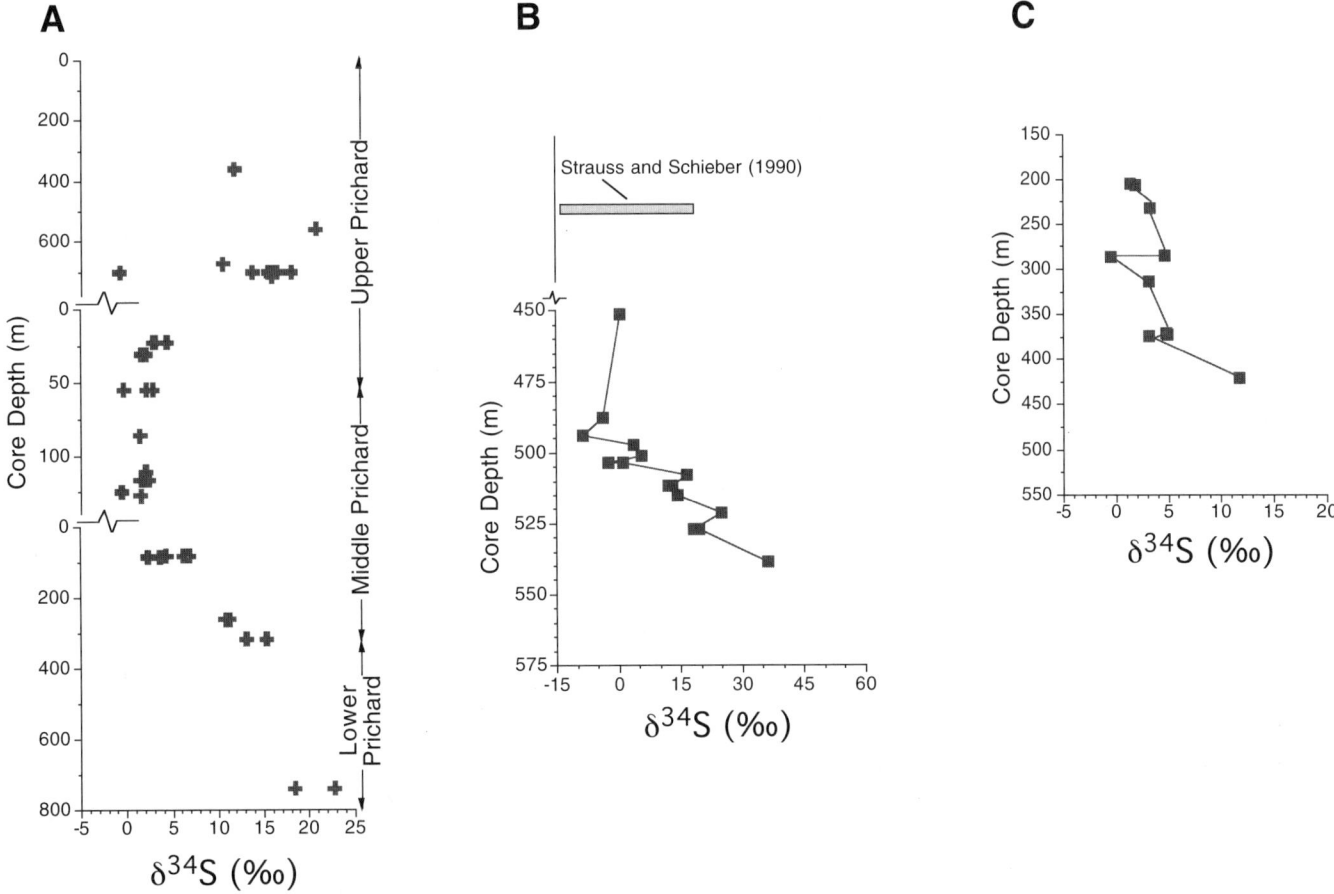

Figure 3. Stratigraphic distributions of $\delta^{34}S$ data for drill core samples of the (A) Prichard and (B and C) Newland Formations (ca. 1.47 Ga) of the Belt Supergroup, northwestern United States. Prichard data are from the "main" Belt Basin and represent fine-scale pyrite/pyrrhotite laminae, lenses, and disseminations within dark-gray to black argillites. The Newland data are from disseminated pyrite within black shales collected within the Helena Embayment—the easternmost extension of the Belt Basin. The Newland shales are host to massive, SEDEX-type sulfide mineralization in stratigraphic and spatial proximity to the data shown here. Nevertheless, the low-temperature, bacteriogenic pyrite in these shales and argillites appears to be unrelated to mineralizing, hydrothermal activity. Further details are available in Lyons et al. (2000) and Luepke and Lyons (2001).

analogous patterns have been found in pyrite and pyrrhotite from the lower Belt Supergroup away from mineralized regions in metamorphosed argillitic equivalents. These data show both decreasing and increasing stratigraphic trends for $\delta^{34}S$ (Fig. 3) (Luepke and Lyons, 2001). Collectively, the isotopic data and the relationships between organic carbon and pyrite sulfur in these sediments were viewed as convincing evidence for marine inputs into the Mesoproterozoic Belt Basin. The extent of marine deposition within the Belt Supergroup has been debated for decades (Winston, 1990; Winston and Link, 1993).

The abundance of ^{34}S-enriched values in the lower Belt Supergroup was assumed to reflect marine inputs into a highly restricted setting that may have been shut off intermittently from the open ocean. At the time of initial publication, the stratigraphic trends reproduced in Figure 3 were viewed as evidence for temporal variation in the strength of the local marine connection—with the lowest $\delta^{34}S$ values reflecting the most open conditions. In subsequent years, we have observed that large and often-systematic $\delta^{34}S$ variations over relatively short stratigraphic intervals within single sedimentary units are not unusual for Proterozoic sulfide accumulations (Ross et al., 1995; Strauss, 1997, 2002), including SEDEX mineralization (Carr and Smith, 1977; Smith et al., 1978; McGoldrick et al., 1999), and that these patterns may have global implications (Kah et al., 2004; Lyons et al., 2004).

The pronounced ^{34}S enrichments that dominate the sulfides of the lower Belt Supergroup are also common throughout the Proterozoic record (Lambert and Donnelly, 1991; Logan et al., 1995; Canfield, 1998; Canfield and Raiswell, 1999; Gorjan et al., 2000; Shen et al., 2002, 2003; Strauss, 2002). Nowhere are these enrichments better expressed than in SEDEX mineralization. In Figure 4, we have compiled a large number of $\delta^{34}S$ data for SEDEX deposits spanning the Proterozoic. This compilation reflects a range of sulfide minerals, including ore-phase galena and sphalerite, as well as sedimentary sulfides (dominantly pyrite) in host sediments away from the hydrothermal mineralization. This selection of data is not intended to be a comprehensive survey of Proterozoic SEDEX ore deposits. Instead, we have tried to assemble a representative suite of the largest data sets, with an additional goal of providing broad temporal coverage. The most striking observations about Figure 4, and the focus of much of the discussion that follows, are the wide ranges of $\delta^{34}S$ values for each of the deposits and the predominance of ^{34}S enrichments in the ore bodies and host shales.

IMPLICATIONS

Stratigraphic Isotope Trends

The patterns of comparatively rapid $\delta^{34}S$ variability described above for Proterozoic metal sulfides—that is, variations of tens of per mil over stratigraphic thicknesses of hundreds of meters or less—are also observed in samples of sulfate-S of Proterozoic age. These patterns have been observed both in gypsum (Fig. 5) (Kah et al., 2001, 2004) and carbonate-associated

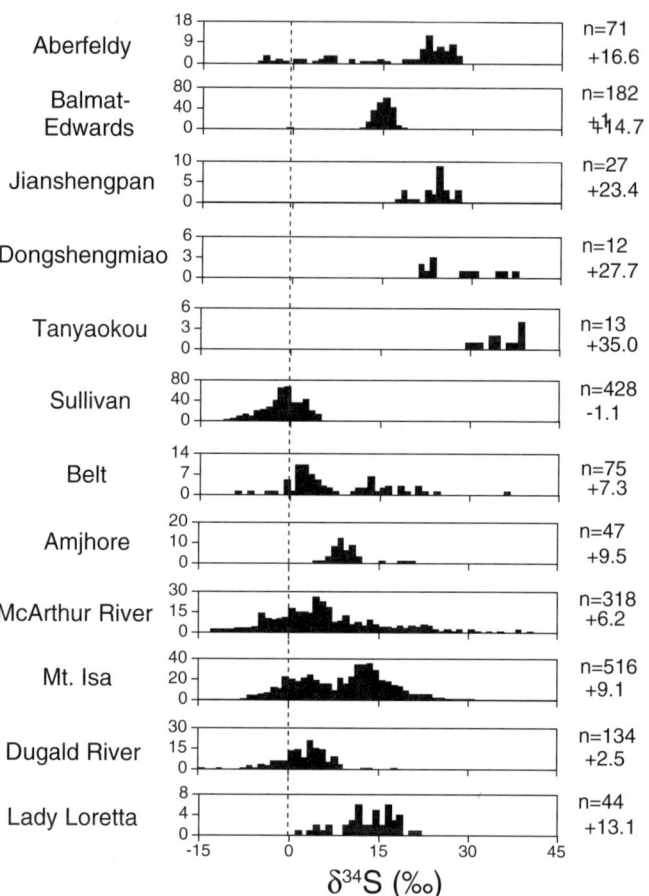

Figure 4. Summary of $\delta^{34}S$ data for sulfides from Proterozoic SEDEX deposits. Data represent pyrite and the full range of base metal sulfides present, as well as disseminated sulfides within the host sediments. References and location information are as follows: Aberfeldy, Scotland (Willan and Coleman, 1983); Balmat-Edwards, United States (Whelan et al., 1984); Jianshengpan, Dongshengmiao, and Tanyaokou, China (Ding and Jiang, 2000); Sullivan, Canada (Campbell et al., 1978, 1980); Belt Supergroup, United States (Lyons et al., 2000; Luepke and Lyons, 2001); Amjhore, India (Guha, 1971; Pandalai et al., 1991); McArthur River, Australia (Smith and Croxford, 1975; Rye and Williams, 1981; Eldridge et al., 1993; Shen et al., 2002); Mount Isa, Australia (Solomon, 1965; Smith et al., 1978; Andrew et al., 1989; Davidson and Dixon, 1992; Painter et al., 1999; McShane, 1996); Dugald River, Australia (Dixon and Davidson, 1996; Davidson and Dixon, 1992); Lady Loretta, Australia, (Carr and Smith, 1977; Scott et al., 1985). With the exception of the Chinese deposits, the histograms are stacked in a generally correct stratigraphic order. The Aberfeldy deposit is late Neoproterozoic. The Balmat-Edwards, Sullivan, Belt, and Amjhore deposits fall in a range of 1.3–1.5 Ga. McArthur River, Mount Isa, Dugald River, and Lady Loretta have ages of 1.6–1.7 Ga. Additional details about age relationships are available in Lyons et al. (2004) and in the primary references. We note the abundance of ^{34}S-enriched sulfides present in these deposits. By contrast, pyrite forming in the modern water columns of the Black Sea and Cariaco Basin have $\delta^{34}S$ values of –30‰ to –40‰, with $\Delta^{34}S$ values of ~–50‰ to –60‰ (Lyons, 1997; Lyons et al., 2003).

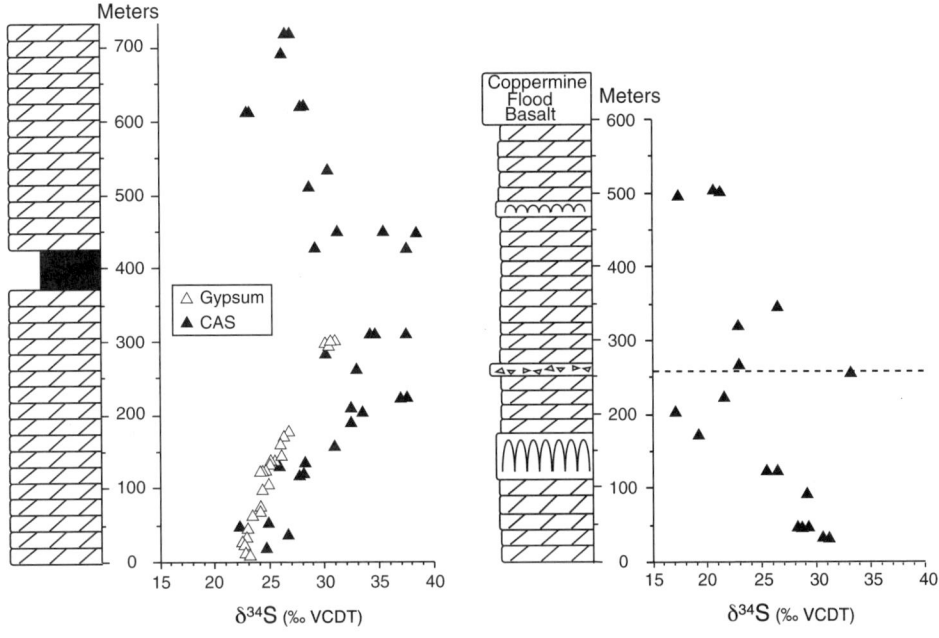

Figure 5. Left: Stratigraphic distribution of $\delta^{34}S$ data for massive, bedded gypsum and carbonate-associated sulfate (CAS) extracted from dolomite of the ca. 1.2 Ga Society Cliffs Formation, Bylot Supergroup, northeastern Arctic Canada. Right: CAS data from dolomite of the 1.3 Ga Dismal Lakes Group, north-central Arctic Canada (after Kah et al., 2001, 2004).

sulfate (CAS) (Fig. 5) (Hurtgen et al., 2002, 2004; Kah et al., 2004; Gellatly and Lyons, 2005). In contrast to earlier arguments for local controls on the patterns observed in the Belt basin (Lyons et al., 2000; Luepke and Lyons, 2001), the spatially and temporally widespread nature of this style of isotopic variability now suggests a global control. Specifically, these $\delta^{34}S$ records preserved in sulfate and sulfide, including SEDEX deposits, may reflect rapid changes in the isotopic composition of seawater sulfate on a global scale. Such patterns of variability are unlike the temporally broad, first-order trend of the Claypool et al. (1980) sulfate curve, which has long defined the Phanerozoic paradigm for seawater isotopic behavior (also Strauss, 1997; Kampschulte and Strauss, 2004). Almost certainly the "Claypool paradigm" misrepresents the extent of short-term $\delta^{34}S$ variability recorded in younger rocks. We are exploring this possibility and see early signs of greater complexity in the Paleozoic record (Gill et al., 2006; also Kampschulte and Strauss, 2004), and marine barite is revealing the finer texture of $\delta^{34}S$ variability during the Mesozoic and Cenozoic (Paytan et al., 1998, 2004). Nevertheless, the shorter-term $\delta^{34}S$ shifts recorded in the barite data are smaller (on the order of 5‰) and longer (roughly 5–10 m.y.) than those of the Proterozoic and likely the Paleozoic.

A Proterozoic ocean with appreciably lower sulfate concentrations, beneath an atmosphere with less oxygen, would be more vulnerable to rapid isotopic variability. A lower mass of sulfate in the ocean equates to a shorter residence time and thus higher sensitivity to $\delta^{34}S$ variation as driven by flux terms such as pyrite burial, weathering inputs, and volcanogenic sulfur. Kah et al. (2004) modeled the rate of $\delta^{34}S$ variability ($d\delta^{34}S/dt$) observed at several Proterozoic localities, using both gypsum and CAS data, and estimated mid-Proterozoic seawater sulfate concentrations that were only 5% to 15% of present-day levels.

Previous sulfur workers have suggested that upsection increases in $\delta^{34}S$ observed over tens to hundreds of meters could, in general, reflect progressive sulfate depletions under closed-system sulfate reduction—that is, isotopic evolution behaving as a Rayleigh distillation. This explanation is particularly attractive within the likely restricted, rifted tectonic settings that characterize most SEDEX deposits; however, it fails to explain the upsection *decreases* in $\delta^{34}S$ that are also observed. Furthermore, analogous trends have been observed in platform carbonate (CAS) and evaporite sequences, where local restriction and progressive ^{34}S enrichment through local pyrite burial are difficult to imagine (Kah et al., 2001, 2004).

Mass balance calculations that weigh sulfate availability against total sulfide burial within shale hosts provide another important constraint (Fig. 6). The large amounts of metal sulfide within SEDEX deposits imply that a substantial fraction of the total basinal sulfate/sulfide reservoir is supplying sulfur to localized sites of enrichment. Consequently, sulfur supplied from single volumes of local pore water and the immediately overlying water column is typically not adequate to explain the large amount of ore sulfur present. In shales unaffected by SEDEX processes, however, we can explore stratigraphic isotopic trends in light of simple assumptions about the sulfur source-sink relationship. If pyrite in the shale is accumulating across the deep basin floor, and mechanisms are not in place for localized enrichment, we can estimate how much pyrite could be generated beneath a given parcel of seawater. We performed this calculation for a 1 m × 1 m area of the water column, assuming closed system

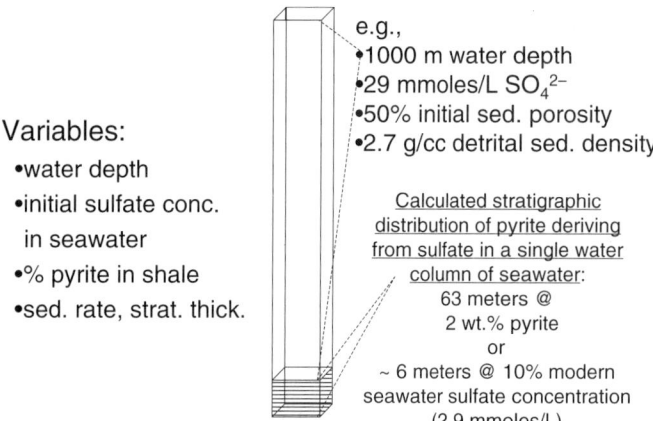

Figure 6. Simple mass balance calculation to assess maximum possible pyrite production from one basin-volume of seawater and specifically the pyrite's stratigraphic distribution in a shale host (unaffected by SEDEX mineralization) as predicted from concentrations in the final rock. Details are provided in the text.

behavior for that parcel and quantitative conversion of sulfate to H_2S and sequestration as FeS_2 (Fig. 6). In other words, we are testing the oft-cited argument that systematic upsection increases in $\delta^{34}S$ within a black shale might reflect progressive evolution via BSR in a closed system.

Our simple model assumes that sufficient Fe is present to capture all of the sulfur in the system over the evolution of the closed system. In the example shown, water depth is assumed to be 1000 m. Seawater sulfate concentration is chosen as today's value (~29 mM). The non-pyrite (detrital sediment) component is assumed to have a density of 2.7 g/cc, and the detrital sediment is assigned an initial porosity of 50%. In this example, all of the sulfide is precipitated as pyrite—other metal sulfides would modify the results only slightly—and the iron sulfide is assumed to precipitate in the original pore space.

Under these model conditions, the stratigraphic distribution of the pyrite resulting from a single water column of seawater can be calculated for any wt.% pyrite in the final sample—taken as a reasonable black shale value of 2% in this example (Fig. 6)—yielding accumulation over a stratigraphic thickness of only 63 m. This thickness is insufficient to explain the stratigraphic extents of many progressive ^{34}S enrichments observed in the shale record (e.g., Fig. 3), and a shallower water column or lower original seawater sulfate concentration could reduce the stratigraphic estimates significantly. For example, our estimate of ~10% modern sulfate concentrations for the mid-Proterozoic would reduce the thickness estimates by a factor of ten, thus adding further credibility to arguments for global rather than local controls on observed $\delta^{34}S$ trends in the shales.

If the estimates for low sulfate in the Proterozoic ocean are correct, the stratigraphic trends preserved within the metal sulfides of shales and ore deposits could be tracking global seawater sulfate values. Under conditions of local isolation, the effects of low oceanic sulfate availability would be exacerbated during sulfate reduction, leading to sulfide $\delta^{34}S$ data that would also track the oceanic trend with little net fractionation between the parent sulfate and product sulfide. We have shown that many of the sulfide data are ^{34}S enriched and that their ranges overlap strongly with those for the sulfate. Rapid isotopic variability in the ocean, in combination with pyrite formation under conditions of limited local and global sulfate availability, would yield the $\delta^{34}S$ ranges summarized in Figure 4.

Controls on ^{34}S Enrichments

Other workers have suggested that widespread ^{34}S enrichments in sedimentary pyrite are attributable to comparatively low sulfate concentrations in the Proterozoic ocean. For example, Shen et al. (2002) estimated sulfate levels of 0.5–2.4 mM based on a model for sulfide accumulation in the late Paleoproterozoic deep waters of the McArthur Basin of northern Australia. Shen et al. (2003), in a focused study of the Mesoproterozoic Roper Group of the McArthur Basin, invoked a sulfate minimum zone—a hypothesized interval within a stratified Proterozoic water column characterized by exaggerated sulfate deficiencies—to explain pronounced ^{34}S enrichments observed within those sediments. A Proterozoic sulfate minimum zone, as first suggested by Logan et al. (1995), could have exacerbated the already low sulfate concentrations in the ocean. Hurtgen et al. (2002, 2004) noted that ^{34}S enrichments in pyrite and large and rapid $\delta^{34}S$ excursions recorded in CAS—specifically in response to the Neoproterozoic glacial episodes—could reflect a suppressed riverine sulfate flux, in combination with nearly complete sulfate reduction within an isolated, anoxic global ocean (cf. Shields et al., 2004). These "snowball Earth" patterns, although generally linked to the overall low sulfate concentrations of the Proterozoic, are a subset of the longer-term controls and patterns described here.

Smaller fractionations during bacterial cycling of S would also facilitate ^{34}S enrichments in Proterozoic sulfides. Specifically, Canfield and Teske (1996), in a now debated model (e.g., Hurtgen et al., 2005), favored reduced fractionations prior to the Neoproterozoic—arguing that the enhanced ^{34}S depletions during disproportionation reactions had minimal or no influence prior to further increases in Earth surface oxidation late in the Proterozoic (cf. Johnston et al., 2005). Regardless, Canfield and Raiswell (1999), Strauss (2002), Shen et al. (2003), and others have shown that fractionations at least as large as those possible by BSR alone, and thus light $\delta^{34}S$ values, are also observed during the mid-Proterozoic, which suggests sulfate concentrations of >200 µM (Habicht et al., 2002).

Some diagenetic models suggest that extreme ^{34}S enrichments in sulfide analogous to those described here (e.g., Fig. 4) may require sulfate concentrations of less than 1 mM (Habicht et al., 2002; see Canfield, 2004). Because most pyrite in marine systems forms early, in the surface sediments or in the water column, $\delta^{34}S$ values tend to be ^{34}S depleted in Phanerozoic sediments.

$\delta^{34}S$ values commonly range between –20‰ and –40‰ in many modern euxinic sediments and Phanerozoic black shales (Sageman and Lyons, 2003). For example, although the modern Black Sea is only weakly linked to the Mediterranean Sea, the balance between rates of net sulfate reduction and sulfate replenishment, in combination with pyrite that forms primarily in the water column, yields typical open-system, ^{34}S-depleted pyrite that is almost 60‰ lighter than the coeval seawater sulfate (Calvert et al., 1996; Lyons, 1997; Wilkin and Arthur, 2001).

The 1 mM sulfate maximum is lower by roughly a factor of two than the mid-Proterozoic predictions summarized here (e.g., Kah et al., 2004). Although not a big difference, these results suggest that SEDEX deposits and Proterozoic sediments in general, through quantitative sulfate reduction and efficient hydrogen sulfide retention as metal sulfides, may exceed typical, more-recent early diagenetic environments in their ability to generate extreme ^{34}S enrichments. In modern marine sediments, as much as 95% of hydrogen sulfide generated is reoxidized (Jørgensen et al., 1990), in part because of limited supplies of reactive iron (Canfield et al., 1992). Under lower sulfate/sulfide conditions, iron limitation becomes less of a factor, particularly if iron and other sulfide-reactive metals are available in great abundance—suggesting anomalously high burial efficiencies for Proterozoic H_2S (Hurtgen et al., 2005). Also, the very high sedimentation rates expected in the isolated rift settings and the lack of Proterozoic bioturbation, regardless of benthic O_2 levels, would have a pronounced effect on sulfate transport, reactive carbon availability, and sulfide retention within the sediments and could enhance ^{34}S enrichment (Hurtgen et al., 2005). Furthermore, rates of BSR associated with SEDEX mineralization may have been anomalously high.

The Possible Role of Methane

Reduced concentrations of sulfate in the early ocean would have favored the production and preservation of methane (Habicht et al., 2002). Sulfate-reducing bacteria compete with methanogens for metabolizable organic compounds, and anaerobic oxidation of methane (AOM) occurs through a consortium of microorganisms, with BSR playing a central role (Boetius et al., 2000). Also, methanogens are anaerobes, and methanotropic bacteria readily oxidize methane in the presence of oxygen. Conditions of low oxygen, in combination with limited sulfate availability, would support a methane-rich ocean-atmosphere system (Pavlov et al., 2000; Habicht et al., 2002), which may have persisted, albeit at a reduced level, throughout most of the Proterozoic (Pavlov et al., 2003). The possibility of high levels of atmospheric methane during the Precambrian was indirectly advanced through the paleosol study of Rye et al. (1995), which has been challenged recently by Ohmoto et al. (2004).

Although a connection to the Proterozoic sulfur geochemistry is highly speculative, we also note that the most extreme ^{34}S-enrichments found in modern marine sediments occur in association with high fluxes of methane. High fluxes of methane within sedimentary systems support AOM and thus anomalously high subsurface rates of BSR and comparatively high associated $\delta^{34}S$ values for hydrogen sulfide and pyrite (Aharon and Fu, 2000, 2003; Arvidson et al., 2004; M. Formolo, 2006, personal commun.). It is interesting to note that pyrite $\delta^{34}S$ values as high as those found commonly in the Proterozoic are best developed today where subsurface, AOM-driven secondary sulfur is overprinting *freshwater* sediments (Jørgensen et al., 2004). Despite a lack of direct evidence, high methane availability may have supported high rates of Proterozoic BSR, which, in combination with lower overall sulfate concentrations, basin restriction, and strong hydrogen sulfide retention, would have favored high $\delta^{34}S$ values for the metal sulfides.

In modern deep-water, cold seep environments, bacterial mats of sulfide-oxidizing *Beggiatoa* are often abundant, reflecting the high levels of methane and associated BSR that typify the near-surface environment immediately below the mats (Larkin et al., 1994; Zhang et al., 2005). Such mats form today chemoautotrophically—deriving energy from oxidation of the copious sulfide formed during AOM. In addition to their abundances of pyrite, modern cold seep and gas hydrate localities are also known for enhanced, often extensive barite deposition (Dickens, 2001; Greinert et al., 2002; Torres et al., 2003).

Organic biomarkers consistent with sulfide-oxidizing bacteria were recently reported from inter-ore beds at the giant Proterozoic McArthur River Zn-Pb SEDEX deposit (Logan and Hinman, 2001), and a diverse microfossil assemblage is preserved in black chert nodules intimately associated with Zn-Pb mineralization (Oehler and Logan, 1977). Additional petrographic evidence for a microbial role in sulfide formation in the Proterozoic deposits comes from a diversity of macro- and micro-textures in iron and base metal sulfides (McGoldrick, 1999). In particular, "crinkly-wavy" laminated pyrite beds, which are ubiquitous in the northern Australian deposits and the eastern Belt Basin, are interpreted as pyritized microbial mats (Schieber, 1986, 1990; McGoldrick, 1999; McGoldrick et al., 1999) (Fig. 7).

Other Evidence for Low Proterozoic Sulfate

Low sulfate concentrations in the Proterozoic ocean are further suggested by other independent evidence: (1) a comparative scarcity of massive bedded gypsum deposits prior to the mid to late Proterozoic (Kah et al., 2004); (2) the generally lower amounts of barite associated with Proterozoic SEDEX deposits relative to Phanerozoic examples (Lydon, 1996)—although barium availability also plays an important role (Cooke et al., 2000); (3) possibly the low CAS concentrations in Proterozoic carbonates (Hurtgen et al., 2002, 2004; Pavlov et al., 2003; Gellatly and Lyons, 2005); and (4) perhaps even the widespread occurrence of dolomite in Precambrian rocks compared to younger sediments. Finally, carbon-sulfur relationships summarized in Strauss (2002) suggest marine deposition during the Paleo- and Mesoproterozoic that was characterized by lower concentrations of seawater sulfate than in the latest Precambrian and Phanerozoic. Lyons et

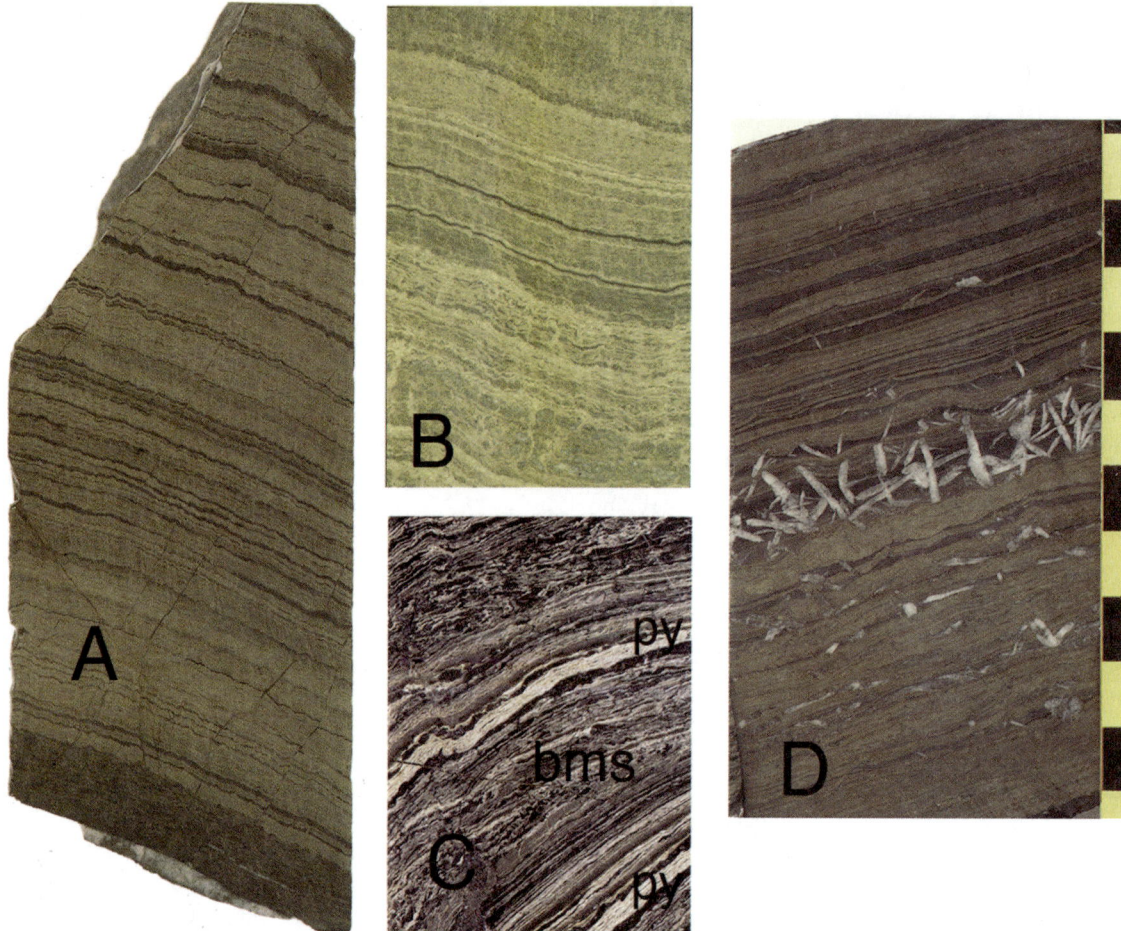

Figure 7. Core photos showing microbial mat-like textures preserved in pyrite and base metal sulfide ore from SEDEX deposits (Schieber, 1990; McGoldrick, 1999). (A) Unpolished slabbed core showing crinkly laminations in bedded pyrite from the late Paleoproterozoic Lady Loretta Zn-Pb-Ag deposit, northwest Queensland (width of core is ~45 mm). (B) Polished slabbed core from late Paleoproterozoic Grevillea Zn-Pb-Ag prospect (northwest Queensland) with crinkly laminated pyrite (width of sample ~20 mm). (C) Polished slabbed core of laminated high-grade Zn-Pb sulfide ore from the Lady Loretta deposit (width of sample ~20 mm); note two prominent crinkly laminated pyrite bands (py) and partly disrupted laminated base metal sulfide and pyrite domains (bms). (D) slabbed core from the Mesoproterozoic Sheep Creek Cu-Co, Montana (width of sample ~55 mm); note compactional draping of pyrite laminae around early-formed barite laths.

al. (2000) described analogous high C/S ratios for the shales of the lower Belt Supergroup (cf. Hurtgen et al., 2005).

Paleoredox

Proterozoic seawater sulfate concentrations, while low, were at least a factor of ten higher than those present during the Archean, as would be expected with the increased oxidation at Earth's surface beginning ca. 2.3 Ga. This extreme sulfur limitation alone may explain the absence of Archean SEDEX mineral deposits. Nonetheless, despite this fundamental shift in redox (Holland, 2002; Farquhar and Wing, 2003; Bekker et al., 2004), a growing body of evidence is suggesting that oxygen deficiency may have persisted in the deep ocean for as long as 1.5 b.y. following the Great Oxidation Event. Canfield (1998) first predicted this from a simple model that required only estimates of oxygen concentrations in the Proterozoic atmosphere and respiratory oxygen demand in the ocean—but without direct evidence. He further suggested that despite the relatively low sulfate conditions, the combination of a growing marine sulfate reservoir in the presence of deep-water anoxia may have led to widespread BSR and thus ocean-scale euxinia. Subsequent to Canfield's initial prediction, Mesoproterozoic euxinia was shown by Shen et al. (2003), Poulton et al. (2004), and Brocks et al. (2005) to have occurred at least on the scale of an individual basin and perhaps more generally throughout the ocean (Arnold et al., 2004).

Canfield (1998) extrapolated his model predictions for euxinia to suggest that BIFs may have disappeared ca. 1.8 Ga not because of comprehensive oxygenation of the ocean but rather through a buildup of H_2S in the water column, which would have modulated Fe solubility through pyrite formation. As such, the 1.8 Ga threshold would represent the time when increasing sulfur availability overcame the Fe fluxes to the ocean, including hydrothermal inputs (Isley and Abbott, 1999; Abbott and Isley, 2001; cf. Lowell and Keller, 2003). Consequently, it may not be fortuitous that SEDEX deposits first appear roughly when BIFs disappear—with both phenomena marking the generally widespread availability of H_2S. Also of relevance, Condie et al. (2001) argued for a prominent maximum in Precambrian black shale deposition during a window from 2000 to 1700 Ma, which they linked at least partially to superplume activity.

An additional and essential benefit of widespread anoxia is the enhanced preservation of sulfide minerals residing on or near the seafloor. It is well known that sulfides deposited on the seafloor today along mid-ocean ridges are rapidly oxidized off axis beneath the oxic water column (Alt et al., 1989; Alt, 1994; Edwards et al., 2003).

Paleotectonics and the Importance of Local Controls

Although tectonic conditions must also be favorable for the production of hydrothermal, mineralizing fluids, high levels of rift-related heat flow are certainly not unique to the sites and time windows of SEDEX ore deposition. Intracratonic and continental margin, fault-controlled basins and troughs are common in rift zones throughout geologic history, and patterns of rifting are often intimately related to supercontinent breakup. It may be relevant, however, that the mid-Proterozoic maximum in SEDEX mineralization coincides with a proposed peak in superplume activity and thus oceanic hydrothermal fluxes (Isley and Abbott, 1999; Condie et al., 2001; see also Barley and Groves, 1992). Isolated, rift basins may also have favored the development of local anoxia, although the importance of this is less clear for the Proterozoic deposits when deep-water oxygen deficiency may have been common. Finally, rift-associated hydrothermal activity and fault conduits for possible methane and other hydrocarbon migration would have been ideal for enhancing metal availability and perhaps hydrogen sulfide production. A close relationship between SEDEX-type mineralization and ancient hydrocarbon seeps or vents has been suggested by others (e.g., Johnson et al., 2004).

Comparisons with Younger Deposits

Temporal selectivity of SEDEX mineralization is, to some extent, biased by a few very large and well-studied deposits. Similarly, the number of SEDEX deposits for a given time interval may be less meaningful than the total mass of mineralization. Nevertheless, peak periods of SEDEX mineralization seem clear and include both Proterozoic and Paleozoic time intervals (Fig. 1) (Lydon, 1996). The most straightforward conclusion from this distribution is that the Proterozoic onset of large-scale SEDEX mineralization almost certainly coincides with the early to mid-Proterozoic accumulation of sulfate and sulfide within the global ocean system. Although sulfate was increasing over this interval, the predominance of ^{34}S-enriched sulfides suggests that sulfate remained well below modern levels.

The biggest complication in an otherwise consistent story is that Phanerozoic deposits are also known for their ^{34}S enrichments—for example, Red Dog, Rammelsberg, and the deposits of the Selwyn Basin, all of which are Paleozoic in age (Anger et al., 1966; Goodfellow, 1987; Eldridge et al., 1988; Jennings and King, 2002) (Fig. 8). Given this, it would be easy to de-emphasize the importance of global controls in favor of local parameters, such as rift-related basin restriction with associated anoxia and local sulfate deficiencies. Although the abundant barite associated with Paleozoic deposits suggests that sulfate was increasing in the ocean, our recent work (Gill et al., 2006) and the studies of Horita et al. (2002), Lowenstein et al. (2003), Brennan et al. (2004), and Canfield (2004) suggest that sulfate increased but remained low relative to the modern ocean well into the Paleo-

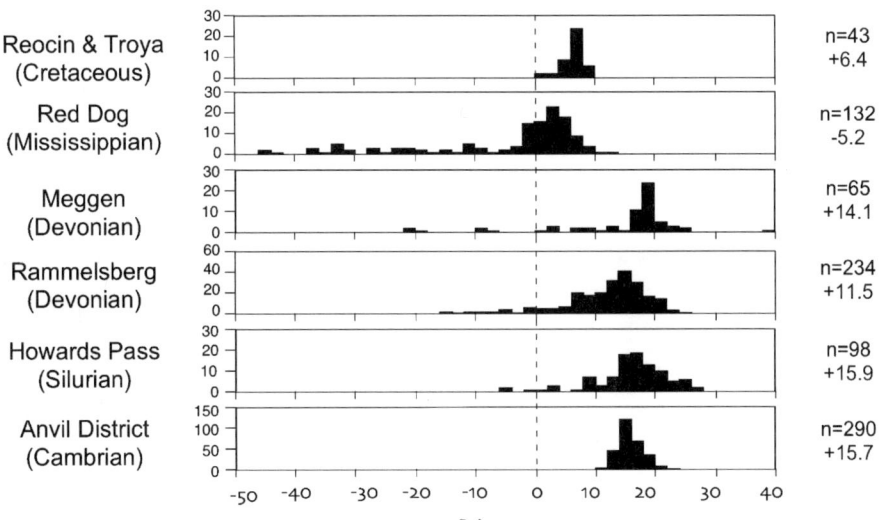

Figure 8. Summary of δ^{34}S data for selected Phanerozoic SEDEX deposits. Data represent pyrite and base metal sulfides. References and location information are as follows: Anvil District, Canada (Shanks et al., 1987); Howards Pass, Canada (Goodfellow and Jonasson, 1987); Rammelsberg, Germany (Anger et al., 1966); Meggen, Germany (Buschendorf et al., 1963); Red Dog, United States (Kelley et al., 2004); Reocin and Troya, Spain (Velasco et al., 1994).

zoic. As for the Proterozoic, restricted, rift basin settings during the Phanerozoic would have intensified the conditions of comparatively low global oceanic sulfate availability.

The predominance of ^{34}S-enriched metal sulfides of Phanerozoic SEDEX deposits highlights even more clearly the necessity for high rates of BSR in combination with anomalously high efficiency in sulfide mineralization. Again, high, fault-controlled fluxes of methane and other hydrocarbons would drive high microbial production of hydrogen sulfide at or below the sediment-water interface, and associated hydrothermal activity would supply the metals necessary for sulfide mineralization. Most Phanerozoic SEDEX deposits are of the "Selwyn-type" (Cooke et al., 2000), and therefore the ore-forming fluids carried metals and reduced S together. Mixing this ore-fluid sulfide with sedimentary sulfide formed at the sites of ore deposition may produce S isotope patterns that are inherently more complex than those seen in "McArthur-type" deposits.

Mechanisms of SEDEX mineralization may intrinsically lead to strong local sulfate deficiencies and thus high ^{34}S enrichments—regardless of the age. Zaback et al. (1993) and Jørgensen et al. (2004) remind us, however, that "closed-system" isotopic behavior often marks settings where sulfate resupply is occurring—but via fluxes that are overwhelmed by high rates of consumption by BSR. This model, for example, helps explain *large amounts* of ^{34}S-enriched sulfide, as are observed in SEDEX deposits, particularly if H$_2$S retention is high (Fig. 4). Today, such high rates of BSR are observed in association with AOM, particularly where other easily metabolized hydrocarbons are also released through seeps (Formolo et al., 2004). One could speculate that the fault control that is central to SEDEX mineralization plays a major role as a conduit for hydrocarbon migration in these organic-rich, shale-dominated settings.

The temporal and spatial distribution of Phanerozoic SEDEX mineralization almost certainly reflects the more strongly oxidizing conditions of the ocean, which would influence both H$_2$S availability and preservation of minerals precipitated at or near the seafloor. As a result, younger deposits may be limited to areas and episodes of basin-scale anoxia (Goodfellow and Jonasson, 1984; Turner, 1992), or they may have required an oxygen-shielding cap, either through barite deposition or mineralization below the sediment-water interface. The drop-off in abundance of Proterozoic deposits following the peak at 1800–1600 Ga is more difficult to explain. Perhaps, in contrast to arguments for persistent euxinia, conditions were already too oxidizing to facilitate widespread bottom-water sulfide production and enhanced metal sulfide preservation. Proterozoic deposits are relatively lacking in barite, as would be expected given the comparatively low sulfate availability in the ocean. Thus, any protection from oxygen offered by a barite cap would be minimal.

SUMMARY

Our goal here was to explore the first-order temporal patterns observed for SEDEX mineralization in light of the most recent models for early ocean-atmosphere chemistry. Despite the many unanswered questions, a number of internally consistent, primary observations have emerged, which provide a framework for further exploring the origins and distributions of SEDEX mineralization:

1. The absence of volumetrically significant SEDEX mineralization during the Archean can be attributed to limitations in sulfide availability, which ultimately reflect seawater sulfate concentrations beneath an oxygen-deficient atmosphere that were perhaps less than 1% of those present today.

2. Following the fundamental shift to a more oxidizing atmosphere ca. 2.3 Ga, sulfate fluxes to the ocean increased significantly through continental weathering of sulfides exposed in the presence of oxygen. Despite increased atmospheric availability, oxygen may have remained scarce in the deep ocean for at least another billion years. This deep ocean condition was at least partially a consequence of atmospheric oxygen levels that remained far below those present today. Increasing sulfate combined with an O$_2$-poor ocean could have spawned widespread euxinia throughout the Proterozoic ocean. The disappearance of banded iron formations ca. 1.8 Ga may mark the attainment of a critical threshold in H$_2$S accumulation in the ocean.

3. Given conditions that were tectonically favorable to hydrothermal sourcing of metals, the appearance of SEDEX mineralization ca. 1.8 Ga likely marks the ubiquity of H$_2$S in the marine system—including oxygen-deficient bottom waters—which favored the production and preservation of metal sulfide mineralization on the seafloor.

4. The common occurrence of ^{34}S enrichments within SEDEX sulfides is viewed as a combined product of globally low seawater sulfate, high rates of bacterial sulfate reduction, and locally enhanced sulfate limitations under the partially closed rifted marine basins in which SEDEX mineralization occurs. Limited sulfate availability may also be recorded in the lower amounts of barite associated with Proterozoic deposits.

5. Despite the possibility of a widely sulfidic deep ocean, the high rates of sulfate reduction expressed in the ^{34}S enrichments must, to an appreciable extent, reflect local bacterial activity. Enhanced local hydrocarbon fluxes and/or globally pervasive methane could have supported these high rates.

6. Following a mid-Proterozoic peak in SEDEX mineralization, the generally waning occurrence of these deposits throughout the latter part of the Proterozoic and only intermittent Phanerozoic occurrences may reflect increasing oxygenation of deep seawater and corresponding decreases in ambient H$_2$S. Under these conditions, basins with restricted circulation and local bottom-water anoxia become increasingly important.

7. Paleozoic SEDEX sulfides commonly bear ^{34}S enrichments analogous to those observed in the Proterozoic. The enrichments are partially a product of the local

reservoir properties—specifically inhibited seawater/sulfate exchange with the open ocean. However, as for the Proterozoic, lower global seawater sulfate availability relative to today and high rates of bacterial sulfate reduction are suggested. It is an oversimplification to imagine that restricted marine conditions a priori favor extreme sulfate limitations. For example, Paleozoic deposits show large amounts of associated barite. Furthermore, purely closed system behavior is inconsistent with large amounts of metal sulfide formation. Instead, we emphasize the importance of anomalously high rates of bacterial reduction and H_2S retention under conditions of limited but not isolated sulfate supply.

ACKNOWLEDGMENTS

Support for this project was provided by NSF grants EAR-9596079 and EAR-9725538 (Lyons and Kah). The authors benefited from many conversations with A. Anbar, D. Canfield, M. Hurtgen, and J. Luepke. D. Winston and G. Zieg are thanked for introducing the authors to the rocks of the Belt Supergroup. Research into the Proterozoic Zn-Pb-Ag deposits of northern Australia has been supported by grants from the Australian Research Council (ARC) and Australian Mineral Industry Research Association (AMIRA) to McGoldrick. Frank Corsetti and Barry Maynard provided insightful reviews, and we thank Steve Kesler and Hiroshi Ohmoto for their many efforts in assembling this volume.

REFERENCES CITED

Abbott, D., and Isley, A., 2001, Oceanic upwelling and mantle-plume activity: Paleomagnetic tests of ideas on the source of the Fe in early Precambrian iron formations, in Ernst, R.E., and Buchan, K.L., eds., Mantle plumes: Their identification through time: Geological Society of America Special Paper 352, p. 323–339.

Aharon, P., and Fu, B., 2000, Microbial sulfate reduction rates and sulfur and oxygen isotope fractionations at oil and gas seeps in deepwater Gulf of Mexico: Geochimica et Cosmochimica Acta, v. 64, p. 233–246, doi: 10.1016/S0016-7037(99)00292-6.

Aharon, P., and Fu, B., 2003, Sulfur and oxygen isotopes of coeval sulfate-sulfide in pore fluids of cold seeps sediments with sharp redox gradients: Chemical Geology, v. 195, p. 201–218, doi: 10.1016/S0009-2541(02)00395-9.

Alt, J.C., 1994, A sulfur isotopic profile through the Troodos ophiolite, Cyprus: Primary composition and the effects of seawater hydrothermal alteration: Geochimica et Cosmochimica Acta, v. 58, p. 1825–1840, doi: 10.1016/0016-7037(94)90539-8.

Alt, J.C., Anderson, T.F., and Bonnell, L., 1989, The geochemistry of sulfur in a 1.3 km section of hydrothermally altered oceanic crust, DSDP Hole 504B: Geochimica et Cosmochimica Acta, v. 53, p. 1011–1023, doi: 10.1016/0016-7037(89)90206-8.

Anbar, A.D., and Knoll, A.H., 2002, Proterozoic ocean chemistry and evolution: A bioinorganic bridge?: Science, v. 297, p. 1137–1142, doi: 10.1126/science.1069651.

Andrew, A.S., Heinrich, C.A., Wilkins, R.W.T., and Patterson, D.J., 1989, Sulfur isotope systematics of copper ore formation at Mount Isa, Australia: Economic Geology and the Bulletin of the Society of Economic Geologists, v. 84, p. 1614–1626.

Anger, G., Nielsen, H., Puchelt, H., and Ricke, W., 1966, Sulfur isotopes in the Rammelsberg ore deposit (Germany): Economic Geology and the Bulletin of the Society of Economic Geologists, v. 61, p. 511–536.

Arnold, G.L., Anbar, A.D., Barling, J., and Lyons, T.W., 2004, Molybdenum isotope evidence for widespread anoxia in mid-Proterozoic oceans: Science, v. 304, p. 87–90, doi: 10.1126/science.1091785.

Arvidson, R.S., Morse, J.W., and Joye, S.B., 2004, The sulfur biogeochemistry of chemosynthetic cold seep communities, Gulf of Mexico, U.S.A: Marine Chemistry, v. 87, p. 97–119, doi: 10.1016/j.marchem.2003.11.004.

Barley, M.E., and Groves, D.I., 1992, Supercontinent cycles and the distribution of metal deposits through time: Geology, v. 20, p. 291–294, doi: 10.1130/0091-7613(1992)020<0291:SCATDO>2.3.CO;2.

Bekker, A., Holland, H.D., Wang, P.-L., Rumble, D., III, Stein, H.J., Hannah, J.L., Coetzee, L.L., and Beukes, N.J., 2004, Dating the rise of atmospheric oxygen: Nature, v. 427, p. 117–120, doi: 10.1038/nature02260.

Boetius, A., Ravenschlag, K., Schubert, C.J., Rickert, D., Widdel, F., Gieseke, A., Amann, R., Jørgensen, B.B., Witte, U., and Pfannkuche, O., 2000, A methane microbial consortium apparently mediating anaerobic oxidation of methane: Nature, v. 407, p. 623–626, doi: 10.1038/35036572.

Brennan, S.T., Lowenstein, T.K., and Horita, J., 2004, Seawater chemistry and the advent of biocalcification: Geology, v. 32, p. 473–476, doi: 10.1130/G20251.1.

Brocks, J.J., Love, G.D., Summons, R.E., Knoll, A.H., Logan, G.A., and Bowden, S.A., 2005, Biomarker evidence for green and purple sulphur bacteria in a stratified Palaeoproterozoic sea: Nature, v. 437, p. 866–870.

Brunner, B., and Bernasconi, S.M., 2005, A revised isotope fractionation model for dissimilatory sulfate reduction in sulfate reducing bacteria: Geochimica et Cosmochimica Acta, v. 69, p. 4759–4771.

Buschendorf, Fr., Nielsen, H., Puchelt, H., and Ricke, W., 1963, Schwefel-isotopen-untersuchungen am pyrit-sphalerit-baryt-lager Meggen/Lenne (Deutschland) und an verschiedenen Devon-evaporiten: Geochimica et Cosmochimica Acta, v. 27, p. 501–523.

Calvert, S.E., Thode, H.D., Yeung, D., and Karlin, R.E., 1996, A stable isotope study of pyrite formation in the Late Pleistocene and Holocene sediments of the Black Sea: Geochimica et Cosmochimica Acta, v. 60, p. 1261–1270, doi: 10.1016/0016-7037(96)00020-8.

Campbell, F.A., Ethier, V.G., Krouse, H.R., and Both, R.A., 1978, Isotopic composition of sulfur in the Sullivan orebody, British Columbia: Economic Geology and the Bulletin of the Society of Economic Geologists, v. 73, p. 246–268.

Campbell, F.A., Ethier, V.G., and Krouse, H.R., 1980, The massive sulfide zone; Sullivan orebody: Economic Geology and the Bulletin of the Society of Economic Geologists, v. 75, p. 916–926.

Canfield, D.E., 1998, A new model for Proterozoic ocean chemistry: Nature, v. 396, p. 450–453, doi: 10.1038/24839.

Canfield, D.E., 2001, Isotope fractionation by natural populations of sulfate-reducing bacteria: Geochimica et Cosmochimica Acta, v. 65, p. 1117–1124, doi: 10.1016/S0016-7037(00)00584-6.

Canfield, D.E., 2004, The evolution of the Earth surface sulfur reservoir: American Journal of Science, v. 304, p. 839–861.

Canfield, D.E., and Raiswell, R., 1999, The evolution of the sulfur cycle: American Journal of Science, v. 299, p. 697–723.

Canfield, D.E., and Teske, A., 1996, Late Proterozoic rise in atmospheric oxygen concentration inferred from phylogenetic and sulphur-isotope studies: Nature, v. 382, p. 127–132, doi: 10.1038/382127a0.

Canfield, D.E., and Thamdrup, B., 1994, The production of ^{34}S-depleted sulfide during bacterial disproportionation of elemental sulfur: Science, v. 266, p. 1973–1975.

Canfield, D.E., Raiswell, R., and Bottrell, S., 1992, The reactivity of sedimentary iron minerals toward sulfide: American Journal of Science, v. 292, p. 659–683.

Carr, G.R., and Smith, J.W., 1977, A comparative isotopic study of the Lady Loretta zinc-lead-silver deposit: Mineralium Deposita, v. 12, p. 105–110, doi: 10.1007/BF00204509.

Chambers, L.A., Trudinger, P.A., Smith, J.W., and Burns, M.S., 1975, Fractionation of sulfur isotopes by continuous cultures of *Desulfovibrio desulfuricans*: Canadian Journal of Microbiology, v. 21, p. 1602–1607.

Claypool, G.E., Holser, W.T., Kaplan, I.R., Sakai, H., and Zak, I., 1980, The age curves of sulfur and oxygen isotopes in marine sulfate and their mutual interpretations: Chemical Geology, v. 28, p. 199–260, doi: 10.1016/0009-2541(80)90047-9.

Condie, K.C., Des Marais, D.J., and Abbott, D., 2001, Precambrian superplumes and supercontinents: A record in black shales, carbon isotopes, and paleoclimates?: Precambrian Research, v. 106, p. 239–260, doi: 10.1016/S0301-9268(00)00097-8.

Cooke, D.R., Bull, S.W., Large, R.R., and McGoldrick, P.J., 2000, The importance of oxidized brines for the formation of Australian Proterozoic stratiform sediment-hosted Pb-Zn (Sedex) deposits: Economic Geology and the Bulletin of the Society of Economic Geologists, v. 95, p. 1–17.

Davidson, G.J., and Dixon, G.H., 1992, Two sulphur isotope provinces deduced from ores in the Mount Isa eastern succession, Australia: Mineralium Deposita, v. 27, p. 30–41, doi: 10.1007/BF00196078.

Des Marais, D.J., Strauss, H., Summons, R.E., and Hayes, J.M., 1992, Carbon isotope evidence for the stepwise oxidation of the Proterozoic environment: Nature, v. 359, p. 605–609, doi: 10.1038/359605a0.

Detmers, J., Brüchert, V., Habicht, K.S., and Kuever, J., 2001, Diversity of sulfur isotope fractionations by sulfate-reducing prokaryotes: Applied and Environmental Microbiology, v. 67, p. 888–894, doi: 10.1128/AEM.67.2.888-894.2001.

Dickens, G.R., 2001, Sulfate profiles and barium fronts in sediment on the Blake Ridge: Present and past methane fluxes through a large gas hydrate reservoir: Geochimica et Cosmochimica Acta, v. 65, p. 529–543, doi: 10.1016/S0016-7037(00)00556-1.

Ding, T.-P., and Jiang, S.-Y., 2000, Stable isotope study of the Langshan polymetallic mineral district, Inner Mongolia, China: Resource Geology, v. 50, p. 25–38.

Dixon, G.H., and Davidson, G.J., 1996, Stable isotope evidence for thermochemical sulfate reduction in the Dugald River (Australia) strata-bound shale-hosted zinc-lead deposit: Chemical Geology, v. 129, p. 227–246, doi: 10.1016/0009-2541(95)00177-8.

Edwards, K.J., McCollom, T.M., Konishi, H., and Buseck, P.R., 2003, Seafloor bioalteration of sulfide minerals: Results from in situ incubation studies: Geochimica et Cosmochimica Acta, v. 67, p. 2843–2856, doi: 10.1016/S0016-7037(03)00089-9.

Eldridge, C.S., Compston, W., Williams, I.S., Both, R.A., Walshe, J.L., and Ohmoto, H., 1988, Sulfur isotope variability in sediment-hosted massive sulfide deposits as determined using the ion microprobe SHRIMP: I. An example from the Rammelsberg orebody: Economic Geology and the Bulletin of the Society of Economic Geologists, v. 83, p. 443–449.

Eldridge, C.S., Williams, N., and Walshe, J.L., 1993, Sulfur isotope variability in sediment-hosted massive sulfide deposits as determined using the ion microprobe SHRIMP: II. A study of the H.Y.C. Deposit at McArthur River, Northern Territory, Australia: Economic Geology and the Bulletin of the Society of Economic Geologists, v. 88, p. 1–26.

Farquhar, J., and Wing, B.A., 2003, Multiple sulfur isotopes and the evolution of the atmosphere: Earth and Planetary Science Letters, v. 213, p. 1–13, doi: 10.1016/S0012-821X(03)00296-6.

Formolo, M.J., Lyons, T.W., Zhang, C., Kelley, C., Sassen, R., Horita, J., and Cole, D.R., 2004, Quantifying carbon sources in the formation of authigenic carbonates at gas hydrate sites in the Gulf of Mexico, v. 205, p. 253–264.

Gellatly, A.M., and Lyons, T.W., 2005, Trace sulfate in mid-Proterozoic carbonates and the sulfur isotope record of biospheric evolution: Geochimica et Cosmochimica Acta, v. 69, p. 3813–3829.

Gill, B.C., Lyons, T.W., and Saltzman, M.R., 2006, Parallel, high-resolution carbon and sulfur isotope records of the evolving Paleozoic marine sulfur reservoir: Palaeogeography, Palaeoclimatology, Palaeoecology (in press).

Goodfellow, W.D., 1987, Anoxic stratified oceans a source of sulphur is sediment-hosted stratiform Zn-Pb deposits (Selwyn Basin, Yukon, Canada): Chemical Geology, v. 65, p. 359–382.

Goodfellow, W.D., and Jonasson, I.R., 1984, Ocean stagnation and ventilation defined by $\delta^{34}S$ secular trends in pyrite and barite, Selwyn Basin, Yukon: Geology, v. 12, p. 583–586, doi: 10.1130/0091-7613(1984)12<583:OSAVDB>2.0.CO;2.

Goodfellow, W.D., and Jonasson, I.R., 1987, Environment of formation of the Howards Pass (XY) Zn-Pb deposit, Selwyn Basin, Yukon, in Morin, J.A., ed., Mineral deposits of Northern Cordillera: Canadian Institute of Mining and Metallurgy, Special Volume 37, p. 19–50.

Goodfellow, W.D., Lydon, J.W., and Turner, R.J.W., 1993, Geology and genesis of stratiform sediment-hosted (SEDEX) zinc-lead-silver sulphide deposits: Geological Association of Canada Special Paper 40, p. 201–251.

Gorjan, P., Veevers, J.J., and Walter, M.R., 2000, Neoproterozoic sulfur-isotope variation in Australia and global implications: Precambrian Research, v. 100, p. 151–179, doi: 10.1016/S0301-9268(99)00073-X.

Greinert, J., Bollwerk, S.M., Derkachev, A., Bohrmann, G., and Suess, E., 2002, Massive barite deposits and carbonate mineralization in the Derugin Basin, Sea of Okhotsk: Precipitation processes at cold seep sites: Earth and Planetary Science Letters, v. 203, p. 165–180, doi: 10.1016/S0012-821X(02)00830-0.

Guha, J., 1971, Sulfur isotope study of the pyrite deposit of Amjhore, Shahbad District, Bihar, India: Economic Geology and the Bulletin of the Society of Economic Geologists, v. 66, p. 326–330.

Habicht, K.S., and Canfield, D.E., 2001, Isotope fractionation by sulfate-reducing natural populations and the isotopic composition of sulfide in marine sediments: Geology, v. 29, p. 555–558, doi: 10.1130/0091-7613(2001)029<0555:IFBSRN>2.0.CO;2.

Habicht, K.S., Gade, M., Thamdrup, B., Berg, P., and Canfield, D., 2002, Calibration of sulfate levels in the Archean Ocean: Science, v. 298, p. 2372–2374, doi: 10.1126/science.1078265.

Hoffman, P.F., Kaufman, A.J., Halverson, G.P., and Schrag, D.P., 1998, A Neoproterozoic snowball Earth: Science, v. 281, p. 1342–1346, doi: 10.1126/science.281.5381.1342.

Holland, H.D., 2002, Volcanic gases, black smokers, and the Great Oxidation Event: Geochimica et Cosmochimica Acta, v. 66, p. 3811–3826, doi: 10.1016/S0016-7037(02)00950-X.

Horita, J., Zimmermann, H., and Holland, H.D., 2002, Chemical evolution of seawater during the Phanerozoic: Implications from the record of marine evaporites: Geochimica et Cosmochimica Acta, v. 66, p. 3733–3756, doi: 10.1016/S0016-7037(01)00884-5.

Hurtgen, M.T., Arthur, M.A., and Halverson, G.P., 2005, Neoproterozoic sulfur isotopes, the evolution of microbial sulfur species, and the burial efficiency of sulfide as sedimentary pyrite: Geology, v. 33, p. 41–44, doi: 10.1130/G20923.1.

Hurtgen, M.T., Arthur, M.A., and Prave, A.R., 2004, The sulfur isotope composition of carbonate-associated sulfate in Mesoproterozoic to Neoproterozoic carbonates from Death Valley, California, in Amend, J.P., Edwards, K.J., and Lyons, T.W., eds., Sulfur biogeochemistry—Past and present: Geological Society of America Special Paper 379, p. 177–194.

Hurtgen, M.T., Arthur, M.A., Suits, N., and Kaufman, A.J., 2002, The sulfur isotopic composition of Neoproterozoic seawater sulfate: Implications for a snowball Earth?: Earth and Planetary Science Letters, v. 203, p. 413–429, doi: 10.1016/S0012-821X(02)00804-X.

Isley, A.E., and Abbott, D.H., 1999, Plume-related mafic volcanism and the deposition of banded iron formation: Journal of Geophysical Research, v. 104, p. 15461–15477, doi: 10.1029/1999JB900066.

Jennings, S., and King, A.R., 2002, Geology, exploration history and future discoveries in the Red Dog district, western Brooks Range, Alaska, in Cooke, D., and Pongratz, J., eds., Giant ore deposits: Characteristics, genesis and exploration: Hobart, Australia, Centre for Ore Deposit Research, CODES Special Publication 4, p. 151–158.

Johnson, C.A., Kelley, K.D., and Leach, D.L., 2004, Sulfur and oxygen isotopes in barite deposits of the western Brooks Range, Alaska, and implications for the origin of the Red Dog massive sulfide deposits: Economic Geology and the Bulletin of the Society of Economic Geologists, v. 99, p. 1435–1448.

Johnston, D.T., Wing, B.A., Farquhar, J., Kaufman, A.J., Strauss, H., Lyons, T.W., Kah, L.C., and Canfield, D.E., 2005, Active microbial sulfur disproportionation in the Mesoproterozoic: Science, v. 310, p. 1477–1479.

Jørgensen, B.B., Bang, M., and Blackburn, T.H., 1990, Anaerobic mineralization in marine sediments from the Baltic Sea-North Sea transition: Marine Ecology Progress Series, v. 59, p. 39–54.

Jørgensen, B.B., Böttcher, M.E., Lüschen, H., Neretin, L.N., and Volkov, I.I., 2004, Anaerobic methane oxidation and a deep H_2S sink generate isotopically heavy sulfide in Black Sea sediments: Geochimica et Cosmochimica Acta, v. 68, p. 2095–2118, doi: 10.1016/j.gca.2003.07.017.

Kah, L.C., Lyons, T.W., and Chesley, J.T., 2001, Geochemistry of a 1.2 Ga carbonate-evaporite succession, northern Baffin and Bylot islands: Implications for Mesoproterozoic marine evolution: Precambrian Research, v. 111, p. 203–234, doi: 10.1016/S0301-9268(01)00161-9.

Kah, L.C., Lyons, T.W., and Frank, T.D., 2004, Low marine sulphate and protracted oxygenation of the Proterozoic biosphere: Nature, v. 431, p. 834–838, doi: 10.1038/nature02974.

Kakegawa, T., and Ohmoto, H., 1999, Sulfur isotope evidence for the origin of 3.4 to 3.1 Ga pyrite at the Princeton gold mine, Barberton Greenstone Belt, South Africa: Precambrian Research, v. 96, p. 209–224, doi: 10.1016/S0301-9268(99)00006-6.

Kakegawa, T., Kawai, H., and Ohmoto, H., 1998, Origins of pyrite in the ~2.5 Ga Mt. McRae Shale, the Hamersley District, Western Australia: Geochimica et Cosmochimica Acta, v. 62, p. 3205–3220, doi: 10.1016/S0016-7037(98)00229-4.

Kampschulte, A., and Strauss, H., 2004, The sulfur isotopic evolution of Phanerozoic seawater based on the analysis of structurally substituted sulfate in carbonates: Chemical Geology, v. 204, p. 255–286, doi: 10.1016/j.chemgeo.2003.11.013.

Kelley, K.D., Leach, D.L., Johnson, C.A., Clark, J.L., Fayek, M., Slack, J.F., Anderson, V.M., Ayuso, R.A., and Ridley, W.I., 2004, Textural, compositional, and sulfur isotope variations of sulfide minerals in the Red Dog Zn-Pb-Ag deposits, Brooks Range, Alaska: Implications for ore formation: Economic Geology and the Bulletin of the Society of Economic Geologists, v. 99, p. 1509–1532.

Lambert, I.B., and Donnelly, T.H., 1991, Atmospheric oxygen levels in the Precambrian: A review of isotopic and geological evidence: Palaeogeography, Palaeoclimatology, Palaeoecology, v. 97, p. 83–91, doi: 10.1016/0031-0182(91)90184-S.

Large, R., McGoldrick, P., Bull, S., and Cooke, D., 2004, Proterozoic stratiform sediment-hosted zinc-lead-silver deposits of northern Australia, in Deb, M., and Goodfellow, W.D., eds., Sediment hosted lead-zinc sulphide deposits; attributes and models of some major deposits in India, Australia and Canada: New Delhi, Narosa Publishing House, p. 1–23.

Larkin, M.J., Aharon, P., and Henk, M.C., 1994, Beggiatoa in microbial mats at hydrocarbon vents in the Gulf of Mexico and Warm Mineral Springs, Florida: Geo-Marine Letters, v. 14, p. 97–103, doi: 10.1007/BF01203720.

Logan, G.A., and Hinman, M.C., 2001, Biogeochemistry of the 1640 Ma McArthur River (HYC) lead-zinc ore and host sediments, Northern Territory, Australia: Geochimica et Cosmochimica Acta, v. 65, p. 2317–2336, doi: 10.1016/S0016-7037(01)00599-3.

Logan, G.A., Hayes, J.M., Hieshima, G.B., and Summons, R.E., 1995, Terminal Proterozoic reorganization of biogeochemical cycles: Nature, v. 376, p. 53–56, doi: 10.1038/376053a0.

Lowell, R.P., and Keller, S.M., 2003, High-temperature seafloor hydrothermal circulation over geologic time and Archean banded iron formations: Geophysical Research Letters, v. 30, no. 7, p. 1391, doi: 10.1029/2002GL016536.

Lowenstein, T.K., Hardie, L.A., Timofeeff, M.N., and Demicco, R.V., 2003, Secular variation in seawater chemistry and the origin of calcium chloride basinal brines: Geology, v. 31, p. 857–860, doi: 10.1130/G19728R1.1.

Luepke, J.J., and Lyons, T.W., 2001, Pre-Rodinian (Mesoproterozoic) supercontinental rifting along the western margin of Laurentia: Geochemical evidence from the Belt-Purcell Supergroup: Precambrian Research, v. 111, p. 79–90, doi: 10.1016/S0301-9268(01)00157-7.

Lydon, J.W., 1996, Sedimentary exhalative sulphides (SEDEX), in Eckstrand, O.R., Sinclair, W.D., and Thorpe, R.I., eds., Geology of Canadian mineral deposit types: Geological Survey of Canada, Geology of Canada, v. 8, p. 130–152.

Lyons, T.W., 1997, Sulfur isotopic trends and pathways of iron sulfide formation in upper Holocene sediments of the anoxic Black Sea: Geochimica et Cosmochimica Acta, v. 61, p. 3367–3382, doi: 10.1016/S0016-7037(97)00174-9.

Lyons, T.W., Werne, J.P., Hollander, D.J., and Murray, R.W., 2003, Contrasting sulfur geochemistry and Fe/Al and Mo/Al ratios across the last oxic-to-anoxic transition in the Cariaco Basin, Venezuela: Chemical Geology, v. 195, p. 131–157, doi: 10.1016/S0009-2541(02)00392-3.

Lyons, T.W., Kah, L.C., and Gellatly, A.M., 2004, The Precambrian sulphur isotope record of evolving atmospheric oxygen, in Eriksson, P.G., et al, eds., The Precambrian Earth: Tempos and events: Developments in Precambrian geology: Amsterdam, Elsevier, p. 421–440.

Lyons, T.W., Luepke, J.J., Schreiber, M.E., and Zieg, G.A., 2000, Sulfur geochemical constraints on Mesoproterozoic restricted marine deposition: Lower Belt Supergroup, northwestern United States: Geochimica et Cosmochimica Acta, v. 64, p. 427–437, doi: 10.1016/S0016-7037(99)00323-3.

McGoldrick, P.J., 1999, Northern Australian SEDEX deposits: Microbial oases in Proterozoic seas, in Mineral deposits: Processes to processing: Proceedings of the 5th Biennial SGA Meeting and the 10th Quadrennial IAGOD Symposium, London, 22–26 August 1999: Rotterdam, A.A. Balkema, v. 2, p. 885–888.

McGoldrick, P.J., Dunster, J., and Aheimer, M., 1999, New sedimentological, geochemical and textural observations from the Lady Loretta deposit: Implications for ore genesis, in Holm, O., Pongratz, J., and McGoldrick, P., eds., Basins, fluids and Zn-Pb ores: Hobart, Australia, Centre for Ore Deposit Research, CODES Special Publication 2, p. 49–58.

McShane, M.B.J., 1996, Stratigraphy and mineralisation, Bernborough, Mount Isa South, northwest Queensland [Postgraduate thesis: Brisbane, University of Queensland, 86 p.

Oehler, J.H., and Logan, R.G., 1977, Microfossils, cherts, and associated mineralization in the Proterozoic McArthur (H.Y.C.) lead-zinc-silver deposit: Economic Geology and the Bulletin of the Society of Economic Geologists, v. 72, p. 1393–1409.

Ohmoto, H., Kakegawa, T., and Lowe, D.R., 1993, 3.4-billion-year-old biogenic pyrites from Barberton, South Africa: Sulfur isotope evidence: Science, v. 262, v. 555–557.

Ohmoto, H., Watanabe, Y., and Kumazawa, K., 2004, Evidence from massive siderite beds for a CO_2-rich atmosphere before ~1.8 billion years ago: Nature, v. 429, p. 395–399, doi: 10.1038/nature02573.

Painter, M.G.M., Golding, S.D., Hannan, K.W., and Neudert, M.K., 1999, Sedimentologic, petrographic and sulfur isotope constraints on fine-grained pyrite formation at Mount Isa Mine and environs, Northwest Queensland, Australia: Economic Geology and the Bulletin of the Society of Economic Geologists, v. 94, p. 883–912.

Pandalai, H.S., Changkakoti, A., Krouse, H.R., and Gunalan, N., 1991, The relationship between carbon, sulfur and pyritic iron in the Amjhore Deposit, Bihar, India: Economic Geology and the Bulletin of the Society of Economic Geologists, v. 86, p. 862–869.

Pavlov, A.A., Hurtgen, M.T., Kasting, J.F., and Arthur, M.A., 2003, Methane-rich Proterozoic atmosphere?: Geology, v. 31, p. 87–90, doi: 10.1130/0091-7613(2003)031<0087:MRPA>2.0.CO;2.

Pavlov, A.A., Kasting, J.F., Brown, L.L., Rages, K.A., and Freedman, R., 2000, Greenhouse warming by CH_4 in the atmosphere of early Earth: Journal of Geophysical Research, v. 105, p. 11981–11990, doi: 10.1029/1999JE001134.

Paytan, A., Kastner, M., Campbell, D., and Thiemens, M.H., 1998, Sulfur isotopic composition of Cenozoic seawater sulfate: Science, v. 282, p. 1459–1462, doi: 10.1126/science.282.5393.1459.

Paytan, A., Kastner, M., Campbell, D., and Thiemens, M.H., 2004, Seawater sulfur isotope fluctuations in the Cretaceous: Science, v. 304, p. 1663–1665, doi: 10.1126/science.1095258.

Poulton, S.W., Fralick, P.W., and Canfield, D.E., 2004, The transition to a sulphidic ocean ~1.84 billion years ago: Nature, v. 431, p. 173–177, doi: 10.1038/nature02912.

Ross, G.M., Bloch, J.D., and Krouse, H.R., 1995, Neoproterozoic strata of the southern Canadian Cordillera and the isotopic evolution of seawater sulfate: Precambrian Research, v. 73, p. 71–99, doi: 10.1016/0301-9268(94)00072-Y.

Rye, D.M., and Williams, N., 1981, Studies of the base metal sulfide deposits at McArthur River, Northern Territory, Australia. III: The stable isotope geochemistry of the H.Y.C., Ridge, and Cooley deposits: Economic Geology and the Bulletin of the Society of Economic Geologists, v. 76, p. 1–26.

Rye, R., Kuo, P.H., and Holland, H.D., 1995, Atmospheric carbon dioxide concentrations before 2.2 billion years ago: Nature, v. 378, p. 603–605, doi: 10.1038/378603a0.

Sageman, B.B., and Lyons, T.W., 2003, Geochemistry of fine-grained sediments and sedimentary rocks, in Mackenzie, F.T., ed., Sediments, diagenesis, and sedimentary rocks: Amsterdam, Elsevier, Treatise on Geochemistry, v. 7: p. 115–158.

Schieber, J., 1986, The possible role of benthic microbial mats during the formation of carbonaceous shales in shallow mid-Proterozoic basins: Sedimentology, v. 33, p. 521–536.

Schieber, J., 1990, Pyritic shales and microbial mats: significant factors in the genesis of stratiform Pb-Zn deposits of the Proterozoic: Mineralium Deposita, v. 25, p. 7–14.

Scott, K.M., Smith, J.W., Sun, S.-S., and Taylor, G.F., 1985, Proterozoic copper deposits in NW Queensland, Australia: Sulfur isotopic data: Mineralium Deposita, v. 20, p. 116–126, doi: 10.1007/BF00204322.

Shanks, W.C., Woodruff, L.G., Jilson, G.A., Jennings, D.S., Modene, J.S., and Ryan, B.D., 1987, Sulfur and lead isotope studies of stratiform Zn-Pb-Ag deposits, Anvil Range, Yukon: Basinal brine exhalation and anoxic bottom-water mixing: Economic Geology and the Bulletin of the Society of Economic Geologists, v. 82, p. 600–634.

Shen, Y., Buick, R., and Canfield, D.E., 2001, Isotopic evidence for microbial sulphate reduction in the early Archaean era: Nature, v. 410, p. 77–81, doi: 10.1038/35065071.

Shen, Y., Canfield, D.E., and Knoll, A.H., 2002, Middle Proterozoic ocean chemistry: Evidence from the McArthur Basin, northern Australia: American Journal of Science, v. 302, p. 81–109.

Shen, Y., Knoll, A.H., and Walter, M.R., 2003, Evidence for low sulphate and anoxia in a mid-Proterozoic marine basin: Nature, v. 423, p. 632–635, doi: 10.1038/nature01651.

Shields, G., Kimura, H., Yang, J., and Gammon, P., 2004, Sulphur isotopic evolution of Neoproterozoic-Cambrian seawater: New francolite-bound sulphate δ^{34}S data and a critical appraisal of the existing record: Chemical Geology, v. 204, p. 163–182, doi: 10.1016/j.chemgeo.2003.12.001.

Smith, J.W., and Croxford, N.J.W., 1975, An isotopic investigation of the environment of deposition of the McArthur mineralization: Mineralium Deposita, v. 10, p. 269–276, doi: 10.1007/BF00207885.

Smith, J.W., Burns, M.S., and Croxford, N.J.W., 1978, Stable isotope studies of the origins of mineralization at Mount Isa. I: Mineralium Deposita, v. 13, p. 369–381, doi: 10.1007/BF00206570.

Solomon, P.J., 1965, Investigations into sulphide mineralization at Mount Isa, Queensland: Economic Geology and the Bulletin of the Society of Economic Geologists, v. 60, p. 737–765.

Strauss, H., 1997, The isotopic composition of sedimentary sulfur through time: Palaeogeography, Palaeoclimatology, Palaeoecology, v. 132, p. 97–118, doi: 10.1016/S0031-0182(97)00067-9.

Strauss, H., 2002, The isotopic composition of Precambrian sulphides—Seawater chemistry and biological evolution, *in* Altermann, W., and Corcoran, P.L., eds., Precambrian sedimentary environments: A modern approach to ancient depositional systems: Oxford, Blackwell Science, International Association of Sedimentologists Special Publication 33, p. 67–105.

Strauss, H., and Schieber, J., 1990, A sulfur isotope study of pyrite genesis: The mid-Proterozoic Newland Formation, Belt Supergroup, Montana: Geochimica et Cosmochimica Acta, v. 54, p. 197–204.

Torres, M.E., Bohrmann, G., Dubé, T.E., and Poole, F.G., 2003, Formation of modern and Paleozoic stratiform barite at cold methane seeps on continental margins: Geology, v. 31, p. 897–900, doi: 10.1130/G19652.1.

Turner, R.J.W., 1992, Formation of Phanerozoic stratiform sediment-hosted zinc-lead deposits: Evidence for the critical role of ocean anoxia: Chemical Geology, v. 99, p. 165–188, doi: 10.1016/0009-2541(92)90037-6.

Velasco, F., Herrero, J.M., Gil, P.P., Alvarez, L., and Yusta, I., 1994, Mississippi Valley-type, sedex, and iron deposits in Lower Cretaceous rocks of the Basque-Cantabrian Basin, northern Spain, *in* Fontbote, L., and Boni, M., eds., Sediment-hosted Zn-Pb Ores: Special Publication 10, Society for Geology Applied to Ore Deposits: Springer-Verlag, Berlin-Heidelberg, p. 246–270.

Whelan, J.F., Rye, R.O., and deLorraine, W., 1984, The Balmat-Edwards zinc-lead deposits-synsedimentary ore from Mississippi Valley-type fluids: Economic Geology and the Bulletin of the Society of Economic Geologists, v. 79, p. 239–265.

Wilkin, R.T., and Arthur, M.A., 2001, Variations in pyrite texture, sulfur isotope composition, and iron systematics in the Black Sea: Evidence for Late Pleistocene and Holocene excursions of the O_2–H_2S redox transition: Geochimica et Cosmochimica Acta, v. 65, p. 1399–1416, doi: 10.1016/S0016-7037(01)00552-X.

Willan, R.C.R., and Coleman, M.L., 1983, Sulfur isotope study of the Aberfeldy barite, zinc, lead deposit and minor sulfide mineralization in the Dalradian metamorphic terrain, Scotland: Economic Geology and the Bulletin of the Society of Economic Geologists, v. 78, p. 1619–1656.

Winston, D., 1990, Evidence for intracratonic, fluvial and lacustrine settings of Middle and Late Proterozoic basins of western U.S.A., *in* Gower, C.F., Rivers, T., and Ryan, B., eds., Mid-Proterozoic Laurentia-Baltica: Geological Association of Canada Special Paper 38, p. 535–564.

Winston, D., and Link, P.K., 1993, Middle Proterozoic rocks of Montana, Idaho and eastern Washington, *in* Reed, J.C., Jr., et al., eds., Precambrian: Conterminous U.S.: Boulder, Colorado, Geological Society of America, Geology of North America , v. C-2, p. 487–517.

Zaback, D.A., Pratt, L.M., and Hayes, J.M., 1993, Transport and reduction of sulfate and immobilization of sulfide in marine black shales: Geology, v. 21, p. 141–144, doi: 10.1130/0091-7613(1993)021<0141:TAROSA>2.3.CO;2.

Zhang, C.L., Huang, Z., Cantu, J., Pancost, R.D., Brigmon, R.L., Lyons, T.W., and Sassen, R., 2005, Lipid biomarkers and carbon-isotope signatures of a microbial (*Beggiatoa*) mat associated with gas hydrates in the Gulf of Mexico: Applied and Environmental Microbiology, v. 71, p. 2106–2112, doi: 10.1128/AEM.71.4.2106-2112.2005.

MANUSCRIPT ACCEPTED BY THE SOCIETY 29 OCTOBER 2005

Geological Society of America
Memoir 198
2006

Precambrian Mississippi Valley–type deposits: Relation to changes in composition of the hydrosphere and atmosphere

Stephen E. Kesler
Martin H. Reich*

Department of Geological Sciences, University of Michigan, Ann Arbor, Michigan 48109, USA

ABSTRACT

We have evaluated the temporal distribution of Mississippi Valley-type (MVT) Zn-Pb deposits with special attention to the nature and number of deposits of Precambrian age. Our evaluation is based on the widely used model for MVT mineralization involving metal-bearing brines that lack reduced S and that deposit sulfides only where they encounter a reservoir of sulfide or where sulfate in the metal-bearing brine is reduced to sulfide. For MVT systems of this type, basins with abundant sulfate would be most favorable for development of MVT mineralization because these would allow transport of metals in sulfate-rich brines and deposition of metals in areas where the sulfate was reduced. Because abundant sulfate requires abundant atmospheric oxygen, the distribution of MVT deposits through time might reflect compositional changes in Earth's atmosphere, especially the suggested Great Oxidation Event (GOE).

A compilation of new data for the Bushy Park-Pering district in the Transvaal Supergroup of South Africa, the world's oldest known MVT province, and published information on other Precambrian MVT deposits in the Ediacara, Berg Aukas/ Abenab, Gayna River, Warrabarty, Nanisivik, Kamarga (Century), McArthur River (Coxco), Ramah, and Esker districts shows that they are generally similar in geologic setting and mineralogy to those in Phanerozoic rocks. Fluid inclusions in some Neoproterozoic deposits, including Berg Aukas/Abenab, Gayna River, Warrabarty, and Nanisivik, record higher temperatures and salinities than found in most Phanerozoic deposits, possibly reflecting igneous activity or a more proximal basinal setting during Precambrian time. Fluid inclusion leachate data for several Precambrian MVT deposits suggest that their parent brines formed by evaporation of seawater, and S isotope compositions indicate that the S was derived largely from coeval seawater sulfate. Comparisons of data from all deposits show no evidence for a gradual increase in temperature or salinity backward through time, such as might be caused by higher heat flow during early stages of Earth history, although the magnitude of this effect might be lost in the uncertainty of most fluid inclusion measurements. These observations confirm that MVT deposits reflect the chemistry of their source basins, which are as old as 2.6 Ga. No MVT deposits or suitable host rocks of an older age are known.

Precambrian MVT deposits do differ from their Phanerozoic analogues in the magnitude of mineralization. Precambrian deposits and districts formed at an

---

*Reich also at: Departamento de Geología, Facultad de Ciencias Físicas y Matemáticas, Universidad de Chile, Santiago, Chile.

Kesler, S.E., and Reich, M.H., 2006, Precambrian Mississippi Valley–type deposits: Relation to changes in composition of the hydrosphere and atmosphere, *in* Kesler, S.E., and Ohmoto, H., eds., Evolution of Early Earth's Atmosphere, Hydrosphere and Biosphere—Constraints from Ore Deposits: Geological Society of America Memoir 198, p. 185–204, doi: 10.1130/2006.1198(11). For permission to copy, contact editing@geosociety.org. ©2006 Geological Society of America. All rights reserved.

estimated rate of 5.5 per billion years versus a significantly larger rate of ~60 per billion years for Phanerozoic deposits, and the Phanerozoic deposits are considerably larger. Furthermore, the transition from low-magnitude, Precambrian-type to high-magnitude, Phanerozoic-type MVT mineralization took place at the beginning of Cambrian time rather than at the 2.3 Ga GOE. This appearance of widespread MVT mineralization is closer to the time at which sulfate concentrations in the world ocean are estimated to have reached present-day levels. Although these conclusions are subject to considerable uncertainty because of the limited number of Precambrian deposits, the lack of an increase in the frequency of MVT mineralization at the GOE suggests that widespread MVT mineralization requires higher levels of sulfate than could have been provided by this event, or that the appearance of sulfate in the ocean was considerably delayed. Finally, the presence of MVT deposits in basins that formed considerably before the GOE suggests that local sulfate concentrations were available at even early points in Earth's history.

Keywords: Mississippi Valley-type deposit, MVT, basin, brine, Pb-Zn.

INTRODUCTION

Mineral deposits have an uneven distribution through time that has been related to long-term changes in global heat flow, tectonism, and compositions of the atmosphere and oceans (Meyer, 1988; Barley and Groves, 1992). This uneven temporal distribution is particularly distinct for deposits that formed at Earth's surface, such as uranium-bearing conglomerates, laterites, evaporites, and iron formations, all of which contain redox-sensitive elements that might reflect increasing oxygen contents in the atmosphere and sulfate contents in the hydrosphere (Holland, 1984).

Ore deposits that formed in the deep subsurface also show temporal variations that might be due, at least in part, to changes in the composition of the atmosphere and hydrosphere. Among subsurface environments, sedimentary basins are most likely to reflect the surface environment because their rocks originated at the surface. The most widespread hydrothermal mineral deposits that formed by sedimentary hydrothermal systems are sandstone and unconformity U deposits, which contain redox-sensitive U, and sedex, Irish-type, and Mississippi Valley-type (MVT) Zn-Pb deposits, which contain redox-sensitive S. Elsewhere in this volume, reviews for two of these deposit types, sandstone-U (Gauthier-Lafaye) and sedex Zn-Pb (Lyons et al.) suggest that their temporal distribution reflects increasing atmospheric oxygen and marine sulfate, respectively. Whereas sedex Zn-Pb deposits form on the seafloor and sandstone-U deposits form in subsurface zones dominated by meteoric water, MVT Zn-Pb deposits form at somewhat greater depths in zones dominated by formation waters. These deposits also show a distinct age distribution, with fewer Precambrian representatives (Fig. 1). Here, we review the characteristics of these Precambrian MVT deposits with the goal of determining whether this pattern reflects a change in the oxygen content of the atmosphere and sulfate content of the oceans.

GEOLOGICAL SETTING AND GEOCHEMICAL CHARACTERITICS OF MVT DEPOSITS

MVT deposits are part of a continuum of Zn-Pb deposits, also including sedex and Irish-type, that form in craton margin and intracratonic sedimentary basins (Sangster, 1990). Although all three deposit types have similar mineralogy, textures, host rocks, and geology settings differ considerably (Fig. 2). At one end of the spectrum are MVT deposits, which consist of coarse-grained sphalerite, galena, and other minerals (Fig. 3A) that fill breccia-type porosity in dolomitized limestone along the margins of sedimentary basins (Anderson and Macqueen, 1988). MVT sulfides contain primary fluid inclusions with temperatures of 75–200 °C and salinities of 10–30 equivalent wt.% NaCl that are interpreted to be basinal brines that were expelled from deeper

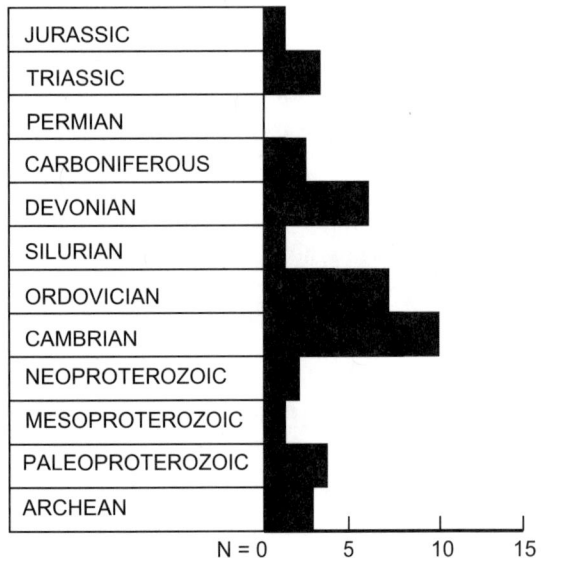

Figure 1. Age distribution for host rocks for MVT deposits, modified from Leach and Sangster (1993) to include additional Precambrian deposits.

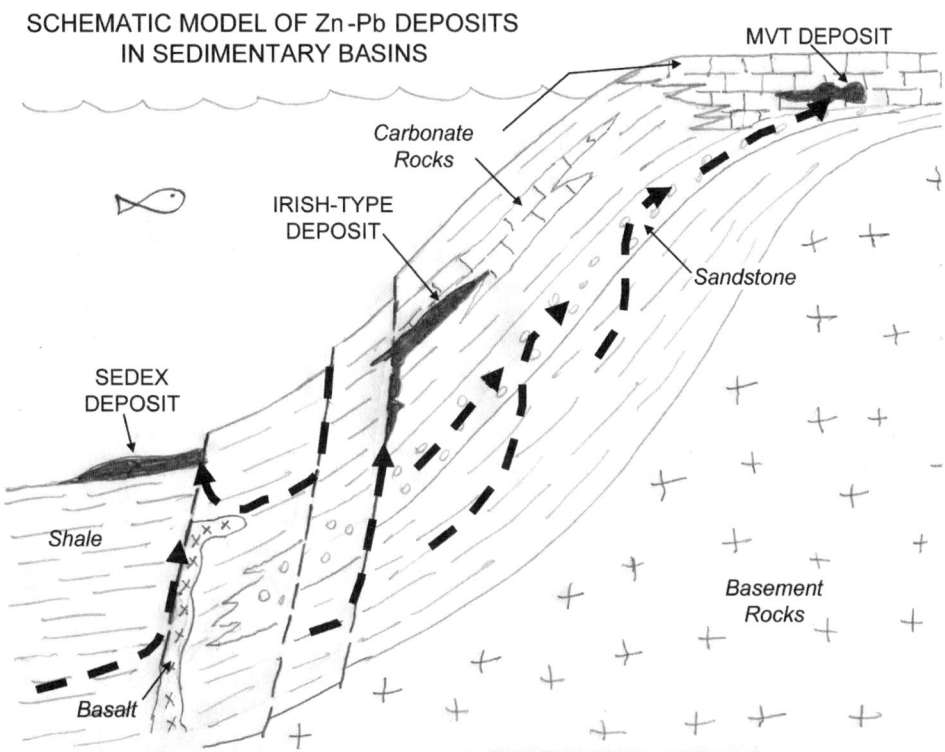

Figure 2. Highly schematic model for formation of MVT, sedex, and Irish-type Zn-Pb deposits in sedimentary basins. Dashed arrows show flow paths for mineralizing solutions. The model is not meant to indicate that these three deposits can form at the same time or that all of them form exclusively in an extensional environment.

parts of adjacent basins (Fig. 2). At the other end of the continuum are sedex deposits, which consist largely of fine-grained, well layered Zn-Pb and other sulfides (Fig. 3B) in black shales along rifted margins of basins (Scott, 1997). Although sedex sulfides do not provide much useful fluid inclusion data, other lines of evidence suggest that their parent fluids were slightly hotter and less saline than MVT fluids, that they were a mixture of basinal brine and seawater, and that they were driven at least partly by heat from mafic dikes that intruded rift-margin faults (Fig. 2). Irish-type deposits contain both layered and coarse-grained ores and appear to have formed largely along faults that cut rocks between the seafloor on which sedex deposits formed and the deeper reservoirs in which MVT deposits formed (Fig. 2).

We have focused on MVT deposits in this study because they form at greatest depth in the subsurface and are most likely to reflect the chemical composition of their host sedimentary basins. Most MVT deposits consist entirely or dominantly of galena and sphalerite with sparry dolomite, calcite, and local quartz and pyrite or marcasite. Some MVT deposits, which consist dominantly of fluorite, have been considered to form a distinct group in recent surveys and are not included in this summary (Sangster, 1990; Leach and Sangster, 1993). Similarly, barite is present in only a few deposits and, where present, it is not coeval

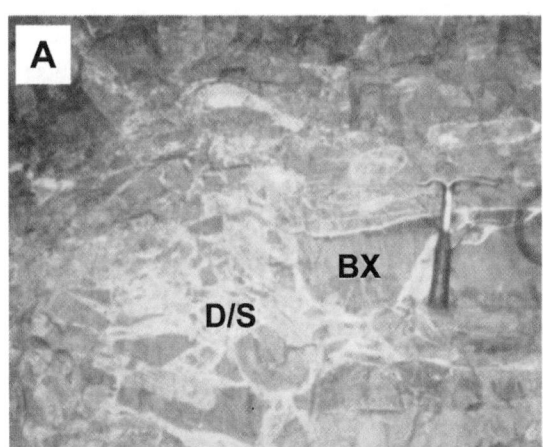

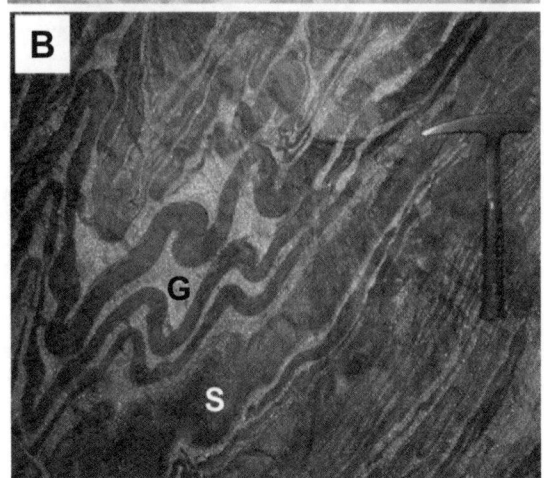

Figure 3. (A) Typical MVT ore showing brecciated blocks of dolostone (BX) surrounded by a matrix of sparry dolomite and sphalerite (D/S) (New Market mine, Tennessee). (B) Typical sedex ore showing deformed layers of sphalerite (S) and galena (G) (Sullivan mine, British Columbia).

with galena and sphalerite and thus does not provide a useful constraint on the composition of the metal-bearing MVT fluid. Other metals that are found in small amounts locally, including copper, cobalt, arsenic, and silver, are not sufficiently systematic in their distribution to classify MVT deposits further, and most were derived largely from nearby wall rocks (Burstein et al., 1992). Thus, insights into the processes that formed most MVT deposits are provided largely by constraints on the behavior of Pb, Zn, and S, and it is these metals that we must look to for information on the redox state of hydrothermal fluids in their host sedimentary basins.

Experimental studies show that chloride complexes are effective ligands for Zn and Pb in low-temperature, basinal brines and that bisulfide and organic complexes are less effective, particularly in the presence of abundant dissolved Ca and Na (Giordano, 1985; Barrett and Anderson, 1988; Sicree and Barnes, 1996). The presence of reduced (sulfide) S in MVT brines greatly reduces the solubility of chloride-complexed Zn and Pb (Anderson, 1983). The most widely applied model for formation of MVT deposits is based on this relation and involves metal-bearing brines that lack reduced S and that deposit MVT sulfides only where they encounter a reservoir of sulfide or where sulfate in the metal-bearing brine is reduced to sulfide (Anderson and Macqueen, 1988; Anderson, 1991; Leach and Sangster, 1993).

According to this "Anderson model," basinal environments containing only reduced S could not form MVT deposits because their low-temperature brines would not transport sufficient metals. Conversely, basinal environments that lack S would allow extensive migration of metals in low-temperature brines, but would not have the capacity to deposit MVT sulfides. The most favorable basinal environments for formation of MVT deposits would be those containing sulfate-bearing evaporites and brines and local accumulations of sulfide formed from this evaporite sulfate. All other things being equal, the MVT-forming capacity of such sedimentary basins should increase with increasing amounts of sulfate that could be reduced locally to cause ore deposition. Possible mechanisms for sulfate reduction in Precambrian sedimentary settings include near-surface bacterial action and deeper thermochemical sulfate reduction, as is the case in modern settings (Ohmoto and Goldhaber, 1997; Habicht et al., 2002). Because basinal brines form largely by evaporation of seawater or dissolution of marine evaporites (Hanor, 1987), their sulfate contents should reflect the composition of their source ocean and the oxygen content of its coeval atmosphere.

Although the Anderson model provides a useful framework for evaluating possible relations between MVT ore formation and compositional evolution of the atmosphere and hydrosphere, it is not without complexities. Ohmoto and Goldhaber (1997) have described eight pathways by which seawater sulfate can be fixed as sulfide in ore deposits. Whereas some of these pathways involve single-stage reduction of sulfate by biogenic, thermochemical, or more direct inorganic processes, others involve multiple steps such as bacterial reduction of seawater sulfate during diagenesis, formation of pyrite, and its subsequent dissolution and transport. In general, biogenic and diagenetic sulfides have lower $\delta^{34}S$ values than sulfides formed by thermochemical reduction of seawater sulfate, and some MVT deposits, such as the large Tri-State district, contain sulfides with low $\delta^{34}S$ values that might be of this origin (Ohmoto and Rye, 1979; Ohmoto and Goldhaber, 1997; Leach and Sangster, 1993). Although the existence of numerous possible pathways makes it more difficult to identify the actual source of S, and the presence of districts such as Tri-State show that alternative pathways operated locally, they do not negate the generalization that high basinal sulfate contents favor formation of MVT deposits.

Other possible complications to the model used here involve solubility constraints. For instance, Ba cannot be transported in the presence of significant dissolved sulfate (Blount, 1977), and the presence of barite in MVT deposits has been cited as evidence that MVT brines lacked sulfate. However, as noted above, barite is actually absent from most MVT deposits and, where present, it was not deposited at the same time as galena and sphalerite. A more important complication involves the possibility that acid MVT brines might carry metals and reduced S together, thus obviating the need for sulfate-rich environments noted above (Sverjensky, 1981). Ohmoto et al. (1990) have suggested that MVT districts such as the Upper Mississippi Valley, where sulfides have uniform $\delta^{34}S$ values and equilibrium fractionation, formed from fluids that contained both metals and sulfide, which requires that the fluid was either unusually hot or acidic. Wall-rock alteration in most MVT deposits does not provide evidence of highly acid brines, however, and maintenance of acidity in brines hosted by carbonate rocks requires high levels of CO_2, which have been documented in only a few MVT deposits (Haynes et al., 1989). If this exception does apply, it is most likely for MVT districts that formed at high temperatures, such as the Upper Mississippi Valley district (McLimans et al., 1980; Ohmoto et al., 1990). As is apparent from this list of complications, the Anderson model probably is less clearly applicable to Irish-type and sedex deposits, and they are not included in this discussion (Sangster, 1990).

In addition to general acceptance of the Anderson model, use of Precambrian MVT deposits as possible sensors of oxygen and sulfate contents of the atmosphere and hydrosphere requires that (1) their geology and geochemistry be similar to Phanerozoic MVT deposits to which the Anderson model was originally applied, and (2) their age be constrained by geologic, paleomagnetic, or isotopic observations. In the following sections, we review Precambrian MVT deposits from these two perspectives.

GEOLOGIC FEATURES OF PRECAMBRIAN MVT DEPOSITS

MVT deposits that are hosted by Precambrian rocks and to which comparisons might be made include, in order of increasing host rock age, Ediacara, Berg Aukas/Abenab, Gayna River, Warrabarty, Nanisivik, McArthur River (Coxco), Kamarga, Esker, Ramah, and Bushy Park-Pering-Zeerust (Fig. 4). Geological fea-

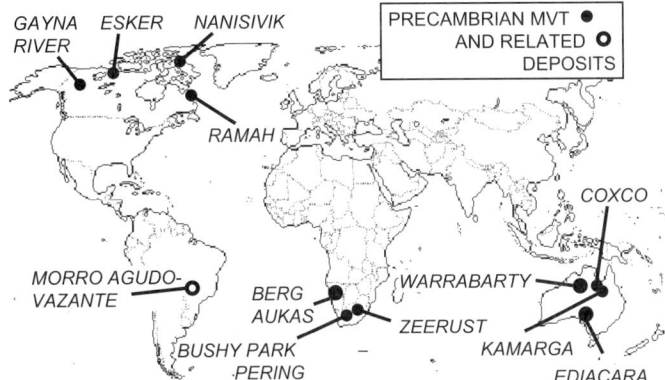

Figure 4. Location of Precambrian and related MVT deposits discussed in this paper.

tures of these deposits and districts are summarized in Table 1 and discussed briefly in the next section.

Some deposits of possible MVT origin are not included in this survey because of inadequate data, because they appear to be transitional in character, or because metamorphic and other overprints obscure their original features. For instance, Neoproterozoic platform carbonate sequences in Brazil, including the Una (Irecê Basin) and Bambui (São Francisco Basin) Groups that originally covered an area of more than 300,000 km^2, host Pb-Zn mineralization of uncertain type. The best-known deposits, including Morro Agudo, Vazante, and related deposits in the São Francisco Basin, have been classified as MVT, sedex, and Irish-type in different studies, but most recent descriptions suggest that they are least like MVTs and most like sedex (Iyer et al., 1992; Hitzman et al., 1995; Kyle and Misi, 1997; Misi et al., 2005). Similarly, the Balmat-Edwards district in New York has a general geologic setting typical of MVT deposits, including association with evaporites, but it is so highly metamorphosed that original textures necessary to determine its origin have been largely obliterated (Whelan et al., 1984, 1990).

MVT Deposits in Neoproterozoic Basins

Two of the four possible Neoproterozoic MVT districts, Ediacara and Berg Aukas/Abenab, have unusual metal contents. The Ediacara district, which is in the central Flinders Ranges of South Australia, is actually hosted by Cambrian shelf carbonates of the Ajax Limestone (Table 1), but is part of a dominantly Proterozoic sedimentary sequence (Christie-Blick et al., 1995). Mineralization consists of galena and pyrite with minor chalcopyrite and sphalerite and rare tetrahedrite and pearceite and has higher silver grades than most Phanerozoic MVTs (Drew and Both, 1984; McFarlane and Bone, 1994). The Berg Aukas/Abenab district, which is in the foreland thrust belt of the Damara orogen of Namibia, consists of galena, sphalerite, and sparry dolomite with trace amounts of pyrite, tetrahedrite, enargite, and chalcopyrite, as well as vanadium, silver, germanium, gallium, and cadmium at levels above those typical of Phanerozoic MVT deposits (Frimmel et al., 1996).

The Gayna River district, which is hosted by passive margin, platform carbonates of the Little Dal Group in Canada, is more typical of Phanerozoic MVTs in both composition and geologic setting (Table 1). The Little Dal Group is distinguished by its extensive stromatolites with internal structures typical of Precambrian reefs and by the presence of evaporites, including local salt casts (Hardy, 1979; Narbonne and Aitken, 1995; Turner et al., 2000). MVT mineralization is hosted by sedimentary and solution-collapse breccias that contain sphalerite and galena with sparry dolomite and local barite with snow-on-the-roof and colloform textures typical of Phanerozoic MVTs (Hardy, 1979; Hewton, 1982).

A final deposit that might belong in this group is Warrabarty in the Patterson orogen of Western Australia (Anderson et al., 2001, 2002). Warrabarty consists of Zn-rich mineralization in lower greenschist facies, metamorphosed carbonaceous dolostones, and limestones of the Broadhurst Formation, a part of the Meso-Neoproterozoic Throssell Group. Mineralization consists largely of sphalerite and galena with lesser pyrite, pyrobitumen, and sparry dolomite in breccias, veins, and disseminations in dolomitized wall rock (Smith, 1996).

MVT Deposits in Mesoproterozoic Basins

MVT districts are scarce in rocks of Mesoproterozoic age, the only good example being the Nanisivik district (Fig. 4). Although commonly classified as MVT, Nanisivik is somewhat unusual in form and mineralogy (Table 1). The ore-hosting Society Cliffs Formation is a platform carbonate sequence consisting largely of stromatolitic to massive dolostone with gypsum horizons that are among the oldest known extensive evaporite deposits (Kah et al., 2001). Solution-collapse and karst breccias are widespread in the upper part of the Society Cliffs Formation, but debate persists about their relation to MVT mineralization (Jackson and Ianelli, 1981; Olson, 1984; Ford, 1986). The ore zone is unusually sulfide-rich and ranges in texture from alternating layers of pyrite, sphalerite, sparry dolomite, and galena that fill open spaces to massive sulfides showing evidence of multiple stages of wall-rock replacement (Olson, 1984; Ghazban et al., 1990). Large iron sulfide bodies surround the Pb-Zn ore bodies and make up most of the district (Arne et al., 1991; Sutherland and Dumka, 1995). Iron-rich MVT mineralization forms from hot brines at temperatures of 200 °C or more, which is consistent with fluid inclusion evidence discussed below and might account for the unusual tube-like form of the main ore zone (St. Marie et al., 2001).

MVT Deposits in Paleoproterozoic Basins

Probable MVT districts in Paleoproterozoic rocks, including Coxco (McArthur River), Kamarga (Century), Ramah, and Esker, differ somewhat from Phanerozoic MVTs, especially in

TABLE 1. GEOLOGICAL CHARACTERISTICS OF PRECAMBRIAN MVT DEPOSITS COMPILED FROM SOURCES IDENTIFIED IN THE TEXT

Deposit	Ediacara	Berg Aukas	Gayna River	Warrabarty	Nanisivik	Coxco (McArthur River)
Host Rocks						
Basin Setting	Passive Margin	Rift Margin	Passive Margin	Transtensional Rift	Rift Margin	Rift Margin
Formation/Subgroup	Ajax	Abenab		Broadhurst	Society Cliffs	Emmeruga, Mara
Group		Otavi	Little Dal	Throssell	Uluksan	McArthur
Supergroup			MacKenzie Mtns.		Bylot	
Age	Cambrian	0.75 Ga	0.78–1.08 Ga	0.71–1.08 Ga	ca. 1.25 Ga	1.60–1.68 Ga
Possible Paleoaquifer	Pound Qtzt	Nosib Gp	Katherine Fm.	Coolbro Sandstone	Eqaluik Gp	Twallalh Gp
Evidence for Evaporites	no	No	Yes	No	Yes	Yes
Deformation	Minor	Minor	Minor	Moderate	Minor	Minor
Metamorphism	Not significant	Not significant	Not significant	Mid-greenschist	Not significant	Not significant
Mineral Deposits						
Area of District (km2)	30	1500	200	Unknown	600	600
Tonnage	29 Mt	small	~50 Mt	Unknown	13.3 Mt	Small
Grade	~1% Pb		5% Pb+ Zn	3–6% Zn+Pb	9.7% Zn, 0.9% Pb	2.5% Zn, 0.5% Pb
Additional Information					~35 Mt Fe sulfide	
Ore Setting	styl, bx	diss, bz	bx, repl	bx, vn, repl	Possible bx	bx
Major Minerals	gal, py, qz, cpy	sph, gal, dol	sph, gal, dol	sph, gal, dol	sph, gal, dol, py/mar	sph, gal, py/mar, dol
Minor Minerals	cpy, sph, tet, pear, dol		bar	cpy		cpy, bn, tet, bar, qz
Unusual Elements	Ag, As	V, Ag, Ge, Ga, As		None		Cu
Bitumen	No		Not reported	Yes	Yes	Yes
Age of Mineralization	Cambrian(?)	0.75 Ga	0.78–1.08 Ga	possible 0.84 Ga	1.095 Ga	1.64–1.68 Ga
Evidence for Age	Geologic	Geologic	Geologic	Model Pb isotope	Paleomagnetic	Geologic

Deposit	Kamarga	Ramah	Esker	Pering	Bushy Park
Host Rocks					
Basin Setting	Rift(?)	Passive Margin	Passive Margin	Passive Margin	Passive Margin
Formation/Subgroup	Gunpowder Creek	Reddick Bight	Rocknest	Campbellrand	Campbellrand
Group	McNamara	Ramah	Coronation	Ghaap	Ghaap
Supergroup				Transvaal	Transvaal
Age	1.653 Ga	ca. 2.0 Ga	ca. 1.9 Ga	2.64 to 2.43 Ga	2.64 to 2.43 Ga
Possible Paleoaquifer	Torpedo Creek Fm. (?)	Roswell Harbour Fm.	Odjick Fm	Makwassie Qz Phry	Makwassie Qz Phry
Evidence for Evaporites	Yes	No	Yes	No	No
Deformation	Minor	Extensive	Extensive	Minor	Minor
Metamorphism	Not significant	Greenschist facies	Low grade	Not significant	Not significant
Mineral Deposits					
Area of District (km2)	2(?)	1000	60	2400	?
Tonnage	50 Mt	No information	80 M	18 Mt	10 Mt
Grade	3% Zn+Pb	No information	4.7% Pb+Zn	3.6% Zn, 0.6% Pb	5% Zn, 0.6% Pb
Additional Information	(Includes 10 Mt 10%)				
Ore Setting	vn, bx, diss	bx, repl	diss, repl, bx	bx	bx
Major Minerals	py, sph, gal	sph, gal, dol	gal, sph, dol	sph, gal, dol	sph, gal, dol
Minor Minerals	cpy, jor, fl	py, qz, fs	cpy	qz, cpy	cpy, py
Unusual Elements	As, Sb	Cu	Cu	Cu	Not reported
Bitumen	Yes	Yes	No	Yes	
Age of Mineralization	Probably 1.6X Ga	Possibly 1.86 Ma	1.90–1.84 Ma	pre 1.98 Ga	Pre 1.98 Ga
Evidence for Age	Geologic	Geologic	Geologic	Geologic	Geologic

Note: Mineralogy: bar—barite, bn—bornite, cpy-chalcopyrite, ccy—chalcedony, dol—dolomite, fs—feldspar, fl—fluorite, gal—galena, jor—jordanite, mar—marcasite, pear—pearcite, po—pyrrhotite, py—pyrite, qz—quartz, sph—sphalerite, tet—tetrahedrite Form of Ore -bx—breccia, diss—disseminated, repl—replacement, styl—stylolite.

their apparently closer association with possibly coeval sedex mineralization. These relations are best displayed in the "Carpentaria zinc belt" of Australia, which contains both the McArthur River-Coxco and Kamarga districts.

MVT mineralization at Coxco is hosted by the latest Paleoproterozoic McArthur (Coxco) and McNamara (Kamarga) Groups in north-central Australia. Both districts contain much better known sedex deposits, HYC in the McArthur Group and Century in the McNamara Group (Plumb et al., 1990). The McArthur River district and immediately adjacent area contains MVT mineralization at Coxco and possibly at Ridge and Cooley (Williams, 1978; Walker et al., 1983; Selley et al., 2001; Garven and Bull, 2000). Overlying middle and upper McArthur Group platform dolostones, which host MVT mineralization, contain layers of acicular, radiating carbonate fans (Coxco needles) thought to have been precipitated as aragonite (Winefield, 2000). Mineralization at Coxco is hosted by a karst system that formed during uplift and weathering of domal stromatolites and fine-grained dolomite of the Mara Dolomite Member, and is very similar to Phanerozoic MVTs (Walker et al., 1983). It contains a first stage of colloform sphalerite, galena, and pyrite-marcasite associated with abundant organic matter, which was covered by detritus from the overlying Lynnot Formation and then followed by a second stage of crosscutting, coarse-grained pyrite-marcasite, sphalerite, galena, sparry dolomite, and minor bitumen in veins and dolomite breccias of probable tectonic origin. Mineralization at Cooley and Ridge is similar to the second stage at Coxco, but includes late chalcopyrite, tetrahedrite, and bornite, which are more abundant than in most Phanerozoic MVT deposits (Williams, 1978).

The McNamara district hosts probable MVT mineralization at Kamarga, which consists of pyrite and sphalerite in veins, breccias, disseminations, and massive replacements in the Gunpowder Creek Formation (Jones, 1986). The Gunpowder Creek Formation contains abundant textures typical of evaporites that have undergone replacement and is thought to have been deposited in a sabkha environment (Jones et al., 1999).

The Esker district near the Arctic coast in Canada is hosted by passive margin shelf sediments of the Rocknest Formation, which include abundant silica pseudomorphs after halite, gypsum, and possibly anhydrite (Grotzinger, 1986a, 1986b, 1986c). MVT mineralization, consisting largely of sparry dolomite, sphalerite, and galena with minor chalcopyrite, is in ore-matrix breccias and disseminations in a regionally extensive stromatolitic reef zone, and several possibly coeval Cu-Co-Pb-Zn sedex showings are in the underlying Odjick Formation (Rhondacorp 2002; Wachowiak, 2001). Paragenetic relations indicate that MVT mineralization followed regional dolomitization, silicification, and sparry dolomitization (Wachowiak, 2001; Wachowiak et al., 1997, 1998).

The Ramah district in northeastern Labrador, Canada, is the most strongly deformed and metamorphosed of the deposits reviewed here, and this limits the certainty with which it can be placed in the MVT class. The district is hosted by shelf carbonates of the Reddick Bight Formation, part of the Paleoproterozoic Ramah Group (Morgan, 1975; Knight and Morgan, 1977, 1981; Mengel et al., 1991). These rocks were deformed and metamorphosed during the 1.86 Ga Torngat orogeny, reaching greenschist facies in the area of the MVT prospects (Korstgard et al., 1987; Mengel et al., 1991; Mengel and Rivers, 1994; Scott and Gauthier, 1996; Hayashi et al., 1997). Mineralization is hosted by breccias that are cemented by sparry dolomite, pyrite, sphalerite, galena, quartz, and calcite, as well as carbonaceous material thought to have been derived from bitumen (Archibald, 1992; Archibald and Wilton, 1994; Wilton et al., 1993, 1994). Feldspar gangue is present locally, although it might have formed during later remobilization, as is indicated by crosscutting veins with MVT-like mineralogy.

MVT Deposits in Archean Basins

With the exception of Gayna River and Nanisivik, the deposits described so far are not particularly compelling examples of MVT mineralization. MVT deposits in the Transvaal Supergroup of South Africa, however, are remarkably similar to Phanerozoic deposits. The Neoarchean Ghaap and Chuniespoort Groups in the lower part of the Transvaal sequence consist largely of carbonate platform sediments with abundant stromatolitic reefs and overlying banded iron formation and contain the Pering, Bushy Park, Zeerust, and other smaller MVT districts (Beukes, 1987; Altermann and Nelson, 1998; Eriksson and Altermann, 1998). These deposits are found in two remnants of the original Transvaal Basin, which are known somewhat confusingly as the Griquatown West and Transvaal Basins.

The Griquatown West Basin contains the Pering and Bushy Park Zn-Pb deposits. The two deposits are sufficiently far apart to be considered separate districts, although each district contains only one important deposit. Pering and surrounding prospects are largely in breccias of probable karst origin that are aligned along fracture systems (Wheatley et al., 1986b; Kruger et al., 2001). Mineralization at Pering includes early fine-grained, colloform sphalerite and subordinate galena and chalcopyrite associated with sparry dolomite and hydrocarbons, and a final stage of coarse-grained sphalerite, galena, sparry dolomite, quartz, and calcite (Greyling et al., 2001). Bushy Park, the more northerly of the two districts, is also hosted by breccias of probable collapse origin and consists largely of coarse-grained sphalerite, sparry dolomite, and galena (Wheatley et al., 1986a; Martini et al., 1995; Baugaard et al., 2001; Schaefer et al., 2001).

The Transvaal Basin contains the Zeerust district and scattered smaller deposits that surround the Bushveld Complex, which intruded the Transvaal sequence ca. 2.06 Ga (Altermann and Nelson, 1998; Eriksson and Altermann, 1998). Mineralization at Zeerust consists dominantly of fluorite in stratabound zones and breccia bodies and the smaller deposits consist largely of sphalerite and galena with variable amounts of fluorite (Martini, 1976; Roberts et al., 1993; Martini et al., 1995; Poetter, 2001). As discussed in the next section these deposits are thought to have formed as part of a large hydrothermal system around the Bushveld Complex. In view of their high fluorine content and igneous association, they are not included in this survey.

Age of Precambrian MVT Deposits

Geological relations provide the only constraints on the age of mineralization in most of the Precambrian MVT districts discussed here. At Ediacara, mineralization took place during formation of the Neocambrian–Cambrian sedimentary sequence (Drew and Both, 1984). At Berg Aukas/Abenab, mineralization is related to 0.75 Ga rifting along the northern margin of the Damara Basin (Frimmel et al., 1996). Age relations for Warrabarty are particularly poorly defined. No measurements are available on the deposit and the Throssell Group, with which mineralization is roughly coeval, is limited only by 1.08 Ga granites that it overlies unconformably and a post-ore 0.71 Ga metamorphic event (Blockley and Myers, 1990; Smith, 1996). Common Pb isotope models suggest that mineralization was synchronous with formation of sediment-replacement copper deposits in the area (Nifty) ca. 0.84 Ga (Smith, 1996).

At Gayna River, main-stage mineralization is younger than the 0.78 Ga dikes that cut it (Hewton, 1982). At Coxco (McArthur River), early mineralization was deposited while karst zones in the Reward Dolomite were being filled by clastic sediment of the overlying Lynott Formation, which constrains it to an age of ca. 1.64 Ga, and second stage mineralization at Coxco took place after deposition of the Lynott Formation (Walker et al.,

1983; Page et al., 2000). No age constraints are recognized for mineralization at Cooley and Ridge, but they and Coxco should be roughly coeval with the 1.69 Ga enclosing sedimentary rocks if MVT mineralization is part of the HYC sedex system (Plumb et al., 1990). The age of mineralization at Kamarga has not been measured; its suggested relation to the Century sedex deposit constrains it to an age only slightly less than that of the 1.67 Ga McNamara Group host rocks (Plumb et al., 1990; Jones et al., 1999). Ages at Esker and Ramah are also not well known. At Esker, MVT mineralization is cut by faults related to the earliest phase of compressional deformation in the Coronation Group, making it probably post-1.90 Ga and pre-1.84 Ga in age (Hoffman and Bowring, 1984; Bowring and Grotzinger, 1992; Wachowiak, 2001). Breccia-hosted ore at Ramah is cut by structures related to the Torngat orogen, which formed during collision of the Nain and Rae terranes ca. 1.86 Ga (Wilton et al., 1993; Archibald, 1992).

Paleomagnetic and isotopic measurements supplement information from geologic relations at Nanisivik, where a diabase dike interpreted to cut ore has been correlated with the 0.72 Ga Franklin dike event (Olson, 1984; Heaman et al., 1992; Pehrsson and Buchan, 1999; Symons et al., 2000). Rb-Sr isotope compositions of sphalerite, dolomite, and leachates from Nanisivik do not fall on an isochron but are consistent with an age between ca. 0.75 and 1.25 Ga (Christensen et al., 1993). A paleomagnetic pole on recrystallized dolomite around the ore zone corresponds to an age of ca. 1.095 Ga on the North American apparent polar wander path (Symons et al., 2000). Although we have accepted a Mesoproterozoic age for Nanisivik here, some organic matter in the deposit appears to reflect a younger age (Gize, 1986), the age of the Franklin dikes has been challenged, and Ar-Ar measurements on MVT-related orthoclase yield Ordovician ages (Sherlock et al., 2004). Until these contradictory observations are resolved, inclusion of Nanisivik in this compilation should be regarded as tentative.

In the Transvaal Supergroup deposits of South Africa, the age of MVT mineralization is constrained by geologic relations and isotopic measurements. Mineralization is found largely in porosity related to an extensive karst system that developed on the Ghaap and Chuniespoort rocks in one or more intervals during Paleoproterozoic time. At Bushy Park-Pering, the age of ore-related karsting limits mineralization to pre-2.1 Ga (Martini et al., 1995). The Kalkdam and Katlani prospects, which are hosted by Ventersdorp lavas that underlie the Transvaal sequence in the Pering area, have similar Pb isotope compositions and yield a Rb-Sr isochron age of 1.977 Ga, which has been interpreted to indicate that mineralizing fluids were expelled during deformation of the 2.0 Ga Kheis Belt on the western edge of the Kaapvaal craton (Duane et al., 1991; Kruger et al., 2001). Relatively high $^{87}Sr/^{86}Sr$ ratios for calcite, dolomite, and sphalerite at Pering and Bushy Park require that the brine contacted evolved, potassium-rich rock outside the ore-hosting carbonate sequence, possibly during the Kheis event (Kruger et al., 2001; Schaefer et al., 2001; Kesler et al., 2003). Ar-Ar measurements on illite thought to coexist with ore minerals at Bushy Park yield an age of 2.145 ± 0.007 Ga, which is the best age estimate available at this time for Bushy Park-Pering mineralization (Schaefer, 2002).

In the Transvaal Basin, some Zn-Pb MVT deposits might have formed ca. 2.35 Ga, as indicated by relations at Genadendal where mineralization in Chuniespoort carbonate rocks forms a feeder for Zn-rich shale at the base of the overlying Pretoria Group (Martini, 1990; Eriksson et al., 2001a). However, fluorite mineralization at Zeerust yields a Sm-Nd isochron age of 2.06 Ga that is essentially the same as the Bushveld Complex, which dominated fluid migration in the Transvaal Basin (Kesler et al., 2003).

GEOCHEMISTRY OF PRECAMBRIAN MVT DEPOSITS

Fluid Inclusion Temperatures and Geochemistry

Temperatures of 75–200 °C are commonly cited for Phanerozoic MVT deposits with most of the variation related to position in the flow path relative to the source basin or possible sources of additional heat (Leach and Sangster, 1993; Rowan et al., 2001). Significantly higher heat flow in Precambrian time probably produced a hotter basinal environment. Estimates based on the history of global heat flow and resulting geotherms (Pollack and Chapman, 1977; Pollack, 1997) indicate that continental temperatures during Paleoproterozoic time were probably ~100 °C higher at depths of ~10 km, a level typical of the base of a thick sedimentary basin. A more conservative estimate of 50 °C is used here, assuming shallower source depths for mineralizing brines, lower temperatures in active sedimentary basins, and published estimates of ocean Precambrian temperatures (Knauth, 2005). This effect is shown in Figure 5 as a gradually increasing increment to the Phanerozoic temperature range. Increased ocean temperatures could have facilitated generation of basinal brines with salinities above the 10%–30% level typical of Phanerozoic deposits, although it is more difficult to estimate this effect because of the range of solution compositions observed in MVT deposits. An approximation based on the increase in salinity with temperature in the $NaCl-H_2O$ system (Sourirajan and Kennedy, 1962) results in salinities of ~33% for late Archean brines with temperatures of 250 °C. This effect is shown as a similar continuous decrease in inclusion salinities though Precambrian time in Figure 6.

Homogenization temperatures for fluid inclusions, largely in sphalerite, from Precambrian MVT deposits are generally more complex, with higher temperatures and more CO_2 and locally CH_4, than their Phanerozoic counterparts (Fig. 5). Ediacara, the youngest deposit, has homogenization temperatures typical of Phanerozoic MVT deposits (Drew and Both, 1984), but temperatures for Berg Aukas (100–210 °C), Gayna River (156–231 °C), Warrabarty (165–245 °C for gray-stage sphalerite), and Nanisivik (87 to above 300 °C) are progressively higher (Misiewicz, 1988; Carriere and Sangster, 1992; Olson,

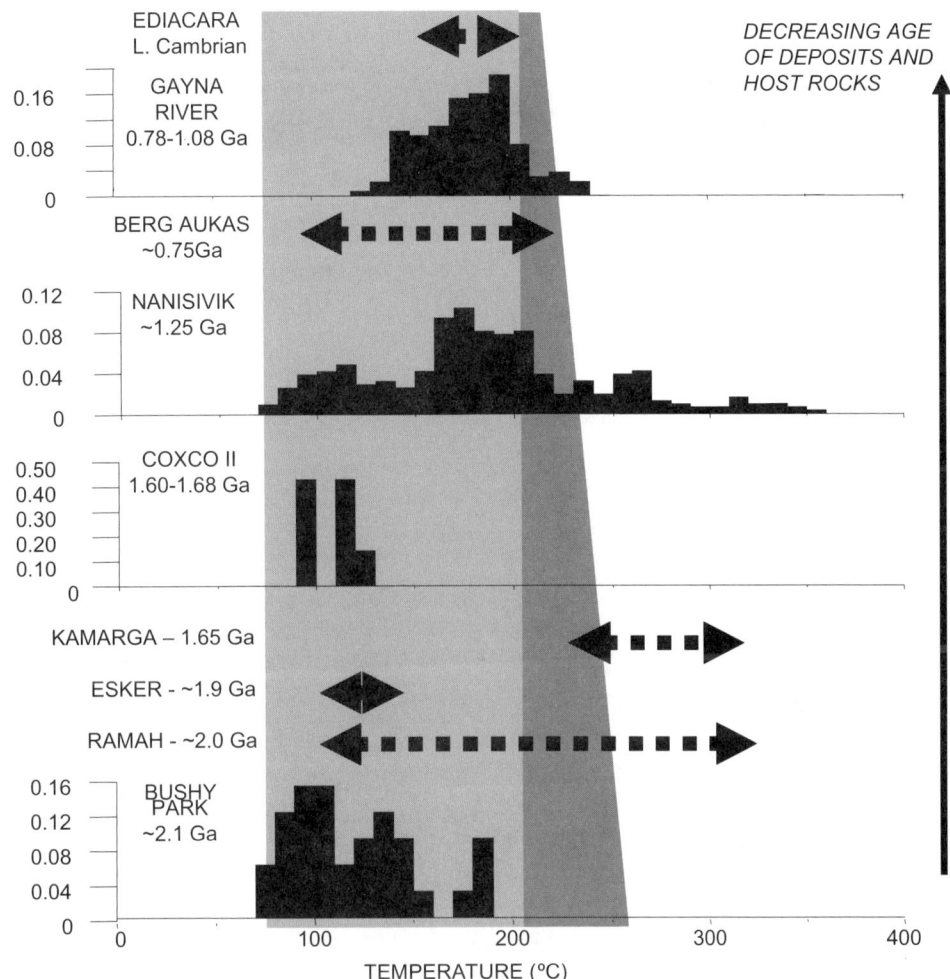

Figure 5. Homogenization temperatures of fluid inclusions in sphalerite from Precambrian MVT deposits compiled from sources discussed in the text. Arrows show range of values where data are not available to plot a histogram. Shaded rectangle in background shows range of temperatures typical of Phanerozoic MVT deposits (Leach and Sangster, 1993), and darker shaded triangle shows increased temperatures that might result from higher Precambrian heat flow as discussed in the text.

1984; McNaughton and Smith, 1986). Highest temperatures at Nanisivik, which are slightly above 300 °C, have been attributed to reheating by the diabase dike that cuts ore, but temperatures of 250 °C have been interpreted as primary, perhaps reflecting higher heat flow for the host rift-margin basin (Olson, 1984; McNaughton and Smith, 1986).

This trend of increasing homogenization temperatures with increasing age does not continue for Mesoproterozoic and older deposits, however. Inclusions in sphalerite from second-stage mineralization at Coxco (McArthur River) homogenize at 100–170 °C and those at Esker homogenize at ~100–150 °C, both of which are typical of Phanerozoic MVT deposits. Homogenization temperatures for sphalerite at Ramah fall in the same range (120–180 °C), but inclusions in quartz and dolomite extend to 320 °C and contain CO_2, probably reflecting a metamorphic overprint that might have corrupted the homogenization temperature record (Archibald, 1992). At Kamarga, fluid inclusions in sphalerite have homogenization temperatures of 270–320 °C and are accompanied by CO_2-rich vapor inclusions that are not as clearly related to a post-ore overprint (Jones et al. 1999).

Fluid inclusions in Archean-hosted Zn-rich MVT deposits are more typical of those in Phanerozoic-hosted MVT deposits with the exception of high-temperature, gas-rich inclusions of uncertain origin that are present in some of the older deposits. Reconnaissance observations that we have made on sphalerite and sparry dolomite from Bushy Park yielded homogenization temperatures of 77–120 °C for primary and pseudosecondary aqueous inclusions and 130–195 °C for secondary aqueous inclusions, all in sphalerite (Table 2). Secondary aqueous inclusions from sparry dolomite homogenized at temperatures of

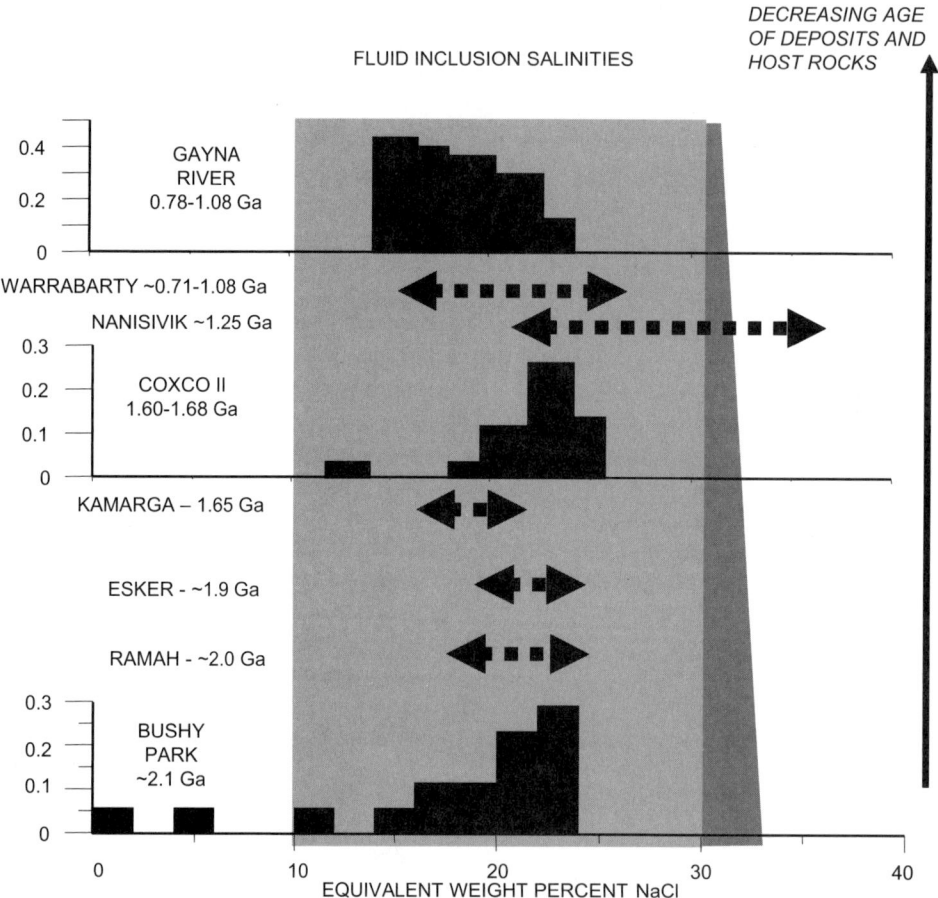

Figure 6. Salinity of fluid inclusions from Precambrian MVT deposits compiled from sources discussed in the text. Arrows show range of values where data are not available to plot a histogram. Shaded rectangle in the background shows the range of salinities typical of Phanerozoic MVT deposits (Leach and Sangster, 1993), and darker shaded triangle shows increased salinities that might result from higher temperatures caused by higher Precambrian heat flow, as discussed in the text.

TABLE 2. SUMMARY OF FLUID INCLUSION MEASUREMENTS
FOR THE BUSHY PARK DISTRICT, SOUTH AFRICA

Sample no.	Mineral	Type	Homogenization temperature (°C)	Freezing temperature (°C)	Equivalent wt% NaCl
BP-01-A	Sphalerite	(L+V)	184–195	–9 to –8	12–13
BP-03	Sphalerite	(L+V)	129–192	–11 to –8	12–15
BP-07	Sphalerite	(L+V)	77–117	–20 to –7	11–22
BP-10	Sphalerite	(L+V)	135–140	–11 to –8	12–15
BP-13	Dolomite	(L+V)	150–187	–11 to –5	8–15
BP-13	Sphalerite	(L+V)	83–188	–20 to –1	2–22

150–190 °C, slightly higher than temperatures of 100–175 °C reported by Wheatley et al. (1986b) for gangue carbonate at Pering. Schaefer et al. (2001) reported a similar range of homogenization temperatures (90–168 °C) for Bushy Park sphalerite and dolomite, with no systematic difference between inclusions in different minerals. They also reported primary vapor-rich inclusions containing variable proportions of CO_2 and CH_4 from Bushy Park, and used these inclusions to determine a pressure correction of ~50 °C for the homogenization temperatures. Greyling et al. (2001) reported temperatures of 157–210 °C for Pering based on intersecting isochors in aqueous and carbonic inclusions.

Freezing temperatures (and salinities estimated from them) for fluid inclusions in sphalerite from some of the Precambrian MVT deposits extend to values in and above the high end of the range typical of Phanerozoic MVT deposits (Fig. 6). Whereas the range of freezing temperatures for inclusions at Gayna (–12 to –24 °C) falls in the center of the Phanerozoic range, those from Coxco II (McArthur River) (–22 to –28 °C) and Esker (–18 to –24 °C) are nearer the high end of the range (Walker et al., 1983; Wachowiak et al., 1997, 1998; Carriere and Sangster, 1992; Smith, 1996). Freezing temperatures from quartz and dolomite at Ediacara (–23 to –28 °C) are also near the top of the Phanerozoic

MVT range. Similar patterns are seen in the deposits for which only salinities are quoted. Inclusions in sphalerite and dolomite from Warrabarty have salinities of 15–26 total salt and contain significant amounts of Ca (Smith, 1996); inclusions in quartz and dolomite from Ramah have salinities of 15–29 equivalent wt% NaCl (D. Wilton, 2002, written commun.), and inclusions from sphalerite and dolomite at Berg Aukas have average salinities of 23 equivalent wt% NaCl (Misiewicz, 1988). All of these are in the upper part of the Phanerozoic MVT range (Fig. 6). Inclusion fluids at Nanisivik have even higher salinities, with first-melting (eutectic) temperatures of –50 °C or more, unusually low last-melting temperatures of –25 to –45 °C, and estimated salinities of 24–35 equivalent wt% NaCl (McNaughton and Smith, 1986).

Our reconnaissance freezing measurements on primary and pseudosecondary inclusions in Bushy Park sphalerite indicate the presence of significant Ca or Mg in the fluids, and yield final-melting temperatures of –1 to –20 °C, corresponding to salinities of up to 22 equivalent wt% NaCl (Fig. 6). Secondary inclusions in sphalerite and sparry dolomite, which are not included in Figure 6, have final melting temperatures of –3 to –14 °C, corresponding to salinities of ~5–17 equivalent weight percent NaCl. Schaefer et al. (2001) reported a similar range of salinities from 27 to 1 equivalent wt% NaCl for inclusions at Bushy Park and suggested that it reflected the presence of a high salinity ore fluid that mixed with meteoric water to form fluids with intermediate salinities, which is in agreement with our observations. Greyling et al. (2001) reported a similar large range of salinities (but not freezing temperatures) for Pering, extending from only a few equivalent wt% NaCl to values as high as 50%, reflecting abundant Ca and Mg in the inclusions.

Na-Cl-Br compositions of fluid inclusion leachates from most Phanerozoic MVT minerals fall on or near the seawater evaporation line in plots of Na/Br versus Cl/Br, suggesting that their source brines formed by evaporation of seawater (Kesler et al., 1995, 1996; Viets et al., 1996; Chi and Savard, 1997; St. Marie and Kesler, 2000). Of the Precambrian MVT deposits and districts included in this study, Na-Cl-Br leachate data are available for Nanisivik (Viets et al., 1996) and Berg Aukas (Chetty and Frimmel, 2000) and are reported here for Bushy Park (Table 3). Two leachates from sphalerite at Nanisivik and one from sparry dolomite at Berg Aukas plot just above the seawater evaporation line, with Nanisivik farther along the evaporation trend (Fig. 7). Leachates from Bushy Park sphalerite and sparry dolomite plot in about the same location as Berg Aukas, also just above the seawater evaporation line (Fig. 7). The similarity in position of Precambrian and Phanerozoic MVT deposits in Figure 7 suggests that the Na-Cl-Br composition of seawater and MVT brine-forming process were similar throughout most of Earth history.

Sulfur Isotope Geochemistry

Sulfides in most of the Precambrian MVT deposits have high $\delta^{34}S$ values that approach those of sulfate in coeval evaporites or seawater, a pattern that is consistent with the Anderson model

TABLE 3. COMPOSITION OF FLUID INCLUSION LEACHATES FROM THE BUSHY PARK DISTRICT, SOUTH AFRICA

Sample no.	Mineral	Na/Br (atomic)	Cl/Br (atomic)	Na/K (atomic)	Ca/Mg (atomic)
BP-1	Sphalerite	196	336	17.3	6.4
	Sphalerite	204	335		
	Dolomite	171	333		
BP-2	Sphalerite	204	331		
BP-3	Sphalerite	216	336		
	Dolomite	188	347		
	Dolomite	192	361		
BP-6	Sphalerite	204	321	16.2	7.3
	Sphalerite	197	303	16.4	7.9
	Sphalerite	188	313		
BP-8	Sphalerite	212	336		
BP-10	Sphalerite	184	305		
	Dolomite	167	308		
BP-13	Sphalerite	163	305		
	Sphalerite	161	308		
	Sphalerite	150	285		

Note: Analyses carried out at University of Michigan using method described by St. Marie et al. (2000).

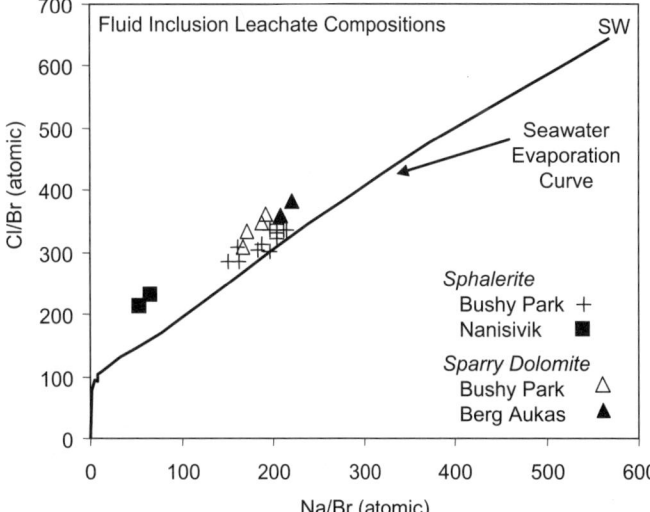

Figure 7. Na/Br versus Cl/Br diagram comparing composition of fluid inclusion leachates from Precambrian MVT deposits with the composition of evaporated modern seawater (line). Data from this study for Bushy Park, from Chetty and Frimmel (2000) for Berg Aukas, and from Viets et al. (1996) for Nanisivik.

applied here (Fig. 8). Best agreement between MVT sulfides and rock sulfate is seen at Nanisivik, where $\delta^{34}S$ values of 21‰–31‰ for MVT sulfides are almost identical to $\delta^{34}S$ values of 22‰–32‰ for evaporite gypsum from the Society Cliffs Formation (Olson, 1984; Ghazban et al., 1990). Ghazban et al. (1990) showed that $\delta^{34}S$ values of the sulfides in Nanisivik ore could be accounted for by deposition from a fluid containing S with $\delta^{34}S$ values of 26 ± 1‰ and suggested that the S was derived from seawater, and more recent data of Kah et al. (2001) show that it could have come directly from Society Cliffs gypsum. Unusually low $\delta^{13}C$ values of 6‰ to –12‰ for sparry dolomite in the Nanisivik ore assemblages suggest that the sulfate was reduced by reaction

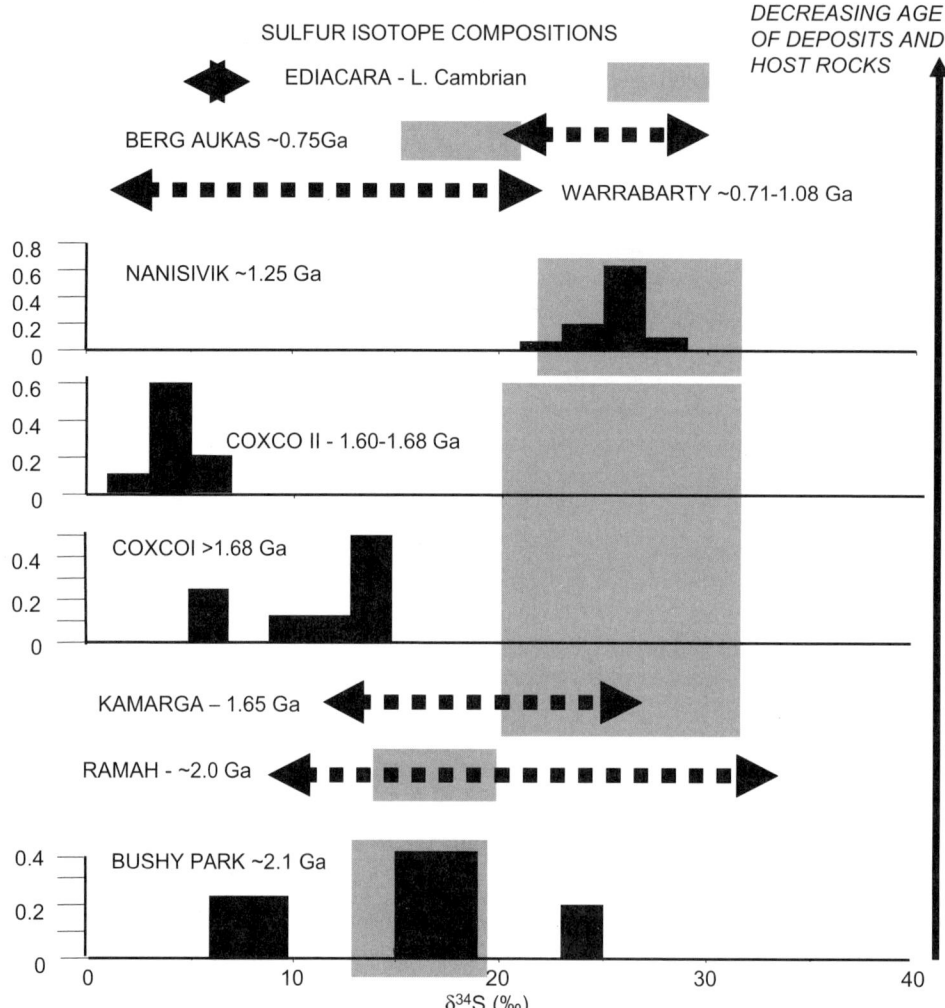

Figure 8. Histograms showing isotopic composition of sulfur in sphalerite and galena from Precambrian MVT deposits. Arrows show range of values where data are not available to plot a histogram. Shaded rectangles show isotopic composition of sulfate in seawater coeval with host rocks for the deposits, based on sources discussed in the text.

with organic matter, providing strong support for the Anderson model (Ghazban et al., 1990).

Sulfides at Coxco (McArthur River) have $\delta^{34}S$ values significantly lower than coeval seawater. $\delta^{34}S$ values of 1.3‰–21.7‰ are lower than the 20‰–32‰ estimated for sulfate in the McArthur Group from analysis of trace sulfate in carbonate rocks and barite that probably replaced sedimentary sulfates (Walker et al., 1983; Bottomley et al., 1992). Second-stage sulfides at Coxco have lower values of 0.9‰–16.1‰, and sphalerite, galena, and pyrite from the Cooley and Ridge deposits have even lower $\delta^{34}S$ values that are similar to those in the HYC sedex deposit (Walker et al., 1983; Rye and Williams, 1981). A possible source of S with the necessary low $\delta^{34}S$ values is pyrite or H_2S with $\delta^{34}S$ values near 0‰ in nearby Tawallah Group black shales (Shen et al., 2002).

Sulfides of possible MVT origin from Ramah have $\delta^{34}S$ values of ~8–34‰ (D. Wilton, 2002, written commun.). No S isotope values are available for sulfate in the sedimentary host rocks, although $\delta^{34}S$ values estimated for seawater sulfate of Paleoproterozoic age range from ~15‰ to 20‰, which falls within the range of MVT sulfide values (Fig. 8). Sulfur isotope data are not available for ore at Esker, but trace sulfate in the Rocknest Formation has $\delta^{34}S$ values of ~12‰–25‰ (Ueda et al., 1991). Sulfur isotope analyses have not been published for the Gayna deposits. $\delta^{34}S$ values for gray-stage sphalerite at Warrabarty range from 1.5‰ to 20.4‰, a somewhat larger range than seen in the other deposits, but have a "distinct mode" at 11‰–14‰ (Smith, 1996). Uncertainty about the age of Warrabarty mineralization limits the degree to which it can be compared with coeval seawater sulfate, although the best comparison is probably with that for Berg Aukas.

$\delta^{34}S$ values obtained in this study for Bushy Park sulfides (Table 4) range from 15.2‰ to 16.6‰ for galena and 15.8‰–19.7‰ with one unusually high value of 24.1‰ for sphalerite

TABLE 4. SULFUR ISOTOPE ANALYSES OF MINERALS
FROM THE BUSHY PARK DISTRICT, SOUTH AFRICA

Sample	Mineral	^{TM34}S (‰)
BP-13-GA	Galena	15.2
BP-UEO-3-GA	Galena	16.6
BP-UEO-1-SP	Sphalerite	15.8
BP-13-SP	Sphalerite	17.3
BP-UEO-3-SP	Sphalerite	17.3
BP-UEO-2-SP	Sphalerite	18.8
BP-UEO-6-SP	Sphalerite	19.7
BP-UEO-8-SP	Sphalerite	24.1

Note: Analyses carried out at Queen's University in the laboratory of T.K. Kyser.

(Fig. 8). Schaefer et al. (2001) reported lower $\delta^{34}S$ values of 4.8‰–8.5‰ for sphalerite and galena and –9.7‰–26.7‰ for diagenetic(?) pyrite from Bushy Park. Sulfate evaporites are not known in the Transvaal sequence, but $\delta^{34}S$ values of 13‰–17‰ have been reported for trace sulfate in carbonate rocks from the Malmani Subgroup (Buchanan and Rouse, 1982, *in* Strauss, 1993), similar to the range estimated by Canfield and Raiswell (1999) for late Archean seawater. This range is very similar to most of our values for sulfides at Bushy Park (Table 4), suggesting that MVT sulfide was derived largely from coeval seawater sulfate. The lower $\delta^{34}S$ values observed for Bushy Park sphalerite by Schaefer et al. (2001) probably reflect additions of S from diagenetic pyrite or related sources.

SIGNIFICANCE OF PRECAMBRIAN MVT DEPOSITS TO EARLY EARTH ATMOSPHERE AND HYDROSPHERE COMPOSITIONS

Comparison of Precambrian and Phanerozoic MVT deposits

Our results show that Precambrian and Phanerozoic MVT deposits share many similarities. Deposits of both ages are along the margins of passive-margin and rifted basins, and are hosted largely by platform carbonate rocks containing extensive reefs. With the exception of Ramah and Esker, most of the Precambrian MVT deposits are in undeformed sequences or foreland thrust belts, as are most of their Phanerozoic counterparts. Both Precambrian and Phanerozoic deposits have generally similar and simple mineralogy. Where ore mineralogy is more complex and includes elements such as copper, silver, and cobalt, these are interpreted to have resulted from local features that contaminated the ore fluid (Frimmel et al., 1996; Wu et al., 1997). Fluid inclusion observations, including homogenization temperatures and salinities, are also generally similar for deposits of both ages, and Precambrian exceptions seem to be just that rather than indications of systematic temporal changes. For instance, only Gayna River and Nanisivik, among the unmetamorphosed Precambrian MVT deposits discussed here, have significantly higher temperatures than those of Phanerozoic deposits (which probably reflect more proximal basinal sources or later dike-related heating), and even older Precambrian deposits such as Coxco and Bushy Park have temperatures typical of Phanerozoic deposits (Fig. 5).

More detailed grade-tonnage and frequency comparisons, however, indicate that the two groups differ. The grade-tonnage plot for Precambrian and Phanerozoic MVT deposits (Fig. 9) can be divided into four quadrants at a grade of 7% Pb+Zn and a tonnage of 10 million that results in a nearly equal number of Phanerozoic deposits in all four quadrants (29%, 24%, 26%, and 21% for NW, NE, SE, and SW quadrants, respectively). Precambrian deposits are distributed much differently among the quadrants, however, with none (0%) in the NW, 13% in the NE, 62% in the SE, and 25% in the SW (Fig. 9). Viewed only from the perspective of grade, only 13% of the Precambrian deposits fall above the 7% division compared with 53% of the Phanerozoic deposits. However, 75% of the Precambrian MVTs have more than 10 million tons compared with only 50% of the Phanerozoic deposits. Thus, Precambrian deposits are generally lower grade but higher tonnage, which should mean that they have generally similar metal contents. This is confirmed by the fact that ~50% of the Precambrian deposits plot above the 1 million ton diagonal in Figure 9, compared with ~41% of the Phanerozoic deposits. Considering all of the uncertainties involved, this suggests that the amount of metal moved by Precambrian and Phanerozoic MVT-forming systems was about the same, but that fewer Precambrian systems made ore.

Frequency comparisons indicate further than there were not many Precambrian systems. Figure 1 shows that ~11 MVT deposits and districts formed during Precambrian time versus at least 30 during Phanerozoic time. In view of the large difference in duration of these time periods, this amounts to a very different indicated "MVT-formation rate" of ~5.5 per billion years for Precambrian deposits and districts versus 60 per billion years for Phanerozoic deposits and districts. If Nanisivik is not

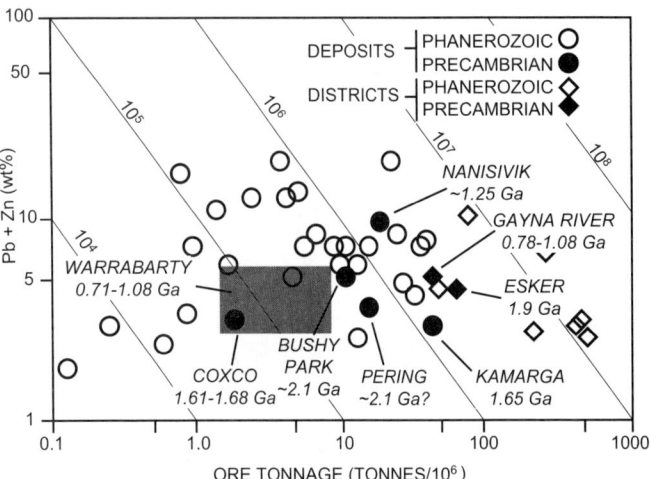

Figure 9. Grade-tonnage plot comparing MVT deposits and districts of Phanerozoic and Precambrian age. Data for Phanerozoic deposits and districts from Leach and Sangster (1993) and data for Precambrian deposits and districts from references cited here. Grade-tonnage for Warrabarty shown as a range estimated from Smith (1996).

Precambrian in age, as noted above, the Precambrian MVT-formation rate would be even lower. The age of ore-hosting sedimentary units varies greatly for both Precambrian and Phanerozoic rocks, although the significance of this variation is complicated by uncertainty about the exact age of mineralization. For instance, Cambrian rocks contain a significant majority of deposits that probably formed later in Paleozoic time (Leach et al., 2001). Within the Precambrian, MVT formation appears to have been greatest during late and early Proterozoic time, with only Nanisivik from the intervening period. These complications can be minimized by comparing the abundance of deposits in 0.5-b.y. groups that bracket the range of possible depositional ages for most of the deposits. As seen in Figure 10, a plot of this type shows a steep rise in MVT formation for the last 0.5 b.y. of Earth history and a much smaller and not greatly changeable rate during earlier (Precambrian) time.

Frequency comparisons of this type should be normalized to the amount of favorable carbonate-bearing sedimentary basins that remain from each of the time periods of interest, thus taking into account both formation of appropriate host rocks and their preservation during later events. For instance, carbonate rocks are scarce in most early Archean greenstone belts and related sedimentary basins; significant volumes of carbonate sediment began to form only by ca. 2.7–2.4 Ga when favorable shelf environments developed for the first time (Eriksson et al., 1998; 2001a, 2001b). Thus, the lower frequency for Precambrian MVT deposits might simply reflect a lack of suitable shelf carbonates in which ore could have formed. Although quantitative data are not available on the change in shelf carbonates through time, they should be related to continental growth rates because shelf environments formed on the margins of flooded continents. Continental growth rates are themselves matters of significant debate (Condie, 2000), but even those based on episodic growth produce relatively smooth cumulative curves for the change in total crust volume with time. As can be seen in Figure 10, growth rates of this type are not similar to our estimate of the formation of MVT deposits through Precambrian time. This is not surprising in view of the other important factors that control shelf carbonate development, especially sea level and climate (Walker et al., 2002).

In the absence of global data, more detailed comparisons of shelf carbonate volumes must be confined to specific areas. One of the largest areas of early Precambrian carbonate sedimentation was the Transvaal-West Griqualand-Kanye Basins of the Kaapvaal Craton in South Africa, where platform sediments covered an area of ~600,000 km^2 (Beukes, 1987). If carbonates of the Kaapvaal Craton correlate with those of the Jeerinah Formation in the Pilbara Craton of Australia to form the Vaalbara terrane, as suggested by Cheney (1996), this platform probably covered at least an additional 100,000 km^2. MVT deposits in this sequence can be compared with two important shelf sequences in North America: (1) Cambrian–Ordovician platform sediments of the Appalachian Basin (USA) with an area of at least 100,000 km^2, and (2) Devonian platform sediments of the Lennard shelf (Australia) with an area of ~50,000 km^2 (Kesler, 1996; Vearncombe et al., 1996). MVT mineralization is widespread in the Vaalbara of South Africa but absent from correlative rocks in Australia (Blockley and Myers, 1990). In the Appalachian province, MVT mineralization is found in all areas, although it is most abundant in the south and progressively less abundant northward (Kesler, 1996). In the Lennard shelf, MVT mineralization is found largely in two areas, at the north and south ends of the belt (Vearncombe et al., 1996).

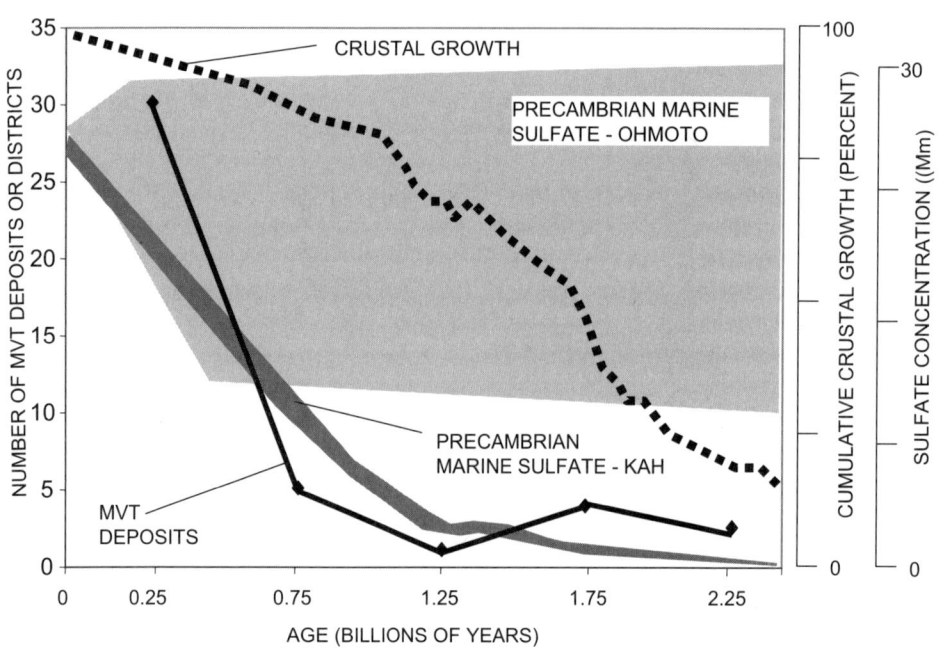

Figure 10. Number of MVT deposits and districts hosted by rocks of Phanerozoic and Precambrian age compiled from data of Leach and Sangster (1993) and this paper, plotted in 0.5-b.y. intervals. Crustal growth curve estimated from data of Condie (2000) and competing estimates of marine sulfate from Kah et al. (2004) and Ohmoto (2004). Fluctuations in sulfate concentration (Lowenstein et al., 2003) of the Paleozoic ocean are not shown.

The density of MVT mineralization in these Precambrian and Phanerozoic basins differs greatly, however. Looking only at reserves and production for deposits and districts that have been or probably will be mined, the Vaalbara province contains ~1 million tons of Zn compared with ~4 million tons for the Appalachian province and 4 million tons for the Lennard shelf. If MVT Zn-Pb and F-Ba districts, rather than deposits and prospects, are counted, the Vaalbara contains approximately six districts compared with at least 25 for the Appalachian province and about six for the much less extensive Lennard shelf. Thus, when viewed in terms of the areas, Phanerozoic shelf sequences clearly contain more MVT ore than their Precambrian analogues.

If the scarcity of Precambrian MVT deposits cannot be attributed to a lack of reef carbonate rocks, perhaps it is the porosity of the carbonates that limited Precambrian MVT mineralization. Low porosity would have limited the volume of MVT minerals that could be deposited by invading fluids, and therefore the grade of the deposit. This might be suspected because Phanerozoic-age carbonate reefs were built by a variety of organisms, whereas Precambrian reefs are almost exclusively stromatolitic (Webb, 2001; Petrov and Semikhatov, 2001; Grotzinger, 1986a, 1986b, 1986c; Hoffman, 1989; Ricketts and Donaldson, 1989). Systematic comparisons of porosity in stromatolitic reefs versus reefs of other types are not available, but anecdotal information that can be gleaned from literature on oil fields does not offer much encouragement for this line of thought. Recent studies of stromatolitic reef rocks of a range of ages include descriptions of oil and MVT mineralization, as well as porosity estimates of 8%–11% (Counter, 1993; Whitesell, 1995; Kuznetzov, 1997; Grotzinger, 2000; Lemon, 2000; Osmond, 2000). Porosities at and below the low end of this range are sufficient to produce a deposit containing 7% Pb+Zn regardless of the relative abundances of galena and sphalerite.

These comparisons argue against the tempting possibility that the scarcity of Precambrian MVTs can be attributed solely to a lack of suitable Precambrian carbonate rocks. Furthermore, by including in this compilation several Precambrian MVT districts that have been explored but not produced, we have also minimized the likelihood that less extensive exploration in Precambrian terranes can account for their lower MVT endowment. Thus, we conclude that Precambrian MVT deposits are indeed less abundant than Phanerozoic deposits and must seek a reason beyond local geologic factors for this difference.

Relation between Atmosphere-Hydrosphere Compositions and MVT Mineralization

With no geologic reasons to account for the global MVT history shown in Figure 10, the pattern likely reflects changes in the composition of the atmosphere and hydrosphere. As noted at the outset of this study, the efficiency of MVT mineralization should be directly related to the sulfate content of basinal fluids and, by extension, the ocean in which the sediments were deposited.

Sulfate in the ocean is linked, in turn, to the composition of the atmosphere and, especially, its oxygen content.

Two competing histories have been suggested for the evolution of the ocean and coexisting atmosphere. One history holds that the oxygen content of the atmosphere increased by as much as an order of magnitude from levels of less than 10^{-2} present atmospheric level (PAL) during the Great Oxidation Event (GOE) ca. 2.3 Ga (Holland, 1999; 2002; Bekker et al., 2004). Prior to the GOE, sulfate concentrations in the ocean are estimated to have been less than 1% of present levels and the main source of sulfate was disproportionation or photochemical oxidation of volcanic SO_2 and oxidation of H_2S by cyanobacteria (Hattori and Cameron, 1986; Canfield and Raiswell, 1999; Farquhar et al., 2000; Habicht et al., 2002; Holland, 2002; Farquhar and Wing, 2003). After the GOE, sulfate concentrations in the ocean are thought to have increased gradually through Proterozoic time, reaching present levels of 10–30 mM only near the end of Proterozoic time (Shen et al., 2002; Kah et al., 2004). The alternative view holds that there was no GOE and the oxygen content of the atmosphere and the sulfate content of the oceans have fluctuated between 10 and 30 mM since at least Neoarchean time (Ohmoto, 1997, 2004).

The two competing marine sulfate histories are shown in Figure 10, where it can be seen that our curve for the change in MVT mineralization through time nearly parallels the GOE-type curve. Similarity of the MVT and GOE-type seawater sulfate curves suggests that sulfate concentrations similar to those in the modern ocean are required for efficient MVT mineralization, and that a scarcity of sulfide-rich traps derived from this sulfate limited the formation of MVT deposits in Precambrian time. Evaporative brines derived from low-sulfate seawater were probably capable of dissolving and transporting metals in much the same amounts observed in Phanerozoic basins. However, low sulfate concentrations in basinal waters would have limited the amount of sulfide that could have formed from such brines (or their rare evaporites), whether by bacterial or thermochemical reduction. Local accumulations of sulfate-rich seawater, such as those proposed for Archean and Neoproterozoic time, might have been the necessary precursors to sulfide-rich traps for early MVT ore formation (Cameron, 1982, 1983; Kasting, 1991). In that context, it is interesting to note that some Precambrian MVT deposits, especially Nanisivik and Kamarga, were close to evaporite sequences, a factor that might have led to higher sulfate and derived sulfide concentrations.

Similarity of these curves does not exclude other possible controls on MVT genesis. Christensen et al. (1997) and Leach et al. (2001) have pointed out that Phanerozoic MVT deposits formed largely during Devonian and Permian time, possibly because of long-distance brine migration during episodic compressive or extensional tectonic events (Bradley and Leach, 2003; Kesler et al., 2004). In addition, the sulfate content of the world ocean during Phanerozoic time varied considerably in response to large-scale continent aggregation and disaggregation, although it did not reach levels as low as those estimated for Paleoproterozoic and Neoarchean time (Horita et al., 2002; Lowenstein et al., 2003). All of these factors operate at timescales considerably smaller than the 500 Ma interval used in Figure 10, however, and it appears that they were masked by a longer-term increase in the sulfate content of the ocean during Precambrian time.

Although the similarity of the MVT and GOE-type sulfate curves agrees with the "GOE-type history" for Earth's early atmosphere and hydrosphere, perplexing complications remain. In particular, the Neoarchean–Paleoproterozoic Transvaal Basin, which is significantly older than the GOE, generated MVT deposits very similar to those in Phanerozoic rocks. Perhaps these deposits, which probably formed ca. 2.0 Ga, took advantage of GOE-related sulfate. Or, maybe they did not form by the Anderson-type model advocated here. Neither of these alternatives seems geologically reasonable, however, and it is much more likely that the basinal brines that formed Bushy Park and Pering were part of the Transvaal sedimentary system and therefore representative of Neoarchean–Paleoproterozoic conditions. Thus, enough sulfate to form MVT deposits was almost certainly available, even at this early stage of Earth history. Whether this was sulfate that formed because of local environmental factors or these deposits are the sole remaining MVT signal of a more sulfate-rich early Earth remains to be determined.

Finally, although basin-hosted deposits are most likely to reflect atmosphere and ocean compositions, our results provide encouragement for study of other subsurface deposits. Deeper deposits related to igneous activity might also be useful. For instance, the oxidation state of magmas controls the form of S in the magma, which controls, in turn, whether metals form immiscible sulfides in the magma or separate into a vapor phase that could form hydrothermal ore deposits. Although current estimates suggest that fO_2 of the mantle has remained within a single log-unit of present values (Canil, 1997; Delano, 2001; Holland, 2002), even these small differences can produce large changes in the SO_2/H_2S ratio of magmas and associated hydrothermal solutions (Core, 2004).

ACKNOWLEDGMENTS

We are grateful to Derek Wilton, Nawojka Wachowiak, Bruce Ehlers, Peter McGoldrick, and Bruce Gemmell for sharing information with us on Ramah, Esker, Bushy Park, Kamarga, and Warrabarty, respectively. The manuscript has been improved by conversations with B.H. Wilkinson and Linda Kah, and challenging reviews by H.L. Barnes, M.B. Goldhaber, and H. Ohmoto.

REFERENCES CITED

Altermann, W., and Nelson, D.R., 1998, Sedimentation rates, basin analysis and regional correlation of three Neoarchean and Paleoproterozoic sub-basins of the Kaapvaal craton, as inferred from precise U-Pb zircon ages from volcaniclastic sediments: Sedimentary Geology, v. 120, p. 225–256, doi: 10.1016/S0037-0738(98)00034-7.

Anderson, G.M., 1983, Some geochemical aspects of sulphide precipitation in carbonate rocks, in Kisvarsanyi, G., Grant, S.K., Pratt, W.P., and Koenig, J.W., eds., Proceedings, International Conference on Mississippi Valley-Type Lead-Zinc Deposits: Rolla, University of Missouri-Rolla, p. 61–76.

Anderson, G.M., 1991, Organic maturation and ore precipitation in Southeast Missouri: Economic Geology and the Bulletin of the Society of Economic Geologists, v. 86, p. 909–926.

Anderson, G.M., and Macqueen, R.W., 1988, Ore deposit models—6: Mississippi Valley-type lead-zinc deposits: Geoscience Canada Reprint Series, v. 3, p. 79–90.

Anderson, B.R., Gemmell, J.B., and Berry, R.F., 2001, The geology of the Nifty copper deposit, Throssell Group, western Australia: Implications for ore genesis: Economic Geology and the Bulletin of the Society of Economic Geologists, v. 96, p. 1535–1565.

Anderson, B.R., Gemmell, J.B., and Nelson, D.R., 2002, Lead isotope evolution of mineral deposits in the Proterozoic Throssell Group, western Australia: Economic Geology and the Bulletin of the Society of Economic Geologists, v. 97, p. 897–911.

Archibald, S., 1992, The geology, mineralization, geochemistry and metallogeny of the Paleoproterozoic Ramah Group, northern Labrador [M.Sc. thesis]: St. John's, Memorial University of Newfoundland, 76 p.

Archibald, S.M., and Wilton, D.H.C., 1994, A lower Proterozoic (ca. 2.0 by) MVT deposits from the Ramah Group, northern Labrador: Geological Association of Canada Program with Abstracts, v. 19, p. 4.

Arne, D.C., Curtis, L.W., and Kissin, S.A., 1991, Internal zonation in a carbonate-hosted Zn-Pb-Ag deposit, Nanisivik, Baffin Island, Canada: Economic Geology and the Bulletin of the Society of Economic Geologists, v. 86, p. 699–717.

Barley, M.E., and Groves, D.I., 1992, Supercontinent cycles and the distribution of metal deposits through time: Geology, v. 20, p. 291–294, doi: 10.1130/0091-7613(1992)020<0291:SCATDO>2.3.CO;2.

Barrett, T.J., and Anderson, G.M., 1988, The solubility of sphalerite and galena in 1–5 m NaCl solutions to 300°C: Geochimica et Cosmochimica Acta, v. 52, p. 813–820, doi: 10.1016/0016-7037(88)90353-5.

Baugaard, W.D., Gregg, J.M., and Ahler, B., 2001, Stratigraphy and diagenesis of the Paleoproterozoic Bushy Park zinc-lean deposit, South Africa: Geological Society of America Abstracts with Programs, v. 33, p. 337.

Bekker, A., Holland, H.D., Wang, P.-L., Rumble, D., III, Stein, J.H., Coetzee, L.L., and Beukes, N.J., 2004, Dating the rise of atmospheric oxygen: Nature, v. 427, p. 117–121, doi: 10.1038/nature02260.

Beukes, N.J., 1987, Facies relations, depositional environments and diagenesis in a major Early Proterozoic stromatolitic carbonate platform to basinal sequence, Campbellrand Subgroup, Transvaal Supergroup, Southern Africa: Sedimentary Geology, v. 54, p. 1–46, doi: 10.1016/0037-0738(87)90002-9.

Blockley, J.G., and Myers, J.S., 1990, Proterozoic rocks of the Western Australian Shield—Geology and mineralization, in Hughes, F.E., ed., Geology of the mineral deposits of Australia and Papua New Guinea: Melbourne, Australasian Institute of Mining and Metallurgy, p. 607–615

Blount, C.W., 1977, Barite solubilities and thermodynamic quantities up to 300°C and 1400 bars: American Mineralogist, v. 62, p. 942–957.

Bottomley, D.J., Veizer, J., Nielsen, H., and Moczydlowska, M., 1992, Isotopic composition of disseminated sulfur in Precambrian sedimentary rocks: Geochimica et Cosmochimica Acta, v. 56, p. 3311–3322, doi: 10.1016/0016-7037(92)90307-5.

Bowring, S.A., and Grotzinger, J.P., 1992, Implication of new chronostratigraphy for the tectonic evolution of Wopmay Orogen, northwest Canadian shield: American Journal of Science, v. 292, p. 1–20.

Bradley, D.C., and Leach, D.L., 2003, Tectonic controls of Mississippi Valley-type lead-zinc mineralization in orogenic forelands: Mineralium Deposita, v. 38, p. 652–667, doi: 10.1007/s00126-003-0355-2.

Burstein, I.B., Shelton, K.L., Hagni, R.D., and Brandom, R.T., 1992, Mobilization of copper by MVT fluids penetrating igneous basement rocks; sulfur isotope studies of the Precambrian Boss-Bixby Fe-Cu deposits, Southeast Missouri: Geological Society of America Abstracts with Programs, v. 24, no.7, p.234.

Cameron, E.M., 1982, Sulphate and sulphate reduction in early Precambrian oceans: Nature, v. 296, p. 145–148, doi: 10.1038/296145a0.

Cameron, E.M., 1983, Evidence from early Proterozoic anhydrite for sulphur isotopic partitioning in Precambrian oceans: Nature, v. 304, p. 54–56, doi: 10.1038/304054a0.

Canfield, D.E., and Raiswell, R., 1999, The evolution of the sulfur cycle: American Journal of Science, v. 299, p. 697–723.

Canil, D., 1997, Vanadium partitioning and the oxidation sate of Archean komatiite magmas: Nature, v. 389, p. 842–845, doi: 10.1038/39860.

Carriere, J.J., and Sangster, D.F., 1992, Preliminary studies of fluid inclusions in sphalerite, quartz, and dolomite from Gayna River MVT deposit, Northwest Territories: Geological Survey of Canada Report 92-01A, p. 47–53.

Cheney, E.S., 1996, Sequence stratigraphy and plate tectonic significant of the Transvaal succession of southern Africa and its equivalent in Western Australia: Precambrian Research, v. 79, p. 3–24, doi: 10.1016/0301-9268(95)00085-2.

Chetty, D., and Frimmel, H.E., 2000, The role of evaporites in the genesis of base metal sulphide mineralization in the Northern Platform of the Pan-African Damara Belt, Namibia: Geochemical and fluid inclusion evidence from carbonate wall-rock alteration: Mineralium Deposita, v. 35, p. 364–376, doi: 10.1007/s001260050247.

Chi, G., and Savard, M.M., 1997, Sources of basinal and Mississippi Valley-type mineralizing brines; mixing of evaporated seawater and halite-dissolution brine: Chemical Geology, v. 143, p. 121–125, doi: 10.1016/S0009-2541(97)00096-X.

Christensen, J.N., Halliday, A.N., Kesler, S.E., and Sangster, D.F., 1993, Further evaluation of the Rb-Sr dating of sphalerite; the Nanisivik Precambrian MVT deposit, Baffin Island, Canada: Geological Society of America Abstracts with Programs, v. 25, p. 471.

Christensen, J.C., Halliday, A.N., and Kesler, S.E., 1997, Rb-Sr dating of sphalerite and the ages of Mississippi Valley-type Pb-Zn deposits: Society of Economic Geologists Special Publication 4, p. 527–535.

Christie-Blick, N., Dyson, I.A., and von der Borch, C.C., 1995, Sequence stratigraphy and the interpretation of Neoproterozoic history: Precambrian Research, v. 73, p. 3–26, doi: 10.1016/0301-9268(94)00096-A.

Condie, K.C., 2000, Episodic continental growth and supercontinents: Afterthoughts and extensions: Tectonophysics, v. 322, p. 153–162, doi: 10.1016/S0040-1951(00)00061-5.

Core, D.P., 2004, Oxygen and sulfur fugacities of granitoids: Implications for ore-forming processes [Ph.D. dissertation]: Ann Arbor, University of Michigan, 396 p.

Counter, K.J., 1993, Diagenesis of the Lower Ordovician Ogdensburg Formation, northern New York and southeastern Ontario; evidence for a hydrothermal "hot spot," in Program with Abstracts, American Association of Petroleum Geologists and Society of Economic Paleontologists and Mineralogists Annual Meeting, New Orleans, p. 88.

Delano, J.W., 2001, Redox history of the Earth's interior since ~3900 Ma: Implications for prebiotic molecules: Origins of Life and Evolution of the Biosphere, v. 31, p. 311–341, doi: 10.1023/A:1011895600380.

Drew, G.J., and Both, R.A., 1984, The carbonate-hosted silver-lead deposits of the Ediacara mineral field, South Australia; petrological, fluid inclusion and sulphur isotope studies: Australian Journal of Earth Sciences, v. 31, p. 177–201.

Duane, M.J., Kruger, F.J., Roberts, P.J., and Smith, C.B., 1991, Pb and Sr isotopes and origin of Proterozoic base metal (fluorite) and gold deposits, Transvaal Sequence, South Africa: Economic Geology and the Bulletin of the Society of Economic Geologists, v. 86, p. 1491–1505.

Eriksson, P.G., Condie, K.C., Tirsgaard, H., Mueller, W.U., Altermann, W., Miall, A.D., Aspler, L.B., Catuneanu, O., and Shiarenzelli, J.R., 1998, Precambrian clastic sedimentation systems: Sedimentary Geology, v. 120, p. 5–53, doi: 10.1016/S0037-0738(98)00026-8.

Eriksson, P.G., and Altermann, W., 1998, An overview of the geology of the Transvaal Supergroup dolomites (South Africa): Environmental Geology, v. 36, p. 179–188, doi: 10.1007/s002540050334.

Eriksson, P.G., Altermann, W., Catuneanu, O., van der Merwe, R., and Bumby, A.J., 2001a, Major influences on the evolution of the 2.67–2.1 Ga Transvaal basin, Kaapvaal craton: Sedimentary Geology, v. 141–142, p. 205–231, doi: 10.1016/S0037-0738(01)00075-6.

Eriksson, P.B., Martins-Neto, M.A., Nelson, D.R., Aspler, L.B., Chiarenzelli, J.R., Catuneanu, O., Sarkar, S., Altermann, W., and Rautenbach, C.J., de W., 2001b, An introduction to Precambrian basins: Their characteristics and genesis: Sedimentary Geology, v. 141–142, p. 1–35.

Farquhar, J., and Wing, B.A., 2003, Multiple sulfur isotopes and the evolution of the atmosphere: Earth and Planetary Science Letters, v. 213, p. 1–13, doi: 10.1016/S0012-821X(03)00296-6.

Farquhar, J., Bao, H., and Thiemens, M., 2000, Atmospheric influence of Earth's earliest sulfur cycle: Science, v. 289, p. 756–758, doi: 10.1126/science.289.5480.756.

Ford, D.C., 1986, Genesis of paleokarst and strata-bound zinc-lead sulfide deposits in Proterozoic dolostone, northern Baffin Island, Canada—A discussion: Economic Geology and the Bulletin of the Society of Economic Geologists, v. 81, p. 1562–11566.

Frimmel, H.D., Deane, J.G., and Chadwick, P.J., 1996, Pan-African tectonism and the genesis of base metal sulfide deposits in the northern foreland of the Damara orogen, Namibia, in Sangster, D.F., ed., Carbonate-hosted lead-zinc deposits: Society of Economic Geologists Special Publication 4, p. 204–217.

Garven, G., and Bull, S.W., 2000, Groundwater flow modeling of sedex-type ore genesis in the McArthur Basin, northern Australia: Geological Society of America Abstracts with Programs, v. 32, no. 7, p. 60–61.

Ghazban, F., Schwarcz, H.P., and Ford, D.C., 1990, Carbon and sulfur isotope evidence for in situ reduction of sulfate, Nanisivik lead-zinc deposits, Northwest Territories, Baffin Island, Canada: Economic Geology and the Bulletin of the Society of Economic Geologists, v. 85, p. 360–375.

Giordano, T.H., 1985, A preliminary evaluation of organic ligands and metal-organic complexing in Mississippi Valley-type ore solutions: Economic Geology and the Bulletin of the Society of Economic Geologists, v. 80, p. 96–106.

Gize, A.P., 1986, The development of a thermal mesophase in organic matter from high-temperature ore deposits, in Dean, W.E., ed., Organics and ore deposits: Proceedings, Denver Regional Exploration Geologists Society Symposium, Wheatridge, Colorado, p. 137–150.

Greyling, L.N., Huizenga, J.M., and Gutzmer, J., 2001, A review of Mississippi Valley-type deposits with emphasis on the Paleoproterozoic Pering Pb-Zn deposit, South Africa: Johannesburg, Economic Geology Research Institute, University of the Witwatersrand, Information Circular no. 356, 32 p.

Grotzinger, J.P., 1986a, Cyclicity and paleoenvironmental dynamics, Rocknest platform, northwest Canada: Geological Society of America Bulletin, v. 97, p. 1208–1231, doi: 10.1130/0016-7606(1986)97<1208:CAPDRP>2.0.CO;2.

Grotzinger, J.P., 1986b, Evolution of early Proterozoic passive-margin carbonate platform, Rocknest Formation, Wopmay Orogen, Northwest Territories, Canada: Journal of Sedimentary Petrology, v. 56, p. 831–847.

Grotzinger, J.P., 1986c, Upward shallowing platform cycles: A response to 2.2 billion years of low-amplitude, high-frequency (Milankovitch band) sea-level oscillations: Paleoceanography, v. 1, p. 403–416.

Grotzinger, J.P., 2000, Facies and paleodevelopmental setting of thrombolite-stromatolite reefs, terminal Proterozoic Nama Group (ca. 550–543 Ma), central and southern Namibia: Communications of the Geological Survey of South West Africa/Namibia, v. 12, p. 221–233.

Habicht, K.S., Gade, M., Thamdrup, B., Berg, P., and Canfield, D.E., 2002, Calibration of sulfate levels in the Archean ocean: Science, v. 298, p. 2372–2375, doi: 10.1126/science.1078265.

Hanor, J.S., 1987, Origin and migration of subsurface sedimentary brines: Tulsa, Oklahoma, Society of Economic Paleontologists and Mineralogists Short Course, v. 21, 247 p.

Hardy, J.L., 1979, Stratigraphy, brecciation and mineralization, Gayna River, N.W.T. [M.Sc. thesis]: Toronto, University of Toronto, 64 p.

Hattori, K., and Cameron, E.M., 1986, Archean magmatic sulphate: Nature, v. 319, p. 45–47, doi: 10.1038/319045a0.

Hayashi, K., Fujisawa, H., Holland, H.D., and Ohmoto, H., 1997, Geochemistry of ~1.9 Ga sedimentary rocks from northeastern Labrador, Canada: Geochimica et Cosmochimica Acta, v. 61, p. 4115–4437, doi: 10.1016/S0016-7037(97)00214-7.

Haynes, F.M., Beane, R.L., and Kesler, S.E., 1989, Simultaneous transport of metal and reduced sulfur, Mascot-Jefferson City zinc district, East Tennessee: Evidence from fluid inclusions: American Journal of Science, v. 289, p. 994–1038.

Heaman, L.M., LeCheminant, A.M., and Rainbird, R.H., 1992, Nature and timing of Franklin igneous events, Canada: Implications for a Late Proterozoic mantle plume and the break-up of Laurentia: Earth and Planetary Science Letters, v. 109, p. 117–131, doi: 10.1016/0012-821X(92)90078-A.

Hewton, R.S., 1982, Gayna River: A Proterozoic Mississippi Valley-type zinc-lead deposit, in Hutchinson, R.W., Spence, C.D., and Franklin, J.M., eds., Precambrian sulphide deposits: H.S. Robinson Memorial Volume: Geological Association of Canada Special Paper 25, p. 667–700.

Hitzman, M.W., Thorman, C.H., Romagna, G., Oliviera, R.F., Dardenne, M.A., and Drew, L.J., 1995, The Morro Agudo Zn-Pb deposits, Minas Gerais, Brazil; a Proterozoic Irish-type carbonate-hosted sedex-replacement deposit: Geological Society of America Abstracts with Programs, v. 27, p. 408.

Hoffman, P.F., 1989, Pethei reef complex (1.9 Ga), Great Slave Lake, N.W.T., in Geldsetzer, H.H., James, N.P., and Tebbutt, G.E., eds., Reefs: Canada and adjacent areas: Canadian Society of Petroleum Geologists Memoir 13, p. 38–48.

Hoffman, P.F., and Bowring, S.A., 1984, Short-lived 1.9 Ga continental margin and its destruction, Wopmay orogen, northwest Canada: Geology, v. 12, p. 68–72, doi: 10.1130/0091-7613(1984)12<68:SGCMAI>2.0.CO;2.

Holland, 1984, The chemical evolution of the atmosphere and oceans: Princeton, New Jersey, Princeton University Press, 582 p.

Holland, H.D., 1999, When did the Earth's atmosphere become oxic? A reply: Geochemical News, v. 100, p. 20–22.

Holland, H.D., 2002, Volcanic gases, black smokers, and the Great Oxidation Event: Geochimica Cosmochimica Acta, v. 66, p. 3811–3826.

Horita, J., Zimmermann, H., and Holland, H.D., 2002, Chemical evolution of seawater during the Phanerozoic; implications from the record of marine evaporites: Geochimica et Cosmochimica Acta, v. 66, p. 3733–3756, doi: 10.1016/S0016-7037(01)00884-5.

Iyer, S.S., Hoefs, J., and Krouse, H.R., 1992, Sulfur and lead isotope geochemistry of galenas from the Bambui Group, Minas Gerais, Brazil; implications for ore genesis: Economic Geology and the Bulletin of the Society of Economic Geologists, v. 87, p. 437–443.

Jackson, G.D., and Ianelli, T.R., 1981, Rift-related cyclic sedimentation in the Neohelikian Borden basin, northern Baffin Island: Canada Geological Survey Paper 81-10, p. 269–302.

Jones, D.A., 1986, The Kamarga deposit: Stratabound zinc-lead mineralisation in the middle Proterozoic McNamara Group, northwest Queensland [Ph. D. dissertation]: Armidale, Australia: University of New England, 147 p.

Jones, D.J., Bull, S., and McGoldrick, P., 1999, The Kamarga deposit: A large, low grade, stratabound zinc resource in the Proterozoic 'Carpentaria Zinc Belt' of northern Australia in Stanley, C.J., ed., Mineral deposits: Processes to processing: Rotterdam, A.A. Balkema, p. 873–876.

Kah, L.C., Lyons, T.W., and Chesley, J.T., 2001, Geochemistry of a 1.2 Ga carbonate-evaporite succession, northern Baffin and Bylot Islands; implications for Mesoproterozoic marine evolution: Precambrian Research, v. 111, p. 203–234, doi: 10.1016/S0301-9268(01)00161-9.

Kah, L.C., Lyons, T.W., and Frank, T.D., 2004, Low marine sulphate and protracted oxygenation of the Proterozoic biosphere: Nature, v. 431, p. 834–838, doi: 10.1038/nature02974.

Kasting, J.E., 1991, Box models for the evolution of atmospheric oxygen: An update: Global and Planetary Change, v. 5, p. 125–131, doi: 10.1016/0921-8181(91)90133-H.

Kesler, S.E., 1996, Appalachian Mississippi Valley-type deposits: Paleoaquifers and brine provinces, in Sangster, D.F., ed., Carbonate-hosted lead-zinc deposits: Society of Economic Geologists Special Publication 4, p. 29–57.

Kesler, S.E., Appold, M.S., Walter, L.M., Martini, A.M., Huston, T.J., and Kyle, J.R., 1995, Na-Cl-Br systematics of Mississippi Valley-type brines: Geology, v. 23, p. 641–644, doi: 10.1130/0091-7613(1995)023<0641:NCBSOM>2.3.CO;2.

Kesler, S., Chesley, J.T., Christensen, J.N., Hagni, R.D., Heijlen, W., Kyle, J.R., Muchez, P., Misra, K.C., and van der Voo, R., 2004, Tectonic controls of Mississippi Valley-type lead-zinc mineralization in organic forelands; discussion: Mineralium Deposita, v. 39, p. 512–514, doi: 10.1007/s00126-004-0422-3.

Kesler, S.E., Gleason, J.D., Smith, C.N., Ahler, B.A., and Taylor, D.R., 2003, Age and provinces of Precambrian MVT mineralization, Transvaal Supergroup, South Africa: Geological Society of America Abstracts with Programs, v. 35, p. 234.

Kesler, S.E., Martini, A.M., Appold, M.S., Walter, L.M., Huston, T.J., and Furman, F.C., 1996, Na-Cl-Br systematics of fluid inclusions from Mississippi Valley-type deposits, Appalachian basin: Constraints on solute origin and migration paths: Geochimica et Cosmochimica Acta, v. 60, p. 225–233, doi: 10.1016/0016-7037(95)00390-8.

Knight, I., and Morgan, W.C., 1977, Stratigraphic subdivision of the Aphebian Ramah Group, northern Labrador: Geological Survey of Canada Paper 77-15, p. 1–31.

Knight, I., and Morgan, W.C., 1981, The Aphebian Ramah Group, northern Labrador: Geological Survey of Canada Paper 81-10, p. 313–330.

Korstgard, J., Ryan, B., and Wardle, R., 1987, The boundary between Proterozoic and Archean crustal blocks in central West Greenland and northern Labrador, in Park, R.G., and Tarney, J., eds., Evolution of the Lewisian and comparable Precambrian high-grade terrains: Geological Society [London] Special Publication 27, p. 247–259.

Knauth, L.P., 2005, Temperature and salinity history of the Precambrian ocean: Implications for the course of microbial evolution: Paleogeography, Paleoclimatology, Paleoecology, v. 219, p. 53–69.

Kruger, F.J., Duane, M.J., Turner, A.M., Whitelaw, H.T., and Verhagen, B.T., 2001, Paleohydrology of c. 2Ga old Mississippi Valley-type Pb-Zn deposit, South Africa: Radiogenic and stable isotope evidence: Johannesburg, Economic Geology Research Institute, University of the Witwatersrand, Information Circular no. 355, 22 p.

Kuznetzov, V.G., 1997, The role of karst in the formation of Yurubchen-Tokhomsk field, Siberian Craton, Russia: American Association of Petroleum Geologists and Society of Economic Paleontologists and Mineralogists, v. 6, p. 65.

Kyle, J.R., and Misi, A., 1997, Origin of Zn-Pb-Ag sulfide mineralization within upper Proterozoic phosphate-rich carbonate strata, Irece basin, Bahia, Brazil: International Geology Review, v. 39, p. 383–399.

Leach, D.L., Bradley, D.C., Lewchuk, M., Symons, D.T.A., Brannon, J., and de Marsily, G., 2001, Mississippi Valley-type lead-zinc deposits through geological time: Implications from recent age-dating research: Mineralium Deposita, v. 36, p. 711–740, doi: 10.1007/s001260100208.

Leach, D.L., and Sangster, D.F., 1993, Mississippi Valley-type deposits, in Kirkham, R.V., Sinclair, W.D., Thorpe, R.I., and Duke, J.M., eds., Mineral deposit modeling: Geological Association of Canada Special Paper 40, p. 289–314.

Lemon, N.M., 2000, A Neoproterozoic fringing stromatolite reef complex, Flinders Ranges, South Australia: Precambrian Research, v. 100, p. 109–120, doi: 10.1016/S0301-9268(99)00071-6.

Lowenstein, T.K., Hardie, L.A., Timofeeff, M.N., and Demicco, R.V., 2003, Secular variation in seawater chemistry and the origin of calcium chloride basinal brines: Geology, v. 31, p. 857–860, doi: 10.1130/G19728R.1.

Martini, J.E.J., 1976, The fluorite deposits in the dolomite series of the Marico district, Transvaal, South Africa: Economic Geology and the Bulletin of the Society of Economic Geologists, v. 71, p. 625–635.

Martini, J.E.J., 1990, The Genadendal Zn-Pb-Mn mineralization, South Africa: A possible Early Proterozoic alkaline hydrothermal system: Economic Geology and the Bulletin of the Society of Economic Geologists, v. 85, p. 1172–1185.

Martini, J.E.J., Eriksson, P.G., and Snyman, C.P., 1995, The Early Proterozoic Mississippi Valley-type Pb-Zn-F deposits of the Campbellrand and Malmani Subgroups, South Africa: Mineralium Deposita, v. 30, p. 135–145, doi: 10.1007/BF00189342.

McFarlane, K., and Bone, Y., 1994, Mississippi Valley-type lead-zinc mineralization at Donkey Bore and Old Wirrealpa Springs, central Flinders Range, South Australia: Geological Society of Australia Abstracts, v. 37, p. 278.

McLimans, R.K., Barnes, H.L., and Ohmoto, H., 1980, Sphalerite stratigraphy of the Upper Mississippi Valley zinc-lead district, southwest Wisconsin: Economic Geology and the Bulletin of the Society of Economic Geologists, v. 75, p. 351–361.

McNaughton, K., and Smith, T.E., 1986, A fluid inclusion study of sphalerite and dolomite from the Nanisivik lead-zinc deposit, Baffin Island, Northwest Territories, Canada: Economic Geology and the Bulletin of the Society of Economic Geologists, v. 81, p. 713–720.

Mengel, F., and Rivers, T., 1994, Metamorphism of pelitic rocks in the Paleoproterozoic Ramah group, Saglek area, northern Labrador: Mineral reactions, P-T conditions and influence of bulk composition: Canadian Mineralogist, v. 32, p. 781–801.

Mengel, R., Rivers, T., and Reynolds, P., 1991, Lithotectonic elements and tectonic evolution of Torngat Orogen, Saglek Fiord, northern Labrador: Canadian Journal of Earth Sciences, v. 28, p. 1407–1423.

Meyer, C. 1988, Ore deposits as guides to the geologic history of the Earth: Annual Review of Earth and Planetary Sciences, v. 16, p. 147–171, doi: 10.1146/annurev.ea.16.050188.001051.

Misi, A., Sundaram, S.S.I., Coelho, C.E.S., Tassinari, C.C.G., Franca-Rocha, J.S., Cunha, I. de A., Gomes, A.S.R., de Oliveira, T.F., Teixeira, J.B.G., and Filho, V.M.C., 2005, Sediment-hosted lead-zinc deposits of the Neoproterozoic Bambuí Group and correlative sequences, São Francisco Craton, Brazil: A review and a possible metallogenic evolution model: Ore Geology Reviews, v. 26, p. 263–304.

Misiewicz, J.E., 1988, The geology and metallogeny of the Otavi Mountain Land, Damara Orogen, SWA/Namibia with particular reference to the Berg Aukas Zn-Pb-V deposit—A model of ore genesis [M.Sc. thesis]: Grahamstown, South Africa, Rhodes University, 143 p.

Morgan, W.C., 1975, Geology of the Precambrian Ramah Group and basement rocks in the Nachvak Fiord-Saglek Fiord area, north Labrador: Geological Survey of Canada Paper 74-54, p. 1–42.

Narbonne, G.M., and Aitken, J.D., 1995, Neoproterozoic of the Mackenzie Mountains, northwestern Canada, in Knoll, A.H., and Malcolm, W., eds., Neoproterozoic stratigraphy and Earth history: Precambrian Research, v. 73, p. 101–121.

Ohmoto, H., 1997, When did the Earth's atmosphere become oxic?: Geochemical News, v. 93, p. 12–27.

Ohmoto, H., 2004, The Archean atmosphere, hydrosphere and biosphere, in Eriksson, P.G., Altermann, W., Nelson, D.R., Mueller, W.U., and Catuneanu, O., eds., The Precambrian Earth: Tempos and events: Developments in Precambrian Geology, v. 12, p. 361–388.

Ohmoto, H., and Goldhaber, M.B., 1997, Sulfur and carbon isotopes, in Barnes, H.L., ed., Geochemistry of hydrothermal ore deposits (3rd edition): New York, John Wiley and Sons, p. 517–612.

Ohmoto, H., Kaiser, C.J., and Geer, K.A., 1990, Systematics of sulphur isotopes in recent marine sediments and ancient sediment-hosted base metal deposits, in Herbert, H.K. and Ho, S.E., eds., Stable isotopes and fluid processes in mineralization: Geology Department and University Extension, University of Western Australia, Publication 23, p. 70–120.

Ohmoto, H., and Rye, R.O., 1979, Isotopes of sulfur and carbon, in Barnes, H.L., ed., Geochemistry of hydrothermal ore deposits (2nd edition): New York, John Wiley and Sons, p. 509–567.

Olson, R.A., 1984, Genesis of paleokarst and strata-bound zinc-lead sulfide deposits in a Proterozoic dolostone, northern Baffin Island, Canada: Economic Geology, v. 79, p. 056–1103.

Osmond, J.C., 2000, West Willow Creek field; first productive lacustrine stromatolite mound in the Eocene Green River Formation, Uinta Basin, Utah: Mountain Geologist, v. 37, p. 157–170.

Page, R.W., Jackson, M.W., and Krassay, A.A., 2000, Constraining sequence stratigraphy in North Australian basins: SHRIMP U-Pb zircon chronology between Mt. Isa and McArthur River: Australian Journal of Earth Sciences, v. 47, p. 431–459, doi: 10.1046/j.1440-0952.2000.00797.x.

Pehrsson, S.J., and Buchan, K.L., 1999, Borden dykes of Baffin Island, N.W.T.: A Franklin U-Pb baddeleyite age and a paleomagnetic reinterpretation: Canadian Journal of Earth Sciences, v. 36, p. 65–73, doi: 10.1139/cjes-36-1-65.

Petrov, P.Y., and Semikhatov, M.A., 2001, Sequence organization and growth patterns of late Mesoproterozoic stromatolites reefs: An example from the Burovaya Formation, Turukhansk Uplift, Siberia: Precambrian Research, v. 111, p. 257–281, doi: 10.1016/S0301-9268(01)00163-2.

Plumb, K.A., Ahmad, M., and Wygralak, A.S., 1990, Mid-Proterozoic basins of north Australian craton—Regional geology and mineralisation, in Hughes, F.E., ed., Geology of the mineral deposits of Australia and Papua New Guinea: Melbourne, Australasian Institute of Mining and Metallurgy, p. 881–902.

Poetter, P.A., 2001, Origin and exploration potential of sedimentary-hosted Mississippi Valley-Type F-Zn-Pb deposits in the Neoarchean platform dolostones of the Transvaal Supergroup, Northwestern Province, South Africa [abs.]: Muenster, WWU Muenster, Diplom thesis, (http://general.rau.ac.za/geology/PPM/Abstracts.html).

Pollack, H.N., 1997, Thermal characteristics of the Archean, in de Wit, M.H., and Ashwal, L.D., eds., Greenstone belts: Oxford, UK, Clarendon Press, p. 223–232.

Pollack, H.N., and Chapman, D.S., 1977, On the regional variation of heat flow, geotherms, and lithospheric thickness: Tectonophysics, v. 38, p. 279–296, doi: 10.1016/0040-1951(77)90215-3.

Rhondacorp 2002, www.rhondacorp.com, 30 June 2002.

Ricketts, B.D., and Donaldson, J.A., 1989, Stromatolite reef development on a mud-dominated platform in the Middle Precambrian Belcher Group of the Hudson Bay, in Geldsetzer, H.H., James, N.P., and Tebbutt, G.E., eds., Reefs: Canada and adjacent areas: Canadian Society of Petroleum Geologists Memoir 13, p. 113–119.

Roberts, P.J., Gize, A.P., Duane, M.J., and Verhagen, B., 1993, Precambrian hydrocarbon residues associated with Mississippi Valley-type mineralization in the Transvaal Sequence, South Africa: South African Journal of Geology, v. 96, p. 57–60.

Rowan, E.L., Goldhaber, M.B., and Hatch, J.R., 2001, The role of regional fluid flow in the Illinois basin's thermal history: Constraints from fluid inclusions and maturity of Pennsylvanian coals: American Association of Petroleum Geologists Bulletin, v. 86, p. 257–277.

Rye, D.M., and Williams, N., 1981, Studies of the base metal sulfide deposits at McArthur River, Northern Territory, Australia III: The stable isotope geochemistry of the H.Y.C., Ridge, and Cooley deposits: Economic Geology and the Bulletin of the Society of Economic Geologists, v. 76, p. 1–26.

St. Marie, J., and Kesler, S.E., 2000, Iron-rich and iron-poor Mississippi Valley-type mineralization, Metaline district, Washington: Economic Geology and the Bulletin of the Society of Economic Geologists, v. 95, p. 1091–1106.

St. Marie, J., Kesler, S.E., and Allen, C.R., 2001, Origin of iron-rich Mississippi Valley-type deposits: Geology, v. 29, p. 59–62, doi: 10.1130/0091-7613(2001)029<0059:OOIRMV>2.0.CO;2.

Sangster, D.F., 1990, Mississippi Valley-type and sedex lead-zinc deposits: A comparative examination: Transactions of the Institution of Mining and Metallurgy, section B, v. 99, p. B21–B42.

Schaefer, M., 2002, Paleoproterozoic Mississippi Valley-Type Pb-Zn Deposits of the Ghaap Group, Transvaal Supergroup in Griqualand West, South Africa [Ph.D. dissertation]: Johannesburg, Rand Afrikaans University, 367 p.

Schaefer, M., Gutzmer, J., and Beukes, N.J., 2001, Genesis of carbonate-hosted Pb-Zn deposits in the Later Archean Transvaal Supergroup, Northern Cape Province, South Africa, in Piestrzykinski, A., ed., Mineral deposits at the beginning of the 21st century: Proceedings of the Biennial SGA-SEG Meeting: Lisse, Swets and Zeitlinger Publishers, p. 177–180.

Scott, S.D., 1997, Submarine hydrothermal systems and deposits, in Barnes, H.L., ed., Geochemistry of hydrothermal ore deposits (3rd edition): New York, John Wiley and Sons, p. 797–876.

Scott, D.J., and Gauthier, G., 1996, Comparison of TIMS (U-Pb) and laser ablation microprobe ICP-MS (Pb) techniques for age determination of detrital zircons from Paleoproterozoic metasedimentary rocks from northeastern Laurentia, Canada, with tectonic implications: Chemical Geology, v. 131, p. 127–142, doi: 10.1016/0009-2541(96)00030-7.

Selley, D., Winefield, P., Bull, S.W., Scott, R.J., and McGoldrick, P.J., 2001, Sub-basins, depositional cycles and the tectono-sedimentary setting of the HYC Zn-Pb-Ag deposit: Geological Society of America Abstracts with Programs, v. 33, p. 270–271.

Shen, Y., Canfield, D.E., and Knoll, A.H., 2002, Middle Proterozoic ocean chemistry: Evidence from the McArthur Basin, northern Australia: American Journal of Science, v. 302, p. 81–109.

Sherlock, R.L., Lee, J.K.W., and Cousens, B.L., 2004, Geologic and geochronologic constraints on the timing of mineralization at the Nanisivik zinc-lead Mississippi Valley-type deposit, northern Baffin Island, Nunavut, Canada: Economic Geology and the Bulletin of the Society of Economic Geologists, v. 99, p. 279–293.

Sicree, A.A., and Barnes, H.L., 1996, Upper Mississippi Valley district ore-fluid model: The role of organic complexes: Ore Geology Reviews, v. 11, p. 105–131, doi: 10.1016/0169-1368(95)00018-6.

Smith, S.G., 1996, Geology and geochemistry of the Warrabarty carbonate-hosted Zn-Pb prospect, Paterson orogen, Western Australia [Ph.D. dissertation]: Hobart, University of Tasmania, 162 p.

Sourirajan, S., and Kennedy, G.S., 1962, The system H_2O-NaCl at elevated temperatures and pressures: American Journal of Science, v. 260, p. 115–141.

Strauss, H., 1993, The sulfur isotope record of Precambrian sulfates: New data and a critical evaluation of the existing record: Precambrian Research, v. 63, p. 225–246, doi: 10.1016/0301-9268(93)90035-Z.

Sutherland, R.A., and Dumka, D., 1995, Geology of the Nanisivik mine, N.W.T., Canada, in Misra, K., ed., Carbonate-hosted lead-zinc-fluorite-barite deposits on North America: Society of Economic Geologists Guidebook Series, v. 22, p.4–18.

Sverjensky, D.A., 1981, The origin of a Mississippi Valley-type deposit in the Viburnum Trend, southeast Missouri: Economic Geology and the Bulletin of the Society of Economic Geologists, v. 76, p. 1848–1872.

Symons, D.T.A., Symons, T.B., and Sangster, D.F., 2000, Paleomagnetism of the Society Cliffs dolostone and the age of the Nanisivik zinc deposits, Baffin Island, Canada: Mineralium Deposita, v. 35, p. 672–682, doi: 10.1007/s001260050270.

Turner, E.C., Narbonne, G.M., and James, N.P., 2000, Framework composition of early Neoproterozoic calcimicrobial reefs and associated microbialites,

Mackenzie Mountains, N.W.T., Canada, *in* Grotzinger, J.P., and James, N.P., eds., Carbonate sedimentation and diagenesis in the evolving Precambrian world: Society for Sedimentary Geology Special Publication 67, p. 179–205.

Ueda, A., Cameron, E.M., and Krouse, H.R., 1991, ^{34}S-enriched sulphate in the Belcher Group, N.W.T., Canada: Evidence for dissimilatory sulphate reduction in the early Proterozoic ocean: Precambrian Research, v. 49, p. 229–233, doi: 10.1016/0301-9268(91)90034-8.

Vearncombe, J.R., Chisnall, A.W., Dentith, M.C., Dorling, S.L., Rayner, M.J., and Holyland, P.W., 1996, Structural controls on Mississippi Valley-type mineralization, the southeast Lennard shelf, Western Australia: Society of Economic Geologists Special Publication 4, p. 74–95.

Viets, J.G., Hofstra, A.H., and Emsbo, P., 1996, Solute compositions of fluid inclusions in sphalerite from North American and European Mississippi Valley-type ore deposits: Ore fluids derived from evaporated seawater, *in* Sangster, D.F., ed., Carbonate-hosted lead-zinc deposits: Society of Economic Geologists Special Publication 4, p. 465–482.

Wachowiak, N.M., 2001, Sediment-hosted Pb-Zn (Esker) /Cu mineralization, ~1.89 Ga Rocknest Platform, Northwest Territories/Nunavut, Canada; characteristics, paragenetic analysis and controls [M.Sc. thesis]: Toronto, University of Toronto, 176 p.

Wachowiak, N.M., Gummer, P.K., Grotzinger, J.P., and Spooner, E.T.C., 1998, Approximately 1.89 Ga MVT/SSC mineralization, Esker deposit, Rocknest carbonate platform, Wopmay orogen, N.W.T., Canada: Geological Society of America Abstracts with Programs, v. 30, p. 369.

Wachowiak, N.M., Gummer, P.K., Pope, M., and Grotzinger, J.P., 1997, Stratabound Zn-Pb mineralization to the Esker horizon, Paleoproterozoic Rocknest Formation, Wopmay orogen, Canada, *in* Program and Abstracts, NWT Geoscience Forum 25th Anniversary, Yellowknife: Yellowknife, Canada, Indian and Northern Affairs Canada, p. 96–98.

Walker, L.J., Wilkinson, B.H., and Ivany, L.C., 2002, Continental drift and Phanerozoic carbonate accumulation in shallow-shelf and deep-marine settings: Journal of Geology, v. 110, p. 75–87, doi: 10.1086/324318.

Walker, R.N., Gulson, B., and Smith, J., 1983, The Coxco deposit—A Proterozoic Mississippi Valley-type deposit in the McArthur River District, Northern Territory, Australia: Economic Geology and the Bulletin of the Society of Economic Geologists, v. 78, p. 214–249.

Webb, G.E., 2001, Biologically induced carbonate precipitation in reefs through time, *in* Stanley, G.D., ed., The history and sedimentology of ancient reef systems: Topics in Geobiology, v. 17, p. 25–37.

Wheatley, C.J., Friggens, P.J., and Dooge, F., 1986a, The Bushy Park carbonate-hosted zinc-lead deposit, Griqualand West, *in* Anhaeusser, C.R., and Maske, S., eds., Mineral deposits of southern Africa, p. 891–900.

Wheatley, C.J.V., Whitfield, G.G., Kenny, K.J., and Birch, A., 1986b, The Pering carbonate-hosted zinc-lead deposit, Griqualand West, *in* Anhaeusser, C.R., and Maske, S., eds., Mineral deposits of southern Africa, p. 867–874.

Whelan, J.F., Rye, R.O., and deLoraine, W.F., 1984, The Balmat-Edwards zinc-lead deposits; synsedimentary ore from Mississippi Valley-type fluids: Economic Geology and the Bulletin of the Society of Economic Geologists, v. 79, p. 239–265.

Whelan, J.F., Rye, R.O., deLoraine, W.F., and Ohmoto, H., 1990, Isotopic geochemistry of the mid-Proterozoic evaporite basin; Balmat, New York: American Journal of Science, v. 290, p. 396–424.

Whitesell, T.C., 1995, Diagenetic features associated with sequence boundaries in upper Miocene carbonate strata, Las Negras, Spain [M.Sc. thesis]: Lawrence, University of Kansas, 292 p.

Williams, N., 1978, Studies of the base metal sulfide deposits at McArthur River, Northern Territory, Australia I: The Cooley and Ridge deposits: Economic Geology and the Bulletin of the Society of Economic Geologists, v. 73, p. 1005–1035.

Wilton, D.H.C., Archibald, S.A., Churchill, R.A., Phillips, R.D., and Saunders, J.K., 1993, Metallogenic potential of the lower Proterozoic supracrustal sequence developed on the Archean Nain Craton, northern Labrador, *in* Davies, J.F., Gibson, H.L., and Whitehead, R.S., eds., Third annual field conference of the Geological Society of CIM, Bathurst, NB, Canada, p. 23–41.

Wilton, D.H.C., Archibald, S.M., Hussey, A.M., and Butler, R.W., Jr., 1994, Metallogeny of the Ramah Group: Discovery of a new Pb-Zn exploration target, northern Labrador: Current Research (1994) Newfoundland Department of Mines and Energy, Geological Survey Branch, Report 94-1, p. 415–428.

Winefield, P.R., 2000, Development of late Paleoproterozoic aragonitic seafloor cements in the McArthur Group, northern Australia, *in* Grotzinger, J.P., and Noel, J.P., eds., Carbonate sedimentation and diagenesis in the evolving Precambrian world: Society for Sedimentary Geology Special Publication 67, p. 145–159.

Wu, Y., Hagni, R.D., and Paarlberg, N.L., 1997, Silver distribution in iron sulfides at the Buick and Brushy Creek mines, Viburnum Trend, Southeast Missouri, *in* Sangster, D.F., ed., Carbonate-hosted lead-zinc deposits: Society of Economic Geologists Special Publication 14, p. 577–586.

MANUSCRIPT ACCEPTED BY THE SOCIETY 29 OCTOBER 2005

Superheavy S isotopes from glacier-associated sediments of the Neoproterozoic of south China: Oceanic anoxia or sulfate limitation?

Liu Tie-bing*
Institute of Geology, Chinese Academy of Sciences, P. O. Box 634, Beijing 100029, China

J. Barry Maynard*
Department of Geology, University of Cincinnati, P. O. Box 210013, Cincinnati Ohio 45221, USA

John Alten
Department of Geology, Miami University, 114 Shideler Hall, Oxford, Ohio 45056, USA

ABSTRACT

Black shales and Mn carbonates interbedded with glacial deposits from the Neoproterozoic of southern China exhibit extremely heavy values of pyrite S isotopes that may reflect the peculiar environment of Earth at this time. $\delta^{34}S$ averages +30‰ at Tanganshan and +44‰ at Xiangtan, compared with typical values of 0‰ to +5‰ found in younger deposits. Furthermore there is no distinction between the shales and the Mn carbonate ores in the Neoproterozoic, unlike the younger deposits, which show much lighter $\delta^{34}S$ in the shales than in the Mn ores (the spread is 25‰). Most other chemical parameters are very similar to both the younger Mn deposits and those from the Paleoproterozoic. The exception is the rare earth elements (REE). All Neoproterozoic Fe ores and most Neoproterozoic Mn ores lack the positive Eu anomaly that characterizes Archean and Paleoproterozoic Fe-Mn accumulations. On the other hand, Neoproterozoic Mn deposits have positive Ce anomalies on North American Shale Composite (NASC) normalized plots, in contrast to other $MnCO_3$ ores. The ΣREE is also higher than in other Mn deposits, but lower than in modern deep-sea crusts.

Sulfide S values in all Neoproterozoic shales tend to be exceptionally variable and to often show much heavier values than can be found in marine strata from the Phanerozoic. Therefore the anomalous $\delta^{34}S$ values we observed reflect peculiar conditions in the world oceans at this time rather than purely local effects. Times of enrichment of seawater sulfate in ^{34}S do not correspond to periods of glaciation, so the likely cause of the S isotope patterns is not worldwide glaciation, but a generally low level of dissolved sulfate S in the Neoproterozoic oceans that allowed modest increases in the amounts of S removed as pyrite to drive down the oceanic S reservoir enough to produce strong Rayleigh reservoir effects. The abundance worldwide of Sturtian-age Mn and Fe deposits indicates an increase in Fe flux to the oceans that would have been sufficient to depress SO_4^{2-} levels severely and to result in residual dissolved S extremely enriched in ^{34}S. REE evidence indicates that most of this enhanced Fe and Mn flux came from diagenetic remobilization of detrital oxides

*E-mail: liutiebinggold@sina.com; maynarjb@uc.edu

rather than from ridge-crest hydrothermal systems, in contrast to the Paleoproterozoic banded iron formations. Rapid introduction of lateritic soil residues to restricted basins by low-latitude glaciation could have provided the needed excess Fe and Mn to drive this system.

Keywords: Neoproterozoic, glaciation, REE, S isotopes.

INTRODUCTION

The Neoproterozoic was a time of profound change in Earth's oceans and biosphere. One peculiar manifestation of these changes is the reappearance of iron formations in abundance in the sedimentary record, for example Rapitan in Canada (Yeo, 1981), the iron-rich beds of the Chuos Formation in Namibia (Breitkopf, 1988), and the Braemar Ironstone of South Australia (Lottermoser and Ashley, 2000). Accompanying these iron formations are some of the world's most important Mn ore deposits, e.g., Urucum in Brazil (Urban et al. 1992), and in all cases the ore bodies are found in close association with glacial deposits (Maynard, 1991). In a study of the relationship of Mn ore genesis to black shale sedimentation, Liu (1988, 1990) compared the whole-rock carbon-sulfur geochemistry of one of these Neoproterozoic deposits, Xiangtan, with younger deposits and found the patterns to be similar. Subsequent work on the sulfur isotopic composition of the Chinese Neoproterozoic deposits, however, has shown them to contain extraordinarily heavy S (Tang and Liu, 1999; Li et al., 1999).

Is this super-heavy sulfur peculiar to the Neoproterozoic of south China? Is it somehow related to the Mn-Fe mineralization? Is it somehow related to glaciation? To address these questions we have undertaken detailed geochemical analysis of two

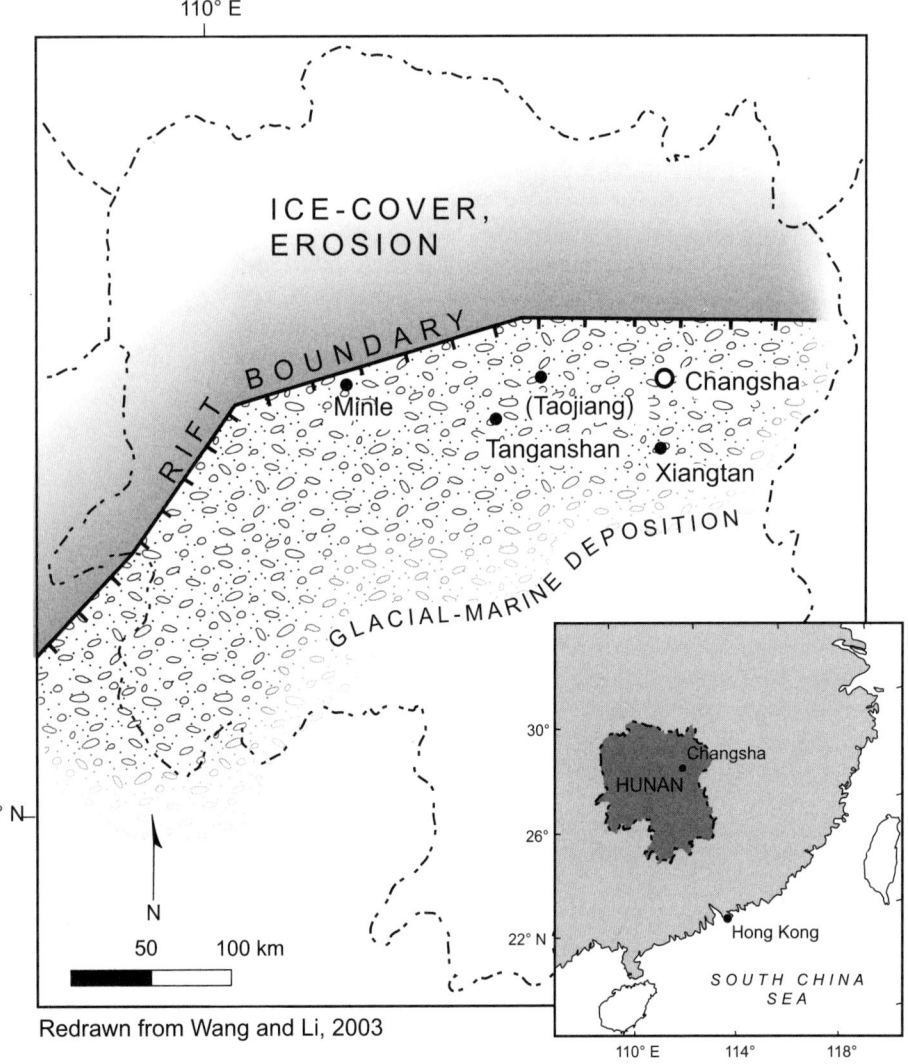

Figure 1. Location of the Xiangtan, Tanganshan, and Minle Neoproterozoic Mn deposits in south China and the Ordovician Taojiang deposit.

of these deposits, Xiangtan and Tanganshan. In this paper we show how isotopic, trace element, and rare earth element (REE) data constrain models for the genesis of Mn-Fe deposits of this time period and help understand the Neoproterozoic atmosphere-ocean system.

Geological Setting

There are 11 large to medium-sized commercial Mn operations in Neoproterozoic host rocks of south China distributed throughout the Hubei, Guizhou, Sichuan, and Hunan Provinces (Xu et al., 1990; Fan et al., 1992; Li et al., 1999). We studied Xiangtan and Tanganshan in Hunan province (Fig. 1). The ores are contained within black shales that are overlain and underlain by glacial deposits (Fig. 2). The regional stratigraphic section comprises, from the base, the Liantuo Formation sandstones, the Gucheng or Chunmu sandstones and diamictites, the Datangpo Formation (Minle Formation of some authors), comprising shales and Mn carbonates, and, at the top, the Nantuo Formation diamictites. The Nantuo is overlain abruptly by carbonates of the Doushantuo Formation and this contact is an important stratigraphic marker in south China. Thus there are two diamictite intervals separated by a black shale with Mn carbonate beds.

Ages are sparse, but a tuff bed in the Liantuo Formation has yielded one date of 748 ± 12 Ma from single-zircon U-Pb dating, and the Datangpo/Minle Formation has been dated by Rb-Sr at 728 ± 27 Ma (Li et al., 2000). These ages and sequence stratigraphic interpretations led Wang and Li (2003, p. 155) to suggest that the Nantuo diamictites are correlative with the Sturtian glacial deposits of Australia. If so, the Datangpo Mn mineralization is comparable to the Braemar ironstone, which is interbedded with the glacial deposits (Gorjan et al., 2000). The overlying Doushantuo carbonates, however, have been dated to 584 ± 26 Ma by Lu-Hf and 599.3 ± 4.2 Ma by Pb-Pb methods (Barfod et al., 2002), which suggests that the Nantuo glacial deposits correspond to the later Marinoan glaciation episode. Dobrzinski et al. (2004), however, argued that both the Chunmu and the Nantuo diamictites are Sturtian, on the basis of the 728 Ma date for the intervening Datangpo Formation. Paleomagnetic data are also ambiguous. Rui and Piper (1997) assigned the Datangpo Formation to the Sturtian, whereas Macouin et al. (2004) favored a Marinoan age. Both Shen (2002) and Macouin et al. (2004) argued for a Marinoan age based on carbon isotope excursions, but either major glacial episode could produce this effect. Thus there are two conflicting age interpretations. If the Marinoan age is correct, then perhaps the Chunmu tillite is a Sturtian equivalent and the Datangpo Formation represents a relatively long time period. Alternatively, the Chunmu and Nantuo tillites could both be phases of Sturtian glaciation and the Datangpo was rapidly deposited during a brief interglacial episode.

The reconstruction of Xu et al. (1990) shows the Datangpo Formation lying within a fault-bounded marine trough between the Yangtze Block to the northwest and the Hunan Block to the southeast. Continental diamictites occupied the margin of the Yangtze Block during Datangpo deposition, whereas glaciomarine strata are found on the Hunan Block. Thus deposition was in a narrow fjord-like rift valley. Urban et al. (1992) proposed a similar ice-covered long, narrow basin as the setting for the Urucum deposits of Brazil. Eyles and Januszczak (2004a, 2004b) have likewise emphasized the rift setting of Neoproterozoic glacial deposits as key to their genesis.

ANALYTICAL METHODS

Samples were collected from mine passageways and were ground in a tungsten carbide ring mill. They were analyzed for major elements by X-ray fluorescence (XRF) and for trace elements by inductively coupled plasma–mass spectrometry (ICP-MS) supplemented by hydride-atomic absorption (AA) for As and Se and by LECO elemental analyzer for C and S. XRF analyses were done at the University of Cincinnati using a Rigaku 3070 spectrometer. Loss-on-ignition residues of the samples were fused with lithium metaborate. The resulting glass beads were then ground and pressed into disks with polyvinyl alcohol binder. Calibration was done by empirical multiple regression

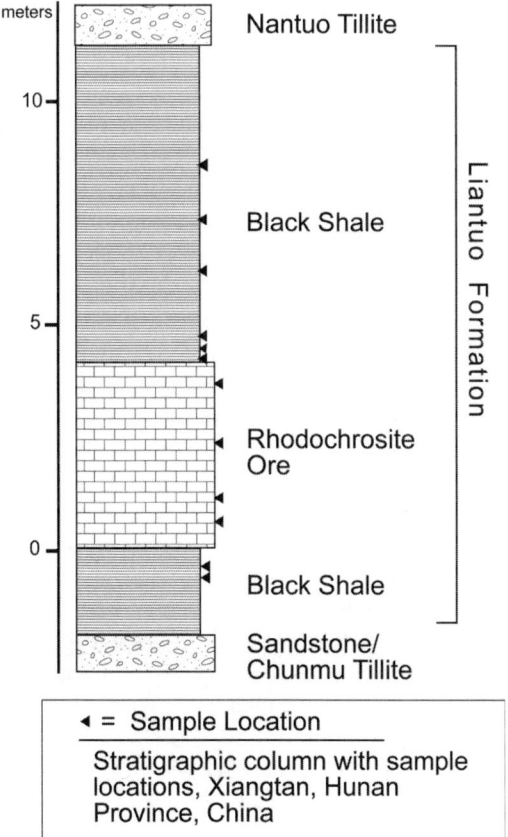

Figure 2. Stratigraphic setting of the Mn mineralization in the Xiangtan deposit (after Liu, 1988).

against a set of U.S. Geological Survey and National Institute of Standards and Technology (NIST) high-Mn standards treated in the same way. Reproducibility is ±2%. ICP-MS and AA were performed by XRAL Laboratories of Toronto, Canada. Precision is about ±5% for most elements with detection limits generally of 0.1–1 ppm for trace elements and 0.05–0.1 for REE. LECO analyses were performed on whole-rock samples and on acid-insoluble residues at Indiana University. Precision was ±1% for C and ±5% for S.

Sulfur was extracted from the samples by a method originally developed by Canfield et al. (1986) and modified by the biogeochemistry laboratory at Indiana University (Bruchert, 1998). First, sample powders are extracted for 24 h with methylene chloride in a soxhlet apparatus. This procedure removes native sulfur and bitumen-bound sulfur. Extracting the bitumen also removes organic coatings from sulfide mineral grains and makes them easier to dissolve in subsequent steps. Any extracted native sulfur is reprecipitated as copper sulfide on copper shot in a receiving flask. The S-coated copper shot is then transferred to a closed reaction system where it is treated with 6N HCl under a nitrogen stream and heated to 90 °C. This process converts the copper sulfide to H_2S, which is bubbled into a $AgNO_3$ trap where the S is precipitated as Ag_2S. Next 0.5–2.0 g of the MeCl-extracted samples are transferred to the same reaction system. Heating the sample to 90 °C in 6N HCl releases S contained in monosulfides (FeS, galena, sphalerite) and also in pyrrhotite. This sulfide converts to H_2S in the acid environment, which is again bubbled into a $AgNO_3$ trap. This S is customarily designated as the acid-volatile fraction (AVS). The sulfur in pyrite and marcasite, FeS_2, occurs in the S^{1-} form and must be reduced to the S^{2-} form to be converted to H_2S gas. This reduction is accomplished using chromous chloride in 6N HCl, again at 90 °C, followed by precipitation of Ag_2S from the H_2S gas. We designate this S as the chrome-reducible fraction (CrRS). The Xiangtan and Tanganshan samples yielded only pyrite S, except for two samples from the ore dump at Tanganshan that had large amounts of native S. We interpret this native S as a weathering product of pyrite and have not investigated surface samples further because of the large impact of weathering, even with relatively brief periods of exposure. Sulfur isotopic ratios were determined at Indiana University using the Finnigan MAT 252 mass spectrometer described by Studley et al. (2002).

RESULTS

Whole Rock Chemistry

The Chinese Mn ores have extreme S isotope values, and might be expected to also have significant anomalies in other chemical properties. However, whole rock data for the Mn ores and for the host black shales are similar to values reported from other Mn deposits and for representative unmineralized black shales. In Table 1 we give major element results for the Xiangtan and Tanganshan ores and shales plus analytical data using the same procedures for a set of NIST standards of Mn and Fe-rich rocks and a representative suite of ore samples from Paleoproterozoic and Phanerozoic Mn deposits. Table 2 gives trace element results and Table 3 gives REE results from the same samples and standards. Many of the NIST standards do not have trace and rare earth elements included in their documentation, so we present values for these standards for the first time. Mamatwan is the largest Paleoproterozoic deposit and comprises a significant fraction of the world's Mn reserves. Molango is the largest Mesozoic deposit, and Taojiang, although much smaller, is representative of Paleozoic deposits. For each deposit, we have selected representative ore samples, either Mn-rich or Fe-rich, and some host rocks. Extensive trace element and REE data for these deposits have not previously been published.

Inspection of the tabulated results for the ores and shales in the Chinese deposits shows a continuum of compositions between high and low Al end members. The data set shows a strong positive correlation between % Mn and % $C_{carbonate}$ ($r^2 = 0.74$) and a negative correlation between % Mn and both % C_{org} ($r^2 = -0.77$) and % Al_2O_3 ($r^2 = -0.71$). No other elements show a positive correlation to % Mn and there are no correlations between any of the whole rock parameters and $\delta^{34}S$.

Thus the highest-grade ore samples from the Chinese deposits are low in Al and C_{org}, but high in C_{carb}. We have used 3% Al_2O_3 as the upper cutoff to separate ore samples from sub-ore grade, but mineralized, samples. In Figure 3 we compare the composition of this end-member ore type to an average $MnCO_3$ ore as calculated from the representative ores given in Tables 1 and 2, giving equal weight to ores of each of the three ages. Two groups of elements show enrichment in the Chinese ores: one group comprising P and Mo, which are probably related to the high organic carbon content of the ores, and another comprising Ti, Y, Zr, and Nb, which are elements typically associated with heavy minerals. Al is identical in the two data sets, so these excess resistate elements are not associated with clays. Pb is also relatively high in the Chinese ores and Ca and Ta are depleted.

Because the Chinese ores are hosted by black shales, we have used the geostandard SDO-1 as a comparison. The Chinese black shales are very similar to this standard (Fig. 4). Compared with SDO-1, Mn is somewhat enriched in the Chinese host shales, so Mn mineralization is distributed at low grade in the host rocks, a pattern that is also seen at Molango. Mo and U are low compared to SDO-1, which is probably related to much higher organic carbon in the standard (10% compared to ~3%–4% for the Chinese shales). Thus, just as for the ores, the host rocks appear to have normal chemistries except for their S isotopic values.

For the REEs, some differences can be seen between the Chinese material and other ores (Fig. 5). Almost all of the older and younger Mn deposits in Table 3 have negative Ce anomalies. Both the Chinese ore deposits by contrast display positive anomalies. Yang et al. (1999) also reported a positive Ce anomaly for one sample from the correlative Songtao deposit. For Eu, the samples from other deposits all have positive

TABLE 1. MAJOR ELEMENTS, C AND S FOR Mn AND Fe ORES AND RELATED STANDARD REFERENCE MATERIALS

Sample name	Analysis unit	Fe % oxide	Mn % oxide	Ti % oxide	Si % oxide	Al % oxide	Ca % oxide	K % oxide	P % oxide	Mg % oxide	Na % oxide	LOI % oxide	total % oxide	C$_{carb}$ %	C$_{org}$ %	S %	δ^{34}S pyrite ‰
Standards																	
NIST-25d	Mn ore	4.16	66.9	0.13	2.52	5.52	0.11	1.40	0.30	0.43	0.03	14.7	96.1			0.36	
NIST-26	Fe ore	79.5	0.14	0.07	6.10	0.96	2.50	0.02	0.09	3.33	0.00	4.46	97.2			0.03	
NIST-27b	Fe ore	93.9	0.17	0.02	1.31	0.62	0.08	0.05	0.09	0.05	0.01	0.39	96.7			0.02	
NIST-28a	Fe ore	93.8	0.53	0.02	1.50	0.76	0.06	0.04	0.07	0.05	0.00	3.00	99.8			0.02	
NIST-69b	Bauxite	7.49	0.11	1.90	13.4	48.8	0.15	0.09	0.14	0.10	0.02	27.2	99.4				
NIST-88b	Dolomite	0.27	0.02	0.02	1.13	0.36	30.1	0.16	0.02	21.0	0.03	47.0	100.1				
NIST-97b	Flint clay	1.13	0.01	2.43	41.2	40.0	0.04	0.91	0.07	0.22	0.09	13.9	100.0			0.01	
NIST-690	Fe ore	93.2	0.23	0.02	3.71	0.18	0.20	0.00	0.05	0.20	0.00	0.06	97.9			0.05	
NIST-1413	Al sand	0.39	0.03	0.10	81.7	9.55	0.77	3.77	0.07	0.07	1.63	1.55	99.6			0.29	
VL-1	Laterite	35.8	0.07	3.15	1.16	37.4	0.03	0.09	0.16	0.03	0.01	22.4	100.3				
Tanganshan																	
TG-3	Mn ore	7.74	27.9	0.56	30.6	6.99	1.90	2.12	0.21	1.93	0.06	19.9	99.9	3.60	3.88	4.45	31.41
TG-4	Mn ore	2.91	59.5	0.05	0.50	0.24	4.70	0.02	0.30	2.51	0.09	31.6	102.4	10.16	2.45	1.32	23.78
TG-5	Mn ore	4.59	56.4	0.30	6.03	2.54	2.90	0.04	0.26	1.12	0.07	26.1	100.4	7.19	3.29	2.60	29.59
TG-6	Shale	6.34	12.5	0.74	50.9	9.68	5.24	0.55	0.24	3.63	0.10	10.9	100.8	1.25	4.02	1.76	
TG-7	Shale	4.02	1.57	0.81	56.2	15.9	0.85	9.18	0.16	1.90	0.13	8.97	99.7	0.00	4.26	3.53	29.70
TG-8	Shale	7.22	6.99	0.68	50.0	11.9	2.03	4.98	0.34	3.64	0.20	12.6	100.5	0.60	4.34	5.24	31.07
TG-9	Shale	9.46	20.6	0.49	33.4	6.82	8.40	0.21	0.61	1.46	0.34	17.7	99.6	3.85	3.48	7.08	30.48
Xiangtan																	
XT-0.3	Shale	5.24	13.2	0.65	42.0	11.6	1.38	3.72	0.37	1.43	0.21	20.2	100.1	1.82	0.94	4.56	48.44
XT-0.4	Mn ore	2.79	26.9	0.47	39.1	8.47	1.26	2.76	0.26	1.28	0.14	17.1	100.5	4.45	0.91	2.46	41.46
XT-0.6	Mn ore	2.18	60.1	0.10	3.13	0.85	1.68	0.52	0.21	1.92	0.04	30.5	101.3	9.43	0.69	0.77	57.58
XT-1.3	Mn ore	2.2	49.0	0.15	17.0	0.80	2.06	0.31	0.12	2.18	0.04	26.4	100.2	8.07	0.55	0.80	47.36
XT-2.4	Mn ore	4.67	39.8	0.54	15.0	5.01	2.05	1.34	0.48	1.43	0.21	29.0	99.5	6.49	2.07	3.14	36.82
XT-3.7	Mn ore	4.25	34.5	0.21	22.8	5.26	2.83	0.65	0.37	1.97	0.21	26.9	99.9	6.41	1.58	2.71	42.86
XT-4.2	Shale	3.35	0.15	0.71	60.0	14.2	0.10	4.23	0.08	1.37	0.04	16.9	101.2	0.00	5.08	3.42	
XT-4.3	Dolomite	0.31	0.29	0.03	20.1	1.33	24.6	0.39	0.11	15.7	0.00	37.2	100.0	10.12	4.27	0.96	15.36
XT-4.5	Mn ore	3.25	25.7	0.47	31.9	8.25	2.54	2.21	0.21	1.61	0.10	23.7	100.0	4.93	2.76	2.40	43.65
XT-6.1	Shale	8.34	16.3	0.99	34.2	9.68	1.33	3.44	0.20	1.50	0.00	23.9	100.0	2.36	2.99	9.09	39.26
XT-7.2	Shale	2.43	0.18	0.72	64.1	15.3	1.47	4.51	0.06	1.65	0.05	11.0	101.5	0.00	4.14	3.34	60.93
XT-8.1	Shale	2.87	0.3	0.72	62.9	14.2	1.45	4.41	0.07	1.56	0.04	13.2	101.7	0.00	3.47	2.94	53.05
Other Mn and Fe Ores																	
SO-5100	Mamatwan Fe formation	37.1	15.7	0.15	18.5	2.58	5.6	0.86	0.04	3.36	0.10	15.0	98.8			0.06	
SO-5101	Mamatwan Mn ore	6.55	42.8	0.01	5.78	0.18	18.9	0.01	0.02	2.96	0.03	20.2	97.4			0.03	
SO-5102	Mamatwan Mn ore	4.93	45.3	0.01	3.46	0.14	20.5	0.00	0.03	3.54	0.02	20.8	98.7			0.04	
SO-5103	Mamatwan Mn ore	7.05	46.4	0.01	6.25	0.25	14.1	0.00	0.03	3.91	0.03	19.6	97.7			0.03	
SO-5104	Mamatwan Mn ore	5.57	53.8	0.01	4.90	0.11	14.7	0.00	0.03	2.43	0.02	15.3	96.9			0.05	
SO-5108	Hotazel Fe formation	35.0	0.16	0.01	67.1	0.17	0.09	0.01	0.04	0.07	0.02	0.86	103.5			0.01	
SO-5206	Molango shale	4.64	2.56	0.29	33.3	6.84	17.2	1.32	0.10	6.70	0.18	27.5	100.5			2.54	
SO-5207	Molango Mn ore	13.5	35.9	0.09	10.8	2.59	2.64	0.01	0.13	8.23	0.01	26.6	100.4			0.27	
SO-5209	Molango Mn ore	11.0	36.3	0.07	12.0	2.79	2.19	0.00	0.14	8.79	0.00	27.4	100.7			0.17	
SO-5301	Taojiang Mn ore	1.02	30.0	0.01	11.2	0.51	23.0	0.22	0.05	1.26	0.02	31.9	99.2			0.33	
SO-5302	Taojiang Mn ore	1.04	38.2	0.02	4.59	0.63	16.9	0.27	0.05	2.34	0.02	34.8	98.8			0.06	

TABLE 2. TRACE ELEMENTS

Sample name	Analysis unit	Ag ppm	As ppm	Ba ppm	Bi ppm	Cs ppm	Cu ppm	Ga ppm	Mo ppm	Nb ppm	Ni ppm	Pb ppm	Rb ppm	Sb ppm	Se ppm	Sn ppm	Sr ppm	Ta ppm	Te ppm	Th ppm	Tl ppm	U ppm	V ppm	Y ppm	Zn ppm	Zr ppm
Standards																										
NIST-25d	Mn ore	4	32	2000	0.7	0.8	401	22	10	4	481	36	19	5.2	0.3	5	163	<0.5	0.2	4.2	5.2	1.3	104	16	706	18
NIST-26	Fe ore	5	2	23.8	<0.1	0.1	5	2	<2	4	7	<5	0.8	0.2	<0.1	118	16	<0.5	0.2	0.8	<0.5	0.6	32	7	<10	4
NIST-27b	Fe ore	4	2.4	49.9	<0.1	0.2	42	2	<2	2	19	21	0.6	1.3	0.1	6	12	1.2	0.3	0.4	<0.5	0.3	39	35	26	2
NIST-28a	Fe ore	4	12	33.2	<0.1	0.6	16	4	2	3	17	25	1.3	0.3	<0.1	3	26	<0.5	0.2	0.4	<0.5	3.6	104	21	18	2
NIST-69b	Bauxite	5	14	94.8	1.2	0.2	19	70	16	918	13	47	3.6	0.9	1.8	13	133	33	0.4	95	<0.5	12	145	63	15	2800
NIST-88b	Dolomite	9	0.8	21.5	<0.1	0.2	<5	<1	2	<2	9	<5	3.3	<0.1	0.1	<1	69	<0.5	0.3	0.5	<0.5	0.2	<5	8.9	<10	<2
NIST-97b	Flint clay	2	3	178	0.9	3.4	24	63	8	45	35	49	34	0.8	0.3	11	91	2.8	0.3	37	<0.5	7.9	254	37	72	450
NIST-690	Fe ore	3	4	21.7	<0.1	<0.1	13	2	<2	3	11	<5	0.4	0.2	<0.1	<1	4.1	<0.5	0.1	0.3	<0.5	0.1	37	3.6	<10	2
NIST-1413	Al sand	3	0.3	1040	<0.1	2.3	6	9	<2	4	7	224	119	0.5	0.2	<1	212	<0.5	0.1	2.5	0.5	1	10	6.5	21	107
VL-1	Laterite	2	2.3	21	<0.1	<0.1	115	46	<2	15	8	10	0.8	0.4	0.2	3	1.8	0.8	<0.1	14	<0.5	1.4	880	3.6	171	173
Tanganshan																										
TG-3	Mn ore	3	121	296	0.5	3.5	83	19	91	24	109	61	56	15	4.2	4	44	1	0.5	11	0.5	2.4	273	28	22	87
TG-4	Mn ore	3	30	48.7	0.3	0.2	<5	3	4	15	27	31	0.9	5.8	1.1	3	93	<0.5	0.2	1	<0.5	0.5	29	33	20	26
TG-5	Mn ore	9	66	51.4	0.7	0.3	16	9	9	35	63	104	1.7	22	2.8	10	46	<0.5	0.2	5.5	<0.5	1.3	131	48	18	40
TG-6	Shale	3	61	100	0.3	4.2	136	18	101	23	149	90	21	46	1.9	8	86	1.5	0.2	18	1.4	5.2	348	59	41	148
TG-7	Shale	3	13	1510	0.1	5.6	43	20	19	18	71	36	215	10	2.4	4	328	1.1	<0.1	17	1.6	3.5	173	59	53	238
TG-8	Shale	3	61	1090	0.2	3.8	62	18	79	19	61	25	94	5	3.5	4	207	1.4	0.3	13	0.7	4	269	51	131	169
TG-9	Shale	3	92	69.8	0.3	1.1	99	10	45	18	77	66	7	18	3.5	5	174	0.7	0.2	12	<0.5	3.3	160	57	76	90
Xiangtan																										
XT-0.3	Shale	5	152	810	0.2	12.2	52	23	10	19	52	111	106	10.8	1	5	111	0.8	0.2	13.7	3.6	3.53	180	46.1	61	157
XT-0.4	Mn ore	3	106	436	0.3	8.5	38	19	14	12	38	44	65.6	7.7	1	4	59.3	0.8	0.2	9.8	2.1	2.45	128	32.1	109	96
XT-0.6	Mn ore	2	37	260	<0.1	1	9	12	5	14	14	8	16.2	2.7	0.5	5	84.4	<0.5	0.2	1.2	<0.5	0.71	53	23.9	45	29
XT-1.3	Shale	5	30.7	257	<0.1	0.7	13	13	8	21	33	8	11	2.9	0.4	3	118	<0.5	0.2	1.2	0.5	0.72	52	41.9	487	30
XT-2.4	Mn ore	3	252	1140	0.4	5.2	55	20	38	23	85	23	35.8	11.1	1.2	4	366	0.6	0.6	12.4	5.2	5.79	137	51.5	38	97
XT-3.7	Mn ore	3	144	5130	0.2	2.1	32	15	25	11	34	17	17.1	4.9	0.6	3	321	0.5	0.3	6.9	2.9	2.58	73	50.6	5	66
XT-4.2	Shale	4	119	1070	0.2	15.1	44	24	3	15	54	71	121	2.2	3.7	3	36.2	0.9	0.4	12.9	2	3.71	65	31.6	69	205
XT-4.3	Dolomite	4	24.1	418	0.1	1.9	34	8	7	2	28	38	8.4	0.9	2.8	<1	1140	<0.5	0.3	1.4	<0.5	3.63	112	25.5	21	134
XT-4.5	Mn ore	2	92.4	15200	0.2	5	39	21	17	12	38	14	53.5	2.6	1.9	3	169	0.5	0.4	8.8	2.1	2.63	65	46.9	44	80
XT-6.1	Shale	3	>500	686	0.4	8.7	91	22	62	49	174	105	88.8	30.8	3	5	95.3	1	1.2	17.1	5.7	5.24	240	42	32	166
XT-7.2	Shale	2	101	522	0.2	12.4	43	21	<2	16	43	22	135	3.1	4.2	4	30.9	1	0.3	14.1	2.7	2.75	64	33.2	39	205
XT-8.1	Shale	2	80.2	481	0.2	4.5	37	21	<2	13	55	189	130	1.3	3.1	<1	39.6	0.8	0.3	14.4	0.7	2.79	68	42.6	51	190
Other Mn Ores																										
SO-5100	Mamatwan Fe fmn	<1		160		1.2	<5	6	<2	2	8	6	23.9			<1	62.8	0.9		2	<0.5	0.18	12	10.1	47	15.2
SO-5101	Mamatwan Mn ore	<1		286		0.2	<5	10	<2	<1	17	6	1			<1	169	1.4		0.4	<0.5	0.16	<5	4.9	9	2.8
SO-5102	Mamatwan Mn ore	<1		467		<0.1	8	8	<2	<1	14	<5	<0.2			<1	226	0.7		0.4	<0.5	0.12	<5	6	27	2.5
SO-5103	Mamatwan Mn ore	<1		89.8		<0.1	<5	8	<2	<1	18	<5	<0.2			<1	62.5	0.7		0.4	<0.5	0.23	6	5.1	159	2.5
SO-5104	Mamatwan Mn ore	<1		598		<0.1	<5	9	<2	<1	25	<5	<0.2			<1	126	0.7		0.3	<0.5	0.12	<5	5	99	1.8
SO-5108	Hotazel Fe fmn	<1		8.9		0.1	<5	2	<2	1	8	10	0.6			<1	3.7	7.4		0.2	<0.5	0.07	13	3.4	<5	2.4
SO-5206	Molango shale	<1		233		4.7	25	9	16	4	131	16	56			<1	331	1.4		3.9	7.9	3.07	67	19.9	287	82.9
SO-5207	Molango Mn ore	<1		65.2		0.1	6	9	2	2	40	6	<0.2			<1	41.2	0.5		1.4	<0.5	1.46	56	8.2	54	12.9
SO-5209	Molango Mn ore	<1		24.3		0.1	8	9	<2	1	79	6	<0.2			<1	38.6	<0.5		1.4	<0.5	1.8	77	9	41	16.3
SO-5301	Taojiang Mn ore	<1		403		0.5	11	6	<2	<1	95	13	7.7			<1	1370	1		0.7	<0.5	0.23	12	11.3	120	3.1
SO-5302	Taojiang Mn ore	<1		465		0.7	31	8	<2	<1	96	16	10.2			<1	974	1.9		0.7	<0.5	0.11	25	9.6	115	2.1

TABLE 3. RARE-EARTH ELEMENTS

Sample name	Analysis unit	La ppm 57	Ce ppm 58	Pr ppm 59	Nd ppm 60	Sm ppm 62	Eu ppm 63	Gd ppm 64	Tb ppm 65	Dy ppm 66	Ho ppm 67	Er ppm 68	Tm ppm 69	Yb ppm 70	Lu ppm 71	SumREE ppm	Ce/Ce* nasc	Eu/Eu* nasc
Standards																		
NIST-25d	Mn ore	18.8	56.5	4.74	18.4	3.8	0.95	3.16	0.49	3.3	0.59	1.67	0.25	1.4	0.21	114	1.30	1.20
NIST-26	Fe ore	4.7	9.7	1.17	4.9	0.8	0.31	1.05	0.17	1.02	0.22	0.63	0.1	0.6	0.07	25	0.90	1.49
NIST-27b	Fe ore	11	25.6	3.7	17.7	5.5	3.42	7.29	1.58	8.8	1.44	3.71	0.55	2.7	0.45	93	0.87	2.37
NIST-28a	Fe ore	10.5	13.7	2.07	9.3	1.9	0.66	2.6	0.45	3.3	0.65	1.97	0.28	1.6	0.25	49	0.64	1.30
NIST-69b	bauxite	74.7	252	10.8	32.7	5.6	0.81	5.45	1.16	7.2	2.09	8.01	1.4	10.5	1.5	414	1.93	0.64
NIST-88b	dolomite	5	4.6	0.85	2.9	0.6	0.13	0.54	0.11	0.74	0.16	0.43	0.05	0.3	0.05	16	0.49	1.00
NIST-97b	flintclay	21.5	37.8	3.99	13.3	2.6	0.68	3.99	0.78	6.4	1.35	4.23	0.68	4.9	0.67	103	0.89	0.93
NIST-690	Fe ore	1.4	1.9	0.26	1.1	0.2	0.1	0.32	0.06	0.45	0.09	0.28	0.05	0.3	<0.05	7	0.69	1.74
NIST-1413	high-Al sand	8.2	13.8	1.68	6.7	1.2	0.47	0.99	0.16	0.95	0.17	0.53	0.09	0.5	0.08	36	0.81	1.89
VL-1	laterite	2.2	13.7	0.48	1.6	0.6	0.19	0.63	0.12	0.75	0.2	0.53	0.11	0.8	0.14	22	2.90	1.36
Tanganshan																		
TG-3	Mn ore	34	82.7	7.35	29.2	5.4	1.5	5.37	0.84	6.4	1.19	3.73	0.55	3.4	0.59	182	1.14	1.22
TG-4	Mn ore	29.1	92.5	7.54	31.1	6.8	2.2	7.19	1.17	6.88	1.42	4.23	0.5	2.8	0.33	194	1.36	1.38
TG-5	Mn ore	34.8	105	8.36	34.5	8.4	2.49	7.85	1.31	9.1	1.67	4.84	0.66	4.2	0.52	224	1.34	1.35
TG-6	shale	61.6	155	14	59.8	9.9	1.95	11.5	2.09	14.5	2.82	8.32	1.08	7.5	1.18	351	1.15	0.80
TG-7	shale	54.9	111	12.8	51.3	10.9	1.52	10	1.55	11.5	2.19	5.59	0.76	5	0.76	280	0.91	0.64
TG-8	shale	42	88.8	10.1	43.1	9.3	2.14	10.2	1.53	10.5	1.81	5.36	0.73	4.7	0.72	231	0.94	0.96
TG-9	shale	54.4	170	14	58.6	12.3	2.68	12.6	1.84	12.9	2.31	6.41	0.87	5.2	0.72	355	1.34	0.95
Xiangtan																		
XT-0.3	shale	54.3	125	12.4	48.4	9.8	1.68	8.49	1.33	8.9	1.71	5.29	0.79	5.4	0.68	284	1.05	0.81
XT-0.4	Mn ore	32.1	75	7.79	32.6	7.3	1.11	6.46	1.13	7.16	1.25	3.7	0.56	3.6	0.48	180	1.03	0.71
XT-0.6	Mn ore	23	69	6	23.7	4.5	0.9	4.89	0.78	5.03	1.15	2.9	0.4	2.6	0.4	145	1.28	0.84
XT-1.3	Mn ore	29	89.8	7.41	31.9	6.5	1.38	6.22	0.97	6.67	1.3	3.65	0.51	2.9	0.36	189	1.33	0.95
XT-2.4	Mn ore	48.2	127	11.5	44.2	8.3	1.58	7.61	1.33	8.06	1.86	5.9	0.86	5.2	0.82	272	1.17	0.87
XT-3.7	Mn ore	43.4	134	11.8	50.1	11.6	2.32	10.2	1.7	11.3	2.19	5.88	0.92	4.8	0.57	291	1.29	0.94
XT-4.2	shale	41.7	69.5	8.76	32.6	7.2	1.35	6.08	0.91	5.8	1.14	3.45	0.5	3.2	0.51	183	0.79	0.90
XT-4.3	dolomite	8.3	15.4	1.71	7.4	2.3	0.57	2.75	0.46	3.39	0.72	2.17	0.33	1.8	0.3	47	0.89	1.00
XT-4.5	Mn ore	39.5	100	9.68	40.7	9	1.2	9.1	1.33	8.53	1.74	5.2	0.72	4.2	0.63	232	1.11	0.58
XT-6.1	shale	81.6	199	15.6	55.7	9.9	1.72	8.56	1.36	8.1	1.62	5.04	0.74	5.5	0.77	395	1.21	0.82
XT-7.2	shale	46	77.9	9.57	36.7	7.4	1.18	6.1	0.91	5.91	1.2	3.95	0.58	3.5	0.49	201	0.81	0.77
XT-8.1	shale	42.5	71	9.28	36.3	7.8	1.65	7.52	1.15	7.12	1.42	4.29	0.61	4.4	0.67	196	0.78	0.95
Other Mn Ores																		
SO-5100	Mamatwan Fe formation	4.1	5.2	0.56	2.5	0.5	0.16	0.74	0.12	1.01	0.24	0.87	0.14	0.9	0.16	17	0.75	1.15
SO-5101	Mamatwan Mn ore	2.5	3	0.41	1.9	0.3	0.13	0.49	0.07	0.5	0.12	0.34	0.06	0.4	0.09	10	0.65	1.49
SO-5102	Mamatwan Mn ore	3.5	3.5	0.52	2.4	0.4	0.2	0.7	0.1	0.63	0.16	0.49	0.06	0.5	0.07	13	0.57	1.66
SO-5103	Mamatwan Mn ore	3.9	3.6	0.51	2.3	0.3	0.12	0.57	0.1	0.51	0.15	0.44	0.06	0.5	0.08	13	0.56	1.27
SO-5104	Mamatwan Mn ore	2.8	3.1	0.42	1.9	0.3	0.18	0.55	0.09	0.57	0.13	0.39	0.05	0.4	0.08	11	0.62	1.95
SO-5108	Hotazel Fe formation	2.3	3.1	0.26	1	0.1	0.07	0.29	0.03	0.38	0.08	0.32	0.03	0.3	0.06	8	0.87	1.80
SO-5206	Molango shale	23.3	37	4.9	20.3	3.8	0.84	4.16	0.62	3.39	0.69	2	0.26	1.8	0.26	103	0.75	0.93
SO-5207	Molango Mn ore	10.4	16.5	2.17	8.7	1.6	0.41	1.72	0.26	1.37	0.28	0.76	0.09	0.7	0.18	45	0.76	1.09
SO-5209	Molango Mn ore	13.2	19.7	2.53	10.3	1.8	0.44	2.02	0.31	1.6	0.33	0.94	0.13	0.8	0.11	54	0.74	1.01
SO-5301	Taojiang Mn ore	11.4	23.9	1.73	7.3	1.6	0.77	2.12	0.32	1.92	0.36	0.94	0.14	0.8	0.11	53	1.17	1.84
SO-5302	Taojiang Mn ore	13.9	20.4	2.06	9.4	1.8	0.86	2.19	0.31	1.8	0.36	1.06	0.14	0.9	0.14	55	0.83	1.90

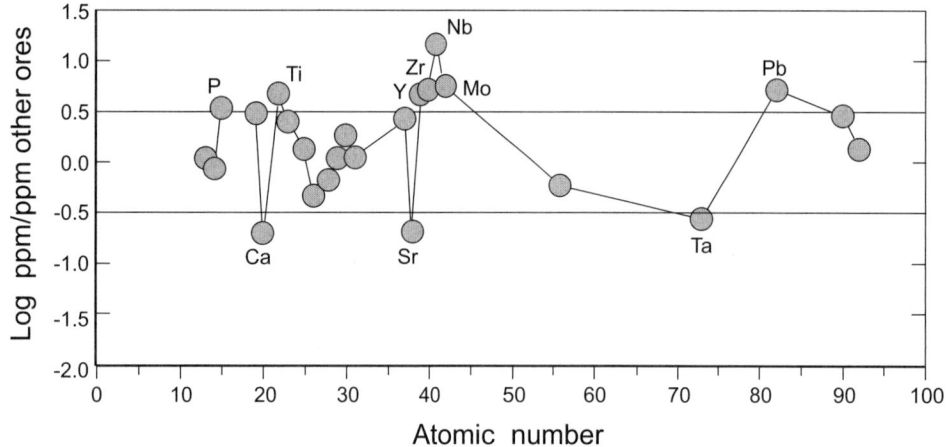

Figure 3. End-member MnCO$_3$ ores from the Chinese Neoproterozoic deposits compared to the average of typical MnCO$_3$ ores from other deposits.

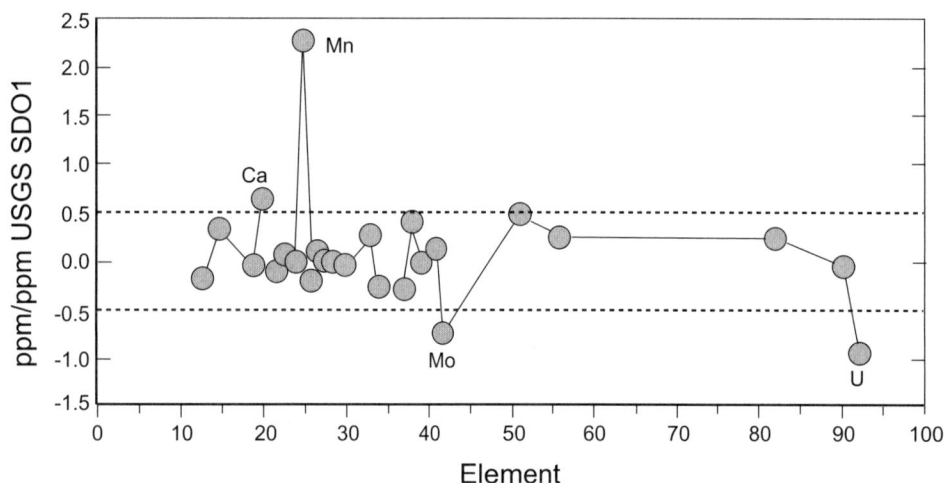

Figure 4. Comparison of compositions of host shales and the black shale standard SDO1.

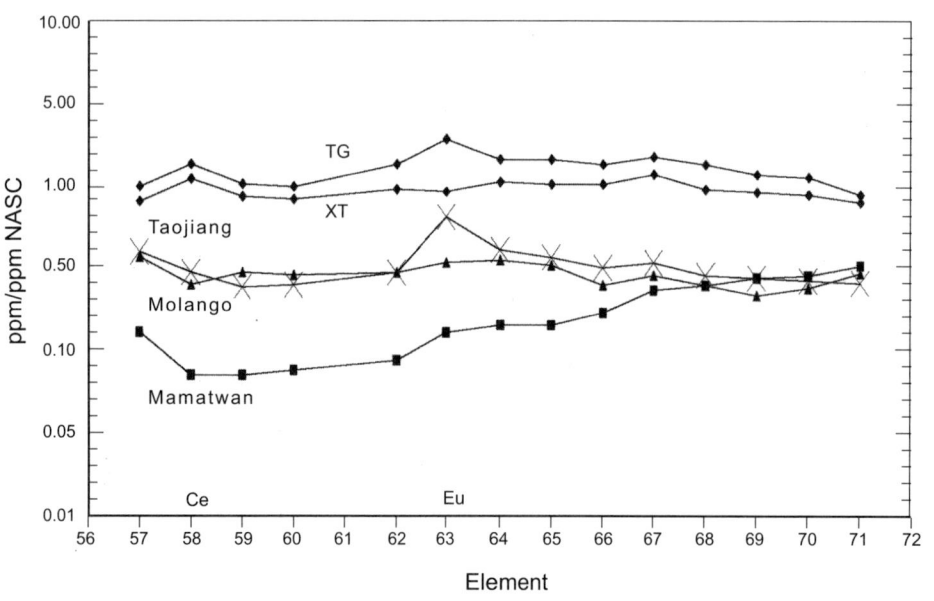

Figure 5. Most Mn carbonate ores exhibit a positive Eu anomaly and a negative Ce anomaly. Both Chinese deposits have prominent positive Ce anomalies and Xiangtan (XT) has a negative Eu anomaly whereas Tanganshan (TG) has the typical positive Eu anomaly.

Eu anomalies, normalized to NASC. For the Chinese deposits, Tanganshan also has a positive anomaly, whereas all of the samples from the Xiangtan deposit have distinct negative Eu anomalies. The La/Yb ratio is less than 1.0 for both of the Chinese deposits, whereas the ratio is substantially greater than 1.0 for the others. On the other hand, the ΣREE is considerably higher in the Neoproterozoic deposits. These differences in REE chemistry are not related to detrital contamination by heavy minerals; there is no correlation between ΣREE and either P_2O_5 or Zr content of the ores, so xenotime or zircon are not controlling the REE chemistry of the ores. Therefore the signature is mostly carried by the Mn minerals themselves and is a fundamental character of the deposits.

In studies of black shales like the host rocks to these deposits, a common question is the degree of anoxia in the basin during deposition. Several ways to estimate this property have been proposed. Among the most common are degree of pyritization (Raiswell et al., 1988), C-S plots, and Mo content (Crusius et al., 1996). Liu (1988) measured degree of pyritization (DOP) in the shales hosting a number of Mn deposits and reported for Xiangtan that footwall shales average 0.70, the ores average 0.49, and the hanging wall shales average 0.71. For comparison, Raiswell et al. (1988) suggested 0.45 as the boundary between fully oxic and suboxic environments and 0.75 as the boundary between suboxic and fully anoxic. The Xiangtan ores, using this scale, were deposited under oxic to suboxic conditions, whereas their host shales were deposited under conditions transitional between suboxic and anoxic. Comparable data are not available for Tanganshan.

In Figure 6 we show a plot of organic C versus total S for Xiangtan and Tanganshan. The data show a pattern typical of anoxic-euxinic conditions, in which there is a poor correlation between S and C_{org}, with S falling mostly between 2% and 4%. Thus the C-S relations indicate more reducing conditions than does DOP.

Mo/Al ratios are commonly used to assess the extent of Mo enrichment in shales (e.g., Warning and Brumsack, 2000). For the Chinese shale samples, the Mo/Al (ppm metal/ % metal) ratio at Tanganshan averages 12, whereas at Xiangtan it averages only 3. For comparison, the ratio is to 16–20 for modern euxinic sediments and <1 for average shale. A comparison of Mo versus C_{org} (Fig. 7) shows no correlation, contrary to the situation seen in most modern sediments, which show an excellent correlation. Furthermore, the Mo contents are almost all much lower than for modern euxinic sediments. The DOP and C/S patterns suggest that the host shales for the Chinese Mn deposits were formed under transitional suboxic-anoxic conditions and that the degree of anoxia was greater at Tanganshan than at Xiangtan. However, Mo levels are somewhat lower in these shales than we would expect from the pattern seen in modern sediments. Euxinic basins sequester large amounts of Mo as MoS_2 in bottom sediments (Barling and Anbar, 2004). Perhaps the low Mo in these Neoproterozoic black shales is related to widespread oceanic anoxia, with a higher flux of Mo to deep-water sediments than today, which resulted in less Mo available for near-shore deposition. Dobrzinski et al. (2004) have performed a similar paleoredox analysis using S/C, U/Th, Cd, Mo, and Ce* as indicators. Their focus was largely on the glacial marine units, which they assigned to oxic depositional conditions. The interglacial interval, however, was more reducing, falling at the border between oxic and suboxic or between suboxic and anoxic, depending on the indicator.

Isotopes

The Chinese Neoproterozoic Mn ores are radically different in S isotope chemistry from other Mn ores. Figure 8 shows the distribution of S isotopes in $MnCO_3$ ores and host black shales from Phanerozoic deposits. Note that the Phanerozoic shales yield broadly negative $\delta^{34}S$ values with a mode at about –25‰.

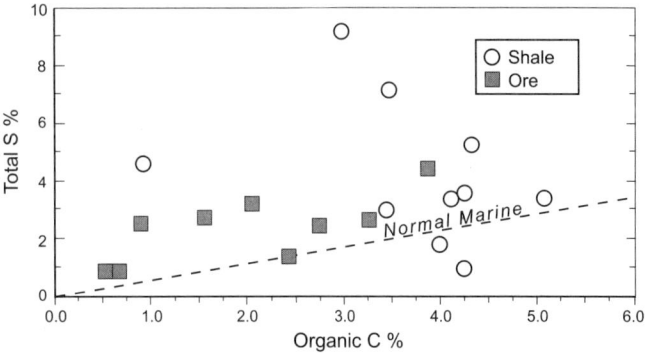

Figure 6. C vs. S plot shows most samples of both ore and shale plot above the normal marine line and have a poor correlation between the two variables. Both features of the plot are consistent with anoxic conditions in the basin of deposition.

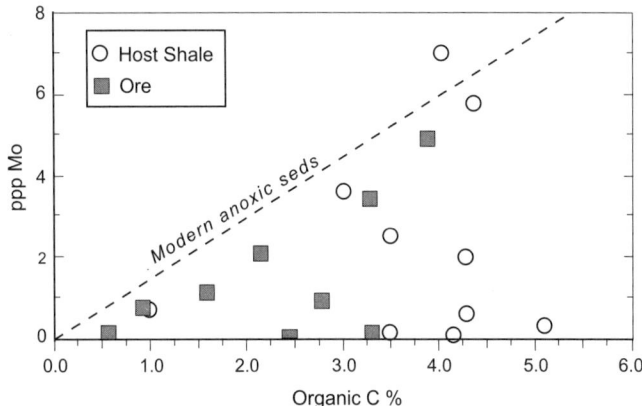

Figure 7. Mo in both ores and host shales does not correlate with the amount of organic carbon, unlike the situation in Phanerozoic shales and modern sediments, and the amount of Mo is less than is found in modern sulfidic sediments.

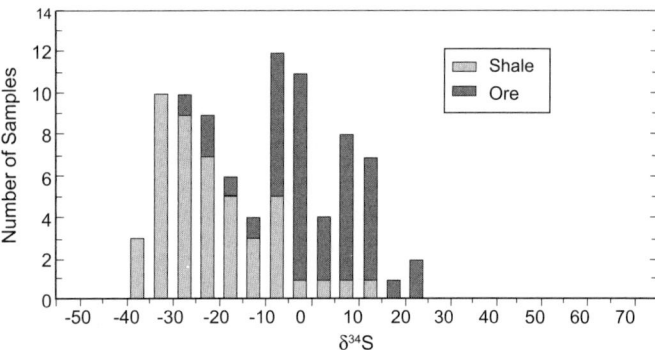

Figure 8. S isotope distribution of pyrite in Phanerozoic Mn ores and in their host shales. Note that the ores are ~30‰ heavier than the shales.

This distribution is typical of syngenetic pyrite formed by bacterial sulfate reduction in fine-grained sediments (e.g., see Ohmoto and Goldhaber, 1997, Fig. 11.17). The $MnCO_3$ ores, on the other hand, have a more irregular distribution with a distinctly higher mode at about +5‰. Okita (1992) explained this heavier S as resulting from formation of pyrite deeper in the sediment than is normally the case during the diagenesis of marine sediments. The large volume of Mn in these sediments poises the Eh at values too high for sulfate reduction. Until all Mn oxide is converted to Mn carbonate, Eh stays high and no sulfides form. Thus pyrite formation takes place farther from the sediment-water interface than usual. This greater depth reduces diffusive exchange with the bulk seawater, resulting in complete reduction of a smaller sulfate reservoir and thus smaller amounts of pyrite with heavier $\delta^{34}S$ than normal.

In Figure 9 we show our data plus previous results from similar Chinese deposits reported by Li et al. (1999) for the Xiangtan and Songtao deposits and by Tang and Liu (1999) for the Minle deposit. The Neoproterozoic ores and their host shales are uniformly shifted to very enriched $\delta^{34}S$ values, with no distinction between the ores and the host shales. Xiangtan and Tanganshan deposits have significantly different $\delta^{34}S$, although both are far more enriched in ^{34}S than other deposits shown in Table 1. The average for Xiangtan is +44.3‰, whereas Tanganshan is somewhat less enriched at +29.3‰. For comparison, Tang and Liu (1999) reported shale $\delta^{34}S$ values that average +53.5‰ and $MnCO_3$ ores that average +49.3‰ for Minle. Li et al. (1999) reported two shale and two ore samples from Xiangtan that average +61.2‰ and +52.1‰ respectively, and two shales and four ore samples from Songtao that average +45.6‰ and +54.4‰, respectively. Thus not only are the Neoproterozoic ores peculiar in their high S values, they are also peculiar in the coincidence of the shale and $MnCO_3$ values, which indicates a difference in ore-forming conditions between the Neoproterozoic and the younger deposits. The Paleoproterozoic Kalahari ore samples were also analyzed for S isotopes, but the amount of S was too low for reliable measurement.

We did not measure C isotopes in our samples, but there are literature data from nearby deposits. Tang and Liu (1999), Li at al. (1999), and Yang et al. (1999) report 24 analyses of carbonate carbon, mostly from the Minle and Songtao deposits. The average $\delta^{13}C$ value is −8.7‰ with a narrow standard deviation of 1.9. This value is similar to the average of −9.1‰ for the Kalahari deposits (Gutzmer, 1996) and −13.1‰ for Molango (Okita, 1987). Thus the Chinese deposits fall within the range for other $MnCO_3$ ores for this parameter.

DISCUSSION

General Model for Mn Ore Genesis

In modern sedimentary basins, Mn shows a strong tendency to be enriched around the margins of areas with deeper, anoxic bottom waters. The Baltic and the Black Seas exhibit this behavior (see, e.g., Sternbeck and Sohlenius, 1997). The cause is the great insolubility of the Fe sulfide, pyrite, compared to the Mn sulfide, alabandite. Fe and Mn are transported into the basin of deposition as coatings of Fe_2O_3 or MnO_2 on detrital particles and are released as soluble Fe^{2+} or Mn^{2+} to the pore waters of the sediment during diagenesis through bacterial reactions (see e.g., Potter et al., 2005, p. 138). The dissolved Fe is quickly incorporated into pyrite, but the Mn diffuses up into the overlying water mass. As a consequence, Fe is vanishingly low in the deep portions of anoxic basins, whereas dissolved Mn is much higher. Both substances form insoluble oxides and so are absent in solution in the shallow, oxygenated water mass. There is a peak in dissolved Mn just beneath the redox interface that reflects the redissolution of MnO_2 particles that form in the shallow water and sink through the interface (Fig. 10).

From this cycling pattern of Mn in modern basins, a general model of Mn ore genesis has been developed in which Mn is solubilized from deep-water sediments in anoxic basins and reprecipitated around the margins of these basins at the point where

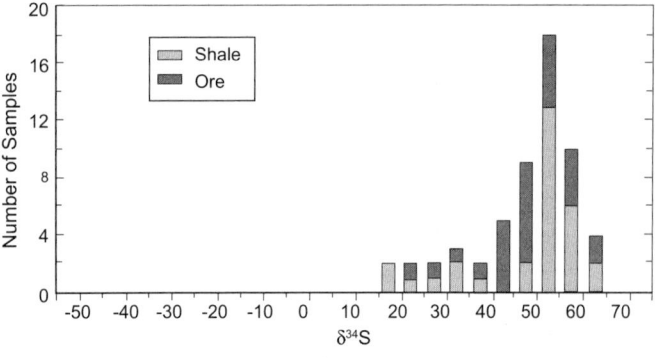

Figure 9. S isotope distribution of pyrite in Neoproterozoic Mn ores of South China and their host shales. Note lack of distinction between the two lithologies and the very heavy values compared to Phanerozoic analogs.

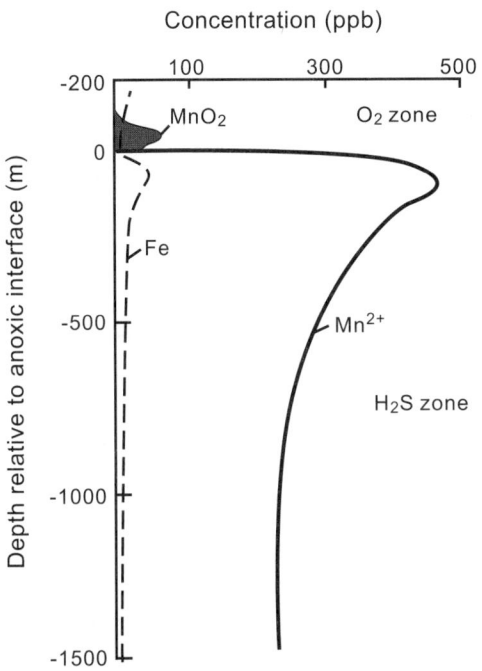

Figure 10. Distribution of Fe and Mn in the Black Sea, a modern anoxic basin that shows Mn concentration at the redox boundary. (Modified from Force and Maynard, 1991, Fig. 11.2)

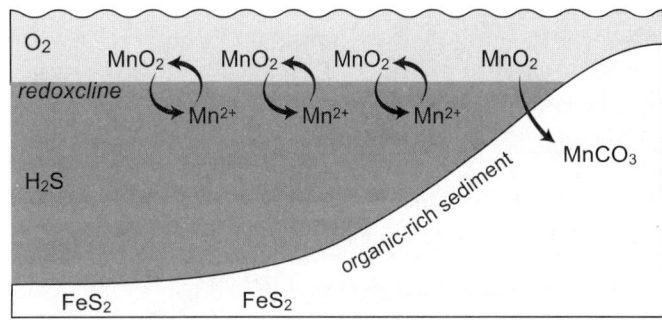

Figure 11. Schematic model for development of $MnCO_3$ by early diagenetic reaction of Mn oxide with sedimentary organic matter.

the redox interface impinges on the seafloor. This model was first articulated by Force and Cannon (1988) from their observations of Mn distributions in modern sediments and facies analysis of several ancient Mn deposits. Subsequently the model has been developed in some detail based on stable isotopic studies of Phanerozoic deposits, particularly Molango in Mexico (Okita, 1987; Okita et al. 1988; Maynard et al., 1990; Okita 1992; Okita and Shanks, 1992). See also reviews by Force and Maynard (1991), who emphasized the ancient record and favored a dominant role for basin geometry, and by Calvert and Pedersen (1996), who emphasized the modern record and argued for a dominant role of surface-water productivity in controlling Mn distribution.

The process of Mn enrichment begins with the precipitation of Mn oxides within the water column at the interface between oxidizing and reducing conditions, usually a halocline. Most of the precipitated Mn simply redissolves as it passes downward through the water column, unless the seafloor is shallow enough to intercept the redox interface (Fig. 11). This phenomenon produces what might be called a "manganese compensation depth." Below this depth, which occurs at about –200 m in the Black Sea, MnO_2 particulates dissolve while settling through the water column and none reach the bottom. Above this depth, solid MnO_2 is stable and does not dissolve as it sinks. Consequently, the deep-water sediments are low in Mn, sediments close to the compensation depth have a strong enrichment in Mn oxide particles, and shallow-water sediments are low in Mn. Thus there is a critical depth for Mn enrichment that produces a "bathtub ring"

around the margins of the basin. Where the clastic sedimentation rate is low, significant Mn accumulations can develop. Although these accumulations start as oxides, they are usually preserved as carbonates in the rock record. Reaction with organic matter in the sediment converts the primary Mn oxide to secondary Mn carbonate, which is depleted in ^{13}C as a result of derivation of a portion of the carbon from decaying organic matter.

A key observation supporting this model is a strong correlation between Mn contents in the rocks and C isotopes (Okita and Shanks, 1992). The production of the $MnCO_3$ mineralization requires the consumption of large amounts of organic matter and most likely occurred during early diagenesis, when bacterial processes are most effective. The process can be represented schematically by the reaction:

$$2MnO_2 + CH_2O + HCO_3^- = 2MnCO_3 + H_2O + OH^-. \quad (1)$$

From this relationship, about one-half the carbon in Mn carbonates is derived from organic matter, one-half from seawater. For marine organic matter $\delta^{13}C$ is –30‰ to –20‰, whereas seawater is close to 0‰. Thus a diagenetic Mn carbonate should have values of –15‰ to –10‰, close to the observed values of –13‰ to –9‰ quoted above.

At the same time that the Mn is oxidizing the organic matter, it also attacks any Fe sulfide in the sediment (Aller and Rude, 1988; Schippers and Jørgensen, 2001). Pyrite is nearly ubiquitous in marine sediments because of the reaction between detrital Fe oxides and seawater sulfate. In Mn-rich sediments, however, this process is blocked because any precursor FeS that forms is quickly destroyed:

$$FeS + 4.5MnO_2 + 4H_2O = FeOOH + 4.5Mn^{2+} \\ + SO_4^{2-} + 7OH^-. \quad (2)$$

Reaction (2) predicts that Mn ore deposits should be very low in S, as is observed in most occurrences other than those in the Neoproterozoic. Also the minor amount of pyrite that does form should be relatively heavy isotopically. As mentioned above, this prediction of heavy S is based on the requirement that any pyrite that forms be relatively late, forming after all of the Mn oxide has been converted to $MnCO_3$. Therefore the degree of contact with the overlying seawater reservoir of sulfate S will be limited, sulfate reduction will go to completion, and the small amount of sulfide that does form will be isotopically close to its parent sulfate, in contrast to normal pyrite in black shale, which is highly depleted in ^{34}S. Subsequent work has shown that this model has broad applicability to Mn ore deposits. See for example Nyame (1998) on the Nsuta deposit of Ghana, and Tsikos (1999) for the Hotazel deposit in the Kalahari Mn field of South Africa.

Ore Genesis in the Neoproterozoic—The Snowball Earth Model

The Neoproterozoic deposits appear chemically to be similar to their older and younger counterparts, except for different S isotope and REE patterns. The similarity in carbon isotopes and trace element chemistries suggests that the $MnCO_3$ formed during early diagenesis from a Mn oxide precursor, just as in other deposits. How was this Mn oxide deposited in such large amounts and what accounts for the super heavy S in the associated pyrite?

An appealing hypothesis has been put forward by Gorjan et al. (2000; see also Gorjan et al., 2003) in which they related isotopic behavior and the abundance of Fe and Mn mineralization to turnover after melting of glacial ice on a "snowball" Earth (Fig. 12). The concept of a frozen Earth in the Neoproterozoic (Kirschvink, 1992) explains many of the biological, sedimentological, and geochemical anomalies seen at this time and accordingly has received much attention. In the Gorjan et al. (2000, 2003) version of this model, the ocean became totally anoxic under its ice cover, which led to a buildup of dissolved Mn and Fe in the bottom water. Because the flux to the oceans of detrital Fe from the continents and hydrothermal Fe from ridge crests is much greater than the flux of dissolved SO_4^{2-} and because the seafloor was everywhere anoxic, pyrite formation in sediments or at the ridges removed most of the sulfur, but left plenty of Fe to accumulate in the water column as soluble Fe^{2+}. This situation is the reverse of the modern oceans, where oxic bottom waters keep most of the detrital Fe insoluble as the oxide and quickly convert any soluble Fe^{2+} in vent fluids into insoluble Fe^{3+}. As a consequence, dissolved S is in great excess over dissolved Fe, and the concentration of SO_4^{2-} is high.

The precipitation of most of the seawater reservoir of sulfur as sulfide resulted in the sequestration of large amounts of ^{32}S in sediments and the accumulation of residual ^{34}S-enriched S in the water column. On melting of the ice, there is overturn of the oceans and this Fe- and Mn-laden deep water wells up onto shallow platforms and deposits the Rapitan-type iron formations characteristic of the Neoproterozoic (Maynard, 1991) and large deposits of Mn oxide. Because this upwelling deep water contains sulfate strongly enriched in ^{34}S, the shallow-water diagenetic pyrite produced from it is similarly very heavy.

Although in broad terms this model provides a satisfactory explanation for a wide array of observations, there are some significant discrepancies:

- The Fe and Mn deposits are interbedded with the glacial deposits rather than succeeding them.
- The ores formed in narrow rifts rather than on open shelves.
- Sulfur isotopic compositions are typically heavy throughout the Proterozoic and excursions appear to be unrelated to glacial episodes.
- The REE evidence from the associated iron formations is incompatible with a totally anoxic deep ocean.

An Ice-covered Rift Model for Ore Genesis in the Neoproterozoic

These aspects of the Mn-Fe mineralization in the Neoproterozoic suggest that the snowball Earth model needs to be refined. The hypothesis that best explains these additional observations is a "partial snowball Earth," with ice-covered continents and marginal seas, but with an ice-free open ocean at low latitudes. See Eyles and Januszczak (2004a, 2004b), Poulsen et al. (2002), and Poulsen (2003) for useful discussions of full versus partial ice cover from a field and from a modeling basis. Three principal lines of evidence support a partial, or "soft" snowball model for mineralization: stratigraphic sequence, stable isotopic values, and REE compositions.

Stratigraphic Setting of Mineralization

The interglacial position of the ores suggests that the snowball Earth model of Fe-Mn deposition needs to be modified to allow the main episode of mineralization to occur during periods of partial melting of the ice cover accompanied by oxidation of surface waters. Multiple glacial advances and retreats are hard to reconcile with a totally frozen Earth. Such fluctuations are, however, compatible with a model of individual ice-covered marine

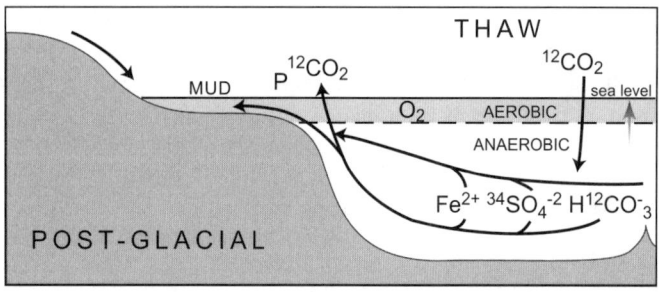

Figure 12. "Snowball Earth" model for Neoproterozoic Fe-Mn deposits. Rapid melting of ice at the end of the event pushes Fe and Mn rich deep water onto shelves to precipitate as oxides. Residual sulfate left in the oceans is very heavy and produces a pulse of heavy S in sediments at the end of glaciation.

basins that could have been time transgressive, as proposed by Eyles and Januszczak (2004a, 2004b).

The dominance of glacial-marine rift environments is also more consistent with partial ice cover. The Neoproterozoic Fe and Mn deposits are uniquely associated with narrow rifts on continental crust (Maynard, 1991); none are found on stable shelves of continental margins. The "hard snowball" explanation for the Fe and Mn mineralization would predict deposition as widespread sheets on continental shelves (see Gorjan et al., 2003, p. 95), much like the famous Pennsylvanian black shales of the U.S. midcontinent described by Heckel (1991), which are related to maximum flooding associated with deglaciation.

Sulfur Isotopic Values Throughout the Neoproterozoic

The recent compilations of sulfide S isotopic data by Strauss (1997) and by Canfield and Raiswell (1999) show that heavy sulfur has been characteristic of both sulfides and sulfates during much of the Proterozoic and is not correlated to the glacial episodes. Another important aspect of the Proterozoic S record is that the spread between sulfate and sulfide values is much less than in Phanerozoic rocks, with several sulfide analyses lying higher than contemporaneous seawater sulfate. Strauss (1997) suggested that these patterns could result from a low-sulfate ocean in which the amount of pyrite formed at the sediment-water interface was limited by the amount of sulfate in the overlying water rather than by the amount of organic carbon, which is the case in modern marine sediments (e.g., Canfield, 2001). A low-sulfate ocean would have been prone to periodic drawdowns of sulfate concentration during times of higher Fe flux to the ocean, which would have resulted in increased pyrite formation. The isotopic composition of the residual sulfate S in seawater would then have spiked to very high values.

Determining a worldwide curve for Neoproterozoic S isotope values comparable to the familiar curves for the Phanerozoic has been difficult because of the scarcity of primary sulfate minerals from this time period and the high variability in sulfide values for rocks of a given age. We have made an approximate sulfide S curve by plotting the medians of the values reported by Strauss (1997) for each time interval. As shown in Figure 13, the median values show large fluctuations, but are invariably heavy compared to Phanerozoic values. There do seem to be periods within the Neoproterozoic that experienced spikes to extremely heavy sulfide S, but these do not match the main glacial episodes. Instead there seems to be a sharp fall in $\delta^{34}S$ following each glaciation, the converse of the prediction from the "hard snowball" model. Perhaps these swings indicate periods of increased deep-water circulation that produced well-oxygenated oceans with higher SO_4^{2-} concentrations.

A finer-scale sulfate S curve is available from the work of Hurtgen et al. (2002), who measured trace sulfate held in carbonate minerals from the Neoproterozoic of Namibia. Their sampling starts above the Sturtian glacial strata and thus does not cover the interval corresponding to the Chinese Mn deposits, but their results do provide important constraints on the mechanisms

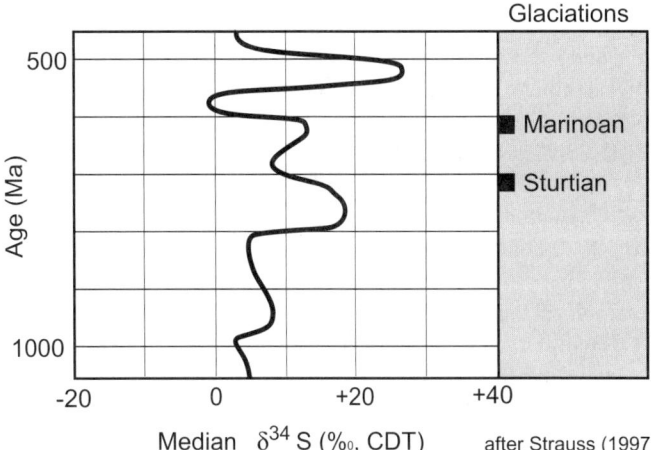

Figure 13. Approximate isotopic curve of sulfide S for the Neoproterozoic. The line shown is the median of a very wide spread of values and so has a higher degree of uncertainty than sulfate S curves for the Phanerozoic. Note the lack of correspondence of heavy S to the end of glaciation. The curves actually suggest a dramatic decrease in δ34S in the post-glacial oceans.

for Neoproterozoic S change. They identified four major excursions to heavy S: one is slightly above the Sturtian glacials and reaches +40‰; the second is 150 m higher in the section and has the greatest departure, to +50‰; the third is midway between the Sturtian and Marinoan glacial episodes and reaches +35‰; and the final excursion is 100 m above the Marinoan equivalents and reaches +40‰. Thus there are several episodes of development of heavy S and only the first one can be convincingly related to glaciation. Their data also shows that the sulfate $\delta^{34}S$ values are comparable to the sulfide values, indicating that sulfate reduction within sediments must have gone essentially to completion during this time interval. Hurtgen et al. (2002) and Canfield (2004) interpreted these patterns to indicate generally low SO_4^{2-} concentrations in seawater at this time. Canfield (2004) has suggested values as low as 200–300 μM, which he attributed to a much greater flux of sedimentary S back to the mantle via subduction of pyrite-rich deep-sea sediments during the Proterozoic than in the modern oceans.

We conclude that the occurrence of extremely heavy $\delta^{34}S$ values of pyrite in the Chinese Mn deposits is not related directly to glaciation but to generally low concentrations of dissolved SO_4^{2-} in the Neoproterozoic world ocean that made seawater subject to rapid and severe swings in its S content. Excess iron, which must have been present judging from the abundance of Rapitan-type iron formations, would have driven SO_4^{2-} concentrations to very low values through formation of more pyrite (Canfield and Raiswell, 1999), and this could account for the uniformly extremely high $\delta^{34}S$ values found. The combination of high Fe flux and a low SO_4^{2-} ocean is what produced the S isotopic signature of the Chinese Mn deposits. We further conclude from our survey of literature data that these very heavy values occurred repeatedly at many times and in many places in the Neoproterozoic and are hence more likely to have resulted from world ocean effects rather than from peculiar chemistry in isolated basins.

REE Evidence for an Oxidizing World Ocean

The nature and source of this Fe flux can be reconstructed from the behavior of REE in the Mn deposits and in their iron formation cousins. Iron formations have distinctive REE patterns for each of the three main periods of iron mineralization (Fig. 14). Both the Archean Algoma-type and Paleoproterozoic Superior-type iron formations have pronounced positive Eu anomalies on NASC normalized plots. Notice also in Table 3 that all of the NIST iron-ore samples, which are from the Lake Superior region, have positive anomalies. This anomaly is conspicuously absent in the Neoproterozoic Rapitan-type deposits (Derry and Jacobsen, 1990; Klein and Beukes, 1993; and Bau and Möller, 1993; Klein and Ladeira, 2004).

Positive Eu anomalies are associated with mid-ocean ridge vent fluids (see, e.g., Cocherie et al., 1994, on Red Sea sediments). Destruction of calcic plagioclase in the oceanic crust leads to a release of excess Eu to hydrothermal solutions. Because of the high temperatures required for the removal of Eu from plagioclase, Eu release is confined to axial vents. The results of Michard and Albarède (1986) suggest that temperatures greater than 350 °C are required. Furthermore, at lower temperatures, there seems to be little release of any of the REE seawater. For example, Wheat et al. (2002) studied low-temperature hydrothermal springs from the Juan de Fuca ridge and found them to be net sinks instead of sources of REE to seawater. It also seems likely that most Mn release is from the axial vents: Murton et al. (1999) were able to account for all of the Mn release from a 50-km segment of the Mid-Atlantic Ridge by flux from the Broken Spur vent field.

Normally, these vent-sourced REE are immediately scavenged by Fe oxides precipitating around the vents (Mitra et al., 1994). If, however, the ridge-crest hydrothermal systems vent into oxygen-free bottom water, this precipitation of Fe oxides and scavenging of REE will not occur, and both the Fe and the vent-signature REE can be concentrated in bottom waters and carried into shallower waters to be precipitated far from their source. The evolution of REE patterns with age suggests that the Archean and Paleoproterozoic iron formations had Fe dominantly sourced from ridge-crest vents discharging into anoxic seawater, whereas the Neoproterozoic deposits received their Fe from seawater with a relatively minor hydrothermal component. The absence of a positive Eu anomaly in the Neoproterozoic deposits indicates that they were deposited during periods of general oxidation of oceanic bottom water. Dobrzinski et al. (2004) came to the same conclusion using a variety of paleoredox indicators.

Neodymium isotopes also suggest a waning hydrothermal influence on Fe deposits through time. Jacobsen and Pimentiel-Klose (1988) reported that Archean and Paleoproterozoic iron formations have ε_{Nd} values similar to the mantle, whereas the Neoproterozoic Urucum iron formations are similar to modern seawater. The average values are +2.7 for Algoma-type IF, +1.0 for Superior-type, and −2.9 for a single determination from Urucum (reported in Derry and Jacobsen, 1990).

Eu anomalies in Mn deposits do not present such a clear picture (Maynard, 2004). Note in Table 3 that virtually all Mn deposits, regardless of age, have positive Eu anomalies. This may indicate that a significant hydrothermal contribution is present in Mn ores throughout geologic time, and furthermore that the bottom water into which the volcanic solutions exhaled was anoxic, as in the Force and Cannon (1988) model. The Chinese deposits are mixed, with the three Tanganshan ore samples each having a positive anomaly whereas all Xiangtan ore samples have a negative anomaly. Neoproterozoic Mn deposits in the Urucum district

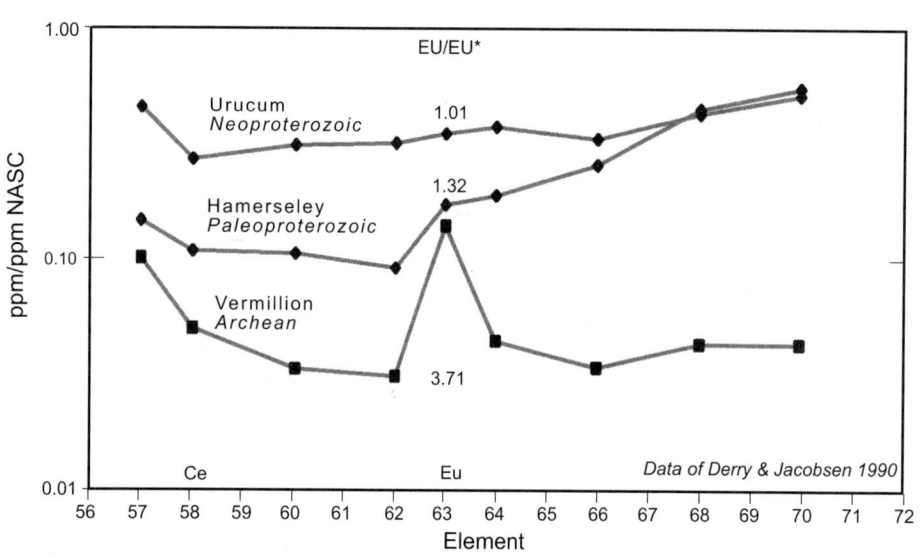

Figure 14. Europium anomalies in iron formations decrease with decreasing age, indicating a diminishing contribution from volcanic-hydrothermal sources to the deposits.

of Brazil have uniformly negative anomalies (Graf et al., 1994), similar to seawater. The Xiangtan and Urucum deposits may have a much lower hydrothermal component than the other Mn ores or perhaps the bottom water in their basins was sufficiently oxic to permit formation of Fe oxides close to the vents, scavenging all of the vent-derived REE.

The behavior of Ce may provide some clues to this situation. Ce shows a much stronger association with the Mn component of Fe-Mn accumulations than to the Fe component (DeCarlo, 1991). Fleet et al. (1983) has suggested that the relative size of the negative Ce anomaly on NASC-normalized plots has a linear relationship to the ratio of hydrothermal- to hydrogenous-sourced material in the Mn accumulation. In the modern oceans, the hydrothermal end member has a strongly negative anomaly with a Ce/Ce* value of ~0.46 (Hein et al., 1996, their table 8 corrected for geometric calculation of Ce*). Hydrogenous Mn accumulations, sourced only from seawater, have a strongly positive Ce anomaly with a value of ~1.6 (data of Usui and Someya, 1997). The rate of deposition accounts for much of this difference (Maynard, 2004). The hydrothermal deposits form relatively quickly and preserve the original REE signature of the water, whereas the hydrogenous Mn nodules form extremely slowly and, in the modern oxygen-rich deep sea, catalyze the oxidation of Ce^{3+} to Ce^{4+}, producing significant Ce enrichments and high Ce/Ce* values. Almost all Fe and Mn ores in Table 3 show negative Ce anomalies, indicating an appreciable hydrothermal contribution into anoxic bottom waters. The Chinese deposits depart from this trend, being uniformly positive like the modern Mn nodules, suggesting at least mildly oxidizing conditions for the Neoproterozoic and slower deposition rates than in other Mn deposits. Slower deposition is also consistent with the higher ΣREE seen in the Chinese deposits (Table 3).

The combination of the Eu and Ce data indicate that Archean and Paleoproterozoic Fe and Mn deposits are dominantly hydrothermal in the source of metals and REE and that the ocean basins worldwide were dominantly anoxic at the time of their formation. The Neoproterozoic Fe deposits, and possibly the Mn deposits, received a minor hydrothermal contribution and were deposited during a time of global oxygenation of the oceans. In the Phanerozoic, iron formations are confined to the immediate vicinity of hydrothermal vents, whereas Mn is far-traveled. Each Mn deposit in Table 3 seems to have had a variable hydrothermal component, but each was associated with a large but isolated basin with anoxic bottom water (e.g., Maynard et al., 1990; Force and Maynard, 1991).

If seafloor hydrothermal processes were supplying only limited amounts of Fe and Mn to the Neoproterozoic ore deposits, what was the source of the high metal fluxes? We suggest that low-latitude glaciation scraped off tropical soils highly enriched in Fe and Mn in the form of oxides rather than silicates and thus highly susceptible to diagenetic remobilization. This lateritic material was then dumped into small ocean basins that had limited communication with the open ocean and so could become anoxic at depth without receiving the mid-ocean ridge signature of REE. Mn was then exported to shallow water to precipitate at the oxic/anoxic interface, while some Fe precipitated in deep water as pyrite, thereby removing most of the dissolved SO_4^{2-} in the basin. This process left residual sulfur strongly enriched in ^{34}S to be incorporated in the shallow-water deposits. Thus the super-heavy values we find are not the direct result of glaciation but are an indirect result through the impact of glaciation on Fe behavior (Fig. 15). Support for a lateritic source of the Fe and Mn comes from the very low Na_2O content of the host shales, which averages 0.14% at Tanganshan and 0.07% at Xiangtan, compared with typical shales, which have ~1% Na_2O (Li, 2000, table VI-4). Dobrzinski et al. (2004) used the CIA index of Nesbitt and Young (1982) to characterize the weathering state of the source area for the Neoproterozoic deposits of south China. They calculated an average CIA of 62 for the lower diamictite, 71 for the interglacial unit, and 65 for the upper glacial unit. For comparison, modern glacial marine sediments of the Scotia Sea average only 55 (data of Diekmann, et al. 2000), whereas modern soils average 72 (Maynard, 1992). Thus the Chinese Neoproterozoic glacial section shows higher degrees of weathering than the modern, and the Datangpo Formation was sourced from deeply weathered material.

Unresolved Problems

This model of Neoproterozoic Fe-Mn mineralization raises several questions that cannot be answered with the data presently available:

1. Why did extensive Fe and Mn mineralization not occur during the second glacial episode of the Neoproterozoic? The available literature indicates that all the sizable occurrences are Sturtian. Are there mis-correlated Marinoan-age deposits or are they truly absent? Could the continents have lacked an extensive lateritic weathering mantle at the time of the Marinoan glaciations? Perhaps the Marinoan glacial episode was truly global whereas the Sturtian was only partial.

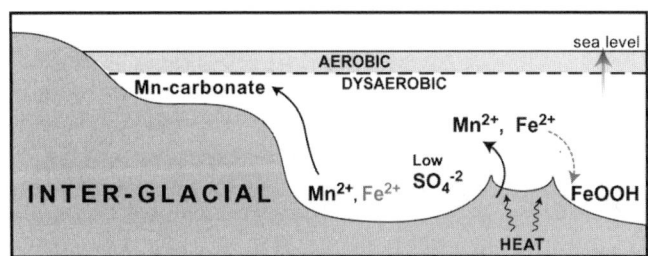

Figure 15. Interglacial model for Fe and Mn mineralization in the Neoproterozoic. Mineralization occurs by upwelling of deep water onto shallow shelves during interglacial transgression. Mid-ocean ridges are intermittently exposed to oxidizing water, which precipitates Fe oxides and REE from vent fluids. The Mn and Fe in the deposits are sourced dominantly from lateritic weathering residues deposited in narrow marine rifts by low-latitude glaciers.

2. What causes the variation in Eu anomaly in Mn ores, but not in the Fe ores? Does it reflect more localized sources for Mn than for Fe? Does it reflect preferential sorption of Eu to Mn oxides under certain conditions?
3. What causes the other S isotope excursions in the Neoproterozoic that are not associated with Fe-Mn deposits? Are these times of higher C_{org} input? Could they be times of evaporite deposition with drawdown of sulfate levels? Could they be times of enhanced Fe flux from mid-ocean ridges? Could they reflect times of greater global anoxia in which more of this mid-ocean ridge flux goes to pyrite rather than to Fe oxide?

Resolving these questions will require detailed geochemical analyses of a range of Neoproterozoic deposits. In particular, the analyses of small amounts of sulfate $\delta^{34}S$ contained in carbonates (e.g., Kampschulte and Strauss, 2004) need to be extended to older rocks and to other sections, and more REE data are needed.

CONCLUSIONS

The Neoproterozoic Mn deposits of China were deposited in a restricted, linear basin during partial retreat of the Sturtian glaciers. Their exceptionally heavy $\delta^{34}S$ values can be explained by an excess of Fe flux over S flux to these restricted basins. Eu anomalies in associated Fe ores indicate that this Fe flux was dominated by diagenetic rather than hydrothermal sources, whereas Ce anomalies in Mn ores suggest relatively oxidizing conditions in the bottom water at the site of Mn deposition. A possible source of enhanced diagenetic flux of Fe and Mn is rapid deposition of lateritic soil residues scraped off by low-latitude glaciers into sub-oxic or anoxic bottom waters of a rift basin. The easily remobilized Fe and Mn oxides in this residue would have accumulated in low-oxygen bottom water and then precipitated under more oxidizing conditions around the edges of the basin during sea-level rise in interglacial periods.

REFERENCES CITED

Aller, R.C., and Rude, P.D., 1988, Complete oxidation of solid phase sulfide by manganese and bacteria in anoxic marine sediments: Geochimica et Cosmochimica Acta, v. 52, p. 751–765, doi: 10.1016/0016-7037(88)90335-3.

Barfod, G.H., Albarède, F., Knoll, A.H., Xiao, S., Télouk, P., Frei, R., and Baker, J., 2002, New Lu-Hf and Pb-Pb age constraints on the earliest animal fossils: Earth and Planetary Science Letters, v. 201, p. 203–212.

Barling, J., and Anbar, A.D., 2004, Molybdenum isotope fractionation during adsorption by manganese oxides: Earth and Planetary Science Letters, v. 217, p. 315–329, doi: 10.1016/S0012-821X(03)00608-3.

Bau M., and Möller, P., 1993) Rare earth element systematics of the chemically precipitated component in Early Precambrian iron formations and the evolution of the terrestrial atmosphere-hydrosphere-lithosphere system: Geochimica et Cosmochimica Acta, v. 57, p, 2239–2249.

Breitkopf, J.H., 1988, Iron formations related to mafic volcanism and ensialic rifting in the Damara Orogen, Namibia: Precambrian Research, v. 38, p. 111–130, doi: 10.1016/0301-9268(88)90087-3.

Bruchert, V., 1998, Early diagenesis of sulfur in estuarine sediments: The role of sedimentary humic and fulvic acids: Geochimica et Cosmochimica Acta, v. 62, p. 1567–1586, doi: 10.1016/S0016-7037(98)00089-1.

Calvert, S.E., and Pedersen, T.F., 1996, Sedimentary geochemistry of manganese: Implications for the environment of formation of manganiferous black shales: Economic Geology and the Bulletin of the Society of Economic Geologists, v. 91, p. 36–47.

Canfield, D.E., 2001, Isotope fractionation by natural populations of sulfate-reducing bacteria: Geochimica et Cosmochimica Acta, v. 65, p. 1117–1124, doi: 10.1016/S0016-7037(00)00584-6.

Canfield, D.E., 2004, The evolution of the Earth surface sulfur reservoir: American Journal of Science, v. 304, p. 839–861.

Canfield, D.E., and Raiswell, R.W., 1999, The evolution of the sulfur cycle: American Journal of Science, v. 299, p. 697–723.

Canfield, D.E., Raiswell, R.W., Westrich, J.T., Reaves, C.M., and Berner, R.A., 1986, The use of chromium reduction in the analysis of reduced inorganic sulfur in sediments and shales: Chemical Geology, v. 54, p. 149–155, doi: 10.1016/0009-2541(86)90078-1.

Cocherie, A., Calvez, J.Y., and Oudin-Dunlop, E., 1994, Hydrothermal activity as recorded by Red Sea sediments: Sr-Nd isotopes and REE signatures: Marine Geology, v. 118, p. 291–302, doi: 10.1016/0025-3227(94)90089-2.

Crusius, J., Calvert, S., Pedersen, T., and Sage, D., 1996, Rhenium and molybdenum enrichments in sediments as indicators of oxic, suboxic, and sulfidic conditions of deposition: Earth and Planetary Science Letters, v. 145, p. 65–78, doi: 10.1016/S0012-821X(96)00204-X.

DeCarlo, E.H., 1991, Paleoceanographic implications of rare earth elements variability within a Fe-Mn crust from the central Pacific Ocean: Marine Geology, v. 98, p. 449–467, doi: 10.1016/0025-3227(91)90116-L.

Derry, L.A., and Jacobsen, S.B., 1990, The chemical evolution of Precambrian seawater: Evidence from REEs in banded iron formations: Geochimica et Cosmochimica Acta, v. 54, p. 2965–2977, doi: 10.1016/0016-7037(90)90114-Z.

Diekmann, B., Kuhn, G., Rachold, V., Abelmann, A., Brathauer, U., Fütterer, D., Gersonde, R., and Grobe, H., 2000, Terrigenous sediment supply in the Scotia Sea (Southern Ocean): Response to Late Quaternary ice dynamics in Patagonia and on the Antarctic Peninsula: Palaeogeography, Palaeoclimatology, Palaeoecology, v. 162, p. 357–387, doi: 10.1016/S0031-0182(00)00138-3.

Dobrzinski, N., Bahlburg, H., Strauss, H., and Zhang, Q.-R., 2004, Geochemical climate proxies applied to the Neoproterozoic glacial succession on the Yangtze platform, south China, in Jenkins, G.S., McMenamin, M.A.S., McKay, C.P., and Sohl, L., eds., The extreme Proterozoic: Geology, geochemistry and climate: Washington D.C., American Geophysical Union, Geophysical Monograph 146, p. 13–32.

Eyles, N., and Januszczak, N., 2004a, 'Zipper-rift': A tectonic model for Neoproterozoic glaciations during breakup of Rodinia after 750 Ma: Earth-Science Reviews, v. 65, p. 1–73, doi: 10.1016/S0012-8252(03)00080-1.

Eyles, N., and Januszczak, N., 2004b, Interpreting the Neoproterozoic glacial record: The importance of tectonics, in Jenkins, G.S., McMenamin, M.A.S., McKay, C. P., and Sohl, L., eds., The extreme Proterozoic: Geology, geochemistry and climate: Washington D.C., American Geophysical Union, Geophysical Monograph 146, p. 125–168.

Fan, D.-L., Liu, T.-B., and Ye, J., 1992, The process of formation of manganese carbonate deposits hosted in black shale series: Economic Geology and the Bulletin of the Society of Economic Geologists, v. 87, p. 1419–1429.

Fleet, A.J., Boström, K., Laubier, L., and Smith, K.L., 1983, Hydrothermal and hydrogenous ferro-manganese deposits; do they form a continuum? The rare earth element evidence, in P.A. Rona, ed., Hydrothermal processes at seafloor spreading centers: New York, Plenum Press, p. 535–555.

Force, E.R., and Cannon, W.F., 1988, A depositional model for shallow-marine manganese deposits around black-shale basins: Economic Geology and the Bulletin of the Society of Economic Geologists, v. 83, p. 83–117.

Force, E.R., and Maynard, J.B., 1991, Manganese: Syngenetic deposits on the margins of anoxic basins, in Force, E.R., Eidel, J.J., and Maynard, J.B., eds., Sedimentary and diagenetic mineral deposits: A basin analysis approach to exploration: El Paso, Texas, Society of Economic Geologists, Reviews in Economic Geology, v. 5, p. 147–159.

Gorjan, P., Veevers, J.J., and Walter, M.R., 2000, Neoproterozoic sulfur-isotope variations in Australia and global implications: Precambrian Research, v. 100, p. 151–179, doi: 10.1016/S0301-9268(99)00073-X.

Gorjan, P., Walter, M.R., and Swart, R., 2003, Global Neoproterozoic (Sturtian) post-glacial sulfide-sulfur isotope anomaly recognized in Namibia: Journal of African Earth Sciences, v. 36, p. 89–98, doi: 10.1016/S0899-5362(03)00002-2.

Graf, J.L., O'Connor, E.A., and Van Leeuwen, P., 1994, Rare earth element evidence of origin and depositional environment of Late Proterozoic ironstone beds and manganese-oxide deposits, SW Brazil and SE Bolivia: Journal of South American Earth Sciences, v. 7, p. 115–133, doi: 10.1016/0895-9811(94)90003-5.

Gutzmer, J., 1996, Genesis and alteration of the Kalhari and Postmasburg manganese deposits, Griqualand West, South Africa [Ph.D. dissertation]: Johannesburg, Rand Afrikaans University.

Heckel, P.H., 1991, Thin, widespread Pennsylvanian black shales of Midcontinent North America, a record a cyclic succession of widespread pycnoclines in a fluctuating epeiric sea, in Dennison, J.M., and Ettensohn, F.R., eds., Tectonic and eustatic controls on sedimentary cycles: Tulsa, Oklahoma, Society for Sedimentary Geology, Concepts in Sedimentology and Paleontology, v. 4, p. 65–87.

Hein, J.R., Gibbs, A.E., Clague, D.A., and Torresan, M., 1996, Hydrothermal mineralization along submarine rift zones, Hawaii: Marine Georesources and Geotechnology, v. 14, p. 177–203.

Hurtgen, M.T., Arthur, M.A., Suits, N.S., and Kaufman, A.J., 2002, The sulfur isotopic composition of Neoproterozoic seawater sulfate: Implications for a snowball Earth?: Earth and Planetary Science Letters, v. 203, p. 413–429, doi: 10.1016/S0012-821X(02)00804-X.

Jacobsen, S.B., and Pimentiel-Klose, M.R., 1988, A Nd isotopic study of the Hamersley and Michipicoten banded iron formations: The source of REE and Fe in Archean oceans: Earth and Planetary Science Letters, v. 87, p. 29–44, doi: 10.1016/0012-821X(88)90062-3.

Kampschulte, A., and Strauss, H., 2004, The sulfur isotopic evolution of Phanerozoic seawater based on the analysis of structurally substituted sulfate in carbonates: Chemical Geology, v. 204, p. 255–286, doi: 10.1016/j.chemgeo.2003.11.013.

Kirschvink, J.L., 1992, Late Proterozoic low-latitude global glaciation: The snowball Earth, in Schopf, J. W., and Klein, C., eds., The Proterozoic biosphere: Cambridge, UK, Cambridge University Press, p. 51–58.

Klein, C., and Beukes, N.J., 1993, Sedimentology and geochemistry of the glaciogenic late Proterozoic Rapitan iron-formation in Canada: Economic Geology and the Bulletin of the Society of Economic Geologists, v. 88, p. 542–565.

Klein, C., and Ladeira, E.A., 2004, Geochemistry and mineralogy of Neoproterozoic banded iron-formations and some selected manganese formations from the Urucum district, Mato Grosso do Sul, Brazil: Economic Geology and the Bulletin of the Society of Economic Geologists, v. 99, p. 1233–1244.

Li, R., Chen, J., Zhang, S., Lei, J., Shen, Y., and Chen, X., 1999, Spatial and temporal variations in carbon and sulfur isotopic compositions of Sinian sedimentary rocks in the Yangtze platform: South China: Precambrian Research, v. 97, p. 59–75, doi: 10.1016/S0301-9268(99)00022-4.

Li, Y-H., 2000, A Compendium of Geochemistry: Princeton University Press, 475 p.

Liu, T.-B., 1988, C-S-Fe correlation of shales hosting sedimentary manganese deposits [Ph.D. dissertation]: Cincinnati, Ohio, University of Cincinnati, 300 p.

Liu, T.-B., 1990, C-S relationships in shales hosting manganese ores from Mexico, China, and Newfoundland; implications for depositional environment and mineralization: Ore Geology Reviews, v. 5, p. 325–340, doi: 10.1016/0169-1368(90)90037-N.

Lottermoser, B.G., and Ashley, P.M., 2000, Geochemistry, petrology and origin of Neoproterozoic ironstones in the eastern part of the Adelaide Geosyncline, South Australia: Precambrian Research, v. 101, p. 49–67, doi: 10.1016/S0301-9268(99)00098-4.

Macouin, M., Besse, J., Ader, M., Gilder, S., Yang, Z., Sun, Z., and Agrinier, P., 2004, Combined paleomagnetic and isotopic data from the Doushantuo carbonates, south China: Implications for the "snowball Earth" hypothesis: Earth and Planetary Science Letters, v. 224, p. 387–398, doi: 10.1016/j.epsl.2004.05.015.

Maynard, J.B., 1991, Iron: Syngenetic deposition controlled by the evolving ocean-atmosphere system, in Force, E., Eidel, J.J., and Maynard, J.B., eds., 1991, Sedimentary and diagenetic mineral deposits: A basin analysis approach to exploration: El Paso, Texas, Society of Economic Geologists, Reviews in Economic Geology, v. 5, p. 141–145.

Maynard, J.B., 1992, Chemistry of modern soils as a guide to interpreting Precambrian paleosols: Journal of Geology, v. 100, p. 279–289.

Maynard, J.B., 2004, Manganiferous sediments, rocks, and ores, in McKenzie, F.T., ed., Sediments, diagenesis, and sedimentary rocks: Treatise on Geochemistry, v. 7: Amsterdam, Elsevier-Pergamon, p.289–308.

Maynard, J.B., Okita, P.M., May, E.D., and Martinez-Vera, A., 1990, Palaeogeographic setting of late Jurassic manganese mineralization in the Molango District, Mexico, in Parnell, J., Ye, L.-J., and Chen, C.-M., eds., Sediment hosted mineral deposits: Boston, Blackwell Scientific, International Association of Sedimentologists Special Publication 11, p. 17–30.

Michard, A., and Albarède, F., 1986, The REE content of some hydrothermal fluids: Chemical Geology, v. 55, p. 51–60, doi: 10.1016/0009-2541(86)90127-0.

Mitra, A., Elderfield, H., and Greaves, M.J., 1994, Rare earth elements in submarine hydrothermal fluids and plumes from the Mid-Atlantic Ridge: Marine Chemistry, v. 46, p. 217–235, doi: 10.1016/0304-4203(94)90079-5.

Murton, B.J., Redbourn, L.J., German, C.R., and Baker, E.T., 1999, Sources and fluxes of hydrothermal heat, chemicals and biology within a segment of the Mid-Atlantic Ridge: Earth and Planetary Science Letters, v. 171, p. 301–317, doi: 10.1016/S0012-821X(99)00157-0.

Nesbitt, H.W., and Young, G.M., 1982, Early Proterozoic climates and plate motions inferred from major element chemistry of lutites: Nature, v. 299, p. 715–717.

Nyame, F.K., 1998, Mineralogy, geochemistry, and genesis of the Nsuta manganese deposit, Ghana [Ph.D. dissertation]: Okayama, Japan, Okayama University.

Ohmoto, H., and Goldhaber, M.B., 1997, Sulfur and carbon isotopes, in Barnes, H.L., ed., Geochemistry of hydrothermal ore deposits: New York, John Wiley and Sons, p. 517–611.

Okita, P.M., 1987, Geochemistry and mineralogy of the Molango manganese orebody, Hidalgo State, Mexico [Ph.D. dissertation]: Cincinnati, Ohio, University of Cincinnati, 362 p.

Okita, P.M., 1992, Manganese carbonate mineralization in the Molango District, Mexico: Economic Geology and the Bulletin of the Society of Economic Geologists, v. 87, p. 1345–1366.

Okita, P.M., and Shanks, W.C., 1992, Origin of stratiform sediment-hosted manganese carbonate ore deposits: Examples from Molango Mexico, and TaoJiang, China: Chemical Geology, v. 99, p. 139–164, doi: 10.1016/0009-2541(92)90036-5.

Okita, P.M., Maynard, J.B., Spiker, E.C., and Force, E.R., 1988, Isotopic evidence for organic matter oxidation by manganese reduction in the formation of stratiform manganese carbonate ore: Geochimica et Cosmochimica Acta, v. 52, p. 2679–2685, doi: 10.1016/0016-7037(88)90036-1.

Potter, P.E., Maynard, J.B., and Depetris, P.J., 2005, Mud and mudstones. Berlin, Springer-Verlag, 297 p.

Poulsen, C.J., 2003, Absence of a runaway ice-albedo feedback in the Neoproterozoic: Geology, v. 31, p. 473–476, doi: 10.1130/0091-7613(2003)031<0473:AOARIF>2.0.CO;2.

Poulsen, C.J., Jacob, R., Pierrehumbert, R.T., and Huynh, T.T., 2002, Testing paleogeographic controls on a Neoproterozoic snowball Earth: Geophysical Research Letters, v. 29, p. 10–1 to 10–4.

Raiswell, R., Buckley, F., Berner, R.A., and Anderson, T.F., 1988, Degree of pyritization of iron as a paleoenvironmental indicator of bottom-water oxygenation: Journal of Sedimentary Petrology, v. 58, p. 812–819.

Rui, Z.Q., and Piper, J.D.A., 1997, Palaeomagnetic study of Neoproterozoic glacial rocks of the Yangzi Block: Palaeolatitude and configuration of South China in the late Proterozoic Supercontinent: Precambrian Research, v. 85, p. 173–199, doi: 10.1016/S0301-9268(97)00031-4.

Schippers, A., and Jørgensen, B.B., 2001, Oxidation of pyrite and iron sulfide by manganese dioxide in marine sediments: Geochimica et Cosmochimica Acta, v. 65, p. 915–922, doi: 10.1016/S0016-7037(00)00589-5.

Shen, Y., 2002, C-isotope variations and paleoceanographic changes during the late Neoproterozoic on the Yangtze Platform, China: Precambrian Research, v. 113, p. 121–133, doi: 10.1016/S0301-9268(01)00205-4.

Sternbeck, J., and Sohlenius, G., 1997, Authigenic sulfide and carbonate mineral formation in Holocene sediments of the Baltic Sea: Chemical Geology, v. 135, p. 55–73, doi: 10.1016/S0009-2541(96)00104-0.

Strauss, H., 1997, The isotopic composition of sedimentary sulfur through time: Palaeogeography, Palaeoclimatology, Palaeoecology, v. 132, p. 97–118, doi: 10.1016/S0031-0182(97)00067-9.

Studley, S.A., Ripley, E.M., Elswick, E.R., Dorais, M.J., Fong, J., Finkelstein, D., and Pratt, L.M., 2002, Analysis of sulfides in whole rock matrices by elemental analyzer-continuous flow isotope ratio mass spectrometry: Chemical Geology, v. 192, p. 141–148, doi: 10.1016/S0009-2541(02)00162-6.

Tang, S.-Y., and Liu, T.-B., 1999, Origin of the early Sinian Minle manganese deposit, Hunan Province, China: Ore Geology Reviews, v. 15, p. 71–78, doi: 10.1016/S0169-1368(99)00015-3.

Tsikos, H., 1999, Petrographic and geochemical constraints on the origin and post-depositional history of the Hotazel iron-manganese deposits, Kalahari Manganese Field, South Africa [Ph.D. dissertation]: Grahamstown, South Africa, Rhodes University, 217 p.

Urban, H., Stribny, B., and Lippolt, H.J., 1992, Iron and manganese deposits of the Urucum district, Mato Grosso do Sul, Brazil: Economic Geology and the Bulletin of the Society of Economic Geologists, v. 87, p. 1375–1392.

Usui, A., and Someya, M., 1997, Distribution and composition of marine hydrogenetic and hydrothermal manganese deposits in the northwest Pacific, in Nicholson, K., Hein, J.R., Bühn, B., and Dasgupta, S., eds., Manganese mineralization: Geochemistry and mineralogy of terrestrial and marine deposits: Geological Society [London] Special Publication 119, p. 177-198.

Wang, J., and Li, Z.-X., 2003, History of Neoproterozoic rift basins in South China: Implications for Rodinia breakup: Precambrian Research, v. 122, p. 141–158, doi: 10.1016/S0301-9268(02)00209-7.

Warning, B., and Brumsack, H.-J., 2000, Trace metal signatures of eastern Mediterranean sapropels: Palaeogeography, Palaeoclimatology, Palaeoecology, v. 158, p. 293–309, doi: 10.1016/S0031-0182(00)00055-9.

Wheat, C.G., Mottl, M.J., and Rudnicki, M., 2002, Trace element and REE composition of a low-temperature ridge-flank hydrothermal spring: Geochimica et Cosmochimica Acta, v. 66, p. 3693–3705, doi: 10.1016/S0016-7037(02)00894-3.

Yang, J.-D., Sun, W.-G., Wang, Z.-Z., Xue, Y.-S., and Tao, X.-C., 1999, Variations in Sr and C isotopes and Ce anomalies in successions from China: Evidence for the oxygenation of Neoproterozoic seawater?: Precambrian Research, v. 93, p. 215–233, doi: 10.1016/S0301-9268(98)00092-8.

Yeo, G.M., 1981, The Late Proterozoic Rapitan glaciation in the northern Cordillera: Geological Survey Canada, Paper 81-10, p. 25- 46.

Xu, X., Huang, H., and Liu, B., 1990, Manganese deposits of the Proterozoic Datangpo Formation, South China: genesis and palaeogeography, in Parnell, J., Ye, L.-J., and Chen, C.-M., eds., Sediment hosted mineral deposits: Boston, Blackwell Scientific, International Association of Sedimentologists Special Publication 11, p. 39–50.

MANUSCRIPT ACCEPTED BY THE SOCIETY 29 OCTOBER 2005

An evaluation of diagenetic recycling as a source of iron for banded iron formations

R. Raiswell*

School of Earth and Environment, University of Leeds, Leeds LS2 9JT, UK

ABSTRACT

REE and Nd isotope data indicate that most of the iron in banded iron formations is derived from hydrothermal sources but do not exclude a significant contribution from terrestrial sources, such as diagenetic recycling. A diagenetic model has been used to estimate the recycling of iron into overlying seawater, due to microbial reduction and dissolution at depth in anoxic sediment pore waters, followed by diffusion upward through a surface layer of sediment that contains oxygenated pore waters. Rates of iron recycling increase with higher pore-water dissolved iron concentrations, decreasing pH and temperature, and smaller thicknesses of the surface oxygenated layer. Iron can be recycled at rates of 1000–5000 μg cm^{-2} yr^{-1} from Proterozoic (pO$_2$ = PAL) pore waters with dissolved Fe^{2+} = 1–5 μg cm^{-3}, pH 6.5 (and T < 65 °C), or pH 7.0 (and T < 40 °C), or pH 7.5 (and T < 20 °C), provided the thickness of the surface oxygenated layer is less than 0.1 cm. Lower pO$_2$ levels and more weakly oxygenated surface layers do not significantly increase the maximum recycling rates but enable these to be achieved at larger thicknesses of the surface layer, for all pH 6.5–7.5 and temperatures from 10 to 65 °C. Rates of iron supplied by diagenetic recycling can be substantially modified by the export efficiency (ε) from the source area and by the ratio (Source Area)/(Sink Area), which can either disperse or concentrate the recycling flux that is delivered to a sink area of banded iron formation. Banded iron formations that require maximum iron delivery rates of 22500 μg cm^{-2} yr^{-1} can be produced only by recycling rates of 5000 μg cm^{-2} yr^{-1} (and ε = 1) from a source area that is at least four times larger than the area of banded iron formation. Modern basins have ratios of shelf area (<200 m water depth) to deep basin area that commonly range from 0.2 to 4. Basins at either extreme have ratios of (Deep Basin Area)/(Shelf Area) or (Shelf Area)/(Deep Basin Area) that exceed 4 and are potentially able to concentrate iron either from a deep basin source area to banded iron formation on the shelf, or from a shelf source area to a banded iron formation depositing in the deep basin. However, these mass balance constraints require the existence of substantial areas of contemporaneous source sediments (or smaller areas of iron-enriched sediments) located either on the shelf or in the deep basin.

Keywords: iron, diagenetic, recycling, sediment enrichment.

---

*r.raiswell@earth.leeds.ac.uk

Raiswell, R., 2006, An evaluation of diagenetic recycling as a source of iron for banded iron formations, *in* Kesler, S.E., and Ohmoto, H., eds., Evolution of Early Earth's Atmosphere, Hydrosphere, and Biosphere—Constraints from Ore Deposits: Geological Society of America Memoir 198, p. 223–238, doi: 10.1130/2006.1198(13). For permission to copy, contact editing@geosociety.org. ©2006 Geological Society of America. All rights reserved.

INTRODUCTION

There is a widespread consensus that Superior-type iron formations occurred in an ocean whose bottom waters contained significant concentrations of dissolved iron. Three major sources for the iron have generally been proposed (e.g., Holland, 1984; Morris, 1993): chemical erosion of the continents, diagenetic mobilization from sediments, and hydrothermal activity. Hydrothermal input to the oceans is currently the favored source for the majority of the iron because of the strong similarities in the rare earth element (REE) patterns of iron formations and seawater-dominated hydrothermal fluids (e.g., Morris and Horowitz, 1983; Klein and Beukes, 1989; Derry and Jacobsen, 1990; Beukes and Klein, 1992; Danielson et al., 1992; Bau and Moller, 1993). The Nd isotope studies also indicate a hydrothermal source for a large proportion of the iron (Jacobsen and Pimental-Klose, 1988), although the isotope signatures indicate perhaps as much as 50% may be derived from the continents (Miller and O'Nions, 1985; Alibert and McCulloch, 1993). However, a hydrothermal source for the REE in seawater does not necessarily indicate that most Fe was hydrothermally sourced.

Holland (1984) has estimated the possible contributions from continental weathering and diagenetic mobilization and concluded that continental weathering was unlikely to have made a significant contribution, because a relatively large riverine discharge is required and it would be difficult to achieve the large-scale separation of dissolved iron from the accompanying sediment load (as is required to produce iron formations that have a low terrigenous content). However, diagenetic mobilization was considered to be a potential source. It is the purpose of this paper to estimate the possible iron contributions from diagenetic mobilization, in the light of recent work on reactive iron supply to modern and ancient euxinic sediments (Wijsman et al., 2001; Anderson and Raiswell, 2004; Raiswell and Anderson, 2005).

Previous estimates of the iron supply to iron formations have generally been based on a time frame supplied by the Dales Gorge Member of the Brockman Iron Formation (Hamersley Basin). The Dales Gorge Member contains alternations of iron-rich and silica-rich bands at three scales, termed macro-, meso-, and microbands. The iron-rich sediments are believed to reflect relatively high levels of hydrothermal activity that are imposed on the more silica-rich conditions resulting from periods of relatively low hydrothermal input (Morris, 1993). The mixing of iron-rich deep waters with silica-rich surface waters occurs on variety of time scales, and the laterally extensive microbands have been interpreted as annual varves (Trendall and Blockley, 1970; Ewers and Morris, 1981; Morris, 1993), indicating sedimentation rates of 20–230 m Myr^{-1} that are also consistent with geochronological evidence (Trendall, 2000). These varves provide a time control that allows the annual flux of iron to the sediments to be estimated as 22.5 mg cm^{-2} yr^{-1} (Trendall and Blockley, 1970). Faster rates of deposition (100–1000 m Myr^{-1}) have been estimated for Hamersley banded iron formations by Barley et al. (1997), assuming that organic shales and sequence boundaries represent periods of low to negligible deposition, respectively.

This paper will be mainly concerned with using models of diagenetic mobilization to estimate iron fluxes and assess their significance in relation to the maximum rates of iron delivery required to form the banded iron formations, but will not be directly concerned with the mechanisms of iron and silica transport and precipitation, the nature of the depositional environment, or post-depositional alteration (see for example Trendall and Blockley, 1970; Ewers and Morris, 1981; Morris, 1993; Krapez et al., 2003; Pickard et al., 2004). A brief consideration is given to riverine fluxes of dissolved and particulate iron but other potential iron sources are considered beyond the present scope.

PREVIOUS WORK

Diagenetic mobilization has previously been invoked by Borchert (1960), Holland (1973, 1984), and Drever (1974), but Holland (1973, 1984) provides the most mechanistic detail. Iron and manganese oxides undergo microbial reduction during diagenesis

$$CH_2O + 4FeOOH + 8H^+ \longrightarrow 4Fe^{2+} + CO_2 + 7H_2O$$

$$CH_2O + 2MnO_2 + 3H^+ \longrightarrow 2Mn^{2+} + HCO_3^- + 2H_2O$$

releasing dissolved metals to the pore waters. Low concentrations of organic C give rise to suboxic diagenesis, where the effects of sulfate reduction are diminished and dissolved iron accumulates in the pore waters rather than being precipitated as iron sulfides. Holland (1984) used data from Aller and Benninger (1981) to show that rates of Mn release from pore waters ranged up to 850 μg cm^{-2} yr^{-1}. He suggested that the fluxes of iron would be comparable and possibly larger, but that relatively little of either metal would reach the overlying water because of oxidation and precipitation as oxides in the uppermost oxygenated layer of sediment. However, this limitation might be less important if bottom waters were depleted in O$_2$, as would be the case if pO$_2$ was less than the present atmospheric level (PAL). Holland (1984) also pointed out that the diagenetic mobilization of iron would be aided by a sulfate-depleted ocean that minimized the precipitation of pore-water dissolved iron as iron sulfides, as a result of microbial sulfate reduction. There thus appear to be two main difficulties in relation to the supply of pore-water dissolved iron by diagenetic mobilization, relating to the extent of removal by the formation and precipitation of iron oxides in surface oxygenated layers of sediment and the competing effects of iron sulfide precipitation due to sulfate reduction at depth in anoxic sediments.

Recent studies of the Black Sea and the Cariaco Basin (Canfield et al. 1996; Raiswell and Canfield 1998; Wijsman et al. 2001; Lyons et al. 2003) have shown that their deep basin sediments, deposited under euxinic bottom waters, are enriched in iron that is highly reactive toward sulfide (Fe$_{HR}$) relative to oxic continental margin and deep-sea sediments. These studies suggested that highly reactive iron enrichment (recognized by a ratio of Fe_{HR} to total Fe or $Fe_{HR}/FeT > 0.4$; see Raiswell and Canfield, 1998) resulted at least partly (and possibly entirely) from the diagenetic mobilization of iron from basin margin sediments into overlying seawater, followed by lateral transport from the basin margin to the deep basin (e.g., Canfield et al. 1996; Lyons 1997; Raiswell and Canfield 1998; Wijsman et al. 2001). Wijsman et al. (2001) also demonstrated that the extent of highly reactive iron enrichment in Black Sea deep basin sediments was within the range of iron release from oxic and dysoxic continental shelf sediments (150–450 µg cm^{-2} yr^{-1}), as estimated from a model of diagenetic iron recycling and in situ measurements of iron fluxes. They proposed that highly reactive iron is first solubilized from iron oxides during anoxic diagenesis of the shelf sediments, and then released as dissolved Fe^{2+} to the overlying seawater. They also proposed that a significant fraction of the dissolved Fe is reoxidized and redeposited into the shelf sediments, but that some escapes and is transferred to the deep basin where the dissolved Fe is precipitated as iron sulfide and deposited (so producing the observed enrichment). However, Anderson and Raiswell (2004) suggested that the iron oxides formed by the oxidation of Fe^{2+} after release to seawater may not all be redeposited into the shelf sediments but may also be transported to the deep basin sediments.

Raiswell and Anderson (2005) have used a simple diagenetic model to show that rates of iron release from typical continental margin sediments can range up to 1000 µg cm^{-2} yr^{-1} (for pore waters containing 1 µg cm^{-3} dissolved iron) but are crucially dependent on pore-water dissolved iron concentrations and the thickness of the surface oxygenated layer. Furthermore the magnitude of the fluxes generated from clastic continental margin sediments is significantly modified by the export efficiency (ε) of recycled iron from the source area, and by the ratio of the source sediment area to the basin sink area. Estimates of ε from the Black Sea average at ~50% and the ratio (Source Area)/(Sink Area) is ~0.27/0.73 (Anderson and Raiswell, 2004), where the source area is defined as those sediments at less than 200 m water depth. Bottom waters at >200 m depth are sulfidic and iron released during diagenesis is immobilized by precipitation as sulfides. In the Black Sea, the ratio (Source Area)/(Sink Area) acts to disperse the flux of recycled iron, but other modern basins have potential ratios of (Source Area)/(Sink Area) that can concentrate recycled iron. The study by Raiswell and Anderson (2005) demonstrates that (1) the flux of iron recycled from clastic sediments during diagenesis may be significant under favorable diagenetic conditions, and (2) the ratio of the (Source Area)/(Sink Area) can significantly enhance or diminish the rates of iron delivered to the sink area sediments.

BENTHIC FLUX MODEL

A general one-dimensional diagenetic equation (Berner, 1980; Boudreau, 1996) for a solute in pore water can be written to describe the effects of diffusion, advection, adsorption, and reaction on variations in solute concentration with time. Boudreau and Scott (1978) have modified this equation to describe the flux of dissolved Mn^{2+} from a reduced pore water through an oxygenated layer of sediment to the sediment-water interface:

$$\frac{\partial(\varphi C)}{\partial t} = \frac{\partial(\varphi D_s \frac{\partial C}{\partial x})}{\partial x} - \frac{\partial(\varphi v C)}{\partial x} + \varphi R \quad (1)$$

where D_s (cm^2 sec^{-1}) = the diffusion coefficient corrected for tortuosity effects and C = the pore-water solute concentration (g cm^{-3}), φ = porosity, t = time (sec^{-1}), v = velocity of pore water relative to the sediment-water interface (cm sec^{-1}), x = distance below the sediment-water interface (cm), and R = rate of oxidation in the oxygenated layer of sediment (g cm^{-3} sec^{-1}). Equation (1) can be simplified by assuming that the advective velocity of the porewater, the diffusion coefficient, and the porosity are independent of time. Boudreau and Scott (1978) also assume that the advective term is small relative to the diffusion and reaction terms and that the reaction term can be expressed as:

$$R = -k_1 (C - C_s) \quad (2)$$

where C_s is the saturation concentration of Fe (III) (g cm^{-3}) with respect to iron oxides and k_1 is the first-order rate constant (sec^{-1}) for the oxidation of Fe(II) and precipitation of Fe(III) oxides. The saturation concentration of iron with respect to iron oxides is assumed to be negligible, which is consistent with the solubilities of $Fe(OH)_3$ (0.7×10^{-9} g cm^{-3}), goethite (0.2×10^{-9} g cm^{-3}), and hematite (0.1×10^{-9} g cm^{-3}) in seawater (de Baar and de Jong, 2001). Furthermore the concentration of Fe(III) in seawater is believed to be $<0.5 \times 10^{-9}$ g cm^{-3} in the absence of organic complexes although estimates are complicated by the formation of hydroxy complexes (Mackey et al., 2002). Concentrations increase by approximately one order of magnitude for each unit decrease in seawater pH (Waite, 2001), but values of C_s are still small enough to be ignored provided $C_p \gg C_s$ (as will normally be the case where pore waters are anoxic and the overlying seawater is in equilibrium with pO_2 values that are able to stabilize iron oxides). For steady-state diagenesis and assuming that iron is removed only by oxidation and precipitation of iron oxides (see below), equation (1) then simplifies to:

$$D_s \frac{d^2 C}{dx^2} - k_1 (C - C_s) = 0 \quad (3)$$

This model (Fig. 1) envisages that transport processes maintain a negligible concentration of Fe^{2+} above the sediment-water interface, below which there exists a thin zone of surface sediment where solid-phase iron oxides precipitate. Porewaters in the surface zone are here termed oxygenated whether oxic (>2 ml O_2 liter^{-1}), dysoxic (0.2–2 ml O_2 liter^{-1}), or suboxic (0–0.2 ml O_2 liter^{-1}), according to the classification of Wignall (1994). There is a gradient in dissolved Fe^{2+} through this zone from the sediment-

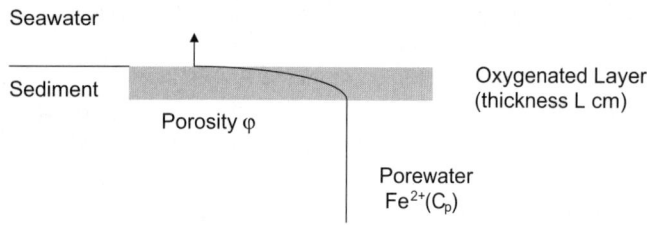

Figure 1. Schematic model for the benthic flux of recycled iron from sediments to the overlying seawater.

water interface to the base of the iron oxide layer. Porewaters below the surface oxygenated layer are reducing and maintain a uniform Fe^{2+} concentration with time. These assumptions produce the boundary conditions that $C = 0$ at $x = 0$ and $C = C_p$ at $x = L$. The benthic flux of Fe^{2+} (Boudreau and Scott, 1978; Boudreau, 1999) is then given by:

$$\text{Flux}(g\ cm^{-3}\ sec^{-1}) = \frac{\varphi(D_s k_1)^{0.5} C_p}{\sinh\left[\left(\frac{k_1}{D_s}\right)^{0.5} L\right]} \quad (4)$$

where C_p is the pore-water concentration of Fe^{2+} (g cm^{-3}) and L (cm) is the depth of the oxygenated layer.

Raiswell and Anderson (2005) have adapted this model to estimate the diagenetic mobilization of iron from clastic sediments. Following their approach, the infinite dilution diffusion coefficient (D^o) for Fe^{2+} is derived from Boudreau (1996) as D^o (cm^2 sec^{-1}) = (3.31 + 0.15 T °C) 10^{-6}. The corrections for tortuosity are derived from Ullman and Aller (1982), who give $D_s = D^o/\varphi F$, where φ is the porosity (here assumed to be 0.85) and the formation factor, F, is approximated as $1/\varphi^m$ and m = 2.5–3.0 for muddy sediments. Simplifying thus approximately produces:

$$D_s = D^o \varphi^{1.7}$$

and hence:

$$D_s = \varphi^{1.7} (3.31 + 0.15\ T°C)\ 10^{-6}. \quad (5)$$

The oxidation of Fe^{2+} can occur chemically or be biologically catalyzed. Chemical oxidation can be inhibited by the presence of humics and other natural organic species (Theis and Singer, 1974) but is very rapid and is believed to predominate over biological oxidation (Nealson, 1997; Santschi et al., 1990). The kinetics of Fe^{2+} oxidation in seawater have been studied by many workers (e.g., Stumm and Morgan, 1980; Davison and Seed, 1983; Roekens and van Grieken, 1983; Millero et al., 1987) and there is general agreement that the rate law has the following form:

$$\frac{d[FeII]}{dt} = -k[FeII][O_2][OH^-]^2 \quad (6)$$

Millero et al. (1987) has the derived the following expression for k:

$$\log k = 21.56 - 1545/T - 3.29 I^{0.5} + 1.52 I$$

where T is temperature in degrees kelvin and I is the ionic strength. Table 1 shows values of log k from 10 to 65 °C, assuming I = 0.723 for Proterozoic seawater. Equation (6) can be modified to derive an apparent first-order rate constant k_1, where

$$\frac{d[FeII]}{dt} = k_1 [FeII]$$

and

$$k_1 = k[O_2][OH^-]^2. \quad (7)$$

Values of k_1 (sec^{-1}) are given in Table 1 for the appropriate oxygen concentrations (Benson and Krause, 1984) to achieve saturation at the specified temperature, and for varying pH (using the apparent K_w for seawater from Millero, 2001). The results in Table 1 show that k_1 for shelf sediments increases by about four orders of magnitude as T increases from 10 to 65 °C, and by two orders of magnitude for pH 6.5–7.5; furthermore k_1 also varies directly with O_2 concentrations. Equation (4) also shows that fluxes of Fe^{2+} from shelf sediments depend on pore-water Fe^{2+} concentrations (C_p) and the thickness of the oxygenated layer (L) in addition to k_1. The values of pH, temperature, and C_p used in the models will be assumed to be those occurring in the pore waters below the surface oxygenated layer, following the approach of Boudreau and Scott (1978). Some reasonable limits are derived below for the temperature and composition of the overlying Proterozoic seawater.

PALEO–MESOPROTEROZOIC SEAWATER CHARACTERISTICS

Most Superior-type iron formations occur over a time range that extends from ca. 2.7 to 1.8 Ga (Paleo–Mesoproterozoic) although both older and younger examples are known (Isley, 1995). The iron formations in the Hamersley Basin (from which rates of iron delivery have been estimated; see earlier) have a rather smaller age range (ca. 2.6–2.4 Ga). The values of seawater temperature and pH over the Paleo–Mesoproterozoic have been the subject of considerable debate and both these parameters are, directly or indirectly, dependent on atmospheric composition. The models of diagenetic iron mobilization additionally require an estimate of pore-water Fe^{2+} concentrations (C_p). However, sulfate reduction exerts an important influence on pore-water dissolved iron profiles in modern marine sediments (e.g., Canfield

TABLE 1. VARIATION IN THE APPARENT FIRST ORDER RATE
CONSTANT (k_1) FOR FEII OXIDATION IN SEAWATER WITH
TEMPERATURE AND pH, ASSUMING SATURATION
WITH RESPECT TO THE ATMOSPHERE

Temp (°C)	log k	pK_w	O_2 μM	k_1 in sec^{-1} at varying pH		
				pH 6.5	pH 7.0	pH 7.5
10	14.4	13.8	275	2.9×10^{-6}	2.9×10^{-5}	2.9×10^{-4}
20	14.59	13.4	226	2.3×10^{-5}	2.3×10^{-4}	2.3×10^{-3}
30	14.76	13.0	191	1.8×10^{-4}	1.8×10^{-3}	1.8×10^{-2}
40	14.92	12.7	164	9.0×10^{-4}	9.0×10^{-3}	9.0×10^{-2}
65	15.3	12.0	31.7	1.1×10^{-2}	1.1×10^{-1}	1.1

Note: Data for pK_w from Millero (2001), log k from Millero et al. (1987) with oxygen saturation concentrations from Benson and Krause (1984).

and Raiswell, 1991), whereas Archean and Paleoproterozoic seawater may have been depleted in sulfate relative to modern seawater (see later). Hence the evidence for seawater sulfate values over this time are also here briefly reviewed. There is a substantial literature relevant to defining the characteristics of Archean and Proterozoic seawater but this review has the limited objective of defining the most appropriate ranges in these characteristics and is thus selective.

Paleo–Mesoproterozoic Seawater Sulfate Concentrations

There are currently conflicting views as to early ocean sulfate concentrations, depending on whether the atmosphere is regarded as oxygen-rich (pO_2 0.5–1.0 PAL) or oxygen-depleted ($pO_2 < 0.01$ PAL). These views on early Earth pO_2 values are summarized in Ohmoto (1997) and Holland (1999) and the references therein provide further detail. Prior to 2.3 Ga the Cloud-Walker-Kasting-Holland (CWKH) model proposes that oxygen was absent or only present at very low concentrations ranging from 10^{-6} atm (Anbar and Holland, 1992) to $10^{-2.5}$ atm (Holland, 1984; Towe, 1996). Concentrations then rose rapidly to a pO_2 of ~10^{-3} atm (Kasting, 1987) or $10^{-1.5}$ atm (Holland, 1984; Lenton, 2003) by 2.25–2.05 Ga. The alternative Dimroth-Kimberley-Ohmoto (DKO) model maintains that pO_2 has remained at PAL ± 0.5 since 4 Ga.

These models directly influence the debate on early seawater sulfate because atmospheric O_2 causes the oxidative weathering of sulfides to sulfate, which is transported by rivers into the oceans. Numerous studies have demonstrated the effects of oxygen and ferric iron (whose stability also depends on pO_2) on pyrite oxidation, and empirical rate laws have been formulated by McKibben and Barnes (1986) and Williamson and Rimstidt (1994). Canfield et al. (2000) have used the rate law of Williamson and Rimstidt (1994) to estimate that the oxidation of pyrite to dissolved sulfate occurs at geologically significant rates even with $pO_2 = 0.04$ PAL. Thus low concentrations of seawater sulfate (compared to present-day) would indicate the existence of pO_2 significantly lower than 0.04 PAL.

It is not intended that this study should contribute to the atmospheric pO_2 debate, as the diagenetic models to be used here do not directly require knowledge of seawater sulfate. However, sulfate concentrations do potentially affect pore-water dissolved Fe^{2+} concentrations, which are an important variable in the benthic flux model. This influence arises because low sulfate concentrations limit the extent of sulfate reduced and fixed as iron sulfides (the low solubility products of which exert an important control on pore-water dissolved Fe^{2+} profiles). Hence the following discussion is confined to establishing the range in Paleo–Mesoproterozoic seawater sulfate estimates. These data will subsequently be used to select appropriate modern sediments from which pore-water dissolved Fe^{2+} concentration ranges can be derived.

The debate on the concentration levels of early seawater sulfate has mainly focused on the Archean (older than 2.5 Ga), and evidence for sulfate levels is usually drawn from $\delta^{34}S$ data on sedimentary sulfides although other approaches can provide useful constraints. For example, Walker and Brimblecombe (1985) have argued that sulfate concentrations up to ~1 mmol can be formed by the oxidation of volcanic SO_2 and H_2S even in the absence of atmospheric oxygen.

The isotopic characteristics of sedimentary sulfides of Archean age show relatively limited variations in $\delta^{34}S$, and only small depletions in ^{34}S (~0 ± 10‰). These isotopic compositions are also only slightly depleted in ^{34}S relative to Archean seawater sulfate (probably +2‰ to +3‰; Ohmoto and Felder, 1987; Canfield et al., 2000). These characteristics have been interpreted by Ohmoto and Felder (1987) as arising from rapid sulfate reduction in a sulfate-rich ocean containing >1 mmol (Ohmoto and Felder, 1987) or ~10 mmol (Ohmoto et al., 1993) sulfate. It was argued that rapid sulfate reduction (1) produces an instantaneous sulfide showing a small kinetic fractionation (0‰–5‰) in favor of ^{32}S, and (2) favors closed system behavior, which produces a cumulative sulfide product that is isotopically similar to the initial sulfate (provided sulfate reduction is rapid enough to minimize sulfate resupply from overlying, sulfate-rich seawater). A warm Archean ocean (30–50 °C) was invoked to produce rapid closed system sulfate reduction with small kinetic fractionations, and hence sulfides with $\delta^{34}S$ values similar to those of contemporaneous sulfate.

The same isotopic characteristics of Archean sulfides have, however, been interpreted by Canfield and Teske (1996) and Habicht and Canfield (1996) as resulting from small kinetic fractionations in favor of ^{32}S that result from reduction in a sulfate-depleted (<1 mmol) ocean. Canfield and Teske (1996) and Canfield et al. (2000) have shown that sulfate reduction rate exerts little effect on the kinetic fractionation factor and that high sulfate reduction rates are associated with large $\delta^{34}S$ depletions (>20‰) in microbial mats and modern marine sediments (Habicht and Canfield, 1996, 2001). Furthermore, temperatures up to 40–45 °C produce significant fractionations (10‰–20‰) with *Desulphovibrio desulphuricans* (Kaplan and Rittenberg, 1964; Chambers and Trudinger, 1979). Consistent with this, mixed microbial populations able to function at higher temperatures (up to 85 °C) showed comparable fractionations of 13‰–28‰ (Canfield et al., 2000). A modeling study (Canfield et al., 2000) also indicated that high sulfate reduction rates result in rapid depletion of organic matter near the sediment surface and thus do not produce closed system behavior. These arguments are difficult to reconcile with the views of Ohmoto and Felder (1987) and Ohmoto et al. (1993) but there are still significant factors to be evaluated. For example, it may be important that sulfate-reducing bacteria show interspecies variations in sulfur isotope fractionation (Detmers et al., 2001). However, these studies are all consistent in suggesting that Paleo–Mesoproterozoic sulfate concentrations were lower than in the present ocean (28 mmol), although the suggested values range from <200 µmol (Canfield et al., 2000) to 1–10 mmol (Ohmoto and Felder, 1987; Ohmoto et al., 1993).

Paleo–Mesoproterozoic Seawater Temperature

Stellar evolution models indicate that solar luminosity increased from ~70% of current values ca. 4.6 Ga to ~85% by 2.0 Ga (Gough, 1981; Newman and Rood, 1977). This weak luminosity would be insufficient to prevent the oceans from freezing solid, and enhanced concentrations of greenhouse gases must therefore have been present to allow the existence of a hydrological cycle (Sagan and Mullen, 1972). The most probable greenhouse gas candidates are assumed to be carbon dioxide, water vapor, and methane, and several studies (see below) have concluded that the enhanced atmospheric concentrations of carbon dioxide required to maintain a hydrological cycle do not indicate variations in ocean chemistry that are inconsistent with the sedimentary rock record. However, a more direct approach is to use evidence of temperature constraints from Phanerozoic glaciations to infer the climatic conditions during the Paleoproterozoic Huronian glaciation, which occurred ca. 2.0–2.5 Ga (Kasting, 1987; Grotzinger and Kasting, 1993). Thus sea-surface temperatures at the height of the most recent glaciation are thought to be ~2 °C (Schneider and Londer, 1984) colder than at present (15 °C), and Kasting (1987) suggested that 5 °C is probably a generous lower limit for seawater temperature during the Paleoproterozoic. An upper limit to seawater temperature is also indicated by the absence of polar ice during the Mesozoic, when surface temperatures are believed to be around 25 °C or ~10 °C warmer than at present. However, Antarctica did not become glaciated until the early Oligocene, when global surface temperatures had declined substantially from the Mesozoic. Thus Kasting (1987) and Grotzinger and Kasting (1993) place Paleoproterozoic surface seawater temperatures at 5–20 °C.

Higher estimates of 30–50 °C have been suggested by Ohmoto and Felder (1987) on the basis of their examination of sedimentary sulfides of Archean age (see earlier). Relatively high Proterozoic temperatures (~50 °C) have also been inferred by Schwartzmann et al. (1993) assuming that the first appearance of cyanobacteria and eukaryotes in the rock record occurred at their upper temperature limit for viable growth (which thus corresponds to the existing surface temperature). Schwartzmann et al. (1993) stressed that this approach would require a re-evaluation of the evidence for Paleoproterozoic glaciation, which is seemingly robust (see earlier).

High seawater temperatures of 55–85 °C have also been deduced by Knauth and Lowe (1978, 2003) from the $\delta^{18}O$ values of Archean cherts from the Swaziland Supergroup. The $\delta^{18}O$ values of these cherts (+15‰ to +22‰) are significantly lower than their Phanerozoic counterparts (+18‰ to +34‰). This difference has hitherto been attributed to formation during late diagenesis or to regional metamorphic, hydrothermal, or long-term resetting of the original values. However, Knauth and Lowe (2003) argued that these explanations are inconsistent with the preservation of $\delta^{18}O$ values through different metamorphic grades and the systematic $\delta^{18}O$ differences between different types of chert, stratigraphic units, and conglomerate clasts. Hence they attributed the low $\delta^{18}O$ values to seawater temperatures of 55–85 °C, rather higher than have been inferred in the studies summarized above. Thus the model discussed here uses a range of 10–65 °C.

Paleo–Mesoproterozoic Seawater pH

Previous studies (Holland, 1973, 1984; Walker, 1983; Grotzinger and Kasting, 1993) have generally concluded that variations in seawater pH since the Archean have been relatively small. These arguments, however, are dependent on assumptions of atmospheric pCO_2, which are in turn constrained by the need for greenhouse gases to maintain a surface temperature that permits the existence of a hydrological cycle (see earlier).

Walker (1983) examined the dissolved equilibria that control the existence of calcite and gypsum in relation to pCO_2. The data showed that higher pCO_2 values produced only relatively small changes in dissolved carbonate, SO_4^{2-}, and Ca^{2+}, which were not inconsistent with the occurrence of gypsum and calcite in the rock record. On this basis pCO_2 values of 1 atm would need to be accompanied by a seawater pH of ~6.2. Grotzinger and Kasting (1993) have also examined the relationship between atmospheric pCO_2 and seawater pH, using the approach of Holland (1984). This approach assumes an ocean saturated with respect to calcite and siderite, with a dissolved Fe^{2+} concentration of 1–3 ppm and dissolved Ca^{2+} concentrations of 3–9 ppm. The carbonate equilibria then allow H^+ to be expressed as a function of pCO_2, producing a pH range of 7.1–7.4 for $pCO_2 = 0.03$ atm, and pH = 6.6–6.9 for $pCO_2 = 0.3$ atm. Grotzinger and Kasting (1993) argued that $pCO_2 = 0.3$ atmosphere is a likely upper limit for the Paleoproterozoic, which equates to a probable pH >6.5.

Seawater pH is clearly sensitive to atmospheric pCO_2, which is not well constrained by greenhouse gas requirements. The main greenhouse gases are CO_2, water vapor, and CH_4 and the same greenhouse conditions could be reached at lower pCO_2 levels with higher concentrations of either water vapor or CH_4. Habicht et al., (2002) have recently pointed out that a low-sulfate ocean (<1 mmol) would result in most anaerobic mineralization of organic carbon occurring through methanogenesis, which is favored when sulfate reduction is suppressed (as in present-day lake sediments; Berner and Raiswell, 1983; Davison et al., 1985). Habicht et al. (2002) found a significant difference between the sulfur isotope fractionations above 200 µmol sulfate (22.6‰ ± 10.3‰) compared with those below 200 µmol (0.7‰ ± 5.2‰). These data suggest that the low kinetic fractionations found in Archean sulfides are attributable to low (200 µmol) concentrations of seawater sulfate. This new estimate of early seawater sulfate is at least five times lower than hitherto suggested (see earlier) and thus permits an enhanced role for methanogenesis, and potentially increased fluxes of methane to the atmosphere. High atmospheric methane concentrations would imply that lower pCO_2 levels would be needed to maintain greenhouse conditions. Paleosol evidence (Rye et al., 1995) also suggests a methane-rich atmosphere accompanied by a relatively low pCO_2 (<0.04 atm), because iron lost from the tops of paleosols is precipitated at depth as silicate rather than carbonate phases. However, the paleosol evidence has recently been reassessed by Ohmoto et al. (2004), who concluded that the data were consistent with an atmosphere containing pCO_2 of 0.02–0.06 atm and without significant concentrations of methane. Overall the evidence on pCO_2 limits indicates that seawater pH can be only loosely constrained to between 6.5 and 7.5.

DIFFUSIVE FLUX MODEL: INPUTS AND RESULTS

Model Inputs

The above discussion has placed some limits on the temperature, pH, and sulfate concentrations in Paleo–Mesoproterozoic seawater, which are here used as a basis to constrain the variables k_1, D_s, and C_p in equation (4). The first-order rate constant k_1 is defined in terms of [OH⁻] and thus a temperature and pH dependence (see eq. 6) arises in relating [OH⁻] to the estimates of pH, because the K_w for seawater is temperature-dependent (see Table 1). The tortuosity-corrected diffusion coefficient D_s is also temperature-dependent as described in equation (5). Hence the models used here will attempt to demonstrate the possible effects of relatively high surface temperatures by considering a range of 10–65 °C for Paleo–Mesoproterozoic seawater (see above), and pore-water temperatures are assumed to fall within the same range.

Literature estimates for the pH of seawater over this time period ranged from 6.5 to 7.5 (see above). By contrast the pH of present-day surface seawater is ~8.1 but the observed pH of pore waters in modern marine shelf sediments undergoing anoxic diagenesis is usually between 7 and 8 (e.g., Boudreau and Canfield, 1988). The decrease in pH is due to the effects of oxic respiration followed by the combined effects of iron reduction and sulfate reduction, which have been modeled by Ben Yaakov (1973), Gardner (1973), and Boudreau and Canfield (1993). The models are consistent with field data indicating that pH varies only between ~7 and 8, with the lower values occurring where greater proportions of sulfide accumulate in the pore waters (as opposed to being precipitated as iron sulfide). An estimation of the pore-water pH of Paleo–Mesoproterozoic clastic marine sediments thus needs to account for differences in both the initial pH of seawater and the changes with depth as a possible consequence of lower seawater sulfate concentrations (see above) and thus less sulfate reduction.

The low Paleo–Mesoproterozoic seawater pH arising from higher atmospheric pCO_2 values makes it unlikely that there could be further significant decreases with depth, for two reasons. First, higher atmospheric pCO_2 produces higher concentrations of dissolved carbonate species, thus providing more effective buffering capacity against further decreases in pH (Gardner, 1973). Second, depth changes in pH in modern sediments arise mainly through the effects of oxic respiration, iron reduction and sulfate reduction (see earlier). Diagenetic models show that the largest pH decreases (to <7) result when sulfide accumulates in solution rather than being precipitated as iron sulfide (Ben Yaakov, 1973; Boudreau and Canfield, 1993. However, the effects of sulfate reduction will be diminished if Paleo–Mesoproterozoic seawater sulfate was significantly lower than present-day. Less sulfate reduction and less production of sulfide makes the accumulation of sulfide in excess of reactive iron oxides less likely. Overall, the above studies suggest that pore-water pH will fall within essentially the same range as the pH of Paleo–Mesoproterozoic seawater (6.5–7.5, see above).

Porewater dissolved iron results from the reduction of iron oxides (e.g., Canfield, 1989) and is closely followed by microbial sulfate reduction in most continental margin sediments (e.g., Canfield and Raiswell, 1991),

$$2CH_2O + SO_4^{2-} \longrightarrow 2HCO_3^- + H_2S$$

with the dissolved sulfide being rapidly precipitated as iron sulfides. The rapid reaction between dissolved sulfide and dissolved Fe^{2+} (plus solid iron oxides) buffers dissolved sulfide at low concentrations until the reactive iron oxides are exhausted. The pore waters are then dominated by dissolved sulfide. Thus a depth profile through modern continental margin sediments typically shows a layer of iron-rich pore waters overlying a layer of dissolved sulfide-rich pore waters (Canfield and Raiswell, 1991). Raiswell and Anderson (2005) have conducted a literature survey of pore-water Fe^{2+} concentrations in anoxic marine shelf sediments, which shows that mean values of 1–5 µg cm⁻³ (Fig. 2A) are obtained by averaging over depths of 10–30 cm. However, a better analogue for a low-sulfate Paleo–Mesoproterozoic pore water may be provided by pore waters from modern, non-acidic lake sediments and Figure 2B shows the range in mean pore-water dissolved Fe^{2+} concentrations in sediments from lakes with sulfate concentrations mostly <200 µmol. Pore-water dissolved iron concentrations in lakes show considerable spatial and seasonal variations, but the range in mean dissolved iron concentrations is a little larger (1–8 µg cm⁻³) than in modern marine sediments, and is sustained over greater depths. Holland (1984) and Grotzinger and Kasting (1993) used values of 1–3 µg cm⁻³ for Fe^{2+} in seawater that is saturated with respect to siderite at pH 6.6–6.9 (pCO_2 = 0.3 atm) and pH 7.1–7.4 (pCO_2 = 0.03 atm). Significantly higher dissolved iron concentrations would be difficult to maintain against removal by siderite (see Ohmoto et al., 2004)

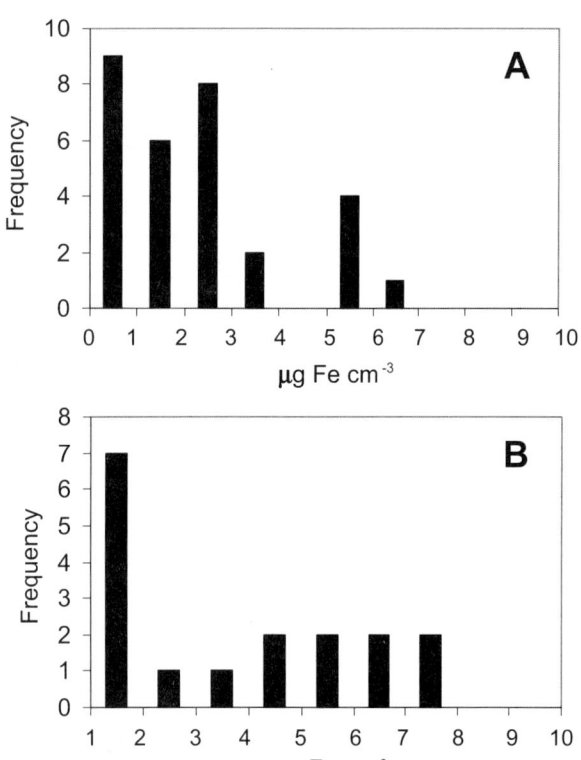

Figure 2. (A) Pore-water dissolved iron concentrations in marine sediments, derived by averaging concentrations over depths of 10–30 cm (from Raiswell and Anderson, 2005). (B) Pore-water dissolved iron concentrations in fresh-water sediments, derived by averaging concentrations over depths of 10–50 cm. Data from Carignan and Lean (1991), Emerson (1976), Feijtel et al. (1988), Matisoff et al. (1980), Nembrini et al. (1983), Nriagu and Dell (1974), and Weiler (1973).

although modern lake pore waters are able to maintain considerable over-saturation with respect to siderite without precipitation occurring (e.g., Emerson, 1976). This argument together with the data in Figure 2 indicates that concentrations of Fe^{2+} of 1–5 μg cm^{-3} represent a reasonable range for pore waters in Paleo–Mesoproterozoic pore waters whatever the sulfate concentrations in seawater. Note also that the pH of the lake pore waters used in Figure 2B in no case falls below 7, and that the changes in pH with depth tend to be small unless the overlying water has a pH >7.5. I therefore conclude that pore-water characteristics can reasonably be assumed to approximate the ranges exhibited in the overlying seawater for temperature and pH.

Model Results

Diffusive fluxes estimated from the model described in equation (4) have been compared (Raiswell and Anderson, 2005) with the flux estimates derived from a more sophisticated diagenetic model (Wijsman et al., 2002) that takes into account all the main carbon mineralization reactions (aerobic respiration, denitrification, manganese reduction, iron reduction, sulfate reduction, and methanogenesis) and the precipitation of iron sulfides. The present model produces diffusive flux estimates that are in some cases three times higher than the model outputs from Wijsman et al. (2002) and the calculated diffusive fluxes at the sites studied by McManus et al. (1997) and Elrod et al. (2004). However, diffusive fluxes derived from equation (4) were consistent with benthic flux measurements from Black Sea shelf sediments by Friedl et al. (1998) and Friedrich et al. (2002) that average 920 μg cm^{-2} yr^{-1} (but see Raiswell and Anderson, 2005). The simple model used here does not reproduce the rapid redox cycling found in more detailed models (e.g., Wang and Van Cappellen, 1996) and hence produces only semiquantitative estimates of diffusive fluxes that are used here to examine iron recycling in a basinal context.

McManus et al. (1997) and Elrod et al. (2004) also report benthic flux measurements of iron from sites on the California continental margin. The data are generally lower than those reported for the Black Sea but show two significant features. First, shallow-water bioirrigated sites (99 m depth) with well-oxygenated bottom waters (101–185 μmol) have benthic flux measurements in the range 2–22 μg cm^{-2} yr^{-1} that are on average 75 times higher than predicted by diffusion flux estimates based on pore-water dissolved iron concentrations (Elrod et al., 2004). Raiswell and Anderson (2005) suggest that bioirrigation decreases pore-water dissolved Fe by transporting Fe to the overlying waters (thus producing low diffusive fluxes but significant bioirrigation fluxes). Enhanced fluxes arising from bioirrigation would not be possible for Paleo–Mesoproterozoic sediments in which benthic activity was absent.

Second, Elrod et al. (2004) show that deeper-water sites with oxygen-depleted bottom waters (<100 μmol) have benthic fluxes in the range 0.2–37 μg cm^{-2} yr^{-1} that are roughly in agreement with estimated diffusive fluxes. Exceptions, however, occur for the highest benthic flux measurements, which are significantly smaller (by a factor of more than ten) than the estimated diffusive fluxes. Elrod et al. (2004) suggest that dissolved Fe was reoxidized in the sediment rather than diffusing into the benthic chamber. Alternatively, Raiswell and Anderson (2005) suggest that iron was released from the sediment but reoxidized and precipitated from seawater in the benthic chamber without being measured. In the absence of the chamber such iron might have been transported away from the site and some fraction exported from the shelf (see later). Overall the benthic flux measurements show that iron fluxes may be significant but are clearly influenced by processes that are currently poorly quantified. Thus the present model estimates should be regarded as no more than semiquantitative.

Figure 3 shows the model estimates of benthic recycled iron fluxes as a function of the oxygenated layer thickness (L = 0.1–10 cm). The model predicts that fluxes continue to increase as L decreases below 0.1 cm. However, values of L <0.1 cm are probably geologically unrealistic and also produce fluxes that require very rapid rates of iron deposition (see later). The three graphs in Figure 3 have been drawn for pH values of 6.5, 7.0, and 7.5, assuming bottom waters saturated with O_2 at the specified temperature (T = 10, 20, 30, 40, and 65 °C) of each plot. The saturation levels of oxygen above 90 μmol (see Table 1) are classified as oxic, whereas a concentration of 31.7 μmol for T = 65 °C is dysoxic (see earlier and Wignall, 1994). All three graphs use a concentration of $C_p = 10^{-6}$ g cm^{-3} or 1 μg cm^{-3} but equation (4) shows that using $C_p = 5$ μg cm^{-3} simply produces a five times larger flux of recycled iron. Figure 3 shows that the diffusive flux of recycled iron is significantly dependent on L and, to a lesser extent, temperature and pH. The highest fluxes of recycled iron are reached for L = 0.1 cm, but values of ~1000 μg cm^{-2} yr^{-1} are only reached at pH 6.5 for T <65 °C, at pH 7 for T <40 °C, and at pH 7.5 for T <20 °C. The maximum diffusive flux represents the case where there is diffusion from the pore waters into the overlying seawater with little or no oxidation in the surface layer. Figure 3 shows that the maximum diffusive fluxes are clearly most sensitive to variations in L for any given C_p, but that increasing temperatures play a subsidiary role in promoting faster oxidation, which becomes more apparent at higher pH.

Figure 4 shows similar plots for the fluxes of recycled iron at lower bottom water concentrations of dissolved O_2, assuming saturation with respect to atmospheric O_2 (PAL = 0.1 and 0.01). The O_2 saturation levels for PAL = 0.1 are all dysoxic except for the T = 65 °C case, where the level is suboxic (see earlier and Wignall, 1994). All O_2 concentrations for PAL = 0.01 are suboxic. There is a marginal increase in the maximum diffusive fluxes (~30%–50%) compared to fluxes at pH 6.5 in Figure 3,

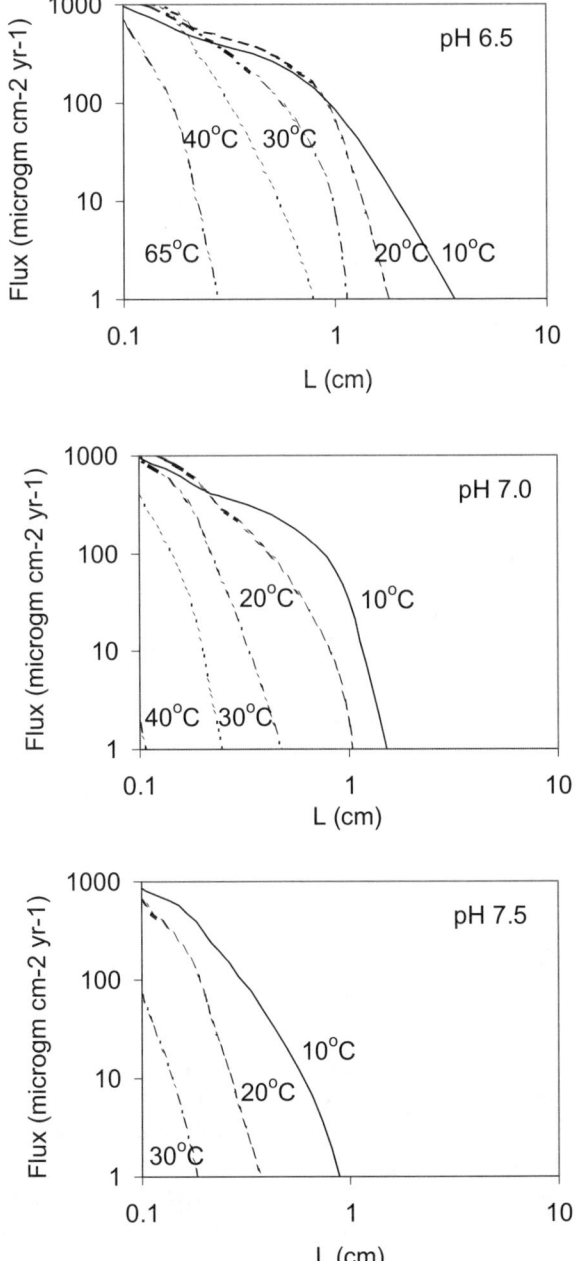

Figure 3. Variations in the benthic flux of recycled iron with oxygenated zone thickness (L cm) for temperatures of 10, 20, 30, 40, and 65 °C at pH 6.5, 7.0, and 7.5, assuming a dissolved iron concentration of 1 μg cm^{-3} and saturation with the atmosphere at pO_2 = PAL.

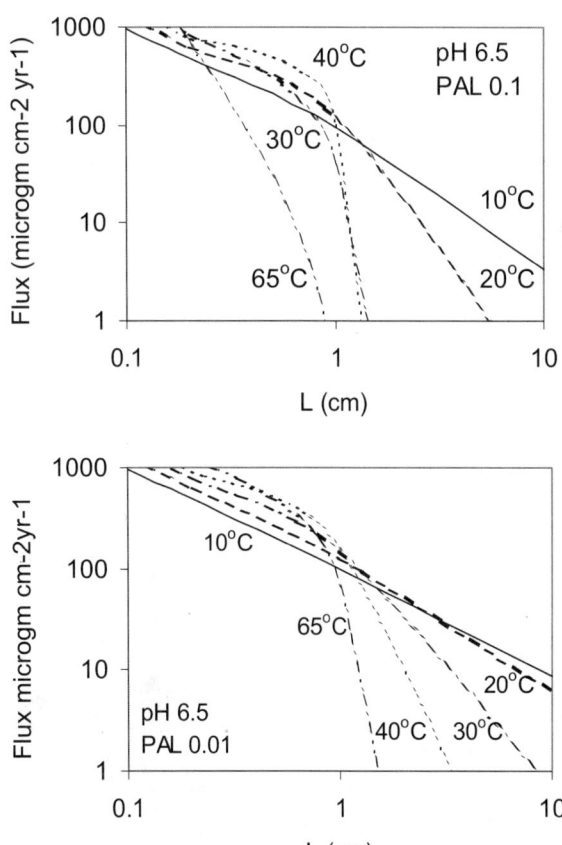

Figure 4. Variations in the benthic flux of recycled iron with oxygenated zone thickness (L cm) for temperatures of 10, 20, 30, 40, and 65 °C at pH 6.5, assuming a dissolved iron concentration of 1 μg cm^{-3} and saturation with the atmosphere at pO_2 = 0.1 PAL (top) and 0.01 PAL (bottom).

but the principal difference is that fluxes of 1000 µg cm^{-2} yr^{-1} are reached at higher values of L (approaching 1 cm.). The effects of increased oxidation in the pore waters at higher temperatures are overcome by lower bottom water O_2 concentrations.

Jorgensen and Boudreau (2001) show that values of L decrease with increasing rates of organic C mineralization (and thus vary seasonally in many modern shelf sediments). Depths of oxygen penetration also increase with increasing bioirrigation. Nonetheless the thickness of the oxygenated zone is often only a few millimeters in shelf sediments. Proterozoic sediments are likely to have relatively low rates of C_{org} mineralization but would be unaffected by bioirrigation as benthic fauna had not then developed. Values of L of around 0.1 cm may thus be reasonable for Paleo–Mesoproterozoic sediments, and the model then indicates that shelf sediments can recycle iron to the overlying seawater at rates of 1000–5000 µg cm^{-2} yr^{-1} (for C_p = 1–5 µg cm^{-3}). These rates are achieved at pO_2 = PAL for smaller values of L when the pH lies toward the lower end of the probable range for Paleo–Mesoproterozoic seawater pH. With pO_2 levels below PAL, near-maximum fluxes are reached even when the oxygenated layer is relatively thick, but the maximum values are similar to the case where pO_2 = PAL.

Note that steady-state conditions require that the addition of reactive iron by sedimentation to the shelf sediments at least equals the loss by diagenetic recycling to the overlying seawater. The mean reactive iron content (Fe_{HR}) of modern continental margin sediments is 0.83‰ ± 0.21% (Raiswell and Canfield, 1998). Sediments with this composition that deposit lithogenous material at rates of at least 0.12 ± 0.03 g cm^{-2} yr^{-1} (equivalent to 0.3 cm yr^{-1} for φ = 0.85) are potentially capable of supplying reactive fluxes of 1000 µg cm^{-2} yr^{-1} provided all the deposited reactive iron is recycled. Higher reactive iron fluxes require proportionately faster sedimentation (fluxes of 5000 µg cm^{-2} yr^{-1} require sedimentation rates of at least 1.5 cm yr^{-1}) or sediments with higher Fe_{HR} contents than modern continental margin sediments (see later). Thus the highest rates of diagenetic iron recycling require relatively high shelf sedimentation rates and/or iron-enriched sediments (see later) and will result in the loss of significant proportions of reactive iron from the shelf sediment source.

Wijsman et al. (2001) found that the Black Sea shelf sediments were depleted in reactive iron relative to mean continental margin sediments, but assessing the extent of depletion strictly requires knowing the composition of the shelf sediments before reactive iron losses have occurred. This may be difficult to achieve (see Anderson and Raiswell, 2004). Kump and Holland (1992) found that Precambrian shales generally exhibited Fe/Ti ratios that were close to the igneous rock trend and were thus neither Fe-enriched nor -depleted. However, some depletions were observed relative to the igneous rock trend and these were attributed to losses by weathering, transport, or diagenesis. This approach would be worth further study.

Rates of iron sourced by diagenetic iron recycling do not, however, represent the rates at which iron accumulates in the receiving sink sediments, which are modified (Raiswell and Anderson, 2005) by the export efficiency (ε) of iron from the source region and the relative sizes of the source-generating and sink depositional areas (and by their spatial relationships). The export efficiency term arises because dissolved iron recycled to overlying seawater may be oxidized to iron oxides that are redeposited on the shelf and not exported to the area of banded iron formation. Raiswell and Anderson (2005) estimated ε to be at least 30% and possibly 80%–90% for the Black Sea. In view of this uncertainty it will be assumed here that ε = 1, which produces maximum iron delivery rates.

SOURCE AND SINK AREA RELATIONSHIPS

The model estimates of recycled iron are clearly inadequate, on their own, to match the required value of ~22500 µg cm^{-2} yr^{-1} (based on the assumption that the microbands are annual varves). However, the relative sizes of the source generating area and the receiving sink area play a crucial role in enhancing or diminishing the delivery rates of recycled iron. Two different cases may be envisaged, depending on whether the sink area of banded iron formation is on the continental shelf (e.g., Morris, 1993) or in the deeper areas of the basin (Krapez et al., 2003; Pickard et al., 2004). The basic assumptions in both cases are as follows:

1. Atmospheric pO_2 values are only loosely constrained but need to be at least sufficient to stabilize iron oxides. Thermodynamic data indicate that iron oxides are stable at pO_2 > 10^{-60} atm, but free O_2 molecules disappear at pO_2 <10^{-21} atm (Ohmoto et al., 2004). The DKO model has pO_2 >10^{-1} atm at the start of the Paleoproterozoic whereas the CWKH model variously estimates pO_2 = 10^{-6} to 10$^{-2.5}$ atm. The value of pO_2 >10^{-6} atm (Fig. 5) used here is arbitrarily chosen but is not in contradiction with either the DKO or the CWKH model and produces weakly suboxic seawater in which iron oxides are stable.
2. A surface ocean layer in equilibrium with atmospheric pO_2 and a deeper Fe^{2+}-bearing water layer that is less-oxygenated (but still weakly suboxic or almost anoxic) in which iron oxides are stable.
3. Clastic shelf sediments deposited under steady-state conditions with sedimentation rates and compositions that are capable of maintaining rates of iron recycling of up to 5000 µg cm^{-2} yr^{-1}. Non-steady-state sedimentation effects may be important but are difficult to quantify in the present context and are here excluded. For example, physical reworking of iron-enriched sediments on the Amazon inner shelf (Aller et al., 1986) suppresses sulfate reduction and produces high concentrations of pore-water dissolved iron (100 µg cm^{-3}) that are periodically mixed into overlying seawater and oxidized to colloidal iron oxides that are exported from the shelf.
4. Porewaters with pH and temperature characteristics that allow maximum iron recycling rates to be achieved. For pO_2 = PAL this requires a pH of 6.5 and T of <65 °C, or

Shelf Deposition

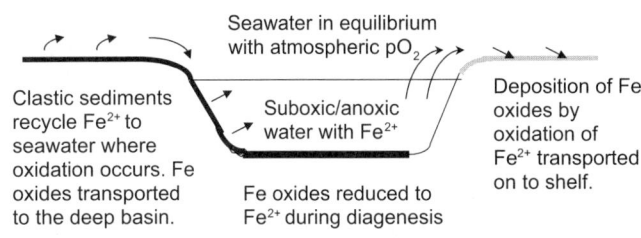

Deep Basin Deposition

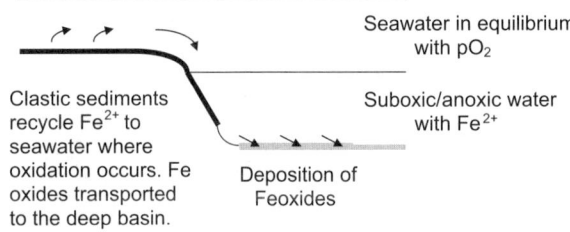

Figure 5. Schematic models of diagenetic recycling to banded iron formations depositing in shelf and deeper basin environments. Heavy dark areas represent source sediments that recycle iron to overlying seawater and heavy light areas represent banded iron formations.

pH 7.0 and T <40 °C, or pH 7.5 and T <20 °C. Lower pO_2 levels (to 0.01 PAL or less) allow the same rates to be achieved at all pH values 6.5–7.5 and temperatures from 10 to 65 °C.

5. Transport of iron oxides can occur from the source areas to the depositional area of banded iron formation. Raiswell and Anderson (2005) point out that iron-enriched sediments ($Fe_{HR}/FeT > 0.4$) commonly occur in modern (the Cariaco Basin, Baltic Sea, and Black Sea) and ancient (e.g., the Jet Rock, the Kimmeridge Clay) euxinic sediments, which suggests that intra-basinal transport is not unusual although the mechanisms may be difficult to identify. In the Black Sea, currents at the depths of the suboxic zone and the sulfide interface have lateral velocities of ~0.5 cm s^{-1} (Buessler et al., 1991) and are able to transport dissolved and particulate iron from the shelf to the deep basin (Kempe et al., 1990; Wijsman et al., 2001). Transport from the shelf to the basin may also involve riverine delivery (see later) into ocean currents or turbidity currents (see Krapez et al., 2003; Pickard et al., 2004), with upwelling generally favored to transport dissolved iron on to the shelf (Ewers and Morris, 1981; Morris, 1993).

The first case (Fig. 5, top) envisages that banded iron formations are deposited below wave base (possibly at water depths of no more than 200 m) on an isolated platform or shelf on the margin of a continent (Morris and Horowitz, 1983; Morris, 1993), and that the source area of clastic sediments is located on the continental shelf or slope and/or in a deep basin area. Part of this source area is envisaged to be in contact with the surface oxygenated waters and part in contact with relatively deep Fe^{2+}-bearing waters that are suboxic/anoxic. Diagenetic iron is recycled as particulate iron oxides from the source area either directly into the deep suboxic/anoxic water or indirectly following downslope transport (Fig. 5). These deep Fe^{2+}-bearing waters are mixed on to the shelf area, where oxidation and deposition of iron oxides occurs to form banded iron formation. Oxidation may occur inorganically, microbially, and/or by photo-oxidation (Cairns-Smith, 1978; Anbar and Holland, 1992; Konhauser et al., 2002).

In this model, the deep ocean acts essentially as a connecting reservoir and, assuming a steady state exists, the rates at which diagenetically recycled iron is exported from the source area ($J_{source} \times \varepsilon$) must match the rates at which iron is deposited into the banded iron formation. The rates of diagenetic iron recycling from the source area in contact with oxygenated bottom waters are a maximum under the conditions defined by assumptions 3 and 4. The rates of iron supply from a source area located under anoxic bottom waters are not modified by passage through a surface oxygenated layer of sediment and equation (4) is thus inappropriate. Under these circumstances the diffusive flux is dependent only on the diffusion coefficient and the concentration gradient between the pore waters and the anoxic overlying seawater. A concentration difference of ~5 µg cm^{-3} between seawater and sediment pore waters <0.1 cm below the sediment-water interface provides a flux of 5000 µg cm^{-2} yr^{-1}.

The flux of iron to the sink area (J_{sink}) is given by:

$$J_{sink} = J_{source} \times \varepsilon \times (\text{Source Area})/(\text{Sink Area})$$

and an iron supply of 5000 µg cm^{-2} yr^{-1} therefore requires that the product of $\varepsilon \times$ (Source Area)/(Sink Area) is at least 4 in order to achieve the delivery of 22,500 µg cm^{-2} yr^{-1} of iron to the banded iron formation. Estimates of (Source Area)/(Sink Area) for the Hamersley Basin can be little more than conjecture, but values of $\varepsilon \times$ (Source Area)/(Sink Area) as high as 5 have been estimated by Raiswell and Anderson (2005) for ancient and modern anoxic sediments. However, the area of the Dales Gorge member is at least 10^5 km^2 (Morris, 1993) and the existence of substantially larger areas of clastic shelf sediment is thought to be unlikely (Morris, 1993).

A rather different situation arises where the source area is located beneath dissolved iron-bearing water in the deep basin, and the shelf area represents the sink area of banded iron formation. Table 2 shows the B/S ratios for some modern basins where S (the shelf area) is defined as the area above 200 m depth (typically the limit of surface water mixing) and B (the deeper basin area) >200 m depth. The data in Table 2 do not imply that these modern basins are capable either of exhibiting source-sink characteristics or of producing iron-enriched sediments. The data also

TABLE 2. RATIOS OF SHELF AREA (S, <200 m DEPTH) TO DEEP BASIN AREA (B, >200 m DEPTH) IN MODERN BASINS

Enclosed/semi-enclosed basins	Basin area (× 10⁶ km²)	S/B	B/S
Gulf of California	0.15	0.89	1.1
Baltic Sea	0.39	99	0.01
Caspian Sea	0.41	2.3	0.43
Black Sea	0.42	0.37	2.7
Red Sea	0.45	0.72	1.4
Sea of Japan	1.01	0.30	3.4
East China Sea	1.20	4.2	0.23
Hudson Bay	1.23	13	0.08
Gulf of Mexico	1.29	0.92	1.1
Sea of Okhotsk	1.39	0.87	2.7
Bering Sea	2.26	0.85	1.2
Marginal Basins			
S. Australian Basin	1.53	0.22	4.6
Argentine Basin	4.40	0.23	4.3
World Oceans	362	0.08	12

Note: After Raiswell and Anderson (2005).

reflect present tectonic, sea-level, and oceanographic conditions and are not necessarily representative of the Proterozoic. A basin source area will deliver iron to a shelf area of banded iron formation at a rate given $J_{source} \times \varepsilon \times$ (Basin Source Area)/(Shelf Sink Area), which would produce an iron delivery rate of 22,500 µg cm⁻² yr⁻¹ from a recycling flux of ~5000 µg cm⁻² yr⁻¹ for an area ratio of ~4. The B/S ratios in modern enclosed, semi-enclosed, and marginal basins typically range up to 4, which suggests that clastic sediments located in the deep basin area have the potential to source most of the iron for banded iron formations.

The second case (Fig. 5, bottom) is based on the deep marine depositional environment inferred by Krapez et al. (2003) and Pickard et al. (2004). Water depth is known only to be below wave base, but a turbiditic origin is deduced from the presence of density current rhythms indicating a depositional environment that is distal from a shelf that supplied the clastics. It is useful to note that currents in the Black Sea disperse clastic sediment uniformly over an area comparable in size to the Hamersley Basin, producing microlaminae that can also be traced for distances of more than 1000 km over the basin floor (Lyons, 1991; Lyons and Berner, 1992). Krapez et al. (2003) and Pickard et al. (2004) offer no evidence for the degree of bottom water oxygenation of the iron formation depositional environment although the existence of iron oxides in the distal area is possible under bottom waters that are almost anoxic. A diagenetic supply of iron in the form of iron oxide colloids generated by recycling might be transported by turbidity currents to a deep marine depositional environment. Typical S/B ratios for modern basins range up to 4 (although values <1 are more common), which suggests that a shelf source area is also viable for the supply of recycled iron to banded iron formation in a deep basin environment.

Better geological constraints, and especially an improved understanding of the basin configuration and the depositional environment of the banded iron formations, are required to reach a definitive view on the contributions of iron from diagenetic recycling. However, the iron mass balance constraints (based on the microbands as varves) suggest that recycling is a potential iron source provided the Hamersley Basin had values of S/B (for deep basin deposition) or B/S (for shelf deposition) of ~4. Both of these extremes imply the existence of large areas of contemporaneous clastic sediment on the shelf or in the deep basin.

RIVERINE IRON FLUXES

Holland (1984) has suggested that riverine particulate and dissolved iron sources might also operate to augment the potential supply by diagenetic recycling. Thus Anderson and Raiswell (2004) have shown that the diagenetic mobilization of recycled iron was only responsible for the supply of 40‰ ± 20% of the reactive iron enrichment in the deep-basin euxinic sediments of the Black Sea. The remaining enrichment was attributed to enhanced reactivity of lithogenous iron, possibly by the fractionation of the riverine particulate flux to produce an iron oxide-enriched component.

Poulton and Raiswell (2002) have shown that the ratio of highly reactive iron to total iron in riverine particulates (0.43 ± 0.03) exceeds that found in modern continental margin and deep sea sediment (overall average 0.26 ± 0.08). It was suggested that highly reactive iron is at present preferentially trapped in inner shore areas (e.g., estuaries, tidal flats) and a limited data set provided some support for this hypothesis. A significant proportion of the highly reactive iron in riverine particulates occurs as iron oxide nanoparticulate spheres that adhere to clay minerals (Poulton and Raiswell, 2005). Adhesion may be diminished on contact with saline estuarine waters owing to collapse of the double layer, and spheres that are detached and sedimented will cause the loss of iron oxides from riverine suspended sediments entering the marine environment. However, detachment without sedimentation could produce an increased export of iron oxides to marine sediments. Such iron oxide-enriched sediments might produce higher concentrations of C_p (pore-water dissolved iron) and corresponding increases in the fluxes of diagenetically recycled iron (see eq. 4).

The long-distance transport of riverine suspended sediment across the shelf to distal environments might also allow sufficient fractionation to produce a component consisting predominantly of iron oxide nanoparticulates. Evidence suggests that riverine particulates can undergo long-range transport into deep basin areas. For example rivers draining the active margins of islands in the East Indies deliver fine particulates into the Equatorial Undercurrent, allowing long-range transport across the Pacific (Gordon et al., 1997; Milliman et al., 1999; Sholkovitz et al., 1999; Mackey et al. 2002). However, this mechanism is problematic because of the large degrees of fractionation required to isolate iron oxides from detrital phases (Holland, 1984).

The flux estimates originating from recycling may also be assisted by the riverine delivery of dissolved iron (see earlier). A rough estimate of the riverine flux into the Hamersley Basin can be made assuming that all silica in the microbands was derived from riverine input, as is suggested by Ge/Si ratios (Hamade et al., 2003). Garrels (1987) estimated that the microbands required an annual flux of 39 mg cm^{-2} yr^{-1}, which would need to be sustained over an area of ~10^5 km^2, giving a total mass of silica of 39 × 10^{15} mg. The maximum silica content of rivers is typically around 10–30 ppm (Holland, 1972), which therefore implies a discharge flux of at least 1–3 × 10^{15} liter yr^{-1}, close to the estimate of 4 × 10^{15} liter yr^{-1} by Holland (1984). This confirmation of Holland's (1984) discharge estimate supports his conclusion that a high riverine dissolved iron concentration (~7 ppm) is needed to provide the iron in banded iron formations. These dissolved iron concentrations require a low atmospheric pO$_2$ (<10^{-60} atm; Ohmoto et al., 2004), which might have existed only prior to 2.3 Ga (if atmospheric evolution followed the CWKH model). Significant riverine dissolved iron contributions to banded iron formations are thus unlikely.

CONCLUSIONS

A benthic recycling model has been used to estimate the rates at which dissolved iron can diffuse through a surface-oxygenated layer of sediment (where precipitation occurs as iron oxides) and into the overlying seawater. The model shows that rates of recycling are controlled mainly by the thickness of the surface-oxygenated layer (L) and the pore-water concentration of dissolved iron (C_p), and to a lesser extent by pH and temperature. Diffusive fluxes of up to 5000 µg cm^{-2} yr^{-1} of iron can be recycled to the overlying seawater when L < 0.1 cm and C_p = 5 µg cm^{-3}, for pH 6.5 (and T < 65 °C), or pH 7.0 (and T < 40 °C), or pH 7.5 (and T < 20 °C). Lower atmospheric pO$_2$ (0.1 and 0.01 PAL) has little effect on the maximum fluxes of iron recycling, although these are achieved at smaller thicknesses of the oxygenated layer for all pH 6.5–7.5 and temperatures 10–65 °C.

Recycled iron fluxes of 5000 µg cm^{-2} yr^{-1} are inadequate compared to the rates of 22,500 µg cm^{-2} yr^{-1} required for banded iron formation. However, the flux of recycled iron derived from the source sediments may be modified substantially by two factors: the export efficiency (ε) from the source area and the ratio of (Source Area)/(Sink Area). The export efficiency term accounts for dissolved iron entering oxygenated overlying waters that is oxidized to iron oxides and redeposited into the source sediments, rather than being exported to banded iron formations. Assuming ε = 1, then the ratio (Source Area)/(Sink Area) can act either to disperse the recycled flux over a relatively large sink area or to focus the flux into a relatively small sink area.

Current depositional models for banded iron formation suggest that deposition occurred either on a shelf or in a distal slope/deep basin environment. Modern basins have (Shelf Area)/(Deep Basin Area) ratios that typically range from 0.2 to 4. Thus basins at either extreme would enable a source shelf supply area to produce the required rates of iron delivery to a deep basin banded iron formation, or a deep basin source area to supply a banded iron formation on the shelf. The iron mass balance constraints indicate that a large area of contemporaneous clastic source sediments (or a smaller area of iron-enriched sediments) would be needed to supply all the iron in banded iron formations.

ACKNOWLEDGMENTS

Lee Kump, Ariel Anbar, and Hiroshi Ohmoto supplied thoughtful and extremely valuable reviews.

REFERENCES CITED

Alibert, C., and McCulloch, M.T., 1993, Rare earth element and neodymium isotopic compositions of the banded iron-formations and associated shales from the Hamersley, Western Australia: Geochimica et Cosmochimica Acta, v. 57, p. 187–204, doi: 10.1016/0016-7037(93)90478-F.

Aller, R.C., and Benninger, L.K., 1981, Spatial and temporal patterns of dissolved ammonium, manganese and silica fluxes from bottom sediments of Long Island Sound: Journal of Marine Research, v. 39, p. 295–314.

Aller, R.C., Mackin, J.E., and Cox, R.T., Jr., 1986, Diagenetics of Fe and S in the Amazon inner shelf muds: Apparent dominance of Fe reduction and implications for the genesis of ironstones: Continental Shelf Research, v. 6, p. 269–289.

Anbar, A.D., and Holland, H.D., 1992, The photochemistry of manganese and the origin of banded iron formations: Geochimica et Cosmochimica Acta, v. 56, p. 2595–2603, doi: 10.1016/0016-7037(92)90346-K.

Anderson, T.F., and Raiswell, R., 2004, Sources and mechanisms for the enrichment of highly reactive iron in euxinic Black Sea sediments: American Journal of Science, v. 304, p. 203–233.

Barley, M.E., Pickard, A.L., and Sylvester, P.J., 1997, Emplacement of a large igneous province as a possible cause of banded iron formation 2.45 billion years ago: Nature, v. 385, p. 55–58, doi: 10.1038/385055a0.

Bau, M., and Moller, P., 1993, Rare earth element systematics of the chemically-precipitated component in Early Precambrian iron formations and the evolution of the terrestrial atmosphere-hydrosphere-lithosphere system: Geochimica et Cosmochimica Acta, v. 57, p. 2239–2249, doi: 10.1016/0016-7037(93)90566-F.

Benson, B.B., and Krause, D.K., Jr., 1984, The concentration and isotopic fractionation of oxygen dissolved in freshwater and seawater in equilibrium with the atmosphere: Limnology and Oceanography, v. 29, p. 620–632.

Ben-Yaakov, S., 1973, pH buffering of pore water of recent anoxic marine Sediments: Limnology and Oceanography, v. 18, p. 86–94.

Berner, R.A., 1980, Early diagenesis: A theoretical approach: Princeton, New Jersey, Princeton University Press, 241 p.

Berner, R.A., and Raiswell, R., 1983, Burial of organic carbon and pyrite sulfur in sediments over Phanerozoic time: A new theory: Geochimica et Cosmochimica Acta, v. 47, p. 855–862, doi: 10.1016/0016-7037(83)90151-5.

Beukes, N.J., and Klein, C., 1992, Models for iron-formation deposition, in Schopf, J.W., and Klein, C., eds., The Proterozoic biosphere: A multidisciplinary study: New York, Cambridge University Press, p. 147–152.

Borchert, H., 1960, Genesis of marine sedimentary iron ores: Transactions of the Institute of Mining and Metallurgy, v. 69, p. 261–277.

Boudreau, B.P., 1996, Diagenetic models and their interpretation: Berlin, Springer-Verlag, 414 p.

Boudreau, P., 1999, Metals and models: Diagenetic modeling in freshwater lacustrine sediments: Journal of Paleolimnology, v. 22, p. 227–251, doi: 10.1023/A:1008144029134.

Boudreau, B.P., and Canfield, D.E., 1988, A provisional diagenetic model for pH in anoxic porewaters: Application to the FOAM site: Journal of Marine Research, v. 46, p. 429–455.

Boudreau, B.P., and Canfield, D.E., 1993, A comparison of closed-system and open-system models for porewater pH and calcite saturation state: Geochimica et Cosmochimica Acta, v. 57, p. 317–334, doi: 10.1016/0016-7037(93)90434-X.

Boudreau, B.P., and Scott, M.R., 1978, A model for the diffusion-controlled growth of deep-sea manganese nodules: American Journal of Science, v. 278, p. 903–929.

Buessler, K.O., Livingston, H.D., and Casso, C., 1991, Mixing between oxic and anoxic waters in the Black Sea as traced by Chernobyl cesium isotopes: Deep-Sea Research, v. 38, p. S725–S745.

Cairns-Smith, A.G., 1978, Precambrian solution photochemistry-inverse segregation and banded iron formations: Nature, v. 276, p. 807–808, doi: 10.1038/276807a0.

Canfield, D.E., 1989, Reactive iron in marine sediments: Geochimica et Cosmochimica Acta, v. 53, p. 619–632, doi: 10.1016/0016-7037(89)90005-7.

Canfield, D.E., and Raiswell, R., 1991, Carbonate precipitation and dissolution: Its relevance to fossil preservation, in Allison, P.A., and Briggs, D.E.G., eds., Taphonomy: Releasing the data locked in the fossil record: New York, Plenum Press, p. 412–453.

Canfield, D.E., and Teske, A., 1996, Late Proterozoic rise in atmospheric oxygen concentration inferred from phylogenetic and sulphur-isotope studies: Nature, v. 382, p. 127–132, doi: 10.1038/382127a0.

Canfield, D.E., Lyons, T.W., and Raiswell, R., 1996, A model for iron deposition to euxinic Black Sea sediments: American Journal of Science, v. 296, p. 818–834.

Canfield, D.E., Habicht, K.S., and Thamdrup, B., 2000, The Archean sulfur cycle and the early history of atmospheric O_2: Science, v. 288, p. 658–661, doi: 10.1126/science.288.5466.658.

Carignan, R., and Lean, D.R.S., 1991, Regeneration of dissolved substances in a seasonally anoxic lake: The relative importance of processes occurring in the water column and in the sediments: Limnology and Oceanography, v. 36, p. 683–707.

Chambers, L.A., and Trudinger, P.A., 1979, Microbiological fractionation of stable sulfur isotopes: A review and critique: Geomicrobiology Journal, v. 1, p. 249–293.

Danielson, A., Moller, P., and Dulski, P., 1992, The europium anomalies in banded iron formations and the thermal history of the ocean crust: Chemical Geology, v. 97, p. 89–100, doi: 10.1016/0009-2541(92)90137-T.

Davison, W., and Seed, G., 1983, The kinetics of the oxidation of ferrous iron in synthetic and natural waters: Geochimica et Cosmochimica Acta, v. 47, p. 67–79, doi: 10.1016/0016-7037(83)90091-1.

Davison, W., Lishman, J.P., and Hilton, J., 1985, Formation of pyrite in freshwater sediments: Implications for C/S ratios: Geochimica et Cosmochimica Acta, v. 49, p. 1615–1620, doi: 10.1016/0016-7037(85)90266-2.

de Baar, H.J.W., and de Jong, J.T.M., 2001, Distributions, sources and sinks of iron in seawater, in Turner, D.R., and Hunter, K.H., eds., The biogeochemistry of iron in seawater: New York, John Wiley and Sons, p. 123–253.

Derry, L.A., and Jacobsen, S.B., 1990, The chemical evolution of Precambrian seawater: Evidence from REEs in banded iron formations: Geochimica et Cosmochimica Acta, v. 54, p. 2965–2977, doi: 10.1016/0016-7037(90)90114-Z.

Detmers, J., Bruchert, V., Habicht, K.S., and Kuever, J., 2001, Diversity of sulfur isotope fractionations by sulfate-reducing prokaryotes: Applied and Environmental Microbiology, v. 67, p. 888–894, doi: 10.1128/AEM.67.2.888-894.2001.

Drever, J.I., 1974, Geochemical model for the origin of Precambrian banded iron formations: Geological Society of America Bulletin, v. 85, p. 1099–1106, doi: 10.1130/0016-7606(1974)85<1099:GMFTOO>2.0.CO;2.

Elrod, V.A., Berelson, W.M., Coale, K.H., and Johnson, K.S., 2004, The flux of iron from continental shelf sediments: A missing source for global budgets: Geophysical Research Letters, v. 31, p. L12307, doi: 10.129/2004GL020216, doi: 10.1029/2004GL020216.

Emerson, S., 1976, Early diagenesis in anaerobic lake sediments: Chemical equilibria in interstitial waters: Geochimica et Cosmochimica Acta, v. 40, p. 925–934, doi: 10.1016/0016-7037(76)90141-1.

Ewers, W.C., and Morris, R.C., 1981, Studies of the Dales Gorge Member of the Brockman Iron Formation, Western Australia: Economic Geology and the Bulletin of the Society of Economic Geologists, v. 76, p. 1929–1953.

Feijtel, T.C., DeLaune, R.D., and Patrick, W.H., Jr., 1988, Biogeochemical control on metal distribution and accumulations in Louisiana sediments: Journal of Environmental Quality, v. 17, p. 88–94.

Friedl, G., Dinkel, C., and Werhli, B., 1998, Benthic fluxes of nutrients in the northwestern Black Sea: Marine Chemistry, v. 62, p. 77–88, doi: 10.1016/S0304-4203(98)00029-2.

Friedrich, J., Dinkel, C., Friedl, G., Pimenov, N., Wijsman, J., Gomoiu, M.-T., Cociasu, A., Popa, L., and Wehrli, B., 2002, Benthic nutrient cycling and diagenetic pathways in the north-western Black Sea: Estuarine, Coastal and Shelf Science, v. 54, p. 369–383, doi: 10.1006/ecss.2000.0653.

Gardner, L.R., 1973, Chemical models for sulfate reduction in closed anaerobic marine environments: Geochimica et Cosmochimica Acta, v. 37, p. 53–68, doi: 10.1016/0016-7037(73)90243-3.

Garrels, R.M., 1987, A model for the deposition of the micro-banded Precambrian iron-formations: American Journal of Science, v. 287, p. 81–106.

Gordon, R.M., Coale, K.H., and Johnson, K.S., 1997, Iron distributions in the Equatorial Pacific: Implications for new production: Limnology and Oceanography, v. 42, p. 419–431.

Gough, D.O., 1981, Solar interior structure and luminosity variations: Solar Physics, v. 74, p. 21–34, doi: 10.1007/BF00151270.

Grotzinger, J.P., and Kasting, J.F., 1993, New constraints on Precambrian ocean chemistry: Journal of Geology, v. 101, p. 235–243.

Habicht, K.S., and Canfield, D.E., 1996, Sulphur isotope fractionation in modern microbial mats and the evolution of the sulphur cycle: Nature, v. 382, p. 342–343, doi: 10.1038/382342a0.

Habicht, K.S., and Canfield, D.E., 2001, Isotope fractionation by sulfate-reducing natural populations and the isotopic composition of sulfide in marine sediments: Geology, v. 29, p. 555–558, doi: 10.1130/0091-7613(2001)029<0555:IFBSRN>2.0.CO;2.

Habicht, K.S., Gade, M., Thamdrup, B., Borg, P., and Canfield, D.E., 2002, Calibration of sulfate levels in the Archean ocean: Science, v. 298, p. 2372–2374, doi: 10.1126/science.1078265.

Hamade, T., Konhauser, K.O., Raiswell, R., Goldsmith, S., and Morris, R.C., 2003, Using Ge/Si ratios to decouple iron and silica fluxes in the Precambrian banded iron formations: Geology, v. 31, p. 35–38, doi: 10.1130/0091-7613(2003)031<0035:UGSRTD>2.0.CO;2.

Holland, H.D., 1972, The chemistry of the atmosphere and oceans: New York, Wiley-Interscience, 351 p.

Holland, H.D., 1973, The oceans: A possible source of iron in iron formations: Economic Geology and the Bulletin of the Society of Economic Geologists, v. 68, p. 545–557.

Holland, H.D., 1984, The chemical evolution of the atmosphere and oceans: Princeton, New Jersey, Princeton University Press, 582 p.

Holland, H.D., 1999, When did the Earth's atmosphere become oxic? A reply: Geochemical News. v. 100, p. 20–22.

Isley, A.E., 1995, Hydrothermal plumes and the delivery of iron to banded iron formations: Journal of Geology, v. 103, p. 169–185.

Jacobsen, S.B., and Pimental-Klose, M.R., 1988, A Nd isotopic study of the Hamersley and Michipicoten banded iron formations: The source of REE and Fe in Archean oceans: Earth and Planetary Science Letters, v. 87, p. 29–44, doi: 10.1016/0012-821X(88)90062-3.

Jorgensen, B.B., and Boudreau, B.P., 2001, Diagenesis and sediment-water exchange, in Boudreau, B.P., and Jorgensen, B.B., eds., The benthic boundary layer: Oxford, UK, Oxford University Press, p. 211–244.

Kaplan, I.R., and Rittenberg, S.C., 1964, Microbiological fractionation of sulphur isotopes: Journal of General Microbiology, v. 34, p. 195–212.

Kasting, J.F., 1987, Theoretical constraints on oxygen and carbon dioxide concentrations in the Precambrian atmosphere: Precambrian Research, v. 34, p. 205–229, doi: 10.1016/0301-9268(87)90001-5.

Kempe, S., Liebezeit, G., Diercks, A.R., and Asper, V., 1990, Water balance in the Black Sea: Nature, v. 346, p. 419, doi: 10.1038/346419a0.

Klein, C., and Beukes, N.J., 1989, Geochemistry and sedimentology of a facies transition from limestone to iron-formation deposition in the Early Proterozoic Transvaal Supergroup, South Africa: Economic Geology and the Bulletin of the Society of Economic Geologists, v. 84, p. 1733–1774.

Knauth, L.P., and Lowe, D.R., 1978, Oxygen isotope geochemistry of cherts from the Onverwacht Group (3.4 billion years), Transvaal, South Africa with implications for secular variations in the isotopic compositions of cherts: Earth and Planetary Science Letters, v. 41, p. 209–222, doi: 10.1016/0012-821X(78)90011-0.

Knauth, L.P., and Lowe, D.R., 2003, High Archean climatic temperatures inferred from oxygen isotope geochemistry of cherts in the 3.5 Ga Swaziland Supergroup, South Africa: Geological Society of America Bulletin, v. 115, p. 566–580, doi: 10.1130/0016-7606(2003)115<0566:HACTIF>2.0.CO;2.

Konhauser, K.O., Hamade, T., Raiswell, R., Morris, R.C., Ferris, F.G., Southam, G., and Canfield, D.E., 2002, Could bacteria have formed Precambrian banded iron formations?: Geology, v. 30, p. 1079–1082, doi:

10.1130/0091-7613(2002)030<1079:CBHFTP>2.0.CO;2.

Krapez, B., Barley, M.E., and Pickard, A.L., 2003, Hydrothermal and resedimented origins of the precursor sediments to banded iron formation: Sedimentological evidence from the Early Palaeoproterozoic Brockman Supersequence of Western Australia: Sedimentology, v. 50, p. 979–1011, doi: 10.1046/j.1365-3091.2003.00594.x.

Kump, L.R., and Holland, H.D., 1992, Iron in Precambrian rocks: Implications for the global oxygen budget of the ancient earth: Geochimica et Cosmochimica Acta, v. 56, p. 3217–3223, doi: 10.1016/0016-7037(92)90299-X.

Lenton, T.M., 2003, The coupled evolution of life and atmospheric oxygen, in Rothschild, L., and Lister, A., eds., Evolution on the planet Earth: The impact of the physical environment: London, Academic Press, p. 35–53.

Lyons, T.W., 1991, Upper Holocene sediments of the Black Sea: Summary of Leg 4 box cores (1988 Black Sea Oceanographic Expedition), in Izdar, E., and Murray, J.W., eds., Black Sea oceanography: Dordrecht, Kluwer, NATO ASI Series, ser. C, v. 351, p. 401–441.

Lyons, T.W., 1997, Sulfur isotopic trends and pathways of iron sulfide formation in upper Holocene sediments of the anoxic Black Sea: Geochimica et Cosmochimica Acta, v. 61, p. 3367–3382, doi: 10.1016/S0016-7037(97)00174-9.

Lyons, T.W., and Berner, R.A., 1992, Carbon-sulfur-iron systematics of the uppermost Holocene sediments of the anoxic Black Sea: Chemical Geology, v. 99, p. 1–27, doi: 10.1016/0009-2541(92)90028-4.

Lyons, T.W., Werne, J.P., Hollander, D.J., and Murray, R.W., 2003, Contrasting sulfur geochemistry and Fe/Al and Mo/Al ratios across the last oxic-to-anoxic transition in the Cariaco Basin, Venezuela: Chemical Geology, v. 195, p. 131–157, doi: 10.1016/S0009-2541(02)00392-3.

Mackey, D.J., O'Sullivan, J.E., and Watson, R.J., 2002, Iron in the western Pacific: A riverine or hydrothermal source for iron in the Equatorial Undercurrent?: Deep-Sea Research, v. 40, p. 877–893.

Matisoff, G., Lindsay, A.H., Matis, S., and Soster, F.M., 1980, Trace metal equilibria in Lake Erie sediments: Journal of Great Lakes Research, v. 6, p. 353–366.

McKibben, M.A., and Barnes, H.L., 1986, Oxidation of pyrite in low temperature acidic solutions: Rate laws and surface textures: Geochimica et Cosmochimica Acta, v. 50, p. 1509–1520, doi: 10.1016/0016-7037(86)90325-X.

McManus, J., Berelson, W.M., Coale, K.H., and Kilgore, T.E., 1997, Phosphorus regeneration in continental margin sediments: Geochimica et Cosmochimica Acta, v. 61, p. 2891–2907, doi: 10.1016/S0016-7037(97)00138-5.

Miller, R.G., and O'Nions, R.K., 1985, Source of Precambrian chemical and clastic sediments: Nature, v. 314, p. 325–330, doi: 10.1038/314325a0.

Millero, F.J., 2001, The physical chemistry of natural waters: New York, Wiley-Interscience, 654 p.

Millero, F.J., Sotolongo, S., and Izaguirre, M., 1987, The oxidation kinetics of Fe(II) in seawater: Geochimica et Cosmochimica Acta, v. 51, p. 793–891, doi: 10.1016/0016-7037(87)90093-7.

Milliman, J.D., Farnsworth, K.L., and Albertin, C.S., 1999, Flux and fate of fluvial sediments leaving large islands in the East Indies: Journal of Sea Research, v. 41, p. 97–107, doi: 10.1016/S1385-1101(98)00040-9.

Morris, R.C., 1993, Genetic modeling for banded iron-formation of the Hamersley Group, Pilbara craton, Western Australia: Precambrian Research, v. 60, p. 243–286, doi: 10.1016/0301-9268(93)90051-3.

Morris, R.C., and Horowitz, R.C., 1983, The origin of the iron-formation-rich Hamersley Group of Western Australia—Deposition on a platform: Precambrian Research, v. 21, p. 273–297, doi: 10.1016/0301-9268(83)90044-X.

Nealson, K.H., 1997, Sediment bacteria: Who's there, what are they doing and what's new?: Annual Review of Earth and Planetary Sciences, v. 25, p. 403–434, doi: 10.1146/annurev.earth.25.1.403.

Nembrini, G.P., Capobianco, J.A., Veil, M., and Williams, A.F., 1983, A Mossbauer and chemical study of the formation of vivianite in sediments of Lago Maggiore (Italy): Geochimica et Cosmochimica Acta, v. 47, p. 1459–1464, doi: 10.1016/0016-7037(83)90304-6.

Newman, M.J., and Rood, R.T., 1977, Implications of solar evolution for the Earth's early atmosphere: Science, v. 198, p. 1035–1067.

Nriagu, J.O., and Dell, C.J., 1974, Diagenetic formation of iron phosphates in Recent lake sediments: American Mineralogist, v. 59, p. 934–946.

Ohmoto, H., 1997, When did the Earth's atmosphere become oxic?: Geochemical News, v. 93, p. 12–27.

Ohmoto, H., and Felder, R.P., 1987, Bacterial activity in the warmer, sulphate-bearing Archean ocean: Nature, v. 328, p. 244–246, doi: 10.1038/328244a0.

Ohmoto, H., Kagegawa, T., and Lowe, D.R., 1993, 3.4 billion-year-old biogenic pyrites from Barberton, South Africa; sulfur isotope evidence: Science, v. 262, p. 555–557.

Ohmoto, H., Watanabe, Y., and Kumazaawa, K., 2004, Evidence from massive siderite beds for a CO_2-rich atmosphere before 1.8 billion years ago: Nature, v. 429, p. 395–399, doi: 10.1038/nature02573.

Pickard, A.L., Barley, M.E., and Krapez, B., 2004, Deep-marine depositional setting of banded iron formation: Sedimentological evidence from interbedded clastic sedimentary rocks in the early Palaeoproterozoic Dales Gorge Member of Western Australia: Sedimentary Geology, v. 170, p. 37–62, doi: 10.1016/j.sedgeo.2004.06.007.

Poulton, S.W., and Raiswell, R., 2002, The low temperature geochemical cycle of iron; from continental fluxes to marine sediment deposition: American Journal of Science, v. 302, p. 774–805.

Poulton, S.W., and Raiswell, R., 2005, Chemical and physical characteristics of iron oxides in riverine and glacial meltwater sediments: Chemical Geology, v. 218, p. 203–221, doi: 10.1016/j.chemgeo.2005.01.007.

Raiswell, R., and Anderson, T.F., 2005, Reactive iron enrichment in sediments deposited beneath euxinic bottom waters: Constraints on supply by shelf recycling, in McDonald, I., Boyce, A.J., Butler, I.B., Herrington, R.J., and Polya, D.A., eds, Mineral deposits and Earth evolution: Geological Society [London] Special Publication 248, p.179–194.

Raiswell, R., and Canfield, D.E., 1998, Sources of iron for pyrite formation: American Journal of Science, v. 298, p. 219–245.

Roekens, E.J., and van Grieken, R.E., 1983, Kinetics of iron oxidation in seawater of various pH: Marine Chemistry, v. 13, p. 195–202, doi: 10.1016/0304-4203(83)90014-2.

Rye, R., Kuo, P.H., and Holland, H.D., 1995, Atmospheric carbon dioxide concentrations before 2.2 billion years ago: Nature, v. 378, p. 603–605, doi: 10.1038/378603a0.

Sagan, C., and Mullen, G., 1972, Earth and Mars: Evolution of atmospheric and surface temperature: Science, v. 177, p. 52–56.

Santschi, P., Hohener, P., Benoit, G., and Bucholtz-ten-Brink, M., 1990, Chemical processes at the sediment-water interface: Marine Chemistry, v. 30, p. 269–315, doi: 10.1016/0304-4203(90)90076-O.

Schneider, S.H., and Londer, R., 1984, The coevolution of climate and life: San Francisco, Sierra Club Books, 543 p.

Schwartzmann, D., McMenamin, M., and Volk, T., 1993, Did surface temperatures constrain microbial evolution?: Bioscience, v. 43, p. 390–393.

Sholkovitz, E.R., Elderfield, H.E., Szymczak, R., and Casey, K., 1999, Island weathering: River sources of rare earth elements to the Western Pacific Ocean: Marine Chemistry, v. 68, p. 39–57, doi: 10.1016/S0304-4203(99)00064-X.

Stumm, W., and Morgan, J.J., 1980, Kinetics and products of ferrous iron oxygenation in aqueous solutions: Environmental Science and Technology, v. 14, p. 561–568.

Theis, T.L., and Singer, P.C., 1974, Complexation of iron (III) by organic matter and its effect on iron (II) oxygenation: Environmental Science and Technology, v. 8, p. 569–573, doi: 10.1021/es60091a008.

Towe, K., 1996, Environmental oxygen conditions during the origin and early evolution of life: Advances in Space Research, v. 18, no. 12, p. 7–15, doi: 10.1016/0273-1177(96)00022-1.

Trendall, A., 2000, The significance of banded iron formation (BIF) in the Precambrian stratigraphic record: Geoscientist, v. 10, p. 4–7.

Trendall, A.F., and Blockley, J.G., 1970, The iron formations of the Precambrian Hamersley Group, Western Australia: With special reference to the associated crocidolite: Western Australia Geological Survey Bulletin, v. 119, p. 336.

Ullman, W.J., and Aller, R.C., 1982, Diffusion coefficients in near shore marine sediments: Limnology and Oceanography, v. 27, p. 552–556.

Waite, T.D., 2001, Thermodynamics of the iron system in seawater, in Turner, D.R., and Hunter, K.H., eds., The biogeochemistry of iron in seawater: New York, John Wiley and Sons, p. 292–342.

Walker, J.C.G., 1983, Possible limits on the composition of the Archean ocean: Nature, v. 302, p. 518–520, doi: 10.1038/302518a0.

Walker, J.C.G., and Brimblecombe, P., 1985, Iron and sulfur in the pre-biologic ocean: Precambrian Research, v. 28, p. 205–222, doi: 10.1016/0301-9268(85)90031-2.

Wang, Y., and Van Cappellen, P., 1996, A multicomponent reactive transport model of early diagenesis: Application to redox cycling in coastal marine sediments: Geochimica et Cosmochimica Acta, v. 60, p. 2993–3014, doi: 10.1016/0016-7037(96)00140-8.

Weiler, R.R., 1973, The interstitial water composition in the sediments of the Great Lakes. I. Western Lake Ontario: Limnology and Oceanography, v. 18, p. 918–931.

Wignall, P.B., 1994, Black shales: Oxford, UK, Clarendon Press, 127 p.

Wijsman, J.W.M., Middelburg, J.J., and Heip, C.H.R., 2001, Reactive iron in Black Sea sediments: Implications for iron cycling: Marine Geology, v. 172, p. 167–180, doi: 10.1016/S0025-3227(00)00122-5.

Wijsman, J.W.M., Herman, P.M.J., Middelburg, J.J., and Soetaert, K., 2002, A model for early diagenetic processes in sediments of the continental shelf of the Black Sea: Estuarine, Coastal and Shelf Science, v. 54, p. 403–421, doi: 10.1006/ecss.2000.0655.

Williamson, M.A., and Rimstidt, D.J., 1994, The kinetics and electrochemical rate-determining step of aqueous pyrite oxidation: Geochimica et Cosmochimica Acta, v. 58, p. 5443–5454, doi: 10.1016/0016-7037(94)90241-0.

MANUSCRIPT ACCEPTED BY THE SOCIETY 29 OCTOBER 2005

Geological Society of America
Memoir 198
2006

Microbial mediation of iron mobilization and deposition in iron formations since the early Precambrian

D. Ann Brown*

Department of Geological Sciences, University of Manitoba, Winnipeg, Manitoba R3T 2N2, Canada

ABSTRACT

It is suggested that sedimentary deposition of iron in sediments from the Archean to the present day can be attributed largely to microbial mediation. Results from laboratory experiments, using a microbial consortium enriched from the source of a biofilm growing on a rock face in an underground research laboratory, are used to advance a plausible explanation for the mobilization, precipitation, transport, and deposition of iron. The consortium produces its own local environments independent of the prevailing atmosphere. It is able to extract iron from minerals such as biotite and magnetite, as well as from a chelated solution; this iron is then metabolized, mainly through dissimulatory iron reduction, to provide cell energy, after which it is immediately precipitated. All organisms require energy for growth and reproduction, but because iron redox reactions are inefficient a large amount of iron must be processed, either directly through metabolism or indirectly due to the local microbial redox microenvironment, with the result that vast quantities accumulate as waste. This precipitate could be the main source of iron in sedimentary iron formations. Siderite and ferrihydrite, the main precipitates, may occur in close juxtaposition within a biofilm. The oxidation state of the iron precipitate is controlled by the nutrient supply, which in turn influences the metabolism of the biofilm organisms and hence their redox. Subsequently, this iron, enmeshed within the biofilm, is either deposited in sediments as fine layers or rolled by wave or current action into particles and granules, which can form structures similar to those found in banded iron formations (BIFs).

Keywords: biofilm, microbial mediation, iron formations, siderite, ferrihydrite.

INTRODUCTION

This paper describes field and laboratory evidence suggesting that microbial activity was important to the genesis of iron formations such as BIFs (banded iron formations). The work is the result of a long-standing dialogue between Gordon A. Gross, a geologist with the Geological Survey of Canada, who has wide experience of iron formations, and me, a microbiologist who was introduced to microbial deposition of iron while investigating a biofilm forming underground in the Canadian Shield. Evidence from experiments using the microbial consortium forming this biofilm is set out to show the significant microbial processes that the consortium can encompass, and a comparison is then made between field and laboratory biofilm textures and some iron formation structures that have been supplied and described by G.A. Gross. Though in general I have used Gross's (1996) nomenclature for stratiform iron deposits, common practice is followed in the use of the

---

*Present address: A&A Research, Old Moorcocks, Rushlake Green, Heathfield, E. Sussex TN21 9PP, UK.

Brown, D.A., 2006, Microbial mediation of iron mobilization and deposition in iron formations since the early Precambrian, *in* Kesler, S.E., and Ohmoto, H., eds., Evolution of Early Earth's Atmosphere, Hydrosphere, and Biosphere—Constraints from Ore Deposits: Geological Society of America Memoir 198, p. 239–256, doi: 10.1130/2006.1198(14). For permission to copy, contact editing@geosociety.org. ©2006 Geological Society of America. All rights reserved.

term BIF for fine-grained iron minerals closely interbanded with quartz, chert, and/or carbonate.

Serious consideration has been given for many years to the processes required to produce the enormous concentrations of iron in BIF sediments (Gross, 1988; Brown et al., 1995). Past studies of the genesis of sedimentary rocks have generally emphasized abiotic physical and chemical processes, and iron in sediments was widely believed to be supplied primarily by inorganic chemical precipitation from hydrothermal waters (Gross, 1983). However, today there is a much better understanding of the influence of biological processes on the geochemistry of the atmosphere and hydrosphere, as well as on sedimentation processes in general. Better data are now also available on the sources of the iron incorporated into sediments, and on the direct microbial reduction of iron as opposed to its oxidation.

The aspects of microbial processes that have been investigated in this study are summarized schematically in Figure 1, and are based on laboratory investigations using microscopic organisms (henceforth called microbes or bacteria) from the biofilm consortium. These include the primary source of the iron; the leaching of this iron into solution where it is utilized through redox reactions by bacteria as their energy source; the consequent precipitation of this iron in either the ferrous or ferric state depending upon the local microenvironment; and the possible diagenesis of these precipitates during their deposition in iron formations. Various particles, granules, layers, and other structures in BIFs are then compared morphologically with textures found in the field and from biofilm processes in the laboratory.

The role of sulfur was not investigated because no iron sulfur compounds were found in the laboratory studies or in the underground biofilm. The status of silica was also not explored, although it is a principal component of many iron formations, in part because silica cannot be metabolized by single-cell organisms known as prokaryotes (Archaea and Bacteria), the only fauna extant at the time of the early BIF deposits, and because some authors (Valley et al., 2002; Knauth et al., 2004) consider that early Archean temperatures on Earth were sufficiently low for marine silica saturation.

The results suggest that both the dissolution of iron from granitic crystalline rocks and its subsequent reprecipitation through the action of natural microbial consortia may offer a plausible and satisfactory alternative explanation for the mobilization, transport, precipitation, and deposition of iron in at least some BIF sediments reaching back in age to the Archean.

History of Geomicrobiological Research

Evidence now seems to indicate that microbes emerged almost as soon as water could form on the surface of Earth. A brief survey shows that the association of microbes with iron deposition has been recognized for nearly two centuries, ever since Ehrenberg (1836) suggested that bacteria were capable of precipitating bog-iron as hydrated iron oxides. Later, when Winogradsky (1888) was investigating microbes in the natural environment, he discovered that bacteria could obtain energy through oxidation of inorganic compounds such as ferrous iron, and was able to show that the bacterium *Leptothrix* requires this dissolved iron for growth. An extensive literature review by Harder (1919)

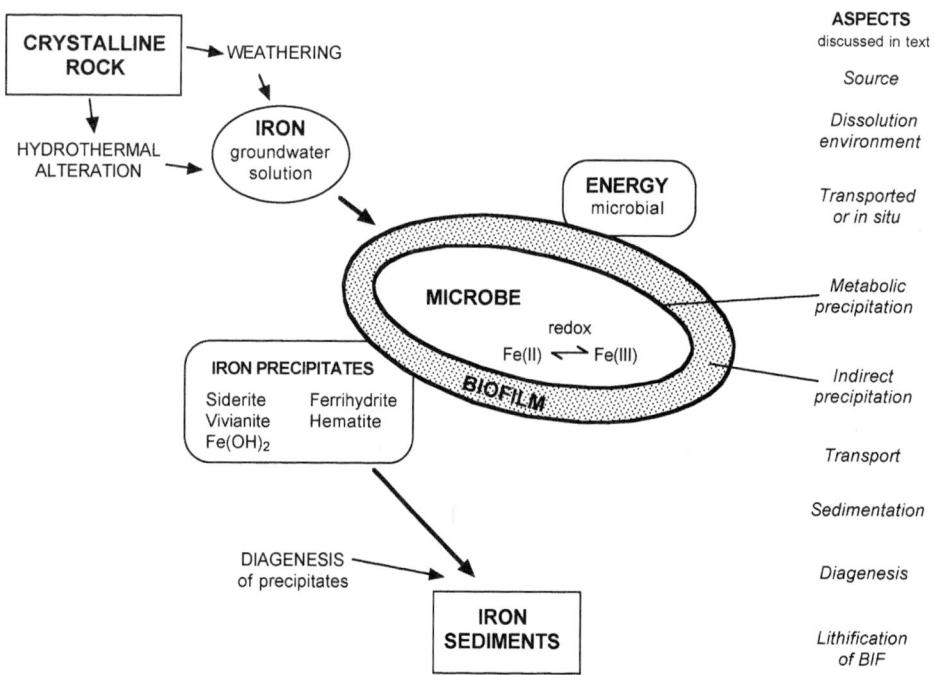

Figure 1. Schematic diagram showing a summary of the various microbial aspects of iron conversion investigated in the laboratory. The dissolution of iron sources (direct mineral/microbe contact not shown) for the provision of cell energy through microbial metabolism results in the precipitation of either ferrous or ferric minerals. These precipitates can become incorporated in sediments where they may be altered by diagenesis and ultimately concentrated in iron formations.

further noted the importance of bacteria for the deposition of sedimentary iron. More recently, it has been recognized that bacteria can grow anaerobically through the coupling of carbon oxidation to the sedimentary reduction of manganese or iron oxides (Nealson and Myers, 1990).

The slow accumulation of knowledge on microbial communities, particularly those in oligotrophic (nutrient-poor) environments, is relevant to the investigation. It was initially thought that microorganisms could only exist at the surface of Earth, but Waksman (1916) demonstrated that the populations and their physiological activities extend throughout the soil profile. Bastin (1926) discovered that bacteria were present in deep subsurface waters of oil field brines, at depths known as the profound groundwater zone, and ZoBell (1952) reported the presence of bacteria in seawater at the bottom of the Philippine Trench, more than 10 km beneath the sea surface. Morita and ZoBell (1955) reported viable bacteria from 8 m below the seafloor of Pacific pelagic sediments, and suggested that they had been incorporated contemporaneously with the million-year-old sediment. ZoBell (1943) proposed that, when nutrients were limited, bacteria formed biofilms that preferentially grew as slime-encased microbes on rock surfaces (sessile) rather than as free-swimming (planktonic) organisms. The importance of such biofilm consortia has been gradually recognized, but progress was slow until it was realized in the 1970s there was a universal interdependence amongst all the organisms found within the surface-associated biofilms. The recent investigation of crystalline rocks for the burial of radioactive waste has shown that the groundwater of the subsurface zone harbors numerous and diverse microbial populations (Fredrickson and Onstott, 1996; Amy and Haldeman, 1997; Pedersen, 1997; Brown and Sherriff, 1999).

The morphologies of filamentous and of domed microfossils, first discovered in the early Proterozoic Gunflint Chert, Canada (1.9 Ga) by Tyler and Barghoorn in 1954, was thought at that time to have cyanobacterial (blue-green algae) origins, and thus possibly to indicate oxygenic photosynthesis. This formation has a marked similarity to other, more iron-rich cherts such as BIFs, and several modern authors (Gross, 1972; Hofmann et al., 1991; Schopf, 1992) have established that autochthonous microbial microfossils and stromatolites were indeed incorporated in Late Archean BIFs. There are now numerous reports of the presence of microbes during deposition of silica-, carbonate-, and iron-rich sedimentary rocks from modern to Precambrian ages (Braga and Martin, 2000; Sumner, 2000; Bonny and Jones, 2003). Even the presence of hydrothermal vent microbial communities during the Middle Archean has been suggested by Rasmussen (2000) in a report of microfossils in a deep-sea volcanogenic massive sulfide deposit from the Pilbara Craton in Australia (3.2 Ga).

Oceanographic research over the last 60 years, especially the discovery of deep-sea thermal vents, has further expanded our knowledge of primitive life. These vents not only create a wide range of local environmental conditions but also discharge large quantities of iron and manganese into the ocean, which are then available to provide the primary energy source for extensive microbial communities (Baross and Deming, 1983). The mass flux of the deeply circulating hydrothermal systems, manifested in part by these vents, is enormous. According to a recent calculation by Harris et al. (2004), the flux from sea mounts equals that from mid-ocean ridges, indicating that it could well have been one of the major sources of iron in marine waters, whether from biologic or abiotic processes. Furthermore, researchers in the Ocean Drilling Program not only have found an incredibly rich diversity of microbes living 1000 m below the seafloor, but also have reported bacteria and remnants of biofilms in waters from an artesian borehole in basalt of the mid-ocean Juan da Fuca Ridge (Cowen et al., 2003). These discoveries suggest that prokaryotic microbes can thrive in seemingly hostile, hot, and anaerobic environments similar to those thought to have been common in shallow Archean marine waters (Baross and Hofmann, 1985). If these microbes were able to exist in such environments, it is highly likely that they would have been involved in the mobilization and deposition of iron.

Sources and Deposition Sites for Iron in BIFs

Though iron formations are known throughout the geological record, from 3.8 Ga (Isua, West Greenland) to contemporary deposition in the Red Sea basins (El Shazly, 1990), the major iron deposits were formed between the Late Archean and the middle Proterozoic (Gross, 1983; Isley, 1995). Many BIFs are extensive (Baur et al., 1985) and comprise alternately banded iron- and silica-rich beds, microbands, and laminations, where magnetite, hematite, and/or siderite are the main primary iron minerals. Nisbet (2000) discussed the evidence for Archean life (possibly as early as 3.7 Ga) and the environments/ecological sites where it may have flourished. These largely coincide with those outlined by Gross (1996, Fig. 3.2) as sites of iron formation deposition. Both authors consider these habitats to be similar to those formed through tectonism today, even though independent cratonic masses may have been smaller and less long lived. The ecological sites are (1) shallow back-arc waters and shores with stromatolite and microbial mat communities with oxygenic photosynthesis at surface and anoxygenic photosynthesis and methanogenesis below; (2) deeper waters and commonly oligotrophic open ocean, with photosynthetic plankton subject to blooms when nutrient supply is above normal; and (3) andesitic island arcs, komatiitic shield volcanoes, and mid-ocean ridges, each driving a hydrothermal circulation supplying nutrients to chemotrophic, thermophilic bacteria, together with some photosynthetic bacteria. Gross (1995, 1996) associates Lake Superior type iron formations with continental shelf and slope; Algoma type with volcanic arcs and spreading ridges; and his Rapitan type with passive margin grabens. The action of the microbial consortium that was used for the laboratory experiments implies an early independence of bacteria from an obligatory photosynthetic basis and, to a certain extent, also from the general atmospheric composition.

The iron that is ultimately deposited in iron formations is derived from terrestrial cratonic and volcanic rocks, as well as from oceanic basalts and from crystallizing magmas. Terrestrial and submarine groundwaters and hydrothermal fluids acquire iron from these sources through both biotic and abiotic reactions with iron-rich oxides and silicates. These waters can transport the iron in ionic or complex organic molecular solution or, as is shown, in microbially reprecipitated particles. Because iron, even in the reduced state, is only marginally soluble in water, the question has been how such vast quantities of iron found in iron formations came to accumulate in the depositional basins. Several investigators, for instance Holland (1973), consider that the periodic upwelling from an ocean reservoir into a restricted basin would have provided sufficient amounts of iron to produce extensive BIFs. Some (Garrels et al., 1973) suggest that the ocean acts as a steady-state system with iron, from whatever source, and silica fluctuating only within narrow limits. Others (Lepp, 1987) think the laterization of cratonic rocks and fluvial transport are sufficient to provide all the necessary iron, and the ocean only acts to homogenize the iron and trace elements. Konhauser et al. (2002) in their study of marine systems suggested that the major oceanic source of the iron in BIFs was hydrothermal, possibly supplemented by continental drainage. Homogenization of the precipitated minerals in the oceanic environment spread them uniformly over large depositional basins. In their model the richer hematite/magnetite bands were formed during major hydrothermal pulses and dependent blooms of iron-precipitating planktonic bacteria, whereas the silica-rich sequences suggest quieter periods when hydrothermal fluids did not reach the basin, or did so only in small amounts. Hamade et al. (2003) have shown that silica and iron sources are decoupled. They suggested that the silica comes from continental landmass weathering and the iron sources are marine hydrothermal.

On the basis of element content and concentration, Gross (1993) suggested that almost all of the iron in iron formations of all ages was volcano-hydrothermal in origin. Beard et al. (2003) summarized the flux and isotope composition of the sources delivering iron to the ocean under present-day levels of atmospheric oxygen: riverine particulate, atmospheric particulate, and hydrothermal loads accounting for 72%, 19%, and 6.5% by weight respectively. The narrow range of $\delta^{56}Fe$ values for terrestrial igneous rocks, clastic sediments, and airborne materials implies that chemical weathering, sedimentary transport, and diagenesis play only a minor role in iron isotope variation. At mid-ocean ridges hot hydrothermal fluids (>300 °C) shift the isotope signature, again in a narrow peak, to a lower value. However, the wider range of values in Archean iron formations (attributed by Beard et al. [2003] to Johnson et al., 2003) implies a mixing of terrestrial and marine hydrothermal sources.

Our leaching experiments show that microbial action can supply terrestrial-source iron to riverine transport. However, from this source it is scattered throughout marine clastic sediments, thus considerably lowering the percentage of riverine particulate flux to iron formation deposition sites. In addition, this work shows how complex the route to the final isotope value of a deposit can be, as it is affected by microbially direct and indirect precipitation, by residual material from the leached rock-forming minerals, as well as their solution and resultant precipitation, and then further by diagenesis during transport and sedimentation. Though Brantley et al. (2001) have shown that fractionation through microbial iron reduction can provide a useful signature, both they and Icopini et al. (2004) have warned of the need to fully understand the mechanisms involved before firm conclusions can be reached. Therefore, the laboratory and field work on the microbial dissolution of iron from rock-forming minerals and its concurrent reprecipitation in a biofilm environment suggests that the iron in BIFs could have been multi-sourced—some from terrestrial weathering and some from marine hydrothermal venting.

The question of whether mediation by microbes could account for the amount of iron deposited in BIFs has been investigated by Konhauser et al. (2002). Their calculations of the number of organisms and of the nutrients required for their growth show that direct metabolic oxidation of iron has the potential to generate most, if not all, of the ferric iron in BIFs, even when no account is taken of any indirect biomineralization that might occur. They conclude that microbes could clearly have been an important contributor to the deposition of iron, and hence microbial mediation must be considered a viable and demonstrable mechanism for BIF development in the Precambrian.

MICROBIAL COMMUNITIES

Microbial Metabolism

Appreciation of the influence of various microorganisms on sedimentary geochemistry has developed slowly. The microbes involved are predominantly Archaea, which tend to inhabit extreme environments, and Bacteria, which prefer more moderate conditions and include the main chemolithotrophic organisms (those that obtain their energy from inorganic sources). Both of these are single-celled prokaryotic microbes that lack a membrane-bound nucleus. The imperative that drives their activities is the need to acquire sufficient energy for individual growth and reproduction. The rate at which this occurs depends on the amount of energy that can be released through redox-coupled microbial metabolism, but the reactions that do occur are those that are thermodynamically possible within a specific local environment. It is noteworthy, however, that in some instances the microbes can adjust their environment to better suit their activities (see niche construction, Laland et al., 2004).

All organisms, whether single or multicellular, have the same basic metabolic pathways, such as those for the biosynthesis of proteins or DNA, and including the energy-producing pathways like the tricarboxylic acid (or Krebs) cycle. Today most organisms, including all eukaryotes (those that contain a defined nucleus), ultimately obtain their energy from the sun, either directly through photosynthesis, or indirectly by using the oxygen so produced as their electron acceptor. Although there is little agreement on the composition of the early terrestrial atmosphere, I concur with Bekker et al. (2004) that it is unlikely to have contained much molecular oxygen before ca. 2.45 Ga. Our present-day atmosphere has been produced through the oxygenic photosynthesis of water, initially by prokaryotic cyanobacteria, but mainly today by eukaryotic green plants. Because the phylogenic record suggests that cyanobacteria were one of the last lineages to diverge from the bacterial tree, it was probably not until after 2.3 Ga that molecular oxygen would actually have begun to accumulate (Blank, 2004).

In the likely absence of available oxygen in the Archean, microbes needed to develop alternative electron acceptors. The chemiosmotic theory (Gottschalk, 1986) describes how chemical energy, in a variety of forms, can be changed into an electrochemical potential across a membrane to produce a proton motive force (pmf) that is utilized to synthesize ATP (used by all living organisms to transfer cell energy). In principle, therefore, any electron potential between electron donor and acceptor can be used to pump protons to the exterior of the cell and thus induce a pmf. Hydrogen carriers are produced, usually by enzymatic oxidation of an organic substrate that interacts with a membrane-bound electron transport chain, which channels the electrons to an oxidant while at the same time pumping protons to the exterior. It should be noted that most of the reactants and products in the sedimentary environment are oxidants or reductants that can be utilized by organisms that follow the chemiosmotic model (Nealson, 1997). Bacteria are thus able to use an extraordinary diversity of ways to obtain energy from either the oxidation or the reduction of inorganic compounds and transition elements. So long as the necessary enzymatic machinery is available, it makes little difference to the organism what that source of energy is, be it nitrate, manganese, uranium, iron, sulfate, or carbon dioxide; these follow a sequence of decreasing redox potential and thermodynamic feasibility (Stumm and Morgan, 1981). It is this ability that enables bacteria to collectively fill virtually every ecological niche found on or under Earth's surface, as long as water is present (Brown and Sherriff, 1999). In the natural environment, however, iron is the most important source of energy because, although the energy it provides is limited, it is the most abundant and available element that undergoes such redox reactions.

Biofilms

Although indigenous Bacteria and Archaea can be present as free-swimming planktonic organisms, in oligotrophic groundwaters they flourish best as microbial consortia that congregate by forming a biofilm at water-rock interfaces. Biofilms are complex communities of independent, but interacting, organisms. They consist of a layer of slime comprising extracellular polymeric substances (EPS) that are usually negatively charged polysaccharides excreted by, and encompassing, the microbes (Costerton et al., 1994). Biofilm development is determined by the velocity and the nutrient content of the local water. Seed organisms adhere to a surface, then multiply and grow to form new colonies while excreting an EPS envelope that provides both a stable environment for the microbes and a medium within which chemical exchange can take place. The negative charges of the EPS and the cell walls of the microbes also "trap" abiotic material, notably positively charged metal ions such as iron, which precipitate and form considerable mineral concentrations (McLean et al., 1996). As biofilms grow mass transfer from bulk liquid to the cells becomes more difficult. The emphasis on a molecule's utilization shifts from its intrinsic metabolic return to its ability to diffuse into the biofilm (Riding, 2002; Battin et al., 2003). As was observed, for instance, in the field at the site of the original biofilm, both the internal runnels that facilitated nutrient transport and the all-enveloping slime provided local and very different microenvironments organized by the microbes themselves. Thus the consortium works together and is able to survive many harsh and extreme external conditions where individual planktonic organisms are not able to exist.

A modern study (Battin et al., 2003) confirms that microbial biofilms can change ecosystem processes to enhance their environment. These changes within the restricted system of a biofilm increase competition between similarly acting species of microbes (in-groups) but enhance cooperation between those of dissimilar function (out-groups), thus building a more efficient consortium (Queller, 2004). These biofilm-forming consortia contain a wide variety of different microbes, but which group of these organisms is active at any particular moment is determined by what nutrients are available and also by the organisms' ability to utilize any accessible energy-producing redox reactions. It is this activity that controls the local microenvironment. Furthermore, because a consortium contains a variety of species, it is able to adapt rapidly to any change in nutrients or the environment. Thus at any one time some organisms will be dormant and will become active only when the environment becomes suitable. Pertinent to some of our experiments, where oxygen is initially present it is first consumed by aerobic organisms, and only when the oxygen is exhausted do the anaerobic microbes become dominant (Nealson, 1997). Metabolic cascades can occur when the waste product from one organism can be used by another as its substrate, thus allowing further energy to be extracted (Nee, 2004). This forms a stepwise degradation of complex organic compounds to simpler ones, and the final product is generally carbon dioxide, which is lost to the atmosphere. Residual organic carbon is therefore found only in sediments where such degradation is incomplete, and thus is unlikely to be found in BIFs, although there may be a signature left from accumulated sediment and minerals.

Biomats

Biofilms are common today on rock and sediment surfaces in terrestrial lakes and streams, as well as on the walls of subsurface groundwater channels and pores. In bodies of calm water they can form free-floating gelatinous clouds several tens of centimeters across. However, both marine and fresh-water biofilms can also become an integral part of planar, soft organic layers and biomats that carpet lake beds and the seafloor over a wide area, trapping sediment as the seasonal supply of nutrients varies. Their resilience to erosion appears to be due in part to filamentous bacteria that stabilize the mat, making it resistant to high-energy environments and allowing the mat to acquire a significant thickness. Biomats of this type have been recognized in South African Archean siliciclastic rocks dated at 2.9 Ga, although it is uncertain what group of bacteria this would have been (Noffke et. al., 2003). Clastic carbonaceous particles of mats, eroded from a shallow-water, euphotic zone, have also been found by Tice and Lowe (2004) in similar South African rocks but dated to 3.4 Ga.

In modern settings, the distribution of microbial species in these mats is mainly determined by the amount of light and oxygen available (Stolz, 2000), with the surface dominated by phototrophic microbes (capable of photosynthesis) such as filamentous cyanobacteria and algal eukaryotes. But in anaerobic environments, even today, consortia can form without phototrophs, as occurs in the catotelm (body) of a bog, where anaerobes are able to cycle electrons through both reduction and oxidation redox reactions (Brown, 1994). In the anaerobic Archean environment similar microbial consortia would also have been present, but they probably were composed of different populations.

In Baja California thick algal mats are formed by several filamentous organisms that contribute to the primary organic deposit (Stolz and Margulis, 1984). Microbes and their EPS within the mats, as well as potentially precipitating iron and/or manganese minerals, trap a layer of fine, inorganic, siliceous sediment that smothers the microbes during flooding. This is then recolonized and anaerobically reworked so no evidence of the original organisms survives, although a recognizable microlayer may be left. The textures of some laminae in BIFs, rather than being the result of settling in quiet water from a bloom of iron-precipitating microbes as suggested by Konhauser et al. (2002), are strongly reminiscent of such sequential biomat growth.

Microbial mats currently forming in Icelandic hot springs show that their ability to bind iron and/or silica helps to prevent the degradation of the organisms by heterotrophic (organic nutrient source) bacteria (Konhauser and Ferris, 1996). The encrusted cells are then incorporated into a matrix of amorphous silica prior to their deposition in the sedimentary record. This suggests that contemporaneous biomineralization, even under a different atmosphere, may have structures similar to those of the geologic past.

Sometimes only small areas of mat are stable, and these can build up into stromatolites, which provide some of the more convincing Precambrian field evidence for microbial mediation in iron deposition. They have been found in rocks as old as the Steep Rock Lake deposit in Canada (3.0 Ga; Blackburn et al., 1991). Stromatolites are small, local, domed or columnar forms of microbial mats, with similar patterns of growth, that commonly occur in shallow seas or pools (Reid et al., 2000). They are dynamic systems affected by tidal and seasonal fluctuations that influence nutritional gradients and thus effectively balance sedimentation (whether mechanical or biochemical) and intermittent lithification to produce millimeter-scale laminations, similar to those described above in Baja California algal mats. Again, sediment trapping smothers the organic surface. This is then recolonized by new organisms that utilize the organic remains of the previous microbes as their carbon source, effectively removing any organic carbon and leaving only a thin layer of sediment as their memorial. An example of this in strongly microlaminated stromatolites from a Lake Superior type iron formation is described later.

On aging, the mats entrap added minerals and become thicker and heavier so that they either break up or slough off excess fragments. These fragments, and the grains of iron and silica minerals they contain, can form the primary granules for sedimentation, whether they are deposited in situ or are transported, possibly fluvially, farther afield.

Interactions between Microbes and Minerals

Microbial metabolic reactions have an extensive influence on the geochemistry of natural groundwaters by cycling iron between oxidized and reduced states (Freeze and Cherry, 1979; Lovley, 1995). This work confirms the ability of bacterial biofilms to accumulate metal ions from their environment and to precipitate them in particulate form. This immobilization of metal ions (Douglas and Beveridge, 1998) and their subsequent transport into aquatic sediments represents a means of concentrating metals in the environment and incorporating them into mineral deposits that are important in shaping our planet.

Two common rock-forming minerals that contain ferrous iron—magnetite and ilmenite—are significant in controlling heterogeneous redox reactions at low temperatures (White and Peterson, 1996). The solid-state mass balance requires that one Fe(II) atom be removed from magnetite for every two Fe(II) atoms oxidized to Fe(III) in the structure to form a new mineral, hematite. On "weathering" at surface, whether biologic or not, both electrons and ferrous iron are released to the aqueous environment, while the surface of the primary mineral is oxidized, possibly in a fashion similar to that of the chemiosmotic model. Much of the iron liberated by weathering is deposited as solid ferric oxyhydroxides such as ferrihydrite, which are characteristic components of young iron oxide accumulations (Schwertmann, 1988). Therefore we investigated the microbial leaching of magnetite and also of biotite (to represent the mafic silicates) in our experiments, although other researchers have investigated hornblende, common in rocks more mafic than granite (Brantley et al. 2001).

In contrast to the oxic zone at the surface, in the subsurface zone below a few meters the groundwater is reducing, a state that is almost entirely due to the microbial reduction of iron (Schwertmann and Taylor, 1989; Lovley, 1995). At one time it was thought that both oxidation and reduction reactions in groundwater were entirely abiotic chemical processes, or, if bacteria were involved, that it was only through their ability to alter the redox and indirectly influence the environment (Halvorson and Starkey, 1927). However, specific enzymatic bacterial reduction reactions have now been clearly demonstrated (Lovley, 1991; Bridge and Johnson, 1998), whereas purely abiotic extraction of iron from rock minerals into surface and subsurface groundwater has not been found to be significant because there appears to be insufficient energy available for its dissolution. Thus, without the "catalysis" of microbial intervention, the rate of iron accumulation in major deposits would be extremely slow if not geologically improbable.

LABORATORY EXPERIMENTS

At the University of Manitoba, together with B.L. Sherriff, I investigated a biofilm-forming consortium that was first discovered growing where mining water from an iron-poor source, typical of the Canadian Shield, flowed over a fresh rock face in the Underground Research Laboratory (URL) excavated by Atomic Energy of Canada Ltd. in the Lac du Bonnet granitic batholith in Manitoba, Canada (Brown et al., 1989). The biofilm formed following an oil spill on the 420 m level, which provided just sufficient carbon for cell mass, but not for cell energy. The biofilm was ~10 mm thick and covered over a square meter in extent (Brown et al., 1994). Iron minerals, both in the oxidized and reduced state, were deposited in close juxtaposition within the biofilm; siderite (identified by Mössbauer spectroscopy) was precipitated against the anaerobic rock face especially around water runnels, and ferrihydrite formed on much of the outer, aerobic surface, while gypsum was precipitated on the outside of the internal water runnels. This precipitation led us to suggest that iron-rich minerals from the granite were being used in energy-producing microbial redox reactions (Brown et al., 1994, 1999). In addition, because both siderite and ferrihydrite were precipitated in such close proximity, it was also suggested that different biofilm microbes within a consortium were able to generate specific microenvironments with a very different redox. Significantly, the overall reaction when using the WATEQ4F modeling program (Ball and Nordstrom, 1991) showed, from chemical considerations alone, that siderite was undersaturated and therefore should not have been formed, suggesting that microbial mediation within the biofilm was necessary for its precipitation (Brown et al., 1994).

This unexpected juxtaposed formation of different iron minerals in the underground led us to investigate the apparent self-organization of aerobic and anaerobic environments produced by the biofilm microbes. A series of laboratory experiments was carried out, utilizing a consortium enriched from the surface water used for mining at the URL, to elucidate the various microbial aspects outlined in Figure 1. A more detailed flow chart of the dissolution of iron from primary sources, through various processes before it is deposited as secondary minerals in sediments, is shown in Figure 2. The boxes in the latter chart indicate the processes that occur during the transport of iron, and the bold type the microbial reactions that were investigated in the laboratory and that are discussed in detail in the following sections.

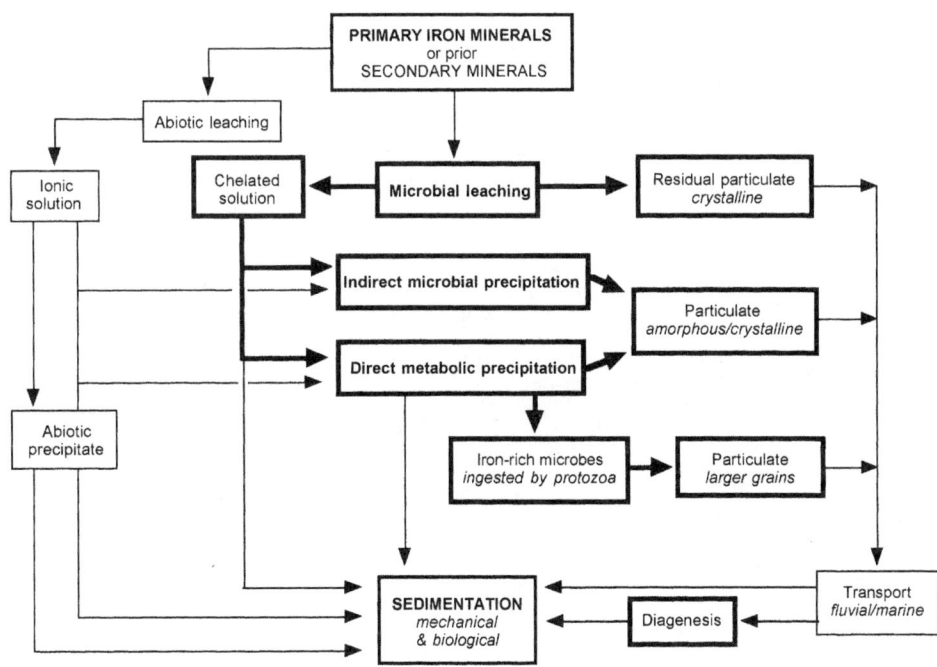

Figure 2. Flow sheet of the various reactions involving iron from the primary mineral to its sedimentation. Bold type—microbially mediated reactions investigated in the laboratory. Heavy boxes—processes investigated.

URL Biofilm

The consortium contains a variety of bacteria with differing morphologies as well as a protozoan (eukaryote), which would not have been present in the Archean (Brown et al., 1998a). The majority are Gram-negative bacteria (that would include *Geobacter metallireducens*), and only a few Gram-positive bacteria are present. Phospholipid fatty acid analysis suggests the presence of *Shewanella* sp., as well as *Desulfovibrio* sp. There are also indications of the presence of *Pseudomonas* and *Actinomycetes*. Transmission electron microscopy (TEM) shows the distinctive morphology of Gram-negative bacteria and confirms the presence of a curved flagellated rod similar to *Desulfovibrio* sp. (Brown et al., 1998b).

Not only does the consortium appear to be very stable, as it has been maintained in the laboratory for nearly 10 yr, but this type of consortium also appears to be widespread. Three seemingly similar consortia were enriched from the URL area. A sample taken from a stream in the English Weald, a historic siderite ironstone mining area, was also found to contain a consortium with comparable reactions (personal observation).

Several methods were tried for growing biofilms by running the culture medium over rock surfaces, but it was difficult both to maintain sterility and to ensure the biofilm remained moist. It tended to dry out, cracking partially into thin gelatinous platelets that morphologically resemble a texture noted in some BIFs. All the experiments were therefore conducted in flasks standing on the laboratory bench, occasionally swirled to aid mixing. Only when a solid substrate, such as mineral chips, was used were the flasks put on an orbital rocker to ensure that all parts of the culture were in contact with the solid. In this latter case the precipitate enmeshed in EPS formed small discrete flocs due to the motion; this becomes important when discussing ovoid granule textures in BIFs.

A bioreactor was maintained to produce inocula for the experiments. The reactor contained granite chips as a source of trace nutrients, in a medium of 5.0 g ferric ammonium citrate, 0.5 g K_2HPO_4, 0.2 g $MgSO_4 \cdot 7H_2O$, and 0.01 g $CaCl_2 \cdot 2H_2O$ per liter at an initial pH of 7.0 (Brown and Hamon, 1994). In laboratory cultures the ferric ammonium citrate used in the bioreactor was replaced with ammonium citrate and ferrous chloride to allow variation in the relative concentrations of iron and carbon. A 10% inoculum was used throughout, and all experiments were carried out at least in triplicate at room temperature (23 °C) with sterile (not inoculated) controls.

Dissolved iron in the URL groundwater has a concentration of less than 1.0 mg L^{-1} (Gascoyne et al., 1989), but in the surrounding granite the iron-bearing minerals biotite, magnetite, ilmenite, and hematite make up 2% of the rock mass, and therefore would seem to be the most probable source of the iron. Silica was also found in the biofilm, but was not investigated because silica cannot be directly metabolized by bacteria; presumably it was adsorbed indirectly onto the EPS slime.

The microenvironments produced by the microbes are quite specific and extremely local. Within three days of inoculation with the consortium, the pH of the medium in stationary laboratory flasks increased from 7.0 to 9.0, and the Eh was reduced to around −300 mV, even though the bungs in these flasks were permeable to air. Thus, although the flasks were in the normal present-day atmosphere, microbial metabolism was sufficient not only to use up all the oxygen in the flask, but also to maintain an anaerobic environment. However, when the iron was present in solid mineral form and the flasks were rocked, the metabolism was slower and the interchange with the atmosphere was much greater, so that an anaerobic environment was not achieved.

Energy

The main source of bacterial energy was found to be dissimulatory iron reduction. This occurs when the iron is not incorporated as a nutrient into the cell, but instead is utilized as an electron acceptor in a redox reaction to generate energy. The iron is reduced from the ferric to the ferrous state with the precipitation of siderite, while at the same time the major electron donor, citrate, is ultimately oxidized to carbon dioxide (Brown et al., 1999). Both citrate and glucose can be fermented, producing an Eh of around −150 mV, and under the experimental conditions citrate was broken down more rapidly than glucose (Table 1). When iron was added to the glucose it initially inhibited the fermentation, but after 40 days the amount of glucose metabolized was similar whether iron was present or not. However, when both citrate and iron were present in the medium, there was a rapid enzymatic reduction of the iron and utilization of citrate, virtually all being consumed within ten days; notably this did not occur when glucose was the carbon source. There was a considerable precipitation of siderite at an Eh below −300 mV, confirming that this lower redox is necessary for its formation (Garrels and Christ, 1990), and also that, here, citrate is the source of the carbonate anion rather than carbon dioxide (although it may have been originally synthesized from carbon dioxide through the tricarboxylic acid cycle). Using citrate as the carbon source in laboratory cultures proved doubly efficient as it provided the electron

TABLE 1. INCUBATION EXPERIMENTS IN TRIPLICATE OVER 40 DAYS

Days	Citrate + iron (% citrate)	Citrate (% citrate)	Glucose + iron (% glucose)	Glucose (% glucose)
0	100	100	100	100
5	3	67	99	80
10	2	66	98	80
15	1	44	92	60
20	–	35	33	54
25	–	29	30	44
30	–	12	19	33
35	–	9	18	29
40	–	4	15	22

Note: Citrate and glucose media are used both with and without iron. Maintained at 23 °C on the bench, with 50 mL of medium in 250 mL flasks and with 10% inoculum. Shown as percentage remaining after a number of days incubation.

donor and also chelated the iron, which by itself is poorly soluble under these pH/Eh conditions.

Although the concentration of sulfate in the groundwater at the URL was considerable (48 mg L^{-1}) no iron sulfides were detected, indicating that the reduction of iron was thermodynamically preferred to that of sulfate.

Dissolution

Direct contact between cells and the mineral involved was once thought to be essential for microbial reduction of iron (Lovley, 1995), but it has now been shown that iron reducing microbes such as *Shewanella* sp. can overcome this problem by releasing soluble quinones into the water (Childers et al., 2002). These act as electron shuttles by transferring electrons from the cell surface to insoluble metal oxides that can be located at a significant distance from the cell. The first organism, *G. metallireducens*, reported by Caccavo et al. (1994) to be capable of completely oxidizing organic compounds to carbon dioxide by utilizing oxidized iron or manganese as electron acceptors uses two methods to extract iron. When grown on relatively insoluble oxides this organism specifically expresses flagella for mobility to reach the mineral and pili for attachment to it, whereas when grown on soluble iron citrate these appendages are not produced, suggesting that the organism is able to detect whether soluble electron acceptors are potentially available (Childers et al., 2002). This confirms that in some cases direct contact with insoluble minerals may indeed be necessary.

Mineral dissolution preferentially occurs at surface dislocation sites. That biogeochemical reactions also occur at these sites can be seen from scanning electron micrograph images, where more bacteria cluster on the mineral surface at fracture edges than on the crystal and cleavage faces (Brown et al., 1997). This implies that the former sites are preferred for the extraction of nutrients, presumably because of the presence of a greater concentration of unsatisfied bonds, or possibly, in the case of a sheet silicate such as biotite, more ready access to the layers of iron atoms in the lattice.

Because the consortium formed in waters draining a granitic batholith, chips of the iron-containing minerals biotite and magnetite were used as the source of iron in the medium for the dissolution experiments. Although the amount of iron in solution was extremely low (~1.0 mg L^{-1}), a week after adding the inoculum the rocked flasks showed a significant amount of orange ferric precipitate enmeshed within the biofilm slime. These results show that the consortium is able to grow on solid iron-containing mineral fragments, suggesting that organisms similar to *Shewanella* and *G. metallireducens* must be present.

In the leaching experiments, after eight weeks the average amount of iron precipitate in the medium from biotite was 54.7 mg L^{-1}, and that from magnetite 6.8 mg L^{-1} (Fig. 3). During this period, precipitation from the biotite dissolution was initially quite low while the consortium organisms became conditioned to extracting iron from a solid substrate. It then increased rap-

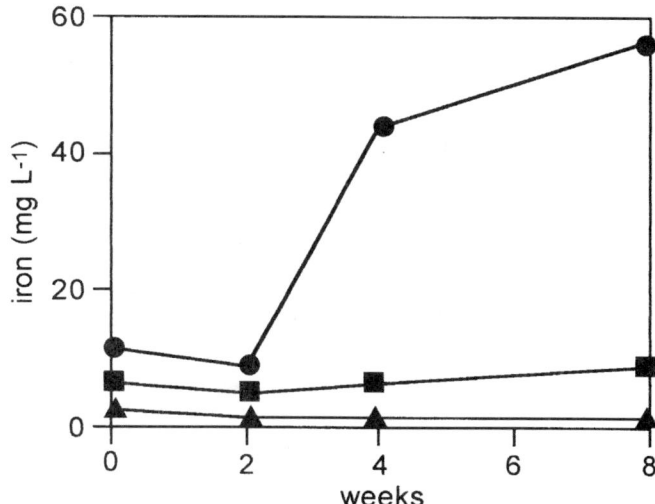

Figure 3. The amount of iron in accumulated precipitate resulting from microbial leaching over time of 5 g each of biotite (●), and magnetite (■) in 50 mL of medium, from an average of 3 flasks for each measurement; iron in solution at each measurement (▲).

idly, but gradually tailed off as the available iron became distanced from the broken cleavage edges. In contrast, the rate of precipitation from the magnetite substrate, although quite low, remained steady while the amount of precipitate increased slowly over time. A very approximate deposition rate of 20 m of hematite per 1 m.y., calculated from the more reliable magnetite experiment, compares well with the lower rate (20–225 m per 1 m.y.) suggested from some sequences at Hamersley, Australia (Trendall, 2000).

So far as is known, iron must be in solution, if only fleetingly, for it to be metabolized by bacteria (Nealson and Little, 1997). Therefore, there has to be a rapid flux of iron from the mineral, through a dilute solute phase, to the microbial sites where metabolism takes place; the iron is then rapidly precipitated and removed from solution. Presumably the small pool of dissolved iron, which remains at a constant low concentration, is sufficient for active metabolism to occur so long as this pool is maintained by the chemical equilibrium. As microbes precipitate the iron, more primary iron must be drawn into solution to preserve this equilibrium, suggesting that the microbial process drives the dissolution of iron from its host minerals. Even when the pH and Eh were adjusted to the same conditions as the cultures and killed organisms were present, no precipitation occurred until viable, metabolizing microbes from the consortium were added (Brown et al., 1999).

Metabolism

In the natural oxygenated environment little iron is found in solution (Beard et al., 2003), except in the extremely acidic and aerobic environments that are produced at massive sulfide exposures or in man-made acid mine drainage. Although simple ionizable ferrous salts are moderately soluble, ferrous iron is much less soluble when bound in complex minerals or organic compounds: a phenomenon exhibited in the reducing groundwaters of the URL. Also, in the laboratory experiments, unless chelated, concentrations of dissolved iron greater than ~1.0 mg L^{-1} were rare. In the natural environment microbes are always present, and if iron is available it will be used as an energy source and be precipitated. Thus it is unlikely that dissolved iron in Archean waters was greater than is found today under similar anaerobic conditions.

Iron is thought to have been one of the earliest, as well as one of the most important, external electron acceptors to have been utilized by microbes (Vargas et al., 1998). Depending upon the local redox conditions, bacteria can obtain energy from either the oxidation or the reduction of iron, although this change of state is less efficient for iron than for manganese (see chemiosmotic model). Iron is almost universally and abundantly present in rocks and so it is the most likely element to have been utilized. Whereas in today's oxic world iron reduction is predominant because much of the iron in soils and minerals is in the oxidized state, in the anaerobic environment of the Archean readily available iron would have been in the ferrous state, and the main reaction would have been its oxidation. This oxidation does not require the presence of free oxygen but can be driven instead by light; anoxygenic phototrophs are bacteria that only have photosystem I and can use ferrous iron as an electron donor. This anaerobic photosynthesis implies that oxygen-independent biological iron oxidation would have been possible before the evolution of two-step oxygenic photosynthesis (Widdel et al., 1993; Ehrenreich and Widdel, 1994), thus making it possible for oxidized iron to be laid down under an anoxic environment. Tice and Lowe (2004) concluded from the shallow-water setting (euphotic zone) and carbon isotopic composition of the Buck Reef Chert, South Africa, that photosynthetic, probably anoxygenic microbes were active in the 3.4 Ga ocean. However, because the modern URL consortium is unable to photosynthesize, oxygen has to be available before it can form ferric iron.

To study the metabolism of the consortium we varied the relative concentrations of both iron and carbon in the medium (Brown et al., 1999). Ammonium citrate was used to provide both carbon and nitrogen for cell growth, and ferrous chloride (mostly oxidized during sterilization) supplied iron for cell energy. The relative efficiency of the incubations is shown, first by the amount of protein produced as a measure of growth and second by the percentage of ferrous iron precipitated. As well, we used the Munsell standard card to define the color of the precipitate because this indicates the balance between the iron valency states within the precipitate (Table 2A). When the ratio of iron to citrate (as molar concentrations) was greater than 1:5, the amount of ferrous iron produced was over 50% of the total iron, and consisted of siderite, ferrous hydroxide, and vivianite (where the phosphate came from the medium) and the color of the precipitate was an olive green. In these cases the reduction of the iron took place in solution while the iron was still chelated (nuclear magnetic resonance data from B.L. Sherriff, 1997, personal commun.). Alternatively, if the ratio was equal to or less than 1:5, this smaller amount of citrate was rapidly consumed by the consortium, removing the chelation. In this latter case the iron was precipitated in the oxic state as very fine grained orange ferrihydrite, which rapidly darkened to brown as it was reduced, but here the average percentage of ferrous iron was lower, and the protein tended to be less. The iron remained reduced so long as there were sufficient carbon metabolites present for the anaerobic environment to be maintained.

Thus the reduction of ferric iron in the laboratory can occur through two different mechanisms, depending on the initial ratio of iron to carbon as citrate (Table 2B). Two populations are found when the percentage of ferrous iron is plotted against protein produced: one, where the precipitate is greenish, ferrous iron averages 56% and protein 16 mg mL^{-1}, and the other, where the precipitate is initially orange, ferrous iron averages 33% and protein 11 mg mL^{-1}. Sterile controls showed no sign of any precipitation. The ability of the consortium to reduce iron is apparently constitutional (the enzyme system is always present and does not need to be induced by the presence of substrate). After 24 serial transfers without iron in the medium, the rate of reduction when iron was reintroduced was the same as that of the original incubations.

This set of experiments shows that it is the relative amount of iron to carbon that determines the state of the iron precipitate. When all the other parameters are kept the same, the larger relative concentration of carbon in the medium produces greater

TABLE 2A. DIFFERENT RATIOS OF IRON (AS FeCl$_2$) TO CARBON (AS CITRATE) IN THE MEDIA

	15 mmol carbon	20 mmol carbon	25 mmol carbon
2.5 mmol iron			
Fe:C ratio	1:6	1:8	1:10
Fe(II) %	50	60	57
Protein (mg mL^{-1})	11	18	20
Munsell standard	5Y 6/4	5Y 5/2	5Y 5/4
5.0 mmol iron			
Fe:C ratio	1:3	1:4	1:5
Fe(II) %	25	34	40
Protein (mg mL^{-1})	8	11	17
Munsell standard	5YR 5/6	5YR 4/4	10YR 4/2
7.5 mmol iron			
Fe:C ratio	1:2	1:2.7	1:3.3
Fe(II) %	26	38	33
Protein (mg mL^{-1})	6	14	12
Munsell standard	5YR 6/6	5YR 3/4	5YR 2/2

Note: This table shows the average results, after eight days incubation in five separate experiments, of the ferrous iron and protein produced, as well as the Munsell standard color of the precipitates.

TABLE 2B. AVERAGES OF THE RESULTS IN TABLE 2A

	Fe(II) %	Protein (mg mL^{-1})	Munsell standard
Fe:C ratio			
1:6 to 1:10	56	16	5Y
1:2 to 1:5	33	11	5 to 10YR

Note: This table shows there are two separate populations.

metabolic activity, hence more rapid iron reduction, so that the iron is directly and rapidly precipitated as siderite. With relatively less carbon, the citrate is first metabolized, thereby removing the chelation, and the iron is precipitated initially as ferrihydrite. This is later reduced, possibly to ferrous hydroxide, which is much less stable than siderite and so is easily reoxidized by ambient oxygen in the air. Thus in this natural consortium only a small difference in the ratio of available nutrients results in completely different iron minerals being precipitated, either a carbonate or an oxide.

The ability of the consortium to define its own redox depending upon the nutrients available has a distinct bearing on the type of iron supplied to sediments. Furthermore, the state of this iron could indicate the changeable environment to which such biofilm consortia are commonly subjected and contribute to the differing laminae found in BIFs. Whether iron is laid down in iron formations as carbonate or as oxide minerals appears to depend upon the local microbial environment at that particular moment rather than on the influence of the general prevailing atmosphere.

Precipitation

The consortium cultures regularly produced both siderite and ferrihydrite, which altered to hematite through the removal of water, and although magnetite was not identified, its formation under similar laboratory conditions has been reported (Lovley et al., 1987). Thus three major iron minerals can all be precipitated through microbial activity, and both siderite and ferrihydrite have been seen to do so in the field within the URL biofilm. Furthermore, even though the consortium did not synthesize magnetite, when utilized as a substrate magnetite was rapidly oxidized to hematite. This clearly demonstrates the consortium's ability, when oxygen is available, to extract energy from iron through either reduction or oxidation reactions. Mössbauer spectroscopy showed that 11% of the magnetite was altered to hematite within three weeks of incubation, but interestingly, maghemite, the usual intermediate in this transformation, was not identified (Brown et al., 1997). Because the reaction is rapid, FeO is probably released directly from the lattice surface and oxidized to hematite, and the lack of maghemite (γFe_2O_3) suggests that once released it is immediately altered to the αFe_2O_3 form. Scanning electron micrographs of the surface of the oxidized magnetite showed hexagonal plates, suggesting the formation of crystals (Brown et al., 1997) similar to those that have been reported in BIFs (LaBerge et al., 1987).

Biomineralization occurs as the consortium microbes extract energy from the iron, precipitating it as siderite, ferrihydrite, and/or magnetite. However, precipitation directly onto the cell wall clogs the active metabolic sites and causes death. The cells then lyse and leave only a ring (i.e., sphere) of iron-rich minerals around an empty cell (Fig. 4), but many cells are able to protect themselves from such a death by excreting a paracrystalline proteinaceous envelope, or S-layer. This layer surrounds the organism and prevents the iron from coming into direct contact with the cell wall, thus allowing normal metabolic activity to continue

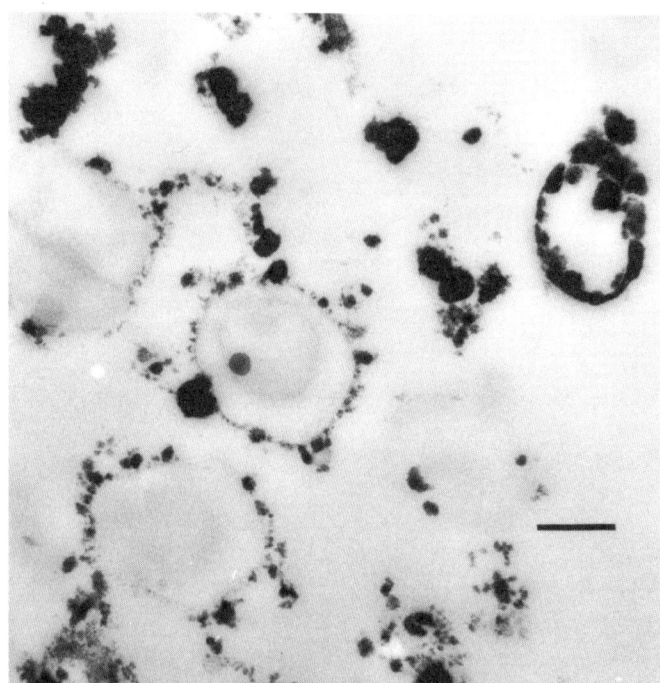

Figure 4. Transmission electron micrograph of an unstained thin section of biofilm bacteria showing iron directly precipitated by metabolic activity. One cell has lysed where the iron precipitate has become too dense for metabolism to occur. Other cells are surrounded by a paracrystalline proteinaceous S-layer on which iron has been metabolically precipitated instead of directly on the cell wall. Scale bar, 0.25 µm.

(Schultze-Lam et al., 1992). Iron is also precipitated indirectly in the EPS and on cell walls owing to their negative charge (Fig. 5).

The spatial distributions of these two types of precipitate have distinct fractal dimensions. Using 12 transmission electron photomicrographs, we measured the density of the iron precipitate for the biofilm matrix and for the cell wall (Fig. 6). Although the clusters do overlap, the means are significantly different; the fractal dimension of the iron mineral spatial distribution is lower for the cell wall environs than that for the more general EPS matrix. This supports the likelihood that they are formed by two different mechanisms, that is, either by direct metabolic activity or indirectly through the microbially induced environment produced within the biofilm itself.

In these experiments the precipitation of iron-bearing minerals, by whatever means, results in its entrapment within the biofilm. In some of the cultures aggressive protozoa actively and preferentially ingested the iron-coated bacteria. When these bacteria had all been consumed, the protozoa encysted, thus efficiently concentrating and stabilizing the iron in even larger particles within the biofilm. Furthermore, because protozoa would not have been extant in the Archean, cultures that did not contain them formed iron microcrystallites of around 8 µm in diameter that were also tightly enmeshed within the EPS (Brown et al.,

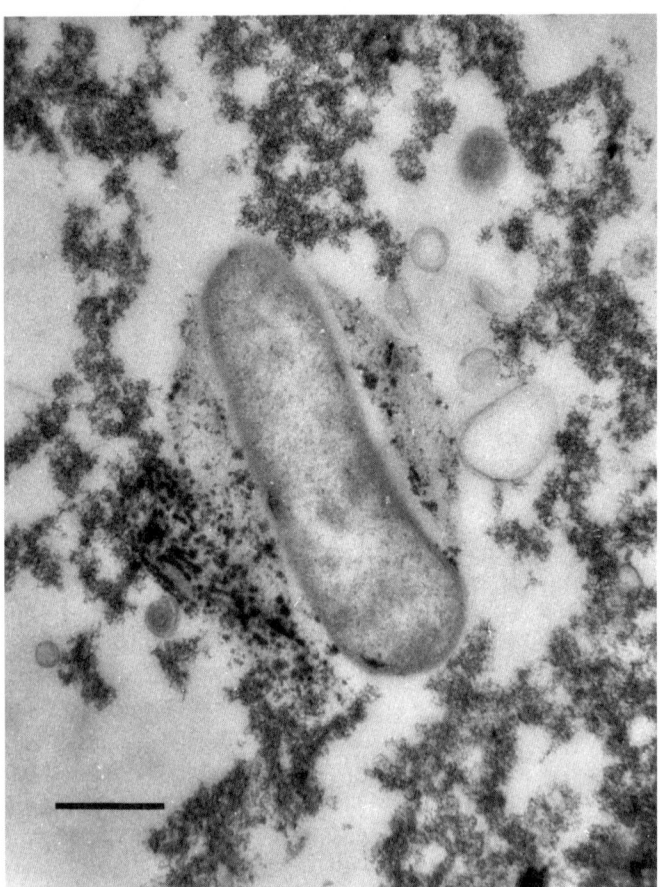

Figure 5. Transmission electron micrograph of an unstained thin section of the biofilm showing the iron indirectly precipitated throughout the EPS matrix due to its negative charge. The S-layer, with metabolically precipitated iron around the central microbe, has been distorted during preparation of the section. Scale bar, 0.25 μm.

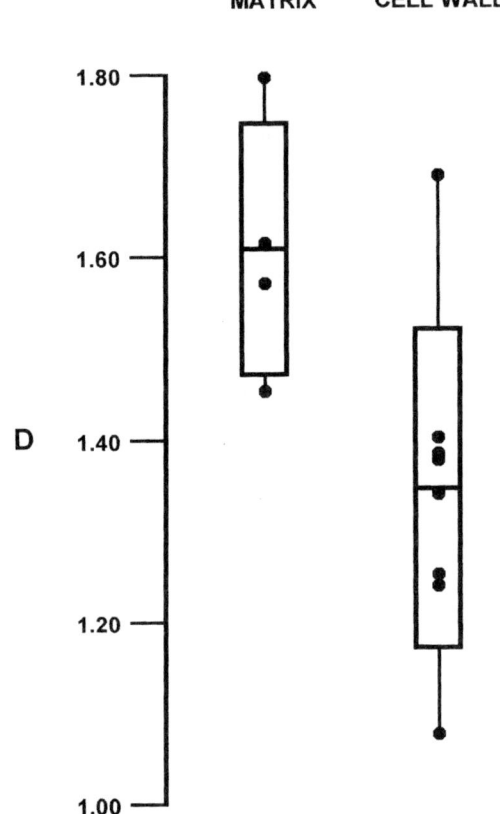

Figure 6. Measurements of fractal dimension, D, from 12 TEM prints showing the different clusters formed by iron precipitates, eight data points for mineralization at the cell wall and four for within the biofilm matrix. The boxes are one standard deviation about the mean.

1998a). Thus, regardless of the probable differences between consortia, the iron would be securely entrapped under Archean conditions, ensuring that it could be transported as particles and deposited during sedimentation. In all the cases tested, almost all the iron involved in the microbial mediation was precipitated, with only a minimal amount remaining in solution.

Finally, it should be noted that because of the influence of pre-glacial marine waters, the URL groundwater composition is quite saline (245 mg L^{-1} of chloride; Gascoyne et al., 1989) indicating that consortia flourishing in fractures underground could also have been active in Archean seas.

Diagenesis

The post-deposition low-temperature diagenesis of the microbially precipitated iron minerals in different oxidation states was also investigated. Experiments at 80 °C for 12 weeks under different headspace gases (air, nitrogen, and hydrogen) showed the ferrous precipitate to be much more stable than the ferric. The ferrous compounds remained reduced except when oxygen was present in the headspace, whereas the ferric compounds, under the same experimental conditions, were not reduced but lost water and were altered to hematite under all headspace gases (Brown et al. 1998b). This suggests that ferrous iron microbially precipitated in sediments as siderite would be more likely to remain in the reduced state, whereas iron deposited as ferrihydrite would tend to form hematite. Notably, the control experiments using chemically precipitated ferric iron formed goethite as well as hematite. Whether ferrihydrite or goethite is formed in the natural environment is under considerable discussion (van der Zee et al., 2003), but this was the only time that goethite was identified in any of the experiments.

Summary Interpretation of Laboratory Experiments

The type of microbial consortium isolated from the URL biofilm is nowadays widespread under both anaerobic and aerobic conditions in the natural environment, indicating that microbial mediation is likely to have an important influence on the geologic deposition of iron, and has probably changed little since the Archean. The various processes performed by the consortium were entirely dependent on microbial activity, given that even under simulated microbial conditions no activity could be detected without the presence of viable organisms. The consortium can utilize iron either from general solution or by close contact with solid minerals, and the very low concentration of iron commonly reported in groundwaters demonstrates its utilization by such microbial activity.

In the natural environment, the microbial imperative is to acquire energy, which drives all the necessary steps for iron to be leached, mobilized, and reprecipitated. This precipitate then provides particulate minerals for transportation to their ultimate depositional sites. Under present global conditions, the consortium microbes are able to manipulate the biofilm microenvironment independently of the surrounding atmosphere. In the experiments the dominant and constitutive energy-producing redox reaction is dissimilatory iron reduction, which has persisted possibly since it evolved under the anoxic atmosphere of the Archean. Where oxygen is available, both ferric oxides, mainly as ferrihydrite, and ferrous carbonate can be precipitated in close juxtaposition. Under suitable conditions this same consortium can also oxidize magnetite to hematite. In the laboratory the ratio of carbon (for cell growth) to iron (for energy) determines the metabolism and thus which mineral is precipitated. Siderite is preferentially formed when the concentration of carbon is greater, and because it is also resistant to oxidation under low temperature diagenesis, it is more likely to be retained in iron sediments. Ferrihydrite is formed when carbon is limited and the environment is less reducing, and although it can be microbially reduced, possibly to ferrous hydroxide, it is more susceptible to oxidation than is siderite, and thus perhaps is the origin of the iron oxides found in BIFs. Ultimately, however, because all these metabolic reactions are inefficient, a large quantity of iron has to be metabolized to produce an adequate supply of energy for bacterial growth, resulting in vast accumulations of iron as a waste product.

IRON FORMATION STRUCTURES

If biofilms mobilized iron from a primary source to its sedimentation site, some structure or texture indicative of this microbial mediation ought to be visible in suitable formations. In fact many rock structures in iron formations either resemble microbially related features observed in the laboratory or can be attributed to biological activities. Some of these features that have been identified in examples of iron formation deposits, from the Archean to the present day, are listed in Table 3.

The observed textures that result from the processes discussed earlier are strong indicators of the depth of water in which the deposition occurs and of whether conditions are generally quiet or are influenced by current and wave action. They may be roughly divided into three: (1) very fine particles of iron minerals originating as precipitates either from within a biofilm or from a bloom (and in addition, silica particles, though these are not discussed here); (2) considerably larger granules of iron minerals and/or jaspilite formed from the breakup of biofilms and biomats or from repeated oolitic growth of biofilm and enmeshed precipitate; and (3) layering at various scales due to successive sedimentation of particles and granules of different composition, including the gradual accumulation of biofilms and biomats.

TABLE 3. COMPOSITION AND GEOMETRY OF GRANULAR MINERALS AND OTHER TEXTURES IN IRON FORMATION DEPOSITS

Description	Fig.	Name	Type	Age (Ga)
Microbanded magnetite with a little silica: relict microovoids of magnetite		Moose Mountain, Canada	Algoma	L. Archean 2.7–2.6
Siderite microovoids and spheroids and pyrite/pyrrhotite in thick siderite beds	8	Wawa, Canada	Algoma	L. Archean 2.7
Magnetite, hematite, and silica in microovoids accumulating on bedding planes in magnetite/jasper lithofacies	9	Timagami, Canada	Algoma	L. Archean 2.7–2.6
Magnetite microovoids in magnetite and siliceous magnetite beds		Anshan, China	?	L. Archean 2.6
Magnetite and hematite microovoids in magnetite, hematite and siliceous magnetite microbands in magnetite/jasper lithofacies		Krivoy-Rog, Ukraine	Lake Superior	E. Proterozoic 2.4–2.2
Hematite and magnetite particles in siliceous oolite and microovoids; hematite in jasper ovoids and larger siliceous granules; hematite micro-particles in jasper and fine-grained magnetite and hematite as rims on siliceous ovoids and oolites	7	Sokoman, Canada	Lake Superior	E. Proterozoic 2.2–1.8
Magnetite (?) microovoids in massive magnetite beds		Bayanobo, China	?	E. to M. Proterozoic 1.8–1
Hematite (?) ovoids and oolites in massive hematite beds and siliceous granules		Rapitan, Canada	Rapitan	L. Proterozoic 0.6
Magnetite (?) microovoids in magnetite micro- and mesobands		Austin Brook, Canada	?	Ordovician
Ovoids and spheroids of Fe and Mn oxides in microbands and beds		Woodstock, Canada	Algoma	Silurian
Microovoids of Fe and Mn oxides in mud		Red Sea		Recent

Note: The global formations range in age from the Archean to Recent. Data provided by G.A. Gross, Geological Survey of Canada (2004).

Fine Particles

The sizes and shapes of the fine particles of iron compounds and their aggregates that formed in the laboratory experiments are comparable to those forming naturally today in biofilms (field observation), and also to those in the least metamorphosed iron formations, which are, in many cases, reported as spheroids (Table 3). Examination of particles in photomicrographs from the consortium (Brown et al., 1998a) suggests that although the smallest particles on microbe walls may be less than 5 nm in average diameter, they are commonly between 5 and 50 nm (Fig. 4), and spheroidal particles within the matrix of the biofilm are 50–100 nm. Encrusted microbes leave iron aggregates 750–1000 nm in diameter, but these together with other precipitates within the biofilm comprise either irregular clumps of spheroids and shapeless particles, 350–450 nm in the longest dimension, or irregular networks of rods comprising three to five spheroids and averaging 250 nm in length by 40 nm in diameter. These particle sizes are comparable to those reported for ferrihydrite in springs (2–5 nm; Carlson and Schwertmann, 1981) or iron hydroxide in microbial mats (100 nm; Konhauser and Ferris, 1996).

Encysted protozoa, appropriate only to post–middle Proterozoic deposits, form much larger secondary particles 2.5–3.5 µm in diameter, whereas secondary (post-microbial action) hexagonal hematite crystals are up to 20 µm in diameter and aggregates of crystallites may be 8 µm in width. Particles from iron formations themselves are reportedly larger (LaBerge et al., 1987; Robbins et al., 1987), but these authors also report submicron hematite "dust" disseminated in jaspilite. They regard their larger grains of spheroidal or crystalline hematite (1–30 µm) as diagenetic-metamorphic recrystallization of the submicron hematite.

Larger Granules

In some deposits thicker (10–300 mm) layers of larger spheroidal to platy granules (0.5–10 mm in diameter) of hematite, siderite, and/or jaspilite interrupt zones of microlaminae. The granules are encountered in depressions on the bedding surface, and show evidence of current action such as imbrication and graded bedding. The granules, especially those of jaspilite, also exhibit repeated growth with outer layers enclosing one or more earlier-formed granules, as shown in specimens from Sokoman, Canada (Fig. 7). This is a common texture developed by biofilms in high energy environments that are intermittently quiescent. These granules resemble the small, spherical flocs of iron-mineral-filled biofilm formed during the orbitally rocked flask experiments described earlier.

All biomats, gelatinous biofilms, and derived flocs must dewater at some stage during sedimentation. Cracks, regarded as syneresis because the outer layers are unbroken, can be observed within jaspilite granules in examples from Sokoman. Fossil biomats are difficult to recognize. The irregular form (viewed from above) of more spheroidal (viewed in cross section) granules, together with the presence of dark, possibly organic material, in coarse siderite layers at Wawa, Canada (Fig. 8) suggests a series of dehydrated biomats that dried in the intertidal zone. Morphologically this strongly resembles the platelets (~0.5–10 mm) of partially dehydrated biofilm grown on granite plaques in the laboratory.

Although plausible abiotic origins have been suggested for the small spheroids and microlaminae, cyclical microbial precipitations, the growth of biofilms and biomats, and near sea-surface blooms seem to be the most likely processes to produce these textures, at least on the finer scale.

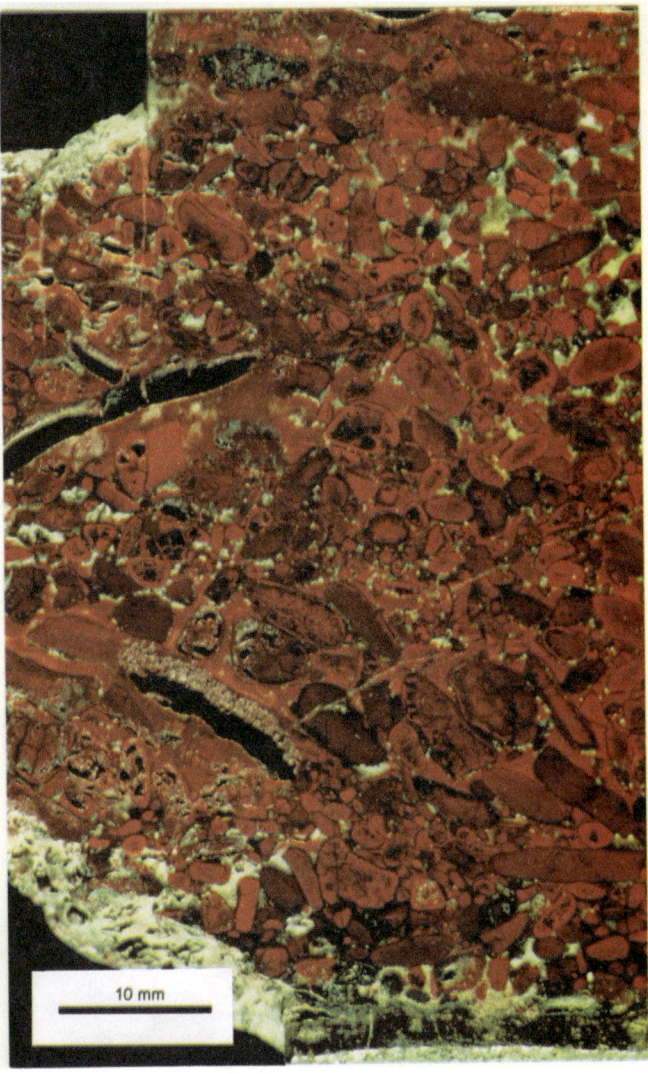

Figure 7. A polished, sectioned hand specimen exhibiting ovoids and irregular granules in a local jasper lithofacies from a Lake Superior type, Proterozoic (2.2–1.8 Ga) iron formation from Sokoman Lake, Labrador, Canada. Some granules show successive layers of biofilm overgrowth and probable internal cracking due to syneresis. Supplied by G.A. Gross, Geological Survey of Canada.

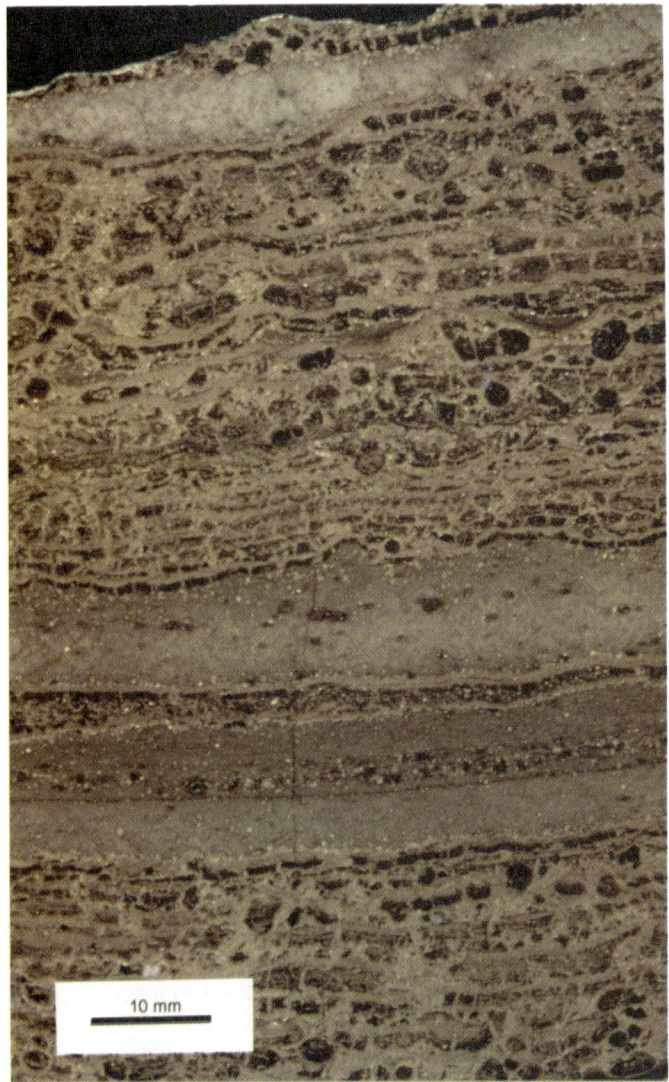

Figure 8. A polished, sectioned hand specimen exhibiting granules and broken layering in siderite lithofacies from an Algoma type, Archean (2.7 Ga) iron formation from Wawa, Ontario, Canada. These probably represent successive, partially dried, microbial mats containing primary microbially precipitated siderite, metamorphically recrystallized. Supplied by G.A. Gross, Geological Survey of Canada.

Figure 9. A polished, sectioned hand specimen exhibiting laminae of magnetite, hematite, and silica in magnetite-jasper lithofacies from an Algoma type, Archean (2.7–2.6 Ga) iron formation from the Timagami area, Ontario, Canada. Iron-rich laminae probably represent the accumulation in quiescent conditions of both direct and indirect fine microbial precipitation. Supplied by G.A. Gross, Geological Survey of Canada.

Layering

Multiple, sequential laminae and banding are common, though not ubiquitous, structures observed in iron formations. Although they range widely in thickness from 10^{-4} m to possibly 10 m, only the thinner layers are pertinent to this study. Quiet-water sedimentation of the finer particles leads to millimeter-scale microlaminae, but some current action is suggested by mild scour-and-fill structures and grains of earlier lithified bands in an example from Timagami, Canada (Fig. 9). I suggest that, before diagenesis and metamorphism, these laminae would comprise the finer particles described earlier.

Because the experiments suggest that a minor change in substrate supply (the iron to carbon ratio) readily changes the state of the iron in microbial precipitates, microlayering would seem to be likely in the natural state where environmental conditions can change rapidly. Seasonality may also cause variations in the supply of iron and sediment or, alternatively, variations could be associated with periods of volcanic activity. Konhauser et al. (2002) have suggested that annual marine blooms of iron-utilizing bacteria are richer when the supply of iron is greater.

A specimen from the Lake Superior Biwabik iron formation, United States (1.9 Ga; Fig. 10), shows small stromatolites coeval with the sediment. A group of these, strongly microlaminated with jaspilite layers, are separated from each other by much less structured, approximately contemporaneous, zones of siliceous

Figure 10. Polished, sectioned hand specimen of stromatolites from the Lake Superior Biwabik iron formation, Minnesota, United States. The complete right-hand stromatolite is 55 mm tall. The specimen is oriented stratigraphically up toward the top of the page. Zones of 10–15 pairs that are more iron-rich than adjacent zones stretch across several stromatolites in the complete hand specimen, probably indicating periods of general increase in the supply of iron.

sediment enclosing magnetite grains. If individual pairs of the iron-rich and iron-poor jaspilite laminae are regarded as annual cycles, then they imply an accumulation rate of ~1000 m per 1 m.y.; this is barely an order of magnitude faster than the average rate (20–225 m per 1 m.y.) suggested over much larger spatial and time scales for BIFs in the Hamersley Group (Trendall, 2000; Isley, 1995). More importantly, there are several zones of especially iron-rich, adjacent microlaminae that stretch at the same stratigraphic level across all the eight stromatolites in this specimen. These groups of 10–15 lamina pairs suggest a possible decadal climatic control on the iron source in the sedimentary basin. However, in this example, although magnetite content in the intervening sediment is weakly correlated with the groups of iron-rich stromatolitic microlaminae, it is not correlated with the individual microlaminae.

CONCLUSIONS

It has been difficult to envisage how iron, although ubiquitous in rocks but to a large extent insoluble in the natural environment, could have been mobilized and transported from primary sources, both terrestrial and submarine, and then redeposited in iron formations. Although abiologic reactions are possible, thermodynamic considerations and deposition rates make it unlikely that they could have been wholly responsible for the amount of iron amassed. The laboratory and field work reported here suggests that the direct and indirect effects of microbial metabolism may have played a significant role in producing the massive accumulations of iron, in geologically credible amounts and time scales necessary for the deposition of stratified sediments, particularly those of early BIFs. Primitive microbial life is thought to have become established very early in the history of Earth, possibly soon after 4.0 Ga. In the anaerobic environment extant at that time, sources of energy for microbial metabolism were limited, since oxygenic photosynthesis had yet to evolve. However, iron is widespread, and redox reactions between the ferrous and ferric states are able to provide a limited but sufficient amount of energy for cell growth. Calculations confirm that many billions of microbes could, by utilizing iron as their energy source, discharge into the environment the many tons of waste iron that is recognized today in formidable iron deposits, such as those of BIFs.

ACKNOWLEDGMENTS

I would like to thank my colleagues, especially G.A. Gross, for the many discussions and their support in this research; B.L. Sherriff, who was instrumental in obtaining my funding and produced the NMR analyses; and the Guelph Regional Scanning Transmission Electron Microscope Facility of the National Sciences and Engineering Research Council of Canada (NSERC) for supplying the TEM figures. I would also like to thank Anton Brown, S. Kesler, H. Ohmoto, K. Towe, and an anonymous reviewer, who have done much work to improve this manuscript.

REFERENCES SITED

Amy, P.S., and Haldeman, D.L., eds., 1997, The microbiology of the terrestrial deep subsurface: New York, CRC Lewis Publishers, 344 p.

Ball, J.W., and Nordstrom, K.D., 1991, User's manual for calculating speciation of major, trace, and redox elements in natural waters: U.S. Geological Survey Open-File Report 91-183, 189 p.

Baross, J.A., and Deming, J.W., 1983, Growth of 'black smoker' bacteria at temperatures of at least 250°C: Nature, v. 303, p. 423–426, doi: 10.1038/303423a0.

Baross, J.A., and Hofmann, S.E., 1985, Submarine hydrothermal vents and associated gradient environments as sites for the origin and evolution of life: Dordrecht, Kluwer Academic Publishers, Origins of Life and Evolution of the Biosphere, v. 15, p. 327–345, doi: 10.1007/BF01808177.

Bastin, E.S., 1926, The presence of sulphate-reducing bacteria in oil-field waters: Science, v. 63, p. 21–24.

Battin, T.J., Kaplan, L.A., Newbold, J.D., and Hansen, C.M.E., 2003, Contributions of microbial biofilms to ecosystem processes in stream mesocosms: Nature, v. 426, p. 439–442, doi: 10.1038/nature02152.

Baur, M.E., Hayes, J.M., Studley, S.A., and Walter, M.R., 1985, Millimeter-scale variation of stable isotope abundances in carbonates from banded iron-formations in the Hamersley group of Western Australia: Economic Geology and the Bulletin of the Society of Economic Geologists, v. 80, p. 270–282.

Beard, B.L., Johnson, C.K., Von Damm, K.L., and Poulson, R.L., 2003, Iron isotope constraints on Fe cycling and mass balance in oxygenated Earth oceans: Geology, v. 31, p. 629–632, doi: 10.1130/0091-7613(2003)031<0629:IICOFC>2.0.CO;2.

Bekker, A., Holland, H.D., Wang, P.-L., Rumble, D., III, Stein, H.J., Hannah, J.L., Coetzee, L.L., and Beukes, N.J., 2004, Dating the rise of atmospheric oxygen: Nature, v. 427, p. 117–120, doi: 10.1038/nature02260.

Blackburn, C.E., Johns, G.W., Ayer, J., and Davis, D.W., 1991, Wabigoon Subprovince, in Thurston, P.C., Williams, H.R., Sutcliffe, R.H., and Stott, G.M., eds., Geology of Ontario: Ontario Geological Survey Special Volume 4, p. 303–381.

Blank, C.E., 2004, Evolutionary timing of the origins of mesophilic sulphate reduction and oxygenic photosynthesis: A phylogenomic dating approach: Geobiology, v. 2, p. 1–20, doi: 10.1111/j.1472-4677.2004.00020.x.

Bonny, S., and Jones, B., 2003, Microbes and mineral precipitation, Miette Hot Springs, Jasper National Park, Alberta, Canada: Canadian Journal of Earth Sciences, v. 40, p. 1483–1500, doi: 10.1139/e03-060.

Braga, J.C., and Martin, J.M., 2000, Subaqueous siliciclastic stromatolites: A case history from late Miocene beach deposits in the Sorbas Basin of SE Spain, in Riding, R.E., and Awramik, S.M., eds., Microbial sediments: Berlin, Springer-Verlag, p. 226–232.

Brantley, S.L., Liemann, L., and Bullen, T.D., 2001, Fractionation of Fe isotopes by soil microbes and organic acids: Geology, v. 29, p. 535–538, doi: 10.1130/0091-7613(2001)029<0535:FOFIBS>2.0.CO;2.

Bridge, T.A.M., and Johnson, D.B., 1998, Reduction of soluble iron and reductive dissolution of ferric iron-containing minerals by moderately thermophilic iron-oxidizing bacteria: Applied and Environmental Microbiology, v. 64, p. 2181–2186.

Brown, A., Soonawala, N.A., Everitt, R.A., and Kamineni, D.C., 1989, Geology and geophysics of the Underground Research Laboratory site, Lac du Bonnet Batholith, Manitoba: Canadian Journal of Earth Sciences, v. 26, p. 404–425.

Brown, D.A., 1995, Carbon cycling in peat and the implications for the rehabilitation of bogs, in Cox, M., Straker, V., and Taylor, D., eds., Wetlands: Archaeology and nature conservation: London, HMSO, p. 99–107.

Brown, D.A., and Hamon, C., 1994, Initial investigation of groundwater microbiology at AECL's Underground Research Laboratory: Pinawa, Atomic Energy of Canada Technical Record, TR-608, 50 p.

Brown, D.A., and Sherriff, B.L., 1999, Evaluation of the effect of microbial subsurface ecosystems on spent nuclear fuel repositories: Environmental Technology, v. 20, p. 469–477.

Brown, D.A., Kamineni, D.C., Sawicki, J.A., and Beveridge, T.J., 1994, Minerals associated with biofilms occurring on exposed rock in a granitic Underground Research Laboratory: Applied and Environmental Microbiology, v. 60, p. 3182–3191.

Brown, D.A., Gross, G.A., and Sawicki, J.A., 1995, A review of the microbial geochemistry of banded iron-formations: Canadian Mineralogist, v. 33, p. 1321–1333.

Brown, D.A., Sherriff, B.L., and Sawicki, J.A., 1997, Microbial transformation of magnetite to hematite: Geochimica et Cosmochimica Acta, v. 61, p. 3341–3348, doi: 10.1016/S0016-7037(97)00159-2.

Brown, D.A., Beveridge, T.J., Keevil, C.W., and Sherriff, B.L., 1998a, Evaluation of microscopic techniques to observe iron precipitation in a natural microbial biofilm: FEMS Microbiology Ecology, v. 26, p. 297–310, doi: 10.1016/S0168-6496(98)00044-0.

Brown, D.A., Sawicki, J.A., and Sherriff, B.L., 1998b, Alteration of microbially precipitated iron oxides and hydroxides: American Mineralogist, v. 83, p. 1419–1425.

Brown, D.A., Sherriff, B.L., Sawicki, J.A., and Sparling, R., 1999, Precipitation of iron minerals by a natural microbial consortium: Geochimica et Cosmochimica Acta, v. 63, p. 2163–2169, doi: 10.1016/S0016-7037(99)00188-X.

Caccavo, F., Lonergan, D.J., Lovely, D.R., Davis, M., Stolz, J.F., and McInerney, M.J., 1994, Geobacter sulfurreducens sp. nov., a hydrogen- and acetate-oxidizing dissimilatory metal-reducing microorganism: Applied and Environmental Microbiology, v. 60, p. 3752–3759.

Carlson, L., and Schwertmann, U., 1981, Natural ferrihydrites in surface deposits from Finland and their association with silica: Geochimica et Cosmochimica Acta, v. 45, p. 421–429, doi: 10.1016/0016-7037(81)90250-7.

Childers, S.E., Ciufo, S., and Lovley, D.R., 2002, Geobacter metallireducens accesses insoluble Fe(III) oxide by chemotaxis: Nature, v. 416, p. 767–769, doi: 10.1038/416767a.

Costerton, J.W., Lewandowski, Z., DeBeer, D., Caldwell, D., Korber, D., and James, G., 1994, Biofilms, the customized microniche: Journal of Bacteriology, v. 176, p. 2137–2142.

Cowen, J.P., Giovannoni, S.J., Kenig, F., Johnson, H.P., Butterfield, D., Rappe, M.S., Hutnak, M., and Lam, P., 2003, Fluids from aging ocean crust that support microbial life: Science, v. 299, p. 120–123, doi: 10.1126/science.1075653.

Douglas, S., and Beveridge, T.J., 1998, Mineral formation by bacteria in natural microbial communities: FEMS Microbiology Ecology, v. 26, p. 79–88, doi: 10.1016/S0168-6496(98)00027-0.

Ehrenberg, D.C.G., 1836, Vorläufige Mittheilungen über das wirkliche Vorkommen fossiler Infusorien und ihre grosse Verbreitung: Annalen der Physik und Chemie, v. 38, p. 213–227.

Ehrenreich, A., and Widdel, F., 1994, Anaerobic oxidation of ferrous iron by purple bacteria, a new type of phototrophic metabolism: Applied and Environmental Microbiology, v. 60, p. 4517–4526.

El Shazly, E.M., 1990, Red Sea deposits, in Chauvel, J.J., Yuqi, C., El Shazly, E.M., eds., Ancient banded iron formations: Athens, Theophrastus Publications, p. 157–222.

Fredrickson, J.K., and Onstott, R.C., 1996, Microbes deep inside the earth: Scientific American, October, p. 68–73.

Freeze, R.A., and Cherry, J.A., 1979, Groundwater: Englewood Cliffs, New Jersey, Prentice-Hall, 524 p.

Garrels, R.M., and Christ, C.L., 1990, Solutions minerals and equilibria: Boston, Jones and Bartlett, 400 p.

Garrels, R.M., Perry, E.A., and Mackenzie, F.T., 1973, Genesis of Precambrian iron-formations and the development of atmospheric oxygen: Economic Geology and the Bulletin of the Society of Economic Geologists, v. 68, p. 1173–1179.

Gascoyne, M., Ross, J.D., Watson, R.L., and Kamineni, D.C., 1989, Soluble salts in a Canadian Shield granite as contributors to groundwater salinity, in Miles, D.L., ed., Proceedings of the 6th International Symposium on Water-Rock Interaction, Malvern, U.K., 1989, p. 247–249.

Gottschalk, G., 1986, Bacterial metabolism (2nd edition): New York, Springer-Verlag, 359 p.

Gross, G.A., 1972, Primary features in cherty iron-formations: Sedimentary Geology, v. 7, p. 241–261, doi: 10.1016/0037-0738(72)90024-3.

Gross, G.A., 1983, Tectonic systems and the deposition of iron-formation: Precambrian Research, v. 20, p. 171–187, doi: 10.1016/0301-9268(83)90072-4.

Gross, G.A., 1988, A comparison of metalliferous sediments, Precambrian to Recent: Krystalinikum, v. 19, p. 59–74.

Gross, G.A., 1993, Element distribution patterns as metallogenic indicators in siliceous metalliferous sediments: Resource Geology Special Issue 17, p. 96–107.

Gross, G.A., 1995, The distribution of rare earth elements in iron-formations: Global Tectonics and Metallogeny, v. 5, p. 63–67.

Gross, G.A., 1996, Stratiform iron, in Eckstrand, O.R, Sinclair, W.D., and Thorpe, R.I., eds., Geology of Canadian mineral deposit types: Geological Survey of Canada, Geology of Canada, no. 8, p. 41–54.

Halvorson, H.O., and Starkey, R.L., 1927, Studies on the transformation of iron in nature. I: Theoretical considerations: Journal of Physical Chemistry, v. 31, p. 626\7–631, doi: 10.1021/j150274a016.

Hamade, T., Konhauser, K.O., Raiswell, R., Goldsmith, S., and Morris, R.C., 2003, Using Ge/Si ratios to decouple iron and silica fluxes in Precambrian banded iron formations: Geology, v. 31, p. 35–38, doi: 10.1130/0091-7613(2003)031<0035:UGSRTD>2.0.CO;2.

Harder, E.C., 1919, Iron-depositing bacteria and their geologic relations: U.S. Geological Survey Professional Paper 113, 89 p.

Harris, R.J., Fisher, A.T., and Chapman, D.S., 2004, Fluid flow through seamounts and implications for global mass fluxes: Geology, v. 32, p. 725–728, doi: 10.1130/G20387.1.

Hofmann, H.J., Sage, R.P., and Berdusco, E.N., 1991, Archean stromatolites in Michipicoten group siderite ore at Wawa, Ontario: Economic Geology and the Bulletin of the Society of Economic Geologists, v. 86, p. 1023–1030.

Holland, H.D., 1973, The oceans: A possible source of iron in iron-formations: Economic Geology and the Bulletin of the Society of Economic Geologists, v. 68, p. 1169–1172.

Icopini, G.A., Anbar, A.D., Ruebush, S.S., Tien, M., and Brantley, S.L., 2004, Iron isotope fractionation during microbial reduction of iron: the importance of adsorption: Geology, v. 32, p. 205–208, doi: 10.1130/G20184.1.

Isley, A.E., 1995, Hydrothermal plumes and the delivery of iron to banded iron formation: Journal of Geology, v. 103, p. 169–185.

Knauth, L., Paul, L., and Lowe, D.R., 2004, High Archean climatic temperature inferred from oxygen isotope geochemistry of cherts in the 3.5 Ga Swaziland Supergroup, South Africa: Nature, v. 42, p. 566–580.

Konhauser, K.O., and Ferris, F.G., 1996, Diversity of iron and silica precipitation by microbial mats in hydrothermal waters, Iceland: Implications for Precambrian iron formations: Geology, v. 24, p. 323–326, doi: 10.1130/0091-7613(1996)024<0323:DOIASP>2.3.CO;2.

Konhauser, K.O., Hamade, T., Raiswell, R., Morris, R.C., Ferris, F.G., Southam, G., and Canfield, D.E., 2002, Could bacteria have formed the Precambrian banded iron formations?: Geology, v. 30, p. 1079–1082, doi: 10.1130/0091-7613(2002)030<1079:CBHFTP>2.0.CO;2.

LaBerge, G.L., Robbins, E.I., and Han, T.-M., 1987, A model for the biological precipitation of Precambrian iron-formations—A: Geological evidence, in Appel, P.W.U., and LaBerge, G.L., eds., Precambrian iron-formations: Athens, Theophrastus Publications, p. 69–95.

Laland, K.N., Odling-Smee, J., and Feldman, M.W., 2004, Causing a commotion: Nature, v. 429, p. 609, doi: 10.1038/429609a.

Lepp, H., 1987, Chemistry and origin of Precambrian iron-formations, in Appel, P.W.U., and LaBerge, G.L., eds., Precambrian iron-formations: Athens, Theophrastus Publications, p. 3–30.

Lovley, D.R., 1991, Dissimilatory Fe(III) and Mn(IV) reduction: Microbiological Reviews, v. 55, p. 259–287.

Lovley, D.R., 1995, Microbial reduction of iron, manganese and other minerals: Advances in Agronomy, v. 54, p. 176–217.

Lovley, D.R., Stolz, J.F., Nord, G.L., and Phillips, E.J.P., 1987, Anaerobic production of magnetite by a dissimilatory iron-reducing microorganism: Nature, v. 330, p. 252–254, doi: 10.1038/330252a0.

McLean, R.J.C., Fortin, D., and Brown, D.A., 1996, Microbial metal-binding mechanisms and their relation to nuclear waste disposal: Canadian Journal of Microbiology, v. 42, p. 392–400.

Morita, R.Y., and ZoBell, C.E., 1955, Occurrence of bacteria in pelagic sediments collected during the Mid-Pacific expedition: Deep-Sea Research, v. 3, p. 66–73.

Nealson, K.H., 1997, Sediment bacteria: who's there, what are they doing, and what's new?: Annual Review of Earth and Planetary Sciences, v. 25, p. 403–434, doi: 10.1146/annurev.earth.25.1.403.

Nealson, K.H., and Myers, C.R., 1990, Iron reduction by bacteria: A potential role in the genesis of banded iron formations: American Journal of Science, v. 290-A, p. 35–45.

Nealson, K.H., and Little, B., 1997, Breathing manganese and iron: Solid-state respiration: Advances in Microbiology, v. 45, p. 213–239.

Nee, S., 2004, More than meets the eye: Nature, v. 429, p. 804–805, doi: 10.1038/429804a.

Nisbet, E., 2000, The realms of Archaean life: Nature, v. 405, p. 625–626, doi: 10.1038/35015187.

Noffke, N., Hazen, R., and Nhleko, N., 2003, Earth's earliest microbial mats in siliciclastic marine environment (2.9 Ga Mozaan Group, South Africa): Geology, v. 31, p. 673–676, doi: 10.1130/G19704.1.

Pedersen, K., 1997, Microbial life in deep granite rock: FEMS Microbiology Reviews, v. 20, p. 399-414, doi: 10.1016/S0168-6445(97)00022-3.

Queller, D.C., 2004, Kinship is relative: Nature, v. 430, p. 975–976, doi: 10.1038/430975a.

Rasmussen, B., 2000, Filamentous microfossils in a 3,235-million-year-old volcanogenic massive sulphide deposit: Nature, v. 405, p. 676–679, doi: 10.1038/35015063.

Reid, R.P., Visscher, P.T., Decho, A.W., Stolz, J.F., Bebout, B.M., Dupraz, C., Macintyre, I.G., Paerl, H.W., Pinckney, J.L., Prufert-Bebout, L., Steppe, T.F., and DesMarais, D.J., 2000, The role of microbes in accretion, lamination and early lithification of modern marine stromatolites: Nature, v. 406, p. 989–992, doi: 10.1038/35023158.

Riding, R., 2002, Biofilm architecture of Phanerozoic cryptic carbonate marine veneers: Geology, v. 30, p. 31–34, doi: 10.1130/0091-7613(2002)030<0031:BAOPCC>2.0.CO;2.

Robbins, E.I., LaBerge, G.L., and Schmidt, R.G., 1987, A model for the biological precipitation of Precambrian iron formations—B: Morphological evidence and modern analogs, in Appel, P.W.U., and LaBerge, G.L., eds., Precambrian iron-formations: Athens, Theophrastus Publications, p. 92–139.

Schopf, J.W., 1992, The oldest fossils and what they mean, in Schopf, J.W., ed., Major events in the history of life: Boston, Jones and Bartlett, p. 29–61.

Schultze-Lam, S., Harauz, G., and Beveridge, T.J., 1992, Participation of a cyanobacterial S-layer in fine-grain mineral formation: Journal of Bacteriology, v. 174, p. 7971–7981.

Schwertmann, U., 1988, Occurrence and formation of iron oxides in various pedoenvironments, in Stucki, J.W., Goodman, B.A., and Schwertmann, U., eds, Iron in soils and clay minerals: Dordrecht, D. Reidel Publishing Co., p. 267–308.

Schwertmann, U., and Taylor, R.M., 1989, Iron oxides, in Boris, J., and Weed, S.B., eds., Minerals in soil environments: Madison, Wisconsin, Soil Science Society of America, p. 379–438.

Stolz, J.F., 2000, Structure of microbial mats and biofilms, in Riding, R.E., and Awramik, S.M., eds., Microbial sediments: Berlin, Springer-Verlag, p. 1–8.

Stolz, J.F., and Margulis, L., 1984, The stratified microbial community at Laguna Figueroa, Baja California, Mexico: A possible model for PrePhanerozoic laminated microbial communities preserved in chert: Origins of Life and Evolution of the Biosphere, v. 14, p. 671–679, doi: 10.1007/BF00933720.

Stumm, W., and Morgan, J.J., 1981, Aquatic chemistry: New York, John Wiley and Sons, 742 p.

Sumner, D.Y., 2000, Microbial vs. environmental influences on the morphology of late Archean fenestrate microbialites, in Riding, R.E., and Awramik, S.M., eds., Microbial sediments: Berlin, Springer-Verlag, p. 307–314.

Tice, M.M., and Lowe, D.R., 2004, Photosynthetic microbial mats in the 3,414-Myr-old ocean: Nature, v. 431, p. 549–552, doi: 10.1038/nature02888.

Trendall, A., 2000, The significance of banded iron formation (BIF) in the Precambrian stratigraphic record: Geoscientist, v. 10, p. 4–7.

Tyler, S.A., and Barghoorn, E.S., 1954, Occurrence of structurally preserved plants in Precambrian rocks of the Canadian Shield: Science, v. 119, p. 606–608.

Valley, J.W., Peck, W.H., King, E.M., and Wilde, S.A., 2002, A cool early Earth: Geology, v. 30, p. 351–354, doi: 10.1130/0091-7613(2002)030<0351:ACEE>2.0.CO;2.

van der Zee, C., Roberts, D.R., Rancourt, D.G., and Slomp, C.P., 2003, Nanogoethite is the dominant reactive oxyhydroxide phase in lake and marine sediments: Geology, v. 31, p. 993–996, doi: 10.1130/G19924.1.

Vargas, M., Kashefi, K., Blunt-Harris, E.L., and Lovley, D.R., 1998, Microbiological evidence for Fe(III) reduction on early earth: Nature, v. 395, p. 65–67, doi: 10.1038/25720.

Waksman, S.A., 1916, Bacterial numbers in soils, at different depths, and in different seasons of the year: Soil Science, v. 2, p. 331–343.

White, A.F., and Peterson, M.L., 1996, Reduction of aqueous transition metal species on the surfaces of Fe(II)-containing oxides: Geochimica et Cosmochimica Acta, v. 60, p. 3799–3814, doi: 10.1016/0016-7037(96)00213-X.

Widdel, F., Schnell, S., Heising, S., Ehrenreich, A., Assmus, B., and Schink, B., 1993, Ferrous iron oxidation by anoxygenic phototrophic bacteria: Nature, v. 362, p. 834–836, doi: 10.1038/362834a0.

Winogradsky, S., 1888, Ueber Eisenbacterien: Botanische Zeitschrift, v. 46, p. 261–270.

ZoBell, C.E., 1943, The effect of solid surfaces upon bacterial activity: Journal of Bacteriology, v. 46, p. 39–56.

ZoBell, C.E., 1952, Bacterial life at the bottom of the Philippine Trench: Science, v. 115, p. 507–508.

MANUSCRIPT ACCEPTED BY THE SOCIETY 29 OCTOBER 2005

Oxygen isotope composition of hematite and genesis of high-grade BIF-hosted iron ores

J. Gutzmer*
Paleoproterozoic Mineralization Research Group, Department of Geology, University of Johannesburg,
P.O. Box 524, Auckland Park 2006, South Africa

J. Mukhopadhyay
Paleoproterozoic Mineralization Research Group, Department of Geology, University of Johannesburg, P.O. Box 524, Auckland Park 2006, South Africa, and Department of Geology, Presidency College, 86/1 College Street, Calcutta, India, 700073

N.J. Beukes
Paleoproterozoic Mineralization Research Group, Department of Geology, University of Johannesburg,
P.O. Box 524, Auckland Park 2006, South Africa

A. Pack
Department of Earth and Planetary Sciences, University of New Mexico, Albuquerque, New Mexico 87131-0001, USA, and CRPG/CNRS Centre de Recherches Petrographiques et Geochimiques, 15 Rue Notre Dame des Pauvres, 54501 Vandoeuvre-les-Nancy, France

K. Hayashi
Department of Mineralogy, Petrology and Economic Geology, Graduate School of Science, Tohoku University, Aoba-ku, Sendai 980-8578 Japan

Z.D. Sharp
Department of Earth and Planetary Sciences, University of New Mexico, Albuquerque, New Mexico 87131-0001, USA

ABSTRACT

High-grade BIF-hosted iron ore deposits are widely believed to have formed by epigenetic residual enrichment of hematite at the expense of other constituents, most notably chert. Processes responsible for the enrichment to high-grade iron ores are, however, only poorly understood and a range of metallogenetic models have been proposed. Field relationships have been used to distinguish three major groups of BIF-hosted high-grade iron ore deposits, namely deposits of ancient supergene, hydrothermal, and supergene-modified hydrothermal origin. Iron ores from all three deposit types are essentially composed of hematite; among different morphological types of hematite, microcystalline platy hematite and martite predominate.

In this contribution, the oxygen isotope geochemistry of ore-forming hematite and martite from several high-grade iron ore deposits is examined, in an attempt to differentiate deposits of hydrothermal origin from those formed in ancient supergene environments. The $\delta^{18}O$ composition of martite and microplaty hematite from depos-

*jg@rau.ac.za

its presumed to be of hydrothermal origin range from +0.9‰ to −7.3‰. **Microcrystalline platy hematite from high-grade ores of ancient supergene origin, in contrast, has $\delta^{18}O$ values ranging between +2.0‰ and −3.9‰. The latter range overlaps with the range that is defined by hematite and magnetite from weakly metamorphosed Archean–Paleoproterozoic BIF (+5‰ to −4‰). The results obtained for ancient supergene deposits developed along the 2.2 Ga Gamagara-Mapedi unconformity strengthen the argument that the Paleoproterozoic atmosphere-hydrosphere-lithosphere system was very similar to that of modern Earth. The marked shift to negative $\delta^{18}O$ values displayed by hematite and martite from hydrothermal iron ore deposits, on the other hand, provides support for the suggestion that aqueous fluids of shallow crustal origin were responsible for the hydrothermal enrichment of banded iron formations to high-grade iron ores.**

Keywords: oxygen isotope geochemistry, hematite, martite, iron ore deposits, BIF.

INTRODUCTION

Most world-class high-grade iron ore deposits (Fig. 1) are the product of epigenetic enrichment of Precambrian banded iron formation (BIF). Despite the economic importance of this deposit type, processes responsible for the enrichment of BIF (containing 30–40 wt% Fe) to high-grade iron ores with 60–67 wt% Fe are poorly understood. Different models, ranging from syngenetic and diagenetic (King, 1989) to hydrothermal (Powell et al., 1999; Barley et al., 1999; Taylor et al., 2001) and ancient supergene (Morris, 1980, 1985; van Schalkwyk and Beukes, 1986) or modern supergene (MacLeod, 1966), have been suggested. Uncertainties about the origin of high-grade hematite deposits are largely due to the monomineralic composition of the ores. They are almost exclusively composed of hematite, a mineral with a wide stability field and simple chemical composition that reveals little about its origin. Furthermore, later deformation and geologically recent deep chemical weathering overprint and obscure many of the primary ore characteristics.

From field geological evidence, Beukes et al. (2002a) recognized three major groups of BIF-hosted high-grade iron ore deposits, namely deposits of ancient supergene, hydrothermal and supergene-modified hydrothermal origin. Type examples for ancient supergene hematite iron ore deposits are those hosted by the ca. 2.45 Ga iron formations of the Asbestos Hills Subgroup of the Transvaal Supergroup in Griqualand West, South Africa (Fig. 1). High-grade iron ores that constitute these deposits are thought to have formed as products of deep lateritic weathering (Gutzmer and Beukes, 1998) immediately below the ca. 2.2 Ga Gamagara-Mapedi unconformity. The high-grade ore grades from the unconformity down into oxidized but not markedly enriched iron formation. The largest and economically most important examples for ancient supergene enriched iron ores are the Sishen and Beeshoek deposits (Fig. 2A). The ores are typically overlain by red beds that contain conglomerates composed of hematite ore and oxidized BIF pebbles (Fig. 2A).

In contrast, high-grade iron ores of hydrothermal origin, of which hard hematite ores of Thabazimbi, South Africa (Fig. 2B),

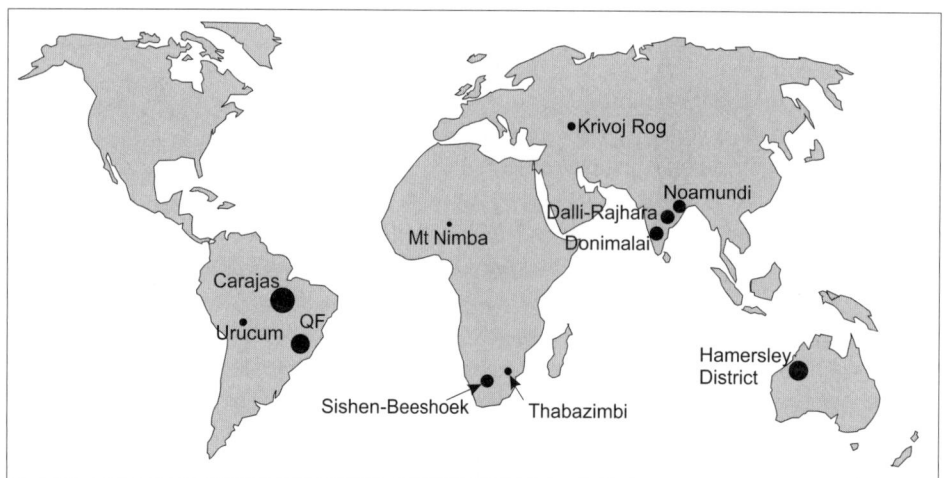

Figure 1. Location map illustrating the distribution of important BIF-hosted high-grade hematite ore deposits. The diameter of the circle shows relative size of the iron ore resource hosted by the deposit/district. The Rooinekke and Zeekoebaart deposits, which are close to the Sishen-Beeshoek deposits in the Northern Cape Province of South Africa, are not displayed on this map because they are currently of no economic significance. Abbreviations: QF—Iron Quadrangle, Minas Gerais, Brazil.

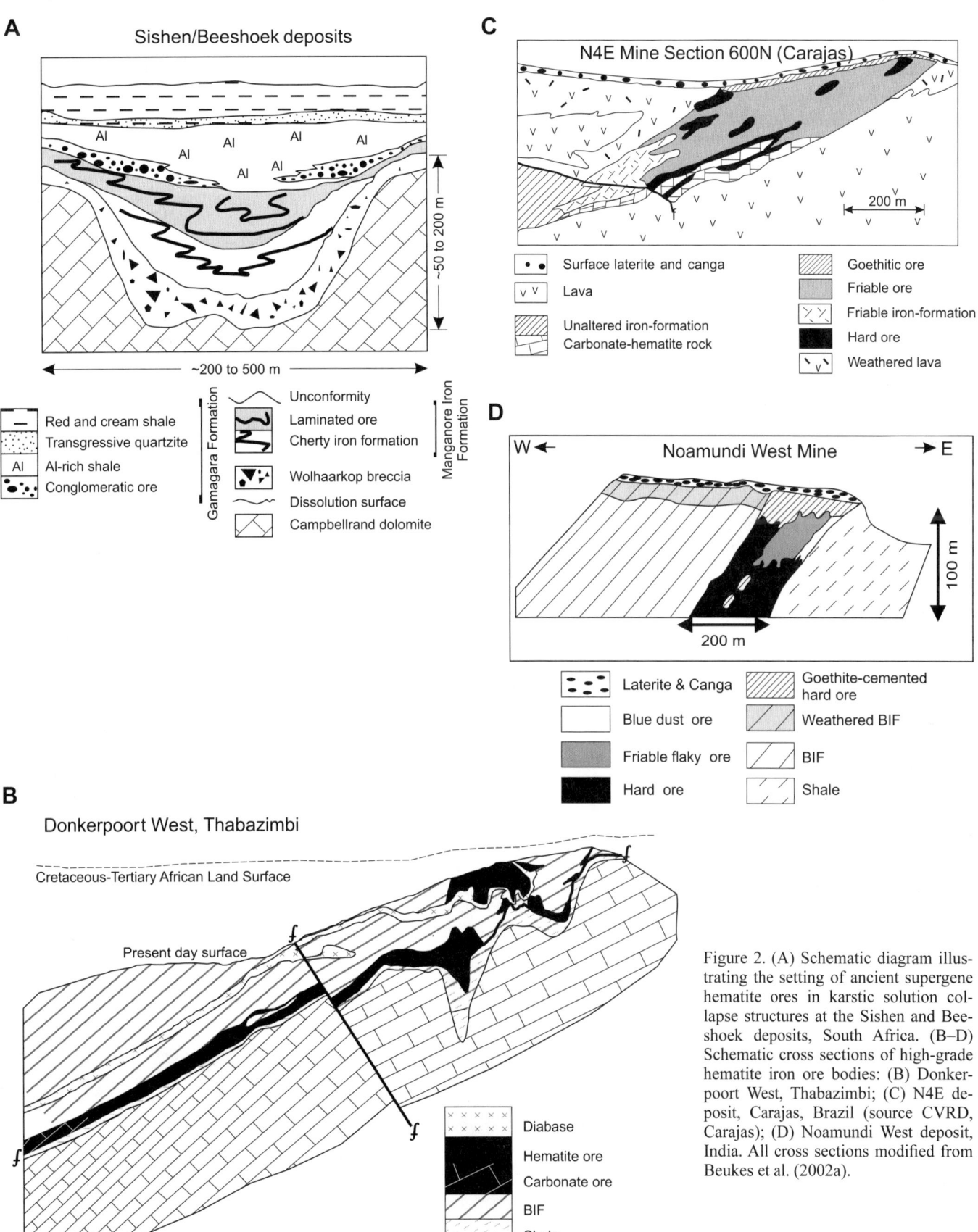

Figure 2. (A) Schematic diagram illustrating the setting of ancient supergene hematite ores in karstic solution collapse structures at the Sishen and Beeshoek deposits, South Africa. (B–D) Schematic cross sections of high-grade hematite iron ore bodies: (B) Donkerpoort West, Thabazimbi; (C) N4E deposit, Carajas, Brazil (source CVRD, Carajas); (D) Noamundi West deposit, India. All cross sections modified from Beukes et al. (2002a).

and Mount Tom Price, Mount Whaleback, Paraburdoo, and Newman in the Hamersley District of Western Australia (Fig. 1) are type examples, are not associated with any erosional unconformity. They are typically located at the base of the hosting iron-formation succession, from which they grade up into unmineralized iron formation. In many of these deposits structural controls on ore distribution, such as faults, sills, or dykes, have been recognized (Taylor et al., 2001; Beukes et al., 2002a).

Supergene-modified hydrothermal iron ores, such as the giant deposits of the Carajas District in Brazil (Fig. 2C) or the Noamundi deposit in India (Fig. 2D), are characterized by the presence of friable hematite ores derived from geologically recent supergene enrichment of carbonate-metasomatized iron formation (Beukes et al., 2002a). Typically, these friable ores are closely associated with hard, non-friable high-grade hematite ore bodies of hydrothermal origin.

Here, we report the results of an investigation into the oxygen isotope geochemistry of hematite from several high-grade iron ore deposits, in an attempt to differentiate deposits of hydrothermal origin from those formed in ancient supergene environments. Samples from eight deposits, hosted by banded iron formations of Archean to Paleoproterozoic age, were studied. Two of these deposits (Sishen, Rooinekke) are considered to be of ancient supergene origin, the other six (Thabazimbi, Mount Tom Price, Paraburdoo, Noamundi, Zeekoebaart, and Carajas) are thought to be of hydrothermal origin (Beukes et al., 2002a). The apparent lack of a significant metamorphic overprint on all of these deposits was an important selection criterion, to avoid the effects of metamorphic recrystallization and resetting of oxygen isotope compositions as described by Hoefs et al. (1982). We restricted our investigation to hard ores that we expect to be either exclusively hydrothermal or supergene in origin, and avoided friable ores that are likely to have a supergene-modified hydrothermal origin.

PETROGRAPHY AND CLASSIFICATION OF THE ORES

The high-grade iron ores from the studied deposits are essentially composed of hematite, with only trace amounts of magnetite, kenomagnetite, goethite, quartz, carbonate (calcite, dolomite), and apatite. Four morphological types of hematite are present in the ores, namely microcrystalline platy hematite (Figs. 3A, 3C), martite (Figs. 3B, 3C), patchy hematite (Fig. 3B), and specular hematite (Figs. 3A, 3D). Microcystalline platy (microplaty) hematite and martite predominate in all deposits. The microplaty hematite occurs as subhedral to euhedral plates, typically less than 25 μm in diameter (Fig. 3C). Martite is closely associated with the microplaty hematite in the form of euhedral to subhedral magnetite octahedra (<100 μm in size) that are partly or completely replaced by a porous scaffold of hematite laths (Fig. 3C) (Ramdohr, 1980). In all of the deposits thought to be of hydrothermal origin, martite is much more abundant than magnetite in the iron-formation protolith. Indeed, martite predominates in all hard iron ores from the six deposits of presumed hydrothermal origin that were investigated during this study. In contrast, high-grade iron ores of the Sishen and Rooinekke deposits, both presumably of ancient supergene origin, are predominantly composed of microplaty hematite, with only minor amounts of martite. Martite from all the studied deposits contains tiny remnants of magnetite and kenomagnetite. Trace amounts of cryptocrystalline goethite are present in samples from the Noamundi, Mount Tom Price and Paraburdoo deposits (Fig. 1).

Patchy hematite (Fig. 3B) and specular hematite (Figs. 3A, 3D) are much less abundant than microcrystalline platy hematite and martite. Patchy hematite, characterized by its anhedral shape and grain sizes not exceeding 50 μm, is thought to have formed by recrystallization of microplaty hematite or martite. Specular hematite is distinctly coarser grained than the other three morphological types of hematite, with grain sizes that may exceed 1 mm (Fig. 3D). Crystals are often developed as thin blades that may be arranged in radial sheaves or massive aggregates. The formation of specular hematite always postdates ore formation; it is restricted to the infill of vugs or veinlets that crosscut preexisting iron ore.

On the basis of mesoscopic appearance (porosity and remnant sedimentary lamination) and petrographic features (presence and abundance of morphological types of hematite), hard iron ores from all studied deposits have been grouped into massive martite, massive martite-microplaty hematite, laminated microplaty hematite, and conglomeratic ore.

Massive martite ore is defined by massively intergrown martite with minor intergranular porosity filled by microplaty hematite. This ore type was identified at Carajas, Thabazimbi, Noamundi, and Paraburdoo. It appears to have originated as large masses of magnetite that were subsequently transformed into martite. Massive martite–microplaty hematite ore is composed of approximately equal amounts of microplaty hematite and martite with conspicuous porosity. It was encountered at Mount Tom Price, Thabazimbi, and Carajas. In contrast to the more massive ores, laminated microplaty hematite ore often displays microbanding defined by the regular intercalation of millimeter-thin laminae of more densely and less densely intergrown microplaty hematite, with open pore space between the hematite platelets. This ore type constitutes a large part of the ore reserves at Sishen, Rooinekke, and Noamundi. Conglomeratic ore is restricted to the Sishen and Beeshoek deposits in South Africa (van Schalkwyk and Beukes, 1986). Lenses of conglomeratic ore are developed along the Gamagara-Mapedi unconformity and are composed of hematite pebbles derived from the underlying laminated microplaty hematite ores. The hematite ore pebbles are hosted by a matrix of microplaty hematite and some specularite.

ANALYTICAL METHODS

Samples of different ore and deposit types were taken either from drill core or deep open-pit exposures, far below the zone of modern weathering. Oxygen isotope analyses were conducted at

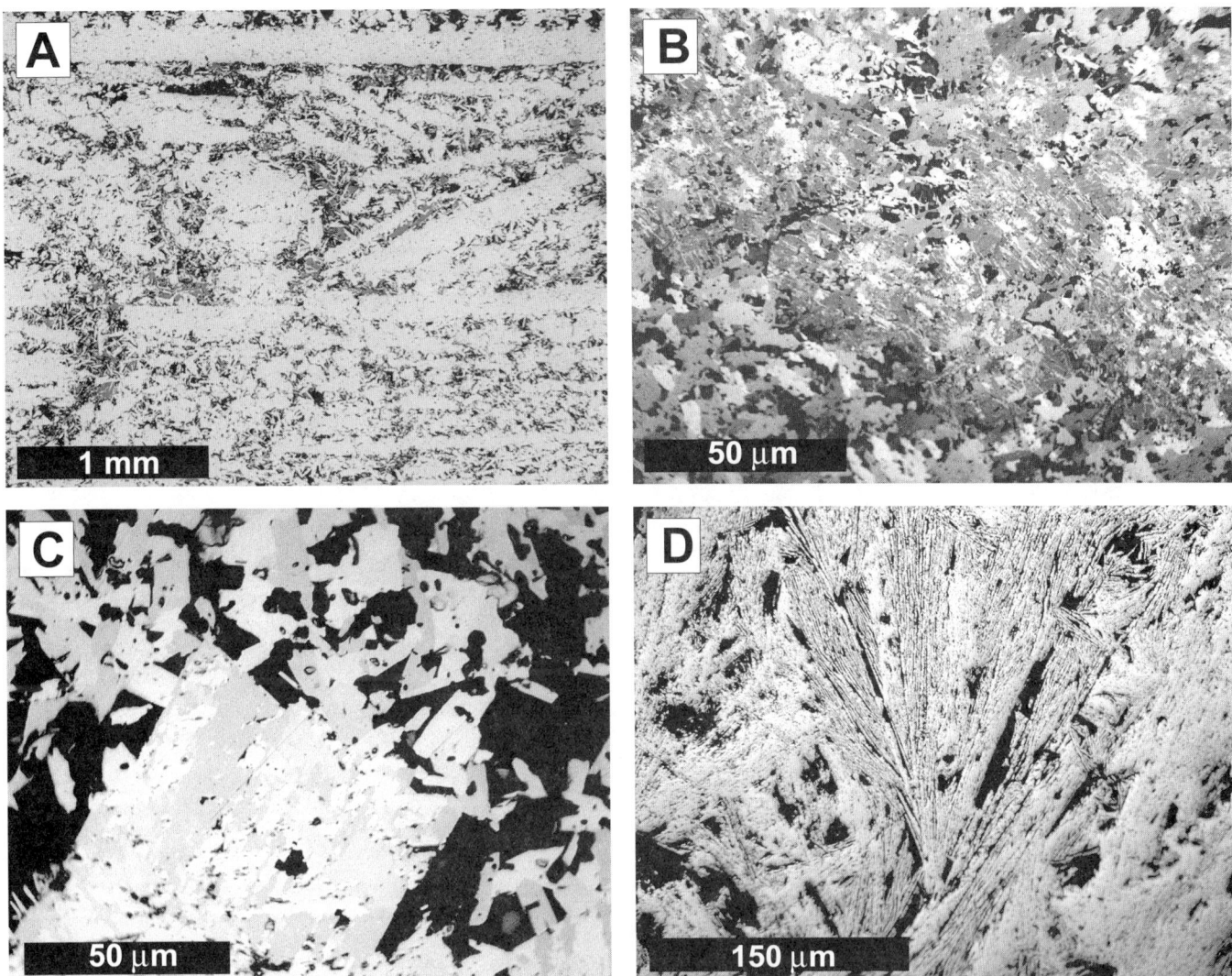

Figure 3. Reflected light photomicrographs illustrating different morphological types of hematite. (A) Laminated microcrystalline platy hematite ore from the Sishen deposit. Note the apparent dissolution collapse brecciation. The breccia clasts are composed of microcrystalline platy hematite and minor hematite; interstices are filled by a porous mass of specularite with trace amounts of quartz (dark gray). (B) Massive martite ore from the Noamundi West deposit (India). Note the characteristic cross-hatch extinction pattern of martite that occurs as poorly defined grains intimately intergrown with patchy hematite (crossed nicols, oil immersion). (C) Martite grain with well-preserved octahedral shape, intergrown with microplaty hematite. Massive martite-microcrystalline platy hematite ore, Noamundi West deposit (note different shades of gray, which are due to anisotropy of hematite under crossed nicols, oil immersion). (D) Coarse crystalline specularite sheaves in vein crosscutting laminated microcrystalline platy hematite ore from the Sishen deposit (South Africa).

the Department of Earth and Planetary Sciences, University of New Mexico, USA, and at the Department of Mineralogy, Petrology and Economic Geology, Tohoku University, Japan.

At the University of New Mexico, oxygen isotope analyses were conducted by the modified fluorination technique as described by Sharp (1990). Monomineralic hematite grains, between 1 mg and 2 mg in weight, were loaded along with standards in a Ni-sample holder and placed in the reaction chamber of the laser fluorination line. The chamber was covered with an IR-transparent BaF_2 window. Samples were pre-fluorinated overnight in order to remove impurities and moisture from the surfaces. Samples were heated and reacted in a BrF_5 atmosphere using a CO_2 laser with IR emission at 10.6 μm. Excess BrF_5 was frozen on a cold trap cooled with liquid nitrogen. Traces of F_2 were eliminated by reaction with hot KCl. Chlorine gas was removed from the analyte O_2 gas by freezing in a cold trap. Sample oxygen gas was collected on two successive 13X molecular sieve cold fingers and released into the dual inlet system of a Finnigan-MAT Delta Plus XL mass spectrometer. All samples are reported relative to standard mean ocean water (VSMOW), defined such that

NBS-28 quartz has a $\delta^{18}O$ value of 9.6‰. Calibrated in-house standards were routinely run with each set of analyses. Reproducibility was found to be better than 0.15‰ (1σ).

At Tohoku University, samples were analyzed both in situ on polished chips and by complete evaporation of 1–2 mg ore chips that were previously separated from a larger specimen. Oxygen was extracted using a 0–17 Watt JLC-200 CO_2-laser (λ = 10.6 μm) and BrF_5 by a fluorination method similar to the technique used at the University of New Mexico. The optical system to introduce the laser beam into the stainless steel reaction chamber through a BaF_2 window is coaxial with a reflected microscope, which enables accurate in situ laser ablation. The diameter of the focused laser beam is ~100 μm; however, the area of thermal halo affected by laser ablation is ~200 μm. Therefore, oxygen extracted by CO_2-laser ablation from a single spot is generated from the area of ~200 μm. Oxygen was converted to CO_2 quantitatively by the reaction with a heated diamond. The $^{18}O/^{16}O$ ratios are measured by a Finnigan-MAT252 mass spectrometer. The accuracy of in situ CO_2-laser fluorination oxygen isotope analyses at Tohoku University was examined by the repeated analyses of working standards (Hayashi et al., 2001). The precision of measured $\delta^{18}O$ of this study is ± 0.3‰ (1σ) per mil.

All data are expressed in per mil (‰) relative to VSMOW. A total of 87 samples of hematite and martite from a total of eight deposits were analyzed (Table 1). Oxygen isotope analyses of specular hematite were obtained only from the Thabazimbi and Sishen deposits, South Africa (Table 1). A summary of the data is presented in Table 2.

The high-grade iron ores studied here typically contain only trace amounts of gangue minerals, including quartz, carbonate, and apatite. These trace minerals generally occur finely disseminated and are too fine-grained to permit isotopic analyses. The occurrence of megaquartz is restricted to rare veins that are either associated with paragenetically late specularite or in an uncertain paragenetic relationship to ore-forming event(s). Coarser-grained, sparitic carbonates (calcite, dolomite) associated with high-grade iron ores at Thabazimbi and Carajas show textural evidence suggesting that they are coeval with microplaty hematite and therefore part of the ore-forming event. Detailed fluid inclusion studies and stable isotope analyses carried out on sparitic carbonates from Thabazimbi, which are not presented here, confirm the cogenetic hydrothermal origin of these minerals (Gutzmer et al., 2001a; Netshiozwi, 2002).

RESULTS

The $\delta^{18}O$ composition of martite-textured hematite (referred to below as martite) from the different deposits ranges from +0.9‰ to −7.3‰, with an average of −4.8‰ (Tables 1, 2). Compared to massive martite ores, ores composed of martite and microplaty hematite are enriched in ^{18}O, with $\delta^{18}O$ ranging from +2.0‰ to −5.1‰, with an average of −1.6‰ (Table 2). Still more enriched in ^{18}O is specularitic hematite hosted in veins that cross-

TABLE 1. COMPILATION OF $\delta^{18}O_{VSMOW}$ SIGNATURES OF ORE-FORMING HEMATITE FROM SELECTED HIGH-GRADE BANDED IRON FORMATION–HOSTED IRON ORE DEPOSITS

Sample no.	Description	$\delta^{18}O_{VSMOW}$	Lab
Paraburdoo (Hamersley District, Australia)			
PB-1-7	Martite	−5.0	1
PB-1-8	Martite	−3.7	1
PB-1-9	Martite	−4.7	1
PB-1-10	Martite	−6.4	1
Mount Tom Price (Hamersley District, Australia)			
MTP-1-1	Martite + Microplaty hematite	−4.4	1
MTP-1-2	Martite + Microplaty hematite	−4.3	1
MTP-1-3	Martite + Microplaty hematite	−3.8	1
MTP-1-4	Martite + Microplaty hematite	−5.1	1
MTP-1-5	Martite + Microplaty hematite	−5.0	1
MTP-1-6	Martite + Microplaty hematite	−4.6	1
MTP-2-11	Martite + Microplaty hematite	−4.7	1
Noamundi (India)			
IO-20-1	Microplaty hematite	−2.2	1
IO-20-2	Microplaty hematite	−2.6	1
IO-20-3	Microplaty hematite	−1.3	1
IO-20-5	Microplaty hematite	−2.4	1
IO-10-1	Martite	−3.9	1
IO-16C-7	Martite	−6.7	1
IO-16C-8	Martite	−5.7	1
IO-16C-10	Martite	−4.6	1
IO-16C-11	Martite	−2.6	1
IO-4A-5	Martite	−4.8	1
IO-4A-6	Martite	−5.0	1
IO-4A-7	Martite	−4.4	1
IO-4A-9	Martite	−5.8	1
N4E Mine, Carajas (Brazil)			
BH529-120-7	Martite	−4.0	1
BH529-120-6	Martite	−5.4	1
BH529-120-5	Martite	−6.0	1
BH529-120-4	Martite	−4.4	1
BH529-90-3	Martite	−5.7	1
BH529-90.0-2	Martite	−7.4	2
BH529-90-1	Martite	−4.4	1
BH-384-12	Martite	−6.5	1
BH384-160	Martite	−6.7	2
BH541-177.5	Martite	−7.0	2
Thabazimbi (RSA)			
BH-607-4	Martite	−4.1	1
BH-607-3	Martite	−3.4	1
BH-607-2	Martite	−4.7	1
BH-607-1	Martite + Microplaty hematite	−1.5	1
BH2099	Martite	−2.7	1
BH-1706-89.5	Martite	−6.0	2
DON-383-159	Specularite	+3.4	2
DON-396-215	Martite	−5.6	2
DKP442-358	Martite	+0.9	2
KW 309-89.92	Martite	−3.4	2
BOB#3	Specularite	−1.4	2
BE#12	Martite	−0.3	2
BW#3	Martite + Microplaty hematite	−1.9	2
VDB#2	Specularite	−2.4	2
KF-83-9	Martite + Microplaty hematite	−3.7	2

(continued)

TABLE 1. COMPILATION OF $\delta^{18}O_{VSMOW}$ SIGNATURES OF ORE-FORMING HEMATITE FROM SELECTED HIGH-GRADE BANDED IRON FORMATION–HOSTED IRON ORE DEPOSITS (continued)

Sample no.	Description	$\delta^{18}O_{VSMOW}$	Lab
Zeekoebaart (RSA)			
Baart-1	Martite	−6.1	2
Rooinekke (RSA)			
Rooinekke#1a	Microplaty hematite	−2.3	2
Rooinekke#1b	Microplaty hematite	−1.2	2
Rooinekke#2	Microplaty hematite	−1.5	2
Sishen (RSA)			
WP-CL-1	Microplaty hematite (conglomerate clast)	−2.3	1
WP-CL-3	Microplaty hematite (conglomerate clast)	−3.5	1
WP-CL-4	Microplaty hematite (conglomerate clast)	−2.4	1
WP-CL-5	Microplaty hematite (conglomerate clast)	−1.5	1
WP-CL-6	Microplaty hematite (conglomerate clast)	−2.2	1
WP-MT-3	Microplaty hematite (conglomerate matrix)	+0.2	1
WP-MT-2	Microplaty hematite (conglomerate matrix)	−1.9	1
WP-MT-1	Microplaty hematite (conglomerate matrix)	+0.2	1
SHV-3-8	Specularite	+2.0	1
SHV-3-7	Specularite	+2.7	1
SHV-PT-3	Specularite	+1.0	1
SHV2-PT-0	Specularite	+3.0	1
SHV-3-9	Microplaty hematite	−3.9	1
1975-175-5	Microplaty hematite	−1.1	1
1975-189-1	Microplaty hematite	−2.3	1
1975-189-2	Microplaty hematite	−2.9	1
1975-199-3	Microplaty hematite	−2.7	1
1975-189-4	Microplaty hematite	−2.1	1
1975-189-8	Microplaty hematite	−2.9	1
1975-238-11	Microplaty hematite	−2.8	1
1975-238-12	Microplaty hematite	−2.5	1
1772#3	Microplaty hematite (conglomerate clast)	+2.0	2
1772#5b	Specularite	+0.5	2
1772#5c	Microplaty hematite (coating around chert pebble)	−0.1	2
1772#7	Microplaty hematite (conglomerate clast)	+0.9	2
ADR#15	Microplaty hematite	−2.1	2
4552#9b	Microplaty hematite	−0.5	2
4552#9d	Microplaty hematite	+0.4	2
4225#9a	Microplaty hematite (coating around chert pebble)	−0.7	2
4225#9e	Microplaty hematite (conglomerate clast)	−0.8	2
3226#3	Microplaty hematite	−0.9	2
3226#4b	Microplaty hematite	+1.3	2
3226#12	Microplaty hematite (conglomerate clast)	−2.3	2

Note: All data are expressed in per mil relative to Vienna standard mean ocean water (VSMOW).
Data source/Lab: 1—Dept. of Mineralogy, Tohoku University, Japan; 2—Dept. of Earth Sciences, University of New Mexico, USA.

TABLE 2 SUMMARY OF OXYGEN ISOTOPE COMPOSITION OF HEMATITE FROM SELECTED HIGH-GRADE BANDED IRON FORMATION–HOSTED IRON ORE DEPOSITS

Deposit/district	Textural type	N	Range	Average
Carajas (Brazil)	Martite	10	−4.0 to −7.3	−5.6
Mt Tom Price (Hamersley)	Martite + Microplaty	7	−3.8 to −5.1	−4.9
Paraburdoo (Hamersley)	Martite	4	−3.7 to −6.4	−4.9
Noamundi (India)	Martite	9	−2.6 to −6.7	−4.3
	Microplaty	4	−1.3 to −2.6	−2.1
Thabazimbi (RSA)	Martite	9	+0.9 to −6.0	−3.9
	Martite + Microplaty	2	−1.5 to −1.9	−1.7
	Specularite	3	+3.4 to −2.4	+0.1
Zeekoebaart (RSA)	Martite	1	−6.1	−6.1
Sishen (RSA)	Microplaty	28	+2.0 to −3.9	−1.5
	Specularite	5	+3.0 to 0.5	+1.9
Rooinekke (RSA)	Microplaty	3	−1.2 to −2.3	−1.7

cut the high-grade ore; the latter values range from +3.4‰ to −2.4‰, with an average of +0.4‰ (Table 2).

There is a very obvious difference in the results obtained for hard high-grade hematite ores from deposits that have been classified by Beukes et al. (2002a) as ancient supergene in origin, i.e., Sishen and Rooinekke, and those presumed to be of hydrothermal origin. Microcrystalline platy hematite from high-grade ores from Sishen and Rooinekke has $\delta^{18}O$ values ranging between +2.0‰ and −3.9‰ (Fig. 4A). Compared to this, samples from deposits of hydrothermal origin are depleted in ^{18}O, with $\delta^{18}O$ values ranging between +0.9‰ and −7.3‰ (Fig. 4A). $\delta^{18}O$ values below −4‰ were observed in all six deposits considered to be of hydrothermal origin, but are conspicuously absent from the two deposits deemed to be of ancient supergene origin. Results obtained in this study are in excellent agreement with similar data reported by Powell et al. (1999) for hematite in microplaty hematite ore from Mount Tom Price (−1‰ to −4‰, $n = 4$) and by Bird et al. (1993) for hard hematite ore (−5.0‰, $n = 1$) and soft hematite ore (−5.0‰ to −6.5‰, $n = 4$) from the N4E deposit in the Carajas District.

DISCUSSION

Introduction

There is growing consensus that high-grade iron ore bodies hosted by banded iron formations form by epigenetic residual enrichment of iron at the expense of other constituents (Barley et al., 1999; Taylor et al., 2001; Beukes et al., 2002a). Residual enrichment of iron is attributed to effective leaching of SiO_2 and other constituents present in the protolith BIF by either meteoric or hydrothermal fluids. The BIF protolith is an intimate and fine-grained mixture of quartz/chert and iron oxides (hematite and magnetite), as well as Fe^{2+}-rich carbonates (ankerite, siderite) and silicates (e.g., greenalite, grunerite). Ferrous iron hosted in magnetite, carbonates, or silicates is oxidized to form hematite, i.e., ferric iron oxide, during the enrichment process. The predominance of martite in samples from some high-grade hematite ore deposits suggests that the oxidation of carbonates and silicates to hematite proceeds via magnetite as an intermediate product.

Little is known about the fluids involved in the enrichment process. For the ancient supergene deposits of Sishen and Rooinekke, descending meteoric water may have driven the alteration process. Hydrothermal fluids that might have been responsible for the formation of high-grade BIF-hosted iron ore have been characterized only for Mount Tom Price (Hagemann et al., 1999; Taylor et al., 2001) and Thabazimbi (Netshiozwi, 2002). Hagemann et al. (1999) identified three compositionally different aqueous fluid inclusion types in megaquartz-hematite veins

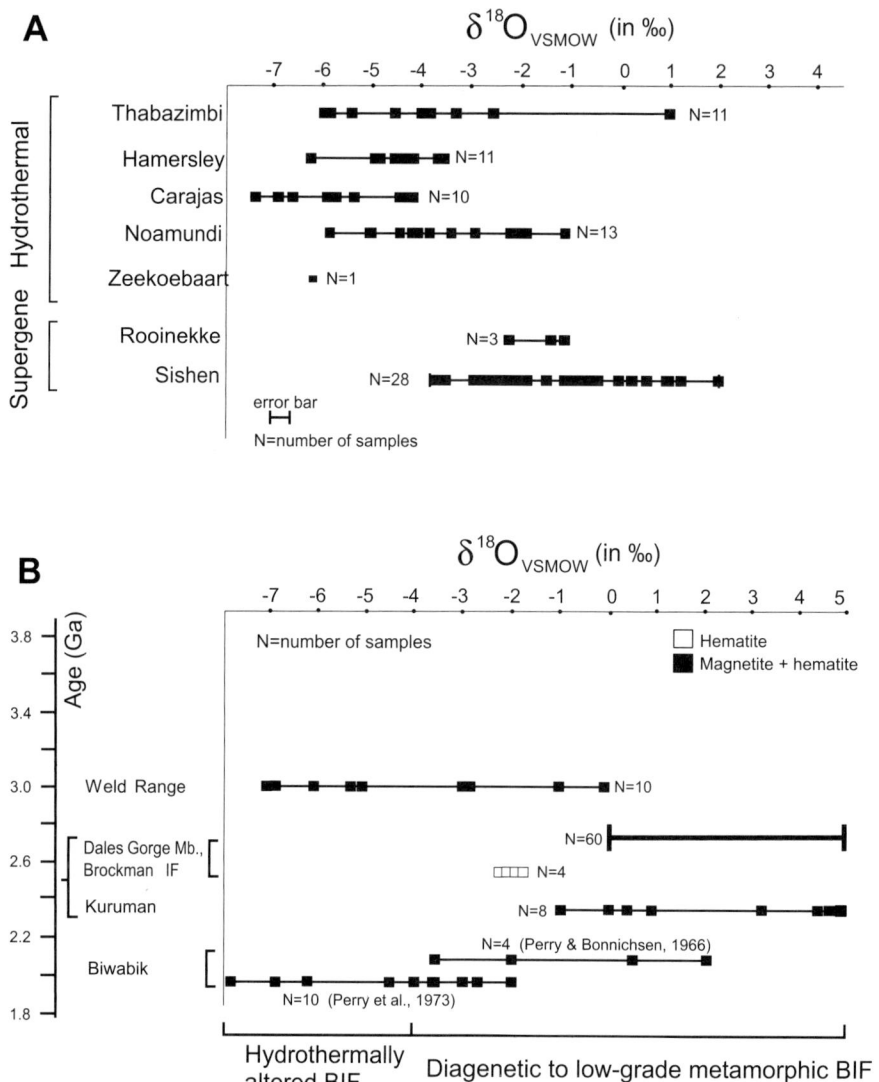

Figure 4. (A) Oxygen isotope composition of hematite from high-grade iron ores in selected deposits. Division of deposits studied into those of presumably hydrothermal origin and those of presumably ancient supergene origin is based on field geological relationships as described by Beukes et al. (2002a). Values obtained for specularitic hematite are not displayed. (B) Oxygen isotope composition of iron oxides (hematite, magnetite) in low-grade metamorphic BIF of Archean-Paleoproterozoic age. Data sources: Weld Range (Perry and Ahmad, 1983); Dales Gorge Member (Becker and Clayton, 1976); Kuruman Iron Formation (Perry and Ahmad, 1983); Biwabik Iron Formation (Perry and Bonnichsen, 1966; Perry et al., 1973). Data for Mount Tom Price and Paraburdoo are combined and displayed as Hamersley.

at the Mount Tom Price deposit, including a low-salinity NaCl-dominated fluid type (salinity [s] = 0.7–3.3 wt% NaCl$_{equivalent}$, homogenization temperature T_H = 138–190 °C), a moderately saline NaCl-dominated fluid (s = 12.3–14.1 wt% NaCl$_{equivalent}$, T_H = 130–209 °C), and a CaCl$_2$-rich brine (s = 18–25 wt% CaCl$_2$). However, megaquartz veins studied by Hagemann et al. (1999) are spatially but not necessarily genetically associated with the formation of the high-grade iron ore. Textural evidence available for sparitic dolomite and calcite intimately associated with high-grade hematite ore at Thabazimbi, on the other hand, suggests that these carbonates formed cogenetic with ore-forming microplaty hematite (Gutzmer et al., 2001a). The study of inclusions within these carbonates should thus characterize the ore-forming fluid. A detailed investigation by Netshiozwi (2002) revealed the presence of two fluid end members, a high-salinity brine (s = 27 wt% NaCl$_{equivalent}$) and a fluid of moderate to low salinity (s = 7–10 wt% NaCl$_{equivalent}$), with homogenization temperatures ranging from 120 °C to 180 °C for both fluids. The results from Thabazimbi compare well with those obtained by Hagemann et al. (1999) at Mount Tom Price, suggesting that hydrothermal enrichment of iron-formation protolith to high-grade hematite iron ore is the result of interaction of iron formation with water-rich fluid mixtures at moderate temperatures (Barley et al., 1999; Hagemann et al., 1999; Netshiozwi, 2002).

It is likely that the oxygen isotope composition of fine-grained ore-forming hematite/martite represents a mixture between the oxygen isotopic composition of iron oxides that were present in the BIF protolith and oxygen introduced by the ore-forming fluid. Evaluating the significance of the $\delta^{18}O$ values of ore-forming hematite thus requires knowledge of the oxygen isotopic composition of iron oxide in the protolith and of the aqueous fluid responsible for enrichment, as well as the temperature of ore-formation and temperature-dependent oxygen isotope fractionation between water and iron oxide minerals.

Oxygen Isotope Geochemistry of Iron Oxides in BIF Protolith

Hoefs (1992) provided a comprehensive compilation of $\delta^{18}O$ values of iron oxides (magnetite, hematite) in banded iron formations. Data are available for only four weakly metamorphosed iron formations of Archean and Paleoproterozoic age, namely the ca. 3.0 Ga Weld Range Iron Formation of the Yilgarn Craton (Perry and Ahmad, 1983), the 2.45 Ga Dales Gorge Member of the Brockman Iron Formation (Becker and Clayton, 1976), the 2.45 Ga Kuruman Iron Formation of the Transvaal Supergroup (Perry and Ahmad, 1983), and the 1.9 Ga Biwabik Iron Formation of the Superior Craton (Perry and Bonnichsen, 1966; Perry et al., 1973). Oxygen isotopic compositions of fine-grained magnetite and hematite of these iron formations range from +5‰ to −8‰ (Fig. 4B). This range, together with equivalent data for coexisting quartz, is thought to represent a disequilibrium array indicative of open-system fluid-rock interaction and recrystallization of iron oxide minerals during diagenesis, and low-grade metamorphism (Gregory, 1986; Hoefs, 1992).

We critically evaluated the original data sets available. From mineralogical and petrographical descriptions of the material analyzed, and original descriptions of the geological setting of the deposits that were sampled, we were able to define the range of $\delta^{18}O$ values of magnetite and hematite in very weakly metamorphosed BIF to −4‰ and +5‰, a smaller range than that reported by Hoefs (1992). Most of the data points in Figure 4B fall within the latter range, with the notable exception of four data points from the Weld Range (Perry and Ahmad, 1983) and three data points from the Biwabik Iron Formation (from the study of Perry et al., 1973). In their contribution, Perry et al. (1973) state explicitly that iron oxides unusually depleted in ^{18}O in the Biwabik Iron Formation are restricted to a part of the succession that had undergone intensive hydrothermal alteration. Similarly, the detailed petrographic descriptions by Gole (1980) of Weld Range BIF suggest widespread hydrothermal alteration as indicated by the presence of, for example, chalcopyrite and arsenopyrite. It thus appears reasonable to attribute the presence of iron oxides with unusually low $\delta^{18}O$ values to hydrothermal alteration, and to exclude them from the expected range of isotopic compositions of magnetite and hematite in weakly metamorphosed BIF of Paleoproterozoic and Archean age.

Temperature-Dependent Isotope Fractionation

The temperature-dependent isotope fractionation between iron oxides and water, especially at low temperature, has been at the center of much debate (see review by Yapp, 2001). However, for our purpose it is important to note that similar, though not identical, temperature-dependent oxygen isotope fractionation curves for the hematite-water system were obtained experimentally by Yapp (1990) and Bao and Koch (1999), and that these curves correspond reasonably well to the curve constructed by Clayton and Epstein (1961) from natural assemblages. In this contribution, we have chosen the fractionation curve experimentally determined by Yapp (1990) in the temperature range 20 °C to 120 °C and extrapolated it to 300 °C (Fig. 5). The resultant curve shows an important crossover point at ~100 °C. At temperatures below this point, hematite precipitating in isotopic equilibrium from an aqueous solution should be enriched, and above this point it should be depleted in ^{18}O, relative to the parent solution (Fig. 5).

Hydrothermal Hematite Formation

$\delta^{18}O$ values as low as −7.3‰ observed for hematite ores of hydrothermal origin (Table 1) suggest a depletion of at least 3‰ in $\delta^{18}O$ compared to iron oxides in the BIF protolith (Fig. 4B). Fluid inclusion constraints available for the Thabazimbi (Netshiozwi, 2002) and Mount Tom Price deposits (Hagemann et al., 1999; Taylor et al., 2001) suggest temperatures of ore formation in the range of 120 °C to 209 °C. No reliable pressure correction can be applied to these values, but because of the apparent lack of any significant regional metamorphic overprint recorded by the host rock strata, it is unlikely that it would exceed 10 °C.

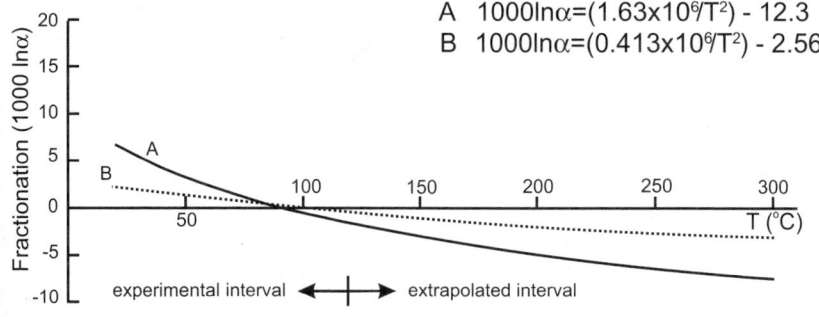

Figure 5. Temperature-dependent oxygen isotope fractionation curves for the system hematite-water. (A) was determined experimentally by Yapp (1990) in the temperature interval 20 °C to 120 °C; (B) was suggested by Clayton and Epstein (1961) on the basis of natural assemblages in the temperature interval 25 °C to 120 °C. Both curves are extrapolated to 300 °C.

If fluid inclusion homogenization temperatures reflect the temperature of ore formation at the Thabazimbi and Mount Tom Price deposits, estimates of the maximum $^{18}O/^{16}O$ ratio of the ore-forming fluid is possible from the most ^{18}O-depleted oxygen isotope compositions in hematite of the two deposits (−5.1‰ for Mount Tom Price, −6.0‰ for Thabazimbi; Table 1). These values are regarded as the most fluid-dominated, because they are farthest removed from the $\delta^{18}O$ values reported for iron oxide in the BIF protolith. However, they are still likely to constitute a mixture of fluid-derived oxygen and oxygen sourced from protore iron oxides; i.e., calculations can be used only to derive a maximum value for $\delta^{18}O$ of the ore-forming fluid. The assessment is based on two further assumptions: (1) there is equilibrium partitioning of oxygen isotopes, and (2) the temperature-dependent oxygen isotope fractionation curve of Yapp (1990) for the hematite-water system remains valid at temperatures exceeding 120 °C (Fig. 5).

Calculations suggest that the ore-forming fluid at Thabazimbi had a maximum $\delta^{18}O$ composition of −4.4‰ at 120 °C, or −1.7‰ at 180 °C. At Mount Tom Price the value was −2.4‰ at 140 °C, and +0.2‰ at 200 °C. Because these values are estimates of the maximum possible $\delta^{18}O$ values of the ore-forming fluid, we conclude that the actual hydrothermal fluid must have been depleted in ^{18}O relative to VSMOW. Such depletion is typical for shallow crustal fluids, i.e., fluids of meteoric origin that have not interacted extensively with silicate rocks (Hoefs, 1997). This is in excellent agreement with the results of a fluid chemistry study at Mount Tom Price indicating that the ore-forming fluids were similar in composition to those of modern basinal brines (Hagemann et al., 1999).

It is also interesting to speculate on possible reasons for the observation that martite-textured hematite is more depleted in ^{18}O than microplaty hematite from the same deposit (Table 2). This apparent difference could be explained by inherent differences in temperature-dependent isotope fractionation between the systems magnetite-H_2O and hematite-H_2O. Limited information available on oxygen isotope fractionation in the magnetite-water system (Chacko et al., 2001) suggests that magnetite should be somewhat depleted in ^{18}O compared to cogenetic hematite. Unfortunately, we have no means to assess the possible effects of martitization on the oxygen isotopic composition nor can we ensure that martite and microplaty hematite are formed cogenetically. From petrographic evidence we do, however, assume that paragenetically older magnetite may have formed at a higher temperature, followed by martitization and microplaty hematite deposition. Martite, formed at the expense of magnetite, may thus have inherited its ^{18}O-depleted oxygen isotope composition from its magnetite precursor.

Ancient Supergene Ore Formation

Bird et al. (1993) proposed that prolonged chemical weathering should lead to resetting of the pre-existing isotopic composition of iron oxide minerals contained in BIF. However, oxidation and ore formation in the ancient supergene Rooinekke and Sishen deposits obviously did not result in a significant shift in oxygen isotopic compositions of iron oxides. This is illustrated by the complete overlap between the range of $\delta^{18}O$ values measured for ore-forming hematite (+2.0‰ and −3.9‰, Fig. 4A) and the reference range for iron oxides from weakly metamorphosed BIF (Fig. 4B). This apparent lack of isotopic resetting is despite indications that isotopic equilibrium is usually attained between meteoric water and iron oxide precipitates in a range of modern and ancient surface environments (Yapp, 2001).

Recent investigations suggest that ancient supergene iron ores along the Paleoproterozoic Gamagara-Mapedi unconformity formed at low latitudes (Evans et al., 2002) and that lateritic weathering conditions prevailed (Gutzmer and Beukes, 1998). Consequently, ore formation is attributed to residual enrichment under lateritic weathering conditions. Assuming that equilibrium isotopic fractionation was attained between meteoric water and neoformed iron oxides, the $\delta^{18}O$ values of Paleoproterozoic meteoric water may be constrained (−3‰ to −8‰) from the isotopic composition of ore-forming hematite. This range corresponds very closely to the isotopic composition of modern precipitation in low latitudes, tropical and subtropical environments (see compilation by Fricke and O'Neil, 1999).

Although only a very crude first-order approximation, this result suggests that the oxygen isotopic composition of Paleoproterozoic meteoric water was similar to modern values, and possibly controlled by very similar processes. It is also in agreement with the conclusions of Godderies and Veizer (2000) and Gutzmer et al. (2001b) that Paleoproterozoic ocean water had an oxygen isotopic composition very similar to modern ocean water.

CONCLUSIONS

The oxygen isotope geochemistry of ore-forming hematite may be regarded as the first geochemical tool to differentiate between high-grade BIF-hosted iron ore deposits of ancient supergene or hydrothermal origin. Compared to the oxygen isotope composition of magnetite and hematite in the BIF protolith, hydrothermal hematite ores are depleted in ^{18}O, a shift that is not evident in hematite ores of ancient supergene origin. The oxygen isotope composition thus has the potential to become a tool to categorize high-grade iron ore deposits of unknown origin, especially in geologically complex or poorly exposed areas.

The ^{18}O depletion of ore-forming hematite in deposits of hydrothermal origin provides support for the hypothesis that fluids of shallow crustal origin, such as brines and meteoric water, are responsible for enrichment of banded iron formations to high-grade iron ores. Results obtained for ancient supergene deposits developed along the Paleoproterozoic Gamagara-Mapedi unconformity in South Africa, in contrast, strengthen the argument that the Paleoproterozoic atmosphere-hydrosphere-lithosphere system at that time was very similar to modern Earth (Beukes et al., 2002b). A more comprehensive evaluation of the oxygen isotope geochemistry is at present hampered by the lack of constraints regarding the temperature of ore formation and the nature of the fluids involved in the enrichment process.

ACKNOWLEDGMENTS

The authors acknowledge the support of geologists of Kumba Resources of South Africa, CVRD of Brazil, Hamersley Iron of Australia, and TISCO of India for permitting access and providing guidance during the visits to deposits. The Indo-South African Inter-Government Science and Technology Cooperation Programme, project no. 5/1/12/15, supported this research project. Mukhopadhyay acknowledges a postdoctoral fellowship grant from the National Research Foundation of South Africa. Pack gratefully acknowledges funding by the Deutscher Akademischer Austauschdienst, and support by Viorel Atudorei and Toti Larson during analytical work at the University of New Mexico. Gutzmer, Mukhopadhyay, and Beukes are grateful to Carlos Rosiere for numerous discussions regarding the origin of high-grade iron ores. Many thanks also to Simon Netshiozwi for the preparation of some of the hematite samples from Thabazimbi and Carajas. Finally, we gratefully acknowledge Steve Kesler and two anonymous reviewers for their constructive criticism, which helped to improve this manuscript.

REFERENCES CITED

Bao, H., and Koch, P.L., 1999, Oxygen isotope fractionation in ferric oxide-water systems: Low temperature synthesis: Geochimica et Cosmochimica Acta, v. 63, p. 599–613, doi: 10.1016/S0016-7037(99)00005-8.

Barley, M.E., Pickard, A.L., Hagemann, S.G., and Folkert, S.L., 1999, Hydrothermal origin for the 2 billion year old Mount Tom Price giant iron ore deposit, Hamersley Province, Western Australia: Mineralium Deposita, v. 34, p. 784–789, doi: 10.1007/s001260050238.

Becker, R.H., and Clayton, R.N., 1976, Oxygen isotope study of a Precambrian banded iron-formation, Hamersley Range, Western Australia: Geochimica et Cosmochimica Acta, v. 40, p. 1153–1165, doi: 10.1016/0016-7037(76)90151-4.

Beukes, N.J., Gutzmer, J., and Mukhopadhyay, J., 2002a, The geology and genesis of high-grade hematite iron ore deposits, in Conference Proceedings, Iron Ore 2002, 9-11 September 2002, Perth: Carlton, Australia, Australasian Institute of Mining and Metallurgy, p. 25–30.

Beukes, N.J., Dorland, H., Gutzmer, J., Nedachi, M., and Ohmoto, H., 2002b, Tropical laterites, life on land, and the history of atmospheric oxygen: Geology, v. 30, p. 491–494.

Bird, M.I., Longstaffe, F.J., Fyfe, W.S., Kronberg, B.I., and Kishida, A., 1993, An oxygen-isotopic study of weathering in the Eastern Amazon Basin, Brazil, in Swart, P.K., Lohman, K.C., McKenzie, J., and Savin, S., eds., Climate change in continental isotopic record: American Geophysical Society Monograph 78, p. 295–307.

Chacko, T., Cole, D.R., and Horita, J., 2001, Equilibrium oxygen, hydrogen and carbon isotope fractionation factors applicable to geologic systems, in Valley, J.W., and Cole, D.R., eds., Stable isotope geochemistry: Washington, D.C., Mineralogical Society of America, Reviews in Mineralogy and Geochemistry, v. 43, p. 1–81.

Clayton, R.N., and Epstein, S., 1961, The use of oxygen isotopes in high-temperature geological thermometry: Journal of Geology, v. 69, p. 447–452.

Evans, D.A.D., Beukes, N.J., Kirschvink, J.L., 2002, Paleomagnetism of a lateritic paleo-weathering horizon and overlying Paleoproterozoic redbeds from South Africa: Implications for the Kaapvaal apparent polar wander path and a confirmation of atmospheric oxygen enrichment. Journal of Geophysical Research-Solid Earth, v. 107, p. 2326–2333.

Fricke, H.C., and O'Neil, J.R., 1999, The correlation between $^{18}O/^{16}O$ ratios of meteoric water and surface temperature: Its use in investigating terrestrial climate change over geologic time: Earth and Planetary Science Letters, v. 170, p. 181–196, doi: 10.1016/S0012-821X(99)00105-3.

Godderies, Y., and Veizer, J., 2000, Tectonic control of chemical and isotopic composition of ancient oceans: The impact of continental growth: American Journal of Science, v. 300, p. 434–461.

Gole, M.J., 1980, Mineralogy and petrology of very-low-metamorphic grade Archaean banded iron-formations, Weld Range, Western Australia: American Mineralogist, v. 65, p. 8–25.

Gregory, R.T., 1986, Oxygen isotope systematics of quartz-magnetite pairs from Precambrian iron formations: Evidence for fluid-rock interaction during diagenesis and metamorphism, in Walther, J.V., and Wood, B.J., eds., Fluid-rock interactions during metamorphism: Heidelberg, Springer-Verlag, Advances in Physical Geochemistry, v. 5, p. 132–153.

Gutzmer, J., and Beukes, N.J., 1998, Earliest laterites and possible evidence for terrestrial vegetation in the Early Proterozoic: Geology, v. 26, p. 263–266, doi: 10.1130/0091-7613(1998)026<0263:ELAPEF>2.3.CO;2.

Gutzmer, J., Beukes, N.J., Netshiozwi, S., and Szabo, A., 2001a, Genesis of world-class, high-grade iron ore deposits—The South African experience, in Piestrzynski, A., ed., Mineral deposits at the beginning of the 21st century: Proceedings of the 6th Biennial Joint SGA-SEG meeting, Krakow, 26-29 August 2001: Lisse, A.A. Balkema, p. 1079–1082.

Gutzmer, J., Pack, A., Lueders, V., Wilkinson, J.J., Beukes, N.J., and Van Niekerk, H.S., 2001b, Formation of hydrothermal jasper and andradite during low-temperature hydrothermal seafloor metamorphism, Ongeluk Formation, Northern Cape Province: Contributions to Mineralogy and Petrology, v. 142, p. 27–42.

Hagemann, S.G., Barley, M.E., Folkert, S.E., Yardley, W.D., and Banks, D.A., 1999, A hydrothermal origin for the giant BIF-hosted Tom Price iron ore deposit, in Stanley et al. Rankin, A.H., Bodnar, R.J., Naden, J., Yardley, B.W.D., Criddle, A.J., Hagni, R.D., Gize, A.P., Pasava, J., Fleet, A.J., Seltmann, R., Halls, C., Stemprok, M., Williamson, B., Herrington, R.J., Hill, R.E.T., Prichard, H.M., Wall, F., eds., Mineral deposits: Processes to processing: Proceedings of the 5th Biennial SGA Meeting and the 10th Quadrennial IAGOD Symposium, London, 22–25 August 1999: Rotterdam, A.A. Balkema, v. 1, p. 41–44.

Hayashi, K., Maruyama, T., and Satoh, H., 2001, Precipitation of gold in a low-sulfidation epithermal gold deposit: Insight from a submillimeter-scale oxygen isotope analysis of vein quartz: Economic Geology and the Bulletin of the Society of Economic Geologists, v. 96, p. 211–216.

Hoefs, J., 1992, The stable isotope composition of sedimentary iron oxides with special reference to banded iron formations, in Clauer, N., and Chauduri, S., eds., Isotopic signatures and sedimentary records: Heidelberg, Springer-Verlag, Lecture Notes in Earth Sciences, v. 43, p. 199–213.

Hoefs, J., 1997, Stable isotope geochemistry (4th edition): Berlin, Springer-Verlag, 201 p.

Hoefs, J., Mueller, G., and Schuster, A.K., 1982, Polymetamorphic relations in iron ores from the Iron Quadrangle, Brazil: The correlation of oxygen isotope variations with deformation history: Contribution to Mineralogy and Petrology, v. 79, p. 241–251, doi: 10.1007/BF00371515.

King, H.F., 1989, The rocks speak: Australasian Institute of Mining and Metallurgy Monograph 15, p. 1–316.

MacLeod, W.N., 1966, The geology and iron deposits of the Hamersley Range area, Western Australia: Perth, Geological Survey of Western Australia Bulletin, v. 117, p. 1–170.

Morris, R.C., 1980, A textural and mineralogical study of the relationship of iron ore to banded iron-formation in the Hamersley iron province of Western Australia: Economic Geology and the Bulletin of the Society of Economic Geologists, v. 75, p. 184–209.

Morris, R.C., 1985, Genesis of iron ore in banded iron-formation by supergene and supergene-metamorphic processes—A conceptual model, in Wolff, K.H., ed., Regional studies and specific deposits: Amsterdam, Elsevier, Handbook of Strata-bound and Stratiform Ore Deposits, v. 13, p. 72–235.

Netshiozwi, S.T., 2002, Origin of high-grade hematite ores at Thabazimbi Mine, Limpopo Province, South Africa [M.Sc. thesis]: Johannesburg, Rand Afrikaans University, 135 p.

Perry, E.C., and Ahmad, S.N., 1983, Oxygen isotope geochemistry of Proterozoic chemical sediments, in Medaris, L.G., Byers, C.W., Mickelson, D.M., and Shanks, W.C., eds., Proterozoic geology: Selected papers from an international Proterozoic symposium: Geological Society of America Memoir 161, p. 253–263.

Perry, E.C., and Bonnichsen, B., 1966, Quartz and magnetite oxygen-18-oxygen-16 fractionation in metamorphosed Biwabik iron formation: Science, v. 153, p. 528–529.

Perry, E.C., Tan, C., and Morey, G.B., 1973, Geology and stable isotope geochemistry of the Biwabik iron formation, Northern Minnesota: Economic Geology and the Bulletin of the Society of Economic Geologists, v. 68, p. 1110–1125.

Powell, McA., Oliver, N.S., Li, Z.X., Martin, D.McB., and Ronaszecki, J., 1999, Synorogenic hydrothermal origin for giant Hamersley iron oxide ore bodies: Geology, v. 27, p. 175–178, doi: 10.1130/0091-7613(1999)027<0175: SHOFGH>2.3.CO;2.

Ramdohr, P., 1980, The ore minerals and their intergrowths. Oxford, UK, Pergamon Press, 2 v., 1207 p.

Sharp, Z.D., 1990, A laser-based microanalytical method for the in-situ-determination of oxygen isotope ratios of silicates and oxides: Geochimica et Cosmochimica Acta, v. 54, p. 1353–1357, doi: 10.1016/0016-7037(90)90160-M.

Taylor, D., Dalstra, H.J., Harding, A.E., Broadbent, G.C., and Barley, M.E., 2001, Genesis of high-grade hematite orebodies of the Hamersley Province, Western Australia: Economic Geology and the Bulletin of the Society of Economic Geologists, v. 96, p. 837–878.

van Schalkwyk, J.F., and Beukes, N.J., 1986, The Sishen iron ore deposit, in Anhaeusser, C.R., and Maske, S.S., eds., Mineral deposits of Southern Africa: Johannesburg, Geological Society of South Africa, v. 1, p. 157–182.

Yapp, C.J., 1990, Oxygen isotopes in iron (III) oxides. I: Mineral-water fractionation factors: Chemical Geology, v. 85, p. 329–335, doi: 10.1016/0009-2541(90)90010-5.

Yapp, C.J., 2001, Rusty relics of Earth History: Iron (III) oxides, isotopes and surficial environments: Annual Review of Earth and Planetary Sciences, v. 29, p. 165–199, doi: 10.1146/annurev.earth.29.1.165.

MANUSCRIPT ACCEPTED BY THE SOCIETY 29 OCTOBER 2005

Rare earth elements in Precambrian banded iron formations: Secular changes of Ce and Eu anomalies and evolution of atmospheric oxygen

Yasuhiro Kato
Department of Geosystem Engineering, University of Tokyo, Bunkyo-Ku, Tokyo 113-8656, Japan

Kosei E. Yamaguchi
Institute for Research on Earth Evolution (IFREE), Japan Agency for Marine-Earth Science and Technology (JAMSTEC), Yokosuka, Kanagawa 237-0061, Japan

Hiroshi Ohmoto
Penn State Astrobiology Research Center of the NASA Astrobiology Institute and the Department of Geosciences, The Pennsylvania State University, University Park, Pennsylvania 16802, USA

ABSTRACT

Rare earth element (REE) analyses of Precambrian banded iron formations (BIFs) show that distinct negative Ce anomalies, although rather weak or moderate (Ce/Ce* = 0.5–0.9), are commonly present in Algoma-type BIFs of the Early and Middle Archean, and even in the 3.8–3.7 Ga Isua iron formation (IF). This indicates that the seawater columns from which the BIFs precipitated were not entirely anoxic and that Ce oxidation mechanisms already existed in the 3.8–3.7 Ga oceans. The presence of pronounced negative Ce anomalies (Ce/Ce* = 0.1–0.5) in Late Archean (2.9–2.7 Ga) Algoma-type BIFs suggests that strongly oxygenated oceanic conditions like today emerged by 2.9–2.7 Ga. This suggestion is consistent with geologic evidence that small but widespread Mn deposits formed during Late Archean time. The Hamersley and Transvaal IFs (2.7–2.4 Ga in age) have less noticeable Ce anomalies (Ce/Ce* = 0.7–1.0). These BIFs were deposited on an evolving rift in a land-locked ocean that became anoxic due to intense hydrothermal activity. The 2.2–2.1 Ga Superior-type IFs exhibit distinct negative Ce anomalies (Ce/Ce* = 0.2–0.7), but the ca. 1.9 Ga IFs and the ca. 0.7 Ga IFs have less distinct Ce anomalies. These variations in Ce/Ce* values of the post-2.7 Ga BIFs may reflect the episodicity in global mantle plume activity that created locally anoxic basins.

The Archean Algoma-type BIFs have distinctly positive but variable Eu anomalies (Eu/Eu* = 0.8–7). Their strong positive Eu anomalies suggest large contributions of hydrothermal fluids to the seawater involved in BIF precipitation. The large variation in Eu anomalies in Algoma-type BIFs reflects the large variation in mixing ratios of hydrothermal fluids and ambient seawaters at various depositional sites. However, the post-2.7 Ga Superior-type BIFs exhibit much lower and constant Eu/Eu* values (0.7–1.9). This implies that the REE chemistry of the basin water that hosted voluminous Superior-type BIFs was influenced by a riverine influx from the surrounding continent that grew rapidly due to global mantle plume activity, besides the intense hydrothermal influx.

Keywords: banded iron formation, Precambrian, rare earth element, Ce anomaly, Eu anomaly, redox condition, oxygen, geochemistry.

INTRODUCTION

Banded iron formations (BIFs) are chemical sedimentary rocks deposited in marine environments mostly during the Precambrian era (James, 1954). They occur mostly in cratons with 3.8–1.9 Ga sedimentary sequences, and are classified into two types based on their formation age, size, and associated lithologies (Gross, 1965). Algoma-type BIFs mainly formed during the Archean and developed in volcanic sequences on a relatively small scale (lateral extents <10 km and thickness <100 m), whereas Superior-type BIFs mostly formed during the Late Archean to Paleoproterozoic period and are associated with clastic sedimentary rocks as well as volcanic rocks. Superior-type BIFs, the dominant iron resource in the world, are typically very large; the largest one exceeds 10^5 km^2 in aerial distribution (Trendall and Blockley, 1970). In addition to these Precambrian iron formations (IFs), many IFs were deposited during the Neoproterozoic (ca. 0.7 Ga) and Phanerozoic periods on a small scale (Ohmoto et al., this volume).

A currently popular theory (e.g., Jacobsen and Pimentel-Klose, 1988) postulates that the ultimate sources of iron and silica for both types of BIFs are submarine hydrothermal emanations, mostly from mid-ocean ridges (MORs). This theory was developed primarily from rare earth element (REE) data on BIFs, which show the predominance of mantle-like Nd isotopes and the ubiquitous existence of positive Eu anomalies (e.g., Jacobsen and Pimentel-Klose, 1988; Bau and Dulski, 1996; Kato et al., 1998). Recently, it has been recognized that submarine volcanic rocks are much more abundantly associated with Superior-type BIFs than previously thought (e.g., Barley et al., 1997). Hence, global magmatism (probably plume-related mafic volcanism) could have played an important role in BIF deposition (e.g., Isley and Abbott, 1999; Kato, 2003).

However, depositional environments for BIFs are still controversial. Many researchers have regarded BIFs as chemical precipitates in shallow-sea environments along continental shelves or near island arcs, because BIFs are often associated with terrigenous clastic rocks (e.g., Jacobsen and Pimentel-Klose, 1988; Manikyamba et al., 1993). According to this scenario, Fe^{2+}-rich deep-sea water from a MOR hydrothermal source upwelled to the continental margin, and mixed with oxygenated surface water to produce deposition of BIFs that are conformable with terrigenous clastic sediments.

In contrast, Kato et al. (1998, 2002a) suggested that some Algoma-type BIFs deposited on the flanks of MORs where hydrothermal Fe-oxyhydroxide particulates precipitated in situ. This suggestion was made because of the recognition of remarkable similarities in geologic settings and REE geochemistry between the BIFs and modern MOR hydrothermal Fe-Mn sediments. The recognition of horizontal shortening by thrust duplication of BIF successions, along with new geochronological data, has led many researchers to propose a new geologic and stratigraphic framework for the Archean greenstone belts in the plate tectonic regime (de Wit et al., 1992; Kusky and Kidd, 1992; Krapez, 1993; Barley, 1993, Kato et al., 1998; Komiya et al., 1999). In this framework, depositional sites of Algoma-type BIFs and terrigenous clastic sediments were completely different; the BIFs are interpreted to have been deposited in deep-sea pelagic environments, far away from continental margins, and to have been placed in their present position as accretionary prism complexes by obduction/subduction.

Although opinions greatly vary on the depositional environments and the depositional sites relative to hydrothermal discharge areas for BIFs, the consensus among researchers is that the geochemical features of BIFs reflect those of the water column from which BIFs precipitated. Among these geochemical features, REE chemistry of BIFs has been well documented. Several researchers have attempted to reconstruct the chemical evolution of the Precambrian oceans from the REE features of BIFs (Fryer, 1977; Derry and Jacobsen, 1990; Danielson et al., 1992; Bau and Möller, 1993; Kato et al., 1998).

Redox conditions of Phanerozoic seawater have been inferred from Ce anomalies in marine sedimentary rocks (e.g., MacLeod and Irving, 1996; Kato et al., 2002b). Unlike other REEs, in the presence of dissolved oxygen Ce^{3+} in seawater is removed as CeO$_2$ or Ce(OH)$_4$, especially with Mn hydroxides (Elderfield et al., 1981). Because of this oxidative removal of Ce, modern oxygenated deep-sea waters are depleted in Ce and show very large negative Ce anomalies. These large negative Ce anomalies are also seen in the hydrothermal Fe-Mn sediments on ridge flanks (Bender et al., 1971; Ruhlin and Owen, 1986). Kato et al. (1998, 2002a) have suggested that some Algoma-type BIFs are ancient counterparts of modern ridge-flank Fe-Mn sediments. Europium anomalies of seawaters are mainly determined by the relative contributions of hydrothermal and riverine influxes (Elderfield, 1988). Therefore, we could monitor the outline of ocean chemistry evolution controlled by these two major influxes by using the secular variation of Eu anomalies preserved in BIFs. In this paper, we examine published and unpublished data on REEs of Precambrian BIFs, especially the Ce and Eu anomalies, in order to constrain the redox conditions and relative importance of hydrothermal versus riverine (continental) influxes in the Precambrian oceans.

REES AS AN INDICATOR FOR PALEOCEAN CHEMISTRY

Total concentrations and relative abundances of REEs in marine sediments are very useful geochemical indicators to determine the origins and depositional environments of marine sediments and to estimate various geologic processes during sedimentation (e.g., Murray et al., 1992a; Kato and Nakamura, 2003). REEs typically exist in a trivalent oxidation state, but Ce and Eu are unique in that they can also exist in tetravalent and divalent oxidation states, respectively, giving rise to their anomalous behaviors. Both are very important elements providing some constraints on paleocean chemistry. The Ce anomaly in ancient submarine sediments is one of the most useful indicators to monitor oceanic redox conditions. The Ce anomaly of seawater corresponds well to varying redox conditions (de Baar et al., 1988; German and Elderfield, 1989; Sholkovitz and Schneider, 1991) owing to rapid reduction or oxidation reactions between Ce^{3+} and Ce^{4+} in response to changes in the redox conditions (Sholkovitz et al., 1992). For example, in the Cariaco Trench, the overlying oxygenated waters exhibit striking negative Ce anomalies, but in the oxic/anoxic transition zone Ce anomalies disappear, and in the anoxic bottom water no or positive Ce anomalies are found (de Baar et al., 1988).

However, there is no linear relationship between the magnitudes of Ce anomalies and the contents of dissolved oxygen in seawater. This is because REEs, and thus Ce anomalies, are also influenced by ocean circulation and biogeochemical cycling of nutrients and other trace metals. The presence of either negative or positive Ce anomalies, however, suggests that the redox-related Ce fractionation took place in the oceans, although qualitatively. Hence, if ancient sediments have mimicked the Ce anomaly of contemporaneous seawater, we can get useful information on the paleoredox condition. Earlier studies have related Ce anomalies of fossil apatites and carbonates in marine sediments to anoxic events in the Phanerozoic, assuming that these minerals record the Ce anomalies of contemporaneous seawater (e.g., Wright et al., 1987; Liu et al., 1988). However, it was reported subsequently that the Ce anomalies in fossil minerals change as a result of diagenetic alteration, and thus these fossil materials do not retain their marine Ce anomalies at the time of sedimentation (German and Elderfield, 1990; Sholkovitz and Schneider, 1991). Because the pore waters in reducing nearshore sediments have high REE concentrations (Elderfield and Sholkovitz, 1987; Sholkovitz et al., 1989) and because these apatite-rich fossils can take up large amounts of REEs, Ce anomalies in small fossil minerals may be easily altered during sediment diagenesis; thus, they cannot be used as a paleoredox proxy. On the other hand, it is unlikely that REE signatures of *bulk* sediments are significantly affected during diagenesis, because the amounts of REEs in *bulk* sediments, such as BIFs, are greater than those in diagenetic fluids by five or six orders of magnitude. In fact, a good correlation of Ce anomalies with other independent indicators of oceanic paleoredox conditions has been reported for bulk limestone samples of Cretaceous age (MacLeod and Irving, 1996). Murray et al. (1992b) and Kato et al. (2002b) have shown that strong negative Ce anomalies in Phanerozoic pelagic cherts were not severely affected by diagenesis, but they were predominantly influenced by the chemistry of the immediately overlying seawater. These results confirm the validity of the Ce anomaly in *bulk* sediments as a paleoredox indicator of the oceans. Similarly, Eu anomalies in sedimentary rocks are likely to reflect Eu anomalies of ancient ocean waters, and the magnitude of the Eu anomaly may be used to evaluate the relative contributions of two major sources of REEs (riverine and hydrothermal) in the ancient seawater (Elderfield, 1988). Mitra et al. (1994) have shown that REEs, including Eu, behave conservatively during mixing of hydrothermal fluids and ambient seawater at venting sites.

It is generally believed that REEs of bulk-rock samples are virtually unaffected by post-depositional processes, including diagenesis, metamorphism, and weathering, unless the fluid/rock ratios were very large and the dissolution/reprecipitation of REE-bearing minerals (as well as major minerals) occurred extensively (e.g., Taylor and McLennan, 1985; McLennan and Taylor, 1991). For example, many investigations (Grauch, 1989, and references therein) have recognized that REEs are basically immobile during high-grade metamorphism under relatively fluid-absent conditions, whereas several studies (e.g., Nyström, 1984) have reported REE mobility under low-grade metamorphic conditions where metamorphic fluids were prevalent. In some instances where carbonate and/or fluoride complexes of REEs could have been important in the metamorphic fluids, REEs were apparently remobilized even during high-grade (e.g., the granulite facies) metamorphism (e.g., Pan and Fleet, 1996). However, carbonate veins and fluorine minerals (fluorapatite and fluorite) containing high REE concentrations are rare in BIFs, and there is no evidence for REE mobility during the metamorphism of BIFs. Because the "distribution coefficients" of REEs between BIFs and metamorphic fluids must have been very large ($\sim 10^5$-10^6), as in the case of diagenesis mentioned above, the REE concentrations and patterns of bulk BIF samples would not have been significantly affected by metamorphic processes, except when the fluid/rock ratios were extremely high (e.g., Elderfield and Sholkovitz, 1987; Michard, 1989).

The mobility and redistribution of REEs by weathering have been reported for lateritic soil profiles (e.g., Braun et al., 1998) and soil formations associated with granitic rocks (e.g., Nesbitt, 1979; Aubert et al., 2001). Intensive and/or extensive weathering has resulted in a complete disappearance of original rock texture, dissolution of the primary REE-bearing minerals (e.g., apatite, sphene, and monazite), as well as feldspars and silicates, in the upper soil sections, and reprecipitation of secondary phosphate minerals (e.g., cerianite, florencite, and rhabdophane) in the lower soil sections (Braun et al., 1998 and references therein). However, the REE data selected for this study were obtained on BIF samples that have well-preserved primary banding, indicating that these samples were not subjected to severe weathering. Therefore, it would be reasonable to conclude that the selected

BIF samples have retained their original REE chemistry, which reflected the chemistry of waters involved in BIF precipitation.

Ce AND Eu ANOMALIES IN MODERN OCEAN WATERS AND SUBMARINE HYDROTHERMAL Fe-Mn SEDIMENTS

REE data of modern seawaters, marine sediments, and post-Archean sedimentary rocks are generally normalized to those of post-Archean average Australian shale (PAAS; Taylor and McLennan, 1985) or North American Shale Composite (NASC; Goldstein and Jacobsen, 1988). The data indicate that the REE concentration ratios of post-Archean shales are remarkably similar, suggesting the similar chemical compositions for the continental crust after the Archean era. Similarly, in order to understand the differences in behaviors of REEs in the Archean oceans, REE data on Archean marine sedimentary rocks should be normalized to the average Archean shale, which represents the average Archean upper continental crust. However, there is no consensus among researchers on the average Archean crustal composition, because there are large variations in the REE ratios among rocks of the same age and there may be an evolutionary trend in the REE ratios of Archean shales (e.g., Taylor and McLennan, 1985; Condie, 1993). Therefore, it is not meaningful to select REE data from a single shale for normalization of various-aged BIFs. In this study, the REE abundances of modern ocean waters, sediments, and ancient BIFs are normalized to those of the Leedey chondrite obtained by Masuda et al. (1973) and Masuda (1975). Cerium and europium anomalies are defined quantitatively as $Ce/Ce^* = Ce_N/[(La_N)(Pr_N)]^{1/2}$ and $Eu/Eu^* = Eu_N/[(Sm_N)(Gd_N)]^{1/2}$, respectively, where Ce^* and Eu^* are the hypothetical concentrations that strictly trivalent Ce and Eu would have (Taylor and McLennan, 1985). The subscript "N" indicates chondrite-normalized values. When Ce and Eu are depleted, Ce/Ce^* and Eu/Eu^* values are below unity and vice versa. Another method of calculation, $[Ce/Ce^* = 2Ce_N/(La_N+Pr_N), Eu/Eu^* = 2Eu_N/(Sm_N+Gd_N)]$, is often used for seawaters and marine sediments (e.g., de Baar et al., 1985), but differences between the values of these two types of calculations are negligible for a wide range of anomalies (Kato et al., 2002a).

Evaluation of the presence or absence of a Ce anomaly is often complicated because of the anomalous behavior of La (Bau and Dulski, 1996). Lanthanum enrichment (i.e., positive La anomalies), together with the well-known Ce depletion, has been identified in modern seawaters (e.g., Zhang et al., 1994; Alibo and Nozaki, 1999). Positive La anomalies have also been reported for hydrothermal Fe-Mn sediments on oceanic crust (Barrett and Jarvis, 1988) and some BIFs (Barrett et al., 1988; Bau and Dulski, 1996). Consequently, anomalous La enrichment can create *false* negative Ce anomalies in some cases. In order to distinguish *true* negative Ce anomalies from *false* ones, plots of Ce/Ce^* against Pr/Pr^* ($Pr/Pr^* = Pr_N/[(Ce_N)(Nd_N)]^{1/2}$), which were suggested by Bau and Dulski (1996), are often used: *true* negative Ce anomalies are represented by negative Ce/Ce^* and positive Pr/Pr^* values, whereas *true* positive Ce anomalies are represented by positive Ce/Ce^* and negative Pr/Pr^* values (see Figures 2 and 4). Slightly positive Gd anomalies have been reported for modern seawaters (de Baar et al., 1985; Alibo and Nozaki, 1999) and ancient sediments (Bau and Dulski, 1996). However, anomalous abundances of Gd are not common; Gd anomalies are generally absent in most BIFs. Thus, in this paper Eu anomalies are calculated as $Eu_N/[(Sm_N)(Gd_N)]^{1/2}$, not by using Tb instead of Gd as suggested by Bau and Dulski (1996).

Chondrite-normalized REE patterns of modern seawaters (deep-sea and surface), black smoker fluids, hydrothermal Fe-Mn sediments on the ridge flank of the East Pacific Rise (EPR), hydrothermal Fe-Mn umbers in the Phanerozoic accretionary complexes, and hydrogenetic Mn crusts are shown in Figure 1. The Ce/Ce^*-Pr/Pr^* diagrams for them are given in Figure 2. It is clearly shown that seawaters plot in the region of *true* negative Ce anomalies. Modern oxygenated deep-sea waters (>2000 m) exhibit very large negative Ce anomalies ($Ce/Ce^* = 0.03$–0.25) and moderate negative Eu anomalies ($Eu/Eu^* = 0.63$–0.72) together with positive La and Gd anomalies (Fig. 1A). Moderate HREE (heavy REE) enrichment is also recognized. The positive La and Gd anomalies may be caused by unusually high stability of La and Gd of the lanthanide series in seawater (e.g., de Baar et al., 1985). Surface seawaters (<200 m) have less striking negative Ce anomalies ($Ce/Ce^* = 0.33$–0.43) and HREE enrichment, but similar Eu anomalies ($Eu/Eu^* = 0.68$–0.73), compared to deep-sea waters. The REE patterns of high-temperature black smoker fluids differ greatly from those of seawater, and are characterized by large positive Eu anomalies ($Eu/Eu^* = 7.2$–15.1), no Ce anomaly, and remarkable LREE (light REE) enrichment (Fig. 1B). In addition, black smoker fluids have higher REE concentrations than seawaters by two or three orders of magnitude.

Hydrogenetic Mn crusts (except for one sample) have considerably positive Ce anomalies ($Ce/Ce^* = 1.3$–3.4; Fig. 1E), which are caused by the oxidative scavenging of Ce and its preferential removal from seawater by Mn hydroxides (Elderfield et al., 1981). The hydrogenetic crusts plot in the region of $Ce/Ce^*>1$ and $Pr/Pr^*<1$ in the Ce/Ce^*-Pr/Pr^* diagram (Fig. 2). The magnitude of positive Ce anomalies is probably related to precipitation rates; a very low growth rate results in a pronounced positive Ce anomaly in hydrogenetic Mn crusts (Kuhn et al., 1998). The Eu/Eu^* values fall in a narrow range (from 0.67 to 0.78), which is the same as (or slightly higher than) that of seawater.

Submarine hydrothermal Fe-Mn sediments, which typically occur on flanks of MORs, generally mimic the striking depletion of Ce and the less striking depletion of Eu in deep-ocean waters because of scavenging of REEs from the deep-ocean waters (Fig. 1C). German et al. (1990) have suggested that hydrothermal plume particles (the precursor of Fe-Mn sediments) scavenge REEs with the proportions of 0.1% hydrothermal fluid and 99.9% seawater. Furthermore, most of the uptake of REEs from seawater occurs after the hydrothermal precipitates have settled on the seafloor, because the REE concentrations of plume par-

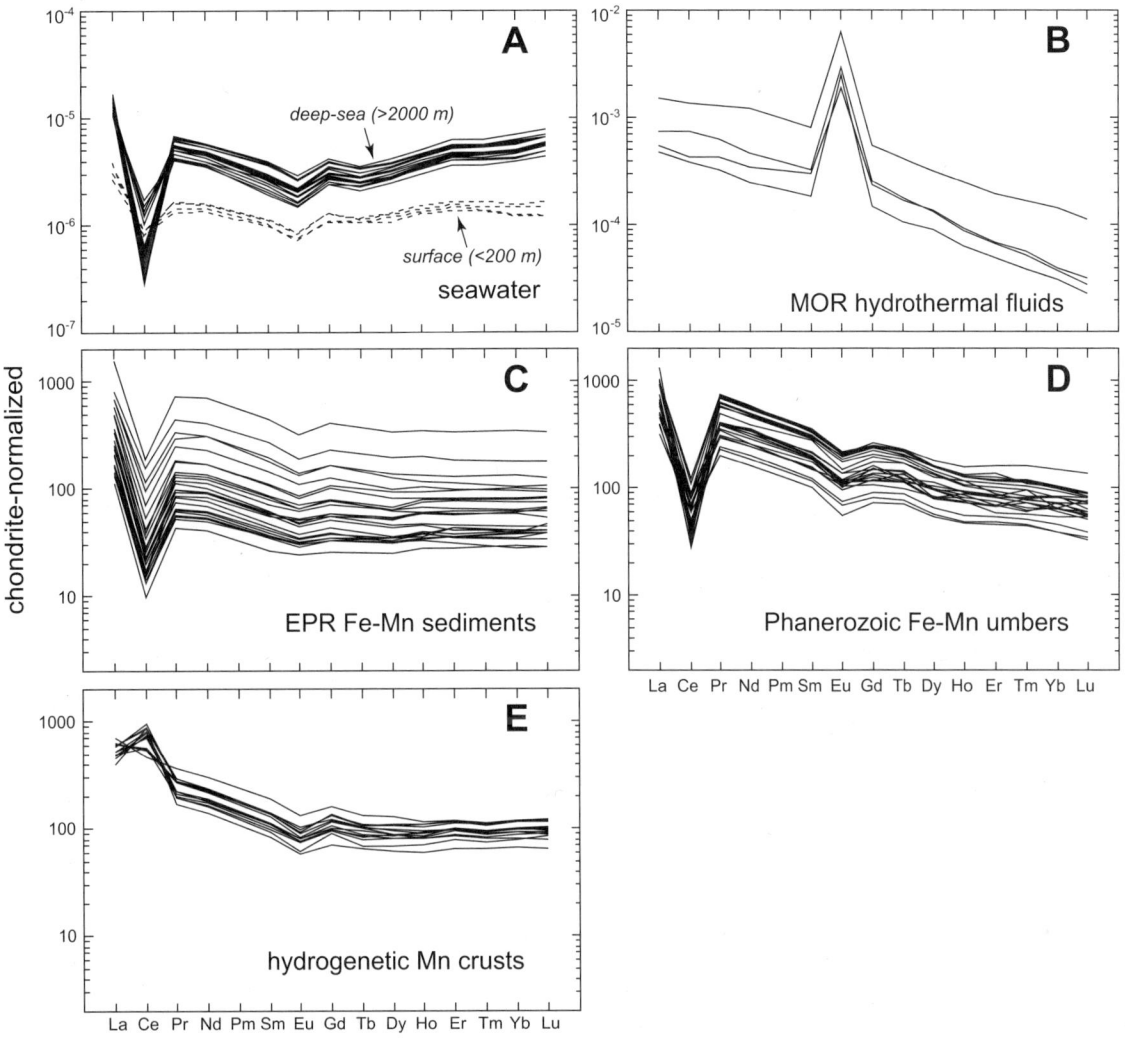

Figure 1. (A) Chondrite-normalized REE patterns of modern seawater. (B) MOR hydrothermal fluids. (C) EPR (East Pacific Rise) hydrothermal Fe-Mn sediments. (D) Phanerozoic hydrothermal Fe-Mn umbers. (E) Hydrogenetic Mn crusts. The data sources are summarized in Table 1.

ticles are found to be one order of magnitude lower than those of sediments. That is, the REEs in hydrothermal Fe-Mn sediments are dominated by seawater components. Consequently, both the Ce/Ce* (0.12–0.26) and Pr/Pr* values of Fe-Mn sediments agree well with those of some deep-sea waters (Fig. 2). On the other hand, the Eu/Eu* values of Fe-Mn sediments (0.75–0.93) are less negative compared to deep-sea waters (0.63–0.72), because of a contribution of submarine hydrothermal fluids with strong positive Eu anomalies. Even a quite trivial contribution from hydrothermal fluids can be detected as Eu anomalies of hydrothermal Fe-Mn sediments, because the hydrothermal fluids have striking positive Eu anomalies and much higher Eu contents than ocean water. In general, however, the subsequent overprinting of REEs from the overlying deep-ocean water with a negative Eu anomaly decreases the Eu anomaly values of the hydrothermal Fe-Mn sediments to the seawater value, with increasing concentrations of REEs; hence the hydrothermal Fe-Mn sediment with the highest ΣREE tends to have the Eu/Eu* value closest to the seawater value (Ruhlin and Owen, 1986; Olivarez and Owen, 1991). This is an important factor to keep in mind when we interpret BIF REE data.

It is well recognized that Phanerozoic Fe-Mn umbers have similar Ce and Eu features of the MOR hydrothermal Fe-Mn sediments (Fig. 1D), as suggested by Ravizza et al. (1999). The umbers that deposited in the oceans but subsequently accreted onto land have wide ranges of Ce/Ce* (0.03–0.24) and Pr/Pr* values, which correspond to the ranges of modern deep-sea waters (Fig. 2). Moreover, the umbers have Eu/Eu* values ranging from 0.60 to 0.90, which are also similar to those of seawaters. These REE features of modern and Phanerozoic submarine hydrothermal Fe-Mn sediments are excellent geochemical proxies of the Ce and Eu anomalies of oxygenated deep-ocean water. Moderate enrichment of HREE is recognized in modern deep-sea waters (Fig. 1A). However, HREE enrichment is absent

274 Y. Kato et al.

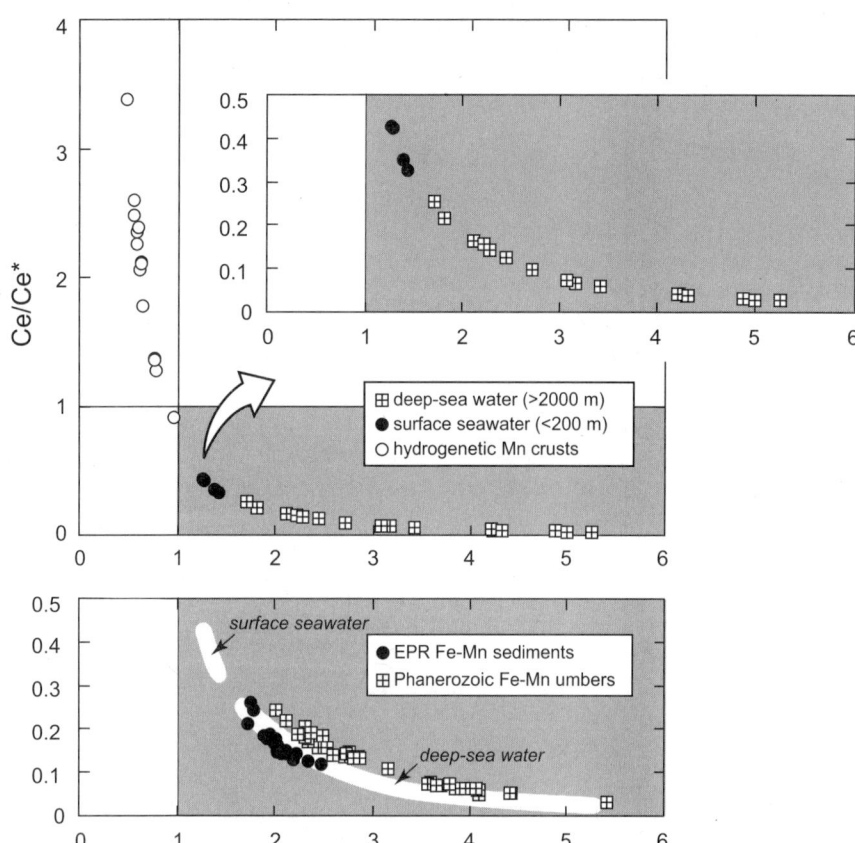

Figure 2. Ce/Ce* versus Pr/Pr* relationships for modern seawater, EPR Fe-Mn sediments, Phanerozoic Fe-Mn umbers, and hydrogenetic Mn crusts. The region of *true* negative Ce anomalies is shown as a shaded area.

in modern and Phanerozoic MOR sediments. This suggests a continuing preferential uptake of LREEs, compared to HREEs, by hydrothermal Fe-Mn sediments; this has been experimentally substantiated by Byrne and Kim (1990) and Koeppenkastrop and De Carlo (1992).

BRIEF DESCRIPTIONS OF SELECTED BIFS

The BIFs selected for this study, excepting the Braemar IF, are chemical sediments free from detritus in order to obtain information about the compositions of contemporaneous seawaters; data on shaly BIFs and ferruginous shales were excluded. Data sources for the selected BIFs are given in Table 1. Reliable data on all REEs (14 elements except for Pm) are necessary in order to discuss Ce and Eu anomalies in detail. However, complete analyses of BIFs using the inductively coupled plasma–mass spectrometry (ICP-MS) method are much fewer than those determined by the ICP or neutron activation analyses. Therefore, where there are no ICP-MS data, we use the ICP, neutron activation, and isotope dilution data: the Marra Mamba, Brockman, Hotazel, Väyrylänkylä, Gunflint, Biwabik, Sokoman, and Rapitan (Braemar) IFs.

Isua, Greenland

The 3.8–3.7 Ga Isua IF, the oldest in the world, is from the Isua supracrustal belt, southern West Greenland. The BIF samples analyzed for REEs by Morinaga (1997) have suffered greenschist-facies metamorphism, as indicated by the mineral assemblage of chlorite, epidote, albite, and quartz (Komiya et al., 1999). The reconstructed lithostratigraphy comprises a simple succession of greenstones with low-K tholeiitic characteristics, BIFs, and turbidites in ascending stratigraphic position (Komiya et al., 1999); this sequence is very similar to the oceanic plate stratigraphy of Phanerozoic age. The Isua IF consists of black, white, gray, green, and red layers that vary in thickness from several millimeters to several centimeters. Black and gray layers are dominant and are composed mainly of magnetite, quartz, amphibole, and chlorite. The red layers contain an unidentified red-colored mineral. Hematite is absent in the Isua IF.

Marble Bar, Australia

The Marble Bar jasper (ferruginous chert) is a member of the Towers Formation of the Warrawoona Group in the Marble Bar greenstone belt, Pilbara, Western Australia (Hickman, 1983; Van Kranendonk et al., 2001). Their depositional age is between 3463 and 3454 Ma (Thorpe et al., 1992; McNaughton et al., 1993). These rocks have undergone relatively low-grade metamorphism of prehnite-pumpellyite to greenschist facies. The Marble Bar jasper/chert formation is composed of red, brown, light green, gray, black, and white layers that vary from 1 mm to 2 cm in thickness. Black, gray, and light green cherts are mainly composed of

TABLE 1. REE DATA SOURCES

	Stratigraphic unit	Age (Ga)	Source of geochemical data	Reference to age
1.	Isua supracrustal belt	3.8 - 3.7	Morinaga, 1997	Nutman et al., 1997
2.	Marble Bar belt	3.46 - 3.45	Kato and Nakamura, 2003	Thorpe et al., 1992
3.	Cleaverville Formation	3.2 - 3.0	Kato et al., 1998	Kiyokawa and Taira, 1998
4.	Sargur belt	3.13 - 2.96	Kato et al., 1996	Nutman et al, 1992
5.	Bababudan belt	2.9 - 2.7	Arora et al., 1995	Chadwick et al., 2000
6.	Kushtagi belt	2.9 - 2.7	Khan et al., 1996	Chadwick et al., 2000
7.	Holenarsipur belt	2.9 - 2.7	Kato et al., 2002a	Chadwick et al., 2000
8.	Sandur belt	2.9 - 2.7	Manikyamba et al., 1993	Chadwick et al., 2000
9.	Marra Mamba IF	2.68 - 2.56	Alibert and McCulloch, 1993	Arndt et al., 1991; Trendall et al., 1998
10.	Brockman IF	2.56 - 2.45	Alibert and McCulloch, 1993	Trendall et al., 1998; Barley et al., 1997
11.	Kuruman and Penge IFs	2.52 - 2.43	Bau and Dulski, 1996	Trendall et al., 1990; Sumner and Bowring, 1996
12.	Boolgeeda IF	2.45 - 2.40	Kato and Kimura, 1999	Barley et al., 1997
13.	Hotazel Formation	2.22 - 2.20	Tsikos and Moore, 1997	Cornell et al., 1996
14.	Väyrylänkylä IF	2.13 - 2.03	Laajoki, 1975; Danielson et al., 1992	Sakko and Laajoki, 1975
15.	Gunflint, Biwabik and Sokoman IFs	1.92 - 1.87	Danielson et al., 1992; Fryer, 1977; Derry and Jacobsen, 1990	Fryer, 1972; Jacobsen and Pimentel-Klose, 1988
16.	Rapitan and Braemar IFs	0.75 - 0.70	Fryer, 1977; Klein and Beukes, 1993; Lottermoser and Ashley, 2000	Klein and Beukes, 1993; Lottermoser and Ashley, 2000
17.	Phanerozoic Fe-Mn umbers	0.08 - 0.07	Ravizza et al., 1999; Fujinaga et al., 1999	Kiminami et al., 1994
18.	EPR hydrothermal Fe-Mn sediments	0.04 - 0.00	Barrett and Jarvis, 1988	
19.	hydrogenetic Mn crusts	0.00	Bau et al., 1996	
20.	seawater	0.00	Zhang and Nozaki, 1996; Alibo and Nozaki, 1999	
21.	MOR hydrothermal fluids	0.00	Mills and Elderfield, 1995; Bau and Dulski, 1999	

microcrystalline quartz, siderite, dolomite, and black aggregates (probably organic compounds). The brown chert layers mostly consist of quartz and goethite, whereas the red chert contains quartz, hematite, and siderite, and the white chert is composed solely of quartz. Goethite is probably a recent weathering product. However, there is no systematic difference between REE features of samples with goethite and those without goethite, suggesting that REEs are virtually unaffected by goethite formation. The average Fe_2O_3* content of the ferruginous cherts is 10.2 wt%, which is significantly lower than in common BIFs. However, field occurrences (e.g., alternation of iron-rich and cherty layers, the association with mid-ocean ridge basalt–like greenstones) and geochemical signatures of the Marble Bar cherts are essentially identical to those of typical Algoma-type BIFs (Kato and Nakamura, 2003).

Cleaverville, Australia

The Middle Archean Cleaverville Formation in the West Pilbara Granite-Greenstone Terrane, Western Australia, is estimated to be between 3.2 and 3.0 Ga in age (Kiyokawa and Taira, 1998). The Formation comprises weakly metamorphosed (greenschist facies) greenstone, BIF, mudstone, sandstone, and minor amounts of conglomerate. Geochemical and petrological characteristics of the underlying greenstones suggest they are Fe-rich low-K tholeiites that formed on a mid-ocean ridge (Ohta et al., 1996). The Cleaverville IF consists of alternating Fe-rich and cherty layers that are several millimeters to several centimeters thick (Kato et al., 1998). The Fe-rich layers are composed mainly of hematite, goethite, and quartz. Chert consists dominantly of quartz with minor amounts of goethite, hematite, and/or kaolinite, and illite.

Sargur, India

The other Middle Archean BIF is the Sargur IF in the Sargur greenstone belt, South India. The Sargur and equivalent greenstone belts occur as small remnants in the Peninsular Gneiss and have been metamorphosed to the upper amphibolite and granulite facies. These belts consist of ultramafic to mafic igneous rocks, fuchsite-bearing quartzites, pelites, calc-silicate rocks, and BIFs. Their depositional age is constrained between 3.13 and 2.96 Ga (Nutman et al., 1992; Chadwick et al., 2000). The Sargur IF is composed of alternating Fe-rich and cherty thin layers, approximately 1–2 mm in thickness (Kato et al., 1996). The Fe-rich layers consist of magnetite and minor amphibole, and the cherty layers consist of quartz. The associated amphibolites have mid-ocean ridge basalt–type geochemical signatures (Kato et al., 1996).

Bababudan, Kushtagi, Northern Holenarsipur, and Sandur, India

Late Archean Algoma-type BIFs widely occur in the Dharwar Supergroup of the Dharwar Craton, South India (Radhakrishna and Naqvi, 1986). The depositional age of the Dharwar Supergroup is broadly bracketed between 2.9 Ga and 2.7 Ga (Nutman et al., 1996; Trendall et al., 1997a, 1997b; Chadwick et al., 2000). The deficiency of precise geochronological data and the absence of good exposure spanning the entire succession preclude geological correlation between the Dharwar greenstone schist belts. The low-grade-metamorphosed (up to amphibolite facies) greenstone belts contain three beds of BIFs interbedded with different types of sedimentary and volcanic rocks (Naqvi et al., 1988). The thickest and most extensive BIF deposition occurred in the Bababudan Group of the lower Dharwar Supergroup. The Bababudan Group is extensively exposed in many greenstone belts, including the Bababudan, Kudremukh, Kushtagi, Chitradurga, Shimoga, and Sandur schist belts. Among these, data sets on BIFs from four schist belts were selected for the present study: the Bababudan belt (Arora et al., 1995), Kushtagi belt (Khan et al., 1996), Sandur belt (Manikyamba et al., 1993), and the northern extension of the Holenarsipur belt (Kato et al., 2002a). Some schist belts (Bababudan, Kudremukh, Kushtagi, northern Holenarsipur) are predominated by oxide facies BIFs and lack stromatolitic carbonate rocks and Mn formations, whereas other belts (Chitradurga, Shimoga, Sandur) are associated with stromatolitic carbonates, Mn formations, and carbonaceous phyllites. These lithological variations suggest that there was a great variation in depositional environments for the Bababudan Group.

The IF samples of the Bababudan belt, collected from the core library at Chickmagalur, are cherty, consisting of magnetite, quartz, grunelite, Mg riebeckite, hornblende, and actinolite (Arora et al., 1995). The cherty IF from the Kushtagi belt is entirely composed of alternating bands of hematite and quartz that were separately analyzed (Khan et al., 1996). The IF collected from the northern Holenarsipur belt consists of two, very thin (usually 1–2 mm thick) alternating lithologies, containing Fe-rich and cherty layers (Kato et al., 2002a). The Fe-rich layers consist dominantly of magnetite and quartz with minor hematite and goethite. The cherty layers are solely composed of quartz with or without small inclusions of magnetite. Cherty IF, ferruginous chert, and chert of the Sandur belt, collected from borehole cores and recent railway and road cuttings, comprise hematite, magnetite, and quartz with lesser amounts of siderite and ankerite (Manikyamba et al., 1993).

Hamersley, Australia

The Hamersley Group, in the Hamersley Basin, Western Australia, includes three BIF groups (Marra Mamba IF, Brockman IF, and Boolgeeda IF, in ascending stratigraphic position). Together they provide a significant proportion of the world's iron ore supply. The depositional ages of these IFs are well determined by U-Pb sensitive high-resolution ion microprobe (SHRIMP) zircon dating (Arndt et al., 1991; Barley et al., 1997; Trendall et al., 1998), although the upper limit of the Boolgeeda IF is not tightly constrained because of a lack of age data for the overlying Turee Creek Group. The Marra Mamba IF and Brockman IF deposited between 2684 and 2561 Ma and between 2561 and 2449 Ma, respectively. The Woongarra Rhyolite underlying the Boolgeeda IF has been precisely dated at 2449 Ma (Barley et al., 1997); the apparent conformity between them shows that the Boolgeeda IF formed soon after 2.45 Ga (Martin et al., 2000). The Marra Mamba and Brockman IFs deposited in a clastic-starved environment and are dominated by either chemical sediments or extrusive volcanic rocks. The Boolgeeda IF is somewhat different from the lower IFs (Marra Mamba and Brockman) and is overlain by turbidites, shallow-marine carbonate rocks, and fluvial and marine siliciclastic rocks of the Turee Creek Group (Martin et al., 2000). The Hamersley Group has suffered low-grade metamorphism varying from the prehnite-pumpellyite facies to greenschist facies (Smith et al., 1982). The analyzed drill core samples of the lower Marra Mamba IF comprise magnetite-rich IF, minnesotaite-chert with subordinate stilpnomelane, and chert-carbonate (Alibert and McCulloch, 1993). Sulfides are ubiquitous as disseminated grains, nodules, and millimeter-thick beds. The upper Marra Mamba IF is characterized by the ubiquitous presence of organic carbon and sulfides, thin laminations, and the presence of both magnetite and relict hematite. The Brockman IF samples are mainly magnetite-rich mesobands and riebeckite-chert. The Boolgeeda IF is composed mostly of magnetite-rich IF and ferruginous chert with millimeter-thick hematite- and chert-rich layers (Kato and Kimura, 1999).

Kuruman and Penge, South Africa

The Kuruman and Penge IFs occur in the Griqualand West sub-basin and the Eastern Transvaal sub-basin, respectively, in South Africa where the Transvaal Supergroup is extensively exposed. The Penge IF is correlative with and lithologically similar to the Kuruman IF (Bau and Dulski, 1996). According to U-Pb ages for zircons from volcanic ash beds, the Kuruman IF is well dated between 2521 and 2432 Ma (Trendall et al., 1990; Sumner and Bowring, 1996). Most of the Transvaal Supergroup was subject to zeolite-facies metamorphism, although metamorphic grades locally reached hornblende-hornfels facies adjacent to the Bushveld Complex. The Kuruman and Penge IFs analyzed for REEs by Bau and Dulski (1996) are drill-core samples and consist of alternating layers of siderite, hematite, and magnetite as well as chert.

Hotazel, South Africa

The Hotazel Formation represents the youngest episode of BIF deposition in the Transvaal Supergroup and overlies andesitic lavas of the Ongeluk Formation, which are dated at 2222 Ma (Cornell et al., 1996). Four distinct BIF units are interbedded with three Mn beds in this formation. The lower Mn bed is extensively mined as the world's largest Mn resource (Kalahari Mn field). The Hotazel IF comprises laminae of chert, iron oxides (magnetite, hematite), iron silicates (greenalite, minnesotaite, stilpnomelane, riebeckite, iron-rich mica), carbonates (calcite, ankerite, siderite), and pyrite (Tsikos and Moore, 1997). There is a stratigraphic change in mineralogy. Oxide-dominated facies characterize the lower parts of the Hotazel IF, whereas silicate- and carbonate-dominated facies increase in upper sections.

Väyrylänkylä, Finland

The Väyrylänkylä IF (Pääkkö and Iso Vuorijärvi IFs) deposited in the Kainuan schist belt, Salmijärvi Basin, Finland. The whole rock Pb-Pb isochron age of the Pääkkö IF is 2080 ± 45 Ma (Sakko and Laajoki, 1975). The IF samples are composed mainly of quartz and magnetite (Laajoki, 1975; Danielson et al., 1992). The Väyrylänkylä IF is characterized by small dimensions, low Fe grade (Fe_{tot} ~26 wt%), and exceptionally high P content (~1.2 wt%), and hence are uneconomic (Laajoki, 1975). Manganiferous rocks locally occur in the formation.

Gunflint, Biwabik, and Sokoman, Canada

Data on the Gunflint, Biwabik, and Sokoman IFs for the present paper are from Fryer (1977), Derry and Jacobsen (1990), and Danielson et al. (1992). The Gunflint IF is exposed along the northern shore of Lake Superior, and the Biwabik IF of the Mesabi Range is its western equivalent. Stille and Clauer (1986) obtained a whole rock Sm-Nd isochron age of 2.08 ± 0.25 Ga from argillites in the Gunflint Formation, but its depositional age are generally estimated to be ca. 1.9 Ga (Jacobsen and Pimentel-Klose, 1988). The Sokoman IF is exposed throughout the Labrador Trough and known as one of the world's largest iron deposits. Fryer (1972) reported a whole rock Rb-Sr isochron age of 1.87 ± 0.05 Ga for a series of slate samples from the Sokoman IF.

Rapitan, Canada, and Braemar, Australia

Two data sets of Neoproterozoic IFs were included in this study, the Rapitan IF in Canada (Fryer, 1977; Klein and Beukes, 1993) and the Braemar IF in South Australia (Lottermoser and Ashley, 2000). The Rapitan IF probably deposited between 755 and 730 Ma (see Klein and Beukes, 1993). Selected samples from the Rapitan IF are banded, nodular, or arenose (Al_2O_3 < 1.2 wt%). Their mineralogy is very simple, consisting of hematite cemented by microquartz. The Braemar IF occurs in the glaciomarine sequences of the Sturtian glaciation (750–700 Ma). The Braemar ironstone facies consists of lenticular laminated and diamictic ironstones. The mineralogy of the laminated ironstones and matrix of diamictic ironstones is simple: magnetite, hematite, and quartz with minor muscovite, chlorite, biotite, carbonate, apatite, plagioclase, and tourmaline. It should be noted that Al_2O_3 contents in the Braemar IF are exceptionally high (2.7–4.0 wt%), reflecting the abundance of aluminous minerals such as muscovite, chlorite, biotite, and plagioclase.

Ce AND Eu ANOMALIES IN BIFS

REE characteristics of the selected BIFs are summarized in Table 2, and chondrite-normalized REE patterns and Ce/Ce*-Pr/Pr* diagrams for them are given in Figures 3 and 4, respectively. Histograms of Ce/Ce* and Eu/Eu* values in the BIFs together with modern seawater, EPR Fe-Mn sediments, and Phanerozoic umbers are also given in Figures 5 and 6, respectively. The REE indices, such as ΣREE, Ce/Ce*, and Eu/Eu* values, of Early Archean IFs (the Isua and Marble Bar IFs) are very similar with each other (Table 2, Figs. 3–6), in spite of their differences in mineralogy. Both BIFs are characterized by distinct positive Eu anomalies and weak but indisputable negative Ce anomalies (Ce/Ce* = 0.6–0.9). Middle Archean IFs (the Cleaverville and Sargur IFs) have distinct positive Eu anomalies, but their magnitudes are not as large as those of the Early Archean IFs (Fig. 6). Negative Ce anomalies are generally weak to moderate (Ce/Ce* = 0.5–0.9; Fig. 5C) in the Cleaverville IF, whereas the Ce anomaly values of the Sargur IF are almost close to one (no Ce anomaly; Fig. 5D).

Generally, positive Eu anomalies in the Late Archean IFs are remarkable and are similar to those of the Early Archean IFs (Fig. 6). Eu anomalies of the Kushtagi and northern Holenarsipur IFs become less striking with increasing ΣREE. This gradual change in Eu anomaly with increasing ΣREE is also observed in

TABLE 2. REE INDICES AND CHARACTERISTICS OF BIFS

Stratigraphic unit	ΣREE	Ce/Ce*	Eu/Eu*	REE characteristics
Isua (3.8-3.7 Ga)	9.41±9.51 0.91-34.3	0.86±0.09 0.61-1.18	2.29±0.47 1.23-3.25	LREE-enriched patterns with pronounced positive Eu anomalies. Some samples exhibit MREE depletion and HREE enrichment. ΣREE varies widely, reflecting two contrasting lithologies (cherty and Fe-rich layers). Weak but true negative Ce anomalies occur in many samples.
Marble Bar (3.5 Ga)	8.12±8.39 0.22-46.2	0.86±0.09 0.60-1.07	2.62±1.21 1.31-6.61	REE patterns varing from HREE- to LREE-enriched, with conspicuous positive Eu anomalies and small but indisputable negative Ce anomalies. The REE patterns of individual samples become more LREE-enriched with increasing ΣREE. Positive Eu anomalies decrease with increasing ΣREE.
Cleaverville (3.2-3.0 Ga)	28.6±25.2 0.63-136	0.93±0.24 0.52-1.90	1.41±0.38 0.79-2.38	Slight LREE-enriched patterns with a large variation in ΣREE. Negative Ce anomalies are weak to moderate; positive Ce anomalies are also identified in some parts of the formation. Positive Eu anomalies are distinct in most samples, and definite negative Eu anomalies are identified in some samples. The lowest Eu/Eu* value (0.79) is identical to the Eu/Eu* values of the Phanerozoic Fe-Mn umbers, suggesting that the Middle Archean seawater was influenced by a riverine flux from the differentiated continental crust bearing negative Eu anomalies.
Sargur (3.1-3.0 Ga)	7.68±2.09 4.58-10.5	0.93±0.04 0.86-1.00	1.70±0.23 1.35-2.17	Gently LREE-enriched patterns with distinct positive Eu anomalies. Several BIF samples have trivial negative Ce anomalies, but most samples have no Ce anomaly.
Bababudan (2.9-2.7 Ga)	8.81±3.57 2.91-19.3	0.76±0.11 0.58-1.10	2.30±0.66 1.27-3.72	MREE-depleted concave patterns with striking positive Eu anomalies. Although false negative Ce anomalies occur in many samples, true negative Ce anomalies are also present in some samples.
Kushtagi (2.9-2.7 Ga)	16.6±12.7 2.30-38.6	0.75±0.18 0.47-0.98	2.04±0.80 1.10-4.17	The REE patterns are classified into two groups with high and low ΣREE, with both groups exhibiting gently LREE-enriched patterns. The cherty samples exhibit a more striking enrichment of Eu and depletion of Ce than the hematite-rich samples.
Holenarsipur (2.9-2.7 Ga)	18.2±14.7 5.18-65.3	0.50±0.19 0.13-0.83	1.60±0.45 0.96-2.45	LREE-enriched patterns with relatively high $(La/Yb)_N$ ratios. Negative Ce anomalies are striking and close to those of modern seawaters. In fact, the Ce/Ce*-Pr/Pr* plots overlap partly those of the modern seawaters. Positive Eu anomalies are noticeable, although the Eu anomalies become less striking with increasing ΣREE.
Sandur (2.9-2.7 Ga)	8.49±8.59 0.97-34.2	1.06±1.13 0.02-5.05	2.65±1.61 0.49-7.13	REE patterns varing from HREE- to LREE-enriched. The Ce anomalies vary from highly positive to greatly negative. True negative Ce anomalies are striking and close to those of modern oxic seawaters. Some parts of samples have very large positive Ce anomalies, which are similar to those of submarine hydrogenetic Mn crusts and nodules. Although the average of Ce/Ce* values is close to one due to very large values (up to 5.05) of several samples, the main cluster of samples shows significantly negative Ce anomalies.
Marra Mamba (2.7-2.6 Ga)	4.28±2.82 1.08-10.3	0.89±0.06 0.80-1.01	1.10±0.18 0.84-1.38	Gently MREE-depleted concave patterns. The Ce anomalies are subtly negative or absent, which contrasts with the strong negative Ce anomalies in the Late Archean Indian BIFs.
Brockman (2.5 Ga)	12.8±7.82 5.58-31.5	0.80±0.08 0.72-1.00	1.07±0.12 0.92-1.37	MREE-depleted concave and gently LREE-enriched patterns. ΣREE of the Brockman IF is higher than that of the Marra Mamba IF. Negative Ce anomalies are more pronounced in the Brockman IF than in the Marra Mamba IF. The Eu/Eu* values fall in a narrow range.
Kuruman & Penge (2.5-2.4 Ga)	10.8±7.42 1.37-30.9	0.86±0.06 0.76-0.96	1.12±0.24 0.74-1.80	Concave REE patterns similar to the Hamersley IFs. LREEs show a wide variation in abundances. The Ce/Ce* values fall in a narrow range which is similar to that of the Hamersley IFs. The Eu/Eu* histograms are slightly shifted toward higher values, compared to the contemporaneous Boolgeeda IF.
Boolgeeda (2.4 Ga)	22.0±16.3 3.01-61.0	0.89±0.06 0.77-1.00	0.91±0.24 0.48-1.89	Gently MREE-depleted concave patterns with a wide range of ΣREE and small but distinct negative Ce anomalies. Negative Eu anomalies are prominent, but positive anomalies are observed in several samples.
Hotazel (2.2 Ga)	16.4±10.8 6.52-38.3	0.73±0.13 0.44-0.90	0.71±0.14 0.50-0.93	MREE-depleted concave and LREE-enriched patterns. Half of the samples show true negative Ce anomalies, but they are not as strong as that of modern seawaters. Significantly low Eu/Eu* values are similar to those of modern seawaters and Phanerozoic Fe-Mn umbers.
Väyrylänkylä (2.1 Ga)	59.8±10.2 42.1-72.2	0.48±0.19 0.28-0.84	0.95±0.09 0.86-1.11	Gently LREE-enriched patterns with a relatively high ΣREE. True negative Ce anomalies are striking. Eu/Eu* values have a narrow range and are higher than those of the Hotazel IF.
Gunflint, Biwabik & Sokoman (1.9 Ga)	33.4±35.9 6.42-130	0.91±0.20 0.56-1.20	0.89±0.15 0.63-1.13	LREE-enriched patterns with a variable ΣREE. Whether clastic-derived REEs are responsible for higher ΣREE in several samples is uncertain, because data on Al_2O_3 and TiO_2 contents are lacking. Negative Ce anomalies are absent or moderate, and moderately negative Ce anomalies are from the Gunflint IF. The Eu/Eu* values are relatively constant.
Rapitan & Braemar (0.7 Ga)	35.7±25.2 4.91-82.1	0.90±0.12 0.66-1.01	0.75±0.08 0.66-0.91	Nearly flat or subtly LREE-enriched patterns. The Ce anomalies of the Braemar IF are absent or weak, whereas those of the Rapitan IF are moderately negative. The magnitudes of negative Eu anomalies are remarkably constant.

Note: ΣREE, Ce/Ce*, and Eu/Eu* values are given as average±1σ (above) and range (below).

the Marble Bar (3.5 Ga) and Cleaverville (3.2–3.0 Ga) IFs (Kato et al., 1998; Kato and Nakamura, 2003). Negative Ce anomalies are distinguished in many Late Archean BIF samples, with the Ce/Ce* values less than 0.6 (Fig. 5). Furthermore, some portions of samples exhibit strong negative Ce anomalies, like modern seawaters, with the Ce/Ce* values less than 0.3. The 2.9–2.7 Ga Sandur IF shows a sizable variation in the REE patterns (Fig. 3H). Although La enrichment is recognizable in some samples, *true* negative Ce anomalies are striking and close to those of modern oxygenated seawaters. The Ce anomalies vary from extremely positive (Ce/Ce* = 5.1 at maximum) to greatly negative (Ce/Ce* = 0.02 at minimum; Figs. 4H, 5H). The Sandur IF shows the greatest variation of Eu/Eu* from 0.5 to 7.1, although positive Eu anomalies are predominant (Fig. 6H).

The 2.7–2.4 Ga Hamersley IFs (the Marra Mamba, Brockman, and Boolgeeda IFs in ascending stratigraphic position) show small or no Ce anomalies (Ce/Ce* = 0.72–1.0; Fig. 5), which contrasts with the strong negative Ce anomalies in the Late Archean Indian BIFs. The Eu anomalies in the Hamersley BIFs vary from negative to positive (Eu/Eu* = 0.84–1.38; dominated by positive anomalies; Figs. 6I, 6J, and 6L). Eu/Eu* values decrease from the Marra Mamba to the Boolgeeda IFs. Generally, the Kuruman and Penge IFs show concave REE patterns that are similar to the Hamersley IFs (Fig. 3K). The Ce/Ce* values fall in a narrow range from 0.76 to 0.96, and there is no significant difference between the Hamersley and Kuruman/Penge IFs (Fig. 5K).

True and significant negative Ce anomalies are well recognized in the 2.2–2.1 Ga IFs (the Hotazel and Väyrylänkylä IFs; Figs. 4, 5). The Hotazel IF has distinctly low Eu/Eu* values from 0.50 to 0.93, whereas the Väyrylänkylä IF has the Eu/Eu* values narrowly ranging from 0.86 to 1.11. Negative Ce anomalies are generally absent or moderate in the ca. 1.9 Ga Gunflint, Biwawik and Sokoman IFs (Figs. 4O, 5O), and their Eu/Eu* values are relatively constant (Fig. 6O; 0.89 ± 0.15). The ca. 0.7 Ga IFs exhibit nearly flat or subtly LREE-enriched patterns (Fig. 3P). The Braemar IF, which has higher Al_2O_3 contents compared to other IFs, has greater REE concentrations (chondrite-normalized La values = 14.3–40.2) compared to the other BIF samples discussed above, clearly reflecting contributions of REEs in clastic sediments. The Ce anomalies of the Braemar ironstones are absent or very weak (Ce/Ce* = ~0.92), whereas those of the Rapitan IF are moderately negative (Ce/Ce* = ~0.66). The contamination of detrital materials may have obscured negative Ce anomalies in the Braemar IF. The magnitudes of negative Eu anomalies in the Braemar IF are very similar to those of the Rapitan IF, with remarkably constant Eu/Eu* values (0.75 ± 0.08; Fig. 6P).

DISCUSSION

The Redox History of the Precambrian Oceans

Figure 7, which shows the secular variation of Ce anomaly in BIFs, gives us new insight into the oxygen scenario for the Precambrian Earth's surface. Dymek and Klein (1988) pointed out the possible presence of slightly negative Ce anomalies in the Isua IF. However, whether or not negative Ce anomalies truly exist in the 3.8–3.7 Ga IF has been ambiguous, because their REE data were determined by instrumental neutron activation analysis and hence were not complete. Especially, the absence of Pr analyses has made it difficult to determine whether the apparent negative Ce anomalies were simple reflections of the presence of positive La anomalies. However, data obtained by the ICP-MS method (Morinaga, 1997) clearly demonstrate the true existence of negative Ce anomalies in the 3.8–3.7 Ga IF. This suggests that some sort of Ce oxidation mechanism already existed in the 3.8–3.7 Ga ocean and that the seawater column from which the Isua IF precipitated was not entirely anoxic. The Ce/Ce* values range from ~0.5 (moderately negative) to ~1.0 (no anomaly) in BIFs of ca. 3.8–3.5 Ga in age, suggesting the development of relatively oxygenated oceanic conditions where aqueous Fe^{2+} precipitated as Fe-oxyhydroxides.

Stronger negative Ce anomalies are observed in many BIFs of 2.9–2.7 Ga age (Late Archean), suggesting that the same oxygenated conditions as those of today existed in the 2.9–2.7 Ga oceans. Even the average Ce/Ce* values of the Late Archean Indian IFs, excepting the Sandur IF, are distinctly negative (0.50 for the Holenarsipur IF) or moderately negative (~0.75 for the Bababudan and Kushtagi IFs) (Table 2), unequivocally demonstrating the widespread presence of negative Ce anomalies in Late Archean seawaters.

The above suggestion is consistent with the presence of small but widespread Mn deposits that formed at the same time as the BIFs in the Dharwar Craton. The geochemical behavior of dissolved Ce is similar to that of dissolved Mn, because the free-energy changes for Ce^{4+} and Mn^{4+} reduction are quite similar (Elderfield, 1988). In modern oceans, oxidation of Ce^{3+} to Ce^{4+}-oxide mostly takes place in the upper water column (Alibo and Nozaki, 1999). Even without additional removal of Ce from the deep water, negative Ce anomalies would be enhanced with time because of the preferential regeneration of REE^{3+} except Ce in the oxygenated deep-sea water (Alibo and Nozaki, 1999). Hence, the contemporaneous Mn deposits in the Dharwar Craton may have formed in shallow environments associated with the oxidative removal of Ce, although positive Ce anomalies have not yet been reported for these Mn deposits (Manikyamba and Naqvi, 1997). The Mn formations, which occur with some BIFs (e.g., the Sandur and Chitradurga IFs), are characterized by an association with stromatolitic carbonate rocks, suggesting their precipitation in shallow environments. Other Indian BIFs (the Bababudan, Kushtagi, and northern Holenarsipur IFs), which lack Mn formations and stromatolitic carbonate rocks, were most likely deposited in deep seas (Kato et al. 2002a). Thus, it is very likely that fully oxygenated seawaters like those today existed in both deep and shallow environments at 2.9–2.7 Ga.

In addition to negative Ce anomalies, huge positive Ce anomalies exist in some parts of the Sandur IF, suggesting that redox-related Ce fractionation took place in the Late Archean ocean. As the mixing of hydrothermal Fe-Mn precipitates with

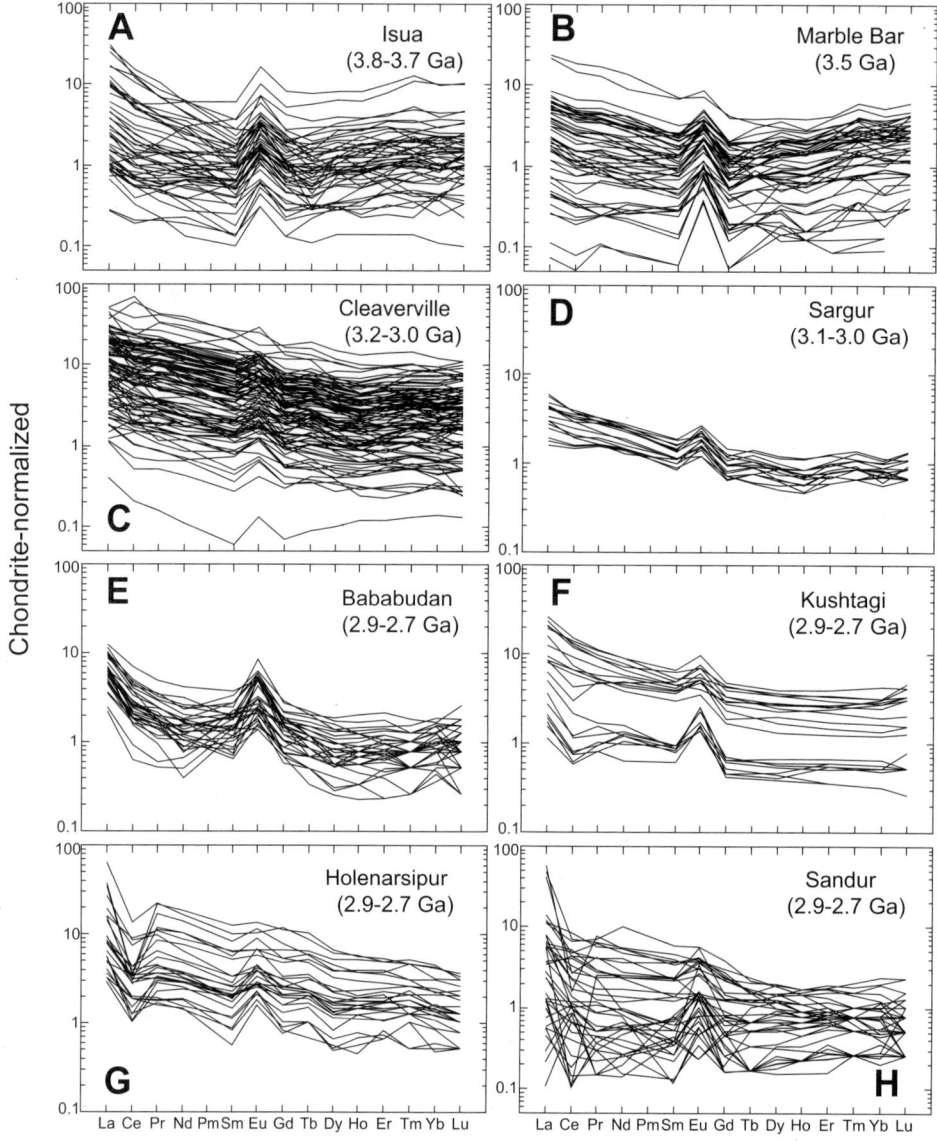

Figure 3 (*on this and following page*). Chondrite-normalized REE patterns of BIFs.

hydrogenetic Mn crusts occurs in the modern ocean system (Kuhn et al., 1998), it seems likely that the Sandur IF precursors were formed by the mixing of hydrothermal Fe-Mn (with negative Ce anomalies) and hydrogenetic Mn (with positive Ce anomalies) precipitates in various proportions. The Sandur IF samples were collected from borehole cores and recent railway and road cuttings (Manikyamba et al., 1993), and hence weathering effects on these samples were probably insignificant. Nevertheless, the fact that the average Ce/Ce* value of the Sandur IF samples is close to one (Table 2) raises the possibility that these positive Ce anomalies, as well as negative Ce anomalies, are products of Ce fractionation by modern weathering. It is well known that positive and negative Ce anomalies are found in some (not all) modern soil profiles: typically, depletions of ΣREE and positive Ce anomalies are found in the upper soil sections, and enrichments of ΣREE and negative Ce anomalies in the lower soil sections (e.g., Koppi et al., 1996; Braun et al., 1998). Detailed investigations on the Sandur IF are necessary to determine the origins of these Ce anomalies.

The Hamersley and Transvaal BIFs, which formed during the period from 2.7 Ga to 2.4 Ga, have weak negative Ce anomalies, varying narrowly from 0.7 to 1.0 in their Ce/Ce* ratios. The magnitudes of their negative Ce anomalies are clearly smaller than those of Early and Middle Archean BIFs. This suggests that the water columns from which these BIFs precipitated were less oxygenated than those for the pre-2.7 Ga BIFs. A plausible explanation is that the depositional environments for the Hamersley and Transvaal BIFs were related to evolving rifts (Alibert and McCulloch, 1993; Ohmoto, 1993). Their depositional settings are quite different from those for the pre-2.7 Ga Algoma-type

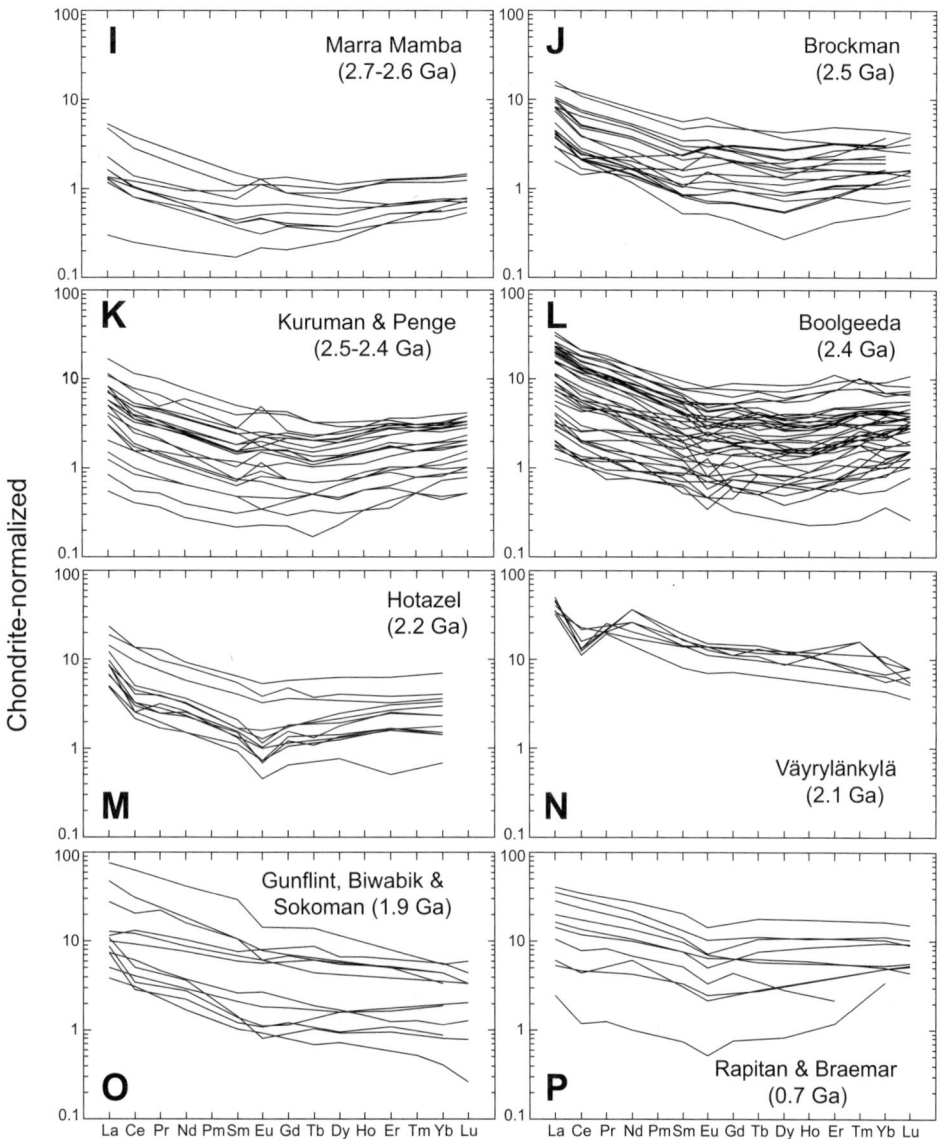

Figure 3 (*continued*).

BIFs that were mostly deposited in the open-ocean environments (Kato et al., 1998). A Fe^{2+}-rich brine pool may have accumulated in land-locked, anoxic basins much like the modern Red Sea brine pools (Ohmoto, 1993; Ohmoto et al., this volume). In this scenario, the geochemical signatures preserved in the post-2.7 Ga BIFs are indicators only of local environments. Globally, the atmospheric and oceanic redox conditions may have been oxygenated persistently since at least 2.9–2.7 Ga.

The Hotazel and Väyrylänkylä IFs with distinct negative Ce anomalies (Ce/Ce* = 0.2–0.7) deposited at 2.2–2.1 Ga. Although their negative anomalies are not as strong as those of the Late Archean Indian BIFs, the close association of these IFs with Mn deposits suggests that oxygenated conditions prevailed in the Hotazel and Väyrylänkylä oceans. The ca. 1.9 Ga Gunflint, Biwabik, and Sokoman IFs exhibit less significant Ce anomalies. The Ce/Ce* fluctuation which shows diminishing negative Ce anomalies at 2.7–2.5 Ga and ca. 1.9 Ga, may have been related to the fluctuation in the global mantle plume activity (Isley and Abott, 1999). The intensive global mantle plume activity at 2.7–2.5 Ga and ca. 1.9 Ga created many rift-related basins (e.g., the Hamersley Basin and the Lake Superior region) and supplied great amounts of Fe^{2+} to the closed basins by submarine hydrothermal fluids. The hydrothermal Fe^{2+} reacted with O$_2$ in the water column to precipitate large amounts of iron oxides (i.e., Superior-type BIFs) and to create anoxic water bodies. Furthermore, the observed Ce/Ce* fluctuation in Late Archean to Paleoproterozoic BIFs may represent the redox conditions only of local seas, rather than of the global oceans. Such explanations can also explain why the ca. 0.7 Ga Rapitan and Braemar IFs exhibit only minor negative Ce anomalies.

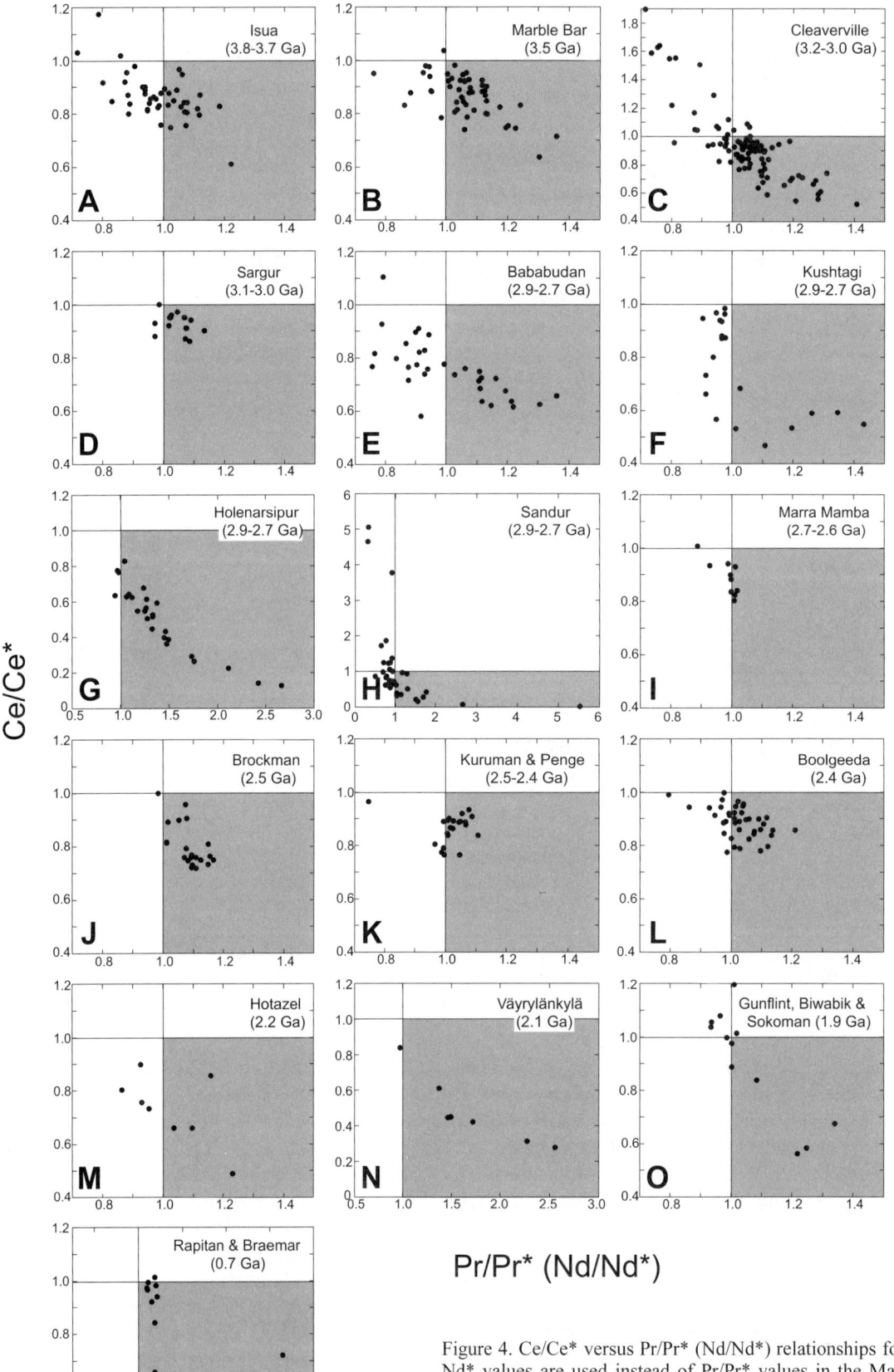

Figure 4. Ce/Ce* versus Pr/Pr* (Nd/Nd*) relationships for BIFs. Nd/Nd* values are used instead of Pr/Pr* values in the Marra Mamba, Brockman, Väyrylänkylä, Lake Superior, and Rapitan/Braemar IFs because of the lack of Pr data for these IFs. The region of *true* negative Ce anomalies is shown as a shaded area.

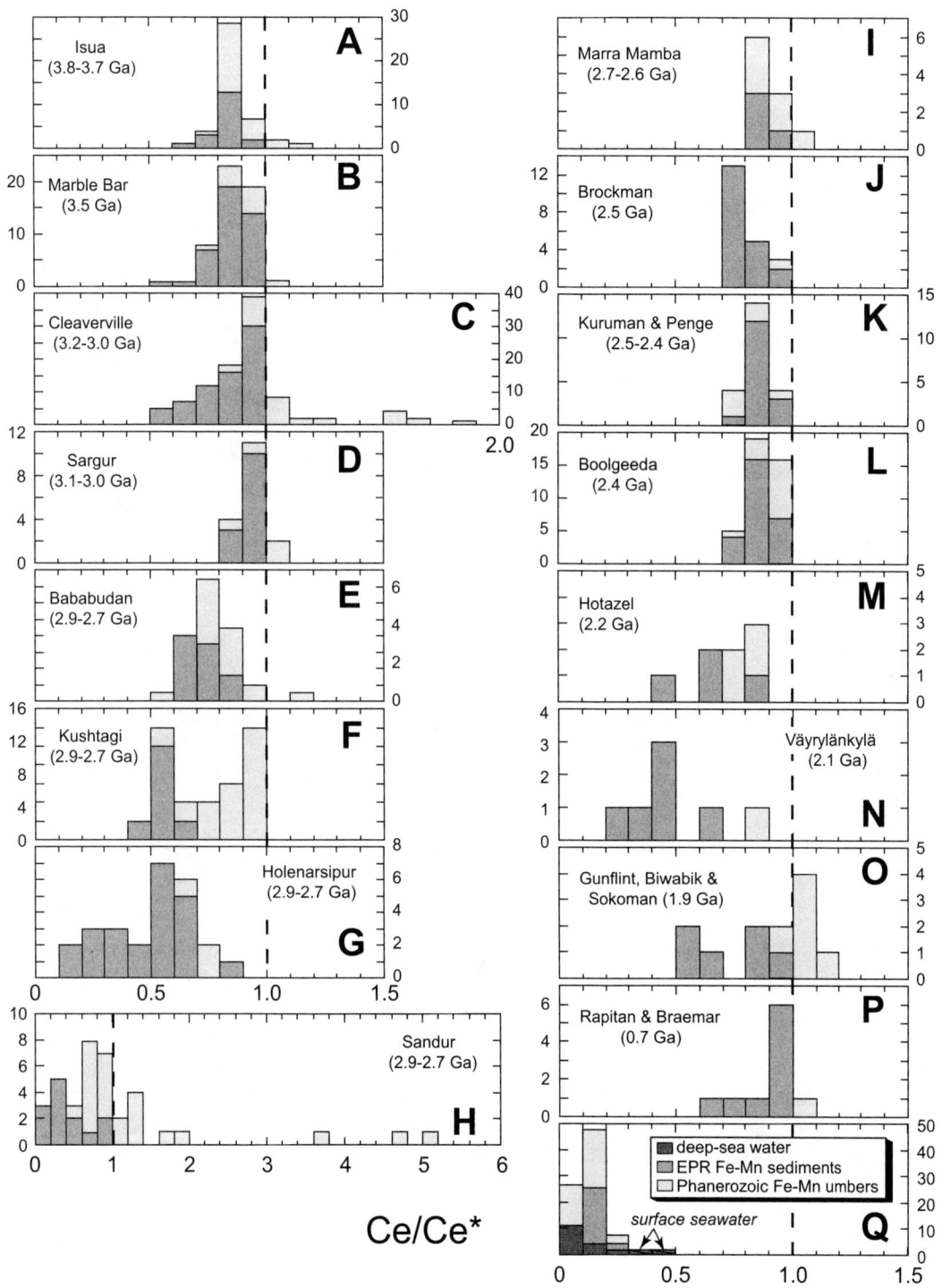

Figure 5. Histograms of Ce/Ce* values in BIFs. For comparison, Ce/Ce* values of modern seawater, EPR Fe-Mn sediments, and Phanerozoic Fe-Mn umbers are presented (Q). *True* negative Ce anomalies, evaluated from the Ce/Ce*-Pr/Pr* diagrams, are shaded in these BIF histograms.

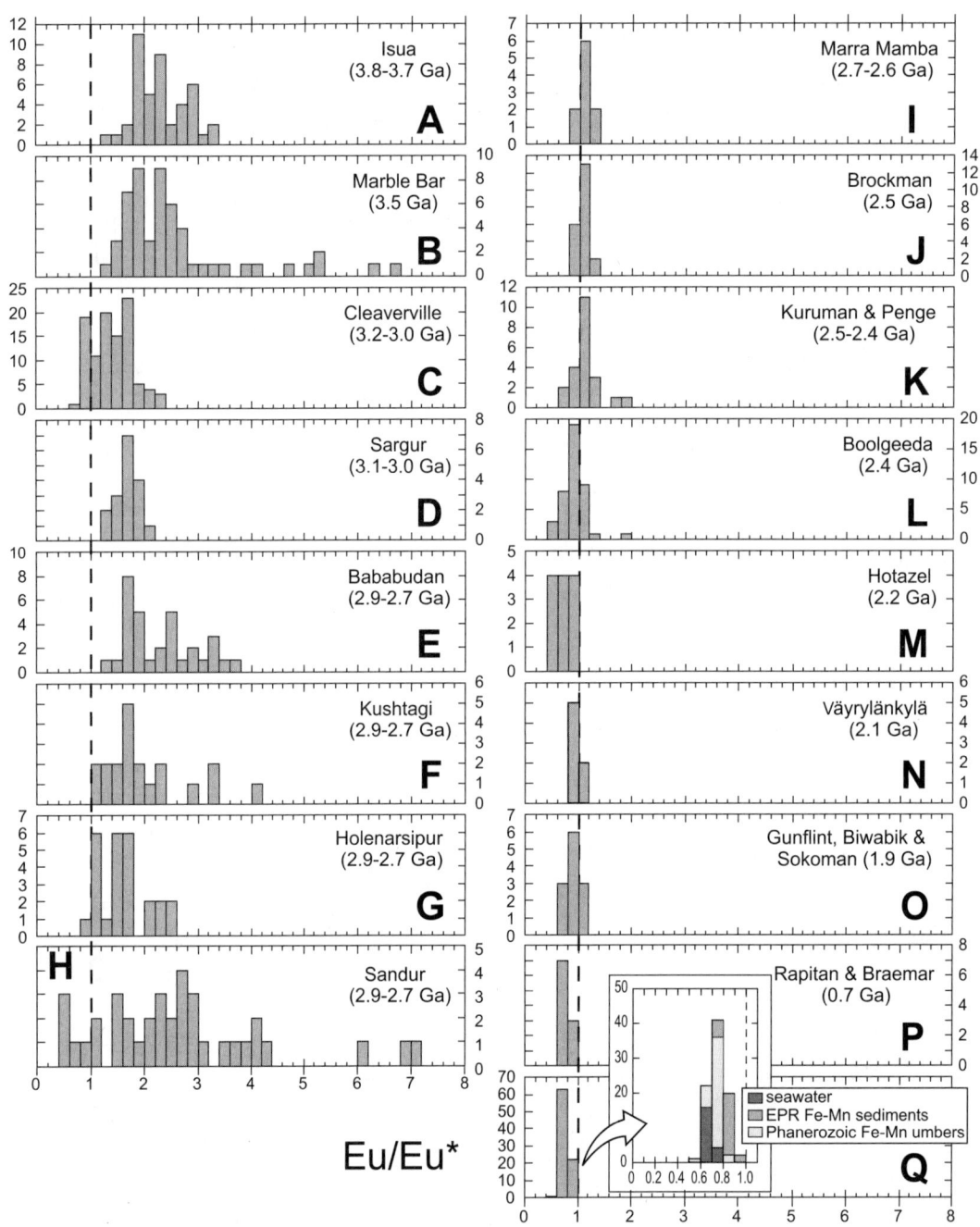

Figure 6. Histograms of Eu/Eu* values in BIFs. The Eu/Eu* values of modern seawater, EPR Fe-Mn sediments, and Phanerozoic Fe-Mn umbers are also presented (Q).

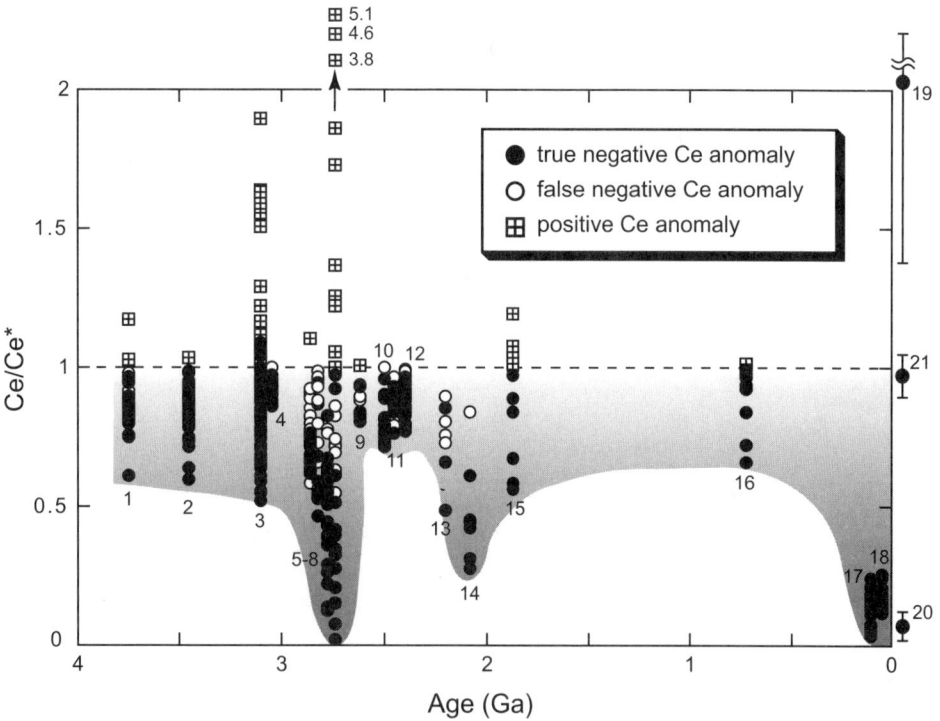

Figure 7. Secular variation of Ce/Ce* values in BIFs. The Ce/Ce* values of modern seawater, MOR hydrothermal fluids, EPR Fe-Mn sediments, Phanerozoic Fe-Mn umbers, and hydrogenetic Mn crusts are also given. Numbers correspond to the stratigraphic units in Table 1.

Secular Variation in Hydrothermal versus Riverine Influxes into the Precambrian Oceans

The secular variation of Eu anomalies in BIFs is summarized in Figure 8. The figure shows that the pre-2.7 Ga Algoma-type BIFs have significantly positive Eu anomalies, although quite variable. Their remarkably positive Eu anomalies suggest that the water column from which the BIFs precipitated had strong positive Eu anomalies, probably caused by larger contributions of hydrothermal fluids. The large variation in the Eu anomalies was probably caused by a large variation in the mixing ratios of hydrothermal fluids and seawaters at various depositional sites in deep-sea pelagic environments (Kato et al., 1998).

Figure 8 shows a drastic change in Eu/Eu* values of BIFs ca. 2.7 Ga. The Superior-type BIFs and the Rapitan-type IFs, which are all younger than 2.7 Ga, exhibit much lower Eu/Eu* values than the pre-2.7 Ga Algoma-type BIFs. Furthermore, their Eu/Eu* range is quite narrow and constant. The significantly lower Eu/Eu* values probably resulted because the BIF-precipitating water columns were dominated by the riverine influx with lower Eu/Eu* values. That is, the riverine influx overwhelmed the hydrothermal influx into the land-locked basins where the post-2.7 Ga Superior-type BIFs formed. The global mantle plume activity was probably responsible for the widespread hydrothermal activity that formed the voluminous Superior-type BIFs at 2.7–2.4 Ga and ca. 1.9 Ga (Isley and Abbott, 1999; Kato, 2003). The global mantle plume activity could also have caused the episodic growth of continental crust and the formation of rift basins (e.g., Condie, 2000), which created the conditions to increase the riverine influx into the land-locked basins and surpass the hydrothermal influx to lower the Eu/Eu* values in the BIF-precipitating water column. The relatively constant Eu/Eu* values in the post-2.7 Ga BIFs may further suggest that the riverine flux from the surrounding continent into the closed basins was nearly constant during the deposition of BIFs. Within the Hamersley Basin, a gradual and slight decrease in the Eu/Eu* values from the 2.7–2.6 Ga Marra Mamba IF (~1.10), to the 2.5 Ga Brockman IF (~1.07), and then to the 2.4 Ga Boolgeeda IF (~0.91) is recognizable (Figs. 6I, 6J, and 6L). This may mean the hydrothermal component decreased and the riverine (continental) component increased comparatively during the period between 2.7 and 2.4 Ga.

CONCLUSIONS

1. Negative Ce anomalies in the 3.8–3.7 Ga Isua BIFs suggest that the water column from which the BIFs precipitated was not entirely anoxic and that a Ce oxidation mechanism already existed in the 3.8–3.7 Ga oceans.
2. The presence of strong negative Ce anomalies in the 2.9–2.7 Ga BIFs indicates that strongly oxygenated oceanic conditions, much like today, already existed at 2.9–2.7 Ga.

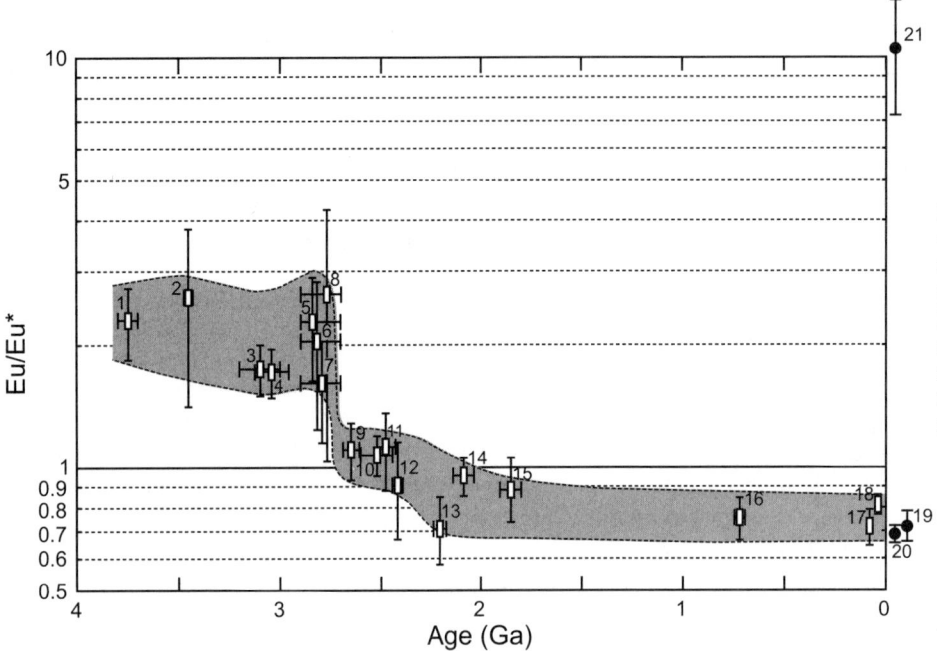

Figure 8. Secular variation of Eu/Eu* values in BIFs. The Eu/Eu* values of modern seawater, MOR hydrothermal fluids, EPR Fe-Mn sediments, Phanerozoic Fe-Mn umbers, and hydrogenetic Mn crusts are given. Numbers correspond to the stratigraphic units in Table 1. Box and bar represent an average Eu/Eu value and ±1σ, respectively.

3. The 2.7–2.4 Ga IFs and the ca. 1.9 Ga IFs have slightly negative Ce anomalies. These BIFs were probably deposited in evolving rifts where less oxygenated water conditions were created by intense hydrothermal activity in closed basins.
4. The presence of strong positive Eu anomalies in the pre-2.7 Ga Algoma-type BIFs suggests large contributions of hydrothermal fluids into the seawater column. The large variation of Eu anomalies in the Algoma-type BIFs reflects a large variation in the mixing ratios of hydrothermal fluids and seawaters at various depositional sites.
5. The post-2.7 Ga BIFs exhibit much lower and constant Eu/Eu* values than the pre-2.7 Ga BIFs, suggesting that water in land-locked basins where the post-2.7 Ga BIFs accumulated was strongly influenced by a riverine influx compared to hydrothermal influx.

ACKNOWLEDGMENTS

This work was supported by the Japanese Ministry of Education, Culture, Sports, Science and Technology through Grant-in Aid Nos. 07238105, 06041038, 08041102, 06740415, and 07740426 to Kato and 07041081 to Ohmoto, and by funds from cooperative programs (No. 40 in 1994, No. 28 in 1995, and No. 32 in 1997) to Kato provided by the Ocean Research Institute, University of Tokyo. Ohmoto is also supported by NASA Astrobiology Institute (NCC2–1057; CAN#NNA04CC06A), NASA Exobiology Program (NAG5–9089), and National Science Foundation (EAR-9706279, EAR-0229556). Yamaguchi acknowledges financial support from IFREE and NASA for his stay at the University of Wisconsin, Madison. Stimulating discussion with Yoshimichi Kajiwara, Shigenori Maruyama, Tsuyoshi Komiya, Michael Bau, Yumiko Watanabe, and the late Yoshiyuki Nozaki are deeply appreciated. We also thank Stephen E. Kesler, Richard W. Murray, and two anonymous reviewers for helpful comments that significantly improved the manuscript.

REFERENCES CITED

Alibert, C., and McCulloch, M.T., 1993, Rare earth element and neodymium isotopic compositions of the banded iron-formations and associated shales from Hamersley, western Australia: Geochimica et Cosmochimica Acta, v. 57, p. 187–204, doi: 10.1016/0016-7037(93)90478-F.

Alibo, D.S., and Nozaki, Y., 1999, Rare earth elements in seawater: Particle association, shale-normalization, and Ce oxidation: Geochimica et Cosmochimica Acta, v. 63, p. 363–372, doi: 10.1016/S0016-7037(98)00279-8.

Arndt, N.T., Nelson, D.R., Compston, W., Trendall, A.F., and Thorne, A.M., 1991, The age of the Fortescue Group, Hamersley Basin, Western Australia, from ion microprobe zircon U-Pb results: Australian Journal of Earth Sciences, v. 38, p. 261–281.

Arora, M., Govil, P.K., Charan, S.N., Uday Raj, B., Balaram, V., Manikyamba, C., Chatterjee, A.K., and Naqvi, S.M., 1995, Geochemistry and origin of Archean banded iron-formation from the Bababudan schist belt, India: Economic Geology and the Bulletin of the Society of Economic Geologists, v. 90, p. 2040–2057.

Aubert, D., Stille, P., and Probst, A., 2001, REE fractionation during granite weathering and removal by waters and suspended loads: Sr and Nd isotopic evidence: Geochimica et Cosmochimica Acta, v. 65, p. 387–406, doi: 10.1016/S0016-7037(00)00546-9.

Barley, M.E., 1993, Volcanic, sedimentary and tectonostratigraphic environments of the ~3.46 Ga Warrawoona Megasequence: A review: Precambrian Research, v. 60, p. 47–67, doi: 10.1016/0301-9268(93)90044-3.

Barley, M.E., Pickard, A.L., and Sylvester, P.J., 1997, Emplacement of a large igneous province as a possible cause of banded iron formation 2.45 billion years ago: Nature, v. 385, p. 55–58, doi: 10.1038/385055a0.

Barrett, T.J., and Jarvis, I., 1988, Rare-earth element geochemistry of metalliferous sediments from DSDP Leg 92: The East Pacific Rise transect: Chemical Geology, v. 67, p. 243–259, doi: 10.1016/0009-2541(88)90131-3.

Barrett, T.J., Fralick, P.W., and Jarvis, I., 1988, Rare-earth-element geochemistry of some Archean iron formations north of Lake Superior, Ontario:

Canadian Journal of Earth Sciences, v. 25, p. 570–580.
Bau, M., and Dulski, P., 1996, Distribution of yttrium and rare-earth elements in the Penge and Kuruman iron-formations, Transvaal Supergroup, South Africa: Precambrian Research, v. 79, p. 37–55, doi: 10.1016/0301-9268(95)00087-9.
Bau, M., and Dulski, P., 1999, Comparing yttrium and rare earths in hydrothermal fluids from the Mid-Atlantic Ridge: Implications for Y and REE behavior during near-vent mixing and for the Y/Ho ratio of Proterozoic seawater: Chemical Geology, v. 155, p. 77–90, doi: 10.1016/S0009-2541(98)00142-9.
Bau, M., and Möller, P., 1993, Rare earth element systematics of the chemically precipitated component in early Precambrian iron formations and the evolution of the terrestrial atmosphere-hydrosphere-lithosphere system: Geochimica et Cosmochimica Acta, v. 57, p. 2239–2249, doi: 10.1016/0016-7037(93)90566-F.
Bau, M., Koschinsky, A., Dulski, P., and Hein, J.R., 1996, Comparison of the partitioning behaviors of yttrium, rare earth elements, and titanium between hydrogenetic marine ferromanganese crusts and seawater: Geochimica et Cosmochimica Acta, v. 60, p. 1709–1725, doi: 10.1016/0016-7037(96)00063-4.
Bender, M., Broecker, W., Gornitz, V., Middle, U., Kay, R., Sun, S., and Biscaye, P., 1971, Geochemistry of three cores from the East Pacific Rise: Earth and Planetary Science Letters, v. 12, p. 425–433, doi: 10.1016/0012-821X(71)90028-8.
Braun, J.J., Viers, J., Dupré, B., Polve, M., Ndam, J., and Muller, J.P., 1998, Solid/liquid REE fractionation in the lateritic system of Goyoum, East Cameroon: The implication for the present dynamics of the soil covers of the humid tropical regions: Geochimica et Cosmochimica Acta, v. 62, p. 273–299, doi: 10.1016/S0016-7037(97)00344-X.
Byrne, R.H., and Kim, K., 1990, Rare earth element scavenging in seawater: Geochimica et Cosmochimica Acta, v. 54, p. 2645–2656, doi: 10.1016/0016-7037(90)90002-3.
Chadwick, B., Vasudev, V.N., and Hedge, G.V., 2000, The Dharwar craton, southern India, interpreted as the result of Late Archaean oblique convergence: Precambrian Research, v. 99, p. 91–111, doi: 10.1016/S0301-9268(99)00055-8.
Condie, K.C., 1993, Chemical composition and evolution of the upper continental crust: Contrasting results from surface samples and shales: Chemical Geology, v. 104, p. 1–37, doi: 10.1016/0009-2541(93)90140-E.
Condie, K.C., 2000, Episodic continental growth models: Afterthoughts and extensions: Tectonophysics, v. 322, p. 153–162, doi: 10.1016/S0040-1951(00)00061-5.
Cornell, D.H., Schutte, S.S., and Eglington, B.L., 1996, The Ongeluk basaltic andesite formation in Griqualand West, South Africa: Submarine alteration in a 2222 Ma Proterozoic sea: Precambrian Research, v. 79, p. 101–123, doi: 10.1016/0301-9268(95)00090-9.
Danielson, A., Möller, P., and Dulski, P., 1992, The europium anomalies in banded iron formations and the thermal history of the oceanic crust: Chemical Geology, v. 97, p. 89–100, doi: 10.1016/0009-2541(92)90137-T.
de Baar, H.J.W., Bacon, M.P., and Brewer, P.G., 1985, Rare earth elements in the Pacific and Atlantic Oceans: Geochimica et Cosmochimica Acta, v. 49, p. 1943–1959, doi: 10.1016/0016-7037(85)90089-4.
de Baar, H.J.W., German, C.R., Elderfield, H., and van Gaans, P., 1988, Rare earth element distributions in anoxic waters of the Cariaco Trench: Geochimica et Cosmochimica Acta, v. 52, p. 1203–1219, doi: 10.1016/0016-7037(88)90275-X.
Derry, L.A., and Jacobsen, S.B., 1990, The chemical evolution of Precambrian seawater: Evidence from REEs in banded iron formations: Geochimica et Cosmochimica Acta, v. 54, p. 2965–2977, doi: 10.1016/0016-7037(90)90114-Z.
de Wit, M.J., Roering, C., Hart, R.J., Armstrong, R.A., de Ronde, C.E.J., Green, R.W.E., Tredoux, M., Peberdy, E., and Hart, R.A., 1992, Formation of an Archaean continent: Nature, v. 357, p. 553–562, doi: 10.1038/357553a0.
Dymek, R.F., and Klein, C., 1988, Chemistry, petrology and origin of banded iron-formation lithologies from the 3800 Ma Isua Supracrustal Belt: West Greenland: Precambrian Research, v. 39, p. 247–302, doi: 10.1016/0301-9268(88)90022-8.
Elderfield, H., 1988, The oceanic chemistry of the rare-earth elements: Royal Society of London Philosophical Transactions, ser. A, v. 325, p. 105–126.
Elderfield, H., and Sholkovitz, E.R., 1987, Rare earth elements in the pore waters of reducing nearshore sediments: Earth and Planetary Science Letters, v. 82, p. 280–288, doi: 10.1016/0012-821X(87)90202-0.
Elderfield, H., Hawkesworth, C.J., Greaves, M.J., and Calvert, S.E., 1981, Rare earth element geochemistry of oceanic ferromanganese nodules and associated sediments: Geochimica et Cosmochimica Acta, v. 45, p. 513–528, doi: 10.1016/0016-7037(81)90184-8.
Fryer, B.J., 1972, Age determinations in the Circum-Ungava Geosyncline and the evolution of Precambrian banded iron-formations: Canadian Journal of Earth Sciences, v. 9, p. 652–663.
Fryer, B.J., 1977, Rare earth evidence in iron-formations for changing Precambrian oxidation states: Geochimica et Cosmochimica Acta, v. 41, p. 361–367, doi: 10.1016/0016-7037(77)90263-0.
Fujinaga, K., Kato, Y., Kiminami, K., Miura, K., and Nakamura, K., 1999, Geochemistry of red shale from the Mugi Formation in the Upper Cretaceous Shimanto Belt, Shikoku, Japan: Memoirs of Geological Society of Japan, no. 52, p. 205–216 (in Japanese with English abstract).
German, C.R., and Elderfield, H., 1989, Rare earth elements in Saanich Inlet, British Columbia, a seasonally anoxic basin: Geochimica et Cosmochimica Acta, v. 53, p. 2561–2571, doi: 10.1016/0016-7037(89)90128-2.
German, C.R., and Elderfield, H., 1990, Application of the Ce anomaly as a paleoredox indicator: The ground rules: Paleoceanography, v. 5, p. 823–833.
German, C.R., Klinkhammer, G.P., Edmond, J.M., Mitra, A., and Elderfield, H., 1990, Hydrothermal scavenging of rare-earth elements in the ocean: Nature, v. 345, p. 516–518, doi: 10.1038/345516a0.
Goldstein, S.J., and Jacobsen, S.B., 1988, Rare earth elements in river waters: Earth and Planetary Science Letters, v. 89, p. 35–47, doi: 10.1016/0012-821X(88)90031-3.
Grauch, R.I., 1989, Rare earth elements in metamorphic rocks, in Lipin, B.R., and McKay, G.A., eds., Geochemistry and mineralogy of rare earth elements: Reviews in Mineralogy, v. 21, p. 147–167.
Gross, G.A., 1965, Geology of iron deposits in Canada: I. General geology and evaluation of iron deposits: Geological Survey of Canada Economic Geology Report 22, 181 p.
Hickman, A.H., 1983, Geology of the Pilbara Block and its environs: Geological Survey of Western Australia Bulletin, v. 127, 268 p.
Isley, A.E., and Abbott, D.H., 1999, Plume-related mafic volcanism and the deposition of banded iron formation: Journal of Geophysical Research, v. 104, p. 15461–15477, doi: 10.1029/1999JB900066.
Jacobsen, S.B., and Pimentel-Klose, M.R., 1988, A Nd isotopic study of the Hamersley and Michipicoten banded iron formations: The source of REE and Fe in Archean oceans: Earth and Planetary Science Letters, v. 87, p. 29–44, doi: 10.1016/0012-821X(88)90062-3.
James, H.L., 1954, Sedimentary facies of iron-formations: Economic Geology and the Bulletin of the Society of Economic Geologists, v. 49, p. 235–293.
Kato, Y., 2003, Banded iron formations: Their genesis and implications for Precambrian Earth: Geochimica et Cosmochimica Acta, v. 67, no. 18S, p. A204.
Kato, Y., and Kimura, S., 1999, Bulk-rock chemical compositions of the Boolgeeda Iron Formation of the Hamersley Group, Western Australia: Its origin and implication from a very low concentration of minor elements: Shigen-Chishitsu, v. 49, p. 175–189.
Kato, Y., and Nakamura, K., 2003, Origin and global tectonic significance of Early Archean cherts from the Marble Bar greenstone belt, Pilbara Craton, Western Australia: Precambrian Research, v. 125, p. 191–243, doi: 10.1016/S0301-9268(03)00043-3.
Kato, Y., Kawakami, T., Kano, T., Kunugiza, K., and Swamy, N.S., 1996, Rare-earth element geochemistry of banded iron formations and associated amphibolite from the Sargur belts, south India: Journal of Southeast Asian Earth Sciences, v. 14, p. 161–164, doi: 10.1016/S0743-9547(96)00054-2.
Kato, Y., Ohta, I., Tsunematsu, T., Watanabe, Y., Isozaki, Y., Maruyama, S., and Imai, N., 1998, Rare earth element variations in mid-Archean banded iron formations: Implications for the chemistry of ocean and continent and plate tectonics: Geochimica et Cosmochimica Acta, v. 62, p. 3475–3497, doi: 10.1016/S0016-7037(98)00253-1.
Kato, Y., Kano, T., and Kunugiza, K., 2002a, Negative Ce anomaly in the Indian banded iron formations: Evidence for the emergence of oxygenated deep-sea at 2.9–2.7 Ga: Resource Geology, v. 52, p. 101–110.
Kato, Y., Nakao, K., and Isozaki, Y., 2002b, Geochemistry of Late Permian to Early Triassic pelagic cherts from southwest Japan: Implications for an oceanic redox change: Chemical Geology, v. 182, p. 15–34, doi: 10.1016/S0009-2541(01)00273-X.

Khan, R.M.K., Das Sharma, S., Patil, D.J., and Naqvi, S.M., 1996, Trace, rare-earth element, and oxygen isotopic systematics for the genesis of banded iron-formations: Evidence from Kushtagi schist belt, Archaean Dharwar Craton, India: Geochimica et Cosmochimica Acta, v. 60, p. 3285–3294, doi: 10.1016/0016-7037(96)00172-X.

Kiminami, K., Miyashita, S., and Kawabata, K., 1994, Ridge collision and in situ greenstones in accretionary complexes: An example from the Late Cretaceous Ryukyu Islands and southwest Japan margin: The Island Arc, v. 3, p. 103–111.

Kiyokawa, S., and Taira, A., 1998, The Cleaverville Group in the West Pilbara Coastal Granitoid-Greenstone Terrain of Western Australia: An example of a Mid-Archaean immature oceanic island-arc succession: Precambrian Research, v. 88, p. 109–142, doi: 10.1016/S0301-9268(97)00066-1.

Klein, C., and Beukes, N.J., 1993, Sedimentology and geochemistry of the glaciogenic Late Proterozoic Rapitan iron-formation in Canada: Economic Geology and the Bulletin of the Society of Economic Geologists, v. 88, p. 542–565.

Koeppenkastrop, D., and De Carlo, E.H., 1992, Sorption of rare-earth elements from seawater onto synthetic mineral particles: An experimental approach: Chemical Geology, v. 95, p. 251–263, doi: 10.1016/0009-2541(92)90015-W.

Komiya, T., Maruyama, S., Masuda, T., Nohda, S., Hayashi, M., and Okamoto, K., 1999, Plate tectonics at 3.8–3.7 Ga: Field evidence from the Isua accretionary complex, southern West Greenland: Journal of Geology, v. 107, p. 515–554, doi: 10.1086/314371.

Koppi, A.J., Edis, R., Field, D.J., Geering, H.R., Klessa, D.A., and Cockayne, D.J.H., 1996, Rare earth element trends and cerium-uranium-manganese associations in weathered rock from Koongarra, Northern Territory, Australia: Geochimica et Cosmochimica Acta, v. 60, p. 1695–1707, doi: 10.1016/0016-7037(96)00047-6.

Krapez, B., 1993, Sequence stratigraphy of the Archean supracrustal belts of the Pilbara Block, Western Australia: Precambrian Research, v. 60, p. 1–45, doi: 10.1016/0301-9268(93)90043-2.

Kuhn, T., Bau, M., Blum, N., and Halbach, P., 1998, Origin of negative Ce anomalies in mixed hydrothermal-hydrogenetic Fe-Mn crusts from the Central Indian Ridge: Earth and Planetary Science Letters, v. 163, p. 207–220, doi: 10.1016/S0012-821X(98)00188-5.

Kusky, T.M., and Kidd, W.S.F., 1992, Remnants of an Archean oceanic plateau, Belingwe greenstone belt, Zimbabwe: Geology, v. 20, p. 43–46, doi: 10.1130/0091-7613(1992)020<0043:ROAAOP>2.3.CO;2.

Laajoki, K., 1975, Rare-earth elements in Precambrian iron formations in Väyrylänkylä, South Puolanka area, Finland: Bulletin of Geological Society of Finland, v. 47, p. 93–107.

Liu, Y.-G., Miah, M.R.U., and Schmitt, R.A., 1988, Cerium: A chemical tracer for paleo-oceanic redox conditions: Geochimica et Cosmochimica Acta, v. 52, p. 1361–1371, doi: 10.1016/0016-7037(88)90207-4.

Lottermoser, B.G., and Ashley, P.M., 2000, Geochemistry, petrology and origin of Neoproterozoic ironstones in the eastern part of the Adelaide Geosyncline, South Australia: Precambrian Research, v. 101, p. 49–67, doi: 10.1016/S0301-9268(99)00098-4.

MacLeod, K.G., and Irving, A.J., 1996, Correlation of cerium anomalies with indicators of paleoenvironment: Journal of Sedimentary Research, v. 66, p. 948–955.

Manikyamba, C., and Naqvi, S.M., 1997, Mineralogy and geochemistry of Archean greenstone belt-hosted Mn formations and deposits of the Dharwar Craton: redox potential of proto-oceans: Geological Society [London] Special Publication 119, p. 91–103.

Manikyamba, C., Balaram, V., and Naqvi, S.M., 1993, Geochemical signatures of polygenetic origin of a banded iron formation (BIF) of the Archaean Sandur greenstone belt (schist belt) Karnataka nucleus, India: Precambrian Research, v. 61, p. 137–164, doi: 10.1016/0301-9268(93)90061-6.

Martin, D.McB., Powell, C.McA., and George, A.D., 2000, Stratigraphic architecture and evolution of the early Paleoproterozoic McGrath Trough, Western Australia: Precambrian Research, v. 99, p. 33–64, doi: 10.1016/S0301-9268(99)00053-4.

Masuda, A., 1975, Abundances of monoisotopic REE, consistent with the Leedey chondrite values: Geochemical Journal, v. 9, p. 183–184.

Masuda, A., Nakamura, N., and Tanaka, T., 1973, Fine structures of mutually normalized rare-earth patterns of chondrites: Geochimica et Cosmochimica Acta, v. 37, p. 239–248, doi: 10.1016/0016-7037(73)90131-2.

McLennan, S.M., and Taylor, S.R., 1991, Sedimentary rocks and crustal evolution: Tectonic setting and secular trends: Journal of Geology, v. 99, p. 1–21.

McNaughton, N.J., Compston, W., and Barley, M.E., 1993, Constraints on the age of the Warrawoona Group, eastern Pilbara Block, Western Australia: Precambrian Research, v. 60, p. 69–98, doi: 10.1016/0301-9268(93)90045-4.

Michard, A., 1989, Rare earth element systematics in hydrothermal fluids: Geochimica et Cosmochimica Acta, v. 53, p. 745–750, doi: 10.1016/0016-7037(89)90017-3.

Mills, R.A., and Elderfield, H., 1995, Rare earth element geochemistry of hydrothermal deposits from the active TAG Mound, 26°N Mid-Atlantic Ridge: Geochimica et Cosmochimica Acta, v. 59, p. 3511–3524, doi: 10.1016/0016-7037(95)00224-N.

Mitra, A., Elderfield, H., and Greaves, M.J., 1994, Rare earth elements in submarine hydrothermal fluids and plumes from the Mid-Atlantic Ridge: Marine Chemistry, v. 46, p. 217–235, doi: 10.1016/0304-4203(94)90079-5.

Morinaga, C., 1997, Geochemical study of the oldest sedimentary rocks from the Isua supracrustal belt [M.S. thesis]: Yamaguchi, Japan, Yamaguchi University, 160 p. (in Japanese with English abstract).

Murray, R.W., Buchholtz ten Brink, M.R., Gerlach, D.C., Russ, G.P., III, and Jones, D.L., 1992a, Interoceanic variation in the rare earth, major, and trace element depositional chemistry of chert: Perspectives gained from the DSDP and ODP record: Geochimica et Cosmochimica Acta, v. 56, p. 1897–1913, doi: 10.1016/0016-7037(92)90319-E.

Murray, R.W., Buchholtz ten Brink, M.R., Gerlach, D.C., Russ, G.P., III, and Jones, D.L., 1992b, Rare-earth, major, and trace element composition of Monterey and DSDP chert and associated host sediment: Assessing the influence of chemical fractionation during diagenesis: Geochimica et Cosmochimica Acta, v. 56, p. 2657–2671, doi: 10.1016/0016-7037(92)90351-I.

Naqvi, S.M., Sawkar, R.H., Subba Rao, D.V., Govil, P.K., and Gnaneswar Rao, T., 1988, Geology, geochemistry and tectonic setting of Archaean greywackes from Karnataka nucleus, India: Precambrian Research, v. 39, p. 193–216, doi: 10.1016/0301-9268(88)90042-3.

Nesbitt, H.W., 1979, Mobility and fractionation of rare earth elements during weathering of a granodiorite: Nature, v. 279, p. 206–210, doi: 10.1038/279206a0.

Nutman, A.P., Chadwick, B., Ramakrishnan, M., and Viswanatha, M.N., 1992, SHRIMP U-Pb ages of detrital zircon in Sargur supracrustal rocks in western Karnataka, southern India: Journal of Geological Society of India, v. 39, p. 367–374.

Nutman, A.P., Chadwick, B., Krishna Rao, B., and Vasudev, V.N., 1996, SHRIMP U/Pb zircon ages of acid volcanic rocks in the Chitradurga and Sandur Groups, and granites adjacent to the Sandur schist belt, Karnataka: Journal of Geological Society of India, v. 47, p. 153–164.

Nutman, A.P., Bennett, V.C., Friend, C.R.L., and Rosing, M.T., 1997, ~3710 and ≥3790 Ma volcanic sequences in the Isua (Greenland) supracrustal belt; structural and Nd isotope implications: Chemical Geology, v. 141, p. 271–287, doi: 10.1016/S0009-2541(97)00084-3.

Nyström, J.O., 1984, Rare earth element mobility in vesicular lava during low-grade metamorphism: Contributions to Mineralogy and Petrology, v. 88, p. 328–331, doi: 10.1007/BF00376757.

Ohmoto, H., 1993, The banded iron formation in the Hamersley Basin, Australia: Products of the oxygen-rich Archean atmosphere?: Geological Society of America Abstracts with Programs, v. 25, no. 6, p. A-89.

Ohmoto, H., Watanabe, Y., Yamaguchi, K.E., Naraoka, H., Haruna, M., Kakegawa, T., Hayashi, K., and Kato, Y., 2006, this volume, Chemical and biological evolution of early Earth: Constraints from banded iron formations, in Kesler, S.E., and Ohmoto, H., eds., Evolution of Early Earth's Atmosphere, Hydrosphere, and Biosphere—Constraints from Ore Deposits: Geological Society of America Memoir 198, doi: 10.1130/2006.1198(17).

Ohta, H., Maruyama, S., Takahashi, E., Watanabe, Y., and Kato, Y., 1996, Field occurrence, geochemistry and petrogenesis of the Archean Mid-Oceanic Ridge Basalts (AMORBs) of the Cleaverville area, Pilbara Craton, Western Australia: Lithos, v. 37, p. 199–221, doi: 10.1016/0024-4937(95)00037-2.

Olivarez, A.M., and Owen, R.M., 1991, The europium anomaly of seawater: Implications for fluvial versus hydrothermal REE inputs to the oceans: Chemical Geology, v. 92, p. 317–328, doi: 10.1016/0009-2541(91)90076-4.

Pan, Y., and Fleet, M.E., 1996, Rare earth element mobility during prograde granulite facies metamorphism: Significance of fluorine: Contributions to Mineralogy and Petrology, v. 123, p. 251–262, doi: 10.1007/s004100050154.

Radhakrishna, B.P., and Naqvi, S.M., 1986, Precambrian continental crust of India and its evolution: Journal of Geology, v. 94, p. 145–166.

Ravizza, G., Sherrell, R.M., Field, M.P., and Pickett, E.A., 1999, Geochemistry of the Margi umbers, Cyprus, and the Os isotope composition of Cretaceous seawater: Geology, v. 27, p. 971–974, doi: 10.1130/0091-7613(1999)027<0971:GOTMUC>2.3.CO;2.

Ruhlin, D.E., and Owen, R.M., 1986, The rare earth element geochemistry of hydrothermal sediments from the East Pacific Rise: Examination of a seawater scavenging mechanism: Geochimica et Cosmochimica Acta, v. 50, p. 393–400, doi: 10.1016/0016-7037(86)90192-4.

Sakko, M., and Laajoki, K., 1975, Whole rock Pb-Pb isochron age for the Pääkkö iron formation in Väyrylänkylä, South Puolanka Area, Finland: Bulletin of Geological Society of Finland, v. 47, p. 113–117.

Sholkovitz, E.R., and Schneider, D.L., 1991, Cerium redox cycles and rare earth elements in the Sargasso Sea: Geochimica et Cosmochimica Acta, v. 55, p. 2737–2743, doi: 10.1016/0016-7037(91)90440-G.

Sholkovitz, E.R., Piepgras, D.J., and Jacobsen, S.B., 1989, The pore water chemistry of rare earth elements in Buzzards Bay sediments: Geochimica et Cosmochimica Acta, v. 53, p. 2847–2856, doi: 10.1016/0016-7037(89)90162-2.

Sholkovitz, E.R., Shaw, T.J., and Schneider, D.L., 1992, The geochemistry of rare earth elements in the seasonally anoxic water column and porewaters of Chesapeake Bay: Geochimica et Cosmochimica Acta, v. 56, p. 3389–3402, doi: 10.1016/0016-7037(92)90386-W.

Smith, R.E., Perdrix, J.L., and Parks, T.C., 1982, Burial metamorphism in the Hamersley Basin, Western Australia: Journal of Petrology, v. 23, p. 75–102.

Stille, P., and Clauer, N., 1986, Sm-Nd isochron-age and provenance of the argillites of the Gunflint Iron Formation in Ontario, Canada: Geochimica et Cosmochimica Acta, v. 50, p. 1141–1146, doi: 10.1016/0016-7037(86)90395-9.

Sumner, D.Y., and Bowring, S.A., 1996, U-Pb geochronologic constraints on deposition of the Campbellrand Subgroup, Transvaal Supergroup, South Africa: Precambrian Research, v. 79, p. 25–35, doi: 10.1016/0301-9268(95)00086-0.

Taylor, S.R., and McLennan, S.M., 1985, The continental crust: Its composition and evolution: Oxford, Blackwell, 312 p.

Thorpe, R.I., Hickman, A.H., Davis, D.W., Mortensen, J.K., and Trendall, A.F., 1992, U-Pb zircon geochronology of Archean felsic units in the Marble Bar region, Pilbara Craton, Western Australia: Precambrian Research, v. 56, p. 169–189, doi: 10.1016/0301-9268(92)90100-3.

Trendall, A.F., and Blockley, J.G., 1970, The iron formations of the Precambrian Hamersley Group, Western Australia, with special reference to the associated crocidolite: Geological Survey of Western Australia Bulletin, v. 199, 366 p.

Trendall, A.F., Compston, W., Williams, I.S., Armstrong, R.A., Arndt, N.T., McNaughton, N.J., Nelson, D.R., Barley, M.E., Beukes, N.J., de Laeter, J.R., Retief, E.A., and Thorne, A.M., 1990, Precise zircon U-Pb chronological comparison of the volcano-sedimentary sequences of the Kaapvaal and Pilbara cratons between about 3.1 and 2.4 Ga, in Glover, J.E., and Ho, S.E., eds., International Archean Symposium, 3rd, Extended Abstract Volume: Perth, Department of Geology, University of Western Australia, p. 81–83.

Trendall, A.F., de Laeter, J.R., Nelson, D.R., and Bhaskar Rao, Y.J., 1997a, Further zircon U-Pb age data for the Daginkatte Formation, Dharwar Supergroup, Karnataka Craton: Journal of Geological Society of India, v. 50, p. 25–30.

Trendall, A.F., de Laeter, J.R., Nelson, D.R., and Mukhopadhway, D., 1997b, A precise zircon U-Pb age for the base of the BIF of the Mulaingiri Formation (Bababudan Group, Dharwar Supergroup) of the Karnataka Craton: Journal of Geological Society of India, v. 50, p. 161–170.

Trendall, A.F., Nelson, D.R., de Laeter, J.R., and Hassler, S.W., 1998, Precise zircon U-Pb ages from the Marra Mamba Iron Formation and Wittenoom Formation, Hamersley Group, Western Australia: Australian Journal of Earth Sciences, v. 45, p. 137–142.

Tsikos, H., and Moore, J.M., 1997, Petrography and geochemistry of the Paleoproterozoic Hotazel iron-formation, Kalahari manganese field, South Africa: Implications for Precambrian manganese metallogenesis: Economic Geology and the Bulletin of the Society of Economic Geologists, v. 92, p. 87–97.

Van Kranendonk, M.J., Hickman, A.H., Williams, I.R., and Nijman, W., 2001, Archaean geology of the East Pilbara Granite-Greenstone Terrane, Western Australia—A field guide: Geological Survey of Western Australia, Record 2001/9, 134 p.

Wright, J., Schrader, H., and Holser, W.T., 1987, Paleoredox variations in ancient oceans recorded by rare earth elements in fossil apatite: Geochimica et Cosmochimica Acta, v. 51, p. 631–644, doi: 10.1016/0016-7037(87)90075-5.

Zhang, J., and Nozaki, Y., 1996, Rare earth elements and yttrium in seawater: ICP-MS determinations in the East Caroline, Coral Sea, and South Fiji basins of the western South Pacific Ocean: Geochimica et Cosmochimica Acta, v. 60, p. 4631–4644, doi: 10.1016/S0016-7037(96)00276-1.

Zhang, J., Amakawa, H., and Nozaki, Y., 1994, The comparative behaviors of yttrium and lanthanides in the seawater of the North Pacific: Geophysical Research Letters, v. 21, p. 2677–2680, doi: 10.1029/94GL02404.

MANUSCRIPT ACCEPTED BY THE SOCIETY 29 OCTOBER 2005

Chemical and biological evolution of early Earth: Constraints from banded iron formations

Hiroshi Ohmoto
Yumiko Watanabe
Penn State Astrobiology Research Center of the NASA Astrobiology Institute and the Department of Geosciences, Pennsylvania State University, University Park, Pennsylvania 16802, USA

Kosei E. Yamaguchi
Institute for Research on Earth Evolution (IFREE), Japan Agency for Marine-Earth Science and Technology (JAMSTEC), Yokosuka, Kanagawa 237-0061, Japan

Hiroshi Naraoka
Department of Chemistry, Okayama University, Okayama, Okayama 700-8530, Japan

Makoto Haruna
Uchida Yoko Co., Ltd., Chuo-ku, Tokyo 104-8282, Japan

Takeshi Kakegawa
Department of Earth Science, Tohoku University, Sendai, Miyagi 980-8578, Japan

Ken-ichiro Hayashi
Department of Earth Evolution Sciences, University of Tsukuba, Tsukuba, Ibaraki 305-8572, Japan

Yasuhiro Kato
Department of Geosystem Engineering, University of Tokyo, Bunkyo-ku, Tokyo 113-8656, Japan

ABSTRACT

Geological and geochemical characteristics of banded iron formations (BIFs) suggest that they formed by mixing locally (or regionally) discharged submarine hydrothermal fluids with local seawater, rather than by upwelling deep ocean water. Submarine hydrothermal fluids typically evolved from local seawater by acquiring heat, metals, and sulfur during deep circulation through a variety of rocks (e.g., volcanics, evaporites) in greenstone terranes that developed under a variety of tectonic settings. In general, when the fluids were heated above ~350 °C, they may have produced Cu- and Zn-rich volcanogenic massive sulfide deposits (VMSDs), whereas those heated less than ~200 °C were generally poor in H_2S and heavy metals, except Fe, and may have subsequently produced BIFs.

Depending on the salinity contrast between discharging hydrothermal fluids (evolved seawater) and local seawater, hydrothermal fluids may (1) mix rapidly with local seawater to form smoker-type BIFs or (2) create a metal- and silica-rich brine pool, mix slowly with the overlying water body, and form brine pool-type BIFs. BIFs associated with VMSDs and volcanic rocks generally belong to smoker-type BIFs; many formed at seawater depths >2.5 km. Large BIFs, including the 2.6–2.4 Ga BIFs in the Hamersley Basin, Australia, the 2.5 Ga Kuruman IF in South Africa, and the 1.87 Ga BIFs in the Lake Superior region, United States-Canada, belong to brine pool-type BIFs. The Hamersley Basin and possibly other large BIF-hosting basins were probably land-locked seas (like the Black Sea) where river waters diluted the surface water zone and the underlying water bodies were anoxic.

During the accumulation of a BIF sequence, the dominant Fe mineralogy frequently changed from ferric (hydr)oxides (oxide BIFs) to siderite (carbonate BIFs) and to pyrite (sulfide BIFs). Such changes were probably caused by changes in the relative amounts of dissolved O_2 (DO), ΣCO_3^{2-}, and ΣS^{2-} in local seawater. From the Fe^{2+}-O_2 mass balance calculations for the formation of iron oxides in smoker-type BIFs, and the relationship between the atmospheric pO_2 and oceanic O_2 depth profile, we conclude that the atmosphere and oceans have been fully oxygenated since ca. 3.8 Ga, except in local anoxic basins. Thermodynamic analyses of the formational conditions of siderite and analyses of the carbon isotopic composition of siderite associated with major BIFs suggest that the pre–1.8 Ga atmosphere was CO_2-rich (pCO_2 >100 PAL) and CH_4-poor ($pCH_4 \approx 10$ ppm); therefore, CO_2, rather than CH_4, was the major greenhouse gas throughout geologic history.

After a decline of hydrothermal fluid flux, BIF-hosting basins generally became euxinic (H_2S-rich) because of the increased activity of sulfate-reducing bacteria (SRB) and SO_4^{2-}-rich seawater, and thereby accumulated organic carbon-rich and pyrite-rich black shales (sulfide-type BIFs). The SO_4^{2-} contents and SRB activity in the oceans have been essentially the same since ca. 3.8 Ga. The Archean oceans were most likely poor in both Fe^{2+} and silica, much like modern oceans. Our study also suggests that diverse communities of organisms, including cyanobacteria, SRB, methanogens, methanotrophs, and eukaryotes, evolved very early in Earth's history, probably by the time the oldest BIFs (ca. 3.8 Ga) formed.

BIFs have been found in rocks of all geologic age. Therefore, they cannot be indicators of an anoxic atmosphere and/or anoxic oceans as suggested by many previous researchers. Instead, BIFs indicate that the atmosphere and ocean chemistry have been regulated at present compositions (except pCO_2) through geologic history by interactions with the biosphere. The general trend of declining size and abundance of BIFs with geologic time reflects the cooling history of Earth's interior.

Keywords: banded iron formation, atmospheric evolution, oceanic evolution, biological evolution, Archean.

INTRODUCTION

"Banded iron formation" (BIF) is a lithological term for a chemical sediment consisting of a thinly layered or laminated rock composed of alternating layers (bands) of chert and primarily iron-rich minerals (hematite, magnetite, siderite, pyrite, and various Fe-rich silicates); its silica content (as SiO_2) ranges from 40 to 50 wt% whereas its iron content (as Fe) typically ranges from 15 to 35 wt% (James, 1954). Large BIF deposits are most abundant in sedimentary sequences older than 1.8 Ga and range from a few centimeters to several hundred meters in thickness and from <~1 km² to >100,000 km² in aerial extent.

The term "iron ore" is typically reserved for a rock containing more than ~35 wt% Fe. Economically, the most important iron ores are those that developed from BIFs by the leaching of silica and conversion of primary/diagenetic iron minerals (hematite, magnetite, and siderite) to "secondary" hematite.

Since Cloud published his model in 1968, theories for the chemical and biological evolutions of both early Earth and BIF formation have influenced each other and continually evolved as new geologic concepts (e.g., plate tectonic and mantle dynamics, submarine hydrothermal processes, and the evolutionary history of microbes) have been developed and new data have become available. Here, we review how BIF models have evolved over the past five decades, and present a new model (synthesis) to explain the latest (unpublished) observations as well as those in literature. In the first section (Background) of this paper, we review the characteristics of BIFs and the evolution of BIF models. In the second section, we review general characteristics of submarine hydrothermal systems, because they are essential in building a new BIF model. In the third section, we present our interpretations of the geological and hydrological processes for the formation of major BIFs, including, but not restricted to ca. 2.7 Ga BIFs in the Abitibi district, Canada; ca. 2.5 Ga BIFs in the Hamersley district, Australia, and in the Kuruman district, South Africa; and ca. 1.9 Ga BIFs in the Lake Superior district, United States-Canada. In the fourth section, we present our models for the precipitation mechanisms of iron oxides, siderite, pyrite, and chert in BIFs and for the conversions of hematite to magnetite and magnetite to hematite. In the final section, we relate BIFs to the evolutionary histories of atmospheric chemistry (pO_2, pCO_2, and pCH_4), oceanic chemistry (Fe^{2+}, silica, O_2, H_2S, SO_4^{2-}, and pH), constituents of the biosphere, and thermal dynamics of the mantle.

BACKGROUND

Over 1300 books, papers, and abstracts have been published on BIFs. Comprehensive reviews on the geology, mineralogy, geochemistry, and models of depositional mechanisms have been assembled in several recent books written or edited by Mel'nik (1982), Trendall and Morris (1983), Appel and LaBerge (1987), Young and Taylor (1989), Farrell (1990), and Eriksson et al. (2004). Other valuable review papers include, but are not restricted to, Dimroth (1976), Cole and Klein (1981), Kimberley (1989a, 1989b), Gross (1991), and Klein and Beukes (1992).

Classifications of Iron Formations

Banded Iron Formations versus Öölitic Ironstones

Although opinions vary as to whether BIFs in specific regions accumulated in open seas, semi-closed seas, lagoons, or lakes (e.g., Eugster and I-Ming Chou, 1973), all researchers agree that BIFs accumulated in large water bodies (i.e., subaqueous). Thus, BIFs contrast with laterites (ferric oxide-rich soils) that also contain >15 wt% Fe but formed subaerially by in situ rock weathering. Öölitic ironstones, in which goethite and/or hematite primarily occur as 1–5 mm öölites and/or pisolites, may have formed mostly (but not always) by the accumulation and reworking of detrital grains of laterites in shallow water (Young, 1989).

In contrast, most iron minerals in BIFs (except some Fe-rich silicates) precipitated by (bio)chemical processes utilizing the dissolved Fe (mostly Fe^{2+}) in overlying water bodies. Fe-precipitation was episodic and resulted in alternating Fe-rich and silica-rich bands. The thickness of an individual iron-rich layer (microband) is typically from ~50 μm to 2 mm, whereas that of an individual chert layer ranges more widely, from ~50 μm to ~1 cm. Because of the variation in thicknesses of Fe-rich and Si-rich microbands, Fe/Si ratios also vary from a ~1 mm to ~10 cm scale (mesobands) to a ~1 m to ~30 m scale (macrobands) (Trendall and Blockley, 1970).

Contrary to a popular notion that BIFs are restricted to rocks of Precambrian age and öölitic ironstones to those of Phanerozoic age, BIFs are also common in Phanerozoic sequences (Tables 1, 2) and öölitic iron ores are not uncommon in Precambrian sequences (Kimberley, 1989b). For example, some BIFs, such as the ca. 1.9 Ga Sokoman BIFs in Canada, contain both banded and öölitic types of iron formations (Klein and Fink, 1976). A large öölitic iron formation occurs in the 2.35 Ga Timeball Hill Formation in South Africa (Beukes, 1983; Bekker et al., 2004). In this paper, we will focus on (bio)chemically precipitated iron formations and exclude from Tables 1 and 2 deposits comprised solely of öölitic iron ores.

Oxide-, Carbonate-, Sulfide-, and Silicate-type BIFs

James (1954) proposed a BIF classification system based on the type of dominant Fe-rich mineral present: oxide (hematite and/or magnetite), carbonate (siderite, ankerite, and/or dolomite), sulfide (pyrite ± pyrrhotite), or silicate (greenalite, minnesotaite, and/or stilpnomelane). This classification scheme is useful because it demonstrates that BIFs formed in several distinct chemical environments and/or through several distinct (bio)chemical reactions.

Jaspers are low-grade oxide-type BIFs. They are composed of thin bands (typically 20 μm to 1 mm thick) of iron oxides (hematite ± magnetite ± siderite ± pyrite) and thicker bands of chert (~1 mm to ~1 cm); their bulk-rock Fe contents are typically ~1–5%, but Fe contents of the iron-rich layers frequently exceed 15%. The iron formations that are closely associated with volcanogenic massive sulfide deposits (VMSDs) are sometimes called "exhalites" or "ferruginous cherts." They should also be included in oxide-type BIFs, because they have essentially the same mineralogical, chemical, and physical characteristics as those not associated with VMSDs.

Most researchers have neglected the importance of sulfide-type BIFs in constraining the environments and processes for the formation of oxide- and carbonate-type BIFs and also the chemical evolution of the atmosphere and oceans. The sulfide-type BIFs in James's (1954) classification refer to pyrite- and organic C-rich black shales with high Fe contents (often >15 wt%) that frequently occur underneath, interbedded with, and/or overlying oxide- or carbonate-type BIFs of similar extent. A well-known example is the Mount McRae Shale that underlies the ca. 2.5 Ga Brockman IF. Pyrite contents in sulfide-type BIFs are variable and may be exceeded by siderite contents in some formations.

James's (1954) sulfide-type BIFs are distinguished from VMSDs of pyrite ± pyrrhotite ± chalcopyrite ± sphalerite ± galena that underlie some oxide-type BIFs. Massive sulfide ore bodies are defined as rocks with sulfide contents more than 50% in volume. Sulfides in the black shales, which typically range from 1 to 10 wt% in S content, most likely formed by utilizing the H_2S generated from local seawater SO_4^{2-} by sulfate-reducing bacteria (SRB), whereas the sulfides in massive

TABLE 1. PARTIAL LIST OF VOLCANIC-ASSOCIATED OXIDE/CARBONATE-TYPE IRON FORMATIONS

	Age	Footwall type	Associated massive sulfide ores	Banded iron formation type	References
Atlantis II Deep, Red Sea	28,000 yr	basalt	Cu, Zn	oxide	Shanks and Bischoff (1980); Pottorf and Barnes (1983)
Fukazawa mine, Japan	12 Ma	rhyolite	Cu, Zn, Pb	oxide	Kalogeropoulos and Scott (1983)
Jalisco, Mexico	Tertiary	tuff		oxide	Zantop (1981)
Iimori, Japan	Jurassic	basalt	Cu	oxide	Kanehira and Tatsumi (1970)
Okuki, Japan	Jurassic	basalt	Cu	oxide	Imai (1978)
Windy Craggy, British Columbia	Triassic	basalt	Cu, Zn	oxide, carb.	Peter and Scott (1998)
Wolverine, Yukon Territory	Mississippian	rhyolite	Zn, Cu	carb., oxide	Piercey et al. (2001)
Nhuelbuta-Queule Mountain, central Chile	Mississippian	basalt	Cu	oxide	Collao et al. (1990)
Avnik, Turkey	Ordovician	rhyolite		oxide	Helvaci (1984)
Lahn-Dill, Germany	Devonian	spilitic tuff		oxide, carb.	Quade (1976)
Altai, western Siberia and eastern Kazakhstan	Devonian	rhyolitic tuff		oxide	Kalugin (1973)
Bathurst, New Brunswick	Ordovician	rhyolite	Cu, Zn, Pb	carb., oxide	van Staal and Williams (1984); Peter and Goodfellow (1996)
Iberian pyrite belt, Spain/Portugal	Ordovician	rhyolite	Cu, Zn, Pb	oxide	Fernández and Moro (1998); Soriano and Marti (1999)
Trondheim, Norway	Ordovician	basalt	Cu	oxide	Grenne and Vokes (1990)
Mount Windsor subprovince, Queensland, Australia	Cambro-Ord.	rhyolite	Cu, Zn, Pb	oxide	Duhig et al. (1992); Davidson et al. (2001)
Mahuilque-Relún area, central Chile	Paleozoic	mica schist	Cu	oxide	Collao et al. (1990)
West Georgia Piedmont, South Carolina	L. Proterozoic/ E. Paleozoic	rhyolite	Cu, Zn	oxide	Abrams and McConnell (1984)
Pilot Knob, Missouri	1.4–1.5 Ga	tuffs		oxide	Rucker et al. (2003)
Starra, Mount Isa, Australia	ca. 1.65 Ga	rhyolite		oxide	Davidson (1992)
Broken Hill, Australia	ca. 1.65 Ga	rhyolite	Zn, Pb	oxide	Haydon and McConachy (1987); Page and Laing (1992)
Prescott-Jerome, Arizona	ca. 1.8 Ga	rhyolite	Cu, Zn	oxide, carb.	DeWitt (1979)
Skellette, Sweden	Proterozoic	basalt	Cu, Zn, Pb	oxide	Parák (1991)
Cuyuna North range, Minnesota	1.87 Ga	basalt (+rhyolite)		oxide, carb.	Morey and Southwick (1995); Schneider et al. (2002)
Iron-River Crystal City, Michigan	1.87 Ga	basalt (+rhyolite)		oxide, carb.	Morey and Southwick (1995); Schneider et al. (2002)
Menominee, Michigan	1.87 Ga	basalt (+rhyolite)		oxide	Morey and Southwick (1995); Schneider et al. (2002)
Bergslagen, Sweden	1.9 Ga	rhyolite	Zn, Pb	oxide, carb.	Allen et al. (1996)
Boolgeeda IF, Hamersley, Australia	2.42 Ga	rhyolite		oxide, carb.	Trendall and Blockley (1970)
Weeli Wolli IF, Hamersley, Australia	2.45 Ga	basalt		oxide, carb.	Trendall and Blockley (1970)
Olary Block, Australia	Proterozoic	rhyolite	Cu	oxide, carb.	Lottermoser and Ashley (1996)
Aggeneys, South Africa	Proterozoic	rhyolite	Cu, Zn, Pb	carb., oxide	Hoffmann (1994)
Otjihase-Matchless Belt, Namibia	Proterozoic	basalt	Cu	oxide	Adamson and Teichmann (1986)
Murmac Bay, Newfoundland	ca. 2.7 Ga	basalt/quartzite		oxide	Hartlaub et al. (2004)
Matagami, Quebec	ca. 2.7 Ga	rhyolite	Cu, Zn, Pb	oxide, carb.	Costa et al. (1983); Liaghat and MacLean (1992)
Hemlo, Ontario	ca. 2.7 Ga	rhyolite	barite	oxide	Lin (2001); Davis and Lin (2003)
Wawa, Ontario	ca. 2.7 Ga	rhyolite		carb.	Morton and Nabel (1984)
Sturgeon Lake, Ontario	ca. 2.7 Ga	rhyolite	Cu, Zn, Pb	oxide	Shegelski (1987); Koopman (1993)
Manitouwage, Ontario	ca. 2.7 Ga	rhyolite	Cu, Zn, Pb	oxide	Zalenski and Peterson (1995)
Temagami, Ontario	ca. 2.7 Ga	rhyolite		oxide	Bowing (1989)
Yellowknife Greenstone Belt	ca. 2.9 Ga	rhyolite	(stockwork)	oxide	Ootes and Lentz (2002)
Isua, Greenland	ca. 3.8 Ga	mafic volcanics	Cu	oxide, carb.	Appel (1987)

TABLE 2. PARTIAL LIST OF SEDIMENT-ASSOCIATED OXIDE/CARBONATE-TYPE IRON FORMATIONS

Deposit/District	Age	Footwall Type	Associated massive sulfide ores	Banded iron formation type	References
Afar Rift, Ethiopia	0.2 Ma	congl./carb.		oxide	Bonatti et al. (1972)
Granada, Spain	Permian-Triassic	carbonate		oxide, carb	Torres-Ruiz, J. (1983)
Tynah, Ireland	Lower Carboniferous	limestone	Zn, Pb	oxide	Russell (1975); Clifford et al. (1986)
Himalaya front, Kashmir–Bhutan	Carboniferous/ Permian	limestone		oxide	O'Rourke (1962)
Urucum, Brazil	L. Proterozoic/ E. Paleozoic	sandstone/congl.		oxide	Dorr and Van (1973)
Damara, Namibia	ca. 0.75 Ga	carbonate/tuff		oxide	Beukes (1973)
Rapitan, NWT, and Yukon, Canada	ca. 0.75 Ga	shale/sandstone		oxide	Klein and Beukes (1993)
Grenville Province, Quebec	ca. 1.2 Ga	limestone	Zn, Pb	oxide	Gauthier et al. (1987)
Sokoman IF, Labrador Trough	1.85 Ga	quartzite/shale		oxide, carb	Klein and Fink (1976)
Mesabi IF, Minnesota	1.87 Ga	quartzite		oxide, carb	Morey and Southwick (1993); Schneider et al. (2002)
Gunflint, Ontario	1.87 Ga	quartzite		oxide, carb	Morey and Southwick (1993); Schneider et al. (2002)
Gamsberg, South Africa	Proterozoic	clastic sed.	Zn, Pb	oxide	Rozendaal (1986)
Hotazel, Kalahar, South Africa	ca. 2.05 Ga	carbonate		oxide, carb	Tsikos et al. (2003)
Kuruman, South Africa	2.5 Ga	carbonate		oxide, carb	Beukes (1973); Klein and Beukes (1992)
Brockman IF, Hamersley, Australia	ca. 2.5 Ga	shale		oxide, carb	Trendall and Blockley (1970); Ewers and Morris (1981)
Marra Mamba IF, Hamersley, Australia	ca. 2.6 Ga	shale		oxide, carb	Trendall and Blockley (1970)
Boquira, Brazil	ca. 2.7 Ga	clastic sed.	Zn, Pb	oxide, carb	Beeson (1990)
Chitrandurga schist belt, India	>2.6 Ga	quartzite/carb.		oxide, carb	Gnaneshwar Rao and Naqvi (1995)

sulfide ores, which typically exceed 50 wt% S, most likely utilized the H$_2$S in locally discharged submarine hydrothermal fluids (Ohmoto, 1996).

Hydrothermal, diagenetic, and metamorphic processes have modified most BIFs. Thus, important questions among researchers concern their primary mineralogy, especially that of hematite (Fe$_2$O$_3$), magnetite (Fe$_3$O$_4$), siderite (FeCO$_3$), and pyrite (FeS$_2$), because these minerals have been used to constrain the chemistry of the atmosphere and oceans. Most researchers agree that the majority of hematite in BIFs formed prior to magnetite *when they occur in the same microbands*. This is because the hematite in BIFs that were subjected only to zeolite-facies metamorphism is typically much finer grained (often <1 μm) than the magnetite (typically >10 μm) (Ahn and Buseck, 1990), and textures showing the replacement of hematite crystals by magnetite crystals are common (e.g., Trendall and Blockley, 1970; LaBerge et al., 1987).

Most researchers have concluded that the primary ferric minerals in BIFs were amorphous and/or crystalline ferric hydroxides, such as goethite (FeOOH), which were transformed to hematite during the early diagenesis of BIFs. This conclusion was based on an *assumption* that the nucleation of primary minerals in BIFs occurred at a temperature where goethite, rather than hematite, is stable. Implications of this assumption, although not explicitly stated, are that the transformation of goethite to hematite occurred when BIFs were buried more than ~2 km in a sediment column where the temperature exceeded the transition temperature (~80 °C) of goethite to hematite (Langmuir, 1997), and that the conversion of hematite to magnetite occurred at even greater depths and higher temperatures. However, this assumption lacks the support of textural evidence (e.g., hematite pseudomorphs after goethite). Some hematite crystals may have directly crystallized from hydrothermal solutions at T ≥~80 °C. Most siderite and pyrite crystals in BIFs formed within the water column of anoxic basins (i.e., syngenetic minerals), but some also formed during the early diagenesis of sediments (i.e., diagenetic minerals) (Ohmoto et al., 2004).

Silicate-type BIFs probably developed from tuff-rich sediments that incorporated abundant Fe-rich carbonates and silica during sediment accumulation. Subsequent diagenetic/metamorphic reactions between the Fe-rich carbonate, silica, and silicates, such as

$$3FeCO_{3\,(siderite)} + 2SiO_2 + 2H_2O \Rightarrow Fe_3Si_2O_5(OH)_{4\,(greenalite)} + 3CO_2$$

were probably responsible for the formation of iron-rich silicates (e.g., French, 1973; Klein, 1983). For this reason, most researchers have not used Fe-silicates to constrain the depositional chemical environments of BIFs.

Algoma- versus Lake Superior–type BIFs

Gross (1965) has recognized that most Archean (>2.5 Ga) BIFs, such as those in the Abitibi Greenstone Belt of Canada, are closely associated with submarine volcanic rocks and are typically small (<100 m thick and <100 km^2 in area); Gross (1965) termed these BIFs Algoma-type. In contrast, some (not all) BIFs from ca. 2.5 to ca. 1.8 Ga in age, such as those in the Lake Superior district of North America and the Hamersley Basin in Australia, appear to have formed in large sedimentary basins with minor volcanism, and are often more than 100 m × 1000 km^2 in size; Gross (1965) termed this BIF group as Lake Superior–type.

Previous Models for the Origin of BIFs

Numerous researchers have presented many different models for the origin of BIFs. Here we review only models that have significantly influenced geologists' perception of the connections between BIFs and the evolution of the atmosphere, hydrosphere, and biosphere.

James's Model (1954)

By the time James (1954) proposed his famous BIF model (Fig. 1A), most geologists had agreed that the dominant form of Fe in BIF-forming waters was ferrous iron (Fe^{2+}), rather than ferric iron (Fe^{3+}), because the concentration of Fe^{3+} in water that is in equilibrium with hematite (Fe_2O_3), or its possible precursor goethite (FeOOH), is much too low (<<10^{-6} m or <<60 ppb at pH >~3) to form vast Fe-rich sediments. It was also known that to transport a sufficient amount of Fe^{2+} at pH >3, the water must be free of molecular O_2, and some oxidation mechanisms must exist in depositional environments to precipitate Fe^{2+} as ferric (hydr)oxides.

James (1954) proposed that the Fe^{2+} was supplied by continental rock weathering under an anoxic atmosphere and transported by rivers to chemically stratified basins. As an Fe^{2+}-rich water mass sank, the Fe^{2+} was successively sequestered as (1) ferric (hydr)oxides by reactions with free O_2 molecules that were generated by cyanobacteria in the surface layer, (2) Fe-rich carbonates by reactions with CO_2 (and HCO_3^-) in intermediate water depths, and (3) Fe-sulfides by reactions with the H_2S (and HS$^-$) generated by SRB in the bottom water (Fig. 1A).

An implication of James's (1954) model is that cyanobacteria and SRB evolved by ca. 3.8 Ga (Isua BIFs). His model also

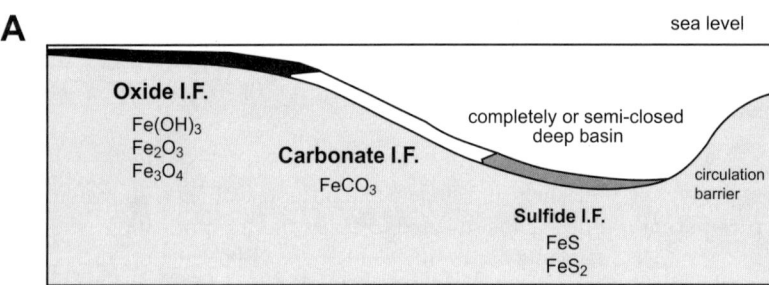

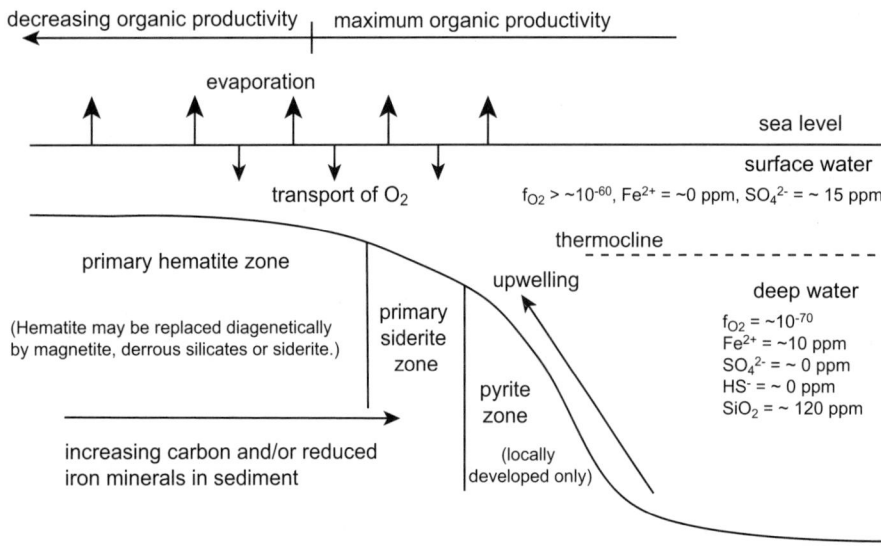

Figure 1. (A) James's (1954) model for the depositional environment of BIFs. (B) Drever's (1974) model for the formation of BIFs.

implies that a BIF bed may change its dominant Fe mineralogy from ferric oxide to siderite to pyrite along a strike, although no documented example of this exists. Typically, changes from one BIF type to another occur in the stratigraphic sequence of a sedimentary basin, indicating that the environments for BIF formation changed (e.g., from oxic to euxinic) over time, rather than spatially, during the history of a sedimentary basin.

James (1954, 1983) did not explicitly relate BIFs to atmospheric and biological evolution. In fact, James (1983) lists ca. 375 Ma BIFs (medium size) in the Altai region of Siberia-Kazakhstan (Table 1) as examples that BIFs are not restricted to Precambrian time.

Cloud's Model (1968)

Cloud (1968) suggested that the Fe^{2+} brought in by rivers was not totally removed as Fe-rich minerals near the coasts, but that instead, most of the Fe^{2+} was carried to the open oceans to create Fe^{2+}-rich and anoxic oceans. According to his model, Fe^{2+} in the deep ocean beneath the photic zone was periodically moved into the photic zone by overturning (instigated by climatic change) to form oxide-type BIFs. He suggested that BIF size was dictated by O_2 availability and that large Superior-type BIFs were evidence for increased O_2 production between ca. 2.5 and 1.8 Ga. By ca. 1.8 Ga, almost all of the Fe^{2+} in the oceans was removed by O_2, causing the ocean waters to become Fe-poor, BIFs to disappear, the atmosphere to become oxic, and red beds to appear. Subsequently, Cloud (1973, 1978) acknowledged the existence of BIFs much younger than ca. 1.8 Ga, including those of Phanerozoic age, but suggested that these formed by different, unspecified mechanisms.

Trendall and Blockley's Model (1970)

The most detailed and extensive BIF investigation was made by Trendall and Blockley (1970) on ca. 2.5 Ga BIFs in the Hamersley Basin, Australia. Their recognition of a basin-wide correlation of individual micro-, meso-, and macrobands and their suggestion that a pair of Fe-rich and silica-rich microbands represents an annual precipitation (Fe precipitation during summers and silica precipitation during winters), led subsequent researchers to suggest that seasonal upwellings of Fe^{2+}-rich water from deep oceans were important in BIF formation. However, their recognition of close temporal associations between volcanic activity and BIFs, and their suggestion of a genetic link between the source of Fe in BIFs and local volcanism, has not been taken seriously by supporters of Cloud's (1968) model. This was because Trendall and Blockley (1970) inferred that the Fe^{2+} was derived by magmatic fluids, but the total mass of Fe in the Hamersley Basin BIF is much greater than that expected from reasonable-sized magmas (e.g., Holland, 1984).

Drever-Holland Model (ca. 1974)

Mass balance calculations of Fe supply in the Hamersley Basin led Holland (1973, 1984) and Drever (1974) to reiterate Cloud's (1968) suggestion that the Fe in BIFs came from upwelling ocean water. However, the Drever-Holland model proposed a different mechanism to transport Fe from land to the oceans. Whereas Cloud suggested fluvial transportation of Fe^{2+}, Drever and Holland suggested that ferric (hydr)oxide minerals were transported to deep oceans as detrital minerals, which were subsequently dissolved in pore fluids as Fe^{2+} in organic-rich sediments to generate Fe^{2+}-rich global oceans.

Holland (1973) and Drever (1974) noted that hematite (or goethite) is stable at $pO_2 > \sim 10^{-60}$ atm and concluded that the generation of a large amount of ferric (hydr)oxide minerals during soil formation could take place even under an anoxic atmosphere. However, such an argument will negate red beds, where hematite crystals are characteristic minerals, as important evidence for an oxygenated atmosphere.

A more serious problem in relating a very low pO_2 value to a real world scenario is that a free-O_2 molecule is not present, even in a volume as large as the entire atmosphere ($V = 4 \times 10^{24}$ cc), when the pO_2 is below $\sim 10^{-43}$ atm because 1 mol of O_2 ($= 6 \times 10^{23}$ molecules; Avogadro's number) occupies 2.24×10^4 cc volume at $pO_2 = 1$ atm and $T = 25$ °C. For a 1L volume of water, a free O_2 molecule disappears at $pO_2 < \sim 10^{-23}$ atm when equilibrium is established between the water and air (Ohmoto et al., 2004). At $pO_2 < 10^{-43}$ (or $< 10^{-23}$) atm, pO_2 is merely a "virtual value" that is calculated from a real pH_2 value (i.e., $pH_2 > 10^{-23}$ atm) by assuming equilibrium in the reaction: $H_2 + 1/2 O_2 = H_2O$ (Ohmoto et al., 2004). In the absence of real O_2 molecules, ferric (hydr)oxides cannot form unless other mechanisms (e.g., photochemical reactions) exist in the formational environments.

Jacobsen-Holland-Klein-Beukes model (ca. 1990)

In the 1970s and 1980s, researchers began to recognize the importance of seawater-rock interactions in the formation of submarine ore deposits (e.g., VMSDs) and in the global geochemical cycles of Na, Mg, Sr, S, and many additional elements (e.g., Ohmoto et al., 1970; Sleep, 1978; Ohmoto and Skinner, 1983; Holland, 1984). Researchers (e.g., Jacobsen and Pimentel-Klose, 1988; Derry and Jacobsen, 1990; Alibert and McCulloch, 1993) also identified the presence of positive Eu anomalies in many BIFs, which are important characteristics of submarine hydrothermal fluids (Kato et al., this volume). These findings have prompted the idea that most Fe^{2+} in the Archean oceans was supplied by submarine hydrothermal fluids that acquired Fe^{2+} from basalts on mid-ocean ridges (MORs), rather than from the weathering of continental rocks (e.g., Jacobsen and Pimentel-Klose, 1988; Holland and Petersen, 1995; Klein and Beukes, 1992). However, these researchers have retained other important elements of Cloud's (1968) model, especially that the global oceans were anoxic and Fe^{2+}-rich (~10 ppm), except possibly in the photic zone, and that periodic upwelling of Fe^{2+}-rich ocean water to nearshore areas produced BIFs under an anoxic atmosphere (Fig. 2A).

Over the years, the proposed date at which the O_2 content of the atmosphere began to rise has moved from ca. 1.8 Ga (Cloud, 1968) to ca. 2.0 Ga (Cloud, 1973) to ca. 2.35 Ga

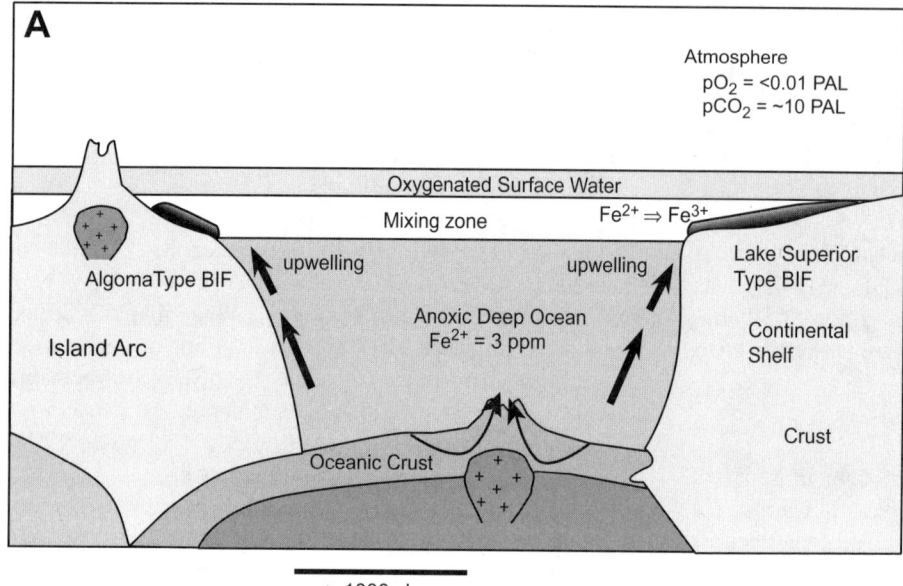

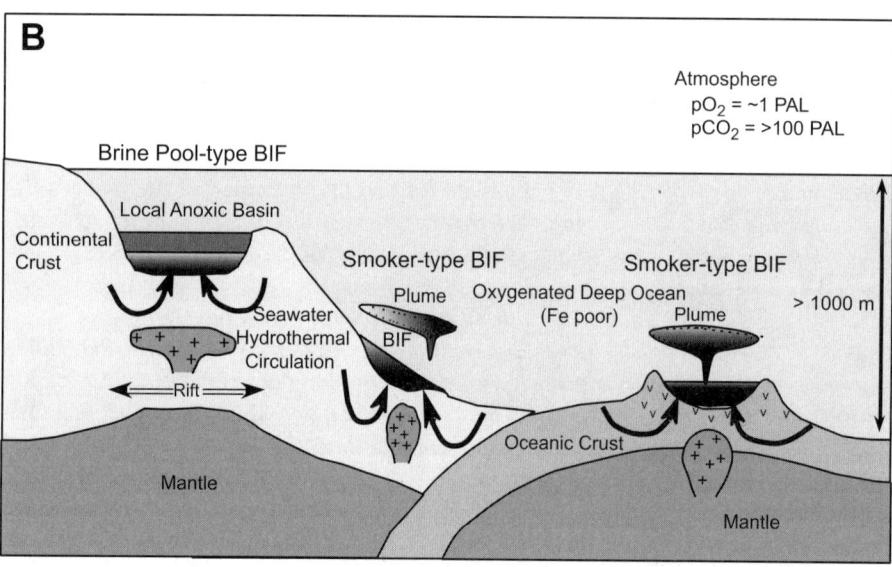

Figure 2. (A) Jacobson-Holland-Klein-Beukes model for the formation of BIFs (Holland and Petersen, 1995). (B) Ohmoto's (1997) model for the formation of BIFs.

(e.g., Kasting, 2001; Holland, 2002; Kasting and Seifert, 2002; Bekker et al., 2004). This has presented a serious contradiction to the Cloud-Walker-Holland-Kasting's atmospheric evolution model, which has linked BIFs to an anoxic atmosphere. Because many BIFs, including those in the Lake Superior region, clearly formed at least 500 m.y. after the 2.35 Ga date for the rise of O_2, BIFs no longer provide evidence for an anoxic atmosphere. Realizing this contradiction, Holland (1999, p. 20) stated, "The formation of BIFs tells us more about the oxidation of the deeper parts of the oceans than about the atmosphere. The cessation of BIF deposition ca. 1.8 Ga may be a signal that the deep ocean basins became oxygenated at that time, and that during the following 1 Ga, the hydrothermal flux of iron was oxidized and precipitated close to the vents, as they are today." In contrast, Canfield and his associates (Canfield et al., 2000; Habicht et al., 2002) have argued that the rise of atmospheric pO_2 around 2.2 Ga changed the ocean from sulfate-poor (SO_4^{2-} content <1/100 of the present level) to slightly sulfate-rich (SO_4^{2-} content ~1/10 of the present level); they also suggested that the ocean became rich in biogenic H_2S, which scavenged Fe^{2+} in the oceans as pyrite to cease the formation of BIFs ca. 1.8 Ga.

Dimroth's Model (1974)

Although most geologists have linked BIFs to an anoxic atmosphere, some geologists (e.g., Dimroth and Kimberley,

1976; Clemmey and Badham, 1982; Ohmoto, 1996; Peter, 2001; Phillips et al., 2001) have argued that BIFs formed throughout geologic history (including the Phanerozoic) under an oxic atmosphere, because the mineralogy and geochemistry of Archean and Proterozoic BIFs are essentially the same as those for Phanerozoic BIFs.

Ohmoto's Model (1997)

Ohmoto (1993) has recognized strong similarities in the geologic settings, mineralogy, and geochemistry of BIFs in many districts with chert-hematite beds that are associated with VMSDs. Subsequently, Ohmoto (1997, 2004) suggested that all BIFs formed by locally discharged submarine hydrothermal fluids under a fully oxygenated atmosphere and oceans (except in local basins) since ca. 3.8 Ga (Fig. 2B). Most of this paper presents supportive evidence for this BIF model.

GENERAL CHARACTERISTICS OF SUBMARINE HYDROTHERMAL SYSTEMS

Origins of Submarine Hydrothermal Fluids

Many geologists have termed BIFs and VMSDs associated with submarine volcanic rocks "exhalites" (e.g., Goodwin, 1973; Spry et al., 2000), implying that locally discharged submarine hydrothermal fluids formed them. Recent geochemical investigations (especially H, O, S, and Sr isotopes and rare earth elements [REEs]) of modern submarine hydrothermal systems (such as black smokers and chimneys on MORs [Barrett and Jambor, 1988] and metalliferous sediments in the Red Sea [Shanks and Bischoff, 1980; Pottorf and Barnes, 1983]) and VMSDs (such as the Japanese Kuroko deposits [e.g., Ohmoto and Skinner, 1983]) have revealed many important characteristics of submarine hydrothermal processes (see reviews by Ohmoto, 1996; Peter and Goodfellow, 1996; Scott, 1997; Spry et al., 2000).

The most important characteristic is that submarine hydrothermal fluids are essentially the product of seawater-rock interactions in regions of submarine igneous activity (i.e., greenstone belts) in a range of tectonic settings (Fig. 2B). Except for some gaseous components (e.g., CO_2, rare gases, and N_2), magmatic contributions to submarine hydrothermal fluids are generally negligible. Seawater that percolated through fractures in the underlying rocks is heated by a local intrusive rock and/or magma and leaches metals (e.g., Fe, Cu, Zn, Pb, Ba), silica, sulfide-S, and sulfate-S from the country rocks, whereas some (or all) of the seawater SO_4^{2-} and country-rock SO_4^{2-} are converted to H_2S. The heated submarine hydrothermal fluids become buoyant and discharge onto the seafloor. Whether the discharged hydrothermal fluids immediately mix with cool local seawater to precipitate minerals, or form brine pools and slowly mix with the overlying seawater to nucleate minerals, depends on the temperature-composition-density relationships between the hydrothermal fluids and local seawater.

Temperature of Hydrothermal Fluids versus Metal Types of Ore Deposits

The temperature of fluid-rock interaction is an important parameter that determines the metal, H_2S, and silica contents of fluids (Ohmoto, 1996; Ohmoto and Goldhaber, 1997) (Fig. 3). In general, fluids with salinity levels like normal seawater may become rich (>1 mM) in H_2S, as well as in metals (Zn, Pb, and Fe), at T >~250 °C; temperatures >~350 °C are required to generate Cu-rich fluids. Fluids rich in both H_2S and metals may precipitate sulfides (ZnS, PbS, FeS_2, and $CuFeS_2$) by cooling, which results when seawater mixes at discharge sites. If hydrothermal fluids are heated only to <~200 °C, they typically remain poor (<1 mM) in H_2S, Zn, Pb, and Cu, but they can be rich (>1 mM) in ΣFe^{2+} (= $FeCl_2^0$ + $FeCl^-$ + Fe^{2+}). (These temperature values become lower with increasing salinity of fluid). Therefore, the temperature of VMSD-forming hydrothermal fluids is generally higher than ~250 °C, whereas BIF-forming fluids are generally lower than ~200 °C.

An important constraint on the formation of Cu-rich VMSDs is seawater depth. If the depth is <~2.5 km (p <~250 atm), hydrothermal fluids of T >~350 °C will separate into liquid and vapor phases (i.e., boiling) before they reach the seafloor, which will cause sulfide minerals to precipitate and form vein- and disseminated-type mineralization, but not massive sulfide ores (Ohmoto,

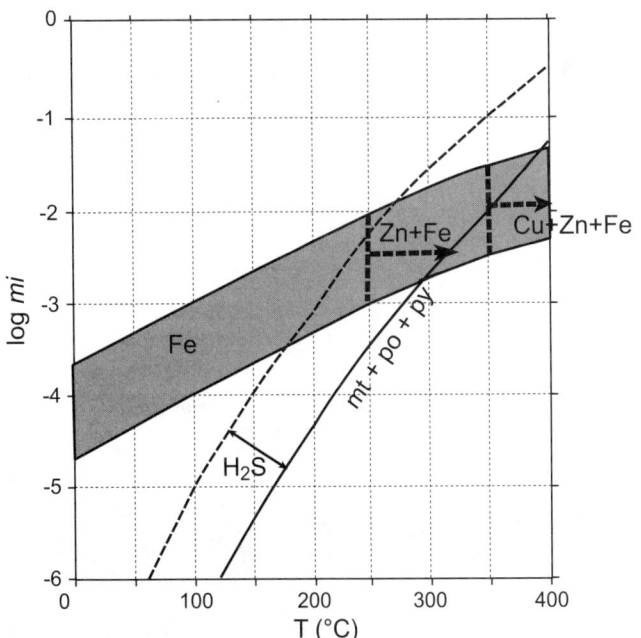

Figure 3. Typical concentration ranges of metals and H_2S in submarine hydrothermal fluids based on thermodynamic calculations (Ohmoto et al., 1983; Ohmoto and Goldhaber, 1997) and analyses of submarine hydrothermal fluids (Scott, 1997). The lower H_2S line represents the stability condition for the assemblage of pyrite+pyrrhotite+magnetite, a common assemblage in basalts.

1996). Therefore, when BIFs are closely associated with Cu-rich massive sulfide deposits, we may conclude that both VMSDs and BIFs formed in a deep (>2.5 km) water body.

Hydrological Characteristics of Submarine Hydrothermal Systems

Circulating seawater acquires heat from rocks. The normal geothermal gradient and heat from an igneous source control the temperature of rocks in the plumbing system; the circulating seawater gradually cools the intrusive source. Therefore, at a discharge site, the temperature of the discharging fluids typically increases with time from near ambient to a maximum temperature, and then decreases back to ambient seawater. Although most sulfide mineralization occurs during the waxing stage, BIF mineralization may occur during both the waxing and waning stages (Cathles, 1983; Ohmoto, 1996). The maximum temperature of the discharging fluids generally decreases with increasing distance from the heat source.

The thermal history (i.e., temperature versus time) and geometry (i.e., sizes and shapes of the total and subsidiary circulation systems, including spacing between subsidiary systems) of a hydrothermal system are largely controlled by the following parameters: (1) the characteristics of the heat source (e.g., depth, size, geometry, and temperature); (2) the geometry of fracture systems; and (3) submarine topography. These three parameters are controlled by (*a*) tectonic setting (e.g., extension versus compression stress regime; ocean ridges versus subduction zone versus backarc rifts versus mantle plumes; and thickness of continental crust under BIF-hosting basins), (*b*) magma type (e.g., basalt versus granite), and (*c*) paleogeography (e.g., paleolatitude, distance from the continent) (Ohmoto, 1996). Three representative cases for the hydrology of submarine hydrothermal systems are described below (Fig. 4A–C).

The first scenario (Fig. 4A) applies to the hydrothermal system of an average VMSD. It is generated by a relatively small (~10 km^3) intrusion (stock, dike) at a shallow depth (~2 km from the seafloor) and has a fluid-circulation of ~5 km in radius (horizontal and vertical dimensions). The fluid temperature at a discharge site above the intrusion may attain a maximum of ~400 °C after ~4000 yr following the intrusion, and gradually decline to ~100 °C after ~10,000 yr (Cathles, 1983).

The second scenario (Fig. 4B) applies to a hydrothermal system associated with a large (>100 km^3), deep-seated (>3 km from the seafloor) basaltic sill (and/or magma), which is periodically replenished by new magmas under an extensional tectonic regime (e.g., MORs, continental rifts, and backarc rifts). Many hydrothermal circulation cells, whose geometries are largely controlled by the major fracture patterns, may be established above the sill. Because new reactive rock surfaces are created by episodic (rather than continuous) tectonic activity and magma injections, the fluid discharge onto the seafloor also becomes episodic; but the overall duration of hydrothermal activity in such an environment may become several orders of magnitude longer (i.e., ~1–100 m.y.) than that associated with a solitary small intrusion.

The third scenario (Fig. 4C) is similar to the second one in tectonic and igneous environments, but here the submarine hydrothermal fluid flow is largely dictated by the large-scale submarine topography and lithology of major aquifers: fluid flow is largely confined to permeable beds (e.g., sandstones, conglomerates, and tuffs) that gently dip toward a large heat source. Such

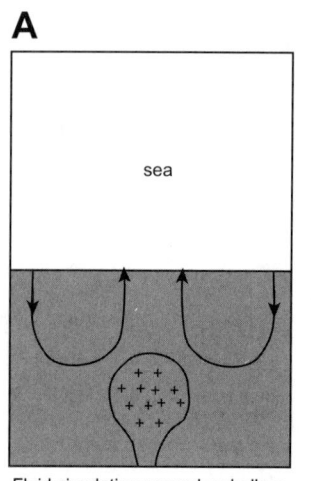
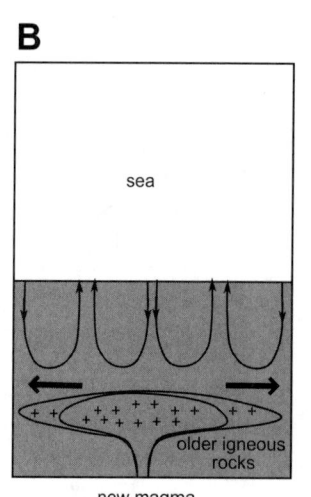
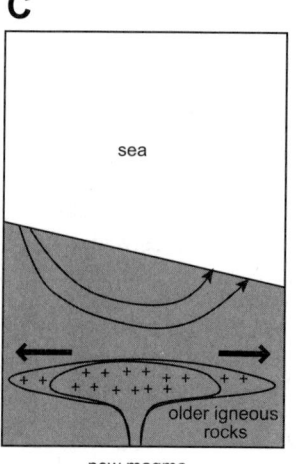

Figure 4. Schematic illustrations of representative sub-seafloor flow patterns of hydrothermal fluids. (A) Around a small, shallow pluton. (B) Over a large, deep sill in a spreading center. (C) Topography- and lithology-controlled.

hydrothermal systems may develop in continental rift systems (e.g., the Red Sea), rather than in MOR systems.

Geometry of Submarine Hydrothermal Deposits

VMSDs are frequently overlain by shales, whose aerial extents are limited essentially to the VMSDs (Ohmoto, 1996). Such a spatial association indicates that submarine depressions (e.g., calderas, troughs, valleys, and basins) are favorable discharging sites for hydrothermal fluids; compared to the zone underneath submarine hills, the head pressure is less (due to a shorter distance to the seafloor) and the permeability is higher (due to a higher fracture density), thus focusing hydrothermal fluid flow to depressions (Ohmoto et al., 1983). Because submarine depressions also provide physical protection (i.e., traps) for the deposited minerals from being dispersed by the bottom currents, they become favorable sites for hydrothermal mineral accumulation. Thus, the shape and size of a hydrothermal deposit may be constrained by the depression in which they form.

More importantly, the geometry of a submarine hydrothermal deposit (VMSD or BIF), whether the precipitated minerals are concentrated around discharging sites as mounds or spread over a large area as a continuous bed, largely depends on how hydrothermal fluids mix with the overlying local water, which is ultimately determined by the density contrast between these fluids. When normal seawater (Σsalts = 3.5 wt%) is heated during water-rock interaction and the amounts of dissolved salts remain basically the same as in normal seawater, the density of the hydrothermal fluids decreases with increasing temperature: e.g., from 1.01 at 25 °C to 0.98 at 100 °C, and 0.75 g/cm^3 at 300 °C at p = 200 bars (Potter and Brown, 1977). When such a hydrothermal fluid is discharged onto the seafloor, it almost instantaneously mixes with local cold seawater at the vents to cause the nucleation of fine-grained minerals (i.e., amorphous and crystalline). These minerals are carried upward as "black smokers" (or "white smokers"; these terms are mineralogy dependent) in a hydrothermal plume until the plume is thermally and chemically homogenized by the cold seawater; the mineral particles continue to settle around the discharge sites as the plume rises though the seawater.

From observations of modern submarine hydrothermal plumes, Peter and Goodfellow (1996) estimated that hydrothermal minerals may be carried as high as ~500 m above the seafloor and may settle as far as 5–10 km from the fluid discharge site. Therefore, a BIF that formed from such a discharge point might extend up to ~10 km away, but more typically between 1 and 5 km (Fig. 5A). If fluids were discharged from multiple sites along a lineament (e.g., trough), the deposits could be much larger (Fig. 5B). Here we term VMSDs and BIFs that formed by mixing low-density, high-temperature fluid with cold seawater near the vents as "smoker-type" deposits; we suggest that most small and linear BIFs (<~100 km in longitudinal direction) belong in this category.

When a hydrothermal fluid acquires high salinity, either by a high degree of evaporation before it circulates through hot rocks or by passage through an evaporite bed, the fluid density may become >1.0 g/cm^3 even at elevated temperatures. For example, if a fluid contains 25 wt% salts, its density at P = 200 bars becomes 1.15 g/cm^3 at 100 °C, 1.08 g/cm^3 at 200 °C, and 0.99 g/cm^3 at 300 °C (Potter and Brown, 1977). Such a fluid will not readily rise through the overlying seawater column and cool quickly, but may instead create a high-temperature brine pool in a depression, such as the hot brine pools in the Red Sea (Pottorf and Barnes, 1983). The nucleation of minerals primarily occurs gently and slowly at and/or near the interface of the brine pool and transition zone, which exhibits thermal and chemical characteristics that are intermediate to the brine and surface water (Fig. 5C). The submarine hydrothermal deposits formed by such processes, termed here "brine pool-type" deposits, may be continuous over a very large area (>100 km × 100 km).

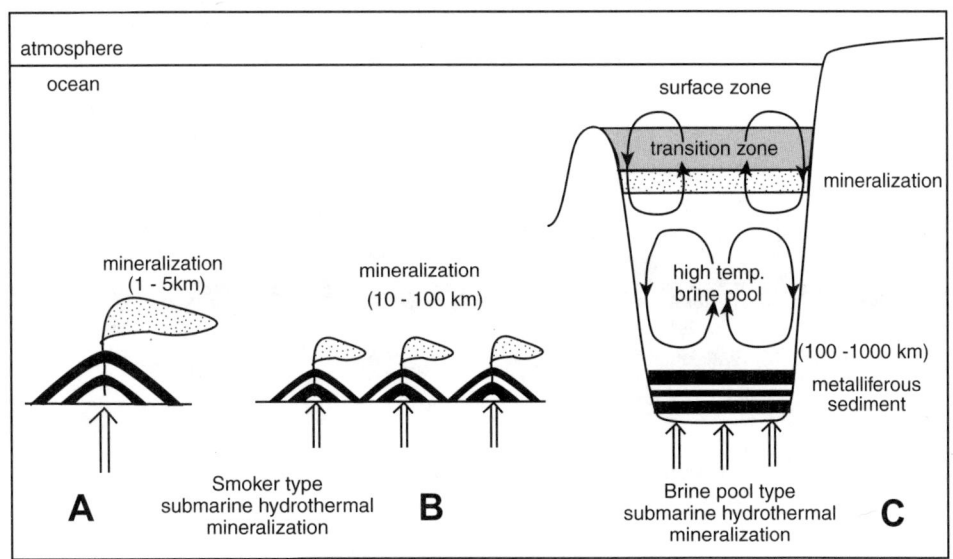

Figure 5. Schematic illustrations of smoker- and brine pool-type submarine mineralization.

GEOLOGIC ENVIRONMENTS FOR BIFS

Classification of Oxide/Carbonate BIFs Based on Footwall Rock Types

The lithology, thickness, and aerial extent of BIF footwall and hanging wall rocks provide the most useful information on the geologic, tectonic, and chemical environments of BIF deposition. Here we define footwalls as rocks that deposited shortly before the BIF, excluding much older rocks that are bounded by a fault or unconformity. The thickness of a footwall rock may range from ~1 m to >1000 m. On an outcrop scale (~1 km), the dominant footwall rock type for oxide- and siderite-type BIFs is either (1) VMSD, (2) submarine volcanic rock (tuffs and lavas), (3) shale (often tuffaceous), (4) dolomite and/or limestone, or (5) conglomerate and/or arkosic sandstone (Fig. 6). Each of the five footwall rocks indicates a specific geologic environment: a deep (>2.5 km) sea for Cu-rich VMSDs, a shallow evaporating water body for dolomite/limestones, and nearshore and shallow water for sandstone/conglomerates. Volcanic- and shale-footwall type BIFs, however, have formed over a wide range of water depths (from the photic zone to >2 km basins).

The type of BIF footwall rock may change over a regional scale. For example, in the Algoma district, Ontario, Canada (Fig. 7), footwall rocks change from felsic volcanics to shales to sandstone-conglomerates over a distance of ~70 km (Goodwin, 1973). The ca. 1.9 Ga BIFs in the Lake Superior region are long thought to have formed on a tectonically stable, igneous-free, continental slope. Although the dominant footwall rock types in the Gunflint-Mesabi districts of the Lake Superior region are quartz-arenites, basalts with minor felsic volcanics (bimodal volcanics) dominate the Menominee-Iron River-Crystal Fall districts (Morey and Southwick, 1995). For these reasons, we avoid the terms Algoma- and Lake Superior-type BIFs and instead use terms such as "VMSD-footwall type BIFs" and "shale-footwall type BIFs."

Examples of each of the BIF footwall types are presented in Tables 1 and 2; contrary to the popular view that all BIFs are older than ca. 1.8 Ga, every BIF type formed throughout geologic history. The tables also reveal an interesting correlation between footwall rock type and metal type of associated sulfide ore bodies: sulfide ore bodies associated with BIFs in volcanic sequence are characteristically Cu-rich, whereas those associated with BIFs in sedimentary sequences are Zn- and Pb-rich. This relationship coincides with the comparison between the metal characteristics of VMSDs versus shale/carbonate-hosted massive sulfide deposits (Ohmoto et al., 1990), suggesting that BIFs in volcanic association were generally formed by higher temperature fluids than those in sedimentary sequences.

VMSD-Footwall Type Oxide/Carbonate BIFs

Although VMSD-footwall type oxide BIFs are generally not familiar to researchers of sediment-hosted BIFs, they are well known among VMSD-exploration geologists. Important examples are (1) the oldest BIFs (ca. 3.8 Ga) in Isua, Greenland, (2) the 2.7 Ga Geco deposits in the Manitowage district, Abitibi, Canada, (3) the 1.8 Ga Jerome deposits, Arizona, (4) the Ordovician Bathurst deposits, New Brunswick, Canada, (5) the Miocene Kuroko deposits, Japan, and (6) the youngest BIFs (25,000 yr old) in the Red Sea. In some cases, such as the Lyon Lake District of the Abitibi district, oxide BIFs occur beneath massive sulfide deposits (Koopman, 1993). The ca. 2.7 Ga oxide BIF in the Hemlo gold deposit in Ontario, Canada, is not associated with massive sulfide ores, but a thick barite ($BaSO_4$) bed occurs in a lower stratigraphic horizon (Lin, 2001; Davis and Lin, 2003), suggesting the 2.7 Ga oceans were not poor in sulfate as previously suggested (e.g., Canfield et al., 2000).

Most VMSDs, such as the ca. 2.7 Ga Kidd Creek deposit in the Abitibi district, Canada (Ohmoto, 1996), are overlain by pyrite- and organic C-rich black shales (e.g., the "sulfide BIFs" of James's [1954] classification) rather than by oxide BIFs. This is possibly because oxide BIF accumulation was interrupted by tuff accumulation and/or seawater conditions in the local submarine depressions became favorable for SRB after the cessation of hydrothermal activity. VMSD-footwall type oxide BIFs undoubtedly formed during the waning stage of local submarine hydrothermal activity because they are associated with well-

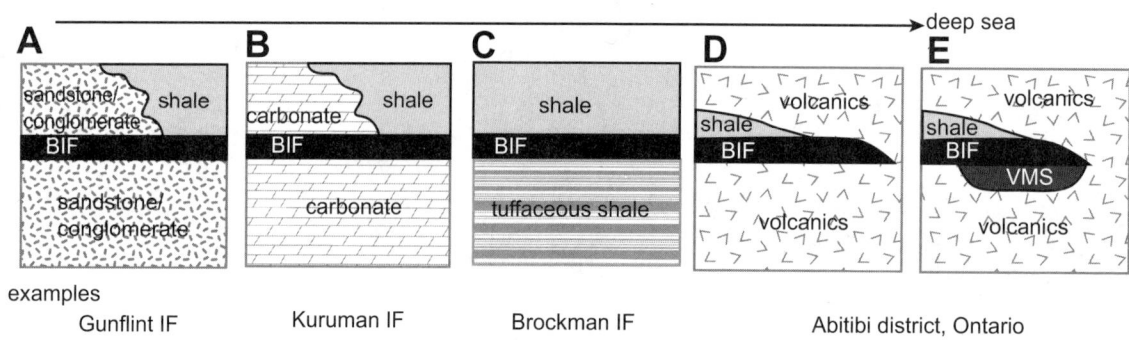

Figure 6. A proposed classification scheme of oxide/carbonate BIFs based on representative footwall rock types.

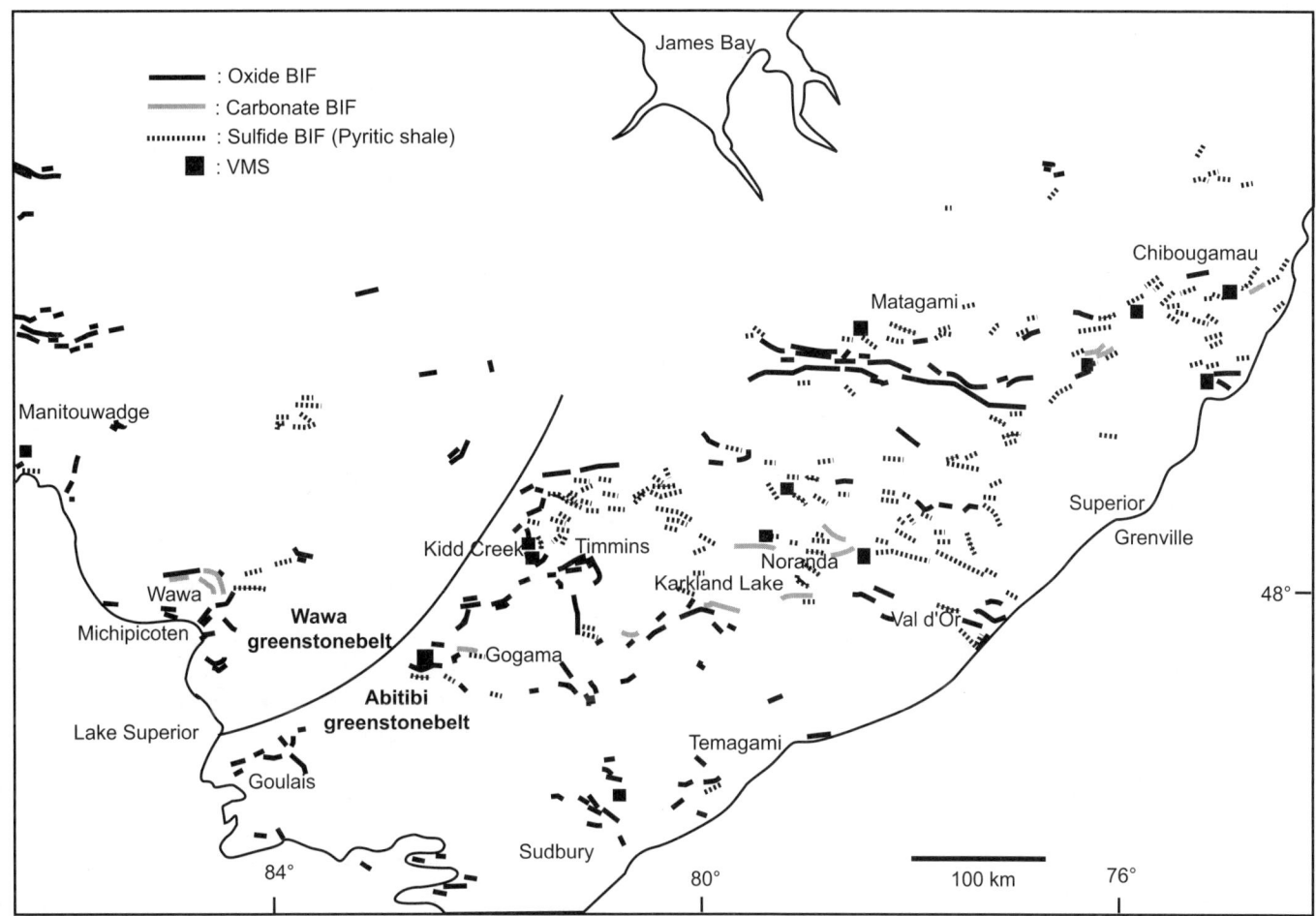

Figure 7. Occurrences of BIFs and major VMSDs in the 2.7 Ga Abitibi and Wawa Greenstone Belts of Ontario, Canada. Compiled from Goodwin (1973) and Fyon et al. (1992).

developed footwall alteration zones (see references in Table 1). Because the associated VMSDs are mostly Cu-rich, these BIFs must have formed in deep (>2.5 km) oceans.

Volcanic-Footwall Type Oxide/Carbonate BIFs

These BIFs are by far the most abundant throughout geologic history (Table 1); they are typically 30–300 m in thickness and extend 2–100 km in longitudinal direction (Goodwin, 1973; Gross, 1990). VMSD-footwall type oxide BIFs are typically ~1 m to ~100 m in thickness and 1–10 km² in aerial extent; they are in fact a subsidiary type of volcanic-footwall type BIFs, because VMSDs are hosted in submarine volcanic rocks, and VMSD-footwall type BIFs generally become volcanic-footwall type in areas outside VMS ore bodies. For this reason, BIFs have been a useful exploration guide for VMSDs (Spry et al., 2000). The best examples of this type of BIF are found in the ca. 2.7 Ga Abitibi Greenstone Belt, Ontario, Canada (Fig. 7).

A previously unnoted important characteristic of volcanic-footwall type BIFs is that they are more commonly associated with felsic volcanic rocks than with mafic volcanic rocks (Table 1). Felsic volcanic rocks may be minor components of bimodal volcanism in some areas (e.g., the ca. 2.3 Ga Menominee, Iron River–Crystal Falls, and Cuyuna North Range of the Lake Superior region) or major volcanic types in other areas. This close association between volcanic-footwall type BIFs and felsic volcanism is the same as that between VMSDs and felsic volcanism throughout the world (Ohmoto, 1996). Such associations suggest that BIF(±VMSD)-forming hydrothermal systems are more frequently developed in continental-rift zones and subduction zones than in MOR regions. From size, shape, and distribution, we may also conclude that most volcanic-footwall type oxide/siderite BIFs belong to the smoker-type mineralization group, but some (including Red Sea BIFs) are clearly brine pool-type.

Shale-Footwall Type Oxide/Carbonate BIFs

The best-known examples of this BIF type are the 2.6–2.4 Ga BIFs in the Hamersley Basin, including the ca. 2.6 Ga Mara Mamba, ca. 2.5 Ga Brockman, and 2.45 Ga Weeli Wolli IFs (Table 2). They range from ~100 to ~500 m in thickness and occur continuously over a > 60,000 km² area. These footwall black shales contain major components of tuffs and basalt sills, and younger BIFs in the Hamersley Basin (i.e., the ca. 2.35 Ga Boolgeeda IF) are volcanic-footwall type BIFs.

Dolomite/Limestone-Footwall Type Oxide/Carbonate BIFs

The best example of this BIF type is the ca. 2.5 Ga Kuruman IF in the Transvaal district, South Africa (Beukes, 1983; Klein and Beukes, 1989). These BIFs (200–400 m thick) overlie the Campbellrand Dolomite (500–2000 m thick) and crop out continuously over a distance of >450 km. They may represent sediments that deposited in shallower and more marginal parts of the Hamersley Basin (see a later section).

An oxide-type BIF at Tynagh, Ireland (Lower Carboniferous), is another example of a dolomite/limestone-hosted BIF (Table 2). It is ~30 m thick and ~2 km in extension, and is closely associated with massive Pb-Zn sulfide deposits (Irish-type massive sulfide) that formed on the seafloor, probably by hydrothermal brines that generated from a deep sedimentary basin (Ohmoto et al., 1990).

Sandstone/Conglomerate-Footwall Type BIFs

Examples of this BIF type include the 1.85 Ga Sokoman IF at the Howell River area (Klein and Fink, 1976) and the 200,000-yr-old Fe-Mn oxide deposits in the Afar Rift, Ethiopia (Table 2). But the best-known examples are the 1.87 Ga BIFs in the Gunflint and Mesabi ranges of the Lake Superior region; they were most likely a continuous formation that extended over a distance of >400 km, but because of the intrusion of the Duluth Gabbro they appear as two separate BIFs. These two BIFs are 100–300 m thick and are underlain by a ~10-m-thick unit of quartz arenites, conglomerates, and carbonates with an algal structure that deposited unconformably on Archean basement rocks (e.g., Bayley and James, 1973; Morey and Southwick, 1995). Abundant geological, petrological, and paleontological evidence (e.g., stromatolites) exists to suggest that these BIFs deposited in shallow waters; granules and öölites are typical features in these BIFs.

BIFs in the Abitibi-Wawa Greenstone Belts, Ontario, Canada

The Abitibi Greenstone Belt, which is ~500 × 1000 km in size and hosts ~150 BIFs and 30 VMSDs (Fig. 7), comprises several autochthonous submarine volcanic terranes that are 2.75–2.67 Ga in age (e.g., Jackson and Fyon, 1991; Thurston, 2002; Ayer et al., 2002). The submarine volcanic rocks consist of bimodal compositions (mostly basalts and rhyolites, and much less, andesites). Recent studies (e.g., Hollings and Kerrich, 2000; Polat and Kerrich, 2001) suggest that these volcanisms were related to both arc basalts (from slab dehydration wedge melting) and dacites/adakites (from later slab melting).

The Abitibi BIFs typically occur in clusters that are spaced at ~20–100 km intervals and the volcanic-footwall type oxide/siderite BIFs typically occur 3–30 km from known VMSDs (Fig. 7). Such a spatial relationship suggests that these BIFs were formed by low-temperature (T < 200 °C) submarine hydrothermal fluids that discharged at peripheral sites of hydrothermal systems, whereas the VMSDs were formed by higher temperature fluids located near the center of heat sources. The close spatial association of BIFs with Cu-rich VMSDs, and the fact that the BIFs are overlain by thick (>2 km) successions of submarine volcanic rocks, without interbedded units of conglomerates and sandstones, suggests that most (if not all) volcanic-footwall type BIFs in the Abitibi Greenstone Belts formed under >2.5 km of seawater.

BIFs in the Michipicoten-Wawa areas of Ontario (Fig. 7) belong to a different greenstone terrane (the 2.75–2.7 Ga Wawa Greenstone Belt) that was autochthonously or allochthonously assembled with the Abitibi belt ca. 2.7 Ga (Thurston, 1991, 2002). These BIFs were probably deposited at shallow (<500 m) water depths, because of the occurrence of stromatolites in some siderite BIFs (Hofmann et al., 1991) and the change in footwall lithology from volcanic rocks to sandstone/conglomerates within a distance of ~70 km (Goodwin, 1973). Although no VMSD has been found within a ~100 km radius of these BIFs (Fig. 7), a distinct footwall alteration zone developed underneath the BIFs in these areas (e.g., Morton and Nebel, 1984; Gross, 1991), suggesting that they also formed from locally discharged hydrothermal fluids. The absence of VMSDs in the Michipicoten-Wawa areas is probably the consequence of a shallow-sea environment.

BIFs in the Hamersley Basin, Western Australia

The prevailing geologic model for the 2.7–2.3 Ga BIFs in the Hamersley Basin, which was developed before the theory of plate tectonics, suggests that they formed in a basin that developed on a stable continental shelf where the basin water was in free communication with the global oceans (e.g., Cloud, 1968; Holland, 1984). However, during the past 20 years or so, the geological and geochemical data accumulated on BIFs and associated rocks suggest that the tectonic, geologic, geochemical, and hydrological processes that took place in the Hamersley Basin, and in other brine pool-type BIFs, were similar to those occurring in the Red and Black Seas during the past ~50 m.y. period (Ohmoto, 1993, Ohmoto et al., 2004). Thus, we name our theory for brine pool-type BIFs as the "Red and Black Seas Hybrid Model." First we will introduce a review of the geologic and hydrologic processes in the Red and Black Seas, and then present geochemical data on the Hamersley BIFs to justify our model.

Mineralization in the Red and Black Seas

Whereas the Red Sea rift system is the site of active tectonic, igneous, and hydrothermal activities, the Black Sea represents a terminal (or resting) stage of a failed rift system with no current igneous or hydrothermal activity, and is filled with a ~15-km-thick sequence of sedimentary and igneous rocks (Fig. 8A). Both the Red and Black Seas host stratified water bodies composed of denser (saline) anoxic bottom water that is overlain by lighter

Deep, is only 5 × 14 km in size, and hosts a ~150-m-deep anoxic, hot brine pool (T = ~60 °C, salinity = ~23%, and $\sum Fe^{2+}$ = ~80 ppm at depths of ~2200 m), which is overlain by a ~50-m-thick transitional zone and a ~2000 m-deep oxic, evaporated seawater (T = ~25 °C, salinity = ~5%) (Pottorf and Barnes, 1983). The submarine hydrothermal fluids feeding the brine pool originated from evaporated seawater and became more saline, hotter, and metal-rich while circulating through Tertiary evaporite beds and underlying basalts before discharging into depressions on the seafloor. A ~4-m-thick bed of metalliferous sediments with average compositions of ~20% Fe, 2.1% Zn, and 0.45% Cu has accumulated in the Atlantis II Deep (Shanks and Bischoff, 1980; Pottorf and Barnes, 1983; El Shazly, 1990).

The temperature of discharging fluids in the Atlantis II Deep brine pool has fluctuated between ~100 and >300 °C during the past ~25,000 yr; iron oxide-rich sediments (i.e., oxide BIFs) developed at T ≈ 100 °C, whereas alternating beds of Fe-Cu-Zn sulfide-rich sediments (i.e., VMSD) precipitated at T > 300 °C (Pottorf and Barnes, 1983; Ohmoto, 1996). High-temperature (T >~250 °C) fluids can be rich in H_2S, as well as Fe, Cu, Zn, and other metals (Fig. 3), thus precipitating metalliferous sulfide-rich sediments when the hot fluids mix with cooler basin bottom water. Lower-temperature hydrothermal fluids (T = ~100 to ~200 °C) are typically Fe^{2+}-rich, but too poor in H_2S to precipitate all the metals as sulfide minerals, thus creating a hot brine pool rich in Fe^{2+}- and silica (Fig. 3). As this hot brine and cooler overlying O_2-rich water mixed, ferric hydroxide and amorphous silica crystals nucleated and settled on the seafloor to form goethite-silica-rich layers (i.e., equivalent to oxide BIFs).

The Black Sea is the world's largest anoxic basin (423,000 km²; ~500 × 1000 km); approximately half of the area is deeper than 2000 m (Fig. 1B). It is land-locked and only connects to the Mediterranean Sea by a narrow waterway (Bosporus Strait). The Black Sea is also stratified; the surface zone (0 to ~50 m in depth) is diluted by river water to ~1.8% salinity, and the bottom water (~100 to ~2000 m in depth) is supplied mostly by oxic Mediterranean seawater (~3.5% salinity) that is diluted by fresh water to ~2.2% salinity. The bottom water is anoxic and rich in H_2S and CH_4 that was produced by sulfate reducers and methanogens, respectively. The transition zone between the oxic and anoxic waters is ~50 m thick (Degens and Ross, 1974; Eremeev, 1992; Murray, 1991).

The nucleation of iron sulfide minerals occurs mostly in the transition zone by utilization of the biogenic H_2S from the bottom water and the particles of Fe^{3+}-bearing minerals (e.g., goethite) that were transported by river water and wind (aerosols) (Muramoto et al., 1991; Wilkin and Arthur, 2001; Jørgensen et al., 2004). The deposition of fine-grained iron sulfide particles was periodically interrupted by the deposition of shales via turbidites to form finely laminated sulfide-rich sediments (e.g., Degens and Ross, 1974). During the active rifting period of the Black Sea (ca. 15 Ma), Mediterranean seawater could have circulated through the hot rocks underneath the Black Sea and discharged onto the seafloor to create a metalliferous brine pool. Such a scenario

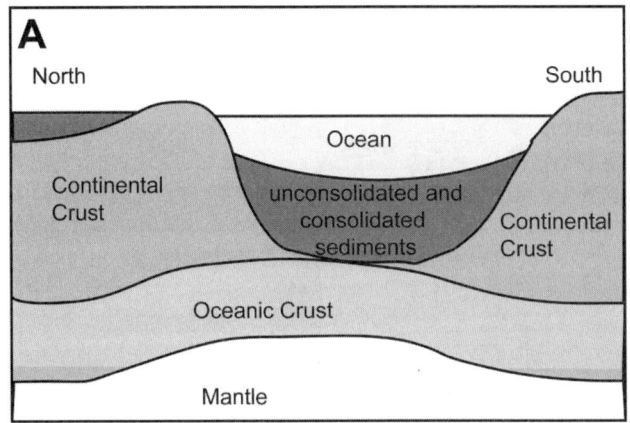

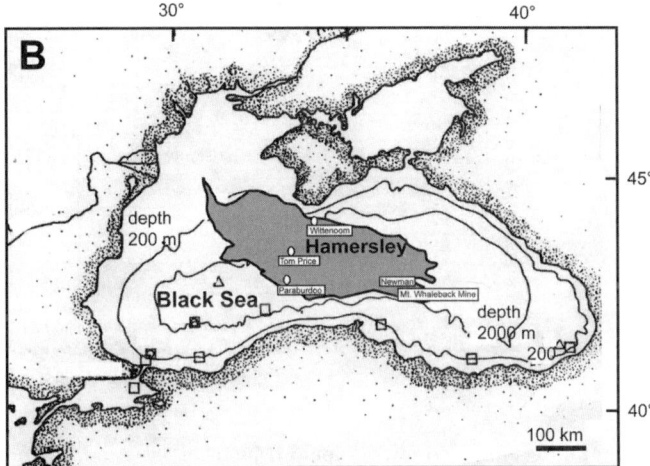

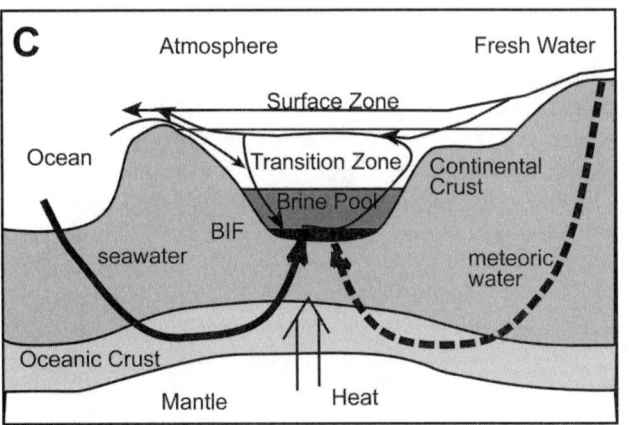

Figure 8. (A) Simplified schematic of the crustal structure beneath the Black Sea (Degens and Ross, 1974). (B) Comparison of the areal dimensions of the Hamersley Basin and the Black Sea. (C) The Red and Black Seas hybrid model for the Hamersley BIFs, highlighting the involvements of seawater and river water in the formation of Superior-type BIFs.

(less saline), cooler, oxic surface water. However, significant differences exist in the origins of these stratified water bodies.

In the Red Sea, many isolated brine pools are created at depths > 2000 m. The most famous brine pool, the Atlantis II is the essence of our Red and Black Seas hybrid model for the Hamersley BIFs.

The Red and Black Seas Hybrid Model for the Hamersley BIFs

The Hamersley Basin, which hosts the largest known BIFs in Earth's history, is only ~60,000 km² in size today (~100,000 km² originally; Trendall and Blockley, 1970), which corresponds to less than one-fourth the size of the Black Sea (Fig. 8B). Our model for the Hamersley BIFs incorporates the following suggestions: (1) The Hamersley Basin was an intracontinental (failed) rift basin, created by mantle upwelling, that experienced large-scale, but intermittent, igneous and hydrothermal activities from ca. 2.8 to ca. 2.4 Ga (Fig. 8C). (2) During BIF formation, the Hamersley Basin was a land-locked sea, connected to the open ocean only by (a) narrow inlet(s), and density stratified like the Black Sea. (3) Large-scale, intermittent submarine hydrothermal systems discharged Fe^{2+} through major fracture systems in (not outside) the Hamersley Basin and created a basin-scale, hot (T = ~50 to ~100 °C) metalliferous brine pool. (4) Ferric (hydr)oxides and silica nucleated primarily in the transition zone between the surface oxic water and underlying hot brine pool, and then settled onto the basin floor. (5) The separation of silica- and iron-rich microbands occurred mostly during the early diagenesis of Fe- and Si-rich sediments. Continuous reactions between the ferric (hydr)oxides and overlying Fe^{2+}-rich brines caused the ferric (hydr)oxides to transform to magnetite. (6) The nucleation of silica also occurred by surface water evaporation, especially during periods of no hydrothermal activity. And (7) when the brine pool temperatures decreased below ~50 °C during the waning periods of hydrothermal activity, SRB became active and produced pyrite- and organic-rich shales (i.e., sulfide BIFs).

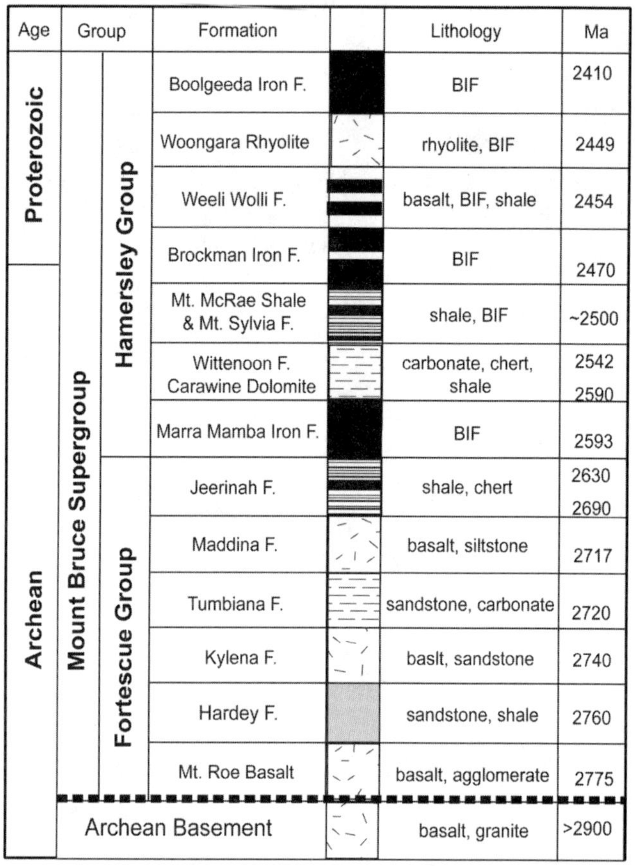

Geologic Evidence for the Red and Black Seas Hybrid Model

Recent geological and petrological investigations of the Hamersley Basin (e.g., Barley and Groves, 1992; Barley, 1993; Barley et al., 1997; Blake, 2001; Eriksson et al., 2002) suggest it was a major igneous province, possibly related to mantle plume(s), during the period between ca. 2.8 and ca. 2.4 Ga. Its stratigraphic record (Fig. 9) suggests the following history: (1) the initial rifting of a thick continental crust (composed mostly of granite batholiths and greenstones older than 2.9 Ga) accompanied by eruptions of large volumes of mostly subaerial flood basalts (e.g., the Mount Roe, Kylena, and Maddina Basalts) and depositions of mostly fluvio-lacustrine sediments (e.g., the Hardey and Tumbiana Formations) from 2775–2680 Ma; no record was preserved from 2680 to 2630 Ma; (2) the deposition of alternating sequences of mostly black shales (e.g., the Jeerinah, Mount Sylvia, and Mount McRae Formations), carbonates (e.g., the Paraburdoo Member of the Wittenoon Dolomite), cherts (e.g., the Bee Gorge Member of the Wittenoon Dolomite), and BIFs (e.g., the Mara Mamba and Brockman IFs), with minor interbedded tuffs, basalt, and rhyolite, from 2630 to 2460 Ma; and (3) eruptions of large volumes of basalts and rhyolites (i.e.,

Age	Group	Formation	Lithology	Ma
Proterozoic	Hamersley Group (Mount Bruce Supergroup)	Boolgeeda Iron F.	BIF	2410
Proterozoic	Hamersley Group (Mount Bruce Supergroup)	Woongara Rhyolite	rhyolite, BIF	2449
Proterozoic	Hamersley Group (Mount Bruce Supergroup)	Weeli Wolli F.	basalt, BIF, shale	2454
Archean	Hamersley Group (Mount Bruce Supergroup)	Brockman Iron F.	BIF	2470
Archean	Hamersley Group (Mount Bruce Supergroup)	Mt. McRae Shale & Mt. Sylvia F.	shale, BIF	~2500
Archean	Hamersley Group (Mount Bruce Supergroup)	Wittenoon F. Carawine Dolomite	carbonate, chert, shale	2542 / 2590
Archean	Hamersley Group (Mount Bruce Supergroup)	Marra Mamba Iron F.	BIF	2593
Archean	Fortescue Group (Mount Bruce Supergroup)	Jeerinah F.	shale, chert	2630 / 2690
Archean	Fortescue Group (Mount Bruce Supergroup)	Maddina F.	basalt, siltstone	2717
Archean	Fortescue Group (Mount Bruce Supergroup)	Tumbiana F.	sandstone, carbonate	2720
Archean	Fortescue Group (Mount Bruce Supergroup)	Kylena F.	baslt, sandstone	2740
Archean	Fortescue Group (Mount Bruce Supergroup)	Hardey F.	sandstone, shale	2760
Archean	Fortescue Group (Mount Bruce Supergroup)	Mt. Roe Basalt	basalt, agglomerate	2775
	Archean Basement		basalt, granite	>2900

Figure 9. A simplified stratigraphic column of the Mount Bruce Supergroup of the Hamersley Basin, Australia. Compiled from data in Trendall and Blockley (1970), Krapez et al. (2003), and Pickard (2003).

bimodal volcanism) and the deposition of BIFs (e.g., the Weeli Wolli and Boolgeeda IFs, and the Woongarra Volcanics) from 2460 to 2410 Ma. Note that bimodal volcanism, including rhyolite that generated by partial melting of lower crustal rocks by mantle-derived basalt magmas, is a characteristic of intracontinental rift-related magmatism.

Thick accumulations of shallow-water-facies carbonate beds of the Tumbiana and Paraburdoo Members of the Wittenoon Dolomite show that the Hamersley Basin was at times highly evaporitic and (semi-)closed into one or more basins. The abundance of Na-rich silicates (e.g., riebeckite, crocidolite), which probably formed by reactions between tuff-rich shale units and alkaline-rich hydrothermal solutions during regional hydrothermal events ca. 2.3–2.2 Ga (Hagemann et al., 1999; Brown et al., 2004), also suggests that rocks enriched in NaCl and KCl (i.e., evaporite beds) existed in the pre-Brockman sequence. Although such evaporite beds are not found in the Hamersley district, probably because they were dissolved away during a long history of fluid-rock interactions, they could have played an important role in the development of hydrothermal brine pools, much like those in the Red Sea.

The sedimentary units between the 2630 Ma Jeerinah Formation and the 2420 Ma Boolgeeda IF total ~2.5 km in thickness and do not exhibit any evidence of subaerial erosion during their deposition. On the basis of sedimentological analyses, Simonson et al. (1993) and Krapez et al. (2003) suggested that carbonates, as well as shales, were transported as turbidites from the northeast margins (shelf) of the Hamersley Basin and deposited at depths > ~400 m (i.e., the depth below the storm wave base). Krapez et al. (2003) also suggested that the precursor sediments to BIFs (hydrothermal muds) were deposited on the flanks of submarine volcanoes, which were located in the deeper, southwestern parts of the Hamersley Basin, and resedimented by density currents.

Geochemical Evidence for the Red and Black Seas Hybrid Model

Discoveries of molecular biomarkers of cyanobacteria (2-methyl hopanes) and eukaryotes (stranes) from shales of the Mount McRae and Jeerinah Formations (Brocks et al., 1999, 2003; Eigenbrode et al., 2004), and the experimentally determined minimum pO_2 requirements for the key enzymes in eukaryotes to produce sterols (Jahnke and Klein, 1983), suggest that the surface water layer of the Hamersley Basin contained free O_2 molecules equivalent to $pO_2 > 0.01$ atm (i.e., >5% of the present atmospheric pO_2 level [PAL]) during the 2.7–2.5 Ga period. This pO_2 value corresponds to 18 µM as the *minimum dissolved oxygen (DO) content* of the surface zone of the Hamersley Basin.

The redox stratification of the Hamersley Basin is suggested by the common occurrence of kerogen with $\delta^{13}C$ values <<–30‰ (Hayes et al., 1983; Brocks et al., 2003). Organic matter with such low $\delta^{13}C$ values is produced in environments where the remnants of primary producers (cyanobacteria and eukaryotes [aerobes]) with $\delta^{13}C$ values ~–30‰ were partially converted to CH_4 ($\delta^{13}C$ values <<–60‰) by methanogens (anaerobes) in an anoxic water body, which was subsequently oxidized to CO_2 ($\delta^{13}C$ values <<–60‰) by methanotrophs (aerobes) that lived at the interface between the anoxic and oxic waters (Hayes, 1994).

One of the characteristic features of BIF-associated black shales is their high content of organic carbon (kerogen). For example, the organic C contents of the 2.5 Ga Mount McRae and 2.7 Ga Jeerinah shales typically range from 1 to 15 wt%, with an average of ~5 wt% (Fig. 10A). The S-C contents of these shales are similar to those of euxinic sediments in the Black Sea (Fig. 10C). The H/C atomic ratio of kerogen in the Hamersley shales are typically reduced to <0.1, compared to a ratio of ~1.5 for fresh organic matter (CH_2O) that deposited onto the seafloor. This reduction in the H/C ratio occurred because of the thermal maturation of organic matter (i.e., losses of H, O, and C atoms as petroleum and natural gases) during a long history of hydrothermal, diagenesis, and metamorphic processes. Our calculations, using an equation that relates the H/C ratio of kerogen to the losses of H, O, and C compounds (Watanabe et al., 1997), suggest that the original contents of organic matter in the Mount McRae and Jeerinah shales was as high as 50–70 vol% of the sediments.

Considering that the average thickness of these shales is ~100 m each, and that they were deposited throughout the Hamersley Basin, we suggest that these two black shales represent the largest accumulations of organic matter in geologic history. By comparison, the average organic C content of Phanerozoic marine shales is only ~1 wt% (Holland, 1984).

The enormous accumulations of organic matter in the McRae and Jeerinah shales indicate that the primary productivity in the surface zone was extremely high and the bottom water of the Hamersley Basin was anoxic to facilitate the preservation of organic matter. The great primary productivity must have been a consequence of very efficient nutrient recycling (nitrate and phosphates) to the surface zone, which would have been possible if the basin was a nearly closed system where river waters supplied nutrients to the basin. Therefore, a land-locked sea like the Black Sea, rather than a wide-open basin, is a better explanation for the organic-rich shales of the Hamersley Basin.

Extensive evidence indicates that large-scale hydrothermal activity most likely occurred intermittently before, during, and after the deposition of BIFs in the Hamersley Basin. The Mount McRae Shale, which forms the footwall of the Brockman IF, exhibits several features typical of feeder zones, especially in areas of large iron deposits. For example, shale samples from the Mount Whaleback mine (DDH186 and RD1) (Fig. 8B) are enriched in Zn (up to ~2000 ppm) and Fe (~20 wt%), whereas some shale samples show depletions of Fe (Fig. 11A). The K_2O contents of the McRae shales decreased from an original value of ~4 wt% to <1 wt% (Fig. 11B). High Zn contents (~2000 ppm) are also reported in samples from the Roy Hill Member of the Jeerinah Formation (Davy, 1985), which is the footwall of the Mara Mamba IF (2590 Ma).

The apparent Rb-Sr age of the Mount McRae shales at the Mount Tom Price Mine decreased from the depositional age of ca. 2500 Ma to 1950 ± 290 Ma by hydrothermal alteration, whereas that at the Mount Whaleback Mine (2460 ± 130 Ma) is essentially the same as the depositional age (Fig. 12A). The apparent Nd-Sm ages of the Mount McRae Shale are 2100 ± 600 Ma for samples from the Mount Tom Price and Mount Whaleback Mines (Fig. 12B). Similarly, Alibert and McCulloch (1993) reported that the Nd-Sm isotope systematics of many BIF samples from the Brockman IF were disturbed from the original ages of ca. 2500 Ma to ca. 2100 Ma by hydrothermal processes. All these data suggest that hydrothermal activity in the Hamersley Basin persisted, probably intermittently, during a ~600 m.y. period from ca. 2.6 to ca. 2.0 Ga.

The Mount McRae Shale contains abundant pyrite nodules that are locally surrounded by a quartz overgrowth layer that is up to ~5 mm in thickness (Haruna et al., 2003). Textural relationships of the nodules with their host shales suggest that these pyrite nodules and the overgrowth quartz crystals grew during the early diagenetic stage (i.e., before solidification) of the Mount McRae Shale, which would correspond to the depositional period of the overlying Brockman IF (Kakegawa et al., 1998; Haruna et al., 2003).

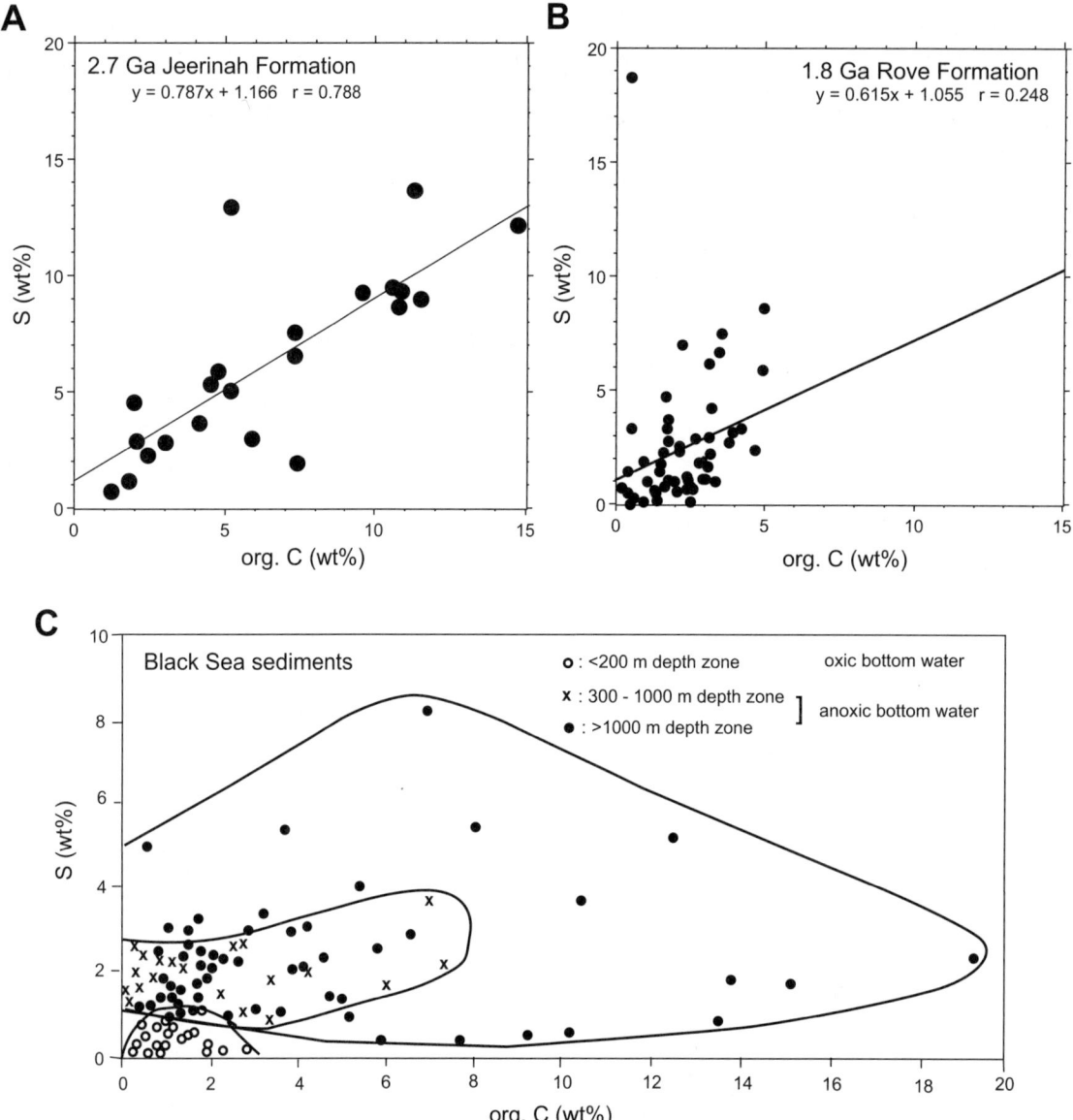

Figure 10. Sulfide-S and organic-C contents of black shales. (A) The 2.7 Ga Roy Hill Member of the Jeerinah Formation, Hamersley Basin, Australia (Davy, 1985). (B) The 1.8 Ga Rove Formation of Ontario, Canada (Poulton et al., 2004). (C) The Black Sea (Ohmoto et al., 1990).

Fluid inclusions are abundant in the quartz overgrowth of pyrite nodules from the Mount Tom Price Mine. Two populations of fluid inclusions are recognized: a high-salinity type (T = 130–220 °C; salinity = 5–12 wt%), and a low-salinity type (T = 150–200 °C; salinity = 0.5–3 wt%) (Fig. 13). These data suggest that the Mount Tom Price Mine was one of the discharging sites for hydrothermal fluids, and that both evaporated seawater (salinity > 3.5%) and meteoric water (salinity < 3.5 wt%) were involved in hydrothermal mineralization. This recognition of two water types in the Hamersley Basin was an important reason for our suggestion that the basin was stratified with a metalliferous hydrothermal brine pool (salinity = 5–12%) and an overlying diluted (salinity < 3.5%) water body (Fig. 8C).

The samples we investigated for alteration and fluid inclusions came from iron ore mines (e.g., the Mount Tom Price and Mount Whaleback Mines). We have not systematically investigated the differences in the footwall alteration in other areas. However, field observations suggest that large pyrite nodules are more abundant in the mine areas than in other areas, which in turn suggests that the present mine sites were also the main fluid discharge locations for BIFs. In many ore districts, major fracture systems in the basement rocks become the main conduits for fluid discharge during and after ore formation (e.g., Ohmoto and Skinner, 1983).

Using an in situ laser ablation method, we have analyzed millimeter-scale variation in the $\delta^{18}O$ values of quartz in the Fe-

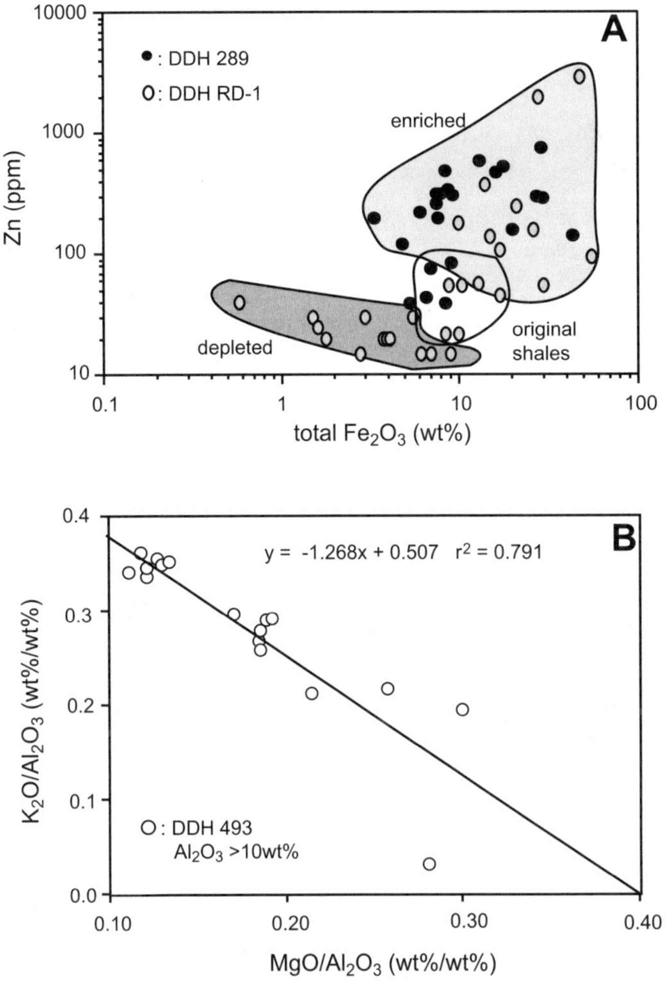

Figure 11. Geochemical data of the 2.55 Ga Mount McRae Shale (Hamersley, Australia), which shows the characteristics of footwall-rock hydrothermal alteration underneath VMSDs (cf. Ohmoto et al., 1983). (A) Some samples are enriched in both Zn and Fe, whereas some are depleted in both Zn and Fe. (B) Depletion of K was accompanied by an increase in Mg (this study).

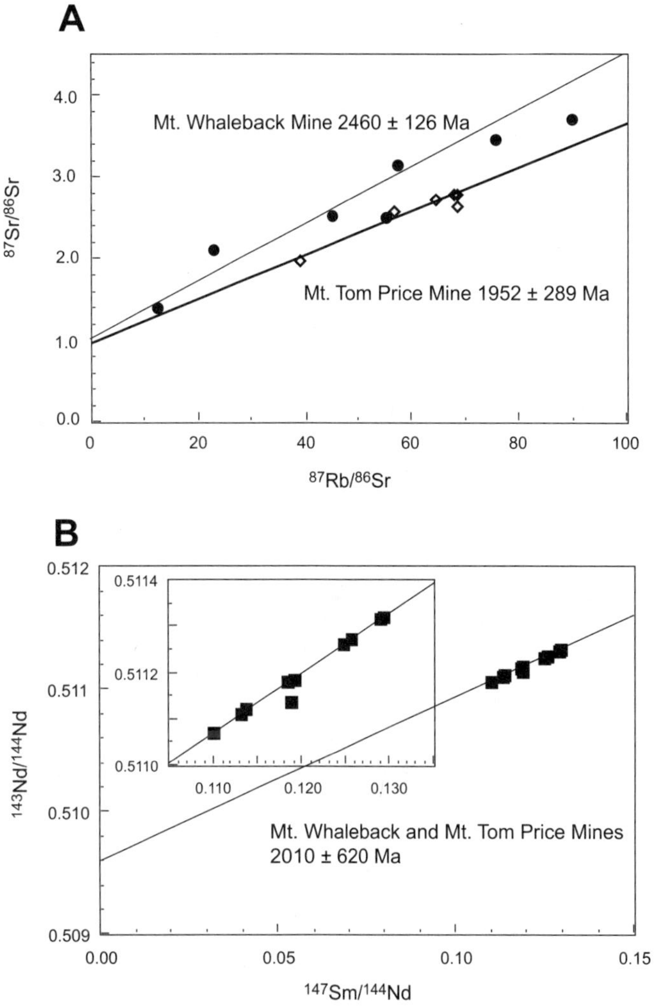

Figure 12. Rb-Sr and Nd-Sm ages of the Mount McRae Shale (Uyeda, 1994).

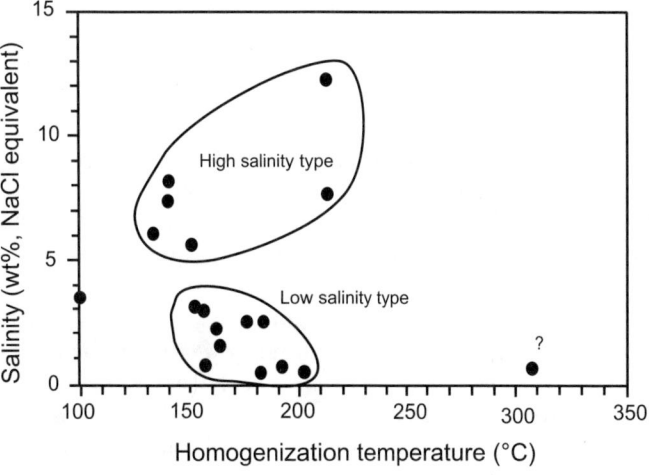

Figure 13. Homogenization and freezing temperatures of fluid inclusions in the quartz overgrowths of pyrite nodules from the Mount McRae Shale. An additional ~30 fluid inclusions were investigated for homogenization temperatures only, most of which were between 120 and 220 °C. Data from Haruna et al. (2003).

poor (silica-rich) and Fe-rich layers of a ~30 cm section of the Brockman IF. Results (Fig. 14A) show a very large variation, from +11.2‰ to +23.0‰, with a general tendency for higher $\delta^{18}O$ values of quartz in the Fe-rich microbands, compared to those in the silica-rich microbands. The $\delta^{18}O$ values of quartz crystals in the silica-overgrowth zone of pyrite nodules in the Mount McRae Shale (~+20‰; Haruna et al., 2003) are similar to those of silica in Fe-rich bands of BIFs.

The $\delta^{18}O$ value of quartz depends on the temperature and $\delta^{18}O$ value of the water involved in quartz formation, whether the quartz and water attained isotopic equilibrium during formation, and whether the $\delta^{18}O$ value of quartz was not modified during

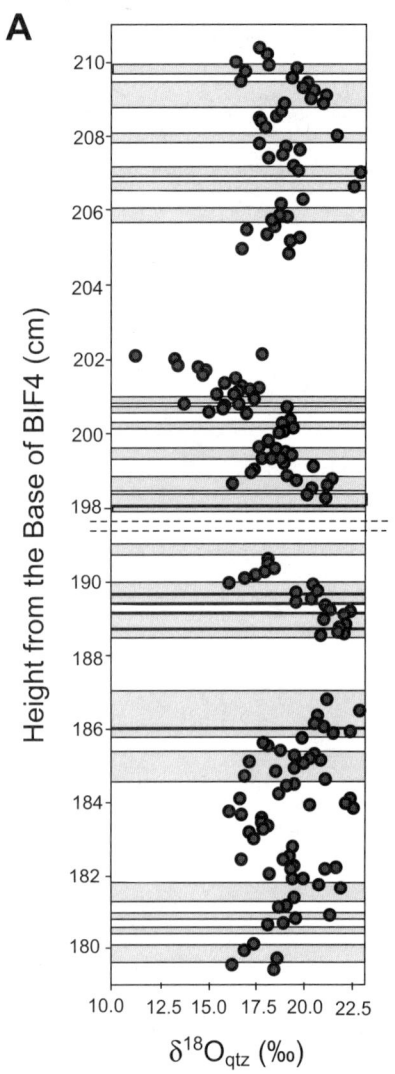
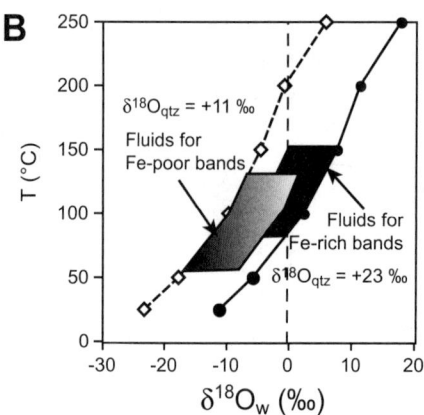

Figure 14. (A) $\delta^{18}O$ profile of a ~30 cm section of a micro-banded BIF sample of the Brockman IF, as determined by an in situ laser ablation method. Gray and white bands represent Fe-rich and Fe-poor cherts, respectively. Note the $\delta^{18}O$ values are generally higher in Fe-rich bands than in Fe-poor bands. (B) T - $\delta^{18}O_{H2O}$ relationships for $\delta^{18}O_{quartz}$ ranging from +11‰ to +23‰. Note that the $\delta^{18}O_{H2O}$ values of ~0‰ (i.e., seawater) and ~−10‰ (i.e., meteoric water) at T ~100 °C can best explain the $\delta^{18}O_{quartz}$ values of Fe-rich versus Fe-poor cherts, respectively (this study).

diagenesis and regional metamorphism. The $\delta^{18}O_{H2O}$ – T relationships are constructed for quartz with a $\delta^{18}O$ range of +11‰ and +23‰ (Fig. 14B) using the quartz-water fractionation factor by Matsuhisa et al. (1978). This figure shows that many combinations of $\delta^{18}O_{H2O}$ and T values can explain the observed $\delta^{18}O$ range for quartz. For example, if the $\delta^{18}O_{H2O}$ value was constant at 0‰ (i.e., today's seawater value), a temperature variation from ~90 to ~200 °C can explain the observed $\delta^{18}O$ variation from +11‰ (T = ~200 °C) for quartz in a silica-rich mesoband to +23‰ (T = ~90 °C) for quartz in a Fe-rich band; however, these temperature relationships between Fe-rich or -poor bands are opposite to what we would expect from a hydrothermal model. Furthermore, brine pool temperatures and quartz nucleation were probably lower than those of the discharging hydrothermal fluids (i.e., T < 200 °C).

If we assume that the probable temperature range of quartz precipitation for these samples was between 50 and 150 °C, a $\delta^{18}O_{H2O}$ range of ~−10‰ to +8‰ can explain the observed $\delta^{18}O$ quartz range of +11‰ to +23‰. In this case, the data can be interpreted as follows: the precipitation of quartz with lower $\delta^{18}O_{qtz}$ values (e.g., +11‰) occurred from meteoric water with an $\delta^{18}O_{H2O}$ value ~−10‰, whereas the precipitation of Fe-oxides and quartz with $\delta^{18}O_{qtz}$ values ~+23‰ occurred from hydrothermal fluids with an $\delta^{18}O_{H2O}$ value ~0 to +8‰; the positive $\delta^{18}O_{H2O}$ values suggest the hydrothermal fluids were seawater modified through evaporation and/or water-rock interaction. Therefore, variations in the mixing ratio of brine-pool water and the freshwater-dominated upper zone (Fig. 8C) best explain our $\delta^{18}O$ data. Note that our oxygen isotope data are not compatible with the conventional BIF model, which postulates an essentially constant $\delta^{18}O_{H2O}$ value of ~0‰ and T <50 °C for BIF mineralization.

Hamade et al. (2003) have found that the Ge/Si ratios of Fe-poor silica bands in the Brockman BIFs are essentially identical to present-day seawater (~0.8 × 10^{-6}), suggesting that the silica in Fe-poor cherts formed from silica derived from the weathering of continental crust (i.e., riverine input), rather than from hydrothermal fluids. The Ge/Si ratios in the Fe-rich bands are as high as 20 × 10^{-6}, which are similar to those in MOR hydrothermal fluids. Such data suggest that the Si, Ge, and Fe in Fe-rich bands were derived from hydrothermal fluids, although Hamade et al. (2003) have concluded that the Ge and Fe were of hydrothermal origin, whereas the Si was from normal seawater.

The neodymium isotopic compositions (ε_{Nd} values) of the Brockman IF are slightly negative (~−1.0), which suggests significant REE contributions (about 1/3) in the Hamersley Basin came from the weathering of old continental crust (i.e., riverine inputs); the other ~2/3 of the REEs were input from submarine hydrothermal fluids (Alibert and McCulloch, 1993). Although many smoker-type BIFs show distinctly positive Eu anomalies (Fig. 15B; also see Kato et al., this volume), the Hamersley BIFs show only weakly positive Eu anomalies (Fig. 15C). This is also consistent with our model that hydrothermal brine-pool water (positive Eu anomalies) was overlain by diluted seawater (no Eu anomalies), and periodic mixing of the two waters resulted in BIF formation.

The absence, or only minor presence, of Ce anomalies in the Brockman IF (Fig. 15C; also see Kato et al., this volume) is also consistent with the suggestion that the BIFs accumulated in an anoxic basin. The Fe-oxides probably had negative Ce anomalies when they formed in the transition zone, but were subsequently modified during and after settling on the seafloor by reactions with anoxic basin water and/or brine-pool water, which were both characterized by the absence of Ce anomalies (Fig. 15A).

The Kuruman IF, South Africa

The Kuruman IF in the Transvaal district of South Africa is another large BIF that has many similarities to the Brockman IF in the Hamersley Basin, Australia, including deposition age (Eriksson et al., 2002; Pickard, 2003). In the Northern Cape Province area of the Transvaal district, BIFs (200–400 m thick) overlie the Campbellrand Dolomite (500–2000 m thick) and crop out continuously over a strike (~N-S) distance >450 km; in the adjacent Transvaal area, the BIFs outcrop discontinuously (due to erosion) over another ~400 km distance (E-W). The exposed parts of these BIFs most likely deposited in shallow water along the basin margins, as suggested from various sedimentological features in the underlying dolomite (e.g., stromatolites) and in the overlying units (e.g., öölites) (e.g., Beukes, 1973, 1983; Klein and Beukes, 1989). However, in the unexposed parts of the sedimentary basin, a thick (~100 m) black shale unit has been found between the BIFs and the Campbellrand Dolomite, indicating that the deposition of oxide BIFs continued in a deep water body.

Recent zircon SHRIMP (sensitive high-resolution ion microprobe) dating of the Kuruman and Brockman IFs shows excellent correlations between several key tuff beds in these two districts (Pickard, 2003). These data suggest that the two BIFs, which are currently separated by the Indian Ocean, formed in the same sedimentary basin ca. 2.5 Ga; the Hamersley district represents deeper parts and the Transvaal district represents the basin margins. Therefore, we classify the Kuruman IF, as well as the Brockman IF, as examples of brine pool-type mineralization.

The isotopic compositions of Fe^{2+} in MOR hydrothermal fluids are very uniform, with $\delta^{56}Fe$ values in the range of –0.5‰ ± 0.2‰, and the isotopic fractionation factors between siderite and Fe^{2+} are ~0‰ (Yamaguchi et al., 2005). Therefore, we would expect the $\delta^{56}Fe$ values of siderite in BIFs to be quite uniform (~−0.5‰), if the Fe in BIFs was supplied from a large Fe^{2+} reservoir, such as Jacobson-Holland-Klein-Beukes's model oceans. However, the $\delta^{56}Fe$ values of siderite in the Kuruman IF, and also those in the 2.60 Ga Carawine Dolomite in the Hamersley Basin (stratigraphically equivalent to the Mara Mamba IF in Fig. 9), range from ~−2‰ to ~+1‰ (Yamaguchi et al., 2005). Such large $\delta^{56}Fe$ variations are, in fact, strong evidence for the brine-pool model, in which the total amount of Fe^{2+} in a brine pool increased by new fluxes of hydrothermal fluids or decreased with time by

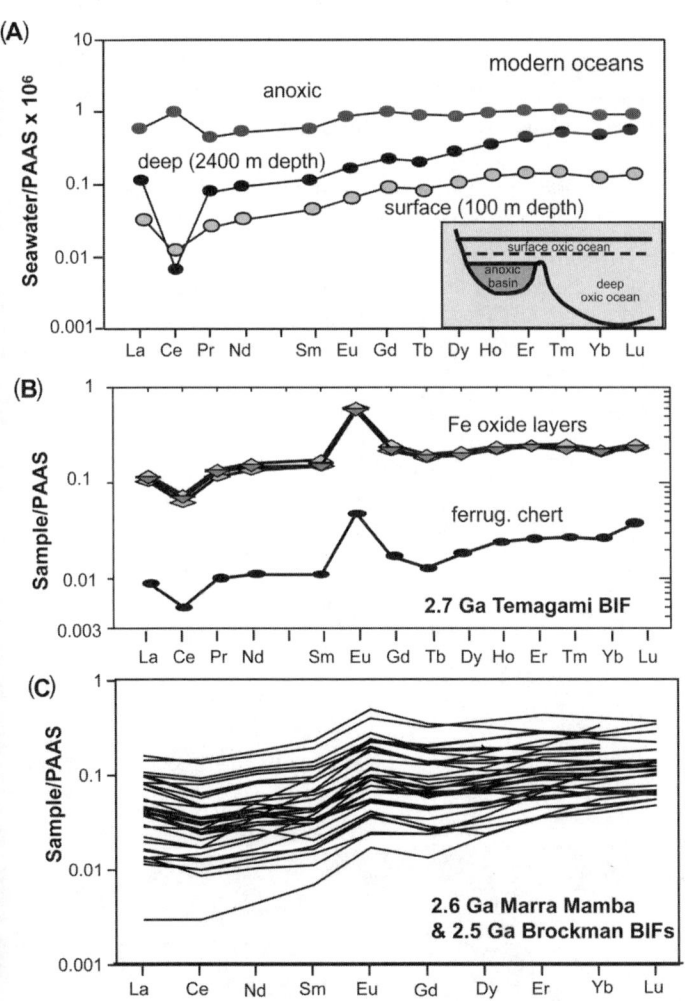

Figure 15. Comparisons of the REE chemistry of (A) modern seawaters; (B) 2.7 Ga Algoma-type BIFs from Temagami, Ontario, Canada (M. Bau, 2003, personal commun.); and (C) 2.5 Ga Brockman IF from the Hamersley Basin, western Australia (Kato et al., this volume).

the precipitation of minerals. For example, the $\delta^{56}Fe$ value of Fe^{2+} in a brine pool may become less than −0.5‰ by the continuous precipitation of ferric hydroxides or greater than −0.5‰ by the continuous precipitation of pyrite and/or siderite, because the isotopic fractionation factors between ferric (hydr)oxide and Fe^+ are positive, whereas those between pyrite (and siderite) and Fe^{2+} are negative (Yamaguchi et al., 2005).

BIFs in the Lake Superior Region, United States-Canada

The Lake Superior region, the type locality of the Lake Superior-type BIFs in Gross's (1965) classification, hosts many major BIFs (Fig. 16). Stratigraphic correlations of BIFs in this region are difficult, because they are highly deformed and metamorphosed. Morey and Southwick (1995) suggested the existence of two distinct BIF groups that formed during different times and under different geologic environments. The first group (e.g., the Menominee, Iron River-Crystal Falls, Marquette, and Cuyuna North Range BIFs), belonging to the volcanic-footwall type of our classification, was suggested to have formed ca. 2.0–1.9 Ga in association with bimodal volcanism. The second group (e.g., the Gunflint, Mesabi, Emily, Gogebic, and Baraga BIFs), belonging to our sandstone/conglomerate-footwall type, were suggested to have formed between ca. 1.9 and 1.85 Ga.

According to the Morey-Southwick (1995) reconstruction of the tectonic and ore-forming histories of the Lake Superior region, BIFs in this region were deposited throughout the progressive growth and ultimate destruction of a rifted continental margin. The suggested sequence of events is as follows: (1) initiation of continental rifting and mafic volcanism (ca. 2.3–2.2 Ga); (2) formation of small BIFs in the intracontinental basin (ca. 2.2–2.0 Ga); (3) continuous rifting, mafic volcanism, and formation of the Menominee, Iron River-Crystal Falls, Marquette, and Cuyuna North Range BIFs (ca. 2.0–1.9 Ga); (4) change from extensional to compressional tectonics, development of an island arc and forearc basin (Animikie Basin), and formation of the Gunflint, Mesabi, Emily, Gogebic, and Baraga BIFs (ca. 1.9–1.85 Ga); (5) continent-arc collision, deepening of the forearc basin, and deposition of the Rove-Virginia black shales (ca. 1.85 Ga); and (6) folding, thrusting, and metamorphism (ca. 1.85 Ga).

However, a recent geochronological and geochemical study of volcanic rocks and iron formations by Schneider et al. (2002) indicates that all iron formations in the Lake Superior district formed at 1874 ± 9 Ma, coeval with arc-related volcanism, in one or more forearc basins during arc accretion from the south.

The lithostratigraphy of the ca. 1.9–1.85 Ga Animikie Group in the Lake Superior region, which includes basal quartzite, BIFs, and an overlying shale formation, suggests that whereas the exposed sections of the Mesabi-Gunflint BIFs most likely deposited in shallow water, possibly the photic zone, the Marquette region was probably under a deep brine pool. Although Morey and Southwick (1995) have not speculated on this, their tectonic model suggests that the Animikie Basin (ca. 1.9–1.85 Ga) was not completely open to the oceans, a setting that would have allowed submarine hydrothermal fluids that discharged from the basin floor to create large brine pool-type BIFs (e.g., the Gunflint and Mesabi BIFs).

The geological and mineralogical characteristics of the ca. 1.9–1.85 Ga BIFs in the Lake Superior region (e.g., the Gunflint and Mesabi BIFs) are more similar to those of the Kuru-

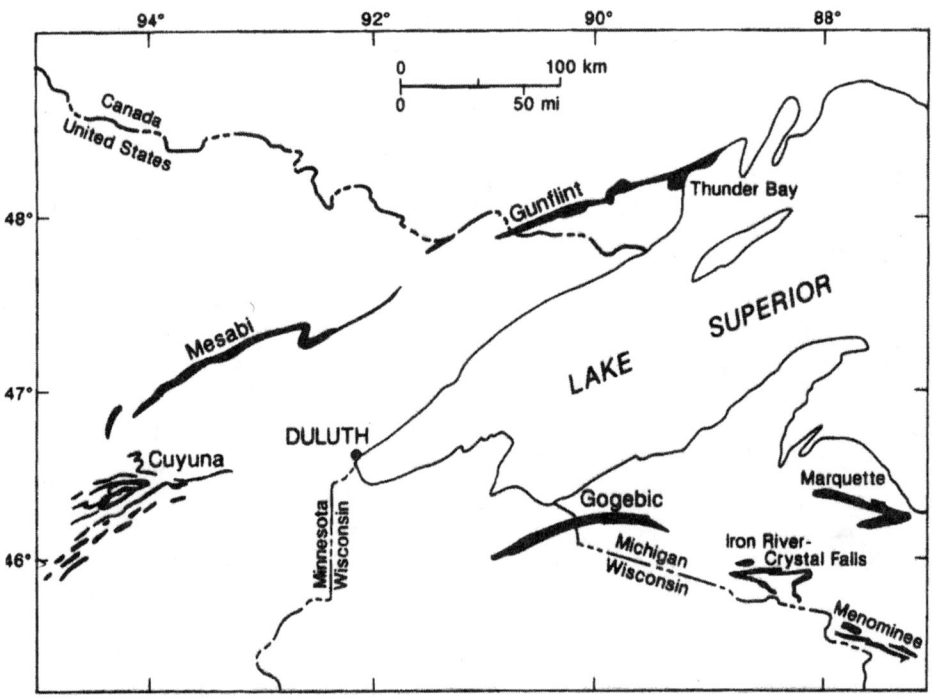

Figure 16. BIFs in the Lake Superior district, United States-Canada (Morey and Southwick, 1995).

man IF (shallow water), than the Brockman IF (deep water). The presence of an anoxic Fe^{2+}-rich, H_2S-poor brine at very shallow depths (<10 m) in the Hamersley and Animikie Basins under an oxygenated atmosphere was not unusual, because the redox boundary in many saline lakes (e.g., the Great Salt Lake) occurs at depths <5 m.

CHEMICAL PROCESSES FOR THE FORMATION OF BIFS AND SECONDARY IRON ORES

Precipitation Mechanisms for Ferric (Hydr)oxides in BIFs

General

Prior to the 1980s, the only known mechanism for the precipitation of ferric (hydr)oxides from aqueous Fe^{2+} was by reaction with molecular O_2, which was generated by the photodissociation of H_2O either with or without biological factors (e.g., cyanobacteria), to facilitate the following reactions:

$$2H_2O + h\nu \Rightarrow 2H_2 + O_2 \qquad (1)$$

and

$$Fe^{2+} + 1/4O_2 + 5/2H_2O \Rightarrow Fe(OH)_3 + 2H^+ \qquad (2)$$

$$\text{or } (Fe^{2+} + 1/4O_2 + H_2O \Rightarrow 1/2Fe_2O_3 + 2H^+). \qquad (2')$$

If the atmosphere was rich in H_2 from volcanic gas and oxygenic photosynthetic organisms (e.g., cyanobacteria) were absent, O_2 production by reaction (1) would become insignificant because of the backward reaction. However, if H_2 is removed through diffusion into space or biosynthesis (i.e., CO_2 reduction),

$$CO_2 + 2H_2 \Rightarrow CH_2O_{(organic\ matter)} + H_2O, \qquad (3)$$

O_2 production from reaction (1) could build an oxygen-rich environment. The combination of reactions (1) and (3), which occurs in cells of oxygenic photoautotrophs (e.g., cyanobacteria), is a well-known overall reaction for oxygenic photosynthesis:

$$CO_2 + H_2O + h\nu \Rightarrow CH_2O_{(organic\ matter)} + O_2. \qquad (4)$$

Therefore, early BIF models (e.g., Cloud, 1968) related the formation of oxide-BIFs to the evolution of cyanobacteria, the size of oxide-BIFs to the amount of O_2 generation, and the precipitation of iron-rich microbands to summer seasons (e.g., Trendall and Blockley, 1970).

Supported by laboratory experiments, Braterman and Cairns-Smith (1987) and Anbar and Holland (1992) proposed that the photolysis of Fe^{2+} by UV lights in the absence of molecular O_2 might have been the cause of ferric hydroxide precipitation in BIFs before the emergence of cyanobacteria:

$$Fe^{2+} + 3H_2O + h\nu \Rightarrow Fe(OH)_3 + 2H^+ + 1/2H_2. \qquad (5)$$

Several recent investigators (e.g., Widdel et al., 1993; Konhauser et al., 2002; Brown, this volume) have suggested that iron-oxidizing bacteria were responsible for the precipitation of ferric hydroxides in BIFs. Two groups of iron-oxidizing bacteria are known: (1) aerobic chemolithoautotrophs (e.g., *Gallionella ferruginea*), and (2) anaerobic anoxygenic photoautotrophs (e.g., *Chromatium* sp.). Aerobic iron-oxidizers basically promote reaction (2), which is the oxidation of Fe^{2+} by dissolved O_2 molecules, whereas anaerobic photoautotrophs promote reaction (5) by utilizing the H_2 generated by reaction (5) in non-oxygenic photosynthesis (such as reaction 3). Therefore, Konhauser et al. (2002) expressed the overall reactions involving iron oxidizers as follows:

$$6Fe^{2+} + 1/2O_2 + CO_2 + 16H_2O \Rightarrow CH_2O + 6Fe(OH)_3 + 12H^+ \qquad (6)$$

$$4Fe^{2+} + CO_2 + 11H_2O \Rightarrow CH_2O + 4Fe(OH)_3 + 8H^+ \qquad (7)$$

Any or all of the above three mechanisms (reactions 5–7) could have caused the nucleation of ferric (hydr)oxides in BIFs, if nucleation was confined to the photic zone and under an anoxic atmosphere. However, if nucleation occurred below the photic zone, the only plausible oxidation mechanism would have been the reaction of hydrothermal Fe^{2+} with molecular O_2 in deep ocean water. The presence of free O_2 molecules below the photic zone can be linked to an oxygenated atmosphere (see below).

Ferric (Hydr)oxides in Smoker-type BIFs

If pre–1.8 Ga oceans were Fe^{2+}-rich, as suggested by many previous investigators (e.g., Cloud, 1968; Drever, 1974; Holland and Petersen, 1995), we would predict pre–1.8 Ga volcanic-associated BIFs (1) to occur as large continuous beds, rather than as small individual bodies and clusters of variable sizes and shapes; (2) to occur in close association with basalts, rather than with less abundant rhyolites; (3) not to occur in close association with hydrothermally altered footwall rocks; (iv) to have uniform mineralogy (e.g., siderite/hematite ratio) and uniform compositions (e.g., Eu and Ce anomalies) among different microbands in a single hand specimen and among different BIF bodies; and (v) to have different mineralogy, textures, and geochemistry than post–1.8 Ga BIFs. However, none of these predictions were found to be true.

In a previous section, we have suggested that many (if not most) VMSD- and volcanic-associated BIFs are the products of smoker-type mineralization, in which the nucleation of minerals occurred by the rapid mixing of hydrothermal fluids with local cold seawater in deep (>2.5 km) oceans, although some volcanic-associated BIFs (e.g., the 2.7 Ga Michipicoten BIFs in the Wawa Greenstone Belt) clearly formed at shallow depths. One may argue that the Fe^{2+} that discharged at a depth >2.5 km did not immediately form ferric oxides, but instead was carried by a hydrothermal plume to the photic zone, where it was converted to ferric hydroxide minerals by photochemical and/or microbial reactions (see reactions 1–7); these minerals then sank through the >2.5-km-deep water column and accumulated on top of a VMS body (i.e., the discharging spot). However, such a scenario is unlikely, because the ferric hydroxide particles would have been widely dispersed by both surface and deep currents and therefore would not have accumulated near fluid discharge sites (i.e., areas of footwall alteration) to form a well-defined BIF body.

Individual grains of iron oxides and carbonates in BIFs show large variations in size and shape (Trendall and Blockley, 1970; Ahn and Buseck, 1990) because of the coagulation of amorphous minerals during settlement through a water column and recrystallization during diagenesis and metamorphism. However, the abundance of very fine grained (<1 μm in size) crystals of hematite and siderite in many BIFs and the confined geometry of both VMSD- and volcanic-footwall type BIFs suggest that the nucleation of iron-bearing minerals and amorphous silica occurred very rapidly at (or near) the discharging sites, much like the black smokers on MOR hydrothermal systems.

The simple cooling of low-temperature (<~250 °C) submarine hydrothermal fluids does not form sulfide-rich ores, because they are poor in H_2S (Fig. 3), or iron oxides, because most hydrothermal fluids do not contain any dissolved O_2 molecules. For example, a typical submarine hydrothermal fluid at ~250 °C may have a f_{O_2} value of ~10^{-40} atm and contain ~10^{-3} moles/kg H_2O of ΣFe^{2+} (e.g., Ohmoto et al., 1983a). According to thermodynamic data, simple cooling may cause this fluid to become supersaturated with respect to magnetite and/or hematite. However, no oxide will form unless mechanisms exist at the depositional site to both continuously dissociate H_2O and remove (oxidize) the dissociated H_2 molecules so that the generated O_2 molecules can react with the Fe^{2+} to form iron oxides.

Anaerobic microbes are one possible mechanism to remove H_2 in submarine environments (see reaction 7). Although such a mechanism may have been important in the formation of some BIFs that formed in the photic zone, it was not significant for most volcanic-associated BIFs that formed in deep oceans. Furthermore, kerogen (a remnant of organisms) is essentially absent in oxide/carbonate BIFs and footwall rocks of most volcanic-associated BIFs, although it is abundant in the overlying sediments (black shales). This suggests that areas of fluid discharge were too hostile for microbes to participate in BIF mineralization, but microbes flourished in the local submarine depressions after the cessation of hydrothermal events.

Another possible mechanism to remove H_2 from the hydrothermal fluid is to react it with the O_2 in local seawater, but this is basically the same as precipitating Fe^{2+} by reactions with O_2 molecules in local seawater.

Ferric (Hydr)oxides in Brine Pool-type BIFs

The current popular model for BIF formation (e.g., Holland, 1984; Klein and Beukes, 1992; Morris, 1993) proposes that a large mass of Fe^{2+}-rich deep ocean water periodically invaded a basin, overturned the entire basin water, mixed with O_2-bearing surface water to nucleate ferric oxides, and returned to the open oceans. Such a fluid circulation scenario, however, would have produced Fe-rich microbands that were very uneven in thickness and distribution but had a general trend of thickening toward shallower parts of the basin. Such large-scale circulation also would have disturbed the sediments, especially in shallow basins. However, two of the important characteristics of BIFs in the Hamersley Basin are that (1) the thicknesses of a Fe-rich microband and mesoband are essentially uniform (averaging ~100 μm and ~1 cm, respectively) throughout the basin, and (2) the sediments were not disturbed by bottom currents (Trendall and Blockley, 1970). These BIF textures suggest that Fe^{2+}-rich water mixed with O_2-rich water gently, uniformly, and simultaneously throughout the Hamersley Basin as suggested in our Red and Black Seas hybrid model (Fig. 8C).

As indicated by the molecular fossils of cyanobacteria and eukaryotes, the surface water of the Hamersley Basin contained at least 18 μM of DO. Reaction (2) and (2)' indicate that 1 mol DO can precipitate 4 mol Fe^{2+} as ferric (hydr)oxide. Therefore, a simple mass balance calculation indicates that mixing a ~2-m-thick layer of this water and a 1.3-cm-thick layer of Fe^{2+}-rich brine (ΣFe^{2+} = 1 mM) will form a microband that is 100 μm thick and 30% Fe_2O_3. For comparison, the thickness of a Fe-rich microband in the Brockman IF—and also in most other BIFs—falls in a range from ~20 μm to ~5 mm with an average of ~100 μm (Trendall and Blockley, 1970) (Fig. 17A). If the DO content of the surface water was 180 μM (i.e., ~50% that of the present ocean surface water), the required thickness of O_2-rich water to precipitate the same amount of iron oxides is reduced to only ~20 cm. Clearly, the thickness of a Fe-rich microband varies with the DO concentration of the surface water, ΣFe^{2+} concentration of the brine, and the thicknesses of the two mixed water bodies. However, our calculations suggest that the microbands in brine-pool type BIFs were produced by the basinwide, gentle disturbance (mixing) of thin layers (centimeters to meters) of brine and surface water.

Such mixing of the bottom layer of the surface water (mostly river water) and the upper layer of the underlying brine pool is expected to occur when the surface water flows from a land-locked basin toward an open ocean (Fig. 8C). The flux of river water to the surface zone of the basin varies seasonally with the climate and geography of the fluvial systems. The depth and thickness of the transition (mixing) zone also depend on the volume of the brine pool, which is influenced by the intensity of submarine hydrothermal activity, sea level, and basin topography. If the basin is completely isolated from an open ocean because of a sea level drop and/or a significant decrease in riverine flux, the thickness of the oxygenated surface water zone may be significantly reduced and the transition zone may become anoxic, ceasing the formation of Fe oxides. If the sea level rises and the basin becomes more open to an ocean, normal seawater may invade the basin and change the characteristics of the O_2-rich surface zone. For all these reasons, we suggest that the mixing of Fe^{2+}-rich brine and O_2-rich surface water occurred periodically, rather than continuously, over time to generate Fe-rich microbands.

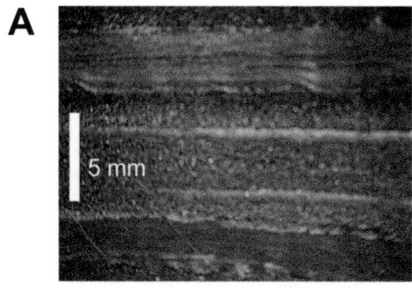

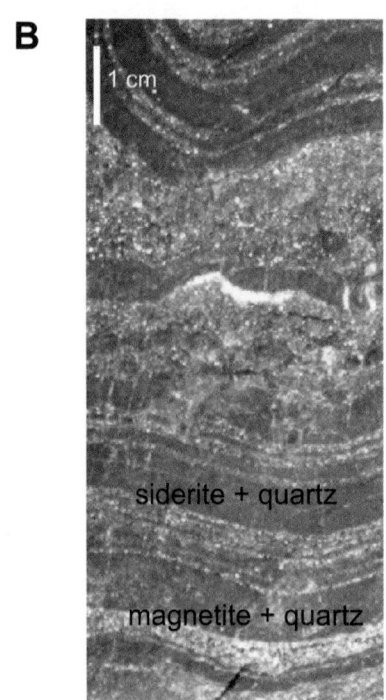

Figure 17. (A) A thin section photo of an oxide BIF from the 2.7 Ga Temagami BIF, Ontario, Canada. Red bands are silica-rich, ΣFe-poor, and high in hematite/magnetite ratio; black bands are silica-poor, ΣFe-rich, and high in magnetite/hematite ratio. (B) A hand specimen showing alternating mesobands of oxide BIFs (magnetite+quartz) and siderite BIFs (siderite+quartz) from the Helen Mine, Wawa, Ontario, Canada.

Formation Mechanisms of Siderite- and Pyrite-Rich BIFs

In every BIF we investigated, the changes in BIF types from oxide-, to carbonate-, to sulfide-type occurred with time (i.e., stratigraphy) during the history of a sedimentary basin, rather than laterally as suggested by James (1954). Alternation of BIF types over a centimeter-vertical scale is very common (Fig. 17B). These features suggest that changes in BIF type resulted from changes in the concentrations of DO, HCO_3^-, and H_2S in local seawater over time.

Whether iron oxide, siderite, or iron sulfide precipitates during the mixing of hydrothermal fluids and local seawater depends on which of the following reactions dominates (Fig. 18):

$$Fe^{2+}_{(hyd)} + 1/4 O_{2(s.w.)} + 5/2 H_2O \Rightarrow Fe(OH)_3 + 2H^+ \quad (2) \text{ at } T < \sim 80 \,°C$$

$$2Fe^{2+}_{(hyd)} + 1/2 O_{2(s.w.)} + 2H_2O \Rightarrow Fe_2O_3 + 4H^+ \quad (2') \text{ at } T > \sim 80 \,°C$$

$$Fe^{2+}_{(hyd)} + HCO_3^-_{(s.w.)} \Rightarrow FeCO_3 + H^+ \quad (8)$$

$$Fe^{2+}_{(hyd)} + 2H_2S_{(s.w.)} \Rightarrow FeS_2 + 2H^+ + H_2 \quad (9)$$

The relative importance of the above reactions is determined by the following three parameters: (1) the chemistry of the hydrothermal fluids, especially the concentrations of ΣFe^{2+} (= Fe^{2+} + $FeCl^+$ + $FeCl_2$ + $FeHCO_3^-$); (2) the chemistry of the local seawater, especially the concentrations of DO, ΣCO_3^{2-}, and ΣS^{2-}; and (3) the kinetics of the above reactions. For example, the O_2 molecules in seawater inside a local depression (basin) may be totally consumed by reactions with hydrothermal ΣFe^{2+} to form iron oxides, whereas the remaining ΣFe^{2+} is consumed by ΣCO_3^{2-} to form siderite. Thus, the abundance ratio of iron oxide to siderite in the precipitates depends on the concentrations of DO and ΣCO_3^{2-} in the seawater, the ΣFe^{2+} concentration of hydrothermal fluids, and the mixing ratio of hydrothermal fluid to seawater to consume all the ΣFe^{2+} in the discharged fluid (Fig. 18).

If the local water becomes anoxic and the temperature falls below ~50 °C, SRB may create an H_2S-rich local environment and mixing of such local seawater with Fe^{2+}-bearing hydrothermal solutions will precipitate iron sulfides. The relationships between the mineralogy of dominant Fe-bearing minerals and the compositions of hydrothermal fluid and local seawater are illustrated in Figures 18 and 19.

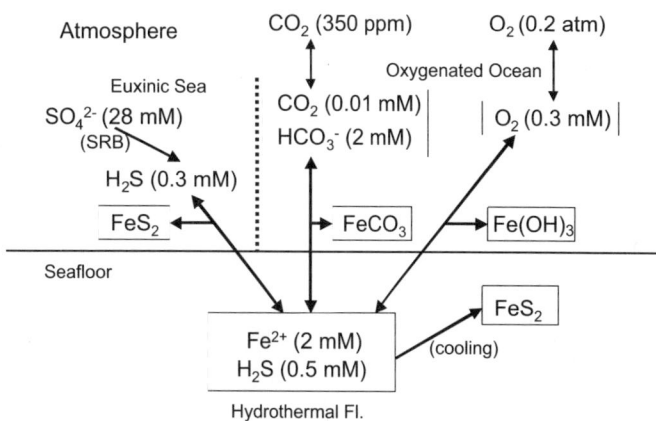

Figure 18. Schematic illustrations of the relationships between local seawater chemistry and BIF Fe-mineralogy: the relative abundance of O_2, ΣCO_3^{2-}, and H_2S in local seawater determines whether iron (hydr)oxide, siderite, or pyrite precipitates when the seawater mixes with Fe^{2+}-rich hydrothermal solutions. The concentration values are of the modern atmosphere and oceans.

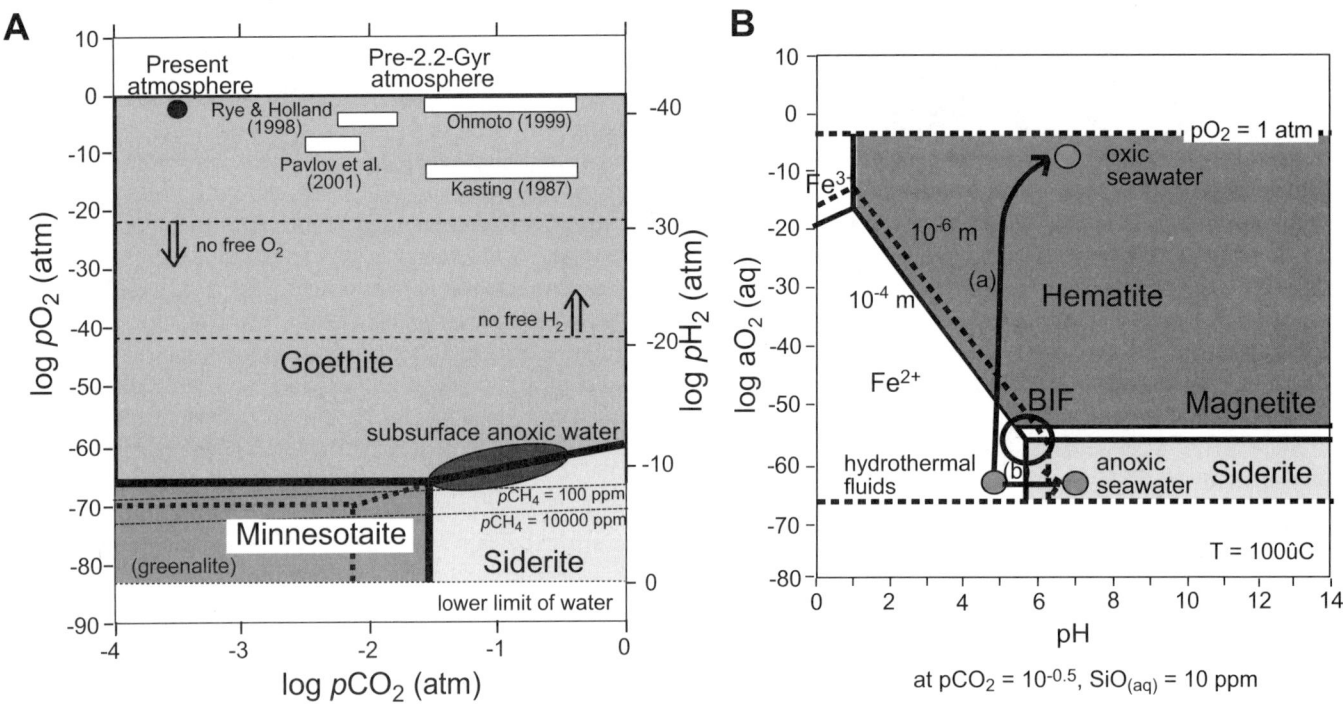

Figure 19. The stability relationships of Fe-bearing minerals. (A) $\log a_{O_2}$ - $\log a_{CO_2}$ diagram at $SiO_{2(aq)} = 10^{-3.4}$ M (24 ppm) at 25 °C. (B) $\log a_{O_2}$ – pH diagram showing the stability fields of Fe-bearing minerals at $pCO_2 = 10^{-0.5}$ atm, $SiO_{2(aq)} = 10$ ppm, and T = 100 °C. Also shown are the changes in fluid chemistry and formational minerals when hydrothermal fluid is mixed with O_2- or HCO_3^--rich local seawater.

The concentration of ΣFe^{2+} in submarine hydrothermal fluids at ~100 to ~200 °C typically ranges from 0.1 mM to 5 mmol per kilogram of H_2O (Fig. 3). The *maximum* concentrations of DO, ΣCO_3^{2-}, and ΣS^{2-} in modern oceans also fall within this range. For example, the average DO concentration in high-latitude ocean surface water is ~0.35 mM. However, the DO content becomes zero in anoxic waters. The concentration of ΣCO_3^{2-} ($\approx HCO_3^-$) in today's oceans averages 2.3 mM but is higher in anoxic basins, such as ~4 mM in the Black Sea (Fry et al., 1991), due to the addition of CO_2 ($+HCO_3^-$) from the oxidation of organic matter in the water column. The ΣS^{2-} content of normal seawater is <<0.01 mM but is much higher in anoxic waters due to bacterial sulfate reduction, such as ~0.3 mM in the anoxic water body of the Black Sea (Ohmoto and Goldhaber, 1997).

The similarities in the concentration of ΣFe^{2+} in hydrothermal fluids and the maximum concentrations of DO, ΣCO_3^{2-}, and ΣS^{2-} in seawater are important because the ΣFe^{2+} in hydrothermal fluids may be quantitatively precipitated as either oxide, carbonate, or sulfide during mixing with a variety of seawaters at ratios ranging from ~1/100 to 100. This is probably the main reason the abundance of Fe-bearing minerals in a BIF is typically unimodal, i.e., oxide-, carbonate-, or sulfide-rich.

The above discussions illustrate that the relative concentrations of DO, ΣCO_3^{2-}, and ΣS^{2-} in local seawater are essential to control the dominant mineralogy (oxide, carbonate, or sulfide) of submarine hydrothermal deposits. A corollary is that if Archean seawater was free of O_2 and H_2S, as suggested by many investigators (e.g., Walker and Brimblecombe, 1985; Canfield and Raiswell, 1999), siderite would have been the only Fe-bearing mineral in volcanic-associated BIFs. However, the abundance of hematite and magnetite indicates that the bottom seawaters near these BIFs contained free O_2 molecules.

Precipitation Mechanisms of Silica in BIFs and Silica Chemistry of the Archean Oceans

Essentially all researchers agree that the cherts in BIFs precipitated as amorphous silica from the aqueous silica in seawater (regardless of whether it was supplied mostly by rivers or by submarine hydrothermal fluids) and were subsequently transformed to opal, cristobalite, and quartz during diagenesis and metamorphism:

$$H_4SiO_{4(aq)} \Rightarrow Si(OH)_{4(amorphous\ silica)} \quad (10)$$

$$Si(OH)_{4(amorphous\ silica)} \Rightarrow SiO_{2(quartz)} + 2H_2O \quad (11)$$

The $H_4SiO_{4(aq)}$ in today's oceans is supplied almost entirely by rivers. The silica concentration of normal seawater is ~5 ppm as SiO_2 (Drever, 1982), which is about the saturation value for quartz at ~4 °C and approximately an order of magnitude less than the saturation value for amorphous silica at ~25 °C (i.e.,

~100 ppm) (Fig. 20). The formation of chert occurs in today's oceans because silica-accreting organisms (e.g., radiolarians and siliceous sponges) extract silica from seawater to form amorphous silica shells; the dead organisms accumulate on the seafloor as amorphous silica-rich sediments, which are subsequently transformed to less soluble minerals (e.g., cristobalite, tridymite, and quartz) during diagenesis.

Silica-accreting organisms probably first appeared ca. 750 Ma (Bengtson, 1994). Knauth and Lowe (2003) have suggested that, in the absence of silica-accreting organisms, the silica concentrations of Precambrian oceans must have been much higher (e.g., ~300 ppm) than today and were controlled by the solubility of amorphous silica, rather than quartz. The precipitation rate of silica is greatly increased when the availability of solid and/or colloid surfaces (e.g., clays, glass, quartz, silicate minerals, and colloids in sediments) for silica nucleation increases (Rimstidt, 1997). Therefore, it is probable that the silica concentrations of the Archean oceans were maintained at levels about the saturation values of quartz, much like modern oceans, although the actual $SiO_{2(aq)}$ values could have been ~20 ppm, instead of ~5 ppm (today's oceans), if the Archean oceans were ~50 °C.

If the Archean oceans were saturated or supersaturated with respect to amorphous silica, silica would have precipitated from high-latitude ocean water during winters and created essentially uniform layers of cherts over a large (>>1000 × 1000 km) area. However, the lateral extents of cherts in greenstone belts are typically ~1 to ~100 km.

Furthermore, the lateral extents of Fe-poor, silica-rich micro- and mesobands are essentially the same as those of Fe-rich, silica-poor bands in most BIFs. These features suggest that most of the SiO_2 in BIFs, as well as Fe, was supplied by local submarine hydrothermal fluids, rather than from normal ocean waters.

The silica concentrations of hydrothermal solutions are typically controlled by the solubility of quartz (Rimstidt, 1997), such as ~250 ppm at ~250 °C (Fig. 20). Simply mixing hydrothermal fluid with normal seawater may cause the precipitation of quartz, but not amorphous silica, if the precipitation of silica-rich minerals is controlled solely by the solubility of quartz and/or amorphous silica (Fig. 20). However, laboratory experiments (e.g., Bazilevskaya, 2004) have demonstrated that the precipitation of amorphous silica occurs from silica-undersaturated solutions when ferric hydroxides are formed, due to the adsorption of amorphous silica on ferric (hydr)oxide particles. Thus, coprecipitation of amorphous silica and ferric (hydr)oxide can be explained. Similarly, mixing brine-pool water that is undersaturated with amorphous silica, but supersaturated with quartz (e.g., SiO_2 = 200 ppm at 100 °C), with cooler, silica-poor surface water (SiO_2 ≈50 ppm) would have caused the precipitation of amorphous silica, if the nucleation of ferric (hydr)oxides occurred.

The separation of amorphous silica from ferric (hydr)oxide particles, and migration and reprecipitation of amorphous silica onto the seafloor, would have occurred during the early diagenesis of Fe-Si-rich sediments, creating Fe-rich (lower) and silica-rich (upper) microbands.

The nucleation of amorphous silica may also occur in the upper zone of a hot brine pool through simple conductive cooling by the overlying cooler water body, because the solubilities of both quartz and amorphous silica decrease with decreasing temperature (Fig. 20). For example, fluid saturated with quartz at 200 °C (SiO_2 = ~200 ppm) may precipitate amorphous silica at ~70 °C during conductive cooling. Such a precipitation mechanism may explain the frequent occurrence of thick (>1 mm), Fe-poor silica bands in the Hamersley BIFs, as well as others. Evaporation of the surface water zone, especially during periods of no brine pool development, would have been another mechanism to cause the precipitation of silica.

Transformation from Hematite-Rich to Magnetite-Rich BIFs

A VMS ore body typically alters its mineralogical and geochemical characteristics as it grows in size because the earlier minerals are continually transformed by reactions with later hydrothermal fluids (Fig. 21A). Similarly, during the growth of a volcanic-associated BIF body, goethite and/or hematite crystals could be subjected to later Fe^{2+}-rich hydrothermal fluids (Fig. 21B). As the crystals react with these "new" hydrothermal fluids, hematite (or goethite) is converted to magnetite through the following reaction (Ohmoto, 2003):

$$Fe_2O_{3(hm)} + Fe^{2+} + H_2O \Rightarrow Fe_3O_{4(mt)} + 2H^+ \quad (12)$$

Reaction (12) is not a redox reaction because there is no change in the valence of any Fe or other atoms; it is simply an acid-base reaction. This proposed mechanism differs from the conventional model for hematite-magnetite transformation (e.g., Perry et al., 1973), which suggests the following redox reaction between the primary ferric (hydr)oxide minerals and organic carbon in BIF sediments to form magnetite during the metamorphic stage:

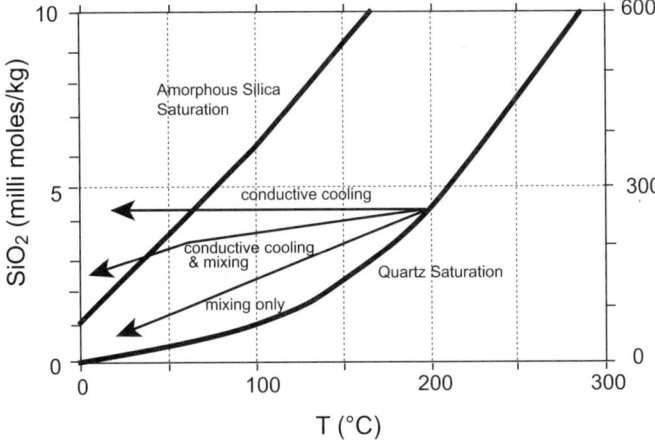

Figure 20. Solubilities of quartz and amorphous silica, and mechanisms of silica precipitation from submarine hydrothermal solutions (modified after Spry et al., 2000).

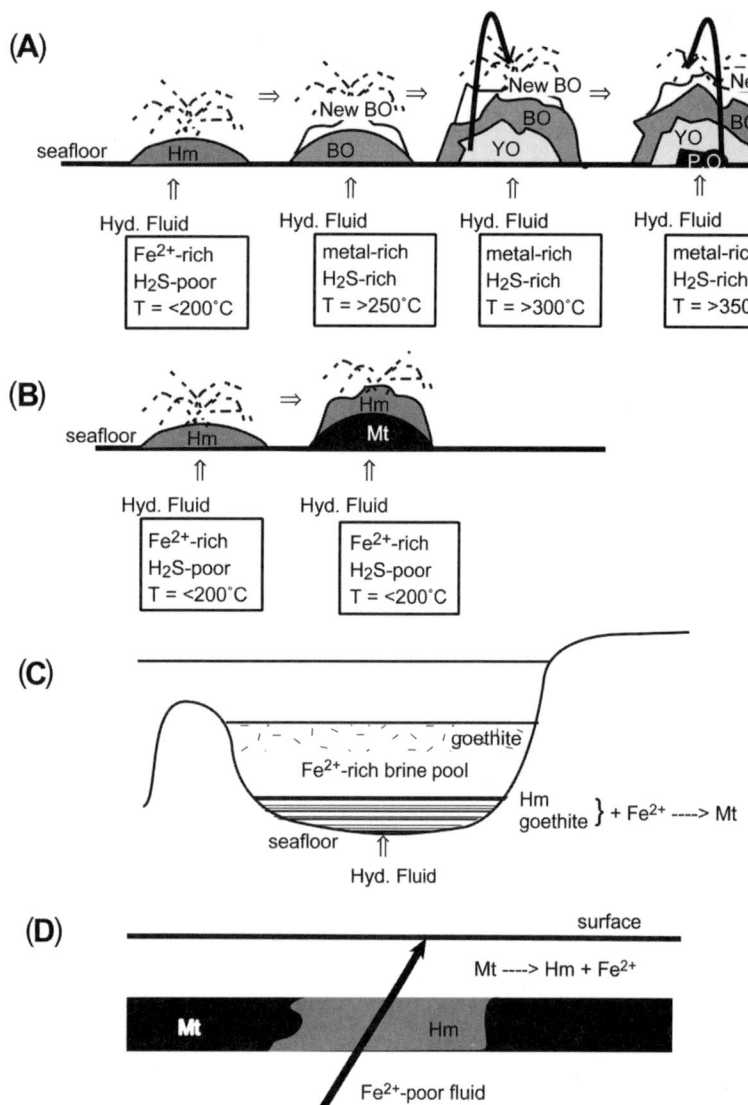

Figure 21. Growth model for submarine hydrothermal deposits. (A) VMSD transformation. (B) BIF transformation. Hm—hematite-chert ore, Mt—magnetite-chert ore, BO—black ore, PO—massive pyrite ore, YO—yellow ore (Ohmoto, 2003). (C) Transformation of hematite to magnetite in brine pool–type BIFs. (D) Transformation of magnetite-rich BIFs to hematite-rich iron ores by subsurface non-redox reactions.

$$6Fe_2O_{3(hm)} + CH_2O_{(org)} \Rightarrow 4Fe_3O_{4(mt)} + CO_2 + H_2O \quad (13)$$

Reaction (13) may explain some (but not all) of the magnetite crystals in BIFs that were subjected to high-grade metamorphism in the Lake Superior region and other regions (French, 1973; Perry et al., 1973). However, reaction (12) best explains the magnetite crystals in most volcanic-associated oxide BIFs, because their metamorphic grades are not high enough to have caused reaction (13) and/or because the kerogen content in BIFs and footwall rocks is too low to have accounted for the high magnetite/hematite ratios (>1) that they contain. Reaction (12) also better explains the general trend that larger BIFs have higher magnetite/hematite ratios. It also explains the well-known feature that low-Fe, silica-rich bands typically exhibit red color due to high hematite/magnetite ratios, whereas Fe-rich bands typically exhibit black color due to high magnetite/hematite ratios (Fig. 17A).

The goethite/hematite crystals in brine pool-type BIFs, which nucleated at the top of brine pools and settled onto the floor of brine pools, would have continuously reacted with the overlying Fe^{2+}-rich brine and partially transformed to magnetite through reaction (12). This explains the high magnetite/hematite ratios (>1) in brine pool-type BIFs.

Formation from Magnetite-Rich BIFs to Hematite-Rich Secondary Iron Ores

The major iron ores in the world are those that developed from BIFs by the loss of silica and transformation of magnetite to hematite; the Fe contents typically increased from ~20 wt% in precursor BIFs to ~50 wt%. A popular model for the formation of these iron ores combines supergene enrichment and metamorphic processes (e.g., Morris, 1985). It postulates that magnetite-rich BIFs were uplifted by a tectonic process and subjected to weathering under an oxygenated atmosphere to form goethite-rich ores:

$$2Fe_3O_{4(mt)} + 1/2O_2 + 3H_2O \Rightarrow 6FeO(OH)_{(gt)} \quad (14),$$

and then subsequently buried to a great depth (>5 km) where they metamorphosed to hematite-rich ores:

$$6FeO(OH)_{(gt)} \Rightarrow 3Fe_2O_{3(hm)} + 3H_2O \quad (15)$$

This paleoweathering-metamorphic model was developed because the principal mechanism for the transformation of magnetite to ferric (hydr)oxides was thought to be the oxidation of Fe^{2+} atoms in magnetite to Fe^{3+} atoms (i.e., reaction 14). Most geologists have also accepted the theory for the rise of atmospheric O_2 ca. 2.2 Ga (e.g., Holland, 1994). For these two reasons, the exploration for iron ores has primarily focused on major unconformities younger than 2.2 Ga. Ohmoto (2003), however, suggested that the more common mechanism for the conversion of magnetite to hematite in most geologic settings is the reverse of reaction (12), i.e., the leaching of Fe^{2+} from magnetite by Fe^{2+}-poor hydrothermal fluids (such as those developed in carbonate- and/or chert-dominated terranes):

$$Fe_3O_{4(mt)} + 2H^+ \Rightarrow Fe_2O_{3(hm)} + Fe^{2+} + H_2O \quad (16)$$

Reaction (16) is not a redox reaction, either. According to this model, the transformation of magnetite-rich BIFs to hematite-rich ores in large ore bodies mostly occurred in subsurface conditions without an oxidant (Fig. 21D). Therefore, the formation of hematite-rich secondary ores cannot be linked to atmospheric O_2 content. Another important implication of this model is that iron ore exploration targets are not confined to paleosurfaces younger than ca. 2.2 Ga; the formation of secondary ores from magnetite-rich BIFs probably occurred throughout geologic time.

Otake et al. (2005) have experimentally confirmed that the non-redox transformations of hematite to magnetite (reaction 12) and magnetite to hematite (reaction 16) indeed occur rapidly (<1 day at 150 °C) in a pH range of ~4 to ~6, even under high pH_2 conditions (up to 50 bars).

CONSTRAINING THE EVOLUTIONARY HISTORIES OF THE ATMOSPHERE, OCEANS, AND BIOSPHERE FROM BIFS

Dissolved O_2 Contents of Deep Ocean Water

REE data of smoker-type BIFs (Fig. 15B; also see Kato et al., this volume) clearly indicate that BIFs are products of the mixing of hydrothermal fluids and seawater, and that Archean global oceans were already oxidized to produce negative Ce anomalies. The DO content of the bottom water in oxide-BIF environments can be estimated from simple mass balance calculations based on reaction (2) or (2′).

According to our studies of several volcanic-associated BIFs, a typical oxide-rich microband is ~1 mm thick and distributed over an area ~5 km in radius with an average composition of ~30% Fe_2O_3 and 70% SiO_2. Thus, the total amount of Fe in an average microband can be calculated as:

$$Fe\ (mol) = (10^{-1} cm) \cdot \pi \cdot (5 \times 10^5 cm)^2 \cdot$$
$$(3.4\ gm/cm^3) \cdot 0.30 \cdot (1/159.7\ g/mol) \cdot 2 = 1.0 \cdot 10^9\ moles$$

where the density of the iron-rich microband was assumed to be 3.4 g/cm³ and the molecular weight of Fe_2O_3 is 159.7 g. The total amount of hydrothermal fluid required to provide 1.0×10^9 mole of Fe is 1.0×10^{12} kg ($= 1.0 \cdot 10^{15}$ cm³ = 1.0 km³), if the concentration of ΣFe^{2+} in the hydrothermal fluid is 1 mM (56 ppm) and its density is 1.0 g/cm³.

Because 1 mole O_2 can precipitate 4 moles Fe^{2+} as Fe_2O_3 (or goethite), 2.5×10^8 moles O_2 are required to precipitate 1.0×10^9 moles Fe^{2+}. If the DO content of local seawater was as low as 10^{-15} moles/kg H_2O, the total volume of seawater required to provide this amount of O_2 was 2.5×10^{23} kg ($= 2.5 \times 10^{11}$ km³). Obviously, this is an unacceptable amount because it far exceeds the total volume of every ocean combined (1.35×10^9 km³). If the DO content of local seawater was 340 μM (the average O_2 content of today's high-latitude surface water), only 7.1×10^{11} kg (= 0.71 km³) of seawater is required to form one microband. This corresponds to a 1.4 mixing ratio (R_{mix}) of hydrothermal fluid to local seawater, where R_{mix} is defined as:

$$R_{mix} = (mass\ of\ hydrothermal\ fluid)/$$
$$(mass\ of\ local\ seawater) \quad (17)$$

The typical R_{mix} values for the formation of volcanic-hosted BIFs are estimated to be between 0.01 and 100, because at R_{mix} <~0.01, positive Eu anomalies would not appear in BIFs, and at R_{mix} > ~100, negative Ce anomalies would not appear. At the minimum R_{mix} value of 0.01, the volume of local seawater required to provide the necessary amount of O_2 (2.5×10^8 moles O_2) to precipitate an average iron-rich microband is 100 km³ and the DO concentration is 2.5 μM (= 2.5×10^{-6} moles/kg H_2O). This DO value is, therefore, the *minimum* value for deep (>2.5 km) ocean water involved in the formation of volcanic-associated BIFs. It compares well with the *minimum* DO value of 18 μM for the photic zone of the ca. 2.7 Ga Hamersley Basin, which was estimated from molecular fossils of eukaryotes.

Atmospheric pO_2

The DO content of today's oceans continuously decreases from ~240 μM (i.e., the amount in equilibrium with an atmosphere of pO_2 = 0.21 atm at 25 °C) to ~50 μM at depths of ~500 m (i.e., the oxygen minimum depth) and then increases to ~170 µM at depths of ~4 km (Fig. 22). The DO decrease in the upper part of the oceans is due to DO consumption during the oxidation of dead organic matter as it sinks through the water column. The increase in DO toward the bottom of the oceans is due to the influx of a cold O_2-rich surface water mass from higher latitudes.

Using the equation by Sarmiento (1992) that relates the DO and PO_4^{3-} contents of the bottom ocean water to the DO and PO_4^{3-} contents of higher latitude water and the solubility of O_2, Lasaga and Ohmoto (2002) derived the following simple equation that relates atmospheric pO_2 to the DO value of deep water (DO_d):

$$DO_d (\mu M) = 340 \cdot (pO_2/ pO_2^0) - 172 \qquad (18)$$

where the pO_2/ pO_2^0 value is the atmospheric pO_2 value relative to the present atmospheric pO_2 level (PAL). Thus, if the atmospheric pO_2 falls below ~0.50 PAL (i.e., 50% of today's level), the DO_d value becomes zero (i.e., anoxic).

If the atmospheric pO_2 was as low as ~0.40 PAL, only the photic zone (areas <~100 m deep) could have been oxygenated (Fig. 22). If Earth's climate in the Archean was much warmer than today (e.g., Karhu and Epstein, 1986; Ohmoto and Felder, 1987), deep ocean circulation may have been the reverse of today's circulation; evaporated seawater may have sunk in the equatorial region and circulated toward higher-latitude areas. If this was the case, an atmospheric pO_2 level much higher than 0.50 PAL was required to maintain an oxygenated bottom water, because the solubility of O_2 in warmer and more saline water is less than that in cold water.

Local "oxygen oases" (e.g., Kasting, 2001) may have been possible in the photic zone, but not in deep oceans because there is no mechanism to generate O_2 below the photic zone, or to transport a small (<100-m-thick) O_2-rich water mass from the photic zone to the bottom waters through a deep anoxic

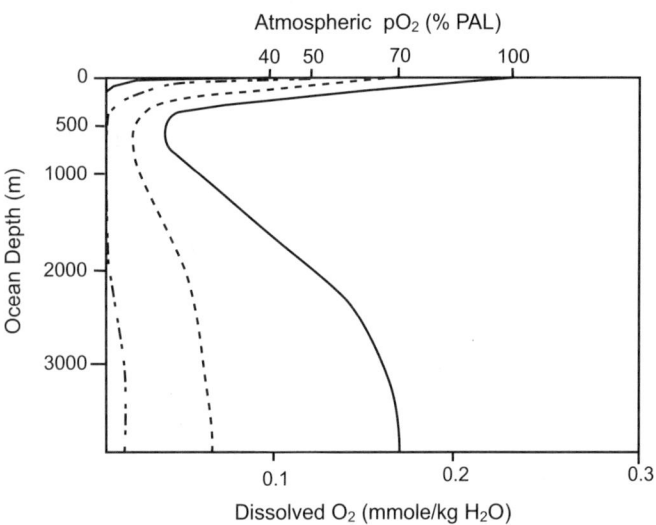

Figure 22. Relationships between the oceanic DO contents and atmospheric pO_2 levels, estimated from equations by Lasaga and Ohmoto (2002).

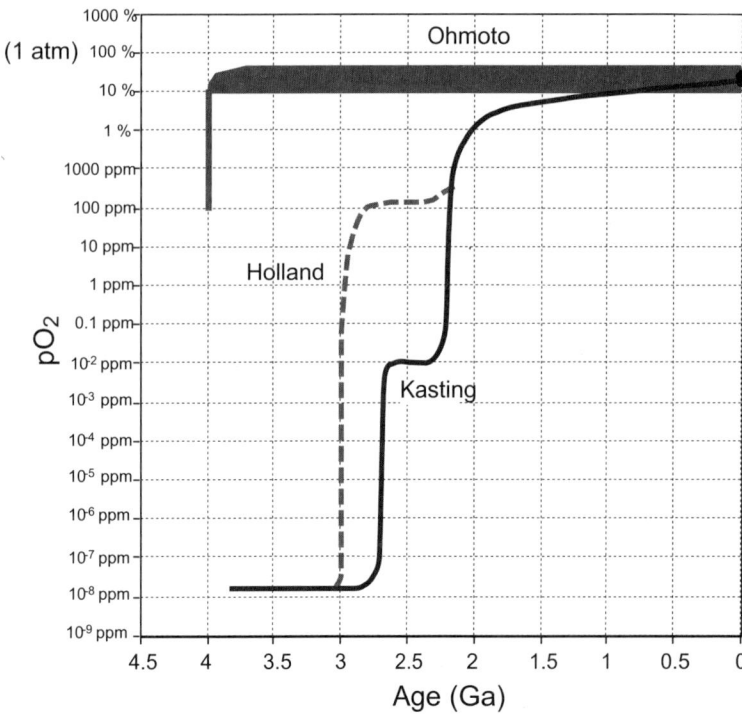

Figure 23. Proposed models for the evolution of atmospheric oxygen by Kasting (Kasting, 2001), Holland and his associates (Rye and Holland, 1998; Holland, 2002; Bekker et al., 2004), and Ohmoto and his associates (Ohmoto, 1997; Lasaga and Ohmoto, 2002).

water column. Therefore, in order to have >2.5 μM of DO below the photic zone to form an average Fe-oxide layer in a volcanic-associated BIF, the atmospheric pO_2 level must have been greater than ~0.50 PAL since ca. 3.8 Ga (i.e., the time of Isua BIF formation) (Fig. 23). Note that the minimum pO_2 requirement becomes higher if the ocean temperatures were higher than today, because the solubility of O_2 decreases with increasing temperature. Therefore, the Archean oceans must have been fully oxygenated, except in local (or regional) anoxic basins, much like today.

An important question that follows our model of a nearly constant atmospheric pO_2 level through geologic time is the O_2-controlling mechanism. According to Holland (1984), the atmospheric pO_2 level has been controlled by the relative fluxes of (1) O_2 production by the burial of organic matter that was produced primarily by oxygenic photoautotrophs, (2) O_2 consumption by reduced volcanic gases (e.g., H_2, CH_4), and (3) O_2 consumption by kerogen, sulfides and Fe^{2+}-bearing minerals during soil formation. Lasaga and Ohmoto (2002) have applied this principle to their numerical simulations of the geochemical cycles of carbon-hydrogen-oxygen-phosphate through the atmosphere-hydrosphere-biosphere-crust-mantle system. Their results suggest that the atmospheric pO_2 rose essentially to the present atmospheric level (1 PAL) within ~10 m.y. since the emergence of oxygenic photoautotrophs (even if the volcanic flux of reduced gas was as much as five times greater than today) and that the atmospheric pO_2 level has remained within ± 50% of the PAL throughout geologic history (Fig. 23). According to Lasaga and Ohmoto (2002), the main reasons that atmospheric pO_2 has maintained this level are two strong negative feedback systems: one between the burial flux of organic matter and atmospheric pO_2 value, and another between the soil O_2 consumption flux and atmospheric pO_2 value.

Atmospheric pCO_2 and pCH_4

Because nuclear fusion in the Sun has been increasing since its formation, the solar flux on early Earth must have been considerably less than today: ~70% of today's value at 4.0 Ga, 85% at 2.5 Ga, and 95% ca. 1 Ga (Kasting, 1987). To compensate for this lower solar luminosity and maintain the liquid oceans, the concentrations of greenhouse gases must have been much higher than today. Assuming CO_2 was the only greenhouse gas besides water vapor, Owen et al. (1979) and Kasting (1987) calculated that the atmospheric pCO_2 was as high as ~1 atm (= 3000 PAL, where 1 PAL = 350 ppm) at 4.5 Ga, then gradually decreased to 0.1 atm (~300 PAL) at 2.5 Ga, and to 0.01 atm (~30 PAL) at 1.0 Ga (curve A, Fig. 24).

Rye et al. (1995) proposed that siderite was absent in pre-2.2 Ga paleosols because the atmospheric pCO_2 was less than the equilibrium pCO_2 value for the greenalite + siderite assemblage. From thermodynamic calculations, they concluded that the atmospheric pCO_2 value must have been lower than that calculated from previous climatic models, and suggested that CH_4, as well as CO_2, was a major greenhouse gas prior to ca. 2.2 Ga. Subsequently, a revised climatic model (Pavlov et al., 2001a, 2001b) suggested values of ~1000 ppm CH_4 and ~2500 ppm CO_2 for the atmosphere at 2.8 Ga (curve B, Fig. 24). By comparison, today's atmosphere contains only ~1 ppm CH_4 and ~350 ppm CO_2.

The atmosphere cannot contain both high O_2 and CH_4 concentrations, because together they react to form CO_2 and H_2O by the following photochemical reaction:

$$2O_2 + CH_4 \Rightarrow CO_2 + 2H_2O \quad (19)$$

Therefore, a methane-rich atmosphere is compatible with an anoxic Archean atmosphere. However, serious flaws exist in Rye et al.'s (1995) assumption that siderite did not form in Archean surface environments, because it is an abundant mineral in BIFs. For the formation of siderite, all of the following conditions must be satisfied: (1) a low pO_2 (<10^{-60} atm); (2) a very high pCO_2 (>$10^{-2.5}$ atm); (3) a slightly acidic to near neutral pH (~5.5 to ~7.5); and (4) a high concentration of ΣFe^{2+} in water (~0.1 to ~100 ppm) (Figs. 19A, 19B). Such a low pO_2 value only occurs when anaerobic organisms produce a substantial amount of H_2 in organic C-rich environments (e.g., subsurface anoxic water) by:

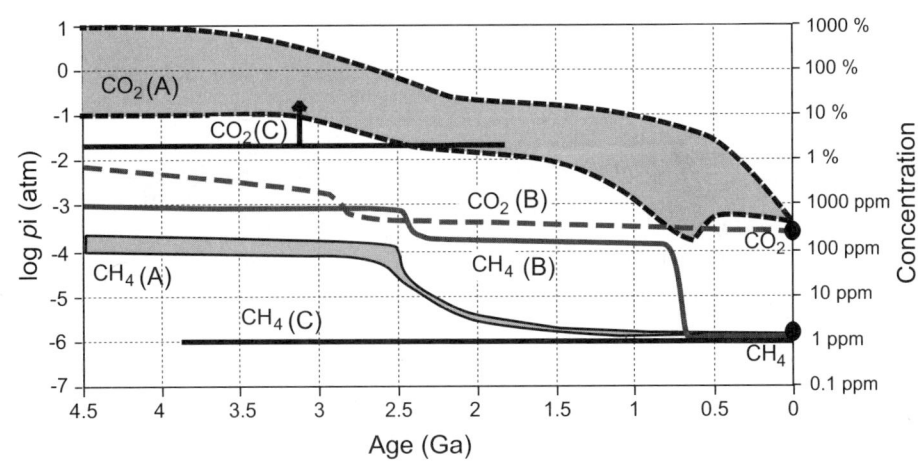

Figure 24. Proposed models for the evolution of atmospheric pCO_2 and pCH_4. Curves (A) are from Kasting (1987), (B) from Pavlov et al. (2001a, 2001b), and (C) from Ohmoto et al. (2004).

$$CH_2O + H_2O \Rightarrow CO_2 + 2H_2 \qquad (20)$$

In fact, siderite appears to have formed only in organic-rich environments throughout geologic history (Ohmoto et al., 2004). But there are distinct differences in the occurrence and carbon isotopic compositions between siderite in Phanerozoic sediments and in pre–1.8 Ga sedimentary sequences. The former typically occurs as minor, disseminated grains with variable $\delta^{13}C$ values (−25‰ to +15‰) (Fig. 25). In contrast, siderite in pre–1.8 Ga sedimentary rocks is very abundant, occurring as an important constituent of BIFs and as Fe-rich massive carbonate beds (e.g., the 2.6 Ga Wittenoon Dolomite in the Hamersley Basin). $\delta^{13}C$ values of these rocks range from +3‰ to −15‰, but are more commonly between −5‰ and 0‰ (Fig. 25), suggesting that 10%–20% of the ΣCO_3^{2-} used to form the siderites came from the decomposition of organic matter by SRB, and 80%–90% of the ΣCO_3^{2-} was from normal seawater that was in equilibrium with the atmosphere. Using these values, together with the thermodynamic evaluation of the conditions for siderite formation, Ohmoto et al. (2004) constrained the pre–1.8 Ga atmospheric pCO_2 level to $>10^{-1.4 \pm 0.2}$ atm (>100 PAL) if T = 25–50 °C and $SiO_{2\,(aq)}$ < 100 ppm (curve C, Fig. 24). With such a high pCO_2, the presence of >10 ppm CH_4 would have made the surface temperatures much warmer than 50 °C. Therefore, the pre–1.8 Ga atmosphere was most likely CO_2-rich and CH_4-poor; a low pCH_4 (<10 ppm) is compatible with a high pO_2 atmosphere.

Oceanic Sulfur Chemistry

The sulfur chemistry of the ocean is closely linked to the evolution of SRB and the atmospheric pO_2 level. Modern oceans are sulfate-rich, with an average concentration of 28 mM SO_4^{2-} (900 ppm S), because 30%–50% of the total SO_4^{2-} flux to the oceans comes from the oxidative weathering of pyrite; the sulfate is in turn removed from the oceans as biogenic sulfides and gypsum. Under an anoxic atmosphere, the SO_4^{2-} content of the ocean water is expected to be much lower than the present value, although there is no consensus as to how low Archean seawater sulfate contents could have been. For example, Walker and Brimblecombe (1985) have estimated that the SO_4^{2-} in Archean oceans was generated only by photochemical reactions of volcanic SO_2 and its concentration was less than ~1/30 of the present value (i.e., <~1 mM); but they also suggested that the SO_4^{2-} content could have been as high as 1/3 of the present value.

A recent proposal by Canfield and associates (Canfield and Teske, 1996; Canfield and Raiswell, 1999; Canfield et al., 2000; Bjerrum and Canfield, 2002; Habicht et al., 2002) suggests that the oceanic SO_4^{2-} content remained ~200 μM (i.e., <1/100 of the present level), except in local evaporitic basins, until ca. 2.2 Ga, then gradually increased to ~10 mM ca. 800 Ma (Fig. 26). This proposal is based on an interpretation that the small difference in $\delta^{34}S$ values between sulfates and biogenic sulfides ($\Delta^{34}S = \delta^{34}S_{sulfate} - \delta^{34}S_{pyrite} = <10‰$), which characterizes Archean sedimentary rocks, only occurs when the sulfate concentration

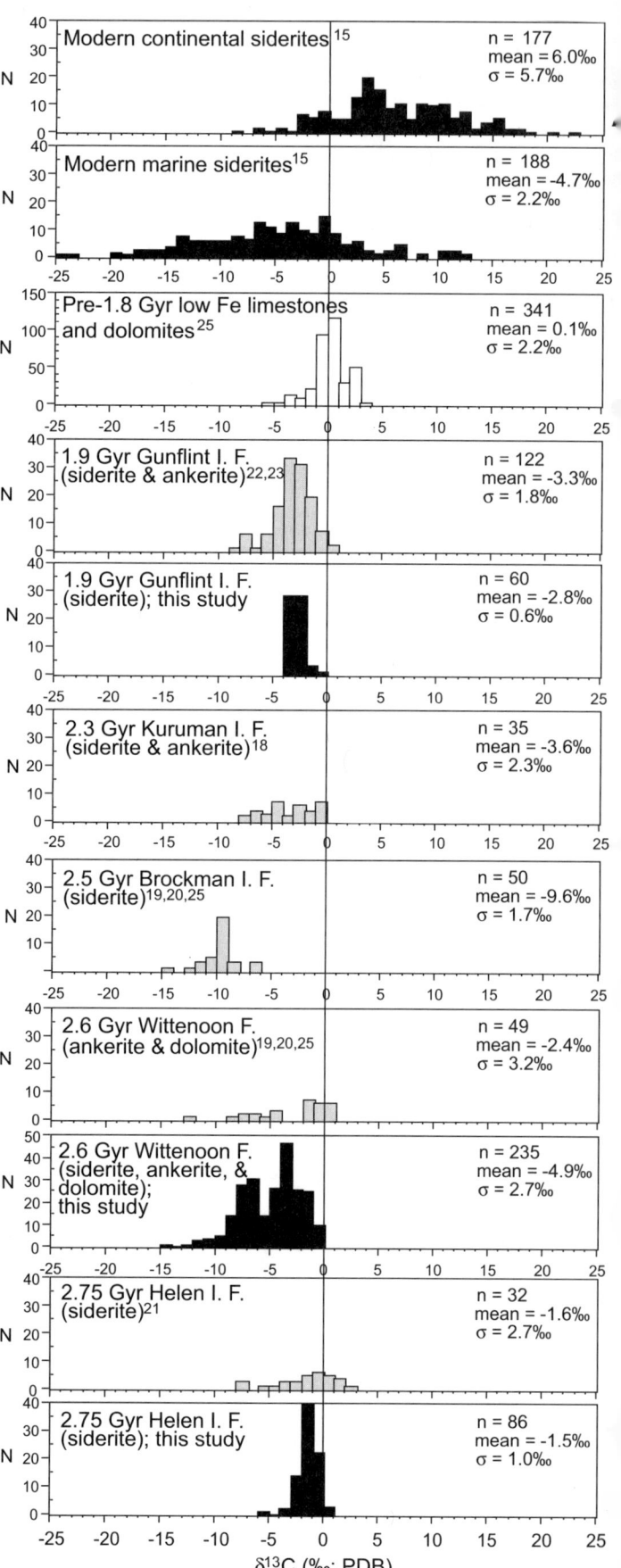

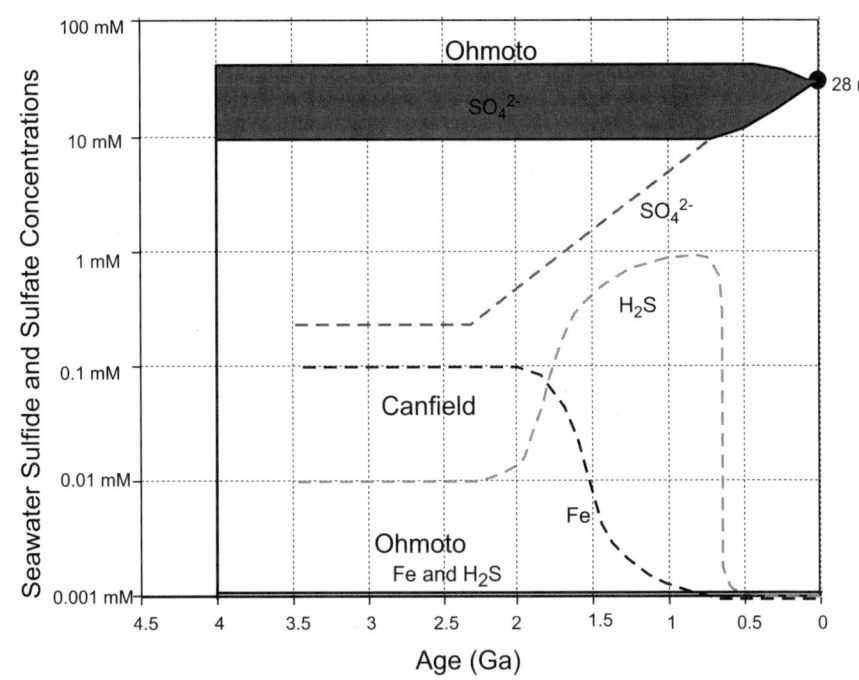

Figure 26. Evolution curves for the sulfate, sulfide, and Fe²⁺ contents of the oceans, as suggested by Canfield and his associates (Canfield and Raiswell, 1999; Bjerrum and Canfield, 2002; Habicht et al., 2002) and Ohmoto (2004).

is <200 µM; $\Delta^{34}S$ values >25‰ (typical values in Phanerozoic sedimentary rocks) occur when the sulfate concentration is > 200 µM. Ohmoto (1992), however, has suggested that the small $\Delta^{34}S$ values are not necessarily an indicator of low-sulfate oceans because $\Delta^{34}S$ values are also influenced by temperature and the relative rates of sulfate supply versus sulfate reduction in a water column and underlying sediments.

Compared to sulfur isotope values, a more reliable indicator of seawater sulfate concentration is the relationship between the sulfide and organic C contents of sediments since SRB produce pyrite through the following reaction:

$$8SO_4^{2-} + 16H^+ + 15C_{NR}\text{-}CH_2O + 4Fe(OH)_3$$
$$\Rightarrow 15C_{NR} + 4FeS_2 + 15CO_2 + 29H_2O \quad (21)$$

where CH_2O refers to the organic molecules metabolizable by SRB and C_{NR} to non-reactive organic carbon molecules (i.e., residual organic C in sediments). Reaction (21) indicates that the pyrite content of a sediment increases with an increasing sulfate content of water and that a positive correlation may exist between the contents of pyrite and organic C in sediments if all the pyrite was formed by bacterial sulfate reduction. However, because the ratio of C_{NR} to CH_2O depends on the environment and type of dominant organic matter (e.g., remnants of cyanobacteria, eukaryotes), the ratio of sulfide S to organic C may vary significantly among sediments (Ohmoto et al., 1990). For example, the sulfide S content and S/C ratio of average normal marine sediments are ~0.2 wt% and 0.4, respectively, but those of freshwater lake sediments are ~0.05 wt% and 0.03, respectively. The S contents and S/C ratios of marine sediments in euxinic basins can be as high as ~10 wt% and ~1, respectively (Ohmoto et al., 1990). Therefore, if the sulfate concentration in the Archean oceans was <200 µM, the sulfide contents in Archean shales are expected to be uniformly <0.1 wt%, and the S/C ratios <<0.1, provided the sulfides were all biogenic.

As mentioned earlier, pyrite-rich black shales (i.e., sulfide BIFs) are the most common hanging wall (i.e., younger) rock type of all oxide/carbonate BIFs. This common temporal relationship between oxide/carbonate BIFs and black shales generates the following questions: (1) Was the deepening of BIF-forming basins to accumulate black shales caused by tectonic subsidence of the region or by a global rise in sea level? (2) Was the development of an euxinic water body the cause or consequence of the cessation of oxide BIF-forming processes? (3) Does the chemistry of BIF-associated black shales reflect global or local seawater chemistry?

For example, Poulton et al. (2004) have concluded that the pyritic black shales (e.g., the Rove Formation) overlying the ca. 1.85 Ga Gunflint IF (Fig. 10B) are excellent evidence that the global oceans became SO_4^{2-}- and H_2S-rich for the first time ca. 1.8 Ga, and that the creation of such oceans was the cause of BIF cessation worldwide. However, if the Rove Formation is evidence for global SO_4^{2-}- and H_2S-rich oceans, then other BIF-associated pyritic black shales, such as the 2.7 Ga Jeerinah Formation

Figure 25. Comparison of the carbon isotopic compositions of carbonate in various rocks. (A, B) Disseminated siderite in modern sediments. (C) Pre–1.8 Ga Fe-poor limestones and dolomites. (D–K) Pre–1.8 Ga Fe-rich carbonates (Ohmoto et al., 2004).

(Fig. 10A) and 2.5 Ga Mount McRae Shale of the Hamersley Basin, must also be evidence for globally SO_4^{2-}- and H_2S-rich oceans. But such an argument would create a serious problem for the Cloud-Holland-Canfield Archean ocean model because the oceans could not become rich in both Fe^{2+} and H_2S at the same time, owing to iron sulfide precipitation.

We interpret the S-C relationships of the BIF-associated Archean black shales (Fig. 10A) as evidence for sulfate-rich oceans, and the occurrences of black shales to the periodic development of local (regional) euxinic basins, rather than globally anoxic oceans. The alternations of oxide- (and/or carbonate)-type BIFs and pyritic black shales in the Hamersley Basin (and other BIF-forming basins) are interpreted as the periodic development of hot brine pools, which created unsuitable environments for SRB and other microbes, much like today's hot (T > 60 °C) brine pool in the Atlantis II Deep, Red Sea (e.g., Kaplan et al., 1969).

Atmospheric Sulfur Cycle

The presence of the mass independent fractionation of sulfur isotopes (MIF-S) in some pre–2.1 Ga sedimentary rocks and the absence of MIF-S in younger rocks, which were first discovered by Farquhar et al. (2000), have been regarded by many recent investigators (Kasting and Siefert 2002; Bekker et al., 2004) as the "smoking gun" for the dramatic change from an anoxic to an oxic atmosphere ca. 2.2 Ga. This is because the MIF-S has been linked to photochemical reactions of volcanic SO_2 in the absence of an ozone shield to produce S^0 with large positive $\Delta^{33}S$ values (>50‰) and SO_4^{2-} with negative $\Delta^{33}S$ values (<–10‰) (where $\Delta^{33}S = \delta^{33}S - 0.515\, \delta^{34}S$); these $\Delta^{33}S$ values are characteristically within the range of 0‰ ± 0.3‰ in Phanerozoic sediments.

Ono et al. (2003) recognized the largest variations in $\Delta^{33}S$ values (–2‰ to +8‰) in pyrite crystals from pyrite-rich black shales (sulfide BIFs) in the Hamersley Basin (e.g., the 2.5 Ga Mount McRae Shale and 2.7 Ga Jeerinah Formation; Fig. 9). However, Bekker et al. (2004) found no evidence of MIF-S (i.e., $\Delta^{33}S = 0‰ \pm 0.3‰$) in the 2.35 Ga Timeball Hill Formation, which hosts pisolitic IFs, in South Africa. These data led Bekker et al. (2004) to conclude that the "Great Oxidation Event" (the change from an anoxic to oxic atmosphere) occurred ca. 2.35 Ga.

However, there are several very serious problems in linking the presence or absence of MIF-S in *geologic* samples to the atmospheric oxygen level: (1) The S^0 produced by the UV radiation of SO_2 in laboratory experiments (Farquhar et al., 2001) has very large positive and variable $\Delta^{33}S$ values, but also very *negative* and variable $\delta^{34}S$ values relative to SO_2. In contrast, geologic samples with positive $\Delta^{33}S$ values show only *positive* $\delta^{34}S$ values. (2) Savarino et al. (2003) reported the discovery of the presence of MIF-S ($\Delta^{33}S = +0.67‰$ and $-0.50‰$) in volcanic ashes associated with the violent eruptions of Mount Pinatubo and an unknown volcano, but also the absence of MIF-S in ashes associated with minor eruptions. (3) Watanabe et al. (2005) found that the MIF-S is absent (i.e., $\Delta^{33}S = 0‰ \pm 0.3‰$) in syngenetic and diagenetic pyrite crystals in all shale samples (>40 analyses) from the 2.76 Ga Hardey Formation (lacustrine) and the 3.0 Ga Mosquito Creek Formation (marine) in the Pilbara district, Australia. (4) Tachibana et al. (2004) have also found no evidence of MIF-S in the 2.45 Ga Livingstone Creek Formation in Elliot Lake, Ontario, Canada. (Note that the Livingstone Creek Formation underlies uraniferous quartz-conglomerate beds that were previously used as evidence for an anoxic atmosphere [e.g., Holland, 1994]).

These recent findings suggest the following possible interpretations: (1) the Archean atmosphere fluctuated between oxic and anoxic; (2) the atmosphere has been oxic since ca. 3.8 Ga, and the MIF-S record indicates periods of violent volcanic eruptions, which ejected large amounts of SO_2 above the stratosphere; or (3) large MIF-S values in some geologic samples were caused by mechanisms other than atmospheric photochemical reactions. Possibilities (1) and (2) should be considered, if atmospheric photochemical reactions are the only cause of MIF-S. Possibility (3) must be considered because of (*a*) the recognition of strong correlations between the MIF-S values and the degree of hydrothermal alteration in samples from the 2.5 Ga Mount McRae Shale and the 2.7 Ga Jeerinah Formation that are highly organic carbon rich (Fig. 27); and (*b*) the creation of significant MIF-S ($\Delta^{33}S = \sim 0.5‰$) during thermochemical reduction of sulfate by amino acids (Watanabe et al., 2006). Clearly, more investigations are needed to understand the causes of MIF-S in geologic samples before linking MIF-S to atmospheric chemistry.

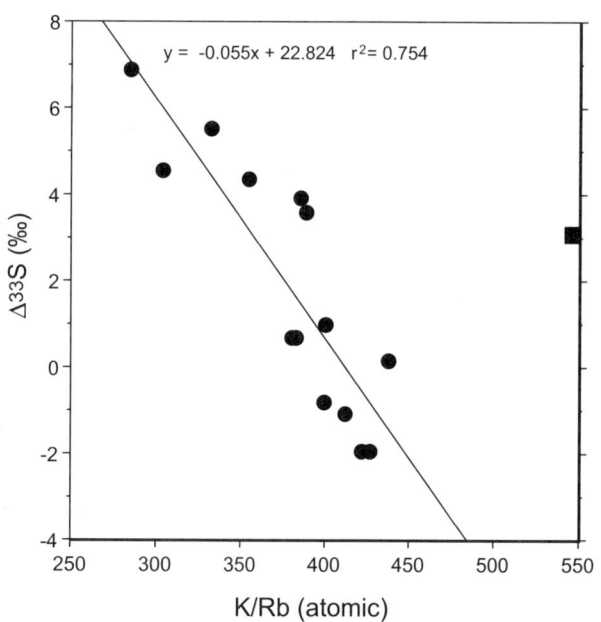

Figure 27. The relationships between the magnitude of MIF-S isotopes ($\Delta^{33}S$) and K/Rb ratios of a 22 m core section of the 2.55 Ga Mount McRae Shale, measured by Ono et al. (2003) and this study, respectively. Other geochemical data (e.g., Zn contents) and mineralogy of the same core sample indicate that trends of increasing $\Delta^{33}S$ value and decreasing K/Rb ratio accompany increasing degrees of hydrothermal alteration.

The Biosphere

Our suggestion of the development of an oxygenated atmosphere and oceans by ca. 3.8 Ga implies that oxygenic photoautotrophs (i.e., cyanobacteria or their ancestors) had already evolved. This suggestion is consistent with others by Schidlowski and Aharon (1992) and Rosing (1999) based on the carbon isotope data of Isua sedimentary rocks. Our suggestion is also consistent with the discovery of Ce anomalies in Isua BIFs (Kato et al., this volume).

The biomarkers, carbon isotopic compositions of kerogen, and S/C relationships of pyritic black shales (sulfide BIFs) in the Hamersley Basin suggest that methanogens and methanotrophs, as well as SRB, cyanobacteria, and eukaryotes, have been active since at least ca. 2.7 Ga. The recent discovery by Buick (2005) of biomarkers of cyanobacteria, methanotrophs, and eukaryotes in 3.2 Ga sedimentary rocks from Pilbara, Australia, indicates the very early divergence of major organisms, and the importance of biological feedbacks in regulating oceanic and atmospheric chemistry.

BIF Distribution through Geologic Time

Because BIFs are found in rocks of all ages (Tables 1, 2), we can unequivocally conclude that BIFs are not indicators of an anoxic atmosphere or anoxic oceans, as many previous researchers suggested. Instead, the formation of oxide- and sulfide-geologic history is an excellent indicator that the atmospheric O_2 and oceanic SO_4^{2-} levels have been essentially constant since ca. 3.8 Ga. The higher abundance of siderite-rich BIFs in pre–1.8 Ga BIFs, however, was probably the result of a higher pCO_2 (>100 PAL) atmosphere.

Isley (1995) and Isley and Abbott (1999) have recognized that the formation of BIFs in geologic history was episodic and coincided with major mantle plume events. The close temporal associations of BIFs with VMSDs, greenstone belts, oceanic rift systems, and mantle plumes (Figs. 28A–D) further suggest that BIFs were related to regional-scale submarine hydrothermal processes. The scarcity of BIFs, VMSDs, mantle plumes, and greenstones during the ca. 1.8–1.0 Ga period was not because the oceans and/or atmosphere became oxygenated, as postulated by the popular model, but because of the change in mantle dynamics. During this period, a supercontinent existed, and large marine rift systems, which were necessary for the formation of VMSDs and BIFs, were scarce.

BIFs became generally smaller and less abundant after ca. 1.8 Ga. These changes probably reflect the fact that mantle plumes became smaller and less frequent, and the temperatures of the mantle and core decreased; such changes would have resulted in smaller and short-lived submarine hydrothermal systems.

In summary, BIFs are excellent indicators that, since ca. 3.8 Ga, (1) the redox structures of the atmosphere (pO_2, pCH_4) and oceans (concentrations of Fe^{2+}, H_2S, and SO_4^{2-}) have been essentially the same; (2) the major microbial constituents in the oceans (e.g., cyanobacteria, sulfate reducers, methanogens) have been essentially the same; (3) the atmospheric pCO_2 and oceanic concentration of ΣCO_3^{2-} have continuously decreased; (4) the pHs of rain- and ocean waters have continuously increased; and (5) the global heat flux from Earth's interiors and submarine volcanic activity has continuously decreased.

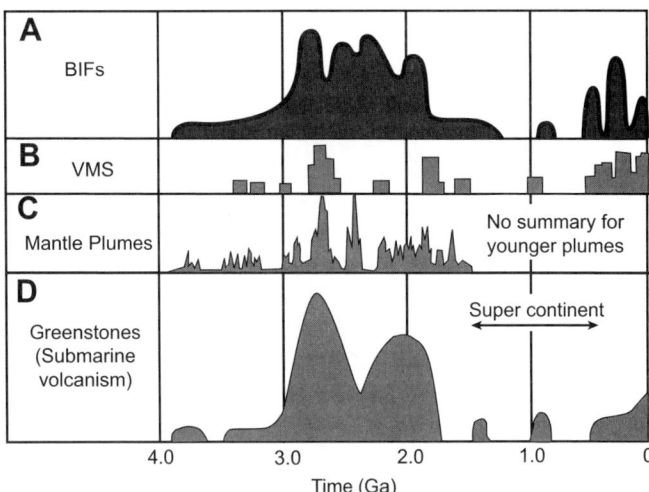

Figure 28. Schematic illustrations of the temporal distributions of (A) BIFs, (B) VMSDs, (C) global plumes, and (D) greenstones. (A) and (B) are modified after Barley and Groves (1992) and Ohmoto (1996) with additional data from literature. (C) is from Isley (1995) and Condie (2001). (D) is from Green (1992) and Condie (2002).

ACKNOWLEDGMENTS

This paper represents our work over the past 15 years or so. We express our sincere appreciation to the following investigators who have helped us in the field and laboratories, and in discussions: Munetomo and Yoko Nedachi of Kagoshima University; Mark Barley, Brian Krapez, and April Pickard of the University of Western Australia; Arthur Hickman, Al Trendall, and Richard Morris of the Geological Survey of Western Australia; Nic Beukes and Jens Gutzmer of Rand Afrikaans University; Kazumasa Kumazawa, Ritsuko Murata, Takahiro Hanamuro, Kaoru Uyeda, Hiroaki Fujimaki, Toshio Nagase, and Shuji Kojima of Tohoku University; Takushi Yokoyama of Kyushu University; and Michael Bau, Denny Walizer, Katya Bazilevskaya, Kate Spangler, Tsubasa Otake, and David Bevacqua of the Pennsylvania State University. This manuscript greatly benefited from critical reviews of an earlier manuscript by Jim Franklin, Bob Rye, Cornelius Klein, Rob Kerrich, Art Rose, and especially Steve Kesler (a perfectionist). This work was supported by the Japanese Ministry of Education (Grant 03102002), the National Science Foundation (EAR9706279, EAR-0229556), the NASA Astrobiology Institute (NCC2-1057; CA#NNA04CC06A), and the NASA Exobiology Program (CA#NNG04GK00G).

REFERENCES CITED

Abrams, C.E., and McConnell, K.I., 1984, Geologic setting of volcanogenic base and precious metal deposits of the West Georgia Piedmont: A multiply deformed metavolcanic terrain: Economic Geology and the Bulletin of the Society of Economic Geologists, v. 79, p. 1521–1539.

Adamson, R.G., and Teichmann, R.F.H., 1986, The Matchless cupreous pyrite deposit, South West Africa/Namibia, in Anhaeusser, C.R., and Maske, S., eds., Mineral deposits of South Africa: Johannesburg, Geological Society of South Africa, v. 2, p. 1755–1760.

Ahn, J.H., and Buseck, P.R., 1990, Hematite nanospheres of possible colloidal origin from a Precambrian banded iron formation: Science, v. 250, p. 111–113.

Alibert, C., and McCulloch, M.T., 1993, Rare earth element and neodymium isotopic compositions of the banded iron-formations and associated shales from Hamersley, western Australia: Geochimica et Cosmochimica Acta, v. 57, p. 187–204, doi: 10.1016/0016-7037(93)90478-F.

Allen, R.L., Lundström, I., Ripa, M., and Christofferson, H., 1996, Facies analysis of a 1.9 Ga, continental margin, back-arc, felsic caldera province with diverse Zn-Pb-Ag-(Cu-Au) sulfide and Fe oxide deposits, Bergslagen region, Sweden: Economic Geology and the Bulletin of the Society of Economic Geologists, v. 91, p. 979–1008.

Anbar, A.D., and Holland, H.D., 1992, The photochemistry of manganese and the origin of banded iron formations: Geochimica et Cosmochimica Acta, v. 56, p. 2595–2603, doi: 10.1016/0016-7037(92)90346-K.

Appel, P.W.U., 1987, Geochemistry of the early Archaean Isua iron-formations, West Greenland, in Appel, P.W.U., and LaBerge, G.L., eds., Precambrian iron-formations: Athens, Theophrastus Publications, p. 31–67.

Appel, P.W.U., and LaBerge, G.L., eds., 1987, Precambrian iron-formations: Athens, Theophrastus Publications, 674 p.

Ayer, J., Amelin, Y., Corfu, F., Kamo, S., Ketchum, J., Kwok, K., and Trowell, N., 2002, Evolution of the southern Abitibi greenstone belt based on U-Pb geochronology: Autochthonous volcanic construction followed by plutonism, regional deformation and sedimentation: Precambrian Research, v. 115, p. 63–95, doi: 10.1016/S0301-9268(02)00006-2.

Bayley, R.W., and James, H.L., 1973, Precambrian iron-formations of the United States: Economic Geology and the Bulletin of the Society of Economic Geologists, v. 68, p. 934–959.

Barley, M.E., 1993, Volcanic, sedimentary and tectonostratigraphic environments of the ~3.46 Ga Warrawoona Megasequence: A review: Precambrian Research, v. 60, p. 47–67, doi: 10.1016/0301-9268(93)90044-3.

Barley, M.E., and Groves, D.I., 1992, Supercontinent cycles and the distribution of metal deposits through time: Geology, v. 20, p. 291–294, doi: 10.1130/0091-7613(1992)020<0291:SCATDO>2.3.CO;2.

Barley, M.E., Pickard, A.L., and Sylvester, P.J., 1997, Emplacement of a large igneous province as a possible cause of banded iron formation 2.45 billion years ago: Nature, v. 385, p. 55–58, doi: 10.1038/385055a0.

Barrett, T.J., and Jambor, J.L., eds., 1988, Seafloor hydrothermal mineralization: Canadian Mineralogist, v. 26, p. 429–888.

Bazilevskaya, E., 2004, Ferric hydroxide formation in silica-rich solutions [M.S. thesis]: University Park, Pennsylvania State University, 105 p.

Beeson, R., 1990, Broken Hill-type lead-zinc deposits—An overview of their occurrence and geological setting: Transactions of the Canadian Institute of Mining and Metallurgy, section B, v. 99, p. 163–175.

Bekker, A., Holland, H.D., Wang, P.-L., Rumble, D., Stein, H.J., Hannah, J.L., Coetzee, L.L., and Beukes, N.J., 2004, Dating the rise of atmospheric oxygen: Nature, v. 427, p. 117–120, doi: 10.1038/nature02260.

Bengtson, S., 1994, The advent of animal skeleton, in Bengtson, S., ed., Early life on Earth: Nobel Symposium 84: New York, Columbia University Press, p. 412–425.

Beukes, N.J., 1973, Precambrian iron-formations of South Africa: Economic Geology and the Bulletin of the Society of Economic Geologists, v. 68, p. 960–1004.

Beukes, N.J., 1983, Paleoenvironmental setting of iron-formations in the depositional basin of the Transvaal Supergroup, South Africa, in Trendall, A.F., and Morris, R.C., Iron-formation: Facts and problems: Amsterdam, Elsevier, p. 131–209.

Bjerrum, C.J., and Canfield, D.E., 2002, Ocean productivity before about 1.9 Gyr ago limited by phosphorus adsorption onto iron oxides: Nature, v. 417, p. 159–162, doi: 10.1038/417159a.

Blake, T.S., 2001, Cyclic continental mafic tuff and flood basalt volcanism in the Late Archaean Nullagine and Mount Jope supersequences in the eastern Pilbara, Western Australia: Precambrian Research, v. 107, p. 139–177, doi: 10.1016/S0301-9268(00)00135-2.

Bonatti, E., Fisher, D.E., Joensuu, O., Rydell, H.S., and Beyth, M., 1972, Iron-manganese-barium deposit from the Northern Afar Rift (Ethiopia): Economic Geology and the Bulletin of the Society of Economic Geologists, v. 67, p. 717–730.

Bowing, R.J., 1989, Rare earth and other geochemical studies of Archean banded iron formation; Sherman and Adams mines, Ontario [Ph.D. thesis]: Hamilton, Ontario, Canada, McMaster University, 399 p.

Braterman, P.S., and Cairns-Smith, A.G., 1987, Iron photoprecipitation and the genesis of the banded iron-formations, in Appel, P.W.U., and LaBerge, G.L., eds., Precambrian iron-formations: Athens, Theophrastus Publications, p. 215–245.

Brocks, J.J., Logan, G.A., Buick, R., and Summons, R.E., 1999, Archean molecular fossils and the early rise of eukaryotes: Science, v. 285, p. 1033–1036, doi: 10.1126/science.285.5430.1033.

Brocks, J.J., Buick, R., Summons, R.E., and Logan, G.A., 2003, A reconstruction of Archean biological diversity based on molecular fossils from the 2.78 to 2.45 billion-year-old Mount Bruce Supergroup, Hamersley Basin, Western Australia: Geochimica et Cosmochimica Acta, v. 67, p. 4321–4335, doi: 10.1016/S0016-7037(03)00209-6.

Brown, M.C., Oliver, N.H.S., and Dickens, G.R., 2004, Veins and hydrothermal fluid flow in the Mt. Whaleback iron ore district, eastern Hamersley Province, Western Australia: Precambrian Research, v. 128, p. 441–474, doi: 10.1016/j.precamres.2003.09.013.

Buick, R., 2005, Paleontological and geochemical records of early life in Archean rocks from the Pilbara Craton, Australia: 2005 Japan Earth and Planetary Science Joint Meeting, Session P074, abstracts 036.

Canfield, D.E., and Teske, A., 1996, Late Proterozoic rise in atmospheric oxygen concentration inferred from phylogenetic and sulphur-isotope studies: Nature, v. 382, p. 127–132, doi: 10.1038/382127a0.

Canfield, D.E., and Raiswell, R., 1999, The evolution of the sulfur cycle: American Journal of Science, v. 299, p. 697–723.

Canfield, D.E., Habicht, K.S., and Thamdrup, B., 2000, The Archean sulfur cycle and the early history of atmospheric oxygen: Science, v. 288, p. 658–661, doi: 10.1126/science.288.5466.658.

Cathles, L.M., 1983, An analysis of the hydrothermal system responsible for massive sulfide deposition in the Hokuroku Basin of Japan, in Ohmoto, H., and Skinner, B.J., eds., The Kuroko and related volcanic massive sulfide deposits: Economic Geology Monograph 5, p. 439–487.

Clemmey, H., and Badham, N., 1982, Oxygen in the Precambrian atmosphere: An evaluation of the geological evidence: Geology, v. 10, p. 141–146, doi: 10.1130/0091-7613(1982)10<141:OITPAA>2.0.CO;2.

Clifford, J.A., Ryan, P., and Kucha, H., 1986, A review of the geological setting of the Tynagh Orebody, Co. Galway, in Andrew, C.J., Crowe, R.W.A., Finlay, S., Pennell, W.M., and Pyne, J.F., eds., Geology and genesis of mineral deposits in Ireland: Dublin, Irish Association for Economic Geology, p. 419–439.

Cloud, P.E., 1968, Atmospheric and hydrospheric evolution on the primitive earth: Science, v. 160, p. 729–736.

Cloud, P.E., 1973, Paleoecological significance of the banded iron-formation: Economic Geology and the Bulletin of the Society of Economic Geologists, v. 68, p. 1135–1143.

Cloud, P.E., 1978, Cosmos, Earth, and man: A short history of the universe: New Haven, Connecticut, Yale University Press, 372 p.

Cole, M.J., and Klein, C., 1981, Banded iron-formations through much of Precambrian time: Journal of Geology, v. 89, p. 169–183.

Collao, S., Alfaro, G., and Hayashi, K., 1990, Banded iron formation and massive sulfide orebodies, south-central Chile: Geologic and isotopic aspects, in Fontboté, L., Amstuz, G.C., Cardozo, M., Cedillo, E., and Frutos, J., eds., Stratabound ore deposits in the Andes: Society for Geology Applied to Mineral Deposits Special Publication 8, p. 209–219.

Condie, K.C., 2001, Mantle plumes and their record in earth history: Cambridge, UK, Cambridge University Press, 306 p.

Condie, K.C., 2002, Continental growth during a 1.9-Ga superplume event: Journal of Geodynamics, v. 34, p. 249–264, doi: 10.1016/S0264-3707(02)00023-6.

Costa, U.R., Barnett, R.L., and Kerrich, R., 1983, The Mattagami Lake Mine Archean Zn-Cu sulfide deposit, Quebec: Hydrothermal coprecipitation of talc and sulfides in a sea-floor brine pool—Evidence from geochemistry, $^{18}O/^{16}O$, and mineral chemistry: Economic Geology and the Bulletin of the Society of Economic Geologists, v. 78, p. 1144–1203.

Davidson, G.J., 1992, Hydrothermal geochemistry and ore genesis of sea-floor volcanogenic copper-bearing oxide ores: Economic Geology and the Bulletin of the Society of Economic Geologists, v. 87, p. 889–912.

Davidson, G.J., Stolz, A.J., and Eggins, S.M., 2001, Geochemical anatomy of silica iron exhalites: Evidence for hydrothermal oxyanion cycling in response to vent fluid redox and thermal evolution (Mt. Windsor Subprovince, Australia): Economic Geology and the Bulletin of the Society of Economic Geologists, v. 96, p. 1201–1226.

Davis, D.W., and Lin, S., 2003, Unraveling the geologic history of the Hemlo Archean gold deposits, Superior province, Canada: A U-Pb geochronological study: Economic Geology and the Bulletin of the Society of Economic Geologists, v. 98, p. 51–67.

Davy, R., 1985, The mineralogy and composition of a core which intersects the Marra Mamba iron formation and the Roy Hill Shale Member: Geological Survey of Western Australia Record, v. 6, 60 p.

Degens, E.T., and Ross, D.A., eds., 1974, The Black Sea—Geology, chemistry, and biology: Tulsa, Oklahoma, American Association of Petroleum Geologists, Memoir 20, 633 p.

Derry, L.A., and Jacobsen, S.B., 1990, The chemical evolution of Precambrian seawater: Evidence from REEs in banded iron formations: Geochimica et Cosmochimica Acta, v. 54, p. 2965–2977, doi: 10.1016/0016-7037(90)90114-Z.

DeWitt, E., 1979, New data concerning Proterozoic volcanic stratigraphy and structure in central Arizona and its importance in massive sulfide exploration: Economic Geology and the Bulletin of the Society of Economic Geologists, v. 74, p. 1371–1382.

Dimroth, E., 1976, Aspects of the sedimentary petrology of cherty iron-formation, in Wolf, K.H. ed., Au, U, Fe, Mn, Hg, Sb, W and P deposits: Amsterdam, Elsevier, Handbook of Strata-bound and Stratiform Ore Deposits, v. 7, p. 203–254.

Dimroth, E., and Kimberley, M.M., 1976, Precambrian atmospheric oxygen: Evidence in the sedimentary distributions of carbon, sulfur, uranium, and iron: Canadian Journal of Earth Sciences, v. 13, p. 1161–1185.

Dorr, J., II, and Van, N., 1973, Iron-formation in South Africa: Economic Geology and the Bulletin of the Society of Economic Geologists, v. 68, p. 1005–1022.

Drever, J.I., 1974, Geochemical model for the origin of Precambrian Banded Iron formations: Geological Society of America Bulletin, v. 85, p. 1099–1106, doi: 10.1130/0016-7606(1974)85<1099:GMFTOO>2.0.CO;2.

Drever, J.I., 1982, The geochemistry of natural waters: Englewood Cliffs, New Jersey, Prentice Hall, 388 p.

Duhig, N.C., Stolz, J., Davidson, G.J., and Large, R.R., 1992, Cambrian microbial and silica gel textures in silica iron exhalites from the mount Windsor Volcanic Belt, Australia: Their petrography, chemistry, and origin: Economic Geology and the Bulletin of the Society of Economic Geologists, v. 87, p. 764–784.

Eigenbrode, J.L., Freeman, K.H., Love, G.D., Snape, C.E., and Summons, R.E., 2004, Hydropyrolytic release of hopane and sterane biomarkers from 2.7 Ga kerogens: Program and Abstracts, 13th Annual V.M. Goldschmidt Conference, Hot Springs, Virginia, p. A240.

El Shazly, E.M., 1990, Red Sea deposits, in Chauvel, J.J., Cheng Yuqi, El Shazly, E.M., Gross, G.A., Laajoki, K., Markov, M.S., Rai, K.L., Stulchikov, V.A., and Augustithis, S.S., eds., Ancient banded iron formations (regional presentations): Athens, Theophrastus Publications, p. 157–222.

Eremeev, V.N., ed., 1992, Complex oceanographic research of the Black Sea: Utrecht, VSP, 130 p.

Eriksson, P.G., Condie, K.C., van der Westhuizen, van der Merwe, R., de Bruiyn, H., Nelson, D.R., Altermann, W., Catuneanu, O., Bumby, A.J., Lindsay, J., and Cunningham, M.J., 2002, Late Archaean superplume events: A Kaapvaal-Pilbara perspective: Journal of Geodynamics, v. 34, p. 207–247.

Eriksson, P.G., Altermann, W., Nelson, D.R., Mueller, W.U., and Catuneanu, O., eds., 2004, The Precambrian Earth: Tempos and events: Amsterdam, Elsevier, Developments in Precambrian Geology, v. 12, 941 p.

Eugster, H.P., and I-Ming Chou, 1973, The depositional environments of Precambrian banded iron-formations: Economic Geology, v. 68, p. 1144–1168.

Ewers, W.E., and Morris, R.C., 1981, Studies of the Dales Gorge Member of the Brockman Iron Formation, Western Australia: Economic Geology and Bulletin of the Society of Economic Geologists, v. 76, p. 1929–1953.

Farquhar, J., Bao, H., and Thiemens, M., 2000, Atmospheric influence of Earth's earliest sulfur cycle: Science, v. 289, p. 756–758, doi: 10.1126/science.289.5480.756.

Farquhar, J., Savarino, J., Airieau, S., and Thiemens, M.H., 2001, Observation of wavelength-sensitive mass-independent sulfur isotope effects during SO2 photolysis; implication for the early atmosphere: Journal of Geophysical Research, ser. E, Planets, v. 106, p. 32829–32839, doi: 10.1029/2000JE001437.

Farrell, L., 1990, Ancient banded iron formations: Athens, Theophrastus Publications, 462 p.

Fernández, A., and Moro, M.C., 1998, Origin and depositional environment of Ordovician stratiform iron mineralization from Zamora (NW Iberian Peninsula): Mineralium Deposita, v. 33, p. 606–619, doi: 10.1007/s001260050176.

French, B.M., 1973, Mineral assemblages in diagenetic and low-grade metamorphic iron-formation: Economic Geology and the Bulletin of the Society of Economic Geologists, v. 68, p. 1063–1074.

Fry, B., Jannasch, H.W., Molyneaux, S.J., Wirsen, C.O., Muramoto, J.A., and King, S., 1991, Stable isotope studies of the carbon, nitrogen and sulfur cycles in the Black Sea and the Cariaco Trench, in Murray, J.W., ed., Black Sea oceanography: Results from the 1988 Black Sea expedition: Deep-Sea Research, part A, Oceanographic Research Papers, v. 38, Suppl. 2A, p. S1003–S1019.

Fyon, J.A., Breaks, F.W., Heather, K.B., Jackson, S.L., Muir, T.L., Stott, G.M., and Thurston, P.C., 1992, Metallogeny of metallic mineral deposits in the Superior Province of Ontario, in Thurston, P.C., Williams, H.R., Sutcliffe, R.H., and Stott, G.M., eds., Geology of Ontario: Ontario Geological Survey Special Volume 4, part 2, p. 1091–1174.

Gauthier, M., Brown, A.C., and Morin, G., 1987, Small iron-formations as a guide to base- and precious-metal deposits in the Grenville Province of southern Quebec, in Appel, P.W.U., and LaBerge, G.L., eds., Precambrian iron-formations: Athens, Theophrastus Publications, p. 297–327.

Gnaneshwar Rao, T., and Naqvi, S.M., 1995, Geochemistry, depositional environment and tectonic setting of the BIF's of the Late Archaean Chitradurga Schist Belt, India: Chemical Geology, v. 121, p. 217–243, doi: 10.1016/0009-2541(94)00116-P.

Goodwin, A.M., 1973, Archean iron-formations and tectonic basins of the Canadian shield: Economic Geology and the Bulletin of the Society of Economic Geologists, v. 68, p. 915–933.

Green, J.C., 1992, Proterozoic rifts, in Condie, K.C., ed., Proterozoic crustal evolution: Amsterdam, Elsevier, p. 97–149.

Grenne, T., and Vokes, F.M., 1990, Sea-floor sulfides at the Høydal volcanogenic deposit, central Norwegian Caledonides: Economic Geology and the Bulletin of the Society of Economic Geologists, v. 85, p. 344–359.

Gross, G.A., 1965, Geology of iron deposits in Canada: General geology and evaluation of iron deposits: Geological Survey of Canada Economic Geology Report 22, v. 1, 181 p.

Gross, G.A., 1990, Geochemistry of iron-formation in Canada, in Chauvel, J.J., Cheng Yuqi, El Shazly, E.M., Gross, G.A., Laajoki, K., Markov, M.S., Rai, K.L., Stulchikov, V.A., and Augustithis, S.S., eds., Ancient banded iron formations (regional presentations): Athens, Theophrastus Publications, p. 3–26.

Gross, G.A., 1991, Genetic concepts for iron-formation and associated metalliferous sediments, in Hutchinson, R.W., and Grauch, R.I., eds., Historical perspectives of genetic concepts and case histories of famous discoveries: Economic Geology Monograph 8, p. 51–81.

Habicht, K.S., Gade, M., Thamdrup, B., Berg, P., and Canfield, D.E., 2002, Calibration of sulfate levels in the Archean ocean: Science, v. 298, p. 2372–2374, doi: 10.1126/science.1078265.

Hagemann, S.G., Barley, M.E., Folkert, S.L., Yardley, W.D., and Banks, D.A., 1999, A hydrothermal origin for the giant BIF-hosted Tom Price iron ore deposit, in Stanley, C.J., et al., Mineral deposits: Processes to processing: Proceedings of the fifth biennial SGA meeting and the tenth quadrennial IAGOD symposium, London: Rotterdam, A.A. Balkema, p. 41–44.

Hamade, T., Konhauser, K.O., Raiswell, R., Goldsmith, S., and Morris, R.C., 2003, Using Ge/Si ratios to decouple iron and silica fluxes in Precambrian banded iron formations: Geology, v. 31, p. 35–38, doi: 10.1130/0091-7613(2003)031<0035:UGSRTD>2.0.CO;2.

Hartlaub, R.P., Heaman, L.M., Ashton, K.E., and Chako, T., 2004, The Archean Murmac Bay Group: Evidence for a giant Archean rift in the Rae Province Canada: Precambrian Research, v. 131, p. 345–372, doi: 10.1016/j.precamres.2004.01.001.

Haruna, M., Hanamuro, T., Uyeda, K., Fujimaki, H., and Ohmoto, H., 2003, Chemical, isotopic, and fluid inclusion evidence for the hydrothermal alteration of the footwall rocks of the BIF-hosted iron ore deposits in the Hamersley district, western Australia: Resource Geology, v. 53, p. 75–88.

Haydon, R.C., and McConachy, G.W., 1987, The stratigraphic setting of Pb-Zn-Ag mineralization at Broken Hill: Economic Geology and the Bulletin of the Society of Economic Geologists, v. 82, p. 826–856.

Hayes, J.M., 1994, Global methanotrophy at the Archean-Proterozoic transition, in Bengtson, S., ed., Early life on earth: Nobel symposium no. 84: New York, Columbia University Press, p. 220–236.

Hayes, J.M., Kaplan, I.R., and Wedeking, K.W., 1983, Precambrian organic geochemistry: Preservation of the record, in Schopf, J.W., ed., Earth's earliest biosphere: Its origin and evolution: Princeton, New Jersey, Princeton University Press, p. 93–134.

Helvaci, C., 1984, Apatite-rich iron deposits of the Avnik (Bingöl) region, southeastern Turkey: Economic Geology and the Bulletin of the Society of Economic Geologists, v. 79, p. 354–371.

Hoffmann, D., 1994, Geochemistry and genesis of manganiferous silicate-rich iron formation bands in the Broken Hill deposit, Aggeneys, South Africa: Exploration and Mining Geology, v. 3, p. 407–417.

Hofmann, H.J., Sage, R.P., and Berdusco, E.N., 1991, Archean stromatolites in Michipicoten Group siderite ore at Wawa, Ontario: Economic Geology and the Bulletin of the Society of Economic Geologists, v. 86, p. 1023–1030.

Holland, H.D., 1973, The oceans: A possible source of iron in iron-formations: Economic Geology and the Bulletin of the Society of Economic Geologists, v. 68, p. 1169–1172.

Holland, H.D., 1984, The chemical evolution of the atmosphere and oceans: Princeton, New Jersey, Princeton University Press, 582 p.

Holland, H.D., 1994, Early Proterozoic atmospheric change, in Bengtson, S., ed., Early life on Earth: Nobel Symposium 84: New York, Columbia University Press, p. 237–244.

Holland, H.D., 1999, When did the Earth's atmosphere become oxic?: Reply: The Geochemical News, v. 100, p. 20–23.

Holland, H.D., 2002, Volcanic gases, black smokers, and the great oxidation event: Geochimica et Cosmochimica Acta, v. 66, p. 3811–3826, doi: 10.1016/S0016-7037(02)00950-X.

Holland, H.D., and Petersen, U., 1995, Living dangerously: Princeton, New Jersey, Princeton University Press, 490 p.

Hollings, P., and Kerrich, R., 2000, An Archean arc basalt-Nb-enriched basalt-adakite association: The 2.7 Ga Confederation assemblage of the Birch-Uchi greenstone belt, Superior Province: Contributions to Mineralogy and Petrology, v. 139, p. 208–226.

Imai, H., 1978, Geology and genesis of the Okuki mine, Ehime Prefecture, and other related cupriferous pyrite deposits in southwest Japan, in Imai, H. ed., Geological studies of the mineral deposits in Japan and East Asia: Tokyo, University of Tokyo Press, p. 233–256.

Isley, A.E., 1995, Hydrothermal plumes and the delivery of iron to banded iron formation: Journal of Geology, v. 103, p. 169–185.

Isley, A.E., and Abbott, D.H., 1999, Plume-related mafic volcanism and the deposition of banded iron formation: Journal of Geophysical Research, ser. B, Solid Earth and Planets, v. 104, p. 15,461–15,477, doi: 10.1029/1999JB900066.

Jackson, S.I., and Fyon, J.A., 1991, The western Abitibi subprovince in Ontario, in Thurston, P.C., Williams, H.R., Sutcliffe, R.H., and Stott, G.M., eds., Geology of Ontario: Toronto, Ontario Geological Survey Special Volume 4, p. 405–482.

Jacobsen, S.B., and Pimentel-Klose, M.R., 1988, A Nd isotopic study of the Hamersley and Michipicoten banded iron formations: The source of REE and Fe in Archean oceans: Earth and Planetary Science Letters, v. 87, p. 29–44, doi: 10.1016/0012-821X(88)90062-3.

Jahnke, L., and Klein, H.P., 1983, Oxygen requirements for formation and activity of the squalene epoxidase in *Saccharomyces cerevisiae*: Journal of Bacteriology, v. 155, p. 488–492.

James, H.L., 1954, Sedimentary facies of iron-formation: Economic Geology and the Bulletin of the Society of Economic Geologists, v. 49, p. 235–293.

James, H.L., 1983, Distribution of banded iron-formation in space and time, in Trendall, A.F., and Morris, R.C., Iron-formation: Facts and problems: Amsterdam, Elsevier, p. 471–490.

Jørgensen, B.B., Böttcher, M.E., Lüschen, H., Neretin, L.N., and Volkov, I.I., 2004, Anaerobic methane oxidation and a deep H_2S sink generate isotopically heavy sulfides in Black Sea sediments: Geochimica et Cosmochimica Acta, v. 68, p. 2095–2118, doi: 10.1016/j.gca.2003.07.017.

Kakegawa, T., Kawai, H., and Ohmoto, H., 1998, Origins of pyrites in the 2.5 Ga Mt. McRae shale, the Hamersley district, western Australia: Geochimica et Cosmochimica Acta, v. 62, p. 3205–3220, doi: 10.1016/S0016-7037(98)00229-4.

Kalogeropoulos, S.I., and Scott, S.D., 1983, Mineralogy and geochemistry of tuffaceous exhalites (tetsusekiei) of the Fukazawa Mine, Hokuroku district, Japan, in Ohmoto, H., and Skinner, B.J., eds., The Kuroko and related volcanic massive sulfide deposits: Economic Geology Monograph, v. 5, p. 412–432.

Kalugin, A.S., 1973, Geology and genesis of the Devonian banded iron-formation in Altai, western Siberia and eastern Kazakhstan, in Genesis of Precambrian iron and manganese deposits, Proceedings, Kiev Symposium, 1970: Paris, UNESCO, Earth Science (Sciences de la Terre), v. 9, p. 159–165.

Kanehira, K., and Tatsumi, T., 1970, Bedded cupriferous iron sulphide deposits in Japan, a review, in Tatsumi, T., ed., Volcanism and ore genesis: Tokyo, University of Tokyo Press, p. 51–76.

Kaplan, I.R., Sweeney, R.E., and Nissenbaum, A., 1969, Sulfur isotope studies on Red Sea geothermal brines and sediments, in Degens, E.T., and Ross, D.A., eds., Hot brines and recent heavy metal deposits in the Red Sea: A geochemical and geophysical account: New York, Springer-Verlag, p. 474–498.

Karhu, J., and Epstein, S., 1986, The implication of the oxygen isotope records in coexisting cherts and phosphates: Geochimica et Cosmochimica Acta, v. 50, p. 1745–1756, doi: 10.1016/0016-7037(86)90136-5.

Kasting, J.F., 1987, Theoretical constraints on oxygen and carbon dioxide concentrations in the Precambrian atmosphere: Precambrian Research, v. 34, p. 205–229, doi: 10.1016/0301-9268(87)90001-5.

Kasting, J.F., 2001, Earth history: The rise of atmospheric oxygen: Science, v. 293, p. 819–820, doi: 10.1126/science.1063811.

Kasting, J.F., and Siefert, J.L., 2002, Life and the evolution of Earth's atmosphere: Science, v. 296, p. 1066–1068, doi: 10.1126/science.1071184.

Kimberley, M.M., 1989a, Nomenclature for iron formations: Ore Geology Reviews, v. 5, p. 1–12, doi: 10.1016/0169-1368(89)90002-4.

Kimberley, M.M., 1989b, Exhalative origins of iron formations: Ore Geology Reviews, v. 5, p. 13–145, doi: 10.1016/0169-1368(89)90003-6.

Klein, C., 1983, Diagenesis and metamorphism of Precambrian banded iron-formations, in Trendall, A.F., and Morris, R.C., Iron-formation: Facts and problems: Amsterdam, Elsevier, p. 417–460.

Klein, C., and Beukes, N.J., 1989, Geochemistry and sedimentology of a facies transition from limestone to iron-formation deposition in the early Proterozoic Transvaal Supergroup, South Africa: Economic Geology and the Bulletin of the Society of Economic Geologists, v. 84, p. 1733–1774.

Klein, C., and Beukes, N.J., 1992, Proterozoic iron formations, in Condie, K.C., ed., Proterozoic crustal evolution: Amsterdam, Elsevier, p. 383–418.

Klein, C., and Beukes, N.J., 1993, Sedimentology and geochemistry of the glaciogenic Late Proterozoic Rapitan Iron-Formation in Canada: Economic Geology and the Bulletin of the Society of Economic Geologists, v. 88, p. 542–565.

Klein, C., and Fink, R.P., 1976, Petrology of the Sokoman Iron Formation in the Howells River area, at the western edge of the Labrador Trough: Economic Geology and the Bulletin of the Society of Economic Geologists, v. 71, p. 453–487.

Knauth, L.P., and Lowe, D.R., 2003, High Archean climatic temperature inferred from oxygen isotope geochemistry of cherts in the 3.5 Ga Swaziland Supergroup, South Africa: Geological Society of America Bulletin, v. 115, p. 566–580, doi: 10.1130/0016-7606(2003)115<0566:HACTIF>2.0.CO;2.

Konhauser, K.O., Hamade, T., Raiswell, R., Morris, R.C., Ferris, F.G., Southam, G., and Canfield, D.E., 2002, Could bacteria have formed the Precambrian banded iron formations?: Geology, v. 30, p. 1079–1082, doi: 10.1130/0091-7613(2002)030<1079:CBHFTP>2.0.CO;2.

Koopman, E.R., 1993, Stratigraphy, structural geology, and structural controls of ore distribution of the Lyon Lake massive sulphide deposit, Sturgeon Lake, Ontario [M.Sc. thesis]: Ottawa, Carleton University, 170 p.

Krapez, B., Barley, M.E., and Pickard, A.L., 2003, Hydrothermal and resedimented origins of the precursor sediments to banded iron formation: Sedimentological evidence from the early Palaeoproterozoic Brockman Supersequence of western Australia: Sedimentology, v. 50, p. 979–1011, doi: 10.1046/j.1365-3091.2003.00594.x.

LaBerge, G.L., Robbins, E.I., and Tsu-Ming Han, 1987, A model for the biological precipitation of Precambrian iron-formations – A: Geological evidence, *in* Appel, P.W.U., and LaBerge, G.L., eds., Precambrian iron-formations: Athens, Theophrastus Publications, p. 69–96.

Langmuir, D., 1997, Aqueous environmental geochemistry: Upper Saddle River, New Jersey, Prentice Hall, 600 p.

Lasaga, A.C., and Ohmoto, H., 2002, The oxygen geochemical cycle: Dynamics and stability: Geochimica et Cosmochimica Acta, v. 66, p. 361–381, doi: 10.1016/S0016-7037(01)00685-8.

Liaghat, S., and MacLean, W.H., 1992, The Key tuffite, Matagami mining district: Origin of the tuff components and mass changes: Exploration and Mining Geology, v. 1, p. 197–207.

Lin, S., 2001, Stratigraphic and structural setting of the Hemlo gold deposits, Ontario, Canada: Economic Geology and the Bulletin of the Society of Economic Geologists, v. 96, p. 477–507.

Lottermoser, B.G., and Ashley, P.M., 1996, Geochemistry and exploration significance of ironstones and barite-rich rocks in the Proterozoic Willyama Supergroup, Olary Block, South Australia: Journal of Geochemical Exploration, v. 57, p. 57–73, doi: 10.1016/S0375-6742(96)00016-7.

Matsuhisa, Y., Goldsmith, J.R., and Clayton, R.N., 1978, Mechanisms of hydrothermal crystallization of quartz at 250 °C and 15 kbar: Geochimica et Cosmochimica Acta, v. 42, p. 173–182, doi: 10.1016/0016-7037(78)90130-8.

Mel'nik, Y.P., 1982, Precambrian banded iron-formations: Physicochemical conditions of formation: Amsterdam, Elsevier, Developments in Precambrian Geology, v. 5, 310 p.

Morey, G.B., and Southwick, D.L., 1995, Allostratigraphic relationships of early Proterozoic iron-formations in the Lake Superior region: Economic Geology and the Bulletin of the Society of Economic Geologists, v. 90, p. 1983–1993.

Morris, R.C., 1985, Genesis of iron ore in banded iron-formation by supergene and supergene-metamorphic processes: A conceptual model, *in* Wolf, K.H., ed., Regional studies and specific deposits: Amsterdam, Elsevier, Handbook of strata-bound and stratiform ore deposits, v. 13, p. 73–235.

Morris, R.C., 1993, Genetic modeling for banded iron-formation of the Hamersley Group, Pilbara Craton, Western Australia: Precambrian Research, v. 60, p. 243–286, doi: 10.1016/0301-9268(93)90051-3.

Morton, R.L., and Nebel, M.L., 1984, Hydrothermal alteration of felsic volcanic rocks at the Helen siderite deposit, Wawa, Ontario: Economic Geology and the Bulletin of the Society of Economic Geologists, v. 79, p. 1319–1333.

Muramoto, J.A., Honjo, S., Fry, B., Hay, B.J., Howarth, R.W., and Cisne, J.L., 1991, Sulfur, iron and organic carbon fluxes in the Black Sea: Sulfur isotope evidence for origin of sulfur fluxes, *in* Murray, J.W., ed., Black Sea oceanography: Results from the 1988 Black Sea expedition: Deep-Sea Research, part A, Oceanographic Research Papers, v. 38, Suppl. 2A, p. S1151–S1187.

Murray, J.W., ed., 1991, Black Sea oceanography, Results from the 1988 Black Sea expedition: Deep-Sea Research, part A, Oceanographic Research Papers, v. 38, Suppl. 2A, p. S655–S1266.

Ohmoto, H., 1992, Biogeochemistry of sulfur and the mechanisms of sulfide-sulfate mineralization in Archean oceans, *in* Schidlowski, M., Golubic, S., Kimberley, M.M., McKirdy, D.M., and Trudinger, P.A., eds., Early organic evolution: Implications for mineral and energy resources: Berlin, Springer-Verlag, p. 378–397.

Ohmoto, H., 1993, The banded iron formations: Products of the oxygen-rich Archean atmosphere?: Geological Society of America Abstracts with Programs, v. 25, no. 6, p. A-89.

Ohmoto, H., 1996, Formation of volcanogenic massive sulfide deposits: The Kuroko perspective: Ore Geology Reviews, v. 10, p. 135–177, doi: 10.1016/0169-1368(95)00021-6.

Ohmoto, H., 1997, When did the Earth's atmosphere become oxic?: Geochemical News, v. 93, p. 12–13 and p. 26–28.

Ohmoto, H., 2003, Non-redox transformation of magnetite-hematite in hydrothermal systems: Economic Geology and the Bulletin of the Society of Economic Geologists, v. 98, p. 157–161.

Ohmoto, H., 2004, Archean atmosphere, hydrosphere, and biosphere, *in* P. Erickson et al., eds. The Precambrian Earth: Tempos and events: Amsterdam, Elsevier, Developments in Precambrian Geology, v. 12, p. 361–368.

Ohmoto, H., and Skinner, B.J., eds., 1983, The Kuroko and related volcanogenic massive sulfide deposits: Economic Geology Monograph 5, 604p.

Ohmoto, H., and Felder, R.P., 1987, Bacterial activity in the warmer, sulphate-bearing, Archaean oceans: Nature, v. 328, p. 244–246, doi: 10.1038/328244a0.

Ohmoto, H., and Goldhaber, M.B., 1997, Sulfur and carbon isotopes, *in* Barnes, H.D., ed., Geochemistry of hydrothermal ore deposits (3rd edition): New York, John Wiley and Sons, p. 517–612.

Ohmoto, H., Kajiwara, Y., and Date, J., 1970, The Kuroko ores in Japan: Products of seawater? Geological Society of America Abstracts with Programs, v. 1, p. 640.

Ohmoto, H., Mizukami, M., Drummond, S.E., Eldridge, C.S., Pisutha-Arnold, V., and Lenagh, T.C., 1983, Chemical processes of Kuroko formation, *in* Ohmoto, H., and Skinner, B.J., eds., The Kuroko and related volcanogenic massive sulfide deposits: Economic Geology Monograph 5, p. 570–604.

Ohmoto, H., Kaiser, C.J., and Geer, K.A., 1990, Systematics of sulphur isotopes in recent marine sediments and ancient sediment-hosted base metal deposits, *in* Herbert, H.K., and Ho, S.E., eds., Proceedings of the conference on stable isotopes and fluid processes in mineralization: Perth, University of Western Australia Special Publication 23, p. 70–120.

Ohmoto, H., Watanabe, Y., and Kumazawa, K., 2004, Evidence from massive siderite beds for a CO_2-rich atmosphere before approximately 1.8 billion years ago: Nature, v. 429, p. 395–399, doi: 10.1038/nature02573.

Ono, S., Eigenbrode, J.L., Pavlov, A.A., Kharecha, P., Rumble, D., Kasting, J.F., and Freeman, K.H., 2003, New insights into Archean sulfur cycle from mass-independent sulfur isotope records from the Hamersley Basin, Australia: Earth and Planetary Science Letters, v. 213, p. 15–30, doi: 10.1016/S0012-821X(03)00295-4.

Ootes, L., and Lentz, D.R., 2002, Occurrence of bleached mafic flows and their association with stockwork sulphides and banded iron-formation in the Crestaurum Formation of the later Archean Yellowknife greenstone belt, Northwest Territories: Ottawa, Geological Survey of Canada, Current Research Report 2002-E5, p. 1–12.

O'Rourke, J.E., 1962, The stratigraphy of Himalayan iron ores: American Journal of Science, v. 260, p. 294–302.

Otake, T., Wesolowski, D.J., Anovitz, L.M., Hayashi, K., and Ohmoto, H., 2005, Magnetite transformation to hematite under high H_2 pressure at 150 °C: Geochimica et Cosmochimica Acta, v. 69, S10, A424.

Owen, T., Cess, R.D., and Ramanathan, V., 1979, Enhanced CO_2 greenhouse to compensate for reduced solar luminosity on early Earth: Nature, v. 277, p. 640–642, doi: 10.1038/277640a0.

Page, R.W., and Laing, W.P., 1992, Felsic metavolcanic rocks related to the Broken Hill Pb-Zn-Ag orebody, Australia: Geology, depositional age, and timing of high-grade metamorphism: Economic Geology and the Bulletin of the Society of Economic Geologists, v. 87, p. 2138–2168.

Parák, T., 1991, Volcanic sedimentary rock-related metallogenesis in the Kiruna-Skellefte belt of northern Sweden, *in* Hutchinson, R.W., and Grauch, R.I., eds., Historical perspectives of genetic concepts and case histories of famous discoveries: Economic Geology Monograph 8, p. 20–50.

Pavlov, A.A., Kasting, J.F., Eigenbrode, J.L., and Freeman, K.H., 2001a, Organic haze in Earth's early atmosphere: Source of low-^{13}C Late Archean kerogens?: Geology, v. 29, p. 1003–1006, doi: 10.1130/0091-7613(2001)029<1003:OHIESE>2.0.CO;2.

Pavlov, A.A., Brown, L.L., and Kasting, J.F., 2001b, UV shielding of NH_3 and O_2 by organic hazes in the Archean atmosphere: Journal of Geophysical Research, E, Planets, v. 106, p. 23267–23287, doi: 10.1029/2000JE001448.

Perry, E.C., Jr., Tan, F.C., and Morey, G.B., 1973, Geology and stable isotope geochemistry of the Biwabik Iron Formation, Northern Minnesota: Economic Geology and the Bulletin of the Society of Economic Geologists, v. 68, p. 1110–1125.

Peter, J.M., 2001, Ancient iron-rich metalliferous sediments (iron formations): Their genesis and use in the exploration for stratiform base metal sulphide deposits, with examples from the Bathurst Mining Camp, *in* Lentz, D. R., ed., Geochemistry of sediments and sedimentary rocks: Secular evolutionary considerations to mineral deposit-forming environments: St. John's, Newfoundland, Geological Association of Canada, p. 1–38.

Peter, J.M., and Goodfellow, W.D., 1996, Mineralogy, bulk and rare earth element geochemistry of massive sulphide-associated hydrothermal sediments of the Brunswick Horizon, Bathurst Mining Camp, New Brunswick: Canadian Journal of Earth Sciences, v. 33, p. 252–283.

Peter, J.M., and Scott, S.D., 1998, Windy Craggy, northwestern British Columbia: The world's largest Besshi-type deposits, *in* Barrie, C.T., and Hannington, M.D., eds., Volcanic-associated massive sulfide deposits: Processes and examples in modern and ancient settings: Reviews in Economic Geology, v. 8, p. 261–295.

Phillips, G. N., Law, J. D. M., and Myers, R. E., 2001, Is the redox state of the Archean atmosphere constrained?: Society of Economic Geologists Newsletter, v. 47, 1, 9–18.

Pickard, A.L., 2003, SHRIMP U-Pb zircon ages for the Palaeoproterozoic Kuruman iron formation, northern Cape Province, South Africa: Evidence for simultaneous BIF deposition on Kaapvaal and Pilbara cratons: Precambrian Research, v. 125, p. 275–315, doi: 10.1016/S0301-9268(03)00113-X.

Piercey, S.J., Paradis, S., Murphy, D.C., and Mortensen, J.K., 2001, Geochemistry and paleotectonic setting of felsic volcanic rocks in the Finlayson Lake volcanic-hosted massive sulfide district, Yukon, Canada: Economic Geology, v. 96, p. 1877–1905.

Polat, A., and Kerrich, R., 2001, Geodynamic processes, continental growth, and mantle evolution in late Archean greenstone belts of the southern Superior Province, Canada: Precambrian Research, v. 112, p. 5–25, doi: 10.1016/S0301-9268(01)00168-1.

Potter, R.W., and Brown, D.L., 1977, The volumetric properties of aqueous sodium chloride solutions from 0° to 500°C at pressures up to 2000 bars based on a regression of available data in the literature: U.S. Geological Survey Bulletin, Report B 1421-C, 36 p.

Pottorf, R.J., and Barnes, H.L., 1983, Mineralogy, geochemistry, and ore genesis of hydrothermal sediments from the Atlantis II Deep, Red Sea, in Ohmoto, H., and Skinner, B.J., eds., The Kuroko and related volcanogenic massive sulfide deposits: Economic Geology Monograph 5, p. 198–223.

Poulton, S.W., Fralick, P.W., and Canfield, D.E., 2004, The transition to a sulphidic ocean approximately 1.84 billion years ago: Nature, v. 431, p. 173–177, doi: 10.1038/nature02912.

Quade, H., 1976, Genetic problems and environmental features of volcano-sedimentary iron-ore deposits of the Lahn-Dill type, in Wolf, K.H., ed., Geochemical studies: Amsterdam Elsevier, Handbook of strata-bound and stratiform ore deposits, v. 2, p. 255–294.

Rimstidt, J.D., 1997, Gangue mineral transport and deposition, in Barnes, H.L., ed., Geochemistry of hydrothermal ore deposits: New York, John Wiley and Sons, p. 487–516.

Rosing, M.T., 1999, ^{13}C-depleted carbon microparticles in >3700-Ma seafloor sedimentary rocks from West Greenland: Science, v. 283, p. 674–676, doi: 10.1126/science.283.5402.674.

Rozendaal, A., 1986, The Gamsberg zinc deposit, Namaqualand district, in Anhaeusser, C.R., and Maske, S., eds., Mineral deposits of South Africa: Johannesburg, Geological Society of South Africa, v. 2, p. 1477–1488.

Rucker, E.A., Dudley, M.A., and Nold, J.L., 2003, Ore textural studies and origin of the Pilot Knob hematite deposit, Iron County, MO: Geological Society of America, North-Central Section, 37th Annual Meeting, Warrensburg, Montana, Abstracts with Programs, v. 35, p. 20.

Russell, M.J., 1975, Lithogeochemical environment of the Tynagh base-metal deposit, Ireland, and its bearing on ore deposition: Transactions of the Canadian Institute of Mining and Metallurgy: Section B: Applied Earth Science, v. 84, p. B128–B133.

Rye, R., and Holland, H.D., 1998, Paleosols and the evolution of atmospheric oxygen: a critical review: American Journal of Science, v. 298, p. 621–672.

Rye, R., Kuo, P., and Holland, H.D., 1995, Atmospheric carbon dioxide concentrations before 2.2 billion years ago: Nature, v. 378, p. 603–605, doi: 10.1038/378603a0.

Sarmiento, J., 1992, Biogeochemical Ocean Models, in Trenberth, K.E., ed., Climate system modeling: Cambridge, UK, Cambridge University Press, p. 519–564.

Savarino, J., Romero, A., Cole-Dai, J., Bekki, S., and Thiemens, M.H., 2003, UV induced mass- independent sulfur isotope fractionation in stratospheric volcanic sulfate: Geophysical Research Letters, v. 30, p. 2131–2134, doi: 10.1029/2003GL018134.

Schidlowski, M., and Aharon, P., 1992, Caron cycle and carbon isotope record: Geochemical impact of life over 3.8 Ga of earth history, in Schidlowski, M., Golubic, S., Kimberley, M.M., McKirdy, D.M., and Trudinger, P.A., eds., Early organic evolution: Implications for mineral and energy resources: Berlin, Springer-Verlag, p. 378–397.

Schneider, D.A., Bickford, M.E., Cannon, W.F., Schulz, K.J., and Hamilton, M.A., 2002, Age of volcanic rocks and syndepositional iron formations, Marquette Range Supergroup: Implications for the tectonic setting of Paleoproterozoic iron formation of the Lake Superior region: Canadian Journal of Earth Sciences, v. 39, p. 999–1012, doi: 10.1139/e02-016.

Scott, S.D., 1997, Submarine hydrothermal systems and deposits, in Barnes, H.L., ed., Geochemistry of hydrothermal ore deposits (3rd edition): New York, John Wiley and Sons, p. 797–875.

Shanks, W.C., III, and Bischoff, J.L., 1980, Geochemistry, sulfur isotope composition, and accumulation rates of Red Sea geothermal deposits: Economic Geology and the Bulletin of the Society of Economic Geologists, v. 75, p. 445–459.

Shegelski, R.J., 1987, The depositional environment of Archean iron formations, Sturgeon-Savant greenstone belt, Ontario, Canada, in Appel, P.W.U., and LaBerge, G.L., eds., Precambrian iron-formations: Athens, Theophrastus Publications, p. 329–344.

Simonson, B.M., Schubel, K.A., and Hassler, S.W., 1993, Carbonate sedimentology of the early Precambrian Hamersley Group of western Australia: Precambrian Research, v. 60, p. 287–335, doi: 10.1016/0301-9268(93)90052-4.

Sleep, N.H., 1978, Thermal structure and kinematics of mid-ocean ridge axis, some implications to basaltic volcanism: Geophysical Research Letters, v. 5, p. 426–428.

Soriano, C., and Marti, J., 1999, Facies analysis of volcano-sedimentary successions hosting massive sulfide deposits in the Iberian pyrite belt, Spain: Economic Geology and the Bulletin of the Society of Economic Geologists, v. 94, p. 867–882.

Spry, P.G., Peter, J.M., and Slack, J.F., 2000, Meta-exhalites as exploration guides to ore, in Spry, P.G., Marshall, B., and Vokes, E.M., eds., Metamorphosed and metamorphogenic ore deposits: Reviews in Economic Geology, v. 11, p. 163–201.

Tachibana, S., Hirai, T., Goto, K., Yamamoto, S., Isozoki, Y., Tada, R., Tajika, E., Shimoda, G., Marishita, Y., and Kita, N.T., 2004, Sulfur isotopic compositions of sulfides from the lower Huronian Supergroup, Ontario, Canada [abs.]: Eos (Transactions, American Geophysical Union), v. 85, p. F1897.

Thurston, P.C., 1991, Archean geology of Ontario: Introduction, in Thurston, P.C., Williams, H.R., Sutcliffe, R.H., and Stott, G.M., eds., Geology of Ontario: Ontario Geological Survey Special Volume 4, p. 73–78.

Thurston, P.C., 2002, Autochthonous development of Superior Province greenstone belts?: Precambrian Research, v. 115, p. 11–36, doi: 10.1016/S0301-9268(02)00004-9.

Torres-Ruiz, J., 1983, Genesis and evolution of the Marquesado and adjacent iron ore deposits, Granada, Spain: Economic Geology and the Bulletin of the Society of Economic Geologists, v. 78, p. 1657–1673.

Trendall, A.F., and Blockley, J.G., 1970, The iron formations of the Precambrian Hamersley group Western Australia: Geological Survey of Western Australia Bulletin, v. 119, 366 p.

Trendall, A.F., and Morris, R.C., eds., 1983, Iron-formation: Facts and problems: Amsterdam, Elsevier, 558 p.

Tsikos, H., Beukes, N.J., Moore, J.M., and Harris, C., 2003, Deposition, diagenesis, and secondary enrichment of metals in the Paleoproterozoic Hotazel Iron Formation, Kalahari manganese Field, South Africa: Economic Geology and the Bulletin of the Society of Economic Geologists, v. 98, p. 1449–1462.

Uyeda, K., 1994, Sr and Nd geochemistry of Mt. McRae Shale in the Hamersley District, Australia [M.S. thesis]: Sendai, Japan, Tohoku University, 64 p.

van Staal, C.R., and Williams, P.F., 1984, Structure, origin, and concentration of the Brunswick 12 and 6 ore bodies: Economic Geology and the Bulletin of the Society of Economic Geologists, v. 79, p. 1669–1692.

Walker, J.C.G., and Brimblecombe, P., 1985, Iron and sulfur in the pre-biologic ocean: Precambrian Research, v. 28, p. 205–222, doi: 10.1016/0301-9268(85)90031-2.

Watanabe, Y., Naraoka, H., Wronkiewicz, D.J., Condie, K.C., and Ohmoto, H., 1997, Carbon, nitrogen, and sulfur geochemistry of Archean and Proterozoic shales from the Kaapvaal Craton, South Africa: Geochimica et Cosmochimica Acta, v. 61, p. 3441–3459, doi: 10.1016/S0016-7037(97)00164-6.

Watanabe, Y., Ikemi, H., and Ohmoto, H., 2005, Absence of mass independent sulfur isotope fractionation in 2.76 Ga freshwater and 3.0 Ga marine sediments: Astrobiology, v. 5, p. 272.

Watanabe, Y., Naraoka, H., and Ohmoto, H., 2006, Mass independent fractionation of sulfur isotopes during thermochemical reduction of native sulfur, sulfite and sulfate by amino acids: American Geophysical Union Spring Meeting, 23–26 May, Baltimore, Maryland (in press).

Widdel, F., Schnell, S., Heising, S., Ehrenreich, A., Assmus, B., and Schink, B., 1993, Ferrous iron oxidation by anoxygenic phototrophic bacteria: Nature, v. 362, p. 834–836, doi: 10.1038/362834a0.

Wilkin, R.T., and Arthur, M.A., 2001, Variations in pyrite texture, sulfur isotope composition, and iron systematics in the Black Sea: Evidence for late Pleistocene to Holocene excursions of the O_2-H_2S redox transition: Geochimica et Cosmochimica Acta, v. 65, p. 1399–1416, doi: 10.1016/S0016-7037(01)00552-X.

Yamaguchi, K.E., Johnson, C.M., Beard, B.L., and Ohmoto, H., 2005, Biogeochemical cycling of iron in the Archean-Paleoproterozoic Earth: Constraints from iron isotope variations in sedimentary rocks from the Kaapvaal and Pilbara Cratons: Chemical Geology, v. 218, p. 135–169, doi: 10.1016/j.chemgeo.2005.01.020.

Young, T.P., 1989, Phanerozoic ironstones: An introduction and review, in Young, T.P., and Taylor, W.E.G., eds., Phanerozoic ironstones: London, Geological Society Special Publication 46, p. ix–xxv.

Young, T.P., and Taylor, W.E.G., eds., 1989, Phanerozoic ironstones: London, The Geological Society, 251 p.

Zalenski, E., and Peterson, V., 1995, Depositional setting and deformation of massive sulfide deposits, iron-formation, and associated alteration in the Manitouwadge greenstone belt, Superior Province, Ontario: Economic Geology and the Bulletin of the Society of Economic Geologists, v. 90, p. 2244–2261.

Zantop, H., 1981, Trace elements in volcanogenic manganese oxides and iron oxides: The San Francisco manganese deposit, Jalisco, Mexico: Economic Geology and the Bulletin of the Society of Economic Geologists, v. 76, p. 545–555.

Manuscript Accepted by the Society 29 October 2005

Index

A

Abitibi Greenstone Belt, 302, 304
Acetate, 3, 11–13, 16, 21, 25
Acetogens, 16, 21
Acetyl-coenzyme A, 16
Acidic springs, 3–4
Active centers, 16–17
Algoma iron formations, 296
Alteration, 97, 126–127
Amazonian Craton, 54
Amino acids, 13, 15, 18, 19, 20
Anderson model, 188, 200
Anoxia (*see also* Oxygen content)
    manganese ore formation and, 213, 216, 218–219, 298
    oceanic sulfur cycle and, 173, 178
    photosynthesis and, 34
    sulfate limitation vs., 206–207, 214–220
Anoxygenic photosynthesis, 34
Antimony, 55, 57, 60–61
Apatite, 77
Arsenic, 152
Atmospheric conditions
    banded iron formations and, 319–322, 324
    diagenetic iron mobilization and, 227, 232
    Franceville Basin and, 157–158
    gold and, 134
    isotopic analysis of, 100–101, 324
    MVT deposits and, 188, 197–200
    nitrogen fixing organisms and, 100–101
    pyrite and, 136, 144–145, 172
    recycling and, 101
    redox conditions and, 121–122, 136, 144
    uranium deposits and, 144–145, 157–158
    Witwatersrand Basin and, 105–106, 121–122
Atmospheric redox state, 122
Autogenesis
    coded polymers and, 19–20
    coenzymes and, 17–18
    differentiation into two domains and, 22–23
    early evolution and, 21
    energy transfer and, 16–17
    enzyme formation and, 15–16
    hydrophobic organic membranes and, 20–21
    hydrothermal mounds as reactors and, 11–13
    initial Earth conditions and, 3–7
    low temperature mounds and, 7–9
    mackinawite and, 9–11
    obduction and photosynthesis and, 23–25
    organic takeover and, 16
    polymerization energy and, 13–15
    RNA synthesis and, 18–19
Autotrophism, 5, 26, 98–99, 325

B

Bababudan belt, 276, 277–279
Bacteria (*see* Microorganisms; Sulfate reducing bacteria)
Banded iron formations (*see also* Superior-type banded iron formations)
    atmospheric carbon dioxide and methane and, 321–322
    biofilms and, 240, 243, 245–251, 251–254
    chemical processes for formation of, 313–319
    classifications of, 293–296
    diagenetic mobilization and, 224–225, 228–234
    distribution of through geologic time, 325
    geologic environments for, 302–313
    hematite and, 258–266
    hydrothermal mounds and, 36–39, 258–266
    isotopic nitrogen analysis and, 98–99
    microorganisms and, 239–241
    models for origin of, 296–299
    overview of, 270, 292–293
    oxygen content and, 172–173, 319–321
    photosynthesis development and, 24
    rare earth elements and, 270–272, 277–279, 279–286
    sulfur cycle and, 172–173, 322–324
Barberton Greenstone Belt
    geology of, 71–72
    granitoid distribution in, 73
    isotopic analysis of, 41–42, 43–46, 47–49
    metabolic analysis of, 34
    mineralization in, 76–77
Barite, 177
Barium, 55, 57, 60–61
Basal brines, 188, 299
Basal paleoplacers, 112, 131–132
Beatrix Reef, 112
Belingwe Greenstone Belt, 43–46, 49
Belt Supergroup, 173–174
Benthic flux model, 225–226, 229–234
Biofilms, 240, 243, 245–251, 251–254
Biogenic sulfides, 34–35
Biomats, 243–244, 252
Biomineralization, 249–250
Biosphere, 2–3, 23, 25–26, 325
Biosynthesis, 18
Biwabik iron formation, 277–279
Black Sea (*see* Red and Black Seas Hybrid model)
Black smokers, 3, 272, 273, 313–314
Blockley (*see* Trendall-Blockley model)
Boolgeeda formation, 276–279
Bouma sequence, 37
Braemar iron formation, 277–279
Brine pool model (*see* Red and Black Seas Hybrid model)
Brines, 188, 299
Brockman iron formation, 224, 276, 277–279, 308–311
Brucite, 3, 5, 7, 11

C

Carbon (*see also* Isotopic carbon analysis; Kerogen), 215
Carbon cycle, 47, 177–178
Carbon dioxide, 26, 321–322
Carbon monoxide dehydrogenase, 13
Carbonate
    banded iron formations and, 293–296, 302–303
    hydrothermal mounds and, 11
    isotopic carbon analysis and, 39
    Lost City spring and, 7
    MVT deposits and, 198–199
Carbonate reefs, 42, 46
Cassiterite, 77–78
Catalysts (*see also* Enzymes), 9–11, 15–16
Cations, 126
Cellular differentiations, 23
Central Rand Group, 107
Cerium
    in banded iron formations, 277–279, 311
    manganese ores and, 219, 270–272
    in modern ocean waters, 272–274
    oceanic redox conditions and, 279–286
Chemical disequilibrium, 3
Chemical tracers, 54, 55–59, 59–65
Chemiosmotic proton potential, 14
Chemiosmotic theory, 243
Chemoautotrophs, 5, 98–99
Chert-iron formations, 36–37, 90, 98–99, 299
Cheshire Formation, 43
Chirality, 15, 20
Chloride, 188
Chlorine, 77
Chondrite, 75, 272
Chromite, 114
Chromium, 55, 57, 60–61
Claypool paradigm, 175
Cleaverville Formation, 275–276
Cloud's model, 297
Cobalt, 18, 145, 152
Coded polymers, 19–20
Coenzymes, 16, 17–18
Communities (*see also* Biofilms; Biomats), 21–23
Compartmentalization, 13
Conglomerate, 40
Convection, 26
Convection cells, 4–6, 25
Copper, 55, 57, 60–61, 77–78
Coxco, 188–192
Cratons (*see also* Individual craton names)
    banded iron formations and, 270
    distribution of in Gondwana, 55–56
    exploration indexes and, 58–59, 62–63
    granitoids of, 70, 71–75, 75–76, 76–77
    isotopic carbon analysis of, 86, 96
    isotopic nitrogen analysis and, 82–85, 86, 89–94
    mineral analysis of, 53–55, 61–62
    MVT deposits and, 198
    overview of, 53–55
    recycling and, 64–65
Crust, 3, 61–62, 63
Cryptozoon structures, 42
Cudapah Basin, 83
Cyanide, 6–7, 11
Cyanobacteria
    biosphere formation and, 325
    formation of, 24
    isotopic fractionation by, 34, 42, 49
    James's model and, 296–297
    Red and Black Seas Hybrid model and, 307, 314–315

D

Dales Gorge Member, 224
Datangpo Formation, 207
Deformation, 161
Dehydrogenases, 13, 15, 16
Detrital mineralogy, 113–115, 127, 144–145, 152–153, 163–164
Dharwar Craton
    banded iron formations in, 276, 277–279
    greenstone belts in, 84

isotopic analysis and, 82–85, 86, 89–94, 96
kerogen and, 87, 89–90
orogenic gold and, 83, 86, 87
rare earth element anomalies and, 279
Diagenetic mobilization (*see also* Recycling)
benthic flux model for, 225–226, 227, 228–234
in Franceville Basin, 161
iron deposits and, 224–225
kerogen generation and, 89–90
microorganisms and, 250
rare earth element composition and, 271
riverine iron fluxes and, 234–235
Diamonds, 54, 87–88, 100
Diffusional effects, 97
Dimroth's model, 298–299
Disproportionation, 46
Dissolution, 247
DNA, 20
Dolomite, 304
Dominion Basin, 107
Drevel (*see* Holland-Drevel model)

E
Early evolution, 21
Electrochemical potentials, 3, 13–14
Electron acceptors
biofilms and, 246–247
deep biosphere establishment and, 23
greigite as, 15
iron as, 248
mackinawite and, 9–11
microbial metabolism and, 242–243
Electron flow, 18
Electron microscopes, 147
Electrons, 17
Elliot Lake Group, 145–146, 147–150, 150–152
Energy
biofilms and, 246–247
hydrothermal mounds and, 13–15, 25
microbial metabolism and, 242–243, 251
peptide nests and, 16–17
Energy content, 13
Entrainment, 108–110
Entropic sinks, 3
Enzymes (*see also* Coenzymes), 15–16, 25, 307
Europium
anomalies of in modern ocean waters, 272–274
banded iron formations and, 218, 270–272, 277–279, 297–298, 311
oceanic redox conditions and, 279–286
Evaporation, 306–307
Evolution, 19–21, 26, 35, 162–163, 279–286
Exhalites, 299
Exploration indexes, 58–59, 62–63
Extracellular polymeric substances, 243

F
Ferredoxins, 16, 17, 20
Ferrihydrite, 249–251
Ferrous hydroxide, 7–9
Fig Tree Group, 41
Fingerprints, 54, 55–59, 59–65
Fission reactors, 164–166
Flavodoxins, 17
Fluid inclusion temperatures, 192–195

Footwall rock types, 302–313
Formaldehyde, 6–7, 13, 18
Franceville Basin
atmospheric oxygen content and, 157–158
description of, 158–161
evolution of, 162–163
isotopic ages of, 161–162
nuclear fission reactors of, 164–166
uranium deposits of, 163–164
Freezing temperatures, 194–195

G
Galena, 170
Geobacter metallireducens, 14
Geomicrobiological research, 240–241
Glaciation, 217, 219
GO-GEOID, 55–56
Gold (*see also* Orogenic gold)
association of with uranium, 138–139
atmospheric conditions and, 134
craton mineral analysis and, 55, 57, 60–61, 83
mineralogy of, 131–133
paragenesis of, 127, 127–128
Re-Os isochron dating of, 137
reduced deposits and, 124
in Witwatersrand Basin, 110–113, 115–117, 122, 123
Gold and platinum group elements (PGE), 54
Gondwana, 54, 55–59, 59–65
Granitoids
in Barberton region, 73
chemical characteristics of, 74–75
classification of, 68
isotopic analysis of, 90–92
in Johannesburg Dome, 72–73, 73–74
Kaapvaal Craton and, 70, 71–75, 75–77
magma and, 68–71
Graphite, 39, 40
Great Oxidation Event, 172–173, 199–200
Green rust, 11
Green sulfur bacteria, 23
Greenhouse gases, 228–229
Greenstone belts (*see also individual greenstone belts*), 54, 70, 84
Greigite, 2, 9–11, 15, 16, 25
Gross National Income, 54, 58–59
Guanine, 18
Gunflint iron formation, 277–279
Gypsum, 177

H
Haloclines, 215
Hamersley Basin
banded iron formations in, 224, 276–279, 304, 306
productivity in, 306–307, 314–315
rare earth element anomalies and, 280–281
Trendall and Blockley model and, 297
Heavy rare earth elements, 75
Heavy sulfur (*see* Isotopic sulfur analysis)
Hematite, 249–250, 258–266, 317–319
Holenarsipur belt, 276, 277–279
Holland-Drever model, 297
Homacetogens, 16
Homeostasis, 22

Hot springs, 244
Hotazel formation, 277–279, 281–282
Hough Lake Group, 145
Huronian Supergroup, 145–146
Hydraulic sorting, 145
Hydrocarbons, 124
Hydrogen, 2, 26
Hydrogen electrodes, 13
Hydrogen sulfide, 3, 23, 177, 178, 180
Hydrological cycle, 228
Hydrophobic organic membranes, 20–21
Hydrosphere, 100–101, 199–200
Hydrosulfide, 6–7
Hydrothermal alteration, 97
Hydrothermal replacement model, 123
Hydrothermal systems
autogenesis and, 11–13
banded iron formations and, 218–219, 258–266, 270, 285
characteristics of, 299–302
as energy source, 3, 25
europium anomalies and, 273, 285
gold and, 115
iron formations and, 224, 242
microorganisms and, 241
serpentinization and, 4–6

I
Ice-covered rift model, 216–218
Icehouses, 3
Ilmenite, 68, 73, 114, 124, 244–245
Income, 54, 58–59
Infrastructure, 54, 58–59
Intrusions, 161
Iron (*see also* Banded iron formations; Chert-iron formations)
amino and nucleic acids in, 20
atmospheric redox state and, 122
cerium and europium anomalies and, 272–274
detrital mineralogy of, 133–134
diagenetic mobilization of, 224–225, 225–226, 229–234
electron flow and, 18
granitoid magmas and, 68–71
in Hadean oceans, 3–4
hematite and magnetite and, 317–319
hydrothermal mounds and, 11, 13, 241
low temperature mounds and, 7, 7–9
manganese and, 214, 217–218
microbial mobilization of, 240, 245–251
particle size and, 251–254
rare earth element behavior and, 218–219
riverine fluxes of, 234–235
Snowball Earth model and, 216
Iron Mountain Formation, 36–37, 38–39
Isobaric interferences, 85
Isotopic analysis
Barberton Greenstone Belt and, 41–42
Belingwe Greenstone Belt and, 43–46
Franceville Basin and, 161–163
Isua Greenstone Belt and, 35–41
mineralization and, 136–137
MVT deposits and, 192
overview of, 34–35
Steep Rock Group and, 42–43

Index

Isotopic carbon analysis
  Barberton Greenstone Belt and, 43–44
  Belingwe Greenstone Belt and, 43–46
  cyanobacteria and, 42
  Dharwar Craton and, 86, 96
  isotopic reservoirs of, 34
  Isua Greenstone Belt and, 39–41
  manganese and, 215–216
  overview of, 34–35
  Steep Rock Group and, 42–43
Isotopic fingerprinting, 34
Isotopic fractionation, 242
Isotopic nitrogen analysis, 16, 82–85, 86, 89–94
Isotopic oxygen analysis, 260–266, 308–311
Isotopic sulfur analysis
  bacterial fractionation and, 171
  Barberton Greenstone Belt and, 43–44
  Belingwe Greenstone Belt and, 43–46
  of Chinese ores, 206–207, 208–-213, 213–214
  Elliot Lake Group and, 147–150
  isotopic reservoirs of, 34
  Isua Greenstone Belt and, 37–39
  MVT deposits and, 195–197
  overview of, 34–35
  Steep Rock Group and, 42–43
  techniques for, 147
  trends in, 173–177
Isua Greenstone Belt
  banded iron formations in, 274, 302
  carbon isotopic analysis of, 39–41
  metabolic analysis of, 34
  overview of, 35–36
  plankton-like organisms in, 47–49
  sedimentary forms sampled, 36–37
  sulfur isotopic analysis of, 37–39

J
Jacobina deposits, 124
Jacobsen-Holland-Klein-Beukes model, 297–298
James's model, 296–297
Jaspers, 293
Jerome VMS deposit, 84
Johannesburg Dome, 72–73, 73–74

K
Kaapvaal Craton (*see also* Barberton Greenstone Belt)
  granitoid genesis in, 75–76
  granitoid mineralization and, 76–77
  metallogenic fingerprinting and, 54
  MVT deposits and, 198
  oxic granitoids of, 70, 71–75
Kerogen
  Cuddapah Basin, 83
  in Dharwar Craton, 87, 89–90
  isotopic analysis of, 86
  isotopic carbon analysis and, 40
  Red and Black Seas Hybrid model and, 307
Kinetics, 2
Kolar Gold Province, 83
Kuruman iron formation, 277–279, 311–312
Kushtagi belt, 276, 277–279

L
Lake Superior-type iron formations, 270, 296, 297, 312–313

Lanthanum enrichment, 272
Laser microprobe analysis, 147
Layering, 252–253
Lead (*see also* Mississippi Valley-type deposits; SEDEX deposits), 55, 57, 60–61, 170
Leader Reef, 127–128
Liantuo Formation, 207
Light rare earth elements, 75
Limestone, 304
Lithophilic elements, 54, 55–59, 59–65
Lost City spring, 7
Low temperature mounds, 7–9, 11–13

M
Mackinawite
  autogenesis and, 9–11, 25
  as basement crystal, 2
  low temperature mounds and, 9
  organic takeover and, 19, 20
  ribose phosphate and, 18
Magma (*see also* Volcagenic Massive Sulfide), 3–9, 54, 68–71, 270
Magnesium, 3, 18
Magnetic susceptibility, 68–71, 73–79
Magnetite
  banded iron formations and, 317–319
  granitoids and, 68, 73, 75
  microbial redox reactions and, 244–245
  reduced deposits and, 124
Manganese
  anoxia and, 213, 216, 218–219
  cerium and, 219, 270–272
  cerium and europium anomalies and, 272–274
  diagenetic mobilization of, 224–225
  Franceville Basin and, 157–158
  isotopic sulfur analysis and, 216–217
  model for ore genesis and, 214–218
  photosynthesis development and, 24
  rare earth element anomalies and, 279–286
  sulfur isotope chemistry and, 213–214
Manganese compensation depth, 215
Manjeri Formation, 43
Mantle degassing, 101
Mantle sulfide, 34–35
Marble Bar Greenstone Belt, 274–275
Marra Mamba formation, 276, 277–279
Mass independent fractionation of sulfur isotopes, 324
Massive martite ore, 260
Matinenda Formation, 145–146
McArthur-type SEDEX deposits, 170–171
Membranes, 9–11, 20–21
Mertite, 260, 262–264
Mesoproterozoic basins, 189
Metabolism, 15, 34–35, 242–243, 247–249
Metal ions, 17–18
Metal sulfides, 16–17
Metallogenesis, 62
Metallogenic fingerprinting, 54, 55–59, 59–65
Metalloproteins, 20
Metamorphism
  isotopic analysis and, 93–94, 99–100
  of Isua Greenstone Belt, 35–36, 40–41
  rare earth element composition and, 271
  weathering and, 317–319
  in Witwatersrand Basin, 126–127

Methane
  banded iron formations and, 321–322
  hydrothermal mounds and, 11
  isotopic carbon fractionation and, 34
  oceanic sulfur cycle and, 177
  serpentinization and, 6–7
  as waste, 3
Methanogens
  Belingwe Greenstone Belt analysis and, 45, 46
  biosphere formation and, 325
  hydrothermal mounds and, 21
  isotopic carbon fractionation by, 34
  oceanic sulfur cycle and, 177
Methanotrophs, 34, 42, 45, 46, 325
Microbial phylogenetic evolution, 35
Microflow and circulation reactors, 26
Microorganisms (*see also* Sulfate reducing bacteria)
  acetate and, 13
  banded iron formations and, 239–241
  Belingwe Greenstone Belt analysis and, 45, 46, 49
  biofilms and, 243
  biomats and, 243–244
  diagenetic mobilization and, 250
  dissolution and, 247
  energy and, 242–243, 251
  ferrihydrite and, 249–251
  geomicrobiological research and, 240–241
  iron and, 240
  mineralization and, 249–250
  minerals and, 244–245
  nitrogen fixing, 100–101
  Red and Black Seas Hybrid model and, 307, 314–315
  silica and, 317
  stromatolites and, 244, 253
Mineral diversity, 54, 55–59, 59–65
Mineralization
  Archean atmosphere and, 122
  controls on, 138–139
  diagenetic iron mobilization and, 230
  granitoids and, 76–78
  manganese ore formation and, 216–217, 219
  microorganisms and, 249–250
  MVT deposits and, 199–200
  Red and Black Seas hybrid model and, 304–306
  SEDEX deposits and, 178–180
  Witwatersrand Basin and, 113–115, 115–117, 127
Mineralogical perspective, 19
Minerals
  craton mineral analysis and, 53–55, 61–62
  microorganisms and, 244–245, 247
  as protective covering, 23
  RNA and, 18
  submarine hydrothermal deposits and, 301
  Witwatersrand Basin and, 108–110
Minor cyanide, 6–7
Mississippi Valley-type (MVT) deposits
  age of, 191–192
  atmospheric conditions and, 188, 197–200
  geochemistry of, 192–197
  geologic features of, 188–191
  overview of, 186–188

Modified paleoplacer model, 115–117, 123
Molybdenum, 213
Mosher Carbonate Formation, 42
Mounds (*see* Hydrothermal systems; Low temperature mounds)
Multienzyme complexes, 22
Muscovite, 92
Mutations, 20

N

Nanocrysts (*see* Mackinawite)
Neodymium, 218–219
Neoproterozoic basins, 189
Neutrons, 165
Ngezi Group, 43
Nickel
  biosynthesis and, 18
  as catalytic metal, 2
  craton mineral analysis and, 55, 57, 60–61
  origins of in pyrite, 152
Nitrogen content (*see* Isotopic nitrogen analysis)
Nitrogen fixing organisms, 100–101
Nuclear fission reactors, 164–166

O

Obduction, 4, 23–25
Oceanic sulfur cycle (*see also* Isotopic sulfur analysis; Sulfate reducing bacteria), 171–173, 174–177, 322–324
Ohmoto's banded iron formation model, 299
Öölitic ironstones, 293
Orebodies, 7, 264
Organic carbon (*see* Kerogen)
Organic membranes, 20–21
Organic takeover, 16, 19, 20
Orogenic gold (*see also* Paleoplacers)
  Dharwar Craton and, 83, 86, 87
  isotopic analysis of, 90–92, 99–100, 110–113
Oryx Goldmine, 112
Osmotic pressure, 13
Oxidation
  diagenetic iron mobilization and, 226
  granitoid magmas and, 68–71
  Great Oxidation Event and, 172–173, 199–200
  mackinawite and, 9
  manganese and, 215
  oceanic iron and, 218–219
  oxidizing atmosphere model and, 122, 136, 218
  in Precambrian, 97–98
  uranium deposits and, 163–164
Oxidative phosphorylation, 14–15
Oxides, 124, 293–296, 302–303
Oxidizing atmosphere model, 122, 136, 218
Oxygen (*see also* Isotopic oxygen analysis)
  banded iron formations and, 319–321
  biofilms and, 243
  evidence for in atmosphere, 24
  evolution of, 279–286
  isotopic analysis of, 260–266
  methane and, 177
  microbial metabolism and, 243
  photosynthesis and, 42, 45, 47
  as waste, 3
Oxygen content (*see also* Anoxia), 26, 172–173, 178, 271, 279–286

P

Paleomagnetic measurements, 192
Paleoplacers, 108, 110–113, 115–117, 123, 131–132
Paleoproterozoic basins, 189–190
Paleoweathering-metamorphic model, 317–319
Paragenesis, 127
Patchy hematite, 260
Pegmatites, 93
Penge iron formation, 277–279
Peptides, 16–17
Peptidyl transferase, 19
pH, 228–230
Phosphates, 16–17
Phosphorylation, 14–15
Photoautotrophs, 325
Photosynthesis
  Barberton Greenstone Belt and, 41
  Belingwe Greenstone Belt and, 45, 46
  biosphere formation and, 325
  isotopic analysis and, 34, 47
  magnesium and, 18
  microbial metabolism and, 243
  obduction and, 4, 23–25
  start date of, 34
  Steep Rock Group and, 42
Phototrophs, 244
Phylogenetic evolution, 35
Placer models, 123, 138–139
Plankton, 40–41
Plate tectonics (*see* Tectonics)
Platinum, 54
Polymerization energy, 13–15
Polymict conglomerate, 40
Porosity, 199
Porphyry, 68, 77–78
Potential energy, 2
Pre-enzymes, 15
Precipitates, 7–9
Precipitation, 249–250
Progressive metamorphism, 93–94
Proterozoic sedimentary exhalative deposits (*see* SEDEX deposits)
Proticity, 14
Protonic potential, 14
Protonmotive force, 15
Protoribosomes, 19, 20
Pyrite
  autogenesis and, 13
  banded iron formations and, 315–316
  detrital mineralogy of, 133–134
  green sandstones and, 160
  isotopic analysis of, 39, 147–150, 217
  Isua Greenstone Belt and, 35
  mackinawite and, 11
  manganese enrichment and, 214–215
  mineralogy of, 128–131
  models for formation of, 152
  origins of, 144–145
  origins of nickel, cobalt, and arsenic in, 152
  origins of sulfur in, 150–152
  oxidizing atmosphere model and, 136, 172
  Re-Os isochron dating of, 137
  redox conditions and, 122
  reduced deposits and, 124
  sulfate reducing bacterial production of, 323
  trace element ratios in, 150
  as waste, 13
  in Witwatersrand Basin, 113, 116, 122, 123
Pyritization, 213
Pyrophosphate, 18, 25
Pyrophosphate hydration, 16
Pyrrhotite, 35, 38, 39

Q

Quinones, 247
Quirke Lake Group, 145

R

Rainbow hydrothermal system, 3
Ranciéite, 24
Rapitan iron formation, 277–279
Rare earth elements
  banded iron formations and, 270–272, 279–286
  Brockman iron formation and, 311
  granitoids and, 75
  as indicators for ocean chemistry, 270–272
  manganese ores and, 218–219
  uranium and, 161
Rare metal pegmatites, 92–93
Rayleigh fractionation, 96
Reactors, 11–13, 164–166
Recycling, 64–65, 101, 230–234, 235
Red and Black Seas Hybrid model, 304–306, 306–307, 307–311, 314–315
Redox state
  atmospheric conditions and, 121–122, 136, 144
  cerium fractionation and, 271
  coenzymes and, 17–18
  electrochemical potentials from, 3
  granitoid magmas and, 68–71
  Great Oxidation Event and, 172–173
  Hamersley Basin and, 307
  microorganisms and, 244–245
  oceanic sulfur cycle and, 178, 179–180
  in Precambrian, 97–98
  proto-Archaea and, 23
  rare earth element anomalies and, 218–219, 279–286
  SEDEX deposits and, 170–171
Reduced carbon, 39
Reduced deposits, 124
Reducing atmosphere model, 122, 136, 144
Reefs, 42, 46
Regulated dynamic systems, 22–23
Ribose phosphate, 18
Riverine transport, 234–235, 242, 285
RNA, 13, 18–19, 19–20
RNA synthesis, 18–19
Roraima Supergroup, 116–117
Rubisco I, 34, 45, 46, 47

S

Sandstones, 158–160, 162–163, 304
Sandur Greenstone Belt, 83–84, 276, 277–279, 279–280
São Francisco Craton, 54
Sargur Greenstone Belt, 276, 277–279
Seawater characteristics, 226–229
Seawater sulfate, 34–35

Index

SEDEX deposits, 170–171, 174–180, 186, 190
Sedimentary organic compounds, 89
Sedimentation
    diagenetic iron mobilization and, 232
    in Franceville Basin, 158–161
    iron and, 133–134
    kerogen formation and, 89–90
    manganese enrichment and, 215
    MVT deposits and, 186–188
    oceanic sulfur cycle and, 177
    pyrite and, 129
    rare earth elements and, 271
    Witwatersrand Basin and, 108–113, 125–126
Selective entrainment, 108–110
Selwyn-type SEDEX deposits, 170–171
Serpentinization, 3, 4–6
Shale footwalls, 303–304
Siderite
    banded iron formations and, 315–316
    formation of, 321–322
    isotopic carbon analysis and, 40
    isotopic sulfur analysis and, 39
    low temperature mounds and, 7–9
    microbial metabolism and, 246, 249–251
Siderophilic elements, 54, 55–59, 59–65
Silica, 293–296, 316–317
Siliclastic sediments, 89–90, 91
Silvermine orebodies, 7
Snowball Earth model, 216
Socioeconomics, 54, 58–59
Sokoman iron formation, 277–279
Solubility constraints, 188
Specular hematite, 260
Sphalerite, 170
Stable isotopes (*see* Isotopic analysis)
Steep Rock Group, 34, 42–43, 49
Steyn paleoplacer, 112–113
Stratification, 158–161, 174–176
Stromatolites, 42–43, 46, 244, 253
Sulfate minimum zone, 176
Sulfate reducing bacteria
    Belingwe Greenstone Belt and, 45, 49
    biosphere formation and, 325
    diagenetic iron mobilization and, 224
    Elliot Lake Group and, 151
    isotopic analysis and, 46, 48
    isotopic carbon fractionation by, 34–35
    James's model and, 296–297
    methane and, 177
    oceanic sulfur cycle and, 179, 180, 322–324
    sulfur isotopes and, 171, 173, 177
Sulfates (*see also* Oceanic sulfur cycle)
    anoxia and, 206–207, 214–220
    banded iron formations and, 172–173
    black smokers and, 3
    diagenetic iron mobilization and, 227–228
    heavy sulfur and, 217
    isotopic reservoirs of, 34–35

isotopic sulfur analysis and, 39, 42, 44
    MVT deposits and, 188, 199–200
Sulfides (*see also* SEDEX deposits; Volcagenic Massive Sulfide (VMS))
    bacterial oxidation of, 177
    banded iron formations and, 293–296
    diagenetic iron mobilization and, 227–228
    energy transfer and, 16–17
    heavy sulfur and, 217
    isotopic analysis and, 34–35, 46
    MVT deposits and, 186–188, 195–197
    Witwatersrand Basin mineralogy and, 127, 134, 136
Sulfur (*see also* Heavy sulfur; Isotopic sulfur analysis; Oceanic sulfur cycle)
    banded iron formations and, 322–324
    isotopic reservoirs of, 34–35
    MVT deposits and, 195–197
    origins of in pyrite, 150–152
Sulfur chemistry, 68
Sulfur cycle, 35, 171–173
Supergene-modified hydrothermal iron ores, 258–266
Superior Province craton, 54, 55–59, 59–65
Superior-type banded iron formations, 270, 296, 297, 312–313
Synthases, 16

T

Tansanshan deposit, 207
Target mine, 111
Tarkwa Basin, 116
Tectonics, 54, 68, 178
Temperature
    diagenetic iron mobilization and, 228, 229
    hydrothermal fluids and, 7–9, 25, 26, 299–300
    isotopic oxygen fractionation and, 265
    MVT deposits and, 192–195
Thermochemical sulfate reduction, 151–152
Thermophiles, 21
Thiolate, 16
Tin, 55, 57, 60–61, 68, 77–78
Titanium, 55, 57, 60–61
Trace elements, 9, 150
Tracers, 54, 55–59, 59–65
Transvaal Supergroup
    banded iron formations and, 258, 277–279, 311–312
    MVT deposits in, 191, 192
    rare earth element anomalies and, 280–281
Trendall-Blockley model, 297
Tungsten, 55, 57, 60–61, 68
Tynagh orebodies, 7, 9

U

Unmodified placer model, 123
Uplift, 124
Uracil, 18

Uraninite
    dating of, 137–138
    origins of, 144–145
    paragenesis of, 127
    reduced deposits and, 124
    reducing atmosphere theory and, 121–122
    as uranium source, 164
    in Witwatersrand Basin, 114, 116, 135-136
Uranium (*see also* Nuclear fission reactors)
    association of with gold, 138–139
    atmospheric conditions and, 144–145, 157–158
    Elliott Lake and, 152
    Franceville Basin and, 157–158, 163–164
    Witwatersrand Basin and, 115, 122, 123

V

Vaal Reef, 137
Väyrylänkylä iron formation, 277–279, 281–282
Vertical tectonics, 54
Volatisphere, 2, 3
Volcagenic Massive Sulfide (VMS)
    characteristics of submarine hydrothermal systems and, 299–302
    chemistry of banded iron formation and, 313–319
    geologic environments and, 302–303
    isotopic analysis of, 83–85, 86, 87, 92–93
Volcanic footwalls, 303, 312

W

Wabigoon Greenstone Belt, 42
Water production, 11
Weathering
    banded iron formations and, 258, 296–297
    iron deposits and, 224
    metamorphism and, 317–319
    rare earth element composition and, 271
    supergene ore formation and, 266
West Rand Group, 107
Witwatersrand Basin
    atmospheric oxygen content and, 105–106
    description of, 106–108, 124–126
    detrital mineralogy of, 113–115, 127
    mineralogy dilemma of, 127–133
    sedimentology of, 108–113

X

Xiangtan deposit, 207

Y

Yellow cake, 115
Yilgarn Craton, 54

Z

Zimbabwe Craton, 54, 59
Zinc (*see also* Mississippi Valley-type (MVT) deposits; SEDEX deposits), 55, 57, 60–61